D1793254

TARRYTOWN TECHNICAL
Information Center
GENERAL FOODS CORP.

Qualitative Analysis of Flavor and Fragrance Volatiles by Glass Capillary Gas Chromatography

Academic Press Rapid Manuscript Reproduction

Qualitative Analysis of Flavor and Fragrance Volatiles by Glass Capillary Gas Chromatography

WALTER JENNINGS
Department of Food Science and Technology

TAKAYUKI SHIBAMOTO
Department of Environmental Toxicology

University of California, Davis, California

1980

ACADEMIC PRESS
A Subsidiary of Harcourt Brace Jovanovich, Publishers

New York   London   Sydney   Toronto   San Francisco

COPYRIGHT © 1980, BY ACADEMIC PRESS, INC.
ALL RIGHTS RESERVED.
NO PART OF THIS PUBLICATION MAY BE REPRODUCED OR
TRANSMITTED IN ANY FORM OR BY ANY MEANS, ELECTRONIC
OR MECHANICAL, INCLUDING PHOTOCOPY, RECORDING, OR ANY
INFORMATION STORAGE AND RETRIEVAL SYSTEM, WITHOUT
PERMISSION IN WRITING FROM THE PUBLISHER.

ACADEMIC PRESS, INC.
111 Fifth Avenue, New York, New York 10003

United Kingdom Edition published by
ACADEMIC PRESS, INC. (LONDON) LTD.
24/28 Oval Road, London NW1 7DX

Library of Congress Cataloging in Publication Data

Jennings, Walter,   Date
    Qualitative analysis of flavor and fragrance
volatiles by glass capillary gas chromatography.

    Includes index.
    1. Flavoring essences—Analysis. 2. Gas chromatography.   I. Shibamoto, Takayuki, joint author.
II. Title.
TP418.J46   664'.52'01544926   79-26034
ISBN 0-12-384250-6

PRINTED IN THE UNITED STATES OF AMERICA

80 81 82 83   9 8 7 6 5 4 3 2 1

CONTENTS

Preface vii

PART A ANALYTICAL CONSIDERATIONS

I	Introduction	1
II	Gas Chromatographic System Requirements	2
	A. Columns	2
	B. Inlets	5
	C. Detectors	8
	D. Temperature Control	8
III	Retention Indices	9
	A. General Considerations	9
	B. Determination of Holdup Time	9
	C. Effect of Liquid Phase and Surface Pretreatment	9
	D. Effect of Temperature	10
	E. Isothermal Procedures	10
	F. Programmed Temperature Procedures	11
IV	Selective Detectors	11
	A. General Considerations	11
	B. Thermal-Conductivity Detector	13
	C. Flame-Ionization Detector	13
	D. Nitrogen–Phosphorus Detector	13
	E. Electron-Capture Detector	14
	F. Flame-Photometric Detector	14
	G. Photoionization Detector	14
V	Ancillary Reactions	15
	A. General Considerations	15
	B. External Reactions	15
	1. Column chromatography	15
	2. High pressure liquid chromatography	16
	3. Simple pre-GC reactions	16
	4. Ozonolysis and hydrogenation	17

	C.	Reaction Gas Chromatography	18
		1. General considerations	18
		2. Precolumn injection	19
		3. Carbon skeleton analysis	19
		4. Hydrogenation	20
		5. Miscellaneous reactions	20
		6. Subtractive reactions	20
VI	Gas Chromatography – Mass Spectrometry		22
	A.	General Considerations	22
	B.	Interfacing	23

PART B   APPENDICES

Appendix I	Compounds and Their Retention Indices	29
Appendix II	Retention Indices in Increasing Order on Methyl Silicone OV101	58
Appendix III	Retention Indices in Increasing Order on Polyethylene Glycol Carbowax 20M	87
Appendix IV	Mass Spectra of Individual Compounds	114

Index   467

PREFACE

In spite of great interest in the analysis of flavor and fragrance volatiles, criteria such as retention indices and mass spectra, which are invaluable aids to their characterizations, are not widely available. A small amount of scattered information appears in the literature, and massive files have been compiled by the research department of several commercial concerns, but these latter data are usually treated as proprietary information and are carefully guarded.

This book represents an effort not only to make such information available to the individual researcher involved in studies of the volatile constituents of fragrances, foods, and natural products, but also to serve as an introduction to ancillary methods that have proved useful in characterization.

The authors are grateful to Yashiharu Ogawa of Ogawa and Co., Ltd., Tokyo, for his encouragement and active support of this undertaking. Many co-workers at Ogawa and Co., Ltd., and at the University of California in Davis have contributed to our efforts. Special mention should be given to K. Harada, who was responsible for the determination of all of the retention indices reported in the appendices, and H. Toda, wo collected the mass spectral data. Sample standards were prepared by K. Kayama, and the glass capillary columns were constructed by O. Mishimura and K. Yamaguchi. Some of the drawings were prepared by M. Sakaguchi and Y. Kamiya; assistance in editing and photographic reproduction was supplied by K. Yamaguchi, S. Mihara and A. Aitoku. N. Lum helped in editing, cropping and arranging the mass spectral data for reproduction and D. Ingraham prepared the computerized printouts of retention indices. To all of these, we express our deep and sincere appreciation.

Walter Jennings
Takayuki Shibamoto

PART A   ANALYTICAL CONSIDERATIONS

I. INTRODUCTION

Samples of interest to the flavor and fragrance analyst may range from relatively simple synthetic mixtures to complex biological samples whose components cover a wide range of volatilities and embrace a challenging array of functional groups. Some are relatively unstable and their analysis requires highly inert analytical systems; others are extremely complex and the separation powers of the column become critically important; a few, including some synthetic mixtures, are relatively simple and high resolving power is unnecessary and may be traded for shorter analysis times. Glass capillary columns are especially suited to all such analyses because, when properly deactivated and installed in a compatible system, their increased inertness can result in the detection of materials that are unable to negotiate a packed or metal system. Columns of fused natural quartz and fused synthetic silica have recently become available, and these are in many cases even more inert (vide infra). In short, glass capillary chromatography can achieve much higher powers of separation, shorter analysis times, and higher sensitivity (see below).

Many procedures have been used as criteria of identification for compounds resolved by gas chromatography, not all of which were (or are) entirely satisfactory. Too much credence was sometimes placed in results from doubtful methods, and these fell into disuse as most investigators sought to utilize more precise and more definitive methods of identification. Provided their limitations are recognized and results are not subjected to overinterpretation, some of those less-rigorous methods can still provide useful information, usually in the form of confirmation. Consider the use of retention behavior as one example.

The use of gas chromatographic retentions as criteria for the identification of volatiles is an old procedure, and one that the more critical analyst has learned to avoid. Establishing the retention of a compound does permit one to rule out by exclusion compounds whose behavior is known to be different, but the possibility of several compounds exhibiting the same retention on a given column under a given set of conditions is very real, particularly in a low-resolution system. With columns capable of achieving upwards of 200,000 effective theoretical plates as compared to the up-to-10,000 effective theoretical plates to which most systems utilizing packed columns are limited, the possibility of co-chromatographing components becomes very much less; it must be emphasized that it does still exist.

Probably the greatest value of retentions is as a complementary criterion. For example, a number of structurally related compounds yield very similar mass spectra, and their differentiation on this basis is sometimes problematical; the sesquiterpene hydrocarbons illustrate this principle well. The individual members of this large class, however, can be readily differentiated by high-resolution gas chromatography. Hence, once the mass spectrometer has established that the compound in question is a sesquiterpene hydrocarbon, a precise retention assignment can establish which sesquiterpene hydrocarbon it must be.

Increased credence can be awarded to identifications assigned on the basis of retentions by resorting to two-dimensional gas chromatography (1–5), i.e., separation on one column followed by a second separation on a dissimular liquid phase. Again, as the resolving power of the chromatographic systems is increased, the possibility of co-chromatographing components becomes less. Hence by utilizing very high resolution systems in two-phase chromatography, a considerable percentage of routine identifications can be established with a reasonable degree of certainty. This approach can be especially valuable when it is coupled with the use of a nondestructive detector and valving systems so that a restricted number of components from one high-resolution gas chromatographic separation can be shunted to a second high-resolution system without the danger of the investigator becoming confused over which peaks from the first separation relate to which peaks in the second separation (6–8).

Complementation of gas-chromatographic retentions with a functional group determination, or a knowledge of the class of compound involved, can reinforce identification assignments. Many workers have employed selective detectors for this purpose or used chemical reactions to separate or modify components prior to gas-chromatographic analysis (e.g., 8–11). Such reactions may be carried out in an isolated system—sometimes in the injection syringe—and may involve, e.g., the subtraction of olefins or carbonyl compounds, or the hydrolysis of esters. Comparison of the original chromatogram with that obtained from the reaction mixture then provides the analyst with additional information. Alternatively, the "reaction chamber" may be incorporated into the gas-chromatographic system, usually as a precolumn, to achieve what has been termed "reaction gas chromatography" (9, 10). Ultimately the retention behavior of the original consitituents and that of the reaction products are compared, sometimes in extremely complex multicomponent systems. Hence the degree of component separation, and the precision with which retentions can be compared, is critically important to such analyses. These "conventional" techniques of subtractive analysis and on-column derivatization can be modified in many cases to render them compatible with glass capillary gas chromatography.

One of the more definitive analytical techniques currently available is offered by the combination of glass capillary gas chromatography and mass spectrometry. Mass spectral fragmentation patterns can be highly characteristic and, when coupled with a precise retention index, offer a powerful tool for component identification.

II. GAS-CHROMATOGRAPHIC SYSTEM REQUIREMENTS

A. Columns

Because of its much higher resolving powers, the wall-coated open tubular column is preferred. Metal capillaries are satisfactory for some types of analyses, but they suffer from several disadvantages: not all compounds can negotiate metal columns; while the separation of some biological compounds, including steroids and drugs, has been demonstrated on both glass and nickel columns (12–14), the latter fail to pass some compounds—e.g., sulfur-containing compounds—(15) that may be critically important to flavor and are detected in the glass capillary system. In addition, metal capillary columns are usually more expensive, and because they coat more uniformly, glass capillary columns are capable of higher resolution. On a "cost-per-plate" basis, glass capillary columns usually represent the greatest value. In addition, it is difficult to locate and remove small segments of column in which imperfections may have developed in metal columns, an operation that is very straightforward in glass capillaries. For these reasons, glass capillary columns, which are readily available today, must be regarded as generally superior and should be used wherever possible.

SCOT or PLOT columns have lower powers of resolution and the possibility of co-chromatographing components is therefore greater. Nevertheless, both because of their larger sample capacities and because the inclusion of small amounts of support material can stabilize some liquid phases that otherwise do not long endure on glass, there may remain good reason to utilize PLOT or SCOT columns. Whisker columns (16, 17) are in reality a special form of PLOT columns.

A system possessing a high degree of inertness is essential; compounds that degrade or rearrange during analysis will produce peaks whose correlation with the original compound is rarely possible. This may require the elimination of metal contact surfaces in the inlet and transfer lines of older systems. Unless the glass capillary column is properly installed in a system designed for or modified to the peculiar special requirements of these columns, prospects of obtaining the degree of separation that it is capable of delivering are indeed dim. Peak tailing, which is normally and sometimes wrongly blamed on the column, can lead to serious problems. Sources of peak tailing (which are also observable with the solvent peak) can be attributed to (a) septum–solvent contact, including absorption of solvent or sample in pieces of the septum that have fallen into the column or inlet chamber; (b) defects in the injection port or flow system related to solvent backflash, polymeric materials in the flow stream, or cracks in inlet liners or other unswept volumes; (c) column defects; and (d) detector (FID) jets, collectors, and electronics, including recorders and integrators.

Column connections should have zero dead volume, and exhibit chemical inertness, ease of use, and thermal stability. Materials used in the connections of glass capillary columns have included Teflon tubing, elastomeric materials, graphite and Vespel ferrules, and platinum–iridium tubing. The use of elastomers or plastics (including heat-shrink Teflon tubing) and "low dead volume" connections usually causes problems and should be avoided. Ideally, the column should be installed so that the inlet end begins inside the heated zone and in an area of high gas velocity; this may involve straightening (and subsequently deactivating; see below) the column ends. The outlet end of the column should terminate as close to the detection point of the detector as possible, and certainly well beyond the point of make-up or auxiliary gas introduction, if used. Plastics, elastomers, or minute dead volumes in the sample-contacting zones can result in an unacceptable degree of tailing and band broadening.

When glass capillaries are heat-straightened, degradation of stationary phase occurs in those parts of the column that were subjected to that high heat treatment. This generates adsorptive sites that can give rise to severe tailing with active compounds; this problem can be corrected (18). Unless these sites are deactivated, the resultant tailing can affect even the precision of the retention index calculation. A major advantage of the fused quartz and fused silica glass capillary columns is that the thin-walled flexible variety is inherently straight, and the column straightening step is unnecessary (18).

Alternatively, several systems have been proposed that permit the installation of unstraightened conventional glass columns. In some cases, this involves a butt joint in a zone of low gas velocity, which must be regarded as undesirable. If the two mating surfaces are precisely flat and closely joined with precise alignment, such a junction would cause no problem, but this is not normally the case. More generally, some slight gap exists between the fixed gas conduit and the column end. This creates a poorly swept volume in a low-gas-velocity zone, and usually also exposes some elastomeric material—which is adsorptive—to the flow stream. Although the volume of this space is small, the gas flow through that zone is very restricted because it is not in the direct flow path. Every peak will possess a tail, equivalent to whatever time is required for the zone to be swept clean by that restricted carrier gas flow.

Figure 1 illustrates a more acceptable method of installing unstraightened columns. In this case, both ends of the column terminate in zones of high gas velocity, eliminating the problem detailed above. If the apparatus incorporates flame ionization detection, the velocity of gas at the detector end can be increased still further by sweeping the end of the column with premixed hydrogen and make-up gas, instead of introducing the hydrogen separately at some point nearer the flame tip. The major problem with the splitter illustrated is that the split point is located in the oven instead of the heated inlet area; at very low split ratios this can lead to discrimination with some samples. The design of various injection devices, and conversion of packed column instruments to glass capillary capability has been discussed (18).

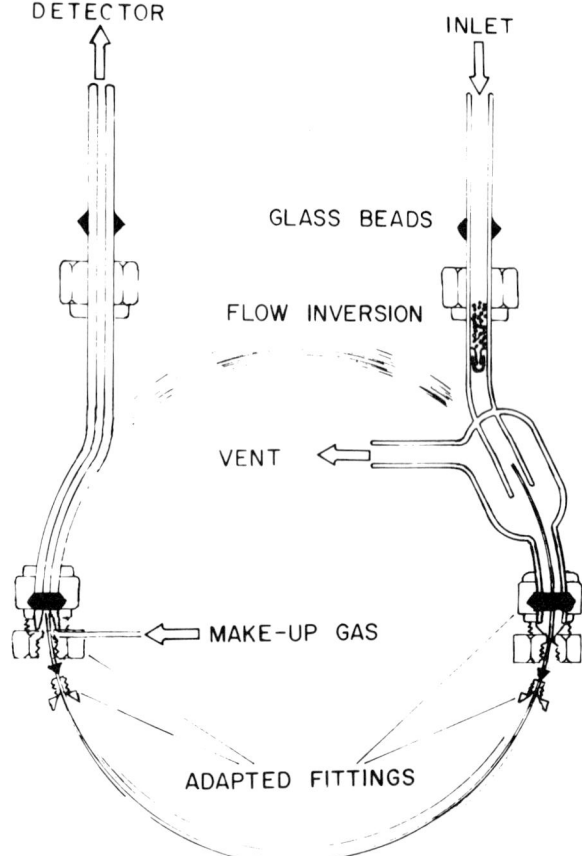

Fig. 1. Schematic of an all-glass inlet splitter and make-up gas adapter designed for use with unstraightened column ends (18).

High-efficiency short columns are becoming increasingly popular. Normally, such columns should have liquid film thicknesses of 0.3–0.5 μm, which is substantially thicker than the 0.05–0.1 μm films supplied by most manufacturers, although one U.S. manufacturer normally supplies columns with coating thicknesses of 0.4–0.5 μm. Thicker films increase sample capacity; capillaries with thin films are normally very long (50–100 m) because partition ratios vary directly with film thickness (18). Hence excessively long columns are necessary to perform an equivalent separation with a thin-film column. Additionally, cryogenic cooling may be required to increase the retention of low-k compounds on thin-film columns. The matter of film thickness in capillary columns has been discussed in detail (19).

B. Inlets

The sample should be introduced onto the column in such a manner that it is restricted to the shortest possible length of the column. Traces of sample remaining in the injection area should be flushed out of the system before the chromatographic process begins; it is critical that the sample plug be short, and that the short sample plug be followed by pure carrier gas rather than be exponentially diluted sample. Otherwise, the peaks on the resultant chromatogram will be broader (i.e., resolution will be poorer) than the column is capable of delivering. Injection techniques have recently been reviewed (20). Four methods of injection are currently receiving wide usage with small-bore capillary columns.

1. *Split Injection*

In this method, most of the high-velocity flow of carrier gas through the inlet is shunted to atmosphere, and a very small portion of that flow stream is placed on the capillary column. In split mode, the sample is generally vaporized in the inlet and a major portion of the sample is vented. Split ratios may vary from 1:5 to 1:1000. Split injections are compatible with the restricted sample capacity of WCOT columns (normally 10–100 ng per component), and because the high flow through the sample injection area rapidly flushes this clean, sample introduction on column is followed with pure carrier gas rather than with exponentially diluted sample. The split injection mode is generally used for the analysis of major components, where these exist at levels of 0.1–10% of the sample mixture. Split-type injections are widely used in the petrochemical field, in the analysis of foods, flavors and essential oils and in the routine organic analysis of simpler complex mixtures. The method can be applied to almost any sample and good results are relatively easy to obtain. Neither solvent volatility nor initial column temperature is crucial (see below). The disadvantages are that with most splitter designs the split may not be linear, the amount of sample that reaches the column is determined only indirectly, and the linearity of the splitter can be affected by changes in the split ratio or the sample size. Some workers utilize 1 ml syringes, but these have generally blunt needles are are quite destructive to septa. Most investigators prefer to use the 10 μl syringe, which has a less destructive needle, and accept the fact that the absolute quantity of sample injected is beyond their control; hence quantitation requires the use of internal standards.

Splitter linearity can be checked quite readily. Probably the best and simplest method is to turn the splitter outlet off (split ratio zero), and remove the capillary and substitute a short packed column, so that everything injected is delivered to the detector (21). The detector should be carefully optimized so that minor changes in flow do not affect its response. A model system, on which the packed column can achieve baseline resolution and preferably representing the range of boiling points and functional groups of the sample, is then injected several times to satisfy any "demand capacity" of the system; the relative detector response for each component is then determined. The packed column is then removed, the capillary is installed, and the splitter is activated. Make-up gas is added to the detector to compensate for the difference between the carrier gas flows of the packed and capillary columns, and the model system is reinjected. A comparison of the two results permits some meaningful decisions as to splitter linearity. It should be noted that while most commercial splitters cite reproducibility as evidence of their excellence, a nonlinear splitter frequently delivers reproducible (and erroneous) results.

A simple concentric-tube inlet splitter is usually sufficient for qualitative analysis, although it must be recognized that a discriminatory split has probably been achieved. The splitter should be provided with a glass liner, not only to ensure a high degree of inertness but also to permit its use as a packed precolumn for subtractive techniques (see below). Precise flow control is critical; otherwise erratic changes in retention times will occur. Figure 2 shows the arrangement generally used to control the pressure drop across (and hence the flow through) a capillary system. Condensation in the needle valve restrictor, or non-steady-state leaks in the column connections or the injector septum, will result in flow variations that render the retention data meaningless. Figure 3 shows an improved design in which non-steady-state leaks or minor changes in split ratio do not affect the pressure drop across the column. This latter configuration should be used wherever possible. A normal pressure regulator is rarely a satisfactory substitute for the more specialized back-pressure regulator in this system. It should be noted that normal and reverse solvent effects normally associated with "splitless" injection (see below) also occur in split mode and can influence the shape of peaks in the vicinity of the solvent peak, even at high split ratios and small injection volumes (22).

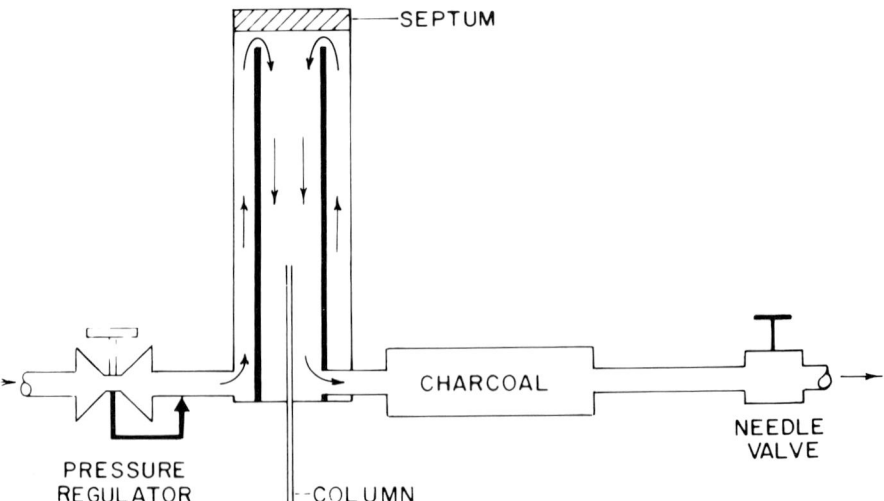

Fig. 2. Restrictor-controlled splitter outlet. Both the pressure regulator and the needle valve restriction affect column head pressure (i.e., pressure drop) and split ratio; non-steady-state leaks (septum, column connections) will also affect column head pressure (and retentions).

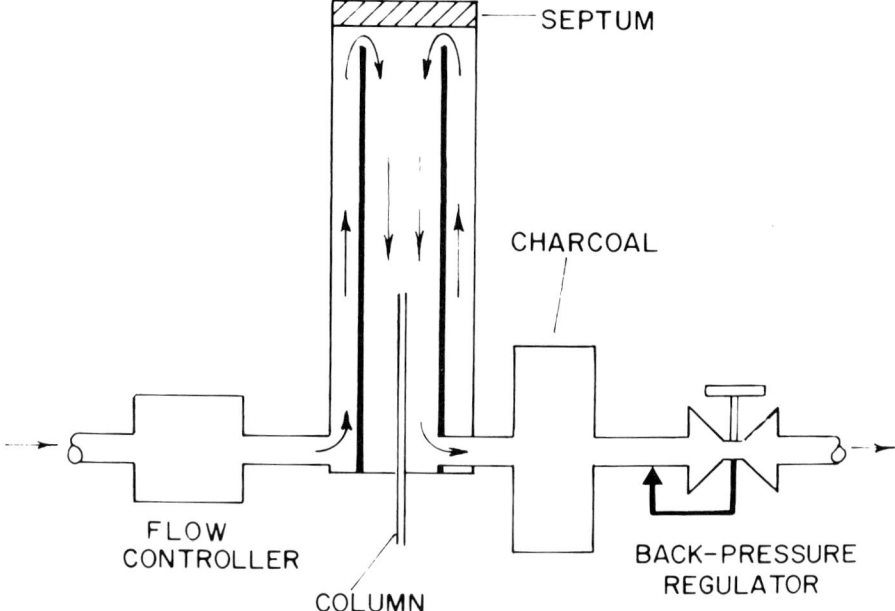

Fig. 3. Back-pressure-regulated splitter outlet. The split ratio is controlled by the flow controller and the column head pressure by the back pressure regulator. Non-steady-state leaks affect the split ratio, but not the column head pressure.

2. *Cold Trapping*

In this method of injection, the column is maintained at a temperature low enough that the vapor pressures of the compounds of interest are negligible; the distribution constants K_D are atypically large, causing an increase in the partition ratios k (18). Hence their progress through the column is infinitely slow. After the injection is completed and any pressure surge accompanying the injection has dissipated, the column temperature is raised, K_D and k assume "normal" values, and the chromatographic process begins. In general, compounds whose boiling points are 30–50° (or more) above the column temperature will have their distribution constants affected to such a degree that they will be concentrated by on-column cold trapping. For the cold trapping of lower-boiling components a U-shaped section formed on the front end of the glass capillary column can be immersed in chilled water or other coolant during the injection period.

3. *"Splitless" Injection*

This technique of sample introduction is based on the injection of the sample together with 0.5–2 μl of a suitable solvent (23–28), while the column is held at a temperature 10–30° below the boiling point of that solvent. The solvent must recondense on the column, forming an area of low phase ratio, which forces an increase in the partition ratios of the sample components (22, 29). Because the phase ratio decreases (and the partition ratios increase) in the direction of carrier gas flow, band-narrowing should be possible on properly executed splitless injection. Cold trapping also plays an important role here, particularly with the higher-boiling sample components.

During the purge activation time, a volume of carrier gas equal to approximately 1.5 times the volume of the inlet vaporization chamber should sweep through the inlet. If the purge activation time is too short, a loss of sample will result; if the purge activation time is too long, a large solvent tail will result. The purge function (30–32) is used to reduce the solvent tail by diverting to atmosphere the last residual sample in the inlet. Hence "splitless" injection does split—or divert—some of the injection in what can be a nonlinear manner.

Advantages claimed for splitless injection include the facts that it permits the analysis of trace components on capillary columns, that the sample size can be determined with reasonable accuracy, and that lower injection port temperatures can be used because a longer vaporization time is permitted. The disadvantages are that the initial column temperature must be compatible with the boiling point of the solvent, the lifetime of the column may be shortened due to solvent overloading, the number of solvents that can be used is restricted, and it is difficult to separate volatile components from the solvent. Additionally, the moment of "injection" cannot be determined, and hence there is a slight error in retention time measurements.

It should be noted that components preceding the solvent but falling within the envelope of the solvent effect can experience a "reverse solvent effect" that leads to peak degradation. Both normal and reverse solvent effects can be demonstrated even with split-mode injection (22).

4. *On-Column Injection*

This method utilizes injection syringes with special (and flexible) 32-gauge needles, which can be inserted into the column. Because these needles are too flexible to penetrate septa, other approaches are utilized. One involves a valved entrance to the injection chamber (32, 33), and another uses a larger needle to serve as a guide through the septum (18). Both the solvent effect and cold trapping play roles in this injection technique.

In a restricted number of highly specialized cases—i.e., the analysis of samples of extreme volatility or oxygen lability—an ampule-type inlet may prove useful.

C. Detectors

Standard ionization-type detectors are most widely used and any that have sufficient sensitivity for the compounds to be detected are usually suitable, provided they have been modified (usually by the addition of an auxiliary or make-up gas) so that residence time between the column outlet and the detection zone (flame) is a minimum. The photoionization detector (PID) offers some special advantages and possesses several disadvantages. Because it is a nondestructive detector, a single fraction eluting from one high-resolution column can be subjected to odor evaluation, trapped and subjected to further chemical characterization, or rerun on a second high-resolution column coated with a dissimular liquid phase (6, 7).

D. Temperature Control

Precise temperature control throughout the oven is, of course, critically important. Various investigators have reported that temperature differences as large as 10° are not uncommon through the ovens of some older chromatographs. The oven should be fitted with a high-velocity fan and the columns

III. RETENTION INDICES

A. General Considerations

Although a number of retention index systems have been proposed (e.g. 34–40), we limit our treatment to the Kovats retention index system (39), in which the retention behavior of a compound is reported relative to that of the n-paraffin hydrocarbons, utilizing a logarithmic scale. A comprehensive discussion of the method has recently been published (41). Each n-paraffin hydrocarbon is assigned by definition an index 100 times its carbon number. The retention index I of a compound can be defined as

$$I^a_b = 100N + 100n \frac{\log t'_{R(A)} - \log t'_{R(N)}}{\log t'_{R(N-n)} - \log t'_{R(N)}}$$

where I is the retention index on liquid phase a at temperature b, and $t'_{R(N)}$ and $t'_{R(N+n)}$ are the adjusted retention times of n-paraffin hydrocarbons of carbon numbers N and $(N-n)$ that are respectively smaller and larger than the adjusted retention time of the unknown, $t'_{R(A)}$.

B. Determination of the Holdup Time t_M

Several methods have been suggested for calculating the gas holdup time of the system, which must be subtracted from the observed retention time t_R to obtain the adjusted retention time t'_R. The method of Peterson and Hirsch (42) is widely used and several other proposals appear interesting (43–47). Various authors (e.g., 47) have commented on the accuracy to be expected in these mathematical determinations. Where feasible, the most direct method is by simple methane injection. Rijks (1) reports that the use of the methane peak to establish t_M is not only simple and direct but is probably the most accurate means.

C. Effect of Liquid Phase and Surface Pretreatment

The retention indices as determined on SCOT or PLOT columns are occasionally slightly different from those obtained on WCOT columns, probably because of a degree of interaction with the support material, but this phenomenon is not wide-spread. Surface pretreatments of a more drastic nature can exercise an effect on retention characteristics, but these very marked effects are generally restricted to coatings that contain strongly acidic or strongly alkaline materials; an example is Carbowax 20M admixed with KOH, which has been used in the separation of amines.

Liquid phases do vary in their selectivity and some changes can be experienced in the retention indices when changing from one to another batch of liquid phase. In general, high-grade stationary phases are to be preferred, e.g., SP 2100 as compared to SF96; both are obtainable as low-viscosity methyl silicones, but one has been carefully refined for gas-chromatographic use. The polarity of a liquid phase may also change with use, which can have a profound effect on its retention behavior. The polyethylene glycol Carbowax 20M in the presence of small amounts of oxygen in the carrier gas will gradually become acidic and its polarity will show a marked increase. A suitable polarity test mixture such as that recommended by Grob and Grob (48) should be used on a routine basis with columns that are to be used for retention measurements. If polarity shifts do occur, they become immediately apparent when checked in this manner. Appendix I lists the retention indices of a number of compounds.

D. Effect of Temperature

The retention index of a given compound, determined at one temperature, will differ from its retention index at another temperature to the extent that the interactions between the liquid phase and the n-paraffin hydrocarbons were affected differently by that temperature shift than were the interactions between the liquid phase and the compound. Hence the magnitude of the shift should be greater with polar compounds, with polar liquid phases, and with compounds of increasing molecular weight (higher retention indices). Ettre (49) pointed out that while this shift is hyperbolic, within a range of perhaps 50°C the departure from linearity is not excessive. He suggested that this temperature dependence of the retention index could be expressed as

$$\Delta I / 10°C = \frac{I_{T_2} - I_{T_1}}{(T_2 - T_1)10}$$

With compounds such as the lower-molecular-weight (i.e., through C_{10} or C_{12}), esters, ketones, and most terpene hydrocarbons, the $\Delta I/10°C$ values are small (usually less than 1.0); short-chain alcohols usually exhibit a negative value (-1.0 to -3.0); with larger-molecular-weight steroids the value may be as large as 40 (18). The magnitude of $\Delta I/10°C$ is greater with polar solutes and with polar liquid phases.

E. Isothermal Procedures

For the greatest precision, retention indices should be determined—as specified—isothermally. Many complex mixtures contain compounds that represent a wide range of volatilities, and several isothermal runs may be required. The first determination is usually made at a relatively low isothermal temperature, and as soon as any degree of peak asymmetry is apparent—which is most easily ascertained by noting that the recorder pen takes perhaps twice as long to rise on the front of the peak trace as it does to fall on the back of that peak—the run is aborted and the oven temperature raised to clear the column. The column is then cooled and equilibrated, and the run is repeated with a mixture of sample plus the appropriate n-paraffin hydrocarbons; again, to save time, the run should be aborted at the same point. The retention indices of the lower-boiling compounds can be calculated from these results; the procedure is then repeated at some intermediate temperature to assign retention indices to the intermediate compounds, and at an elevated temperature for the high-boiling components. Maximum usable sensitivity should be employed, so that a minimum amount of sample can be used.

Otherwise, there is a danger that minor components will undergo separation in a liquid phase that is, in fact, a mixture of liquid phase and major components of the sample; this would, of course, seriously affect retention behavior.

F. Programmed Temperature Procedures

Giddings (50) pointed out that as long as the program rate is truly linear, the correlation between the retention times of higher members of an homologous series and the number of carbon atoms is almost linear. Particularly where the retention index is used to confirm assignments derived from other evidence (e.g., the mass spectrum), even an approximate value may be sufficient, especially if the four or five possible assignments suggested by the mass spectrometer differ widely in the value of their retention indices. Several workers have proposed the use of linearly programmed temperature for the determination of retention indices in complex systems that represent a wide range of volatilities (e.g., 38, 51, 52). A cubic spline interpolation procedure for calculating programmed retention indices has been suggested (53). Some authors suggest that results from programmed runs are in closer agreement with those from isothermal runs if unadjusted retention times are used in the calculation; others prefer to use adjusted retention times in the normal manner. As the rate of temperature programming is increased, the discrepancy between the programmed result and the isothermal retention index will be magnified. Hence, program rates should be kept low (one to two degrees per minute). In most cases, the logarithms of the adjusted retention times of the n-paraffin hydrocarbons are a linear function of the chain length (or the retention index) over a range of at least several members of the series under isothermal conditions. With programmed temperatures, this is not the case, and curvature becomes apparent. Provided the program rate is low, the degree of curvature is slight, and can be approximated by a straight line between two consecutive members of the series. Particularly in programmed runs, it is important to use those two consecutive members of the n-paraffin hydrocarbons that bracket the compound in question in assigning the retention index; interpolation between two points is required, but extrapolation beyond the range bracketed by the standards should be avoided.

IV. SELECTIVE DETECTORS

A. General Considerations

Selective detectors offer another route toward qualitative analysis. While the information gleaned from their use is rarely definitive enough that a structure can be assigned on that evidence alone, it is one more piece of information that, in combination with others, can be useful in making assignments. The normal detector may be bypassed or removed, and the selective detector substituted. Alternatively, and provided that each detector can be connected to its own recorder pen through a dedicated electrometer, two dissimilar detectors can be operated either in parallel or in series to provide simultaneous traces, one from the selective detector and the other from the nonselective detector. In practice, the effluent flow stream can either be split so that it feeds two or more dissimilar detectors (e.g., FID and EC) as illustrated in Fig. 4, or the flow stream can first be passed through a nondestructive detector (e.g., PID) and thence to the second (e.g., FID). In this latter case, a split or leak may be required to narrow the band that is delivered to the second detector (6, 7).

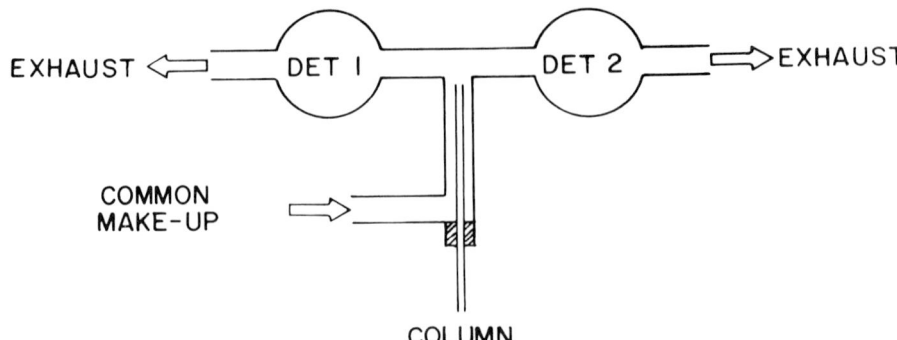

Fig. 4. Schematic of a column outlet split into two detectors that are serviced by a common make-up gas.

Several types of effluent splitters have been suggested; it is, of course, desirable that the sample-contacting surfaces be inert, that the splitter possess minimum volume, and that the gas velocity through the splitter be fairly high. Anderson and Bertsch (54) recently described a variety of multiple-effluent splitters constructed of platinum–iridium tubing and glass, and several designs utilizing glass-lined stainless steel have been used. Some commercial experimentation now in progress seems aimed at producing a variety of inert effluent splitters.

For detectors that can tolerate roughly equivalent amounts of the same make-up gas (e.g., FID, NP), a design similar to that shown in Figure 4 is sometimes satisfactory for the combination splitter–make-up gas adapter. Because the flow streams are identical in composition, both experience the same changes in flow resistance as the temperature is changed, and the unit can usually be programmed without causing changes in the effluent split ratio. Where different types or amounts of make-up gas are required by the two detectors (e.g., FID, EC), separate make-up lines may have to be utilized as shown in Fig. 5. These offer one advantage, in that the effluent split ratio can be varied by changing the pressure of one or the other make-up lines, much as with a Deans switch (55). If the split is too

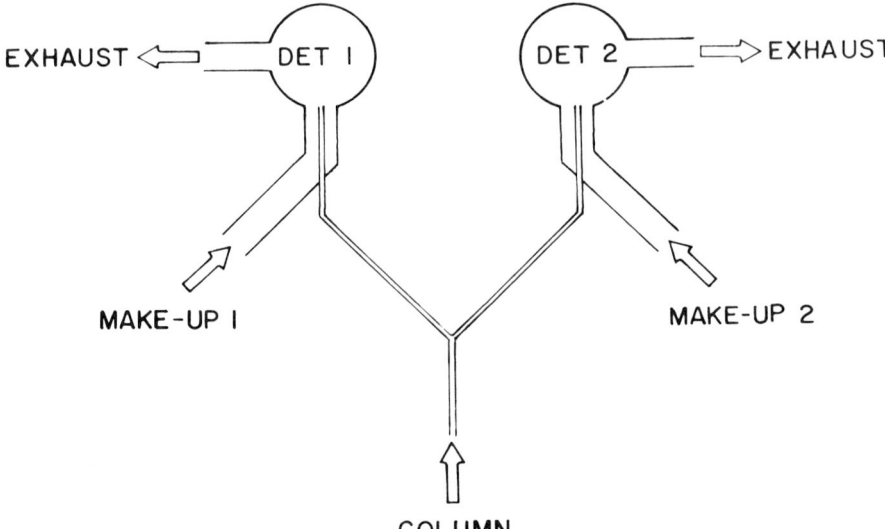

Fig. 5. Schematic of a column outlet split into two detectors that are serviced by dissimilar make-up gases.

much in favor of the FID, a slight increase in the FID make-up, or a decrease in the ECD make-up, will cause the split to shift in favor of the ECD. A disadvantage is that as the temperature is changed, the flow resistances in the two legs change by unequal amounts and the split ratio is affected. This can be a real problem in temperature-programmed runs, as the effluent split ratio can shift to deliver the entire column flow to one detector and exclude the other. In most cases, improved performance will be noted if the make-up lines are insulated so that their flow resistances remain more nearly independent of oven temperature. In special cases, it may be advisable to heat these lines so the make-up gas is hot, and so discourage the condensation of higher-boiling components.

B. Thermal Conductivity Detector

The thermal conductivity detector (TCD) is usually insufficiently sensitive to permit its operation with the low flow rates and small amounts of material that are eluted from analytical (i.e., small-bore) glass capillary columns. Larger-bore (0.75 mm) capillaries can be used to drive some of the small-internal-volume thermistorized TCD units, and the performance of these combinations can be further improved by the use of hydrogen carrier gas, both because its greater thermal capacity yields an increased sensitivity and because it can be utilized at higher linear velocities, and hence higher flow rates. Thermistors, however, are metallic oxides and over a period of time even glass-enveloped thermistors suffer reduction and lose sensitivity in the hydrogen flow stream. Under normal conditions, lifetimes of up to several months have been reported.

Obviously, the separation efficiency of the large-bore columns will be much lower than that which can be achieved with small-bore columns, inasmuch as the theoretical maximum efficiency is inversely proportional to the radius of the column (18). Nevertheless, some interesting work has been accomplished with larger diameter columns (56). The TCD is relatively nonselective and has been used as a standard for comparison with the results of selective detectors. Because it is nondestructive, the flow stream can be passed sequentially through the TCD to a second detector.

C. Flame Ionization Detector

The flame ionization detector (FID) is currently probably the most widely used detector because it is relatively nonselective, has an extended linear dynamic range, and is very stable and simple. Most commercial models of the FID require a make-up gas when operated with the restricted carrier gas flows that are tolerated by capillary columns; they exhibit their highest sensitivity when that make-up gas is nitrogen. The FID is a destructive detector, and for multiple detection it must be operated in parallel with the other detector(s), utilizing an effluent splitter, or be positioned as the final detector in a series operation.

D. Nitrogen–Phosphorus Detector

The nitrogen–phosphorus detector (NPD) is also known as a thermionic detector. Early models were merely modified FIDs, produced by suspending a material such as rhubidium sulfate in the hydrogen flame, which resulted in an enhanced response to nitrogen- or phosphorus-containing compounds, and a decreased response to hydrocarbons; the mechanism of detection is not well understood. Later models utilize electrical heating of the alkali bead. Some models operate with the hydrogen jet unlighted, and depend instead on the generation of a low-temperature plasma. The NPD is highly selective, responding to nitrogen-, phosphorus-, and vicinal carbonyl-containing compounds, and produces little or no response to other substances.

E. Electron Capture Detector

The electron capture detector (ECD) is, again, a highly selective detector. Halogen-containing compounds produce very high responses, and the response increases in the order chlorine < bromine < iodine < fluorine. Molecules possessing multiple-capture sites, as are exhibited by butanedione and carbon tetrachloride, have extremely large response factors. The ECD is particularly useful for monitoring chlorine-containing compounds such as pesticides, bromine-containing compounds such as fire retardants, and the degradation products of these materials. The more modern pulsed systems will function with nitrogen as the carrier (or in our case, as make-up gas), but the sensitivity is approximately doubled when argon containing 5% methane is substituted; a satisfactory compromise for applications of the type with which we are concerned is to use helium as the carrier and argon/methane as make-up. The make-up flow should be large enough to give narrow peaks and fast recovery, but not so large as to cause too great a loss in detector sensitivity.

F. Flame Photometric Detector

The flame photometric detector (FPD) measures the fluorescence emission of hetero atoms—in our case phosphorus and sulfur are of particular interest—of organic molecules subjected to combustion. The light-emitting processes of interest occur above the flame, where the light is collected and amplified via a photomultiplier tube. The photomultiplier operates through a restricted field of vision to eliminate light produced by the combustion of carbon atoms, and an optical filter that is varied depending on which element is to be monitored. The principal sulfur emission is at 394 nm, and that for phosphorus is at 526. Unfortunately, both elements emit at both wavelengths, but the sulfur emission at 526 is limited to 5–10% of that at 394 nm.

G. Photoionization Detector

This nondestructive detector can be utilized either in parallel or in series with other detectors. It requires only carrier (and, in our case, make-up) gas, which can be either nitrogen or helium. The PID has good sensitivity, and is useful in the nanogram to microgram range. It exhibits a wide dynamic range, is sensitive to some inorganic compounds (e.g., I_2, NO_2) and most organics, but the response factors differ on the basis of their ionization potentials. Hence it is a relatively selective detector and has been used, for example, to distinguish which compounds of a hydrocarbon mixture are aromatic, olefinic, or paraffins (57). Ionization potentials of a number of compounds are readily available in standard chemical handbooks.

V. ANCILLARY REACTIONS

A. General Considerations

While gas chromatography is our most powerful tool for separating the components of a volatile mixture, this capability can be exploited more fully if the analyst has some additional information about each of the components separated. A wide range of functional group reactions, based largely on classical techniques of qualitative analysis, has been adapted and used for this purpose. Mixtures have been examined before and after acid (or alkaline) extraction, hydrolysis, treatment with Girard's T reagent or 2,4-dinitrophenylhydrazine, hydrogenation, ozonolysis, and a variety of other reactions. Changes that occurred in the chromatographic pattern could then be correlated with the disappearance of compounds and/or the production of new ones to yield additional information about specific components of the original mixture. Some have viewed the ingenuity of the analyst as the only limitation to these myriad approaches. Final decisions are usually made on the basis of the retention behavior of resolved compounds; consequently, systems capable of higher resolution and more precise retention assignments should make such techniques even more useful. A great deal of the pioneering work that developed ancillary techniques for application to packed-column gas chromatography takes on new meaning when complemented with high-resolution gas chromatography. A number of appropriate reactions have been reviewed by Ettre and McFadden (58); at this writing, a series of coordinated papers on chemical derivatizations in chromatography has been recently published. The subjects include trialkyl ether derivatives (other than TMS)(59), derivatization techniques for pesticide analysis (60), pyrolytic methylation reactions (conversion to and in-injector decomposition of the N-methylammonium salt of an acidic compound)(61), and the use of derivatives in the chromatographic analysis of food additives (62). An excellent general discussion of precolumns in gas chromatography has also appeared (63).

B. External Reactions

This includes those reactions that are performed external to and entirely independent of the chromatographic system. Once the process is completed, the products are then examined by gas chromatography and may be compared to the original.

In many cases, these are merely separation techniques used to remove interfering substances, or to increase the relative concentrations of trace compounds to a point where they can be studied. The individual interested in studying the occurrence of minor constituents of hop oil in beer, for example, is plagued by the relatively massive concentrations of fermentation products that mask these minor components; a preliminary fractionation that produces a sample in which the relative concentration of hop oil constituents is much higher can greatly facilitate his work.

1. Column Chromatography

Solid–liquid adsorption chromatography utilizing a variety of granulated or powdered adsorbents and a wide selection of solvent systems can achieve very efficient preliminary fractionation on a number of complex mixtures (e.g., 64). Because the eluates are usually recovered as dilute solutions, and because the developing solvents are in many cases unsuitable for direct injection into the gas-chromatographic system, it is sometimes desirable to leave the separated materials on the column rather than elute them with the solvent system used for separation. After a certain period of development, the column is washed with a nonpolar solvent such as pentane or isopentane. Selected areas of the column are then removed and washed with a minimum quantity of a polar elutant such as ethyl ether, and the solutions examined by gas chromatography. The eluant should be selected not only on the basis of its extraction ability, but also with attention to its compatibility with the liquid phase to be used in the analysis. Acetone, for example is an undesirable solvent for injections on WCOT columns that are coated with methyl silicone liquid phases, as it encourages their degradation.

An even simpler approach is sometimes fruitful. Terpenoid mixtures can be fractionated quite readily in a microcolumn using a 1 mm × 10 cm piece of thin-walled Teflon tubing that has been packed with silica gel. One to 2 μl of the mixture is placed on one end of the column, which is placed, loaded end down, in a stoppered test tube containing 1 ml of dichloromethane. When the solvent has reached the top, the column is removed and cut into 1 cm sections. The packing from each section is placed in a separate micro-test-tube and washed with a minimum quantity of ethyl ether or dichloromethane, which is injected for analysis. The top section of the column will yield monoterpene hydrocarbons, sesquiterpene hydrocarbons will be centrally located, and the oxygenated terpenes will be found on the bottom tube sections. Various modifications of this technique can be applied to the analysis of a number of samples.

2. High-Pressure Liquid Chromatography

While HPLC is normally viewed as the method of choice for the separation of nonvolatile compounds, it can also be utilized to good advantage as a prefractionation technique. In this case the fraction(s) of interest must be eluted from the column, and extracted from the developing solution with a solvent compatible with the gas chromatographic separation if necessary. The method is beginning to receive wide application from studies concerned with flavor components in alcoholic beverages to polynuclear aromatic hydrocarbons in smoke and air samples. A commercial unit coupling HPLC and GC has been described recently (65, 66). One avenue that has not been adequately investigated is the use of very low boiling solvents—e.g., the Freons—in the preliminary HPLC system. Although this approach would require that the system, including the detector, be operated at either low temperature or elevated pressure, these are not insurmountable difficulties. Because of its low boiling point, the solvent could be evaporated at subzero temperatures, yielding a concentrated fraction for gas-chromatographic analysis.

3. Simple Pre-GC Reactions

Simple procedures utilizing acidic or basic extractions, or treatment of the sample with anionic or cationic exchange resins, may also serve as a preliminary fractionation technique. Organic acids can be isolated from complex mixtures, and examined per se or subjected to esterification. Alternatively, the selective removal of acids or amines, which tend to obliterate regions of the chromatogram by tailing on general-purpose columns, can result in samples that produce much improved chromatograms, which can be subjected to a much higher degree of interpretation. 2,4-Dinitrophenylhyrazine is a useful reagent that has been widely used for the selective removal of carbonyl compounds. In several cases, the derivatives have been isolated and characterized per se by GC (67–69). Innes *et al*. used a number of different chemical absorbers to subtract various functional groups from hydrocarbon gas mixtures, which were subsequently analyzed by GC (70). The *n*-paraffin content of various kerosene fractions was quantitated by Greeley (71), who first passed the material through a molecular sieve, and recovered the sorbed paraffins by cold-trapping the backflushed eluant on the GC column.

A number of materials are normally converted to more volatile derivatives prior to their GC analysis (see below). Among the esterification procedures, one that seems particularly useful involves the derivatization of fatty acids with phenyltrimethylammonium hydroxide (70).

Alkaline hydrolysis has been applied to studies such as fruit flavor analysis, where comparison of the original chromatogram with an extract of the alkaline extract of the hydrolysate established which peaks were esters, and the nature of their alcoholic moieties (72, 73). Following exhaustive extraction, the alkaline hydrolysate can be acidified and reextracted to allow the isolation of the acidic moieties. These can be analyzed on an FFAP column, or converted to methyl esters for analysis on other liquid phases. Establishing (a) the presence in the original essence, and (b) the absence in the hydrolysate of peaks whose retentions correspond to those of esters composed of (c) acid and (d) alcohol moieties whose retentions agree with those of new peaks in the respective hydrolysate fractions can be taken as a reasonable criterion of identification.

Colorimetric reactions have been applied to individual GC fractions to establish the presence of specific functional groups. Methods described by Cronin and Gilbert (74) are reasonably complete, and the use of recently suggested trapping procedures (75) may permit the extension of these techniques to small-bore glass capillary eluates. Cochrane (76) discussed the application of several derivatization reactions to confirm specific organochlorine and organophosphorus herbicides and pesticides. Precolumn techniques were used by Bloch *et al.* (77) to remove olefins from gasoline samples produced from methanol prior to their GC analysis.

4. *Ozonolysis and Hydrogenation*

A variety of methods can be used to establish the existence of carbon–carbon double bonds in chromatographic fractions. One of the more exacting is the comparison of chromatograms before and after hydrogenation (see below). Once the presence of unsaturation is established, the next question is where in the molecule is that unsaturation located. Ozonolysis has been combined with gas chromatography by a number of workers (e.g., 9, 10). The procedure can be accomplished on a microscale quite readily. Approximately 1 mg of sample in 0.5 ml pentane is placed in a 2 ml conical test tube, and ozone (Fig. 6) is bubbled through the mixture for two to three minutes. One drop of 5% acetic

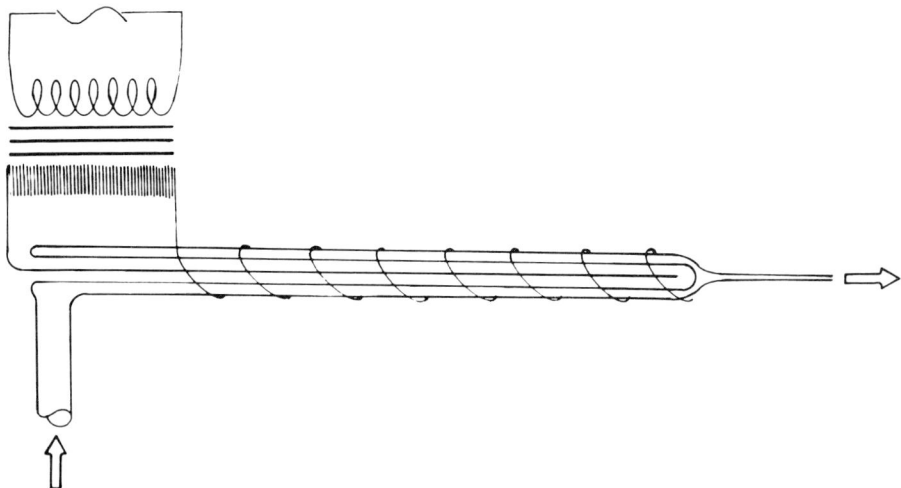

Fig. 6. Schematic of a microozonizer. A field of 20–30 kV can be produced with an automotive ignition coil whose primary is powered with 8–14 V from a variable transformer. When air (or oxygen) is passed through the device, a portion of the oxygen is converted to ozone. Because of the toxicity of ozone, the apparatus should be contained within a fume hood.

acid is added, and the tube shaken vigorously. The pentane layer is removed via pipet or syringe, concentrated under a nitrogen stream, and injected for product analysis. From a knowledge of the chain lengths of the aldehydic fragments, the position of the double bond can usually be deduced:

$$\text{\textcircled{R}}-CH=CH-R' \xrightarrow{O_3} \underset{\underset{\underset{RCHO + R'CHO}{\downarrow}}{O-O-O}}{R-CH\quad CH-R'}$$

Davison and Dutton (78) utilized a modified precolumn technique to accomplish ozonolysis. A short piece of metal tubing with an injection port and carrier gas inlet on one end and a needle on the other was packed with solid support, bent to a U-tube configuration, and fitted to a soldering gun. Ozone was introduced into the unheated U tube to convert unsaturates to ozonides, and flushed out with carrier gas. The products were then injected via the needle while the U tube was heated to 250°C by direct passage of the electric current. A short plug of zinc oxide was inserted in the injection port liner to retain carboxylic acids that were produced in the pyrolytic cleavage. Cronin and Gilbert (74) subjected GC eluates to microozonolysis and hydrogenation, followed by GC analysis.

C. Reaction Gas Chromatography

1. *General Considerations*

This includes those reactions that are performed as an integral part of the injection and chromatographic process. It is usually necessary to construct ancillary apparatus or slightly modify the inlet or carrier gas systems to utilize this type of analysis. With some splitter systems it is simpler to substitute a liner, packed with the appropriate material, to serve as the precolumn on which the reaction occurs.

Frequently a better approach is to prepare a separate "precolumn" (Fig. 7), which is inserted into the normal splitter inlet. In this case, a separate carrier gas source should be utilized to permit the use of a slightly higher pressure, so that the linear velocity through the column remains the same in either mode. Otherwise the comparison of complex chromatograms and the interpretation of results can be rather difficult. The best procedure is usually to inject a mixture of two or three hydrocarbons through the splitter in the normal mode and observe their retention times. The needle of the precolumn injector is then inserted through the regular septum, the normal carrier gas source is turned off, and the

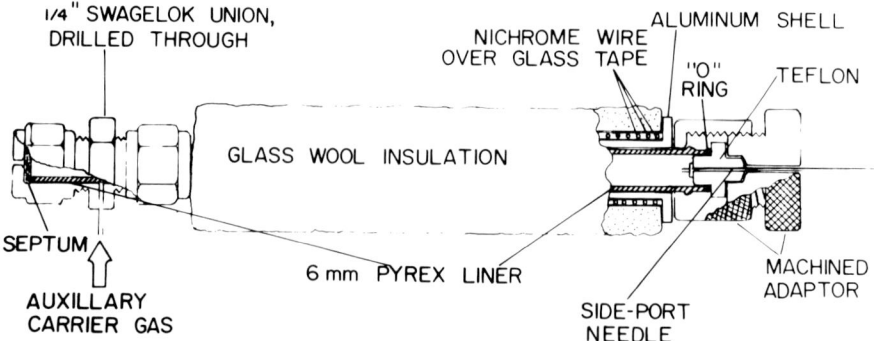

Fig. 7. Schematic of an external injector for reaction gas chromatography. A thermocouple (not shown) should be included to provide accurate temperature readout.

V. ANCILLARY REACTIONS

auxillary carrier gas source supplying the precolumn injector is turned on. Lever-type valves preceding the regulators are ideal for this application. The pressure of the auxiliary carrier gas source is then adjusted so that the test hydrocarbons exhibit the same retention times in this mode as they did in the normal mode.

The reaction chamber, whether it is a packed splitter insert or a separate precolumn injector, must be maintained at a temperature sufficiently high so the injected materials volatize, react, and proceed to the chromatographic column without undue delay. At the same time, the reaction mixture blended with the solid support and housed in the reaction chamber must be stable at that temperature. For these reasons, reactions requiring aqueous solutions of the reactants are rarely satisfactory for reaction gas chromatography.

2. Precolumn Injection

Most inlet splitters are fitted with a removable glass insert that can be packed with a solid support impregnated with one of the catalysts or reactants described below. This can be accomplished more conveniently with, but is not limited to, splitters designed so that the insert can be removed from the septum (outer) end of the splitter. The insert should be contained in an independently heated zone whose temperature can be controlled within fairly close limits. In most instruments, the housing and temperature controls of the standard heated injector designed for on-column injection with ¼ in. packed columns can be used, with its controls and read-out, without change. For carbon skeleton analysis or hydrogenation it is necessary to use hydrogen as the carrier gas. Some instruments utilize a common inlet for the carrier and make-up (or auxiliary) gas: the common gas stream is divided into its carrier and make-up functions within the instrument housing. Some minor plumbing changes may be required in the adaptation of such instruments, but these can be readily accomplished. Hydrogen–air mixtures are explosive, and great care should be exercised to ensure that hydrogen leakage into confined areas where explosive concentrations could accumulate cannot occur.

It is frequently more convenient to construct an external precolumn injector that circumvents many of these difficulties (e.g., 78–83). In either case, it is important to remember that when hydrogen is used as the carrier gas the splitter discharge is flammable, and its rapid dissipation should be encouraged. This can be accomplished by directing the blast of a small centrifugal fan toward the splitter outlet, although some users argue that hydrogen diffuses so rapidly that the chances of accumulating an explosive concentration from limited open-air discharges are almost nonexistent.

3. Carbon Skeleton Analysis

This process catalytically strips from each compound in the mixture injected all functional groups containing oxygen, nitrogen, sulfur, or halogen, yielding the equivalent hydrocarbon and/or its next lower homolog. It utilizes a palladium catalyst and hydrogen carrier gas; reactions included hydrogenation, dehydrogenation, and hydrogenolysis. The catalyst may be prepared by depositing palladium chloride on Chromosorb P (acid-washed) and reducing it under hydrogen flow at an elevated temperature. Sufficient sodium carbonate must be premixed with the support to react with the HCl produced by activation; otherwise cracking of carbon–carbon bonds may occur (79). Alternatively, it is possible to purchase 5% palladium predeposited on calcium carbonate; this is then carefully mixed at a ratio of 1:5 with Chromosorb W or Chromosorb P (both acid-washed) and placed in the reactor for activation. The external injector is normally activated before attachment to the chromatograph.

The activation process usually requires a hydrogen flow of 30–60 cm^3/min. The reactor is adjusted first to 125° for 60 min when, without interrupting the hydrogen flow, the temperature of the reactor is raised to 200° and held for another 30–60 min, after which it is increased to 280–300° for another 30–

60 min period to complete the activation process. The life of the catalyst is variable, and may range from two or three days to two weeks or longer. The loss of activity is related to catalyst temperature, the cleanliness (and dryness) of the hydrogen gas, and the nature of the samples injected. Hydrogen that is quite suitable for normal FID operation, where it is conducted more or less directly to the burner, may be insufficiently pure for carbon skeleton analysis or on-column hydrogenation reactions, in which it passes over the catalytic reactor and then serves as the carrier gas. Attempts to regenerate an exhausted catalyst usually yield unsatisfactory results. In most cases, it is wiser to substitute new catalyst, which is subjected to the normal activation procedure.

The method has been applied to the analysis of halogenated environmental contaminants (84); the analyst should be aware of the fact that the hydrogenolysis of halogen-containing compounds in a sample produces halogen acids, which may cause cracking of carbon–carbon bonds of sample components.

Carbon skeleton analysis has also been used for the characterization of fatty acids, fatty acid esters, and long-chain compounds (85), short-chain hydrocarbons (78), and compounds containing other functional groups (85). It has also been applied to trimethylsilyl derivatives (87). Care must be exercised in interpretation, because of the difficulty in correlating peaks from one complex chromatogram with another; it is advisable to run test mixtures at regular intervals, to ascertain the degree of conversion and the chain length of compound for which the system is effective.

4. *Hydrogenation*

The same apparatus and techniques described above for carbon skeleton analysis, applied under milder conditions (i.e., lower temperatures), can achieve rapid and quantitative hydrogenation of many unsaturated compounds. Several investigators (e.g., 80–86) have studied a variety of solid supports, catalysts, and conditions, but for general use the apparatus previously described, operated at a temperature of 140–175°C, will achieve excellent results. The activation procedure is the same as that described for carbon skeleton analysis; following activation, the temperature of the reactor is dropped to the range specified above. Again, the life of the catalyst is related to the temperature of the reactor, the nature of the samples injected, and the purity (or cleanliness) of the hydrogen carrier gas. Some interesting applications of on-line hydrogenation of fatty acid methyl esters using capillary GC were reported by Kuzmenico *et al.* (87). Complicated hydrocarbon mixtures have been analyzed by reaction–GC hydrogenations (88), and Mounts and Dutton (89) applied microvapor phase hydrogenation to fatty acid esters. Ragaini (90) used a palladium black catalyst to force the isomerization of *n*-butenes in the absence of molecular hydrogen.

5. *Miscellaneous Reactions*

Getty *et al.* used a precolumn for on-column generation of trimethylsilyl derivatives (91). Amine salt precursors have been used to produce nitrosamines by direct injection (92). By treating the column with phosphoric acid, Gaede and Meloan (93) were able to apply GC to the analysis of long-chain amides. It has been reported that the addition of formic acid to the carrier gas allows for the direct GC analysis of barbiturates as their solium salts (94).

6. *Subtractive Reactions*

A variety of chemical and physical reactions have been used to remove specific types of compounds from complex mixtures; comparison of the before- and after-reaction chromatograms yields additional information on those compounds that have been removed. Removal of carbonyl compounds with Girard's T reagent or 2,4-dinitrophenylhydrazine are examples of reactions that have long been used

V. ANCILLARY REACTIONS

to selectively remove specific compounds from complex mixtures. Molecular sieves have been widely used to selectively remove alkane and olefinic hydrocarbons from their branched, cyclic, and aromatic counterparts; many such examples have been cited above. Molecular sieve 5A has been especially popular for this purpose. However, under some conditions a lack of selectivity characterized by loss of some lower-molecular-weight acids, aldehydes, and alcohols has been reported. This lack of specificity has limited the utility of molecular sieves in such reactions, and necessitates the use of test mixtures as a check on the type of compounds removed. Silver, copper, or mercury salts or concentrated sulfuric acid, blended with the solid support in the precolumn, can be used to selectively remove olefins from hydrocarbon mixtures. Chromosorb G, wetted with concentrated sulfuric acid and placed in a reaction chamber maintained at 150–200°C, also removes olefins from hydrocarbon mixtures quite effectively.

Chromosorb-type solid supports blended with boric acid and maintained at 75–200°C have been reported to subtract primary and secondary alcohols (95); tertiary alcohols are dehydrated and converted to the corresponding olefin. Beroza and Sarmiento (86) reported that the retention times of phenols were also affected, that salicylaldehyde was subtracted, and that the subtracted alcohols did gradually bleed off as very broad peaks, indicating that they had been probably converted to the borate esters and were then slowly released over a period of time. The implication of these reports is that such precolumns should be removed and replaced frequently.

Zinc oxide has been used for subtraction of carboxylic acids. Alpha-substituted acids tailed under these conditions, and their retention times were increased; alcohols and phenols are partially subtracted.

There have been many reports that the liquid-phase FFAP (an acronym for "free fatty acid phase" applied to a product made by refluxing PEG Carbowax 20M with terephthalic acid) can function as an aldehyde subtractor, but in most cases it appears to have a variable and unpredictable "demand capacity" for aldehydes. Columns that have been used for some time have apparently satisfied this demand capacity and pass aldehydes quite readily; hence the use of FFAP as an aldehyde subtractor is at best a risky procedure that would require frequent controls. Polyethylene glycol (PEG) 1000 impregnated with 20% hydroxylamine hydrochloride and 6% solium hydroxide and coated on a firebrick support has been reported to subtract aldehydes and ketones. The use of o-dianisidine coated on a solid support placed in the last portion of the reaction chamber has been recommended for subtraction of aldehydes, including alpha-substituted aldehydes; ketones are reportedly unaffected, except for cyclohexanone, which is 50% subtracted. C_{12} and higher epoxides were partially or completely subtracted, while smaller epoxides, ethers, esters, alcohols, phenols, olefins, and hydrocarbons were not affected.

Chromosorb P impregnated with 20% benzidine has also been used at 100–175°C to remove aldehydes, most ketones, and epoxides (96). However, unreactive compounds have been reported to experience some degree of delay in this type of reactor; this would make comparisons of chromatograms rather risky. Polyalkyl amines have been reported to serve as specific abstractors for carbonyl compounds (97).

It would be difficult to overemphasize just how tempting it can be to overinterpret the results of precolumn and on-column reactions. The activity of the reactor is influenced by many variables, including the age, activation procedure, contaminants in the hydrogen gas, size, and the composition and number of samples injected. The degree to which a given compound undergoes reaction can be influenced not only by the activity and temperature of the catalyst, but also by its residence time in the reactor (which is lower for more volatile or less-retained compounds), its concentration in the sample injected, and the nature and amount of other sample components. Some compounds undergo complete reaction, some go beyond this and undergo degradation, and others barely react. The use of controls that duplicate as closely as possible the qualitative and quantitative composition of the sample can help to raise the level of confidence in results from these procedures.

VI. GAS CHROMATOGRAPHY–MASS SPECTROMETRY

A. General Considerations

Gas chromatography is the most effective technique currently available for separating the components of volatile mixtures. Recent developments in improved columns, inlets, and related paraphernalia combine to make high-resolution separation of individual compounds more readily attainable. Mass spectrometry is a powerful tool for structural elucidation of volatiles, and the reliability of the identification is highest when the spectroscopist deals with pure, individual compounds. Hence the combination of the two to provide GC/MS is very logical (98).

At the same time, this can be a stormy marriage. Not only are there a variety of opinions as to the best method of incorporating the two techniques into a single instrument (see below), but a serious and fundamental problem soon becomes apparent: good gas chromatography is not always compatible with good mass spectrometry.

The chromatographer strives for clean, needle-sharp peaks; broad peaks—i.e., those that persevere for an extended time interval—usually testify to poor chromatography, particularly if they occur in the early portion of the chromatogram. The mass spectrometer requires a finite period of time to scan the required mass range and, from the standpoint of consistent spectral results, it is desirable that the concentration of the component remains constant during the scanning period. With any acceptable degree of chromatography such constancy is rarely possible, and component concentration usually changes during the course of the scan.

If the mass spectrometer scans from the lower limit to the higher limit of the mass range and the scanning period happens to coincide with the first appearance of a peak and conclude at the peak maximum, the spectrum is distorted; high-mass fragments are maximized, and low-mass fragments minimized. A scanning period that begins at the peak maximum and concludes with the disappearance of the peak would accentuate the low-mass fragments, and deemphasize the high-mass fragments; the two spectra would be grossly different, although they reflect the same compounds.

For these reasons, must spectroscopists prefer to secure several scans on each component and average the individual scans, i.e., they want reasonably broad peaks, which is just the opposite of what the chromatographer strives to attain. Because of these considerations, the individual interested in lower-boiling compounds finds that it is not always possible to use the full resolving potential of a high-resolution capillary in a GC/MS unit; the peaks of these low-k compounds may be too narrow to permit good mass spectrometry. The use of thick film (low phase ratio) columns to produce higher partition ratios and broader peaks is generally a better solution to this problem that overloaded injections or operation at lower-than-optimum carrier gas flows. Another approach is to substitute nitrogen carrier gas, which can produce equivalent separations at lower flow velocities, giving broader peaks and longer analysis times. Midrange and higher boiling compounds that emerge later in the chromatogram invariably broaden sufficiently that the problem becomes more one of resolution and clean separation.

B. Interfacing

The subject of interfacing has recently been thoroughly reviewed (99). The classical interfaces—ballistic, diffusion, and membrane—were developed to permit coupling of the reduced pressure inlet section of the mass spectrometer and the outlet of a packed gas chromatographic column. The suppliers of most GC/MS systems have worked out satisfactory means of interfacing glass capillary columns to their units; older units not designed for this purpose, and some special-purpose applications may require innovative modification. Several approaches have been used; in some cases the column terminates directly in the ion source, providing what has been termed a "tight source." In such cases, the column serves as a restrictor. Because the outlet end of the column is under vacuum, a lower head pressure is required for the same average linear carrier gas velocity and some loss in efficiency can be expected. This is because the reduced pressure results in displacing the van Deemter curve upward (a higher value for h_{min}, hence a smaller n), but as the curve is much flatter, the effect of velocity is less k-specific. Hence short columns (e.g., 10 m) are rarely coupled directly to the ion source, because a larger fraction of the column length must operate at reduced pressure, and a larger proportion of its theoretical plates are lost under these conditions. The mass spectrometer must, of course, possess sufficient pumping capacity to maintain a suitable vacuum in spite of the direct introduction of carrier gas. Glass, glass-lined stainless steel (GLT) (both via standard Swagelok or similar connectors), and platinum–iridium tubing (100,101) are all used for glass capillary column–mass spectrometer connections. For the first two, Vespel or graphite ferrules are usually substituted for the glass seals. It is important that these are not exposed to the flow stream, because the elastomeric materials do contribute to tailing problems. Platinum–iridium is probably the most reliable connection, provided the seal is accomplished on clean surfaces. As a microtorch, a 3–5 cm length of 0.03 in. i.d., 1/16 in. o.d. stainless steel tubing, attached to a flexible (nylon) hydrogen line via a reducing union, is quite satisfactory. A pressure of 5–10 psi hydrogen produces a suitable flame. Much of the platinum tubing used in making connections is 0.3 mm o.d. With capillary columns that have an i.d. of 0.32 mm, this poses no problem, but those with an i.d. of 0.25 mm must be slightly enlarged (see below).

The platinum should be conditioned by gently heating to a dull red down its entire length. Some authorities prefer to have an oxygen flow through the platinum during this treatment. After cooling, the platinum or platinum–iridium tubing can be cut using X-acto knife or scalpel. The cut should not go clear through but should score the column deeply, perhaps to the half-way mark; the tubing is then bent sharply back away from the cut and broken open. Both pieces of the tubing should be examined under magnification to make sure that they are open and have not been closed by the cutting operation. A small file can be used to dress the end if necessary, and a common sewing needle serves as a handy reamer to reshape out-of-round ends. It is advisable to have a high-velocity gas flow through the tube during any filing or reaming, to minimize the possibility of depositing dust or debris in the tube.

For connecting to the larger-bore (0.32 mm i.d.) columns, the procedure is relatively straightforward. With an inert gas flowing through the column, a 1–2 cm length at the end of the glass capillary is heated to remove all traces of liquid phase, deactivation chemicals, etc. The end of the platinum–iridium tubing should also have been heated to remove drawing lubricants or other surface contaminants. After the two have cooled, the tubing is inserted into the glass capillary column (5–10 mm), and the glass is heated carefully with the microtorch. It should appear to wet the platinum–iridium completely and uniformly over the entire contact surface. The formation of bubbles (which will cause leaks) is indicative of contamination such as residual liquid phase or drawing lubricant.

Small-diameter columns usually require enlarging before insertion of the platinum–iridium tubing. This can be done by heating the tip of a needle or fine sharp tungsten rod in a microflame while it is inserted in the end of the capillary column and rotated to form a funnel; care must be exercised to ensure that the column does not overheat and distort. Alternatively, the end of the column can be blown under gas pressure. The platinum is then inserted into this flared end and a seal is made in the usual fashion by heating down the length of the platinum until contact is made. Again, the seal should be inspected under magnification.

Where a length of platinum–iridium, glass-lined stainless steel, or glass is used between the end of the column and the inlet to the mass spectrometer, it is desirable that this line contain minimum volume; care should be taken to ensure that the entire line is heated and no condensation occurs. Schmid *et al.* (102) described a heated all-glass interface that appears to possess a number of advantages. Fused silica columns are sufficiently flexible that they can be directly coupled to the mass spectrometer per se.

References

1. Rijks, J. A. "Characterization of Hydrocarbons by Gas Chromatography. Means of Improving Accuracy," Ph.D. thesis, Technical University of Eindhoven, The Netherlands, 1973.
2. Rijks, J. A., van den Berg, J. H. M., and Diephendaal, J. P. *J. Chromatogr.* **91,** 63, 1974.
3. Schomburg, G., Husmann, H., and Weeke, F. *J. Chromatogr.* **99,** 63, 1974; **112,** 205, 1975.
4. Schomburg, G., and Husmann, H. *Chromatographia* **8,** 517, 1975.
5. Kaiser, R. E., *J. Chromatogr. Sci.* **12,** 36, 1974.
6. Jennings, W. G., Wyllie, S. G., and Alves, S. *Chromatographia* **10,** 426, 1977.
7. Alves, S. and Jennings, W. G. *Food Chem.* **4,** 149, 1979.
8. Bertsch, W., *HRC&CC* **1,** 187, 289, 1978.
9. Beroza, M., and Inscoe, M. N., *In* "Ancillary Techniques of Gas Chromatography" (L. S. Ettre and W. H. McFadden, eds.), p. 89. Wiley, New York, 1969.
10. Beroza, M., and Coad, R. A., *J. Gas Chromatogr.* **4,** 199, 1966.
11. Klesment, I. *J. Chromatogr.* **69,** 37, 1972.
12. Fenimore, D. C., Whitford, J. H., Davis, C. M., and Zlatkis, A. *J. Chromatogr.* **140,** 9, 1977.
13. Davis, C. M., Meyer, C. J., and Fenimore, D. C. *Clinica Chim. Acta* **78,** 71, 1977.
14. Pfaffenberger, C. *In* "Applications of Glass Capillary Gas Chromatography" (W. Jennings, ed.) (in press).
15. Farwell, S. O., Gluck, S., Blamesberger, W. L., Schutte, T. M., and Adam, D. F. *Anal. Chem.* **51,** 609, 1979.
16. Sheike, J. D., Comins, N. R., and Pretorius, V. *Chromatographia* **8,** 354, 1975; *J. Chromatogr.* **112,** 97, 1975.
17. Onuska, F. I., and Comba, M. E., *J. Chromatogr.* **126,** 133, 1976.
18. Jennings, W. "Gas Chromatography with Glass Capillary Columns," 2nd Ed. Academic Press, New York, 1980.
19. Grob, K. Jr., and Grob, K. *Chromatographia* **10,** 250, 1977.
20. Schomburg, G., Behlau, H., Dielmann, R. Weeke, F., and Husmann, H *J. Chromatogr.* **142,** 87, 1977.
21. Jennings, W. *J. Chromatogr. Sci.* **13,** 185, 1975.
22. Miller, R. J., and Jennings, W. G. *H&C&CC* **2,** 72, 1979.
23. Grob, K., and Grob, G. *J. Chromatogr. Sci.* **7,** 584, 1969.
24. Grob, K., and Grob, G. *J. Chromatogr. Sci.* **7,** 587, 1969.
25. Grob, K., and Grob, G. *Chromatographia,* **5,** 3, 1972.
26. Grob, K. *J. Chromatogr.* **90,** 303, 1974.
27. Grob, K., and Grob, K. Jr. *J. Chromatogr.* **94,** 53, 1974.
28. Grob, K., and Grob, K. Jr. *HRC&CC* **1,** 57, 1978.
29. Jennings, W. G., Freeman, R. R., and Rooney, T. A. *HRC&CC* **1,** 275, 1978.
30. Schulte, E., and Acker, L. *Z. Anal Chem.* **268,** 260, 1974.
31. Schulte, E. *Chromatographia* **7,** 138, 1974.
32. Grob, K., and Grob, K. Jr. *J. Chromatogr.* **151,** 311, 1978.

REFERENCES

33. Grob, K. *HRC&CC* **1,** 263, 1978.
34. Göbler, A. *J. Chromatogr. Sci.* **10,** 128, 1972.
35. Miwa, T. K., Micolajczak, K. L., Earle, F. R., and Wolff, I. A. *Anal. Chem.* **32,** 1739, 1960.
36. Woodfard, E. P., and van Gent, C. M. *J. Lipid Res.* **1,** 88, 1960.
37. Ackman, R. G. *J. Chromotogr. Sci.* **10,** 535, 1972.
38. van den Dool, H. and Fratz, P. D. *J. Chromatogr.* **11,** 463, 1963.
39. Kováts, E. *Adv. Chromatogr.* **1,** 229, 1965.
40. Said, A. S., and Hussein, F. H. *HRC&CC* **1,** 257, 1978.
41. Ettre, L. S. *Chromatographia* **6,** 489, 1973; **7,** 39, 1974.
42. Peterson, M. L., and Hirsch, J. *J. Lipid Res.* **1,** 132, 1959.
43. Kaiser, R. *Chromatographia* **2,** 215, 1969.
44. Ebel, S., and Kaiser, R. E. *Chromatographia* **7,** 696, 1974.
45. Guardino, X., Albaiges, J. Firpo, G., Rodriques-Viñals, R., and GAssiot, M. *J. Chromatogr.* **188,** 13, 1976.
46. Garcia-Dominguez, J. A., Garcia-Munoz, J., Sanchez, E. F., and Molera, M. J. *J. Chromatogr. Sci.* **15,** 520, 1977.
47. Goedert, M., and Guichon, G. *Anal. Chem.* **45,** 1180, 1188, 1973.
48. Grob, K., and Grob, G. *Chromatographia* **4,** 422, 1971.
49. Ettre, L. S., *Anal. Chem.* **36**(8), 31A, 1964.
50. Giddings, J. C., *J. Chromatogr.* **4,** 11, 1960.
51. Watts, R. B., and Kekwick, R. G. O. *J. Chromatogr.* **88,** 165, 1974.
52. Majlát, P., Erdos, Z., and Takáos, J. *J. Chromatogr.* **91,** 89, 1974.
53. Halang, W. A., Langlais, R., and Kugler, E. *Anal. Chem.* **50,** 1829, 1978.
54. Anderson, E. L., and Bertsch, W. *HRC&CC* **1,** 13, 1978.
55. Deans, D. R. *Chromatographia* **1,** 18, 1968.
56. Walradt, J. Paper No. 130, *Pittsburgh Conf. Anal. Chem. Applied Spectrosc.*, Cleveland, Ohio, 1978.
57. Driscoll, J. N., Jaramillo, L. F., and Becker, J. H. Paper No. 406, *Pittsburgh Conf. Anal. Chem. Appl. Spectrosc.*, Cleveland, Ohio, March 5–9, 1979.
58. Ettre, L. S., and McFadden, W. J. "Ancillary Techniques of Gas Chromatography," Wiley, New York, 1969.
59. Poole, C. F., and Zlatkis, A. *J. Chromatogr. Sci.* **17,** 115, 1979.
60. Cochrane, W. P. *J. Chromatogr. Sci.* **17,** 124, 1979.
61. Kossa, W. C., and MacGee, J. *J. Chromatogr. Sci.* **17,** 177, 1979.
62. Conacher, H. B. S., and Page, B. D. *J. Chromatogr. Sci.* **17,** 188, 1979.
63. Kaiser, R. E. *HRC&CC* **2,** 95, 1979.
64. Scheffer, J. J. C., Koedam, A., Schüsler, M. Th. I. W., and Svendensen, A. B. *Chromatographia* **10,** 669, 1977.
65. Cram, S. P., Brown, A. C., Freitas, E., Majors, R. E., and Johnson, E. L. Paper No. 115, *Pittsburgh Conf. Anal. Chem. Appl. Spectrosc.*, Cleveland, Ohio, 1979.
66. Majors, R. E., Johnson, E. L., Cram, S. P., Brown, A. P. III, and Freitas, E. Paper No. 116, *Pittsburgh Conf. Anal. Chem. Appl. Spectrosc.*, Cleveland, Ohio, 1979.
67. Hoshika, Y., and Takata, Y. *J. Chromatogr.* **120,** 379, 1976.
68. Flückiger, R. *Chromatographia* **8,** 435, 1975.
69. Pías, J. B., and Gascó, L. *Chromatogrphia* **8,** 270, 1975.
70. Innes, W. B., Bambrick, W. E., and Andreatch, A. *J. Anal. chem.* **35,** 1198, 1963.
71. Greeley, R. H. *J. Chromatogr.* **8,** 229, 1974.
72. Jennings, W. G. *J. Food Sci.* **26,** 1, 1961.
73. Jennings, W. G., and Creveling, R. K. *Food Sci.* **28,** 91, 1963.
74. Cronin, D. A., and Gilbert, J. *J. Chromatogr.* **71,** 251, 1972; **87,** 387, 1973.
75. Jennings, W. *HRC&CC* **2,** 221, 1979.
76. Cochrane, W. P. *J. Chromatogr. Sci.* **13,** 246, 1975.
77. Bloch, M. G., Callen, R. B., and Stockinger, J. H. *J. Chromatogr. Sci.* **15,** 504, 1977.
78. Davison, V. L., and Dutton, H. J. *Anal. Chem.* **38,** 1302, 1966.
79. Pacáková, V., and Kozlík, *Chromatographia* **11,** 266, 1978.
80. Beroza, M. *Anal. Chem.* **34,** 1801, 1962.
81. Beroza, M. *Nature* **196,** 768, 1962.
82. Beroza, M., and Inscoe, M. N. *In* "Ancillary Techniques of Gas Chromatography" (L. S. Ettre and W. H. McFadden, eds.), pp. 89–144. Wiley, New York, 1969.
83. Harris, W. E. *J. Chromatogr. Sci.* **13,** 514, 1975.
84. Zimmerli, B. *J. Chromatogr.* **88,** 65, 1974.

85. Beroza, M., and Sarmiento, R. *Anal. Chem.* **37,** 1040, 1965.
86. Beroza, M., and Sarmiento, R. *Anal. Chem.* **36,** 1744, 1964.
87. Kuzmenko, T. E., Samusenko, A. L., Uralets, V. P., and Golovnya, R. V.*HRC&CC* **2,** 43, 1979.
88. Iwanow, A., and Eisen, O. *J. Chromatogr.* **69,** 53, 1972.
89. Mounts, T. L., and Dutton, H. *J. Anal. Chem.* **37,** 641, 1965.
90. Ragaini, V. *J. Catalysis* **34,** 1, 1974.
91. Getty, R. H., Stone, J., and Hanson, R. H. *Anal. Chem.* **49,** 1086, 1977.
92. Freed, D. J., and Mujsce, A. M. *Anal. Chem.* **49,** 1544, 1977.
93. Gaede, D., and Meloan, C. E. *Anal. Lett.* **6**(1) 71, 1973.
94. Greenwood, N. D., Guppy, I. W., and Simmons, H. P. *J. Chromatogr. Sci.* **13,** 349, 1975.
95. Ikeda, R. M., Simmons, D. E., and Grossman, J. D. *Anal. Chem.* **36,** 2189, 1964.
96. McKeag, R. G., and Hougen, F. W. *Anal. Chem.* **49,** 1978, 1977.
97. Appleyard, J., and Haken, J. K. *J. Chromatogr.* **99,** 319, 1974.
98. McFadden, W. H. "Techniques of Combined Gas Chromatography–Mass Spectrometry," Wiley, New York, London, 1969.
99. McFadden, W. H. *J. Chromatogr. Sci.* **17,** 2, 1979.
100. Neuner-Jehler, N., Etzweiler, F., and Sarske, G. *Chromatographia* **6,** 211, 1973.
101. Neurer-Jehler, N., and Etzweiler, F. *Chromatographia* **6,** 503, 1973.
102. Schmid, P. P., Müller, M. D., and Simon, W. *HRC&CC* **2,** 225, 1979.

NOTES ON THE APPENDICES

A. Nomenclature

The great bulk of flavor and fragrance volatiles are natural products, and man has been involved in efforts to characterize these substances from the times of our earliest historical records. Many were isolated, described, and given trivial names long before their chemical structures had been elucidated; these names have persevered. For example, the physical constants of 3,7-dimethyl-2,6-octadiene-1-ol are listed in the "Handbook of Chemistry and Physics" under the trivial name "geraniol." In the interest of consistency, common industrial usage has been followed and trivial names—e.g., vanillin, geranyl acetate, methyl ethyl ketone, amyl valerate—have been used throughout these appendices. Other commonly used synonyms appear in the index.

B. Retention Data

In most cases involving the analysis of complex mixtures, isothermal retention indices become impractical because of the range of boiling points (or retentions) embraced by the components. Most practitioners instead resort to temperature programming. While the programmed temperature retention index may vary from the true isothermal value, the variation is usually less than 1%, for the type of compounds considered here, provided that program rates do not exceed one to two degrees per minute. Variations between retention indices determined under isothermal and temperature programmed conditions are usually smaller on nonpolar liquid phases. In addition, McReynolds constants are usually less affected by the nature of the column—glass capillary or packed—when those columns contain a nonpolar liquid phase. Hence retention indices obtained on a methyl silicone liquid phase will usually be in closer agreement with literature values. Data obtained on Carbowax 20M glass capillaries may deviate slightly from literature values, but should be in reasonable agreement with the values in these tables.

The retention indices reported in the following table were determined on smooth-bore wall-coated open-tubular glass capillary columns. One, 0.28 mm × 50 m, was coated with methyl silicone OV-101, admixed with 1% Carbowax 20M as an antitailing additive, and programmed from 80 to 200° at 2°/min. McReynolds constants of these columns are detailed below. The second column, 0.2 mm × 80 m, was coated with polyethyleneglycol Carbowax 20M, and was programmed from 70 to 170° at 2°/min.

Compound	OV-101	Carbowax 20M
Benzene	20	294
Butanol	58	499
2-Pentanone	46	338
Nitropropane	64	530
Pyridine	41	470

C. Mass Spectra

Mass spectra were determined on a Hitachi RMU-6M mass spectrometer equipped with the Hitachi Datalizer 002B.

PART B   APPENDICES

Appendix I

COMPOUNDS AND THEIR RETENTION INDICES

COMPOUNDS	OV 101	CBWX 20M
Acetaldehyde	363	690
Acetaldehyde citronellyl methyl acetal	1374	1596
Acetaldehyde di-(cis-3-hexenyl) acetal	1465	1700
Acetaldehyde ethyl cis-3-hexenyl acetal	1095	1297
Acetaldehyde linalyl ethyl acetal 1	1361	1558
Acetaldehyde linalyl ethyl acetal 2	1362	1564
Acetaldehyde phenylethyl n-propyl acetal	1424	1836
Acetaldehyde styleneglycol acetal 1	1226	1780
Acetaldehyde styleneglycol acetal 2	1333	1793
Acetone	530	810
Acetone 1-phenyl-1,2-ethanediol ketal	1260	1766
Acetonyl acetone	894	1500
Acetophenone	1048	1627
4-Acetyl-6-tert-butyl-1,1-dimethylindan	1706	2145
Acetylcedrene	1768	2213
Acetylcymene	1381	1928
Acetyleugenol	1541	2277
2-Acetylfuran	892	1491
5-Acetyl-2-methyl-8-isopropyl-(2,2,2) bicyclo-2-octene 1-epitone 1	1452	1858
5-Acetyl-2-methyl-8-isopropyl-(2,2,2) bicyclo-2-octene 2-epitone 2	1467	1883
2-Acetylpyridine	1014	1600
2-Acetylpyrrole	1050	1935
2-Acetylthiazone	995	1639
2-acethylthiophene	1069	1760
Alloaromadendrene	1478	1662
Allyl acetate	675	1010
Allyl anthranilate 1	1328	2196
Allyl anthranilate 2	1493	2381
Allyl benzoate	1239	1800
Allyl n-butyrate	867	1161
Allyl isobutyrate	820	1090
Allyl cinnamate 1	1363	2054
Allyl cinnamate 2	1537	2258
Allyl-3-cyclohexyl propionate	1405	1800
Allyl n-decanoate	1462	1755
Allyl 2-ethylbutyrate	995	1254
Allyl formate	586	957
Allyl 2-furoate	1113	1748
Allyl n-heptanoate	1164	1454
Allyl ethyl ether	586	767
Allyl n-hexanoate	1065	1356
Allylionone 1	1605	1993
Allylionone 2	1679	2014
Allylionone 3	1689	2146

COMPOUNDS	OV 101	CBWX 20M
Allylionone 4	1734	2199
Allyl isothiocyanate	860	1352
Allyl levulinate	1114	1743
Allyl n-nonanoate	1365	1655
Allyl n-octanoate	1265	1554
Allyl phenoxyacetate	1429	2176
Allyl phenylacetate	1422	2175
Allyl n-propionate	777	1090
Allyl salicylate	1339	1946
Allyl tiglate	1002	1370
Allyl n-valerate	965	1256
Allyl isovalerate	920	1190
o-Aminoacetophenone	1288	2181
m-Aminoacetophenone	1409	2181
p-Aminoacetophenone	1506	2181
n-Amyl acetate	895	1161
Isoamyl acetate	860	1110
2-n-Amyl-3-acetonyl-1-cyclopentanone	1600	2259
n-Amyl alcohol	756	1213
Isoamyl alcohol	719	1184
sec-Amyl alcohol	685	1091
tert-Amyl alcohol	631	987
n-Amyl anisate	1732	2405
Isoamyl anisate	1686	2333
n-Amyl anthranilate	1700	2510
Isoamyl anthranilate	1656	2447
n-Amyl benzoate	1454	1940
Isoamyl benzoate	1421	1894
Isoamyl benzyl ether	1297	1668
Amyl butyl carbinol	1175	1550
n-Amyl n-butyrate	1078	1305
n-Amyl isobutyrate	1035	1237
Isoamyl n-butyrate	1042	1259
Isoamyl isobutyrate	997	1187
Isoamyl cinnamate	1719	2355
alpha-n-Amylcinnamic aldehyde	1631	2211
alpha-n-Amylcinnamyl acetate	1757	2318
n-Amylcyclopentenone 1	1247	1689
n-Amylcyclopentenone 2	1113	1748
p-tert-Amylcyclohexanone	1113	1800
Isoamyl n-decanoate	1633	1848
Isoamyl n-dodecanoate	1829	2048
Diisoamyl ether	1000	1064
n-Amyl formate	810	1107
Isoamyl formate	775	1058
n-Amyl furfurylacrylate	1544	2153
Isoamyl 2-furoate	1287	1840

COMPOUNDS	OV 101	CBWX 20M
n-Amyl furylpropionate	1453	1947
Isoamyl furylpropionate	1415	1894
n-Amyl n-heptanoate	1375	1600
Isoamyl n-heptanoate	1334	1548
n-Amyl n-hexanoate	1279	1500
Isoamyl n-hexanoate	1238	1451
n-Amyl levulinate	1325	1860
Isoamyl levulinate	1284	1807
n-Amyl 2-methylbutyrate	1126	1324
Isoamyl 2-methylbutyrate	1087	1273
Isoamyl n-nonanoate	1533	1748
n-Amyl n-octanoate	1471	1700
Isoamyl n-octanoate	1433	1648
n-Amyl phenylacetate	1506	2047
Isoamyl phenylacetate	1468	1991
Isoamyl phenylethyl ether	1384	1741
n-Amyl pivalate	1063	1228
n-Amyl n-propionate	955	1180
Isoamyl n-propionate	954	1180
6-Amyl-alpha-pyrone	1434	2166
Isoamyl n-valerate	1138	1354
Isoamyl pyruvate	1712	1910
n-Amyl salicylate	1557	2077
Isoamyl salicylate	1528	2021
n-Amyl tiglate	1216	1519
Isoamyl tiglate	1178	1469
Isoamyl n-undecanoate	1733	1948
n-Amyl 10-undecenoate	1760	2053
Isoamyl 10-undecenoate	1717	2000
n-Amyl n-valerate	1185	1401
n-Amyl isovalerate	1135	1337
Isoamyl n-valerate	1138	1354
Isoamyl isovalerate	1092	1287
Amyl vinyl carbinol	968	1426
Amyl vinyl carbinyl acetate	1094	1365
Anethole	1270	1809
Angelica lactone	1000	1430
Anis alcohol	1267	2210
Anis aldehyde	1234	1982
Anis aldehyde propyleneglycol acetal 1	1508	2231
Anis aldehyde propyleneglycol acetal 2	1512	2234
Anisole	900	1327
Anisyl acetate	1390	2199
Anisyl n-butyrate	1569	2274
Anisyl isobutyrate	1520	
Anisyl formate	1307	1710
Anisyl n-heptanoate	1862	

COMPOUNDS	OV 101	CBWX 20M
Anisyl n-hexanoate	1763	
Anisyl n-propionate	1482	2205
Anisyl n-valerate	1665	
Benzal acetone	1337	2065
Benzaldehyde	947	1502
Benzonitrile	965	1583
Benzophenone	1604	2410
Benzyl acetate	1144	1697
Benzyl acetone	1218	1849
Benzyl alcohol	1033	1822
Benzyl benzoate	1741	
Benzyl n-butyrate	1322	1856
Benzyl isobutyrate	1277	1771
Benzyl n-decanoate	1923	2460
Benzyl ethyl carbionl	1222	1882
Benzyl ethyl ether	1046	1439
Benzyl formate	1058	1675
Benzyl n-heptanoate	1620	2158
Benzyl n-hexanoate	1521	2057
Benzyl n-nonanoate	1823	2362
Benzyl n-octanoate	1720	2260
Benzyl n-propionate	1234	1679
Benzyl n-propyl carbinol	1310	1955
Benzyl isopropyl carbinol	1292	1912
Benzyl tiglate	1474	2075
Benzyl n-valerate	1421	1956
Benzyl isovalerate	1374	1880
Borneol	1164	1698
Isoborneol	1157	1660
Isobornyl acetate	1279	1584
Bornyl benzoate	1749	
Bornyl butyrate	1473	1760
Bornyl formate	1239	1610
Isobornyl formate	1228	1596
Isobornyl n-propionate	1376	1676
Bornyl isovalerate	1512	1774
Bromostyrol	1197	1778
beta-Bourbonene	1406	1546
2,3-Butanedione	575	963
1,3-Butanediol	941	1692
1,4-Butanediol	1046	1861
2,3-Butanedione diacetyl	606	963
n-Butanol	655	1113
sec-Butanol	591	975
tert-Butanol	512	830
2-Butanone	579	908
2-Butanoylfuran	1078	1644

COMPOUNDS	OV 101	CBWX 20M
2-Butanoyl-5-methylfuran	1192	1748
2-Butanoylthiophene	1252	1894
2-Butene-1-ol	650	1193
3-Buten-1-ol	638	1137
3-Buten-2-ol	811	1022
2-Buten-1,4-diol	1104	1983
n-Butyl acetate	793	1059
Isobutyl acetate	758	1000
tert-Butyl acetate	676	893
Butyl acetoacetate	1104	1798
sec-Butyl acetate	746	982
Butyl acrylate	892	1189
Isobutyl acrylate	781	1107
Isobutyl aldehyde	500	800
Isobutyl aldehyde 1-phenyl-1,2-ethanediol acetal	1384	1900
Isobutyl aldehyde propyleneglycol acetal	831	1060
Isobutyl aldehyde propyleneglycol acetal	840	1074
n-Butyl angelate	1116	1417
Isobutyl angelate	1033	1289
n-Butyl anisate	1632	2305
Isobutyl anisate	1586	2233
n-Butyl anthranilate	1600	2419
Isobutyl anthranilate	1556	2347
tert-Butyl benzene	988	1247
sec-Butyl benzene	1011	
n-Butyl benzoate	1354	1841
Isobutyl benzoate	1310	1771
Isobutyl benzyl carbinol	1366	1983
n-Butyl benzyl ether	1238	1613
n-Butyl n-butyrate	979	1207
n-Butyl isobutyrate	939	1139
Isobutyl n-butyrate	941	1152
Isobutyl isobutyrate	900	1084
n-Butyl n-butyryllactate	1331	1733
Isobutyl cinnamate	1605	2228
alpha-n-Butyl cinnamic aldehyde	1535	2160
o-tert-Butylcyclohexyl acetate 1	1286	1565
o-tert-Butylcyclohexyl acetate 2	1304	1596
trans-p-tert-Butylcyclohexyl acetate	1322	1628
cis-p-tert-Butylcyclohexyl acetate	1360	1675
p-tert-Butylcyclohexanone	1208	1645
sec-Butylcyclohexanone 1	1194	1564
sec-Butylcyclohexanone 2	1196	1566
n-Butyl n-decanoate	1575	1798
2-Isobutyl-4,5-dimethyloxazole	1044	1330
2-n-Butyl-4,5-dimethylthiazole	1251	1600

COMPOUNDS	OV 101	CBWX 20M
2-Isobutyl-4,5-dimethylthiazole	1193	1517
Isobutyl disulfide	1205	1435
Butyl disulfide	1295	1580
n-Butyl n-dodecanoate	1772	2000
Di-n-butyl ether	876	965
n-Butyl ethyl ether	684	788
1,3-Butylene glycol acetal	786	1072
1,3-Butylene glycol butyral	967	1232
1,3-Butylene glycol formal	777	1128
1,4-Butylene glycol formal	829	1176
2,3-Butylene glycol formal	740	1037
sec-Butyl ethyl formal	826	992
Isobutyl 2-ethyl-n-hexanoate	1248	1400
2-n-Butyl-4-ethyl-5-methyloxazole	1159	1441
2-Isobutyl-4-ethyl-5-methyloxazole	1106	1359
n-Butyl ethyl ether	876	965
n-Butyl ethyl sulfide	893	1090
n-Butyl formate	696	996
Isobutyl formate	673	955
2-n-Butylfuran	883	1130
Isobutyl beta-2-furylacrylate	1435	2006
n-Butyl 3-furyl-n-propionate	1354	1728
n-Butyl n-heptanoate	1276	1500
Isobutyl n-heptanoate	1239	1448
n-Butyl n-hexanoate	2180	2435
n-Butyl n-hexanoate	1177	1402
Isobutyl n-hexanoate	1140	1350
Isobutyl 2-hexenoate	1180	1461
n-Butyl p-hydroxybenzoate	1738	
Di-n-butyl ketone	1058	1330
n-Butyl lactate	997	1508
n-Butyl levulinate	1225	1760
Isobutyl levulinate	1183	1696
n-Butyl mercaptan	892	944
Isobutyl mercaptan	660	889
n-Butyl N-methylanthranilate	1660	2266
Isobutyl N-methylanthranilate	1617	2174
n-Butyl 2-methylbutyrate	1029	1226
Isobutyl 2-methylbutyrate	991	1171
Butyl methyl ether	615	755
tert-Butyl methyl ether	563	688
Citronellyl ethyl acetal	1423	1626
p-tert-Butyl-alpha-methylhydrocinnamic aldehyde	1506	2039
Isobutyl 2-methylpentanoate	1155	1417
n-Butyl methylphenylglycidate 1	1591	2200
n-Butyl methylphenylglycidate 2	1700	2346

COMPOUNDS	OV 101	CBWX 20M
n-Butyl methyl sulfide	813	1043
2-n-Butyl-4-methylthiazole	1141	1500
2-Isobutyl-4-methylthiazole	1086	1420
n-Butyl n-nonanoate	1475	1700
n-Butyl n-octanoate	1373	1600
Isobutyl n-octanoate	1338	1543
n-Butyl n-pentadecanoate	2080	2330
n-Butyl propyl sulfide	972	1173
n-Butyl phenylacetate	1408	1932
Isobutyl phenylacetate	1371	1864
p-tert-Butylphenylpropyl aldehyde 1	1212	2030
n-Butyl pivalate	963	1128
Isobutyl pivalate	933	1085
n-Butyl n-propionate	889	1130
Isobutyl n-propionate	852	1071
6-sec-Butylquinoline	1592	2235
n-Butyl salicylate	1457	1976
Isobutyl salicylate	1410	1896
n-Butyl n-tetradecanoate	1977	2229
2-n-Butylthiazole	1070	1480
2-Isobutylthiazole	1020	1404
2-n-Butylthiophene	1052	1353
Ethyl angelate	920	1228
n-Butyl tiglate	1116	1419
Isobutyl tiglate	1076	1357
n-Butyl n-tridecanoate	1880	2118
n-Butyl n-undecanoate	1674	1900
n-Butyl 10-undecenoate	1660	1954
Isobutyl 10-undecenoate	1617	1900
n-Butyl n-valerate	1078	1305
n-Butyl isovalerate	1021	1242
Isobutyl n-valerate	1040	1252
Isobutyl isovalerate	992	1184
gamma-Butyrolactone	885	1632
Isobutyrophenone	1185	1685
delta-Cadinene	1524	1761
gamma-Cadinene	1518	1766
Calamenene	1518	1842
Camphene	954	1083
Camphor	1136	1518
Carvacrol	1297	2159
Carveol 1	1209	1790
Carveol 2	1222	1820
Carvomenthene	1022	1150
Carvone	1228	1715
Carvone oxide	1261	1805
Carvyl propionate	1440	1833

COMPOUNDS	OV 101	CBWX 20M
Caryophyllene	1428	1617
Dihydrocuminyl acetate	1418	1900
alpha-Cedrene	1436	1600
beta-Cedrene	1446	1633
alpha-Cedrene epoxide	1585	1961
Cedrol	1609	2100
Cedryl acetate 1	1427	1591
Cedryl acetate 2	1766	2173
Chavicol	1238	2300
1,4-Cineole	1010	1185
1,8-Cineole	1027	1228
Cinnamic aldehyde	1250	1996
Cinnamyl acetate	1419	2103
Cinnamyl alcohol	1300	2207
Cinnamyl n-butyrate	1604	2247
Cinnamyl isobutyrate	1555	2171
Cinnamyl cinnamate	2055	
Cinnamyl formate	1330	2094
Cinnamyl cinnamate	2055	
Cinnamyl n-heptanoate	1905	2545
Cinnamyl n-hexanoate	1805	2445
Cinnamyl n-propionate	1515	2169
Cinnamyl n-valerate	1705	2347
Cinnamyl isovalerate	1655	2271
Citral 1	1222	1661
Citral 2	1249	1706
Citronellal	1137	1465
Citronellol	1215	1722
Citronellyl acetate	1335	1645
Citronellyl n-butyrate	1511	1786
Citronellyl isobutyrate	1469	1705
Citronellyl crotonate	1558	1929
Citronellyl isocrotonate	1526	1833
Citronellyl formate	1261	1600
Citronellyl n-propionate	1427	1700
Citronellyl n-valerate	1608	1880
Citronellyl isovalerate	1563	1800
Citronellyl vinyl ether	1235	1445
alpha-Copaene	1398	1519
beta-Copaene	1445	1626
Coumarin	1418	2361
p-Cresol	1051	2003
p-Cresol methyl ether	1005	1415
p-Cresyl acetate	1150	1700
p-Cresyl benzoate	1764	
p-Cresyl ethylcarbonate	1304	1919
p-Cresyl isobutyrate	1291	1763

COMPOUNDS	OV 101	CBWX 20M
p-Cresyl n-octanoate	1755	2264
p-Cresyl phenylacetate	1827	
p-Cresyl salicylate	1850	
p-Cresyl isotiglate	1482	2011
p-Cresyl isovalerate	1389	1898
Cuminyl acetate	1401	1952
Cuminalcohol	1283	2045
Cuminaldehyde	1227	1766
Cyclohexyl glycidate	1344	1875
Cycohexadecanone	1731	2392
Cyclohexane	677	765
Cyclohexanol	880	1375
Cyclohexanone	875	1306
Cyclohexanone 1,3-butylene ketal	1186	1500
Cyclohexyl acetate	1027	1343
Cyclohexyl n-butyrate	1209	1492
Cyclohexyl isobutyrate	1164	1427
Cyclohexenylcyclohexanone	1553	2128
2-Cyclohexylcyclohexanone	1496	1975
2-Cyclohexyl ethanol	1098	1668
2-Cyclohexylethyl acetate	1233	1591
Cyclohexyl formate	951	1305
Cyclohexyl n-hexanoate	1411	1695
Cyclohexyl n-propionate	1120	1408
Cyclohexyl n-valerate	1310	1595
Cyclohexyl isovalerate	1264	1527
Cyclooctanol	1133	1700
Cyclooctanyl acetate	1280	1645
Cyclopentanol	802	1283
Cyclopentanone	805	1238
p-Cymene	1020	1272
n-Decanal	1188	1485
n-Decanal diethyl acetal	1473	1613
n-Decanal dimethyl acetal	1366	1567
trans-2-trans-4-Decadienol	1310	1938
5-Decanol	1175	1550
Decahydro-beta-naphthol 1	1280	1883
Decahydro-beta-naphthol 2	1323	1995
gamma-Decalactone	1437	2101
delta-Decalactone	1463	2144
n-Decanal	1188	1485
n-Decanal diethyl acetal	1473	1613
n-Decanal dimethyl acetal	1366	1567
n-Decane	1000	1000
Decanol	1263	1723
n-Decan-2-ol	1190	1585
2-Decanoylfuran	1689	2264

COMPOUNDS	OV 101	CBWX 20M
2-Decenal	1449	1842
cis-4-Decenal	1177	1523
1-Decene	991	1043
2-Decen-1-ol	1257	1794
9-Decen-1-ol	1251	1775
9-Decen-1-yl acetate	1383	1722
omega-Decenyl butyrate	1558	1870
cis-3-Decenyl butyrate	1563	1866
trans-2-Decenyl butyrate	1563	1866
cis-3-Decenol	1245	1765
trans-2-Decenol	1257	1792
cis-3-Decenyl acetate	1376	1701
Decyl acetate	1393	1662
n-Decyl n-propionate	1486	1729
Diallyl sulfide	854	1150
Diisoamyl ether	1000	1064
Di-n-amyl fumarate	1558	2006
Di-n-amyl ketone	1258	1528
Diisoamyl ketone	1258	1528
Dibenzyl ether	1631	2323
Di-n-butyl adipate	1658	2087
Dibutyl butyrolactone	1531	2141
Di-n-butyl ether	876	965
Di-n-butyl fumarate	1558	2006
Diisobutyl acetal	999	1072
Di-n-butyl ketone	1058	1330
Diisobutyl ketone	983	1207
Di-n-butyl succinate	1534	2000
Di-n-butyl sulfide	1073	1270
Diisobutyl sulfide	969	1149
1,1-Diethoxycyclohexane		1265
1,1-Diethyoxyethane	710	880
1,1-Diethyoxyhexane	1080	1228
Diethyl adipate	1358	1858
m-Diethylbenzene	1070	1339
o-Diethylbenzene	1088	1372
p-Diethylbenzene	1080	1353
Diethyleneglycol diethyl ether	1058	1572
Diethyleneglycol dimethyl ether	924	1396
Diethyleneglycol monoethyl ether	986	1583
Diethyleneglycol monomethyl ether	920	1321
Diethyl ether	572	590
Diethyl fumarate	1160	1632
Diethyl ketone	681	980
Diethyl malate	994	1638
Diethyl malonate	1043	1542
2,4-Diethyl-5-methyloxazole	983	1274

COMPOUNDS	OV 101	CBWX 20M
Diethyl phthalate	1565	2303
Diethyl sebacate	1752	2272
Diethyl suberate	1553	2065
Diethyl succinate	1153	1642
Diethyl sulfide	690	904
Diethyl trisulfide	1125	1535
Di-n-heptyl ether	1458	1550
D-n-hexyl ether	1269	1349
Di-n-hexyl ketone	1458	1728
Dihydroanethole	1193	1600
Dihydrocarveol	1188	1713
Dihydrocarvone	1183	1600
Dihydrocuminalcohol	1286	1981
Dihydrocarvyl acetate	1319	1670
Dihydrocoumarin	1359	2286
Dihydrocinnamic aldehyde	1139	1783
Dihydro-nor-dicyclopentadienyl acetate	1406	1881
Dihydrojasmone	1400	1892
Dihydroisojasmone	1374	1842
Dihydrolinalool	1122	1512
Dihydrolinalyl acetaldehyde	1315	1659
Dihydromyrcenol	1063	1438
Dihydromyrcenyl acetate	1202	1431
cis-Dihydroocimene	982	1088
trans-Dihydroocimene	995	1110
Dihydropseudo ionone	1473	1885
3,4-Dihydropyran	705	982
Dihydrosafrole	1286	1822
Dihydroterpinyl acetate	1282	1561
3,4-Dimethoxyacetophenone	1532	2393
1,3-Dimethoxy benzene	1143	1740
1,4-Dimethoxybenzene	1145	1705
1,1-Dimethoxy-n-butane	770	969
1,4-Dimethoxy-2-tert-butylbenzene	1398	1870
1,1-Dimethoxy-n-heptane	1063	1261
1,1-Dimethoxy-n-hexane	964	1156
1,1-Dimethoxy-n-pentane	868	1057
1,1-Dimethoxy-n-propane	650	880
2,4-Dimethylacetophenone	1233	1809
2,4-Dimethyl-5-acetylthiazole	1217	1835
Dimethyl adipate	1212	1779
2,4-Dimethyl benzylacetate	1348	1916
2,4-Dimethyl benzyl alcohol	1226	2032
Dimethyl benzyl carbinol	1147	1715
Dimethyl benzyl carbinyl acetate	1302	1755
Dimethyl benzyl carbinyl n-butyrate	1476	1889
Dimethyl benzyl carbinyl n-propionate	1391	1810

COMPOUNDS	OV 101	CBWX 20M
2,6-Dimethylcyclohexanone	985	1322
2,5-Dimethyl-4-ethyloxazole	900	1231
2,5-Dimethyl-4-ethylthiazole	1050	1398
Dimethyl fumarate	997	1530
2,5-Dimethylfuran	697	951
2,6-Dimethyl-n-heptan-2-ol	983	1300
2,6-Dimethyl-hept-5-en-1-al	1039	1358
2,6-Dimethylheptyl-4-acetate	1092	1265
Dimethyl malonate	896	1472
2,4-Dimethyl-n-pentan-3-ol	828	1157
2,2-Dimethyl-n-pentanol	874	1405
2,6-Dimethyl phenol	1087	1883
Dimethyl phenyl ethyl carbinol	1282	1916
Dimethyl phenyl ethyl carbinyl acetate	1428	1908
2,3-Dimethylpyrazine	900	1330
2,5-Dimethylpyrazine	893	1306
2,6-Dimethylpyrazine	895	1325
Dimethyl sebacate	1616	2195
Dimethyl suberate	1414	1985
Dimethyl succinate	1002	1558
2,4-Dimethylthiazole	869	1271
4,5-Dimethylthiazole	917	1359
2,5-Dimethylthiophene	866	1161
Dimethyl trisulfide	952	1400
alpha-p-Dimethylstyrene	1080	1278
Diphenyl	1369	1981
Diphenyl oxide	1386	1991
Dipropyleneglycol 1	1017	1767
Dipropyleneglycol 2	1039	1817
Dipropyleneglycol 3	1043	1823
Dipropyleneglycol 4	1044	1848
Dipropyleneglycol 5	1075	1892
Di-n-Propyl fumarate	1360	1806
Di-n-Propyl ketone	857	1131
Diisopropyl ketone	783	1007
Di-n-Propyl malonate	1227	1700
gamma-Dodecalactone	1647	2317
delta-Dodecalactone	1675	2358
n-Dodecanal	1392	1695
n-Dodecanal diethyl acetal	1671	1811
n-Dodecanal dimethyl acetal	1565	1769
n-Dodecane	1200	1200
n-Dodecanol	1468	1925
n-Eicosane	2000	2000
Elemol	805	1318
7,8-Epoxy-1,3,3,11-tetramethyl-(5,4,0,0(2.4))-undecane	1568	1912

COMPOUNDS	OV 101	CBWX 20M
Estragole	1183	1652
p-Ethoxybenzaldehyde	1308	2017
2-Ethoxythiazole	943	1380
Ethyl acetate	595	872
Ethyl acetylacetate	907	1427
Ethyl acrylate	681	980
Ethyl alcohol	500	900
Ethyl n-amyl ketone	928	1190
Ethyl anisate	1426	2110
Ethyl anthranilate	1396	2232
Ethyl benzoate	1154	1647
2-Ethylbutyl acetate	959	1205
Ethyl n-butyl ketone	869	1155
Ethyl n-butyrate	784	1025
Ethyl isobutyrate	746	956
Ethyl chloroacetate	810	1281
Ethyl cinnamate	1443	2095
Ethyl crotonate	823	1161
Ethyl n-decanoate	1379	1624
Ethyl 2,4-dimethyl-1,3-dioxolylacetate 1	1147	1603
Ethyl 2,4-dimethyl-1,3-dioxolylacetate 2	1154	1616
2-Ethyl-4,5-dimethyloxazole	914	1243
2-Ethyl-4,5-dimethylthiazole	1065	1429
Ethyl disulfide	910	1232
Ethyl n-dodecanoate	1579	1826
Ethyleneglycol monophenyl ether	1213	2100
Ethyl formate	530	806
2-Ethylfuran	694	951
Ethyl 2-furoate	1029	1599
4-Ethylguaiacol	1265	2011
Ethyl n-heptanoate	1082	1321
2-Ethylhexanal	959	1216
2-Ethyl-2-n-hexanal	1007	1336
Ethyl isohexanoate	951	1181
Ethyl n-hexanoate	983	1223
4-Ethyl isohexanoate	951	1181
2-Ethyl-2-n-hexene-1-ol	1051	1518
2-Ethyl-1-hexyl acrylate	1224	1494
2-Ethyl-1-hexyl propionate	1231	1450
2-Ethyl-1-hexyl vinyl ether	1038	1165
Ethyl levulinate	1029	1567
Ethyl 2-methylbutyrate	837	1049
Ethyl 2-methylpentanoate	1005	1302
2-Ethyl-4-methylthiazole	955	1331
4-Ethyl-5-methylthiazole	991	1400
Ethyl n-octanoate	2180	2460
n-Ethyl n-nonanoate	1280	1523

COMPOUNDS	OV 101	CBWX 20M
Ethyl n-octanoate	1180	1423
Ethyl oleate	2155	2489
3-Ethyl-3-pentanol	866	1183
2-Ethyl phenol	1113	2028
3-Ethyl phenol	1146	2150
Ethyl phenylacetate	1219	1773
Ethyl phenyl ethyl acetal	1332	1770
Ethyl pyruvate	785	1253
Ethyl pivalate	776	947
Ethyl n-propionate	691	944
Ethyl n-propyl ketone	767	1055
Ethyl salicylate	1257	1787
Ethyl sorbate	1075	1505
Ethyl n-tetradecanoate	1780	2027
2-Ethylthiazole	879	1300
2-Ethylthiophene	861	1179
Ethyl tiglate	922	1234
Ethyl n-undecanoate	1479	1725
Ethyl 10-undecenoate	1469	1775
Ethyl n-valerate	884	1124
Ethyl isovalerate	840	1060
Ethylvanillin	1448	2414
Eugenol	1351	2103
Isoeugenol 1	1394	2186
Isoeugenol 2	1438	2269
Fenchone	1080	1410
Fenchyl acetate	1220	1473
alpha-Fenchyl alcohol	1110	1574
Furfuryl disulfide	1660	2600
Furan	500	786
Furfural	815	1449
Furfuryl acetate	969	1518
Furfuryl n-butyrate	1148	1655
Furfuryl n-heptanoate	1443	1950
Furfuryl n-hexanoate	1343	1850
Furfuryl n-propionate	1059	1587
Furfuryl n-valerate	1245	1752
Galaxolide 1	1837	2299
Galaxolide 2	1837	2307
Geranial	1252	1730
Geraniol	1243	1797
Geranonitrile 1	1199	1680
Geranonitrile 2	1236	1723
Geranonitrile 3	1249	1792
Geranyl acetate	1364	1735
Geranyl benzoate	1949	
Geranyl n-butyrate	1532	1872

COMPOUNDS	OV 101	CBWX 20M
Geranyl isobutyrate	1493	1795
Geranyl ethyl ether 1	1255	1476
Geranyl ethyl ether 2	1275	1506
Geranyl formate	1282	1684
Geranyl n-heptanoate	1831	2157
Geranyl 2-methylbutyrate	1574	1886
Geranyl n-propionate	1424	1799
Geranyl tiglate	1650	1985
Geranyl n-valerate	1632	1960
Geranyl isovalerate	1593	1895
Guaiacol	1071	1840
Guaiacyl n-caproate	1681	2296
Helional	1543	2383
Heliotropine	1318	2171
Heliotropyl acetate	1485	2325
n-Heneicosane	2100	2100
n-Heptadecane	1700	1700
n-gamma-Heptalactone	1126	1860
n-Heptanal	883	1186
n-Heptanal diethyl acetal	1179	1319
n-Heptanal dimethyl acetal	1069	1265
n-Heptanal ethyleneglycol acetal	1144	1460
n-Heptane	700	700
2,3-Heptanedione	816	1138
n-Heptanol	957	1419
n-Heptan-2-ol	888	1284
n-Heptan-4-ol	879	1250
2-Heptanoylfuran	1384	1956
2-Heptanoylthiophene	1565	2209
1-Hepten-3-ol	868	1322
cis-3-Heptenyl acetate	1084	1400
trans-3-Heptenyl acetate	1080	1388
cis-3-Heptenyl butyrate	1255	1545
trans-2-Heptenyl butyrate	1275	1568
cis-3-Heptenyl propionate	1171	1472
trans-2-Heptenyl propionate	1182	1497
n-Heptyl acetate	1095	1361
n-Heptyl n-butyrate	1275	1503
n-Heptyl isobutyrate	1233	1433
alpha-n-Heptylcinnamic aldehyde	1827	2409
n-Heptyl formate	1012	1310
2-n-Heptylfuran	1184	1429
n-Heptyl mercaptan	1011	1256
n-Heptyl 2-methylbutyrate	1324	1518
n-Heptyl n-octanoate	1666	1892
n-Heptyl phenylacetate	1717	2265
n-Heptyl pivalate	1263	1428

COMPOUNDS	OV 101	CBWX 20M
n-Heptyl salicylate	1790	2332
2-n-Heptylthiophene	1359	1670
n-Heptyl valerate	1372	1614
n-Hexadecane	1600	1600
n-Hexanal	780	1084
n-Hexanal diethyl acetal	1082	1223
n-Hexane	600	600
3,4-Hexanedione	777	1123
n-Hexanol	858	1316
n-Hexan-2-ol	786	1192
2-Hexanoylfuran	1281	1850
2-Hexanoylthiophene	1459	2104
trans-2-Hexenal	832	1207
1-Hexen-3-ol	770	1225
cis-3-Hexenol	847	1351
trans-2-Hexenol	854	1368
cis-3-Hexenyl acetate	987	1300
trans-2-Hexenyl acetate	997	1315
cis-3-Hexenyl anthranilate	1807	
cis-3-Hexenyl benzoate	1553	2122
cis-3-Hexenyl n-butyrate	1170	1450
trans-2-Hexenyl n-butyrate	1180	1461
cis-3-Hexenyl isobutyrate	1129	1377
cis-3-Hexenyl decanoate	1760	2038
cis-3-Hexenyl ethyl acetal	1094	1298
cis-3-Hexenyl formate	902	1252
cis-3-Hexenyl n-heptanoate	1465	1743
trans-2-Hexenyl n-heptanoate	1474	1755
cis-3-Hexenyl n-hexanoate	1370	1654
trans-2-Hexenyl n-hexanoate	1375	1656
cis-3-Hexenyl lactate	1187	1727
cis-3-Hexenyl methoxyformate	1073	1475
cis-3-Hexenyl methyl acetal	1035	1267
cis-3-Hexenyl n-nonanoate	1664	1938
trans-2-Hexenyl n-nonanoate	1673	1953
cis-3-Hexenyl n-octanoate	1564	1838
trans-2-Hexenyl n-octanoate	1573	1853
cis-3-Hexenyl phenylacetate	1610	2220
cis-3-Hexenyl n-propionate	1083	1371
trans-2-Hexenyl n-propionate	1085	1370
cis-3-Hexenyl salicylate	1654	2227
cis-3-Hexenyl n-valerate	1270	1584
trans-2-Hexenyl n-valerate	1275	1560
cis-3-Hexenyl isovalerate	1223	1480
2-Hexoxyacetaldehyde dimethyl acetal	1234	1528
n-Hexyl acetate	1012	1307
n-Hexyl angelate	788	1621

COMPOUNDS	OV 101	CBWX 20M
n-Hexyl benzoate	1558	2066
n-Hexyl n-butyrate	1176	1398
n-Hexyl isobutyrate	1135	1333
alpha-n-Hexylcinnamic aldehyde	1727	2309
2-Hexylcyclopentanone acetic acid methyl ether	1610	2200
trans-Linalool oxide	1082	1451
n-Hexyl formate	994	1258
2-n-Hexyl furan	1083	1326
n-Hexyl n-heptanoate	1470	1699
Hexyl n-hexanoate	1371	1599
n-Hexyl mercaptan	910	1145
n-Hexyl 2-methylbutyrate	1224	1418
n-Hexyl nonanoate	1668	1900
n-Hexyl octanoate	1564	1805
n-Hexyl phenylacetate	1607	2148
n-Hexyl pivalate	1163	1328
n-Hexyl n-propionate	1088	1326
n-Hexyl salicylate	1664	2175
2-n-Hexylthiophene	1256	1564
n-Hexyltiglate	1310	1621
n-Hexyl n-valerate	1275	1498
n-Hexyl isovalerate	1228	1433
alpha-Humulene	1465	1682
p-Hydroxyacetophenone	1144	1790
Hydroxycitronellal	1269	1882
Hydroxycitronellol	1347	2143
Indole	1304	2351
alpha-Ionone	1416	1833
beta-Ionone	1474	1918
gamma-Ionone	1347	1882
Isolongiforanone 1	1610	2072
Isolongiforanone 2	1622	2112
Jasmal	1459	1879
cis-Jasmone	1378	1914
Lavandulol	1154	1662
Lavandulyl acetate	1274	1597
Limonene	1030	1206
Linalool	1092	1506
cis-Linalool oxide	1068	1423
trans-Linalool oxide	1082	1451
Linalyl acetate	1246	1538
Linalyl n-butyrate	1420	1680
Linalyl isobutyrate	1366	1597
Linalyl formate	1206	1570
Linalyl n-heptanoate	1670	1930
Linalyl n-hexanoate	1582	1843

COMPOUNDS	OV 101	CBWX 20M
Linalyl 2-methylbutyrate	1450	1695
Linalyl n-propionate	1324	1596
Linalyl n-valerate	1500	1765
Linalyl isovalerate	1461	1698
Maltol	1105	2030
Menthol	1171	1612
2-Mercaptobenzothiazole	1944	
Methyl acetate	513	813
Menthone	1143	1478
o-Methoxyacetophenone	1269	1975
m-Methoxyacetophenone	1279	2011
p-Methoxyacetophenone	1327	2115
p-Methoxybenzylacetone	1473	2236
o-Methoxy cinnamic aldehyde	1504	2430
o-Methylacetophenone	1118	1679
m-Methylacetophenone	1156	1749
p-Methylacetophenone	1166	1750
n-Methyl-2-acetylpyrrole	1055	1653
2-Methyl-5-acetylthiophene	1185	1836
Methyl n-amyl ketone	872	1172
Methyl isoamyl ketone	825	1100
Methyl anisate	1354	2071
Methyl anthranilate	1332	2181
Methyl acrylate		938
o-Methyl anisole	1000	1432
p-Methyl anisole	1000	1432
Methyl benzoate	1078	1600
p-Methyl benzyl alcohol	1122	1956
Methyl benzyl ether	981	1391
3-Methylbutan-2-ol	671	1052
Methyl n-butyl ketone	772	1070
Methyl isobutyl carbinol	748	1142
Methyl isobutyl ketone	725	1000
Methyl n-butyrate	705	975
Methyl isobutyrate	673	913
Methyl isovalerate	1505	1734
Methyl chavicol	1182	1670
Methyl cinnamate	1365	2051
alpha-Methylcinnamyl acetate	1484	2158
alpha-Methylcinnamyl alcohol	1343	2252
beta-Methylcinnamyl alcohol	1365	2283
alpha-Methylcinnamic aldehyde	1309	1992
alpha-Methylcitronellol	1220	1540
3-Methylcoumarin	1490	2424
6-Methylcoumarin	1545	2630
7-Methylcoumarin	1553	2620
Methyl crotonate	745	1100

COMPOUNDS	OV 101	CBWX 20M
Methyl n-decanoate	1307	1581
Methyl n-decyl ketone	1377	1688
Methyl dihydrocinnamate	1258	1842
Methyl disulfide	730	1081
Methyl n-dodecanoate	1507	1785
Methyl n-dodecyl ketone	1580	1893
Methyl ether	350	524
2-Methyl-5-ethylfuran	791	1024
2-Methyl-3-ethylpyrazine	987	1381
2-Methyl-5-ethylthiophene	957	1245
Methylisoeugenol 1	1427	2044
Methylisoeugenol 2	1468	2126
2-Methylfuran	614	866
5-Methylfurfural	942	1563
Methyl furoate	956	1561
Methyl isogeranylacetate	1362	1765
Methyl n-heptanoate	1006	1276
1-Methyl 4-hepten-1-ol	982	1433
6-Methyl-hept-5-en-2-one	968	1335
Methyl n-heptyl ketone	1074	1377
Methyl hexadecanoate	1911	2204
Methyl n-hexanoate	906	1177
Methyl n-hexyl ketone	972	1275
Methyl n-hexyl ketone 1-phenyl-1,2-ethanediol ketal 1	1739	2224
Methyl n-hexyl ketone 1-phenyl-1,2-ethanediol ketal 2	1758	2248
Methyl hydrojasmonate		2229
Methyl 3-hydroxybutyrate	1318	1454
Methyl 2-hydroxyisobutyrate	1116	2054
Methyl hexyl acetaldehyde	1048	1306
4-Methyl-5-hydroxy phenylacetate	1263	1767
Methyl isohexyl carbinyl acetate	1080	1300
Methyl hexyl ether	824	960
Methylionone 1	1471	1836
Methylionone 2	1506	1897
Methylionone 3	1530	1930
Methylionone 4	1564	1981
Methyllavender ketone	1341	2067
Methyl levulinate	956	1534
Methyl mercaptoacetate	802	1346
Methyl methacrylate	699	1008
Methylmethacrylate	699	1008
2-Methyl-2-(4-methyl n-amyl) tetrahydrofuran	1159	1325
Methyl N-methylanthranilate	1389	2042
Methyl 2-methylbenzoate	1165	1709

COMPOUNDS	OV 101	CBWX 20M
Methyl 3-methylbenzoate	1190	1744
Methyl 4-methylbenzoate	1199	1755
Methyl 2-methylbutyrate	765	1000
Methyl naphthyl ketone	1592	2471
Methyl nicotinate	1116	1779
Methyl n-nonanoate	1207	1479
Methyl nonyl acetaldehyde	1352	1621
Methyl n-nonyl ketone	1276	1585
Methyl octadecanoate	2101	2418
Methyl n-octanoate	1107	1378
2-Methyl-1-octanol	1119	1573
Methyl octyl acetaldehyde	1254	1521
Methyl octyl ether	1016	1152
Methyl n-octyl ketone	1176	1480
2-Methyl-1-pentanol	837	1268
2-Methyl-2-pentanol	797	1160
2-Methyl-3-pentanol	777	1121
3-Methyl-1-pentanol	852	1297
3-Methyl-2-pentanol	797	1160
3-Methyl-3-pentanol	757	1080
4-Methyl-1-pentanol	838	1282
4-Methyl-2-pentanol	758	1124
3-Methyl-2-pentanol	755	1054
Methyl phenylacetate	1154	1747
Methyl phenyl carbinol	1051	1765
o-Methylphenyl ethyl alcohol	1216	2012
2-Methyl-2-phenylhexan-4-one	1405	1938
2-Methyl-1-propanol	616	1054
2-Methyl-n-propan-2-ol	500	871
Methyl propionate	611	896
2-Methyl-5-isopropylacetophenone	1358	1876
N-Methyl-2-pyrrolaldehyde	986	1616
2-Methyl-5-propionylthiophene	1280	1900
Methyl N-propylanthranilate	1678	2575
Methyl n-propyl ketone	672	969
beta-Methyl-p-isopropylphenyl-propionaldehyde	1444	1954
2-Methyl-3-n-propylpyrazine	1074	1462
2-Methyl-3-isopropylpyrazine	1028	1387
2-Methylpyrazine	805	1251
N-Methylpyrrole	715	1139
Methyl pyruvate	680	1217
Methyl salicylate	1181	1754
Methyl sorbate	998	1448
Methyl sulfoxide	840	1554
Methyl n-tetradecanoate	1707	1990
2-Methyltetrahydrofuran	674	901

COMPOUNDS	OV 101	CBWX 20M
Methyl tiglate	850	1188
2-Methylthiazole	780	1256
4-Methylthiazole	800	1263
4-Methyl-5-thiazoleethanol	1283	2216
4-Methyl-5-thiazoleethanol acetate	1368	2077
2-Methylthiophene	775	1123
Methyl p-toluate	1194	1725
Methyl n-tridecanoate	1612	1895
2-Methylundecanal	1353	1609
2-Methyl-1-undecanol	1422	1875
Methyl 10-undecenoate	1396	1733
omega-Methyl undecylenate	1400	1747
Methyl undecyl ether	1318	1453
Methyl n-undecyl ketone	1479	1792
Methyl n-valerate	806	1076
Methyl isovalerate	764	1008
Methyl vinyl ketone	550	995
4-Methyl-5-vinylthiazole	1011	1500
Methylzingerone	1660	2640
Muscone	1831	2281
Musk xylol	1506	2475
gamma-Muurolene	1475	
alpha-Muurolene	1500	1730
Myrcene	986	1156
Myrcenol	1103	1585
Myrcenyl acetate	1247	1574
Myrcenyl propionate	1327	1625
Neral	1227	1680
Nerol	1218	1757
alpha-Nerolidol	1524	1961
beta-Nerolidol	1553	2000
Nerolidylethanol	1851	2462
Neryl acetate	1345	1699
Neryl n-butyrate	1519	1838
Neryl isobutyrate	1474	1764
Neryl formate	1267	1663
Neryl n-heptanoate	1808	2120
Neryl n-hexanoate	1709	2021
Neryl n-propionate	1436	1771
Neryl n-valerate	1610	1930
Neryl isovalerate	1574	1864
n-Nonadecane	1900	1900
gamma-Nonalactone	1328	1991
delta-Nonalactone	1356	2038
n-Nonanal	1087	1382
n-Nonanal diethyl acetal	1374	1514
n-Nonanal dimethyl acetal	1267	1465

COMPOUNDS	OV 101	CBWX 20M
n-Nonane	900	900
n-Nonanoic acid	1286	2110
n-Nonanol	1161	1624
n-Nonan-2-ol	1089	1484
2-Nonanone	1093	1420
5-Nonanone	1074	1360
2-Nonanoylfuran	1588	2163
trans-2-Nonenal	1146	1540
trans-2-nonenol	1157	1691
1-Nonen-3-ol	1068	1520
n-Nonyl acetate	1292	1560
2-Nonyn-1-al dimethyl acetal	1300	1666
Nopyl acetate	1412	1777
Nootketone	1802	2250
Norbornyl acetate	1112	1476
cis-Ocimene	1025	1228
trans-Ocimene	1038	1250
Nopylacetate	1412	1777
gamma-Octalactone	1225	1883
delta-Octalactone	1252	1929
n-Octanal	985	1278
n-Octanal diethyl acetal	1276	1417
n-Octanal dimethyl acetal	1167	1366
n-Octane	800	800
n-Octanol	1061	1519
2-Octanone	991	1304
n-Octan-2-ol	988	1385
2-Octanoylfuran	1487	2062
2-Octanoylthiophene	1667	2313
trans-2-Octenal	1045	1427
1-Octene	797	830
2-Octene	811	880
1-octene-3-ol	968	1420
1-Octene-3-yl propionate	1180	1432
trans-3-Octenol	1036	1541
cis-3-Octenol	1041	1563
n-Octyl acetate	1193	1459
Isooctyl acetate	1154	1419
n-Octyl n-butyrate	1373	1597
n-Octyl isobutyrate	1332	1529
Octyl formate	1117	1426
2-n-Octylfuran	1285	1530
n-Octyl 2-methylbutyrate	1422	1615
Octyl propionate	1283	1536
Octyl salicylate	1895	2435
2-n-Octylthiophene	1463	1780
Octyl-n-valerate	1474	1719

COMPOUNDS	OV 101	CBWX 20M
Paraldehyde	763	1069
Patchouli alcohol	1667	2156
n-Pentadecane	1500	1500
n-Pentadecanol	1373	2252
Pentalide	1823	2255
n-Pentane	500	500
2,3-Pentanedione	681	1044
n-Pentan-2-ol	685	1091
2-Pentanoylfuran	1180	1747
2-Pentanoylthiophene	1355	1993
1-Penten-3-ol	673	1130
Isopentyl ether	999	1067
n-Pentyl ether	1065	1165
2-n-Pentylfuran	983	1229
2-n-Pentyl-3-methyl-2-cyclopenten-1-one	1400	1892
2-n-Pentylthiophene	1153	1462
alpha-Phellandrene	1002	1177
beta-Phellandrene	1025	1216
beta-Phenoxyethyl isobutyrate	1493	2100
Phenylethyl benzoate	1841	
Phenylethyl butyrate	1422	1930
Phenol	1002	1932
Phenoxyethyl isobutyrate	1486	2106
Phenoxyethyl propionate	1447	2126
Phenylacetaldehyde	1024	1646
Phenylacetaldehyde dimethyl acetal	1200	1665
Phenyl isobutyl methyl carbinyl acetate	1459	1943
Phenylethyl acetate	1233	1785
Phenylethyl alcohol	1104	1859
Phenylethyl anthranilate	2091	
Phenylethyl benzoate	1841	
Phenylethyl butyrate	1422	1930
Phenylethyl isobutyrate	1374	1855
Phenylethyl cinnamate	2147	
Phenylethyl n-decanoate	2022	2540
Phenylethyl formate	1156	1752
Phenylethyl n-heptanoate	1718	2233
Phenylethyl n-hexanoate	1618	2134
Phenylethyl 2-methylbutyrate	1472	1945
Phenylethyl n-nonanoate	1921	2439
Phenylethyl n-octanoate	1819	2337
Phenylethyl pivalate	1400	1832
Phenylethyl n-propionate	1328	1855
Phenylethyl tiglate	1562	2154
Phenylethyl n-valerate	1517	2034
Phenylethyl isovalerate	1474	1955
Phenylpropyl acetate	1347	1926

COMPOUNDS	OV 101	CBWX 20M
3-Phenylpropyl alcohol	1218	1993
Phenylpropyl n-butyrate	1535	2083
Phenylpropyl isobutyrate	1490	1996
Phenylpropyl n-propionate	1445	1994
Phenylpropyl n-valerate	1635	2183
Phenylpropyl isovalerate	1590	2100
Phenyl salicylate	1742	
Pinacol	843	
Pinacolone	695	960
cis-Pinane	987	1075
trans-Pinane	981	1062
alpha-Pinene	942	1039
beta-Pinene	981	1124
Piperidine	750	1042
Piperitone	1247	1739
Prenyl acetate	900	1243
n-Propanol	535	1002
2-Propanol	500	884
Propionaldehyde	481	784
Propionaldehyde 1-phenyl-1,2-ethanediol acetal 1	1327	1871
Propionaldehyde 1-phenyl-1,2-ethanediol acetal 2	1333	1880
2-Propionylfuran	988	1563
n-Propionyl methylanthranilate	1673	2350
2-Propionyl-5-methylfuran	1106	1672
2-Propionylpyrrol	1145	1990
2-Propionylthiophene	1164	1821
n-Propyl acetate	694	962
Isopropyl acetate	645	883
p-Isopropylacetophenone	1332	1912
n-Propyl anisate	1527	2205
n-Propyl anthranilate	1500	2320
n-Propyl benzoate	1254	1745
Isopropyl benzoate	1189	1639
Isopropyl benzyl carbinol	1291	1914
n-Propyl n-butyrate	881	1110
n-Propyl isobutyrate	842	1044
Isopropyl n-butyrate	825	1030
Isopropyl cinnamate	1485	2097
n-Propyl cinnamic aldehyde	1531	2111
n-Propyl n-decanoate	1476	1697
Isopropyl n-decanoate	1417	1615
2-n-Propyl-4,5-dimethyloxazole	996	1310
2-Isopropyl-4,5-dimethyloxazole	960	1249
2-n-Propyl-4,5-dimethylthiazole	1151	1500
2-Isopropyl-4,5-dimethylthiazole	1109	1439

COMPOUNDS	OV 101	CBWX 20M
Propyl disulfide	1096	1358
Isopropyl disulfide	1014	1264
n-Propyl n-dodecanoate	1676	1897
Isopropyl n-dodecanoate	1814	1814
Propyl ether	676	766
Isopropyl ether	590	649
Propylene glycol	750	1561
2-n-Propyl-4-ethyl-5methyloxazole	1064	1345
2-Isopropyl-4-ethyl-5-methyloxazole	1021	1279
n-Propyl formate	606	907
Isopropyl formate	573	838
2-n-Propylfuran	782	1083
Propyl furoate	1125	1700
n-Propyl n-heptanoate	1177	1398
Isopropyl n-heptanoate	1120	1317
n-Propyl n-hexanoate	1079	1298
Isopropyl n-hexanoate	1021	1223
n-Propyl levulinate	1125	1663
Isopropyl levulinate	1068	1575
Di-n-Propyl malonate	1227	1700
n-Propyl N-methylanthranilate	1560	2166
Isopropyl N-methylanthranilate	1491	2058
n-Propyl 2-methylbutyrate	933	1134
Propyl methyl ether	512	644
2-n-Propyl-4-methylthiazole	1040	1400
2-Isopropyl-4-methylthiazole	1000	1339
n-Propyl n-nonanoate	1377	1598
Isopropyl n-nonanaote	1318	1516
n-Propyl n-octanoate	1277	1498
Isopropyl n-octanoate	1219	1419
p-Isopropylphhenol	1200	2178
n-Propyl phenylacetate	1300	1848
n-Propyl pivalate	863	1028
Isopropyl pivalate	810	956
n-Propyl n-propionate	785	1010
Isopropyl n-propionate	738	950
n-Propyl salicylate	1357	1878
Isopropyl salicylate	1292	1773
Isopropyl n-tetradecanoate	1811	2017
2-n-Propylthiazole	970	1380
2-n-Propylthiophene	951	1259
n-Propyl tiglate	1020	1320
Isopropyl tiglate	959	1238
n-Propyl n-undecanoate	1576	1797
Isopropyl n-undecanoate	1516	1715
n-Propyl 10-undecenaote	1565	1860
n-Propyl n-valerate	981	1200

COMPOUNDS	OV 101	CBWX 20M
n-Propyl isovalerate	924	1144
Isopropyl n-valerate	924	1125
Isopropyl isovalerate	883	1034
Isopulegol	1145	1573
Isopulegyl acetate	1258	1585
Pulegone	1230	1662
Pyrazine	739	1194
Pyridine	695	1180
Pyrrol-2-carboxaldehyde	1005	1976
cis-Rose oxide	1087	1354
trans-Rose oxide	1100	1370
Sabinene	976	1130
trans-Sabinene hydrate	1078	
Sabinol	1135	1683
Sabinyl acetate	1262	1651
Safrole	1278	1876
Isosafrole	1360	2029
Salicaldehyde	1027	1705
Rosephenone	1538	2172
Styrallyl acetate	1173	1673
Tangerinal 1	1543	1857
Tangerinal 2	1566	1892
gamma-Terpinene	1057	1251
Terpinene-4-ol	1175	1628
alpha-Terpineol	1185	1661
beta-Terpinenol	1137	1616
delta-Terpinenol	1160	1655
Terpinolene		1287
Terpinyl acetate	1333	1687
Terpinyl n-butyrate	1514	1828
Terpinyl isobutyrate	1467	1748
Triethyl citrate	1627	2386
Terpinyl formate	1333	1666
Terpinyl n-propionate	1426	1747
Terpinyl n-valerate	1614	1928
Terpinyl isovalerate	1565	1858
n-Tetradecane	1400	1400
Tetrahydrofuran	636	898
Tetrahydrofurfuryl acetate	1055	1585
Tetrahydrofurfuryl alcohol	884	1494
Tetrahydrofurfuryl propionate	1153	1632
Tetrahydrogeraniol	1185	1626
o-Toluyl thiol	1067	1631
Tetrahydrolinalool	1087	1397
Tetrahydromyrcenol	1090	1414
Tetrahydropyran	690	930
Tetrahydrothiophene	801	1130

COMPOUNDS	OV 101	CBWX 20M
2,3,5,6-Tetramethylpyrazine	1069	1458
Thiazole	715	1246
Thiophene	650	1035
o-Tolualdehyde	1054	1632
m-Tolualdehyde	1053	1632
p-Tolualdehyde	1067	1652
alpha-Thujene	938	1036
Thujopsene	1451	1660
Thymol	1287	2100
Tonalid	1849	2373
Triacetin	1563	2029
n-Tridecane	1300	1300
Triethyl citrate	1627	2386
3,5,5-Trimethyl-n-hexanal	963	1200
3,5,5-Trimethyl-n-hexanol	1041	1480
2,4,5-Trimethyloxazole	829	1179
2,3,5-Trimethylpyrazine	985	1387
4-(2,4,6-Trimethyl-3-cyclohexen-1-yl)-3-buten-2-one 1	1433	1881
4-(2,4,6-Trimethyl-3-cyclohexen-1-yl)-3-buten-2-one 2	1435	1892
Trimethylthiazole	981	1367
gamma-Undecalactone	1542	2210
delta-Undecalactone	1579	2251
n-Undecanal	1290	1589
n-Undecanal diethyl acetal	1572	1712
n-Undecanal dimethyl acetal	1466	1668
n-Undecane	1100	1100
n-Undecanol	1364	1822
6-Undecanol	1281	1640
10-Undecen-1-al	1280	1642
Undecenol	1350	1889
n-Undecyl acetate	1487	1775
Valencene	1487	1751
Valeraldehyde	694	1002
Isovaleraldehyde	649	937
Isovaleraldehyde propyleneglycol acetal 1	938	1167
Isovaleraldehyde propyleneglycol acetal 2	945	1181
gamma-Valerolactone	921	1617
Vanillin	1392	2449
Verbenone	1195	1730

COMPOUNDS	OV 101	CBWX 20M
Vinyl acetate	562	878
Vinyl butyrate	745	1045
Vinyl propionate	650	960
o-Xylene	884	1191
m-Xylene	863	1147
p-Xylene	860	1140
Zingerone	1625	

Appendix II

RETENTION INDICES IN INCREASING ORDER ON METHYL SILICONE OV 101

COMPOUNDS	OV 101	CBWX 20M
Methyl ether	350	524
Acetaldehyde	363	690
Propionaldehyde	481	784
Isobutyl aldehyde	500	800
Ethyl alcohol	500	900
Furan	500	786
2-Methyl-n-propan-2-ol	500	871
n-Pentane	500	500
2-Propanol	500	884
tert-Butanol	512	830
Propyl methyl ether	512	644
Methyl acetate	513	813
Acetone	530	810
Ethyl formate	530	806
n-Propanol	535	1002
Methyl vinyl ketone	550	995
Vinyl acetate	562	878
tert-Butyl methyl ether	563	688
Diethyl ether	572	590
Isopropyl formate	573	838
2,3-Butanedione	575	963
2-Butanone	579	908
Allyl formate	586	957
Allyl ethyl ether	586	767
Isopropyl ether	590	649
sec-Butanol	591	975
Ethyl acetate	595	872
n-Hexane	600	600
2,3-Butanedione diacetyl	606	963
n-Propyl formate	606	907
Methyl propionate	611	896
2-Methylfuran	614	866
Butyl methyl ether	615	755
2-Methyl-1-propanol	616	1054
tert-Amyl alcohol	631	987
Tetrahydrofuran	636	898
3-Buten-1-ol	638	1137
Isopropyl acetate	645	883
Isovaleraldehyde	649	937
2-Butene-1-ol	650	1193
1,1-Dimethoxy-n-propane	650	880
Thiophene	650	1035
Vinyl propionate	650	960
n-Butanol	655	1113
Isobutyl mercaptan	660	889
3-Methylbutan-2-ol	671	1052

COMPOUNDS	OV 101	CBWX 20M
Methyl n-propyl ketone	672	969
Isobutyl formate	673	955
Methyl isobutyrate	673	913
1-Penten-3-ol	673	1130
2-Methyltetrahydrofuran	674	901
Allyl acetate	675	1010
tert-Butyl acetate	676	893
Propyl ether	676	766
Cyclohexane	677	765
Methyl pyruvate	680	1217
Diethyl ketone	681	980
Ethyl acrylate	681	980
2,3-Pentanedione	681	1044
n-Butyl ethyl ether	684	788
sec-Amyl alcohol	685	1091
n-Pentan-2-ol	685	1091
Diethyl sulfide	690	904
Tetrahydropyran	690	930
Ethyl n-propionate	691	944
2-Ethylfuran	694	951
n-Propyl acetate	694	962
Valeraldehyde	694	1002
Pinacolone	695	960
Pyridine	695	1180
n-Butyl formate	696	996
2,5-Dimethylfuran	697	951
Methyl methacrylate	699	1008
Methylmethacrylate	699	1008
n-Heptane	700	700
3,4-Dihydropyran	705	982
Methyl n-butyrate	705	975
1,1-Diethoxyethane	710	880
N-Methylpyrrole	715	1139
Thiazole	715	1246
Isoamyl alcohol	719	1184
Methyl isobutyl ketone	725	1000
Methyl disulfide	730	1081
Isopropyl n-propionate	738	950
Pyrazine	739	1194
2,3-Butylene glycol formal	740	1037
Methyl crotonate	745	1100
Vinyl butyrate	745	1045
sec-Butyl acetate	746	982
Ethyl isobutyrate	746	956
Methyl isobutyl carbinol	748	1142
Piperidine	750	1042
Propylene glycol	750	1561

COMPOUNDS	OV 101	CBWX 20M
3-Methyl-2-pentanol	755	1054
n-Amyl alcohol	756	1213
3-Methyl-3-pentanol	757	1080
Isobutyl acetate	758	1000
4-Methyl-2-pentanol	758	1124
Paraldehyde	763	1069
Methyl isovalerate	764	1008
Methyl 2-methylbutyrate	765	1000
Ethyl n-propyl ketone	767	1055
1,1-Dimethoxy-n-butane	770	969
1-Hexen-3-ol	770	1225
Methyl n-butyl ketone	772	1070
Isoamyl formate	775	1058
2-Methylthiophene	775	1123
Ethyl pivalate	776	947
Allyl n-propionate	777	1090
1,3-Butylene glycol formal	777	1128
3,4-Hexanedione	777	1123
2-Methyl-3-pentanol	777	1121
n-Hexanal	780	1084
2-Methylthiazole	780	1256
Isobutyl acrylate	781	1107
2-n-Propylfuran	782	1083
Diisopropyl ketone	783	1007
Ethyl n-butyrate	784	1025
Ethyl pyruvate	785	1253
n-Propyl n-propionate	785	1010
1,3-Butylene glycol acetal	786	1072
n-Hexan-2-ol	786	1192
n-Hexyl angelate	788	1621
2-Methyl-5-ethylfuran	791	1024
n-Butyl acetate	793	1059
2-Methyl-2-pentanol	797	1160
3-Methyl-2-pentanol	797	1160
1-Octene	797	830
4-Methylthiazole	800	1263
n-Octane	800	800
Tetrahydrothiophene	801	1130
Cyclopentanol	802	1283
Methyl mercaptoacetate	802	1346
Cyclopentanone	805	1238
Elemol	805	1318
2-Methylpyrazine	805	1251
Methyl n-valerate	806	1076
n-Amyl formate	810	1107
Isopropyl pivalate	810	956
Ethyl chloroacetate	810	1281

COMPOUNDS	OV 101	CBWX 20M
3-Buten-2-ol	811	1022
2-Octene	811	880
n-Butyl methyl sulfide	813	1043
Furfural	815	1449
2,3-Heptanedione	816	1138
Allyl isobutyrate	820	1090
Ethyl crotonate	823	1161
Methyl hexyl ether	824	960
Methyl isoamyl ketone	825	1100
Isopropyl n-butyrate	825	1030
sec-Butyl ethyl formal	826	992
2,4-Dimethyl-n-pentan-3-ol	828	1157
1,4-Butylene glycol formal	829	1176
2,4,5-Trimethyloxazole	829	1179
Isobutyl aldehyde propyleneglycol acetal	831	1060
trans-2-Hexenal	832	1207
Ethyl 2-methylbutyrate	837	1049
2-Methyl-1-pentanol	837	1268
4-Methyl-1-pentanol	838	1282
Isobutyl aldehyde propyleneglycol acetal	840	1074
Ethyl isovalerate	840	1060
Methyl sulfoxide	840	1554
n-Propyl isobutyrate	842	1044
Pinacol	843	
cis-3-Hexenol	847	1351
Methyl tiglate	850	1188
Isobutyl n-propionate	852	1071
3-Methyl-1-pentanol	852	1297
Diallyl sulfide	854	1150
trans-2-Hexenol	854	1368
Di-n-Propyl ketone	857	1131
n-Hexanol	858	1316
Allyl isothiocyanate	860	1352
Isoamyl acetate	860	1110
p-Xylene	860	1140
2-Ethylthiophene	861	1179
n-Propyl pivalate	863	1028
m-Xylene	863	1147
2,5-Dimethylthiophene	866	1161
3-Ethyl-3-pentanol	866	1183
Allyl n-butyrate	867	1161
1,1-Dimethoxy-n-pentane	868	1057
1-Hepten-3-ol	868	1322
2,4-Dimethylthiazole	869	1271
Ethyl n-butyl ketone	869	1155
Methyl n-amyl ketone	872	1172
2,2-Dimethyl-n-pentanol	874	1405

COMPOUNDS	OV 101	CBWX 20M
Cyclohexanone	875	1306
Di-n-butyl ether	876	965
n-Butyl ethyl ether	876	965
Di-n-butyl ether	876	965
2-Ethylthiazole	879	1300
n-Heptan-4-ol	879	1250
Cyclohexanol	880	1375
n-Propyl n-butyrate	881	1110
2-n-Butylfuran	883	1130
n-Heptanal	883	1186
Isopropyl isovalerate	883	1034
Ethyl n-valerate	884	1124
Tetrahydrofurfuryl alcohol	884	1494
o-Xylene	884	1191
gamma-Butyrolactone	885	1632
n-Heptan-2-ol	888	1284
n-Butyl n-propionate	889	1130
2-Acetylfuran	892	1491
Butyl acrylate	892	1189
n-Butyl mercaptan	892	944
n-Butyl ethyl sulfide	893	1090
2,5-Dimethylpyrazine	893	1306
Acetonyl acetone	894	1500
n-Amyl acetate	895	1161
2,6-Dimethylpyrazine	895	1325
Dimethyl malonate	896	1472
Anisole	900	1327
Isobutyl isobutyrate	900	1084
2,5-Dimethyl-4-ethyloxazole	900	1231
2,3-Dimethylpyrazine	900	1330
n-Nonane	900	900
Prenyl acetate	900	1243
cis-3-Hexenyl formate	902	1252
Methyl n-hexanoate	906	1177
Ethyl acetylacetate	907	1427
Ethyl disulfide	910	1232
n-Hexyl mercaptan	910	1145
2-Ethyl-4,5-dimethyloxazole	914	1243
4,5-Dimethylthiazole	917	1359
Allyl isovalerate	920	1190
Ethyl angelate	920	1228
Diethyleneglycol monomethyl ether	920	1321
gamma-Valerolactone	921	1617
Ethyl tiglate	922	1234
Diethyleneglycol dimethyl ether	924	1396
n-Propyl isovalerate	924	1144
Isopropyl n-valerate	924	1125

COMPOUNDS	OV 101	CBWX 20M
Ethyl n-amyl ketone	928	1190
Isobutyl pivalate	933	1085
n-Propyl 2-methylbutyrate	933	1134
alpha-Thujene	938	1036
Isovaleraldehyde propyleneglycol acetal 1	938	1167
n-Butyl isobutyrate	939	1139
1,3-Butanediol	941	1692
Isobutyl n-butyrate	941	1152
5-Methylfurfural	942	1563
alpha-Pinene	942	1039
2-Ethoxythiazole	943	1380
Isovaleraldehyde propyleneglycol acetal 2	945	1181
Benzaldehyde	947	1502
Cyclohexyl formate	951	1305
Ethyl isohexanoate	951	1181
4-Ethyl isohexanoate	951	1181
2-n-Propylthiophene	951	1259
Dimethyl trisulfide	952	1400
Isoamyl n-propionate	954	1180
Camphene	954	1083
n-Amyl n-propionate	955	1180
2-Ethyl-4-methylthiazole	955	1331
Methyl furoate	956	1561
Methyl levulinate	956	1534
n-Heptanol	957	1419
2-Methyl-5-ethylthiophene	957	1245
2-Ethylbutyl acetate	959	1205
2-Ethylhexanal	959	1216
Isopropyl tiglate	959	1238
2-Isopropyl-4,5-dimethyloxazole	960	1249
n-Butyl pivalate	963	1128
3,5,5-Trimethyl-n-hexanal	963	1200
1,1-Dimethoxy-n-hexane	964	1156
Allyl n-valerate	965	1256
Benzonitrile	965	1583
1,3-Butylene glycol butyral	967	1232
Amyl vinyl carbinol	968	1426
6-Methyl-hept-5-en-2-one	968	1335
1-octene-3-ol	968	1420
Diisobutyl sulfide	969	1149
Furfuryl acetate	969	1518
2-n-Propylthiazole	970	1380
n-Butyl propyl sulfide	972	1173
Methyl n-hexyl ketone	972	1275
Sabinene	976	1130
n-Butyl n-butyrate	979	1207
Methyl benzyl ether	981	1391

COMPOUNDS	OV 101	CBWX 20M
trans-Pinane	981	1062
beta-Pinene	981	1124
n-Propyl n-valerate	981	1200
Trimethylthiazole	981	1367
cis-Dihydroocimene	982	1088
1-Methyl 4-hepten-1-ol	982	1433
Diisobutyl ketone	983	1207
2,4-Diethyl-5-methyloxazole	983	1274
2,6-Dimethyl-n-heptan-2-ol	983	1300
Ethyl n-hexanoate	983	1223
2-n-Pentylfuran	983	1229
2,6-Dimethylcyclohexanone	985	1322
n-Octanal	985	1278
2,3,5-Trimethylpyrazine	985	1387
Diethyleneglycol monoethyl ether	986	1583
N-Methyl-2-pyrrolaldehyde	986	1616
Myrcene	986	1156
cis-3-Hexenyl acetate	987	1300
2-Methyl-3-ethylpyrazine	987	1381
cis-Pinane	987	1075
tert-Butyl benzene	988	1247
n-Octan-2-ol	988	1385
2-Propionylfuran	988	1563
Isobutyl 2-methylbutyrate	991	1171
1-Decene	991	1043
4-Ethyl-5-methylthiazole	991	1400
2-Octanone	991	1304
Isobutyl isovalerate	992	1184
Diethyl malate	994	1638
n-Hexyl formate	994	1258
2-Acetylthiazone	995	1639
Allyl 2-ethylbutyrate	995	1254
trans-Dihydroocimene	995	1110
2-n-Propyl-4,5-dimethyloxazole	996	1310
Isoamyl isobutyrate	997	1187
n-Butyl lactate	997	1508
Dimethyl fumarate	997	1530
trans-2-Hexenyl acetate	997	1315
Methyl sorbate	998	1448
Diisobutyl acetal	999	1072
Isopentyl ether	999	1067
Diisoamyl ether	1000	1064
Angelica lactone	1000	1430
n-Decane	1000	1000
Diisoamyl ether	1000	1064
o-Methyl anisole	1000	1432
p-Methyl anisole	1000	1432

COMPOUNDS	OV 101	CBWX 20M
2-Isopropyl-4-methylthiazole	1000	1339
Allyl tiglate	1002	1370
Dimethyl succinate	1002	1558
alpha-Phellandrene	1002	1177
Phenol	1002	1932
p-Cresol methyl ether	1005	1415
Ethyl 2-methylpentanoate	1005	1302
Pyrrol-2-carboxaldehyde	1005	1976
Methyl n-heptanoate	1006	1276
2-Ethyl-2-n-hexanal	1007	1336
1,4-Cineole	1010	1185
sec-Butyl benzene	1011	
n-Heptyl mercaptan	1011	1256
4-Methyl-5-vinylthiazole	1011	1500
n-Heptyl formate	1012	1310
n-Hexyl acetate	1012	1307
2-Acetylpyridine	1014	1600
Isopropyl disulfide	1014	1264
Methyl octyl ether	1016	1152
Dipropyleneglycol 1	1017	1767
2-Isobutylthiazole	1020	1404
p-Cymene	1020	1272
n-Propyl tiglate	1020	1320
n-Butyl isovalerate	1021	1242
2-Isopropyl-4-ethyl-5-methyloxazole	1021	1279
Isopropyl n-hexanoate	1021	1223
Carvomenthene	1022	1150
Phenylacetaldehyde	1024	1646
cis-Ocimene	1025	1228
beta-Phellandrene	1025	1216
1,8-Cineole	1027	1228
Cyclohexyl acetate	1027	1343
Salicaldehyde	1027	1705
2-Methyl-3-isopropylpyrazine	1028	1387
n-Butyl 2-methylbutyrate	1029	1226
Ethyl 2-furoate	1029	1599
Ethyl levulinate	1029	1567
Limonene	1030	1206
Benzyl alcohol	1033	1822
Isobutyl angelate	1033	1289
n-Amyl isobutyrate	1035	1237
cis-3-Hexenyl methyl acetal	1035	1267
trans-3-Octenol	1036	1541
2-Ethyl-1-hexyl vinyl ether	1038	1165
trans-Ocimene	1038	1250
2,6-Dimethyl-hept-5-en-1-al	1039	1358
Dipropyleneglycol 2	1039	1817

COMPOUNDS	OV 101	CBWX 20M
Isobutyl n-valerate	1040	1252
2-n-Propyl-4-methylthiazole	1040	1400
cis-3-Octenol	1041	1563
3,5,5-Trimethyl-n-hexanol	1041	1480
Isoamyl n-butyrate	1042	1259
Diethyl malonate	1043	1542
Dipropyleneglycol 3	1043	1823
2-Isobutyl-4,5-dimethyloxazole	1044	1330
Dipropyleneglycol 4	1044	1848
trans-2-Octenal	1045	1427
Benzyl ethyl ether	1046	1439
1,4-Butanediol	1046	1861
Acetophenone	1048	1627
Methyl hexyl acetaldehyde	1048	1306
2-Acetylpyrrole	1050	1935
2,5-Dimethyl-4-ethylthiazole	1050	1398
p-Cresol	1051	2003
2-Ethyl-2-n-hexene-1-ol	1051	1518
Methyl phenyl carbinol	1051	1765
2-n-Butylthiophene	1052	1353
m-Tolualdehyde	1053	1632
o-Tolualdehyde	1054	1632
n-Methyl-2-acetylpyrrole	1055	1653
Tetrahydrofurfuryl acetate	1055	1585
gamma-Terpinene	1057	1251
Benzyl formate	1058	1675
Di-n-butyl ketone	1058	1330
Di-n-butyl ketone	1058	1330
Diethyleneglycol diethyl ether	1058	1572
Furfuryl n-propionate	1059	1587
n-Octanol	1061	1519
n-Amyl pivalate	1063	1228
Dihydromyrcenol	1063	1438
1,1-Dimethoxy-n-heptane	1063	1261
2-n-Propyl-4-ethyl-5methyloxazole	1064	1345
Allyl n-hexanoate	1065	1356
2-Ethyl-4,5-dimethylthiazole	1065	1429
n-Pentyl ether	1065	1165
o-Toluyl thiol	1067	1631
p-Tolualdehyde	1067	1652
cis-Linalool oxide	1068	1423
1-Nonen-3-ol	1068	1520
Isopropyl levulinate	1068	1575
2-acethylthiophene	1069	1760
n-Heptanal dimethyl acetal	1069	1265
2,3,5,6-Tetramethylpyrazine	1069	1458
2-n-Butylthiazole	1070	1480

COMPOUNDS	OV 101	CBWX 20M
m-Diethylbenzene	1070	1339
Guaiacol	1071	1840
Di-n-butyl sulfide	1073	1270
cis-3-Hexenyl methoxyformate	1073	1475
Methyl n-heptyl ketone	1074	1377
2-Methyl-3-n-propylpyrazine	1074	1462
5-Nonanone	1074	1360
Dipropyleneglycol 5	1075	1892
Ethyl sorbate	1075	1505
Isobutyl tiglate	1076	1357
n-Amyl n-butyrate	1078	1305
2-Butanoylfuran	1078	1644
n-Butyl n-valerate	1078	1305
Methyl benzoate	1078	1600
trans-Sabinene hydrate	1078	
n-Propyl n-hexanoate	1079	1298
1,1-Diethyoxyhexane	1080	1228
p-Diethylbenzene	1080	1353
alpha-p-Dimethylstyrene	1080	1278
Fenchone	1080	1410
trans-3-Heptenyl acetate	1080	1388
Methyl isohexyl carbinyl acetate	1080	1300
Ethyl n-heptanoate	1082	1321
n-Hexanal diethyl acetal	1082	1223
trans-Linalool oxide	1082	1451
trans-Linalool oxide	1082	1451
cis-3-Hexenyl n-propionate	1083	1371
2-n-Hexyl furan	1083	1326
cis-3-Heptenyl acetate	1084	1400
trans-2-Hexenyl n-propionate	1085	1370
2-Isobutyl-4-methylthiazole	1086	1420
Isoamyl 2-methylbutyrate	1087	1273
2,6-Dimethyl phenol	1087	1883
n-Nonanal	1087	1382
cis-Rose oxide	1087	1354
Tetrahydrolinalool	1087	1397
o-Diethylbenzene	1088	1372
n-Hexyl n-propionate	1088	1326
n-Nonan-2-ol	1089	1484
Tetrahydromyrcenol	1090	1414
Isoamyl isovalerate	1092	1287
2,6-Dimethylheptyl-4-acetate	1092	1265
Linalool	1092	1506
2-Nonanone	1093	1420
Amyl vinyl carbinyl acetate	1094	1365
cis-3-Hexenyl ethyl acetal	1094	1298
Acetaldehyde ethyl cis-3-hexenyl acetal	1095	1297

COMPOUNDS	OV 101	CBWX 20M
n-Heptyl acetate	1095	1361
Propyl disulfide	1096	1358
2-Cyclohexyl ethanol	1098	1668
trans-Rose oxide	1100	1370
n-Undecane	1100	1100 ?
Myrcenol	1103	1585
2-Buten-1,4-diol	1104	1983
Butyl acetoacetate	1104	1798
Phenylethyl alcohol	1104	1859
Maltol	1105	2030
2-Isobutyl-4-ethyl-5-methyloxazole	1106	1359
2-Propionyl-5-methylfuran	1106	1672
Methyl n-octanoate	1107	1378
2-Isopropyl-4,5-dimethylthiazole	1109	1439
alpha-Fenchyl alcohol	1110	1574
Norbornyl acetate	1112	1476
Allyl 2-furoate	1113	1748
n-Amylcyclopentenone 2	1113	1748
p-tert-Amylcyclohexanone	1113	1800
2-Ethyl phenol	1113	2028
Allyl levulinate	1114	1743
n-Butyl angelate	1116	1417
n-Butyl tiglate	1116	1419
Methyl 2-hydroxyisobutyrate	1116	2054
Methyl nicotinate	1116	1779
Octyl formate	1117	1426
o-Methylacetophenone	1118	1679
2-Methyl-1-octanol	1119	1573
Cyclohexyl n-propionate	1120	1408
Isopropyl n-heptanoate	1120	1317
Dihydrolinalool	1122	1512
p-Methyl benzyl alcohol	1122	1956
Diethyl trisulfide	1125	1535
Propyl furoate	1125	1700
n-Propyl levulinate	1125	1663
n-Amyl 2-methylbutyrate	1126	1324
n-gamma-Heptalactone	1126	1860
cis-3-Hexenyl isobutyrate	1129	1377
Cyclooctanol	1133	1700
n-Amyl isovalerate	1135	1337
n-Hexyl isobutyrate	1135	1333
Sabinol	1135	1683
Camphor	1136	1518
Citronellal	1137	1465
beta-Terpinenol	1137	1616
Isoamyl n-valerate	1138	1354
Isoamyl n-valerate	1138	1354

COMPOUNDS	OV 101	CBWX 20M
Dihydrocinnamic aldehyde	1139	1783
Isobutyl n-hexanoate	1140	1350
2-n-Butyl-4-methylthiazole	1141	1500
1,3-Dimethoxy benzene	1143	1740
Menthone	1143	1478
Benzyl acetate	1144	1697
n-Heptanal ethyleneglycol acetal	1144	1460
p-Hydroxyacetophenone	1144	1790
1,4-Dimethoxybenzene	1145	1705
2-Propionylpyrrol	1145	1990
Isopulegol	1145	1573
3-Ethyl phenol	1146	2150
trans-2-Nonenal	1146	1540
Dimethyl benzyl carbinol	1147	1715
Ethyl 2,4-dimethyl-1,3-dioxolylacetate 1	1147	1603
Furfuryl n-butyrate	1148	1655
p-Cresyl acetate	1150	1700
2-n-Propyl-4,5-dimethylthiazole	1151	1500
Diethyl succinate	1153	1642
2-n-Pentylthiophene	1153	1462
Tetrahydrofurfuryl propionate	1153	1632
Ethyl benzoate	1154	1647
Ethyl 2,4-dimethyl-1,3-dioxolylacetate 2	1154	1616
Lavandulol	1154	1662
Methyl phenylacetate	1154	1747
Isooctyl acetate	1154	1419
Isobutyl 2-methylpentanoate	1155	1417
m-Methylacetophenone	1156	1749
Phenylethyl formate	1156	1752
Isoborneol	1157	1660
trans-2-nonenol	1157	1691
2-n-Butyl-4-ethyl-5-methyloxazole	1159	1441
2-Methyl-2-(4-methyl n-amyl) tetrahydrofuran	1159	1325
Diethyl fumarate	1160	1632
delta-Terpinenol	1160	1655
n-Nonanol	1161	1624
n-Hexyl pivalate	1163	1328
Allyl n-heptanoate	1164	1454
Borneol	1164	1698
Cyclohexyl isobutyrate	1164	1427
2-Propionylthiophene	1164	1821
Methyl 2-methylbenzoate	1165	1709
p-Methylacetophenone	1166	1750
n-Octanal dimethyl acetal	1167	1366
cis-3-Hexenyl n-butyrate	1170	1450
cis-3-Heptenyl propionate	1171	1472

COMPOUNDS	OV 101	CBWX 20M
Menthol	1171	1612
Styrallyl acetate	1173	1673
Amyl butyl carbinol	1175	1550
5-Decanol	1175	1550
Terpinene-4-ol	1175	1628
n-Hexyl n-butyrate	1176	1398
Methyl n-octyl ketone	1176	1480
n-Butyl n-hexanoate	1177	1402
cis-4-Decenal	1177	1523
n-Propyl n-heptanoate	1177	1398
Isoamyl tiglate	1178	1469
n-Heptanal diethyl acetal	1179	1319
Isobutyl 2-hexenoate	1180	1461
Ethyl n-octanoate	1180	1423
trans-2-Hexenyl n-butyrate	1180	1461
1-Octene-3-yl propionate	1180	1432
2-Pentanoylfuran	1180	1747
Methyl salicylate	1181	1754
trans-2-Heptenyl propionate	1182	1497
Methyl chavicol	1182	1670
Isobutyl levulinate	1183	1696
Dihydrocarvone	1183	1600
Estragole	1183	1652
2-n-Heptylfuran	1184	1429
n-Amyl n-valerate	1185	1401
Isobutyrophenone	1185	1685
2-Methyl-5-acetylthiophene	1185	1836
alpha-Terpineol	1185	1661
Tetrahydrogeraniol	1185	1626
Cyclohexanone 1,3-butylene ketal	1186	1500
cis-3-Hexenyl lactate	1187	1727
n-Decanal	1188	1485
n-Decanal	1188	1485
Dihydrocarveol	1188	1713
Isopropyl benzoate	1189	1639
n-Decan-2-ol	1190	1585
Methyl 3-methylbenzoate	1190	1744
2-Butanoyl-5-methylfuran	1192	1748
2-Isobutyl-4,5-dimethylthiazole	1193	1517
Dihydroanethole	1193	1600
n-Octyl acetate	1193	1459
sec-Butylcyclohexanone 1	1194	1564
Methyl p-toluate	1194	1725
Verbenone	1195	1730
sec-Butylcyclohexanone 2	1196	1566
Bromostyrol	1197	1778
Geranonitrile 1	1199	1680

COMPOUNDS	OV 101	CBWX 20M
Methyl 4-methylbenzoate	1199	1755
n-Dodecane	1200	1200
Phenylacetaldehyde dimethyl acetal	1200	1665
p-Isopropylphhenol	1200	2178
Dihydromyrcenyl acetate	1202	1431
Isobutyl disulfide	1205	1435
Linalyl formate	1206	1570
Methyl n-nonanoate	1207	1479
p-tert-Butylcyclohexanone	1208	1645
Carveol 1	1209	1790
Cyclohexyl n-butyrate	1209	1492
p-tert-Butylphenylpropyl aldehyde 1	1212	2030
Dimethyl adipate	1212	1779
Ethyleneglycol monophenyl ether	1213	2100
Citronellol	1215	1722
n-Amyl tiglate	1216	1519
o-Methylphenyl ethyl alcohol	1216	2012
2,4-Dimethyl-5-acetylthiazole	1217	1835
Benzyl acetone	1218	1849
Nerol	1218	1757
3-Phenylpropyl alcohol	1218	1993
Ethyl phenylacetate	1219	1773
Isopropyl n-octanoate	1219	1419
Fenchyl acetate	1220	1473
alpha-Methylcitronellol	1220	1540
Benzyl ethyl carbionl	1222	1882
Carveol 2	1222	1820
Citral 1	1222	1661
cis-3-Hexenyl isovalerate	1223	1480
2-Ethyl-1-hexyl acrylate	1224	1494
n-Hexyl 2-methylbutyrate	1224	1418
n-Butyl levulinate	1225	1760
gamma-Octalactone	1225	1883
Acetaldehyde styleneglycol acetal 1	1226	1780
2,4-Dimethyl benzyl alcohol	1226	2032
Cuminaldehyde	1227	1766
Di-n-Propyl malonate	1227	1700
Neral	1227	1680
Di-n-Propyl malonate	1227	1700
Isobornyl formate	1228	1596
Carvone	1228	1715
n-Hexyl isovalerate	1228	1433
Pulegone	1230	1662
2-Ethyl-1-hexyl propionate	1231	1450
2-Cyclohexylethyl acetate	1233	1591
2,4-Dimethylacetophenone	1233	1809
n-Heptyl isobutyrate	1233	1433

COMPOUNDS	OV 101	CBWX 20M
Phenylethyl acetate	1233	1785
Anis aldehyde	1234	1982
Benzyl n-propionate	1234	1679
2-Hexoxyacetaldehyde dimethyl acetal	1234	1528
Citronellyl vinyl ether	1235	1445
Geranonitrile 2	1236	1723
Isoamyl n-hexanoate	1238	1451
n-Butyl benzyl ether	1238	1613
Chavicol	1238	2300
Allyl benzoate	1239	1800
Bornyl formate	1239	1610
Isobutyl n-heptanoate	1239	1448
Geraniol	1243	1797
cis-3-Decenol	1245	1765
Furfuryl n-valerate	1245	1752
Linalyl acetate	1246	1538
n-Amylcyclopentenone 1	1247	1689
Myrcenyl acetate	1247	1574
Piperitone	1247	1739
Isobutyl 2-ethyl-n-hexanoate	1248	1400
Citral 2	1249	1706
Geranonitrile 3	1249	1792
Cinnamic aldehyde	1250	1996
2-n-Butyl-4,5-dimethylthiazole	1251	1600
9-Decen-1-ol	1251	1775
2-Butanoylthiophene	1252	1894
Geranial	1252	1730
delta-Octalactone	1252	1929
Methyl octyl acetaldehyde	1254	1521
n-Propyl benzoate	1254	1745
Geranyl ethyl ether 1	1255	1476
cis-3-Heptenyl butyrate	1255	1545
2-n-Hexylthiophene	1256	1564
2-Decen-1-ol	1257	1794
trans-2-Decenol	1257	1792
Ethyl salicylate	1257	1787
Di-n-amyl ketone	1258	1528
Diisoamyl ketone	1258	1528
Methyl dihydrocinnamate	1258	1842
Isopulegyl acetate	1258	1585
Acetone 1-phenyl-1,2-ethanediol ketal	1260	1766
Carvone oxide	1261	1805
Citronellyl formate	1261	1600
Sabinyl acetate	1262	1651
Decanol	1263	1723
n-Heptyl pivalate	1263	1428
4-Methyl-5-hydroxy phenylacetate	1263	1767

COMPOUNDS	OV 101	CBWX 20M
Cyclohexyl isovalerate	1264	1527
Allyl n-octanoate	1265	1554
4-Ethylguaiacol	1265	2011
Anis alcohol	1267	2210
Neryl formate	1267	1663
n-Nonanal dimethyl acetal	1267	1465
D-n-hexyl ether	1269	1349
Hydroxycitronellal	1269	1882
o-Methoxyacetophenone	1269	1975
Anethole	1270	1809
cis-3-Hexenyl n-valerate	1270	1584
Lavandulyl acetate	1274	1597
Geranyl ethyl ether 2	1275	1506
trans-2-Heptenyl butyrate	1275	1568
n-Heptyl n-butyrate	1275	1503
trans-2-Hexenyl n-valerate	1275	1560
n-Hexyl n-valerate	1275	1498
n-Butyl n-heptanoate	1276	1500
Methyl n-nonyl ketone	1276	1585
n-Octanal diethyl acetal	1276	1417
Benzyl isobutyrate	1277	1771
n-Propyl n-octanoate	1277	1498
Safrole	1278	1876
n-Amyl n-hexanoate	1279	1500
Isobornyl acetate	1279	1584
m-Methoxyacetophenone	1279	2011
Cyclooctanyl acetate	1280	1645
Decahydro-beta-naphthol 1	1280	1883
n-Ethyl n-nonanoate	1280	1523
2-Methyl-5-propionylthiophene	1280	1900
10-Undecen-1-al	1280	1642
2-Hexanoylfuran	1281	1850
6-Undecanol	1281	1640
Dihydroterpinyl acetate	1282	1561
Dimethyl phenyl ethyl carbinol	1282	1916
Geranyl formate	1282	1684
Cuminalcohol	1283	2045
4-Methyl-5-thiazoleethanol	1283	2216
Octyl propionate	1283	1536
Isoamyl levulinate	1284	1807
2-n-Octylfuran	1285	1530
o-tert-Butylcyclohexyl acetate 1	1286	1565
Dihydrocuminalcohol	1286	1981
Dihydrosafrole	1286	1822
n-Nonanoic acid	1286	2110
Isoamyl 2-furoate	1287	1840
Thymol	1287	2100

COMPOUNDS	OV 101	CBWX 20M
o-Aminoacetophenone	1288	2181
n-Undecanal	1290	1589
p-Cresyl isobutyrate	1291	1763
Isopropyl benzyl carbinol	1291	1914
Benzyl isopropyl carbinol	1292	1912
n-Nonyl acetate	1292	1560
Isopropyl salicylate	1292	1773
Butyl disulfide	1295	1580
Isoamyl benzyl ether	1297	1668
Carvacrol	1297	2159
Cinnamyl alcohol	1300	2207
2-Nonyn-1-al dimethyl acetal	1300	1666
n-Propyl phenylacetate	1300	1848
n-Tridecane	1300	1300
Dimethyl benzyl carbinyl acetate	1302	1755
o-tert-Butylcyclohexyl acetate 2	1304	1596
p-Cresyl ethylcarbonate	1304	1919
Indole	1304	2351
Anisyl formate	1307	1710
Methyl n-decanoate	1307	1581
p-Ethoxybenzaldehyde	1308	2017
alpha-Methylcinnamic aldehyde	1309	1992
Benzyl n-propyl carbinol	1310	1955
Isobutyl benzoate	1310	1771
Cyclohexyl n-valerate	1310	1595
trans-2-trans-4-Decadienol	1310	1938
n-Hexyltiglate	1310	1621
Dihydrolinalyl acetaldehyde	1315	1659
Heliotropine	1318	2171
Methyl 3-hydroxybutyrate	1318	1454
Methyl undecyl ether	1318	1453
Isopropyl n-nonanaote	1318	1516
Dihydrocarvyl acetate	1319	1670
Benzyl n-butyrate	1322	1856
trans-p-tert-Butylcyclohexyl acetate	1322	1628
Decahydro-beta-naphthol 2	1323	1995
n-Heptyl 2-methylbutyrate	1324	1518
Linalyl n-propionate	1324	1596
n-Amyl levulinate	1325	1860
p-Methoxyacetophenone	1327	2115
Myrcenyl propionate	1327	1625
Propionaldehyde 1-phenyl-1,2-ethanediol acetal 1	1327	1871
Allyl anthranilate 1	1328	2196
gamma-Nonalactone	1328	1991
Phenylethyl n-propionate	1328	1855
Cinnamyl formate	1330	2094

COMPOUNDS	OV 101	CBWX 20M
n-Butyl n-butyryllactate	1331	1733
Ethyl phenyl ethyl acetal	1332	1770
Methyl anthranilate	1332	2181
n-Octyl isobutyrate	1332	1529
p-Isopropylacetophenone	1332	1912
Acetaldehyde styleneglycol acetal 2	1333	1793
Propionaldehyde 1-phenyl-1,2-ethanediol acetal 2	1333	1880
Terpinyl acetate	1333	1687
Terpinyl formate	1333	1666
Isoamyl n-heptanoate	1334	1548
Citronellyl acetate	1335	1645
Benzal acetone	1337	2065
Isobutyl n-octanoate	1338	1543
Allyl salicylate	1339	1946
Methyllavender ketone	1341	2067
Furfuryl n-hexanoate	1343	1850
alpha-Methylcinnamyl alcohol	1343	2252
Cyclohexyl glycidate	1344	1875
Neryl acetate	1345	1699
Hydroxycitronellol	1347	2143
gamma-Ionone	1347	1882
Phenylpropyl acetate	1347	1926
2,4-Dimethyl benzylacetate	1348	1916
Undecenol	1350	1889
Eugenol	1351	2103
Methyl nonyl acetaldehyde	1352	1621
2-Methylundecanal	1353	1609
n-Butyl benzoate	1354	1841
n-Butyl 3-furyl-n-propionate	1354	1728
Methyl anisate	1354	2071
2-Pentanoylthiophene	1355	1993
delta-Nonalactone	1356	2038
n-Propyl salicylate	1357	1878
Diethyl adipate	1358	1858
2-Methyl-5-isopropylacetophenone	1358	1876
Dihydrocoumarin	1359	2286
2-n-Heptylthiophene	1359	1670
cis-p-tert-Butylcyclohexyl acetate	1360	1675
Di-n-Propyl fumarate	1360	1806
Isosafrole	1360	2029
Acetaldehyde linalyl ethyl acetal 1	1361	1558
Acetaldehyde linalyl ethyl acetal 2	1362	1564
Methyl isogeranylacetate	1362	1765
Allyl cinnamate 1	1363	2054
Geranyl acetate	1364	1735
n-Undecanol	1364	1822

COMPOUNDS	OV 101	CBWX 20M
Allyl n-nonanoate	1365	1655
Methyl cinnamate	1365	2051
beta-Methylcinnamyl alcohol	1365	2283
Isobutyl benzyl carbinol	1366	1983
n-Decanal dimethyl acetal	1366	1567
n-Decanal dimethyl acetal	1366	1567
Linalyl isobutyrate	1366	1597
4-Methyl-5-thiazoleethanol acetate	1368	2077
Diphenyl	1369	1981
cis-3-Hexenyl n-hexanoate	1370	1654
Isobutyl phenylacetate	1371	1864
Hexyl n-hexanoate	1371	1599
n-Heptyl valerate	1372	1614
n-Butyl n-octanoate	1373	1600
n-Octyl n-butyrate	1373	1597
n-Pentadecanol	1373	2252
Acetaldehyde citronellyl methyl acetal	1374	1596
Benzyl isovalerate	1374	1880
Dihydroisojasmone	1374	1842
n-Nonanal diethyl acetal	1374	1514
Phenylethyl isobutyrate	1374	1855
n-Amyl n-heptanoate	1375	1600
trans-2-Hexenyl n-hexanoate	1375	1656
Isobornyl n-propionate	1376	1676
cis-3-Decenyl acetate	1376	1701
Methyl n-decyl ketone	1377	1688
n-Propyl n-nonanoate	1377	1598
cis-Jasmone	1378	1914
Ethyl n-decanoate	1379	1624
Acetylcymene	1381	1928
9-Decen-1-yl acetate	1383	1722
Isoamyl phenylethyl ether	1384	1741
Isobutyl aldehyde 1-phenyl-1,2-ethanediol acetal	1384	1900
2-Heptanoylfuran	1384	1956
Diphenyl oxide	1386	1991
p-Cresyl isovalerate	1389	1898
Methyl N-methylanthranilate	1389	2042
Anisyl acetate	1390	2199
Dimethyl benzyl carbinyl n-propionate	1391	1810
n-Dodecanal	1392	1695
Vanillin	1392	2449
Decyl acetate	1393	1662
Isoeugenol 1	1394	2186
Ethyl anthranilate	1396	2232
Methyl 10-undecenoate	1396	1733
alpha-Copaene	1398	1519

COMPOUNDS	OV 101	CBWX 20M
1,4-Dimethoxy-2-tert-butylbenzene	1398	1870
Dihydrojasmone	1400	1892
omega-Methyl undecylenate	1400	1747
2-n-Pentyl-3-methyl-2-cyclopenten-1-one	1400	1892
Phenylethyl pivalate	1400	1832
n-Tetradecane	1400	1400
Cuminyl acetate	1401	1952
Allyl-3-cyclohexyl propionate	1405	1800
2-Methyl-2-phenylhexan-4-one	1405	1938
beta-Bourbonene	1406	1546
Dihydro-nor-dicyclopentadienyl acetate	1406	1881
n-Butyl phenylacetate	1408	1932
m-Aminoacetophenone	1409	2181
Isobutyl salicylate	1410	1896
Cyclohexyl n-hexanoate	1411	1695
Nopyl acetate	1412	1777
Nopylacetate	1412	1777
Dimethyl suberate	1414	1985
Isoamyl furylpropionate	1415	1894
alpha-Ionone	1416	1833
Isopropyl n-decanoate	1417	1615
Dihydrocuminyl acetate	1418	1900
Coumarin	1418	2361
Cinnamyl acetate	1419	2103
Linalyl n-butyrate	1420	1680
Isoamyl benzoate	1421	1894
Benzyl n-valerate	1421	1956
Allyl phenylacetate	1422	2175
2-Methyl-1-undecanol	1422	1875
n-Octyl 2-methylbutyrate	1422	1615
Phenylethyl butyrate	1422	1930
Phenylethyl butyrate	1422	1930
Citronellyl ethyl acetal	1423	1626
Acetaldehyde phenylethyl n-propyl acetal	1424	1836
Geranyl n-propionate	1424	1799
Ethyl anisate	1426	2110
Terpinyl n-propionate	1426	1747
Cedryl acetate 1	1427	1591
Citronellyl n-propionate	1427	1700
Methylisoeugenol 1	1427	2044
Caryophyllene	1428	1617
Dimethyl phenyl ethyl carbinyl acetate	1428	1908
Allyl phenoxyacetate	1429	2176
Isoamyl n-octanoate	1433	1648
4-(2,4,6-Trimethyl-3-cyclohexen-1-yl)-3-buten-2-one 1	1433	1881
6-Amyl-alpha-pyrone	1434	2166

COMPOUNDS	OV 101	CBWX 20M
Isobutyl beta-2-furylacrylate	1435	2006
4-(2,4,6-Trimethyl-3-cyclohexen-1-yl)-3-buten-2-one 2	1435	1892
alpha-Cedrene	1436	1600
Neryl n-propionate	1436	1771
gamma-Decalactone	1437	2101
Isoeugenol 2	1438	2269
Carvyl propionate	1440	1833
Ethyl cinnamate	1443	2095
Furfuryl n-heptanoate	1443	1950
beta-Methyl-p-isopropylphenyl-propionaldehyde	1444	1954
beta-Copaene	1445	1626
Phenylpropyl n-propionate	1445	1994
beta-Cedrene	1446	1633
Phenoxyethyl propionate	1447	2126
Ethylvanillin	1448	2414
2-Decenal	1449	1842
Linalyl 2-methylbutyrate	1450	1695
Thujopsene	1451	1660
5-Acetyl-2-methyl-8-isopropyl-(2,2,2) bicyclo-2-octene 1-epitone 1	1452	1858
n-Amyl furylpropionate	1453	1947
n-Amyl benzoate	1454	1940
n-Butyl salicylate	1457	1976
Di-n-heptyl ether	1458	1550
Di-n-hexyl ketone	1458	1728
2-Hexanoylthiophene	1459	2104
Jasmal	1459	1879
Phenyl isobutyl methyl carbinyl acetate	1459	1943
Linalyl isovalerate	1461	1698
Allyl n-decanoate	1462	1755
delta-Decalactone	1463	2144
2-n-Octylthiophene	1463	1780
Acetaldehyde di-(cis-3-hexenyl) acetal	1465	1700
cis-3-Hexenyl n-heptanoate	1465	1743
alpha-Humulene	1465	1682
n-Undecanal dimethyl acetal	1466	1668
5-Acetyl-2-methyl-8-isopropyl-(2,2,2) bicyclo-2-octene 2-epitone 2	1467	1883
Terpinyl isobutyrate	1467	1748
Isoamyl phenylacetate	1468	1991
n-Dodecanol	1468	1925
Methylisoeugenol 2	1468	2126
Citronellyl isobutyrate	1469	1705
Ethyl 10-undecenoate	1469	1775
n-Hexyl n-heptanoate	1470	1699

COMPOUNDS	OV 101	CBWX 20M
n-Amyl n-octanoate	1471	1700
Methylionone 1	1471	1836
Phenylethyl 2-methylbutyrate	1472	1945
Bornyl butyrate	1473	1760
n-Decanal diethyl acetal	1473	1613
n-Decanal diethyl acetal	1473	1613
Dihydropseudo ionone	1473	1885
p-Methoxybenzylacetone	1473	2236
Benzyl tiglate	1474	2075
trans-2-Hexenyl n-heptanoate	1474	1755
beta-Ionone	1474	1918
Neryl isobutyrate	1474	1764
Octyl-n-valerate	1474	1719
Phenylethyl isovalerate	1474	1955
n-Butyl n-nonanoate	1475	1700
gamma-Muurolene	1475	
Dimethyl benzyl carbinyl n-butyrate	1476	1889
n-Propyl n-decanoate	1476	1697
Alloaromadendrene	1478	1662
Ethyl n-undecanoate	1479	1725
Methyl n-undecyl ketone	1479	1792
Anisyl n-propionate	1482	2205
p-Cresyl isotiglate	1482	2011
alpha-Methylcinnamyl acetate	1484	2158
Heliotropyl acetate	1485	2325
Isopropyl cinnamate	1485	2097
n-Decyl n-propionate	1486	1729
Phenoxyethyl isobutyrate	1486	2106
2-Octanoylfuran	1487	2062
n-Undecyl acetate	1487	1775
Valencene	1487	1751
3-Methylcoumarin	1490	2424
Phenylpropyl isobutyrate	1490	1996
Isopropyl N-methylanthranilate	1491	2058
Allyl anthranilate 2	1493	2381
Geranyl isobutyrate	1493	1795
beta-Phenoxyethyl isobutyrate	1493	2100
2-Cyclohexylcyclohexanone	1496	1975
Linalyl n-valerate	1500	1765
alpha-Muurolene	1500	1730
n-Pentadecane	1500	1500
n-Propyl anthranilate	1500	2320
o-Methoxy cinnamic aldehyde	1504	2430
Methyl isovalerate	1505	1734
p-Aminoacetophenone	1506	2181
n-Amyl phenylacetate	1506	2047
Methylionone 2	1506	1897

COMPOUNDS	OV 101	CBWX 20M
p-tert-Butyl-alpha-methylhydrocinnamic aldehyde	1506	2039
Musk xylol	1506	2475
Methyl n-dodecanoate	1507	1785
Anis aldehyde propyleneglycol acetal 1	1508	2231
Citronellyl n-butyrate	1511	1786
Anis aldehyde propyleneglycol acetal 2	1512	2234
Bornyl isovalerate	1512	1774
Terpinyl n-butyrate	1514	1828
Cinnamyl n-propionate	1515	2169
Isopropyl n-undecanoate	1516	1715
Phenylethyl n-valerate	1517	2034
gamma-Cadinene	1518	1766
Calamenene	1518	1842
Neryl n-butyrate	1519	1838
Anisyl isobutyrate	1520	
Benzyl n-hexanoate	1521	2057
delta-Cadinene	1524	1761
alpha-Nerolidol	1524	1961
Citronellyl isocrotonate	1526	1833
n-Propyl anisate	1527	2205
Isoamyl salicylate	1528	2021
Methylionone 3	1530	1930
Dibutyl butyrolactone	1531	2141
n-Propyl cinnamic aldehyde	1531	2111
3,4-Dimethoxyacetophenone	1532	2393
Geranyl n-butyrate	1532	1872
Isoamyl n-nonanoate	1533	1748
Di-n-butyl succinate	1534	2000
alpha-n-Butyl cinnamic aldehyde	1535	2160
Phenylpropyl n-butyrate	1535	2083
Allyl cinnamate 2	1537	2258
Rosephenone	1538	2172
Acetyleugenol	1541	2277
gamma-Undecalactone	1542	2210
Helional	1543	2383
Tangerinal 1	1543	1857
n-Amyl furfurylacrylate	1544	2153
6-Methylcoumarin	1545	2630
Cyclohexenylcyclohexanone	1553	2128
Diethyl suberate	1553	2065
cis-3-Hexenyl benzoate	1553	2122
7-Methylcoumarin	1553	2620
beta-Nerolidol	1553	2000
Cinnamyl isobutyrate	1555	2171
Isobutyl anthranilate	1556	2347
n-Amyl salicylate	1557	2077

COMPOUNDS	OV 101	CBWX 20M
Citronellyl crotonate	1558	1929
omega-Decenyl butyrate	1558	1870
Di-n-amyl fumarate	1558	2006
Di-n-butyl fumarate	1558	2006
n-Hexyl benzoate	1558	2066
n-Propyl N-methylanthranilate	1560	2166
Phenylethyl tiglate	1562	2154
Citronellyl isovalerate	1563	1800
cis-3-Decenyl butyrate	1563	1866
trans-2-Decenyl butyrate	1563	1866
Triacetin	1563	2029
cis-3-Hexenyl n-octanoate	1564	1838
n-Hexyl octanoate	1564	1805
Methylionone 4	1564	1981
Diethyl phthalate	1565	2303
n-Dodecanal dimethyl acetal	1565	1769
2-Heptanoylthiophene	1565	2209
n-Propyl 10-undecenaote	1565	1860
Terpinyl isovalerate	1565	1858
Tangerinal 2	1566	1892
7,8-Epoxy-1,3,3,11-tetramethyl-(5,4,0,0(2.4))-undecane	1568	1912
Anisyl n-butyrate	1569	2274
n-Undecanal diethyl acetal	1572	1712
trans-2-Hexenyl n-octanoate	1573	1853
Geranyl 2-methylbutyrate	1574	1886
Neryl isovalerate	1574	1864
n-Butyl n-decanoate	1575	1798
n-Propyl n-undecanoate	1576	1797
Ethyl n-dodecanoate	1579	1826
delta-Undecalactone	1579	2251
Methyl n-dodecyl ketone	1580	1893
Linalyl n-hexanoate	1582	1843
alpha-Cedrene epoxide	1585	1961
Isobutyl anisate	1586	2233
2-Nonanoylfuran	1588	2163
Phenylpropyl isovalerate	1590	2100
n-Butyl methylphenylglycidate 1	1591	2200
6-sec-Butylquinoline	1592	2235
Methyl naphthyl ketone	1592	2471
Geranyl isovalerate	1593	1895
2-n-Amyl-3-acetonyl-1-cyclopentanone	1600	2259
n-Butyl anthranilate	1600	2419
n-Hexadecane	1600	1600
Benzophenone	1604	2410
Cinnamyl n-butyrate	1604	2247
Allylionone 1	1605	1993

COMPOUNDS	OV 101	CBWX 20M
Isobutyl cinnamate	1605	2228
n-Hexyl phenylacetate	1607	2148
Citronellyl n-valerate	1608	1880
Cedrol	1609	2100
cis-3-Hexenyl phenylacetate	1610	2220
2-Hexylcyclopentanone acetic acid methyl ether	1610	2200
Isolongiforanone 1	1610	2072
Neryl n-valerate	1610	1930
Methyl n-tridecanoate	1612	1895
Terpinyl n-valerate	1614	1928
Dimethyl sebacate	1616	2195
Isobutyl N-methylanthranilate	1617	2174
Isobutyl 10-undecenoate	1617	1900
Phenylethyl n-hexanoate	1618	2134
Benzyl n-heptanoate	1620	2158
Isolongiforanone 2	1622	2112
Zingerone	1625	
Triethyl citrate	1627	2386
Triethyl citrate	1627	2386
alpha-n-Amylcinnamic aldehyde	1631	2211
Dibenzyl ether	1631	2323
n-Butyl anisate	1632	2305
Geranyl n-valerate	1632	1960
Isoamyl n-decanoate	1633	1848
Phenylpropyl n-valerate	1635	2183
gamma-Dodecalactone	1647	2317
Geranyl tiglate	1650	1985
cis-3-Hexenyl salicylate	1654	2227
Cinnamyl isovalerate	1655	2271
Isoamyl anthranilate	1656	2447
Di-n-butyl adipate	1658	2087
n-Butyl N-methylanthranilate	1660	2266
n-Butyl 10-undecenoate	1660	1954
Furfuryl disulfide	1660	2600
Methylzingerone	1660	2640
cis-3-Hexenyl n-nonanoate	1664	1938
n-Hexyl salicylate	1664	2175
Anisyl n-valerate	1665	
n-Heptyl n-octanoate	1666	1892
2-Octanoylthiophene	1667	2313
Patchouli alcohol	1667	2156
n-Hexyl nonanoate	1668	1900
Linalyl n-heptanoate	1670	1930
n-Dodecanal diethyl acetal	1671	1811
trans-2-Hexenyl n-nonanoate	1673	1953
n-Propionyl methylanthranilate	1673	2350

COMPOUNDS	OV 101	CBWX 20M
n-Butyl n-undecanoate	1674	1900
delta-Dodecalactone	1675	2358
n-Propyl n-dodecanoate	1676	1897
Methyl N-propylanthranilate	1678	2575
Allylionone 2	1679	2014
Guaiacyl n-caproate	1681	2296
Isoamyl anisate	1686	2333
Allylionone 3	1689	2146
2-Decanoylfuran	1689	2264
n-Amyl anthranilate	1700	2510
n-Butyl methylphenylglycidate 2	1700	2346
n-Heptadecane	1700	1700
Cinnamyl n-valerate	1705	2347
4-Acetyl-6-tert-butyl-1,1-dimethylindan	1706	2145
Methyl n-tetradecanoate	1707	1990
Neryl n-hexanoate	1709	2021
Isoamyl pyruvate	1712	1910
Isoamyl 10-undecenoate	1717	2000
n-Heptyl phenylacetate	1717	2265
Phenylethyl n-heptanoate	1718	2233
Isoamyl cinnamate	1719	2355
Benzyl n-octanoate	1720	2260
alpha-n-Hexylcinnamic aldehyde	1727	2309
Cycohexadecanone	1731	2392
n-Amyl anisate	1732	2405
Isoamyl n-undecanoate	1733	1948
Allylionone 4	1734	2199
n-Butyl p-hydroxybenzoate	1738	
Methyl n-hexyl ketone 1-phenyl-1,2- ethanediol ketal 1	1739	2224
Benzyl benzoate	1741	
Phenyl salicylate	1742	
Bornyl benzoate	1749	
Diethyl sebacate	1752	2272
p-Cresyl n-octanoate	1755	2264
Methyl n-hexyl ketone 1-phenyl-1,2- alpha-n-Amylcinnamyl acetate	1757	2318
Methyl n-hexyl ketone 1-phenyl-1,2- ethanediol ketal 2	1758	2248
n-Amyl 10-undecenoate	1760	2053
cis-3-Hexenyl decanoate	1760	2038
Anisyl n-hexanoate	1763	
p-Cresyl benzoate	1764	
Cedryl acetate 2	1766	2173
Acetylcedrene	1768	2213
n-Butyl n-dodecanoate	1772	2000
Ethyl n-tetradecanoate	1780	2027
n-Heptyl salicylate	1790	2332

COMPOUNDS	OV 101	CBWX 20M
Nootketone	1802	2250
Cinnamyl n-hexanoate	1805	2445
cis-3-Hexenyl anthranilate	1807	
Neryl n-heptanoate	1808	2120
Isopropyl n-tetradecanoate	1811	2017
Isopropyl n-dodecanoate	1814	1814
Phenylethyl n-octanoate	1819	2337
Benzyl n-nonanoate	1823	2362
Pentalide	1823	2255
p-Cresyl phenylacetate	1827	
alpha-n-Heptylcinnamic aldehyde	1827	2409
Isoamyl n-dodecanoate	1829	2048
Geranyl n-heptanoate	1831	2157
Muscone	1831	2281
Galaxolide 1	1837	2299
Galaxolide 2	1837	2307
Phenylethyl benzoate	1841	
Phenylethyl benzoate	1841	
Tonalid	1849	2373
p-Cresyl salicylate	1850	
Nerolidylethanol	1851	2462
Anisyl n-heptanoate	1862	
n-Butyl n-tridecanoate	1880	2118
Octyl salicylate	1895	2435
n-Nonadecane	1900	1900
Cinnamyl n-heptanoate	1905	2545
Methyl hexadecanoate	1911	2204
Phenylethyl n-nonanoate	1921	2439
Benzyl n-decanoate	1923	2460
2-Mercaptobenzothiazole	1944	
Geranyl benzoate	1949	
n-Butyl n-tetradecanoate	1977	2229
n-Eicosane	2000	2000
Phenylethyl n-decanoate	2022	2540
Cinnamyl cinnamate	2055	
Cinnamyl cinnamate	2055	
n-Butyl n-pentadecanoate	2080	2330
Phenylethyl anthranilate	2091	
n-Heneicosane	2100	2100
Methyl octadecanoate	2101	2418
Phenylethyl cinnamate	2147	
Ethyl oleate	2155	2489
n-Butyl n-hexanoate	2180	2435
Ethyl n-octanoate	2180	2460

Appendix III

RETENTION INDICES IN INCREASING ORDER ON POLYETHYLENE GLYCOL CARBOWAX 20M

COMPOUNDS	OV 101	CBWX 20M
n-Pentane	500	500
Methyl ether	350	524
Diethyl ether	572	590
n-Hexane	600	600
Propyl methyl ether	512	644
Isopropyl ether	590	649
tert-Butyl methyl ether	563	688
Acetaldehyde	363	690
n-Heptane	700	700
Butyl methyl ether	615	755
Cyclohexane	677	765
Propyl ether	676	766
Allyl ethyl ether	586	767
Propionaldehyde	481	784
Furan	500	786
n-Butyl ethyl ether	684	788
Isobutyl aldehyde	500	800
n-Octane	800	800
Ethyl formate	530	806
Acetone	530	810
Methyl acetate	513	813
tert-Butanol	512	830
1-Octene	797	830
Isopropyl formate	573	838
2-Methylfuran	614	866
2-Methyl-n-propan-2-ol	500	871
Ethyl acetate	595	872
Vinyl acetate	562	878
1,1-Diethyoxyethane	710	880
1,1-Dimethoxy-n-propane	650	880
2-Octene	811	880
Isopropyl acetate	645	883
2-Propanol	500	884
Isobutyl mercaptan	660	889
tert-Butyl acetate	676	893
Methyl propionate	611	896
Tetrahydrofuran	636	898
Ethyl alcohol	500	900
n-Nonane	900	900
2-Methyltetrahydrofuran	674	901
Diethyl sulfide	690	904
n-Propyl formate	606	907
2-Butanone	579	908
Methyl isobutyrate	673	913
Tetrahydropyran	690	930
Isovaleraldehyde	649	937

COMPOUNDS	OV 101	CBWX 20M
Methyl acrylate		938
n-Butyl mercaptan	892	944
Ethyl n-propionate	691	944
Ethyl pivalate	776	947
Isopropyl n-propionate	738	950
2,5-Dimethylfuran	697	951
2-Ethylfuran	694	951
Isobutyl formate	673	955
Ethyl isobutyrate	746	956
Isopropyl pivalate	810	956
Allyl formate	586	957
Methyl hexyl ether	824	960
Pinacolone	695	960
Vinyl propionate	650	960
n-Propyl acetate	694	962
2,3-Butanedione	575	963
2,3-Butanedione diacetyl	606	963
Di-n-butyl ether	876	965
n-Butyl ethyl ether	876	965
Di-n-butyl ether	876	965
1,1-Dimethoxy-n-butane	770	969
Methyl n-propyl ketone	672	969
sec-Butanol	591	975
Methyl n-butyrate	705	975
Diethyl ketone	681	980
Ethyl acrylate	681	980
sec-Butyl acetate	746	982
3,4-Dihydropyran	705	982
tert-Amyl alcohol	631	987
sec-Butyl ethyl formal	826	992
Methyl vinyl ketone	550	995
n-Butyl formate	696	996
Isobutyl acetate	758	1000
n-Decane	1000	1000
Methyl isobutyl ketone	725	1000
Methyl 2-methylbutyrate	765	1000
n-Propanol	535	1002
Valeraldehyde	694	1002
Diisopropyl ketone	783	1007
Methyl methacrylate	699	1008
Methylmethacrylate	699	1008
Methyl isovalerate	764	1008
Allyl acetate	675	1010
n-Propyl n-propionate	785	1010
3-Buten-2-ol	811	1022
2-Methyl-5-ethylfuran	791	1024
Ethyl n-butyrate	784	1025

COMPOUNDS	OV 101	CBWX 20M
n-Propyl pivalate	863	1028
Isopropyl n-butyrate	825	1030
Isopropyl isovalerate	883	1034
Thiophene	650	1035
alpha-Thujene	938	1036
2,3-Butylene glycol formal	740	1037
alpha-Pinene	942	1039
Piperidine	750	1042
n-Butyl methyl sulfide	813	1043
1-Decene	991	1043
2,3-Pentanedione	681	1044
n-Propyl isobutyrate	842	1044
Vinyl butyrate	745	1045
Ethyl 2-methylbutyrate	837	1049
3-Methylbutan-2-ol	671	1052
3-Methyl-2-pentanol	755	1054
2-Methyl-1-propanol	616	1054
Ethyl n-propyl ketone	767	1055
1,1-Dimethoxy-n-pentane	868	1057
Isoamyl formate	775	1058
n-Butyl acetate	793	1059
Isobutyl aldehyde propyleneglycol acetal	831	1060
Ethyl isovalerate	840	1060
trans-Pinane	981	1062
Diisoamyl ether	1000	1064
Diisoamyl ether	1000	1064
Isopentyl ether	999	1067
Paraldehyde	763	1069
Methyl n-butyl ketone	772	1070
Isobutyl n-propionate	852	1071
1,3-Butylene glycol acetal	786	1072
Diisobutyl acetal	999	1072
Isobutyl aldehyde propyleneglycol acetal	840	1074
cis-Pinane	987	1075
Methyl n-valerate	806	1076
3-Methyl-3-pentanol	757	1080
Methyl disulfide	730	1081
Camphene	954	1083
2-n-Propylfuran	782	1083
Isobutyl isobutyrate	900	1084
n-Hexanal	780	1084
Isobutyl pivalate	933	1085
cis-Dihydroocimene	982	1088
Allyl isobutyrate	820	1090
Allyl n-propionate	777	1090
n-Butyl ethyl sulfide	893	1090
sec-Amyl alcohol	685	1091

COMPOUNDS	OV 101	CBWX 20M
n-Pentan-2-ol	685	1091
Methyl isoamyl ketone	825	1100
Methyl crotonate	745	1100
n-Undecane	1100	1100
n-Amyl formate	810	1107
Isobutyl acrylate	781	1107
Isoamyl acetate	860	1110
trans-Dihydroocimene	995	1110
n-Propyl n-butyrate	881	1110
n-Butanol	655	1113
2-Methyl-3-pentanol	777	1121
3,4-Hexanedione	777	1123
2-Methylthiophene	775	1123
Ethyl n-valerate	884	1124
4-Methyl-2-pentanol	758	1124
beta-Pinene	981	1124
Isopropyl n-valerate	924	1125
1,3-Butylene glycol formal	777	1128
n-Butyl pivalate	963	1128
2-n-Butylfuran	883	1130
n-Butyl n-propionate	889	1130
1-Penten-3-ol	673	1130
Sabinene	976	1130
Tetrahydrothiophene	801	1130
Di-n-Propyl ketone	857	1131
n-Propyl 2-methylbutyrate	933	1134
3-Buten-1-ol	638	1137
2,3-Heptanedione	816	1138
n-Butyl isobutyrate	939	1139
N-Methylpyrrole	715	1139
p-Xylene	860	1140
Methyl isobutyl carbinol	748	1142
n-Propyl isovalerate	924	1144
n-Hexyl mercaptan	910	1145
m-Xylene	863	1147
Diisobutyl sulfide	969	1149
Carvomenthene	1022	1150
Diallyl sulfide	854	1150
Isobutyl n-butyrate	941	1152
Methyl octyl ether	1016	1152
Ethyl n-butyl ketone	869	1155
1,1-Dimethoxy-n-hexane	964	1156
Myrcene	986	1156
2,4-Dimethyl-n-pentan-3-ol	828	1157
2-Methyl-2-pentanol	797	1160
3-Methyl-2-pentanol	797	1160
Allyl n-butyrate	857	1161

COMPOUNDS	OV 101	CBWX 20M
n-Amyl acetate	895	1161
2,5-Dimethylthiophene	856	1161
Ethyl crotonate	823	1161
2-Ethyl-1-hexyl vinyl ether	1038	1165
n-Pentyl ether	1065	1165
Isovaleraldehyde propyleneglycol acetal 1	938	1167
Isobutyl 2-methylbutyrate	991	1171
Methyl n-amyl ketone	872	1172
n-Butyl propyl sulfide	972	1173
1,4-Butylene glycol formal	829	1176
Methyl n-hexanoate	906	1177
alpha-Phellandrene	1002	1177
2-Ethylthiophene	861	1179
2,4,5-Trimethyloxazole	829	1179
n-Amyl n-propionate	955	1180
Isoamyl n-propionate	954	1180
Pyridine	695	1180
Ethyl isohexanoate	951	1181
4-Ethyl isohexanoate	951	1181
Isovaleraldehyde propyleneglycol acetal 2	945	1181
3-Ethyl-3-pentanol	866	1183
Isoamyl alcohol	719	1184
Isobutyl isovalerate	992	1184
1,4-Cineole	1010	1185
n-Heptanal	883	1186
Isoamyl isobutyrate	997	1187
Methyl tiglate	850	1188
Butyl acrylate	892	1189
Allyl isovalerate	920	1190
Ethyl n-amyl ketone	928	1190
o-Xylene	884	1191
n-Hexan-2-ol	786	1192
2-Butene-1-ol	650	1193
Pyrazine	739	1194
n-Dodecane	1200	1200
n-Propyl n-valerate	981	1200
3,5,5-Trimethyl-n-hexanal	963	1200
2-Ethylbutyl acetate	959	1205
Limonene	1030	1206
n-Butyl n-butyrate	979	1207
Diisobutyl ketone	983	1207
trans-2-Hexenal	832	1207
n-Amyl alcohol	756	1213
2-Ethylhexanal	959	1216
beta-Phellandrene	1025	1216
Methyl pyruvate	680	1217
Ethyl n-hexanoate	983	1223

COMPOUNDS	OV 101	CBWX 20M
n-Hexanal diethyl acetal	1082	1223
Isopropyl n-hexanoate	1021	1223
1-Hexen-3-ol	770	1225
n-Butyl 2-methylbutyrate	1029	1226
n-Amyl pivalate	1063	1228
Ethyl angelate	920	1228
1,8-Cineole	1027	1228
1,1-Diethyoxyhexane	1080	1228
cis-Ocimene	1025	1228
2-n-Pentylfuran	983	1229
2,5-Dimethyl-4-ethyloxazole	900	1231
1,3-Butylene glycol butyral	957	1232
Ethyl disulfide	910	1232
Ethyl tiglate	922	1234
n-Amyl isobutyrate	1035	1237
Cyclopentanone	805	1238
Isopropyl tiglate	959	1238
n-Butyl isovalerate	1021	1242
2-Ethyl-4,5-dimethyloxazole	914	1243
Prenyl acetate	900	1243
2-Methyl-5-ethylthiophene	957	1245
Thiazole	715	1246
tert-Butyl benzene	988	1247
2-Isopropyl-4,5-dimethyloxazole	960	1249
n-Heptan-4-ol	879	1250
trans-Ocimene	1038	1250
2-Methylpyrazine	805	1251
gamma-Terpinene	1057	1251
Isobutyl n-valerate	1040	1252
cis-3-Hexenyl formate	902	1252
Ethyl pyruvate	785	1253
Allyl 2-ethylbutyrate	995	1254
Allyl n-valerate	965	1256
n-Heptyl mercaptan	1011	1256
2-Methylthiazole	780	1256
n-Hexyl formate	994	1258
Isoamyl n-butyrate	1042	1259
2-n-Propylthiophene	951	1259
1,1-Dimethoxy-n-heptane	1063	1261
4-Methylthiazole	800	1263
Isopropyl disulfide	1014	1264
1,1-Diethoxycyclohexane		1265
2,6-Dimethylheptyl-4-acetate	1092	1265
n-Heptanal dimethyl acetal	1069	1265
cis-3-Hexenyl methyl acetal	1035	1267
2-Methyl-1-pentanol	837	1268
Di-n-butyl sulfide	1073	1270

COMPOUNDS	OV 101	CBWX 20M
2,4-Dimethylthiazole	869	1271
p-Cymene	1020	1272
Isoamyl 2-methylbutyrate	1087	1273
2,4-Diethyl-5-methyloxazole	983	1274
Methyl n-hexyl ketone	972	1275
Methyl n-heptanoate	1006	1276
alpha-p-Dimethylstyrene	1080	1278
n-Octanal	985	1278
2-Isopropyl-4-ethyl-5-methyloxazole	1021	1279
Ethyl chloroacetate	810	1281
4-Methyl-1-pentanol	838	1282
Cyclopentanol	802	1283
n-Heptan-2-ol	888	1284
Isoamyl isovalerate	1092	1287
Terpinolene		1287
Isobutyl angelate	1033	1289
Acetaldehyde ethyl cis-3-hexenyl acetal	1095	1297
3-Methyl-1-pentanol	852	1297
cis-3-Hexenyl ethyl acetal	1094	1298
n-Propyl n-hexanoate	1079	1298
2,6-Dimethyl-n-heptan-2-ol	983	1300
2-Ethylthiazole	879	1300
cis-3-Hexenyl acetate	987	1300
Methyl isohexyl carbinyl acetate	1080	1300
n-Tridecane	1300	1300
Ethyl 2-methylpentanoate	1005	1302
2-Octanone	991	1304
n-Amyl n-butyrate	1078	1305
n-Butyl n-valerate	1078	1305
Cyclohexyl formate	951	1305
Cyclohexanone	875	1306
2,5-Dimethylpyrazine	893	1306
Methyl hexyl acetaldehyde	1048	1306
n-Hexyl acetate	1012	1307
n-Heptyl formate	1012	1310
2-n-Propyl-4,5-dimethyloxazole	996	1310
trans-2-Hexenyl acetate	997	1315
n-Hexanol	858	1316
Isopropyl n-heptanoate	1120	1317
Elemol	805	1318
n-Heptanal diethyl acetal	1179	1319
n-Propyl tiglate	1020	1320
Diethyleneglycol monomethyl ether	920	1321
Ethyl n-heptanoate	1082	1321
2,6-Dimethylcyclohexanone	985	1322
1-Hepten-3-ol	868	1322
n-Amyl 2-methylbutyrate	1126	1324

COMPOUNDS	OV 101	CBWX 20M
2,6-Dimethylpyrazine	895	1325
2-Methyl-2-(4-methyl n-amyl) tetrahydrofuran	1159	1325
2-n-Hexyl furan	1083	1326
n-Hexyl n-propionate	1088	1326
Anisole	900	1327
n-Hexyl pivalate	1163	1328
2-Isobutyl-4,5-dimethyloxazole	1044	1330
Di-n-butyl ketone	1058	1330
Di-n-butyl ketone	1058	1330
2,3-Dimethylpyrazine	900	1330
2-Ethyl-4-methylthiazole	955	1331
n-Hexyl isobutyrate	1135	1333
6-Methyl-hept-5-en-2-one	968	1335
2-Ethyl-2-n-hexanal	1007	1336
n-Amyl isovalerate	1135	1337
m-Diethylbenzene	1070	1339
2-Isopropyl-4-methylthiazole	1000	1339
Cyclohexyl acetate	1027	1343
2-n-Propyl-4-ethyl-5methyloxazole	1064	1345
Methyl mercaptoacetate	802	1346
D-n-hexyl ether	1269	1349
Isobutyl n-hexanoate	1140	1350
cis-3-Hexenol	847	1351
Allyl isothiocyanate	860	1352
2-n-Butylthiophene	1052	1353
p-Diethylbenzene	1080	1353
Isoamyl n-valerate	1138	1354
Isoamyl n-valerate	1138	1354
cis-Rose oxide	1087	1354
Allyl n-hexanoate	1065	1356
Isobutyl tiglate	1076	1357
2,6-Dimethyl-hept-5-en-1-al	1039	1358
Propyl disulfide	1096	1358
2-Isobutyl-4-ethyl-5-methyloxazole	1106	1359
4,5-Dimethylthiazole	917	1359
5-Nonanone	1074	1360
n-Heptyl acetate	1095	1361
Amyl vinyl carbinyl acetate	1094	1365
n-Octanal dimethyl acetal	1167	1366
Trimethylthiazole	981	1367
trans-2-Hexenol	854	1368
Allyl tiglate	1002	1370
trans-2-Hexenyl n-propionate	1085	1370
trans-Rose oxide	1100	1370
cis-3-Hexenyl n-propionate	1083	1371
o-Diethylbenzene	1088	1372

COMPOUNDS	OV 101	CBWX 20M
Cyclohexanol	880	1375
cis-3-Hexenyl isobutyrate	1129	1377
Methyl n-heptyl ketone	1074	1377
Methyl n-octanoate	1107	1378
2-Ethoxythiazole	943	1380
2-n-Propylthiazole	970	1380
2-Methyl-3-ethylpyrazine	937	1381
n-Nonanal	1037	1382
n-Octan-2-ol	938	1385
2-Methyl-3-isopropylpyrazine	1028	1387
2,3,5-Trimethylpyrazine	935	1387
trans-3-Heptenyl acetate	1030	1388
Methyl benzyl ether	931	1391
Diethyleneglycol dimethyl ether	924	1396
Tetrahydrolinalool	1087	1397
2,5-Dimethyl-4-ethylthiazole	1050	1398
n-Hexyl n-butyrate	1176	1398
n-Propyl n-heptanoate	1177	1398
Isobutyl 2-ethyl-n-hexanoate	1248	1400
Dimethyl trisulfide	952	1400
4-Ethyl-5-methylthiazole	991	1400
cis-3-Heptenyl acetate	1084	1400
2-n-Propyl-4-methylthiazole	1040	1400
n-Tetradecane	1400	1400
n-Amyl n-valerate	1185	1401
n-Butyl n-hexanoate	1177	1402
2-Isobutylthiazole	1020	1404
2,2-Dimethyl-n-pentanol	874	1405
Cyclohexyl n-propionate	1120	1408
Fenchone	1080	1410
Tetrahydromyrcenol	1090	1414
p-Cresol methyl ether	1005	1415
n-Butyl angelate	1116	1417
Isobutyl 2-methylpentanoate	1155	1417
n-Octanal diethyl acetal	1276	1417
n-Hexyl 2-methylbutyrate	1224	1418
n-Butyl tiglate	1116	1419
n-Heptanol	957	1419
Isooctyl acetate	1154	1419
Isopropyl n-octanoate	1219	1419
2-Isobutyl-4-methylthiazole	1086	1420
2-Nonanone	1093	1420
1-octene-3-ol	968	1420
Ethyl n-octanoate	1180	1423
cis-Linalool oxide	1068	1423
Amyl vinyl carbinol	968	1426
Octyl formate	1117	1426

COMPOUNDS	OV 101	CBWX 20M
Cyclohexyl isobutyrate	1164	1427
Ethyl acetylacetate	907	1427
trans-2-Octenal	1045	1427
n-Heptyl pivalate	1263	1428
2-Ethyl-4,5-dimethylthiazole	1065	1429
2-n-Heptylfuran	1184	1429
Angelica lactone	1000	1430
Dihydromyrcenyl acetate	1202	1431
o-Methyl anisole	1000	1432
p-Methyl anisole	1000	1432
1-Octene-3-yl propionate	1180	1432
n-Heptyl isobutyrate	1233	1433
n-Hexyl isovalerate	1228	1433
1-Methyl 4-hepten-1-ol	982	1433
Isobutyl disulfide	1205	1435
Dihydromyrcenol	1063	1438
Benzyl ethyl ether	1046	1439
2-Isopropyl-4,5-dimethylthiazole	1109	1439
2-n-Butyl-4-ethyl-5-methyloxazole	1159	1441
Citronellyl vinyl ether	1235	1445
Isobutyl n-heptanoate	1239	1448
Methyl sorbate	998	1448
Furfural	815	1449
2-Ethyl-1-hexyl propionate	1231	1450
cis-3-Hexenyl n-butyrate	1170	1450
Isoamyl n-hexanoate	1238	1451
trans-Linalool oxide	1082	1451
trans-Linalool oxide	1082	1451
Methyl undecyl ether	1318	1453
Allyl n-heptanoate	1164	1454
Methyl 3-hydroxybutyrate	1318	1454
2,3,5,6-Tetramethylpyrazine	1069	1458
n-Octyl acetate	1193	1459
n-Heptanal ethyleneglycol acetal	1144	1460
Isobutyl 2-hexenoate	1180	1461
trans-2-Hexenyl n-butyrate	1180	1461
2-Methyl-3-n-propylpyrazine	1074	1462
2-n-Pentylthiophene	1153	1462
Citronellal	1137	1465
n-Nonanal dimethyl acetal	1267	1465
Isoamyl tiglate	1178	1469
Dimethyl malonate	896	1472
cis-3-Heptenyl propionate	1171	1472
Fenchyl acetate	1220	1473
cis-3-Hexenyl methoxyformate	1073	1475
Geranyl ethyl ether 1	1255	1476
Norbornyl acetate	1112	1476

COMPOUNDS	OV 101	CBWX 20M
Menthone	1143	1478
Methyl n-nonanoate	1207	1479
2-n-Butylthiazole	1070	1480
cis-3-Hexenyl isovalerate	1223	1480
Methyl n-octyl ketone	1176	1480
3,5,5-Trimethyl-n-hexanol	1041	1480
n-Nonan-2-ol	1089	1484
n-Decanal	1188	1485
n-Decanal	1188	1485
2-Acetylfuran	892	1491
Cyclohexyl n-butyrate	1209	1492
2-Ethyl-1-hexyl acrylate	1224	1494
Tetrahydrofurfuryl alcohol	884	1494
trans-2-Heptenyl propionate	1182	1497
n-Hexyl n-valerate	1275	1498
n-Propyl n-octanoate	1277	1498
Acetonyl acetone	894	1500
n-Amyl n-hexanoate	1279	1500
n-Butyl n-heptanoate	1276	1500
2-n-Butyl-4-methylthiazole	1141	1500
Cyclohexanone 1,3-butylene ketal	1186	1500
4-Methyl-5-vinylthiazole	1011	1500
n-Pentadecane	1500	1500
2-n-Propyl-4,5-dimethylthiazole	1151	1500
Benzaldehyde	947	1502
n-Heptyl n-butyrate	1275	1503
Ethyl sorbate	1075	1505
Geranyl ethyl ether 2	1275	1506
Linalool	1092	1506
n-Butyl lactate	997	1508
Dihydrolinalool	1122	1512
n-Nonanal diethyl acetal	1374	1514
Isopropyl n-nonanaote	1318	1516
2-Isobutyl-4,5-dimethylthiazole	1193	1517
Camphor	1136	1518
2-Ethyl-2-n-hexene-1-ol	1051	1518
Furfuryl acetate	969	1518
n-Heptyl 2-methylbutyrate	1324	1518
n-Amyl tiglate	1216	1519
alpha-Copaene	1398	1519
n-Octanol	1061	1519
1-Nonen-3-ol	1068	1520
Methyl octyl acetaldehyde	1254	1521
cis-4-Decenal	1177	1523
n-Ethyl n-nonanoate	1280	1523
Cyclohexyl isovalerate	1264	1527
Di-n-amyl ketone	1258	1528

COMPOUNDS	OV 101	CBWX 20M
Diisoamyl ketone	1258	1528
2-Hexoxyacetaldehyde dimethyl acetal	1234	1528
n-Octyl isobutyrate	1332	1529
Dimethyl fumarate	997	1530
2-n-Octylfuran	1285	1530
Methyl levulinate	956	1534
Diethyl trisulfide	1125	1535
Octyl propionate	1283	1536
Linalyl acetate	1246	1538
alpha-Methylcitronellol	1220	1540
trans-2-Nonenal	1146	1540
trans-3-Octenol	1036	1541
Diethyl malonate	1043	1542
Isobutyl n-octanoate	1338	1543
cis-3-Heptenyl butyrate	1255	1545
beta-Bourbonene	1406	1546
Isoamyl n-heptanoate	1334	1548
Amyl butyl carbinol	1175	1550
5-Decanol	1175	1550
Di-n-heptyl ether	1458	1550
Allyl n-octanoate	1265	1554
Methyl sulfoxide	840	1554
Acetaldehyde linalyl ethyl acetal 1	1361	1558
Dimethyl succinate	1002	1558
trans-2-Hexenyl n-valerate	1275	1560
n-Nonyl acetate	1292	1560
Dihydroterpinyl acetate	1282	1561
Methyl furoate	956	1561
Propylene glycol	750	1561
5-Methylfurfural	942	1563
cis-3-Octenol	1041	1563
2-Propionylfuran	988	1563
Acetaldehyde linalyl ethyl acetal 2	1362	1564
sec-Butylcyclohexanone 1	1194	1564
2-n-Hexylthiophene	1256	1564
o-tert-Butylcyclohexyl acetate 1	1286	1565
sec-Butylcyclohexanone 2	1196	1566
n-Decanal dimethyl acetal	1366	1567
n-Decanal dimethyl acetal	1366	1567
Ethyl levulinate	1029	1567
trans-2-Heptenyl butyrate	1275	1568
Linalyl formate	1206	1570
Diethyleneglycol diethyl ether	1058	1572
2-Methyl-1-octanol	1119	1573
Isopulegol	1145	1573
alpha-Fenchyl alcohol	1110	1574
Myrcenyl acetate	1247	1574

COMPOUNDS	OV 101	CBWX 20M
Isopropyl levulinate	1068	1575
Butyl disulfide	1295	1580
Methyl n-decanoate	1307	1581
Benzonitrile	965	1583
Diethyleneglycol monoethyl ether	986	1583
Isobornyl acetate	1279	1584
cis-3-Hexenyl n-valerate	1270	1584
n-Decan-2-ol	1190	1585
Methyl n-nonyl ketone	1276	1585
Myrcenol	1103	1585
Isopulegyl acetate	1258	1585
Tetrahydrofurfuryl acetate	1055	1585
Furfuryl n-propionate	1059	1587
n-Undecanal	1290	1589
Cedryl acetate 1	1427	1591
2-Cyclohexylethyl acetate	1233	1591
Cyclohexyl n-valerate	1310	1595
Acetaldehyde citronellyl methyl acetal	1374	1596
Isobornyl formate	1228	1596
o-tert-Butylcyclohexyl acetate 2	1304	1596
Linalyl n-propionate	1324	1596
Lavandulyl acetate	1274	1597
Linalyl isobutyrate	1366	1597
n-Octyl n-butyrate	1373	1597
n-Propyl n-nonanoate	1377	1598
Ethyl 2-furoate	1029	1599
Hexyl n-hexanoate	1371	1599
2-Acetylpyridine	1014	1600
n-Amyl n-heptanoate	1375	1600
2-n-Butyl-4,5-dimethylthiazole	1251	1600
n-Butyl n-octanoate	1373	1600
alpha-Cedrene	1436	1600
Citronellyl formate	1261	1600
Dihydroanethole	1193	1600
Dihydrocarvone	1183	1600
n-Hexadecane	1600	1600
Methyl benzoate	1078	1600
Ethyl 2,4-dimethyl-1,3-dioxolylacetate 1	1147	1603
2-Methylundecanal	1353	1609
Bornyl formate	1239	1610
Menthol	1171	1612
n-Butyl benzyl ether	1238	1613
n-Decanal diethyl acetal	1473	1613
n-Decanal diethyl acetal	1473	1613
n-Heptyl valerate	1372	1614
n-Octyl 2-methylbutyrate	1422	1615
Isopropyl n-decanoate	1417	1615

COMPOUNDS	OV 101	CBWX 20M
Ethyl 2,4-dimethyl-1,3-dioxolylacetate 2	1154	1616
N-Methyl-2-pyrrolaldehyde	986	1616
beta-Terpinenol	1137	1616
Caryophyllene	1428	1617
gamma-Valerolactone	921	1617
n-Hexyl angelate	788	1621
n-Hexyltiglate	1310	1621
Methyl nonyl acetaldehyde	1352	1621
Ethyl n-decanoate	1379	1624
n-Nonanol	1161	1624
Myrcenyl propionate	1327	1625
Citronellyl ethyl acetal	1423	1626
beta-Copaene	1445	1626
Tetrahydrogeraniol	1185	1626
Acetophenone	1048	1627
trans-p-tert-Butylcyclohexyl acetate	1322	1628
Terpinene-4-ol	1175	1628
o-Toluyl thiol	1067	1631
gamma-Butyrolactone	885	1632
Diethyl fumarate	1160	1632
Tetrahydrofurfuryl propionate	1153	1632
o-Tolualdehyde	1054	1632
m-Tolualdehyde	1053	1632
beta-Cedrene	1446	1633
Diethyl malate	994	1638
2-Acetylthiazone	995	1639
Isopropyl benzoate	1189	1639
6-Undecanol	1281	1640
Diethyl succinate	1153	1642
10-Undecen-1-al	1280	1642
2-Butanoylfuran	1078	1644
p-tert-Butylcyclohexanone	1208	1645
Citronellyl acetate	1335	1645
Cyclooctanyl acetate	1280	1645
Phenylacetaldehyde	1024	1646
Ethyl benzoate	1154	1647
Isoamyl n-octanoate	1433	1648
Sabinyl acetate	1262	1651
Estragole	1183	1652
p-Tolualdehyde	1067	1652
n-Methyl-2-acetylpyrrole	1055	1653
cis-3-Hexenyl n-hexanoate	1370	1654
Allyl n-nonanoate	1365	1655
Furfuryl n-butyrate	1148	1655
delta-Terpinenol	1160	1655
trans-2-Hexenyl n-hexanoate	1375	1656
Dihydrolinalyl acetaldehyde	1315	1659

COMPOUNDS	OV 101	CBWX 20M
Isoborneol	1157	1660
Thujopsene	1451	1660
Citral 1	1222	1661
alpha-Terpineol	1185	1661
Alloaromadendrene	1478	1662
Decyl acetate	1393	1662
Lavandulol	1154	1662
Pulegone	1230	1662
Neryl formate	1267	1663
n-Propyl levulinate	1125	1663
Phenylacetaldehyde dimethyl acetal	1200	1665
2-Nonyn-1-al dimethyl acetal	1300	1666
Terpinyl formate	1333	1666
Isoamyl benzyl ether	1297	1668
2-Cyclohexyl ethanol	1098	1668
n-Undecanal dimethyl acetal	1466	1668
Dihydrocarvyl acetate	1319	1670
2-n-Heptylthiophene	1359	1670
Methyl chavicol	1182	1670
2-Propionyl-5-methylfuran	1106	1672
Styrallyl acetate	1173	1673
Benzyl formate	1058	1675
cis-p-tert-Butylcyclohexyl acetate	1360	1675
Isobornyl n-propionate	1376	1676
Benzyl n-propionate	1234	1679
o-Methylacetophenone	1118	1679
Geranonitrile 1	1199	1680
Linalyl n-butyrate	1420	1680
Neral	1227	1680
alpha-Humulene	1465	1682
Sabinol	1135	1683
Geranyl formate	1282	1684
Isobutyrophenone	1185	1685
Terpinyl acetate	1333	1687
Methyl n-decyl ketone	1377	1688
n-Amylcyclopentenone 1	1247	1689
trans-2-nonenol	1157	1591
1,3-Butanediol	941	1692
Cyclohexyl n-hexanoate	1411	1695
n-Dodecanal	1392	1695
Linalyl 2-methylbutyrate	1450	1695
Isobutyl levulinate	1183	1696
Benzyl acetate	1144	1697
n-Propyl n-decanoate	1476	1697
Borneol	1164	1698
Linalyl isovalerate	1461	1698
n-Hexyl n-heptanoate	1470	1699

COMPOUNDS	OV 101	CBWX 20M
Neryl acetate	1345	1699
Acetaldehyde di-(cis-3-hexenyl) acetal	1465	1700
n-Amyl n-octanoate	1471	1700
n-Butyl n-nonanoate	1475	1700
Citronellyl n-propionate	1427	1700
p-Cresyl acetate	1150	1700
Cyclooctanol	1133	1700
Di-n-Propyl malonate	1227	1700
n-Heptadecane	1700	1700
Propyl furoate	1125	1700
Di-n-Propyl malonate	1227	1700
cis-3-Decenyl acetate	1376	1701
Citronellyl isobutyrate	1469	1705
1,4-Dimethoxybenzene	1145	1705
Salicaldehyde	1027	1705
Citral 2	1249	1706
Methyl 2-methylbenzoate	1165	1709
Anisyl formate	1307	1710
n-Undecanal diethyl acetal	1572	1712
Dihydrocarveol	1188	1713
Carvone	1228	1715
Dimethyl benzyl carbinol	1147	1715
Isopropyl n-undecanoate	1516	1715
Octyl-n-valerate	1474	1719
Citronellol	1215	1722
9-Decen-1-yl acetate	1383	1722
Decanol	1263	1723
Geranonitrile 2	1236	1723
Ethyl n-undecanoate	1479	1725
Methyl p-toluate	1194	1725
cis-3-Hexenyl lactate	1187	1727
n-Butyl 3-furyl-n-propionate	1354	1728
Di-n-hexyl ketone	1458	1728
n-Decyl n-propionate	1486	1729
Geranial	1252	1730
alpha-Muurolene	1500	1730
Verbenone	1195	1730
n-Butyl n-butyryllactate	1331	1733
Methyl 10-undecenoate	1396	1733
Methyl isovalerate	1505	1734
Geranyl acetate	1364	1735
Piperitone	1247	1739
1,3-Dimethoxy benzene	1143	1740
Isoamyl phenylethyl ether	1384	1741
Allyl levulinate	1114	1743
cis-3-Hexenyl n-heptanoate	1465	1743
Methyl 3-methylbenzoate	1190	1744

COMPOUNDS	OV 101	CBWX 20M
n-Propyl benzoate	1254	1745
Methyl phenylacetate	1154	1747
omega-Methyl undecylenate	1400	1747
2-Pentanoylfuran	1180	1747
Terpinyl n-propionate	1426	1747
Allyl 2-furoate	1113	1748
n-Amylcyclopentenone 2	1113	1748
Isoamyl n-nonanoate	1533	1748
2-Butanoyl-5-methylfuran	1192	1748
Terpinyl isobutyrate	1467	1748
m-Methylacetophenone	1156	1749
p-Methylacetophenone	1166	1750
Valencene	1487	1751
Furfuryl n-valerate	1245	1752
Phenylethyl formate	1156	1752
Methyl salicylate	1181	1754
Allyl n-decanoate	1462	1755
Dimethyl benzyl carbinyl acetate	1302	1755
trans-2-Hexenyl n-heptanoate	1474	1755
Methyl 4-methylbenzoate	1199	1755
Nerol	1218	1757
2-acethylthiophene	1069	1760
Bornyl butyrate	1473	1760
n-Butyl levulinate	1225	1760
delta-Cadinene	1524	1761
p-Cresyl isobutyrate	1291	1763
Neryl isobutyrate	1474	1764
cis-3-Decenol	1245	1765
Linalyl n-valerate	1500	1765
Methyl isogeranylacetate	1362	1765
Methyl phenyl carbinol	1051	1765
Acetone 1-phenyl-1,2-ethanediol ketal	1260	1766
gamma-Cadinene	1518	1766
Cuminaldehyde	1227	1766
Dipropyleneglycol 1	1017	1767
4-Methyl-5-hydroxy phenylacetate	1263	1767
n-Dodecanal dimethyl acetal	1565	1769
Ethyl phenyl ethyl acetal	1332	1770
Benzyl isobutyrate	1277	1771
Isobutyl benzoate	1310	1771
Neryl n-propionate	1436	1771
Ethyl phenylacetate	1219	1773
Isopropyl salicylate	1292	1773
Bornyl isovalerate	1512	1774
9-Decen-1-ol	1251	1775
Ethyl 10-undecenoate	1469	1775
n-Undecyl acetate	1487	1775

COMPOUNDS	OV 101	CBWX 20M
Nopyl acetate	1412	1777
Nopylacetate	1412	1777
Bromostyrol	1197	1778
Dimethyl adipate	1212	1779
Methyl nicotinate	1116	1779
Acetaldehyde styleneglycol acetal 1	1226	1780
2-n-Octylthiophene	1463	1780
Dihydrocinnamic aldehyde	1139	1783
Methyl n-dodecanoate	1507	1785
Phenylethyl acetate	1233	1785
Citronellyl n-butyrate	1511	1786
Ethyl salicylate	1257	1787
Carveol 1	1209	1790
p-Hydroxyacetophenone	1144	1790
trans-2-Decenol	1257	1792
Geranonitrile 3	1249	1792
Methyl n-undecyl ketone	1479	1792
Acetaldehyde styleneglycol acetal 2	1333	1793
2-Decen-1-ol	1257	1794
Geranyl isobutyrate	1493	1795
Geraniol	1243	1797
n-Propyl n-undecanoate	1576	1797
Butyl acetoacetate	1104	1798
n-Butyl n-decanoate	1575	1798
Geranyl n-propionate	1424	1799
Allyl benzoate	1239	1800
Allyl-3-cyclohexyl propionate	1405	1800
p-tert-Amylcyclohexanone	1113	1800
Citronellyl isovalerate	1563	1800
Carvone oxide	1261	1805
n-Hexyl octanoate	1564	1805
Di-n-Propyl fumarate	1360	1806
Isoamyl levulinate	1284	1807
Anethole	1270	1809
2,4-Dimethylacetophenone	1233	1809
Dimethyl benzyl carbinyl n-propionate	1391	1810
n-Dodecanal diethyl acetal	1671	1811
Isopropyl n-dodecanoate	1814	1814
Dipropyleneglycol 2	1039	1817
Carveol 2	1222	1820
2-Propionylthiophene	1164	1821
Benzyl alcohol	1033	1822
Dihydrosafrole	1286	1822
n-Undecanol	1364	1822
Dipropyleneglycol 3	1043	1823
Ethyl n-dodecanoate	1579	1826
Terpinyl n-butyrate	1514	1828

COMPOUNDS	OV 101	CBWX 20M
Phenylethyl pivalate	1400	1832
Carvyl propionate	1440	1833
Citronellyl isocrotonate	1526	1833
alpha-Ionone	1416	1833
2,4-Dimethyl-5-acetylthiazole	1217	1835
Acetaldehyde phenylethyl n-propyl acetal	1424	1836
2-Methyl-5-acetylthiophene	1185	1836
Methylionone 1	1471	1836
cis-3-Hexenyl n-octanoate	1564	1838
Neryl n-butyrate	1519	1838
Isoamyl 2-furoate	1287	1840
Guaiacol	1071	1840
n-Butyl benzoate	1354	1841
Calamenene	1518	1842
2-Decenal	1449	1842
Dihydroisojasmone	1374	1842
Methyl dihydrocinnamate	1258	1842
Linalyl n-hexanoate	1582	1843
Isoamyl n-decanoate	1633	1848
Dipropyleneglycol 4	1044	1848
n-Propyl phenylacetate	1300	1848
Benzyl acetone	1218	1849
Furfuryl n-hexanoate	1343	1850
2-Hexanoylfuran	1281	1850
trans-2-Hexenyl n-octanoate	1573	1853
Phenylethyl isobutyrate	1374	1855
Phenylethyl n-propionate	1328	1855
Benzyl n-butyrate	1322	1856
Tangerinal 1	1543	1857
5-Acetyl-2-methyl-8-isopropyl-(2,2,2) bicyclo-2-octene 1-epitone 1	1452	1858
Diethyl adipate	1358	1858
Terpinyl isovalerate	1565	1858
Phenylethyl alcohol	1104	1859
n-Amyl levulinate	1325	1860
n-gamma-Heptalactone	1126	1860
n-Propyl 10-undecenaote	1565	1860
1,4-Butanediol	1046	1861
Isobutyl phenylacetate	1371	1864
Neryl isovalerate	1574	1864
cis-3-Decenyl butyrate	1563	1866
trans-2-Decenyl butyrate	1563	1866
omega-Decenyl butyrate	1558	1870
1,4-Dimethoxy-2-tert-butylbenzene	1398	1870
Propionaldehyde 1-phenyl-1,2-ethanediol acetal 1	1327	1871
Geranyl n-butyrate	1532	1872

COMPOUNDS	OV 101	CBWX 20M
Cyclohexyl glycidate	1344	1875
2-Methyl-1-undecanol	1422	1875
2-Methyl-5-isopropylacetophenone	1358	1876
Safrole	1278	1876
n-Propyl salicylate	1357	1878
Jasmal	1459	1879
Benzyl isovalerate	1374	1880
Citronellyl n-valerate	1608	1880
Propionaldehyde 1-phenyl-1,2-ethanediol acetal 2	1333	1880
Dihydro-nor-dicyclopentadienyl acetate	1406	1881
4-(2,4,6-Trimethyl-3-cyclohexen-1-yl)-3-buten-2-one 1	1433	1881
Benzyl ethyl carbionl	1222	1882
Hydroxycitronellal	1269	1882
gamma-Ionone	1347	1882
5-Acetyl-2-methyl-8-isopropyl-(2,2,2) bicyclo-2-octene 2-epitone 2	1467	1883
Decahydro-beta-naphthol 1	1280	1883
2,6-Dimethyl phenol	1087	1883
gamma-Octalactone	1225	1883
Dihydropseudo ionone	1473	1885
Geranyl 2-methylbutyrate	1574	1886
Dimethyl benzyl carbinyl n-butyrate	1476	1889
Undecenol	1350	1889
Dihydrojasmone	1400	1892
Dipropyleneglycol 5	1075	1892
n-Heptyl n-octanoate	1666	1892
2-n-Pentyl-3-methyl-2-cyclopenten-1-one	1400	1892
Tangerinal 2	1566	1892
4-(2,4,6-Trimethyl-3-cyclohexen-1-yl)-3-buten-2-one 2	1435	1892
Methyl n-dodecyl ketone	1580	1893
Isoamyl benzoate	1421	1894
Isoamyl furylpropionate	1415	1894
2-Butanoylthiophene	1252	1894
Geranyl isovalerate	1593	1895
Methyl n-tridecanoate	1612	1895
Isobutyl salicylate	1410	1896
Methylionone 2	1506	1897
n-Propyl n-dodecanoate	1676	1897
p-Cresyl isovalerate	1389	1898
Isobutyl aldehyde 1-phenyl-1,2-ethanediol acetal	1384	1900
n-Butyl n-undecanoate	1674	1900
Isobutyl 10-undecenoate	1617	1900
Dihydrocuminyl acetate	1418	1900

COMPOUNDS	OV 101	CBWX 20M
n-Hexyl nonanoate	1668	1900
2-Methyl-5-propionylthiophene	1280	1900
n-Nonadecane	1900	1900
Dimethyl phenyl ethyl carbinyl acetate	1428	1908
Isoamyl pyruvate	1712	1910
Benzyl isopropyl carbinol	1292	1912
7,8-Epoxy-1,3,3,11-tetramethyl-(5,4,0,0(2.4))-undecane	1568	1912
p-Isopropylacetophenone	1332	1912
cis-Jasmone	1378	1914
Isopropyl benzyl carbinol	1291	1914
2,4-Dimethyl benzylacetate	1348	1916
Dimethyl phenyl ethyl carbinol	1282	1916
beta-Ionone	1474	1918
p-Cresyl ethylcarbonate	1304	1919
n-Dodecanol	1468	1925
Phenylpropyl acetate	1347	1926
Acetylcymene	1381	1928
Terpinyl n-valerate	1614	1928
Citronellyl crotonate	1558	1929
delta-Octalactone	1252	1929
Linalyl n-heptanoate	1670	1930
Methylionone 3	1530	1930
Neryl n-valerate	1610	1930
Phenylethyl butyrate	1422	1930
Phenylethyl butyrate	1422	1930
n-Butyl phenylacetate	1408	1932
Phenol	1002	1932
2-Acetylpyrrole	1050	1935
trans-2-trans-4-Decadienol	1310	1938
cis-3-Hexenyl n-nonanoate	1664	1938
2-Methyl-2-phenylhexan-4-one	1405	1938
n-Amyl benzoate	1454	1940
Phenyl isobutyl methyl carbinyl acetate	1459	1943
Phenylethyl 2-methylbutyrate	1472	1945
Allyl salicylate	1339	1946
n-Amyl furylpropionate	1453	1947
Isoamyl n-undecanoate	1733	1948
Furfuryl n-heptanoate	1443	1950
Cuminyl acetate	1401	1952
trans-2-Hexenyl n-nonanoate	1673	1953
n-Butyl 10-undecenoate	1660	1954
beta-Methyl-p-isopropylphenyl-propionaldehyde	1444	1954
Benzyl n-propyl carbinol	1310	1955
Phenylethyl isovalerate	1474	1955
Benzyl n-valerate	1421	1956

COMPOUNDS	OV 101	CBWX 20M
2-Heptanoylfuran	1384	1956
p-Methyl benzyl alcohol	1122	1956
Geranyl n-valerate	1632	1960
alpha-Cedrene epoxide	1585	1961
alpha-Nerolidol	1524	1961
2-Cyclohexylcyclohexanone	1496	1975
o-Methoxyacetophenone	1269	1975
n-Butyl salicylate	1457	1976
Pyrrol-2-carboxaldehyde	1005	1976
Dihydrocuminalcohol	1286	1981
Diphenyl	1369	1981
Methylionone 4	1564	1981
Anis aldehyde	1234	1982
2-Buten-1,4-diol	1104	1983
Isobutyl benzyl carbinol	1366	1983
Dimethyl suberate	1414	1985
Geranyl tiglate	1650	1985
Methyl n-tetradecanoate	1707	1990
2-Propionylpyrrol	1145	1990
Isoamyl phenylacetate	1468	1991
Diphenyl oxide	1386	1991
gamma-Nonalactone	1328	1991
alpha-Methylcinnamic aldehyde	1309	1992
Allylionone 1	1605	1993
2-Pentanoylthiophene	1355	1993
3-Phenylpropyl alcohol	1218	1993
Phenylpropyl n-propionate	1445	1994
Decahydro-beta-naphthol 2	1323	1995
Cinnamic aldehyde	1250	1996
Phenylpropyl isobutyrate	1490	1996
Isoamyl 10-undecenoate	1717	2000
n-Butyl n-dodecanoate	1772	2000
Di-n-butyl succinate	1534	2000
n-Eicosane	2000	2000
beta-Nerolidol	1553	2000
p-Cresol	1051	2003
Isobutyl beta-2-furylacrylate	1435	2006
Di-n-amyl fumarate	1558	2006
Di-n-butyl fumarate	1558	2006
p-Cresyl isotiglate	1482	2011
4-Ethylguaiacol	1265	2011
m-Methoxyacetophenone	1279	2011
o-Methylphenyl ethyl alcohol	1216	2012
Allylionone 2	1679	2014
p-Ethoxybenzaldehyde	1308	2017
Isopropyl n-tetradecanoate	1811	2017
Isoamyl salicylate	1528	2021

COMPOUNDS	OV 101	CBWX 20M
Neryl n-hexanoate	1709	2021
Ethyl n-tetradecanoate	1730	2027
2-Ethyl phenol	1113	2028
Isosafrole	1360	2029
Triacetin	1563	2029
p-tert-Butylphenylpropyl aldehyde 1	1212	2030
Maltol	1105	2030
2,4-Dimethyl benzyl alcohol	1226	2032
Phenylethyl n-valerate	1517	2034
cis-3-Hexenyl decanoate	1760	2038
delta-Nonalactone	1356	2038
p-tert-Butyl-alpha-methylhydrocinnamic aldehyde	1506	2039
Methyl N-methylanthranilate	1389	2042
Methylisoeugenol 1	1427	2044
Cuminalcohol	1283	2045
n-Amyl phenylacetate	1506	2047
Isoamyl n-dodecanoate	1829	2048
Methyl cinnamate	1365	2051
n-Amyl 10-undecenoate	1760	2053
Allyl cinnamate 1	1363	2054
Methyl 2-hydroxyisobutyrate	1116	2054
Benzyl n-hexanoate	1521	2057
Isopropyl N-methylanthranilate	1491	2058
2-Octanoylfuran	1487	2062
Benzal acetone	1337	2065
Diethyl suberate	1553	2065
n-Hexyl benzoate	1558	2066
Methyllavender ketone	1341	2067
Methyl anisate	1354	2071
Isolongiforanone 1	1610	2072
Benzyl tiglate	1474	2075
n-Amyl salicylate	1557	2077
4-Methyl-5-thiazoleethanol acetate	1368	2077
Phenylpropyl n-butyrate	1535	2083
Di-n-butyl adipate	1658	2087
Cinnamyl formate	1330	2094
Ethyl cinnamate	1443	2095
Isopropyl cinnamate	1485	2097
Cedrol	1609	2100
Ethyleneglycol monophenyl ether	1213	2100
n-Heneicosane	2100	2100
beta-Phenoxyethyl isobutyrate	1493	2100
Phenylpropyl isovalerate	1590	2100
Thymol	1287	2100
gamma-Decalactone	1437	2101
Cinnamyl acetate	1419	2103

COMPOUNDS	OV 101	CBWX 20M
Eugenol	1351	2103
2-Hexanoylthiophene	1459	2104
Phenoxyethyl isobutyrate	1486	2106
Ethyl anisate	1426	2110
n-Nonanoic acid	1286	2110
n-Propyl cinnamic aldehyde	1531	2111
Isolongiforanone 2	1622	2112
p-Methoxyacetophenone	1327	2115
n-Butyl n-tridecanoate	1880	2118
Neryl n-heptanoate	1808	2120
cis-3-Hexenyl benzoate	1553	2122
Methylisoeugenol 2	1468	2126
Phenoxyethyl propionate	1447	2126
Cyclohexenylcyclohexanone	1553	2128
Phenylethyl n-hexanoate	1618	2134
Dibutyl butyrolactone	1531	2141
Hydroxycitronellol	1347	2143
delta-Decalactone	1463	2144
4-Acetyl-6-tert-butyl-1,1-dimethylindan	1706	2145
Allylionone 3	1689	2146
n-Hexyl phenylacetate	1607	2148
3-Ethyl phenol	1146	2150
n-Amyl furfurylacrylate	1544	2153
Phenylethyl tiglate	1562	2154
Patchouli alcohol	1667	2156
Geranyl n-heptanoate	1831	2157
Benzyl n-heptanoate	1620	2158
alpha-Methylcinnamyl acetate	1484	2158
Carvacrol	1297	2159
alpha-n-Butyl cinnamic aldehyde	1535	2160
2-Nonanoylfuran	1588	2163
6-Amyl-alpha-pyrone	1434	2166
n-Propyl N-methylanthranilate	1560	2166
Cinnamyl n-propionate	1515	2169
Cinnamyl isobutyrate	1555	2171
Heliotropine	1318	2171
Rosephenone	1538	2172
Cedryl acetate 2	1766	2173
Isobutyl N-methylanthranilate	1617	2174
Allyl phenylacetate	1422	2175
n-Hexyl salicylate	1664	2175
Allyl phenoxyacetate	1429	2176
p-Isopropylphhenol	1200	2178
o-Aminoacetophenone	1288	2181
m-Aminoacetophenone	1409	2181
p-Aminoacetophenone	1506	2181
Methyl anthranilate	1332	2181

COMPOUNDS	OV 101	CBWX 20M
Phenylpropyl n-valerate	1635	2183
Isoeugenol 1	1394	2186
Dimethyl sebacate	1616	2195
Allyl anthranilate 1	1328	2196
Allylionone 4	1734	2199
Anisyl acetate	1390	2199
n-Butyl methylphenylglycidate 1	1591	2200
2-Hexylcyclopentanone acetic acid methyl ether	1610	2200
Methyl hexadecanoate	1911	2204
Anisyl n-propionate	1482	2205
n-Propyl anisate	1527	2205
Cinnamyl alcohol	1300	2207
2-Heptanoylthiophene	1565	2209
Anis alcohol	1267	2210
gamma-Undecalactone	1542	2210
alpha-n-Amylcinnamic aldehyde	1631	2211
Acetylcedrene	1768	2213
4-Methyl-5-thiazoleethanol	1283	2216
cis-3-Hexenyl phenylacetate	1610	2220
Methyl n-hexyl ketone 1-phenyl-1,2-ethanediol ketal 1	1739	2224
cis-3-Hexenyl salicylate	1654	2227
Isobutyl cinnamate	1605	2228
n-Butyl n-tetradecanoate	1977	2229
Methyl hydrojasmonate		2229
Anis aldehyde propyleneglycol acetal 1	1508	2231
Ethyl anthranilate	1396	2232
Isobutyl anisate	1586	2233
Phenylethyl n-heptanoate	1718	2233
Anis aldehyde propyleneglycol acetal 2	1512	2234
6-sec-Butylquinoline	1592	2235
p-Methoxybenzylacetone	1473	2236
Cinnamyl n-butyrate	1604	2247
Methyl n-hexyl ketone 1-phenyl-1,2-ethanediol ketal 2	1758	2248
Nootketone	1802	2250
delta-Undecalactone	1579	2251
alpha-Methylcinnamyl alcohol	1343	2252
n-Pentadecanol	1373	2252
Pentalide	1823	2255
Allyl cinnamate 2	1537	2258
2-n-Amyl-3-acetonyl-1-cyclopentanone	1600	2259
Benzyl n-octanoate	1720	2260
p-Cresyl n-octanoate	1755	2264
2-Decanoylfuran	1689	2264
n-Heptyl phenylacetate	1717	2265

COMPOUNDS	OV 101	CBWX 20M
n-Butyl N-methylanthranilate	1660	2266
Isoeugenol 2	1438	2269
Cinnamyl isovalerate	1655	2271
Diethyl sebacate	1752	2272
Anisyl n-butyrate	1569	2274
Acetyleugenol	1541	2277
Muscone	1831	2281
beta-Methylcinnamyl alcohol	1365	2283
Dihydrocoumarin	1359	2286
Guaiacyl n-caproate	1681	2296
Galaxolide 1	1837	2299
Chavicol	1238	2300
Diethyl phthalate	1565	2303
n-Butyl anisate	1632	2305
Galaxolide 2	1837	2307
alpha-n-Hexylcinnamic aldehyde	1727	2309
2-Octanoylthiophene	1667	2313
gamma-Dodecalactone	1647	2317
alpha-n-Amylcinnamyl acetate	1757	2318
n-Propyl anthranilate	1500	2320
Dibenzyl ether	1631	2323
Heliotropyl acetate	1485	2325
n-Butyl n-pentadecanoate	2080	2330
n-Heptyl salicylate	1790	2332
Isoamyl anisate	1686	2333
Phenylethyl n-octanoate	1819	2337
n-Butyl methylphenylglycidate 2	1700	2346
Isobutyl anthranilate	1556	2347
Cinnamyl n-valerate	1705	2347
n-Propionyl methylanthranilate	1673	2350
Indole	1304	2351
Isoamyl cinnamate	1719	2355
delta-Dodecalactone	1675	2358
Coumarin	1418	2361
Benzyl n-nonanoate	1823	2362
Tonalid	1849	2373
Allyl anthranilate 2	1493	2381
Helional	1543	2383
Triethyl citrate	1627	2386
Triethyl citrate	1627	2386
Cycohexadecanone	1731	2392
3,4-Dimethoxyacetophenone	1532	2393
n-Amyl anisate	1732	2405
alpha-n-Heptylcinnamic aldehyde	1827	2409
Benzophenone	1604	2410
Ethylvanillin	1448	2414
Methyl octadecanoate	2101	2418

COMPOUNDS	OV 101	CBWX 20M
n-Butyl anthranilate	1600	2419
3-Methylcoumarin	1490	2424
o-Methoxy cinnamic aldehyde	1504	2430
n-Butyl n-hexanoate	2180	2435
Octyl salicylate	1895	2435
Phenylethyl n-nonanoate	1921	2439
Cinnamyl n-hexanoate	1805	2445
Isoamyl anthranilate	1656	2447
Vanillin	1392	2449
Benzyl n-decanoate	1923	2460
Ethyl n-octanoate	2180	2460
Nerolidylethanol	1851	2462
Methyl naphthyl ketone	1592	2471
Musk xylol	1506	2475
Ethyl oleate	2155	2489
n-Amyl anthranilate	1700	2510
Phenylethyl n-decanoate	2022	2540
Cinnamyl n-heptanoate	1905	2545
Methyl N-propylanthranilate	1678	2575
Furfuryl disulfide	1660	2600
7-Methylcoumarin	1553	2620
6-Methylcoumarin	1545	2630
Methylzingerone	1660	2640

Appendix IV

MASS SPECTRA OF INDIVIDUAL COMPOUNDS

APPENDIX IV MASS SPECTRA OF INDIVIDUAL COMPOUNDS

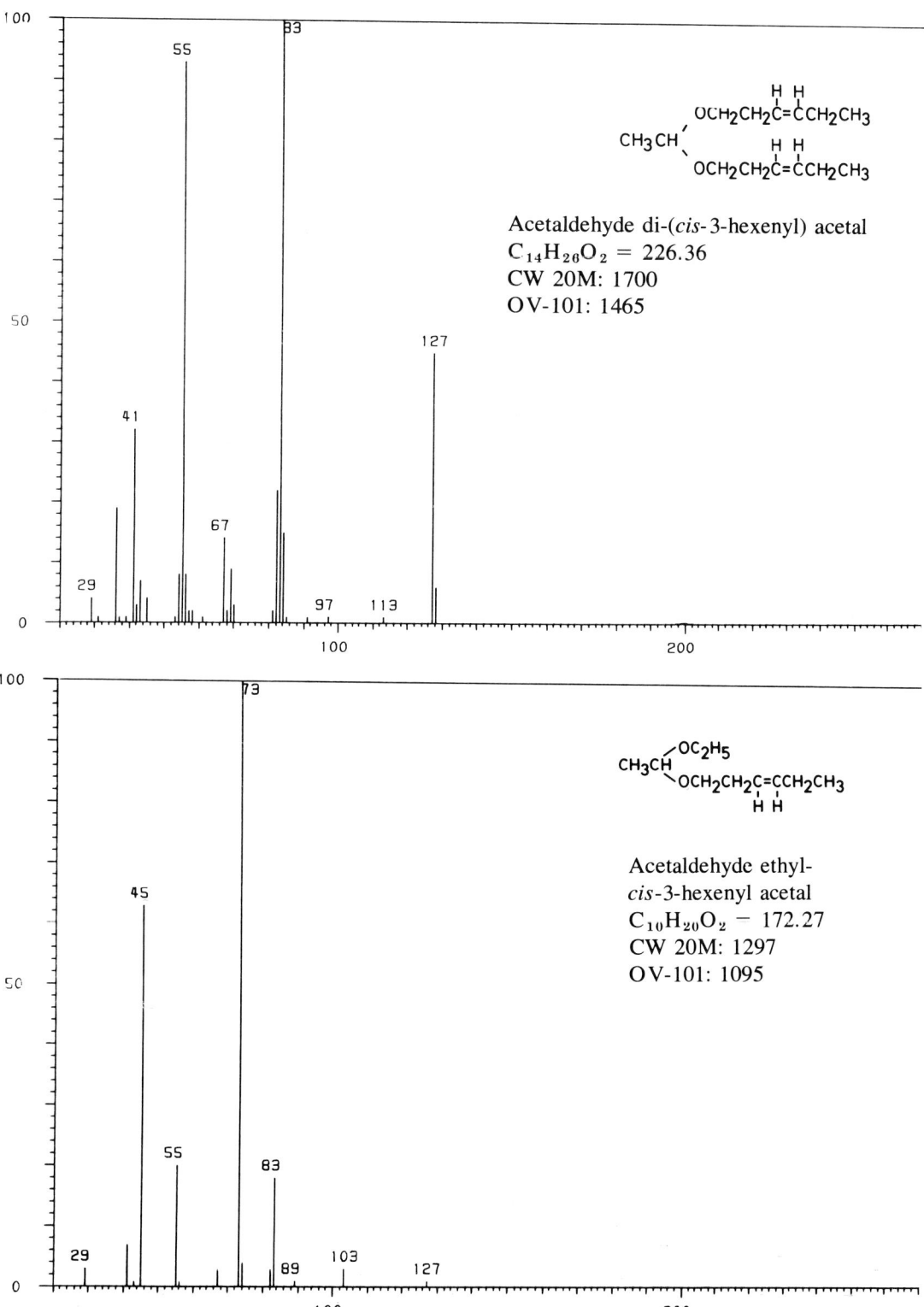

Acetaldehyde di-(*cis*-3-hexenyl) acetal
$C_{14}H_{26}O_2 = 226.36$
CW 20M: 1700
OV-101: 1465

Acetaldehyde ethyl-*cis*-3-hexenyl acetal
$C_{10}H_{20}O_2 = 172.27$
CW 20M: 1297
OV-101: 1095

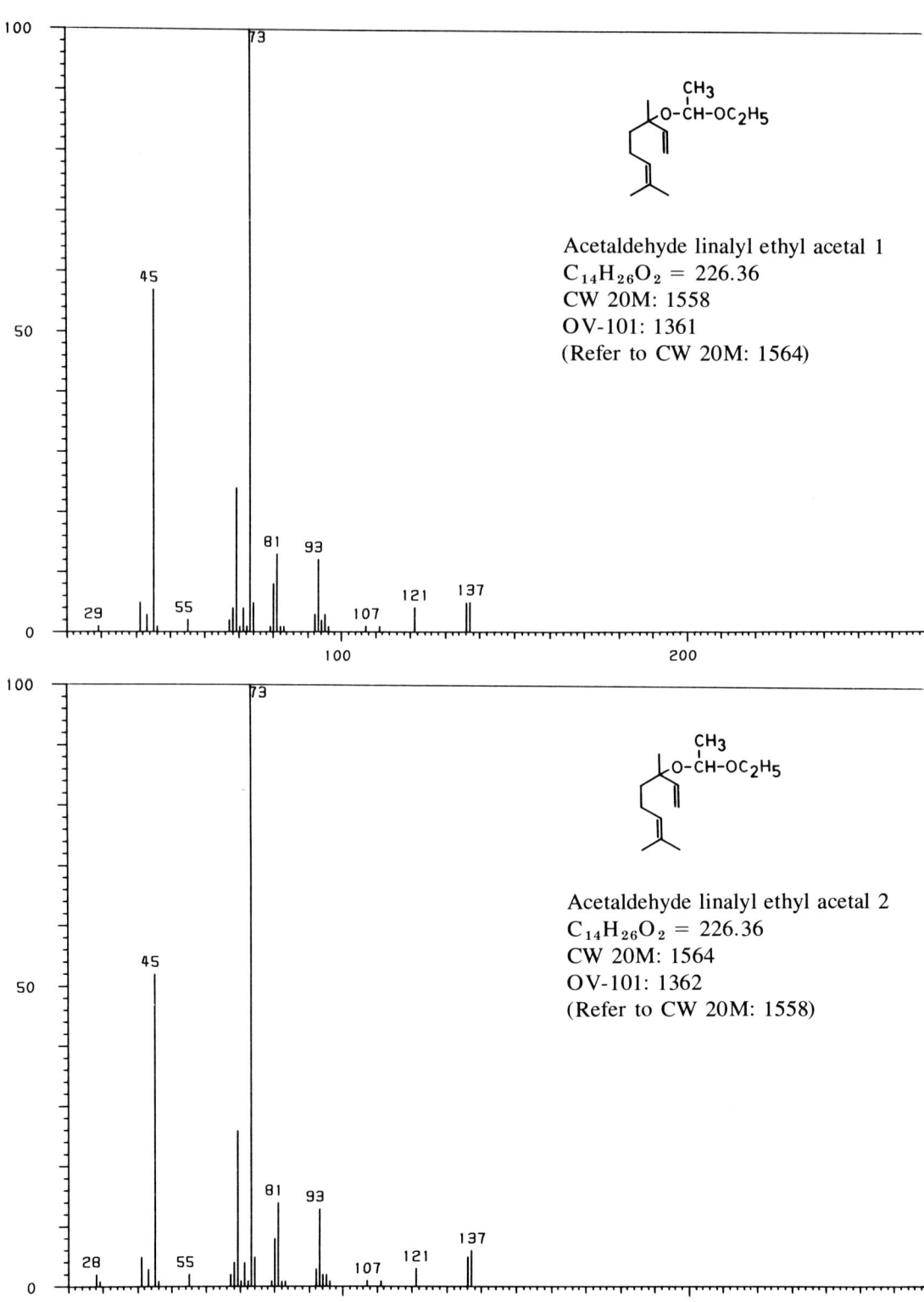

APPENDIX IV   MASS SPECTRA OF INDIVIDUAL COMPOUNDS

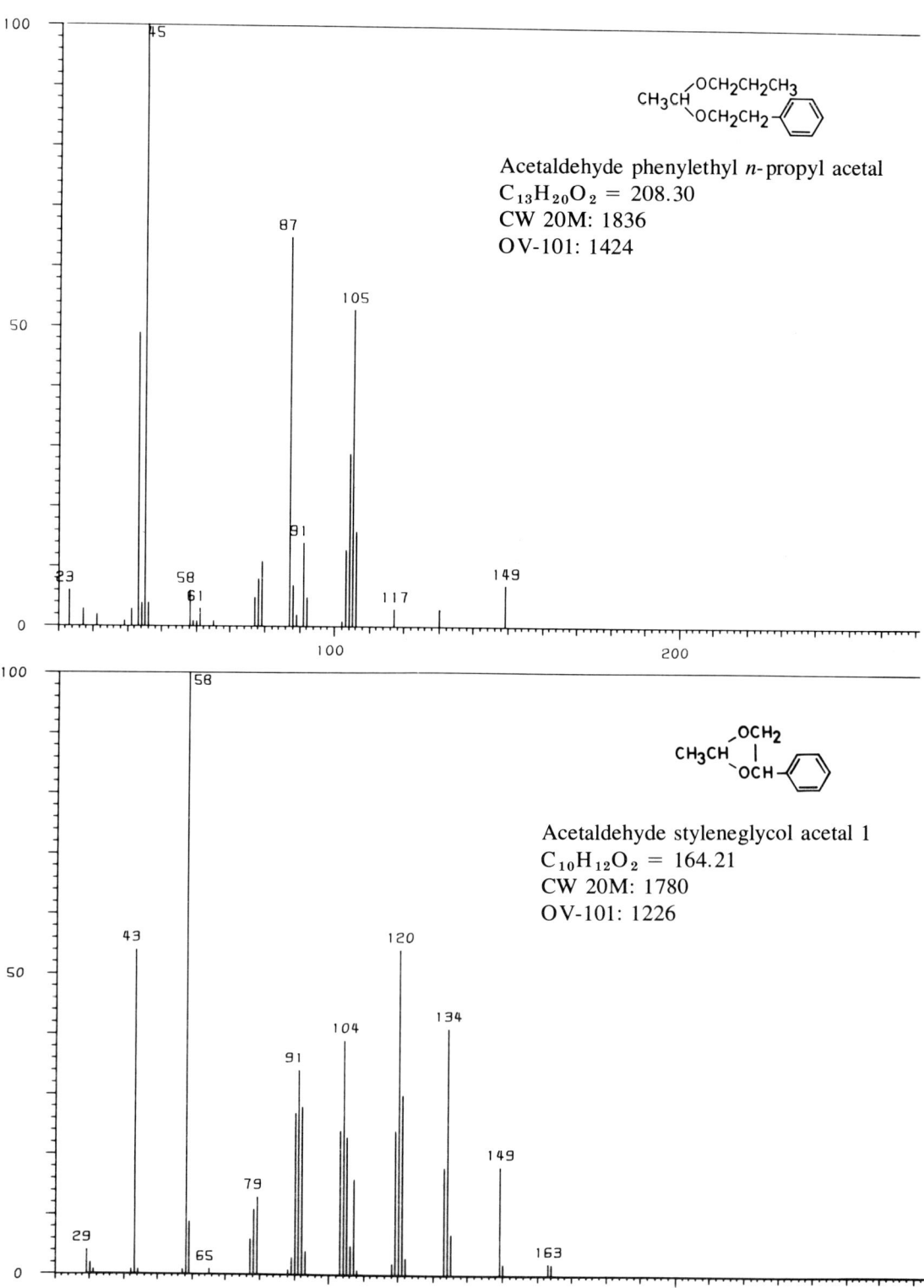

Acetaldehyde phenylethyl *n*-propyl acetal
$C_{13}H_{20}O_2 = 208.30$
CW 20M: 1836
OV-101: 1424

Acetaldehyde styleneglycol acetal 1
$C_{10}H_{12}O_2 = 164.21$
CW 20M: 1780
OV-101: 1226

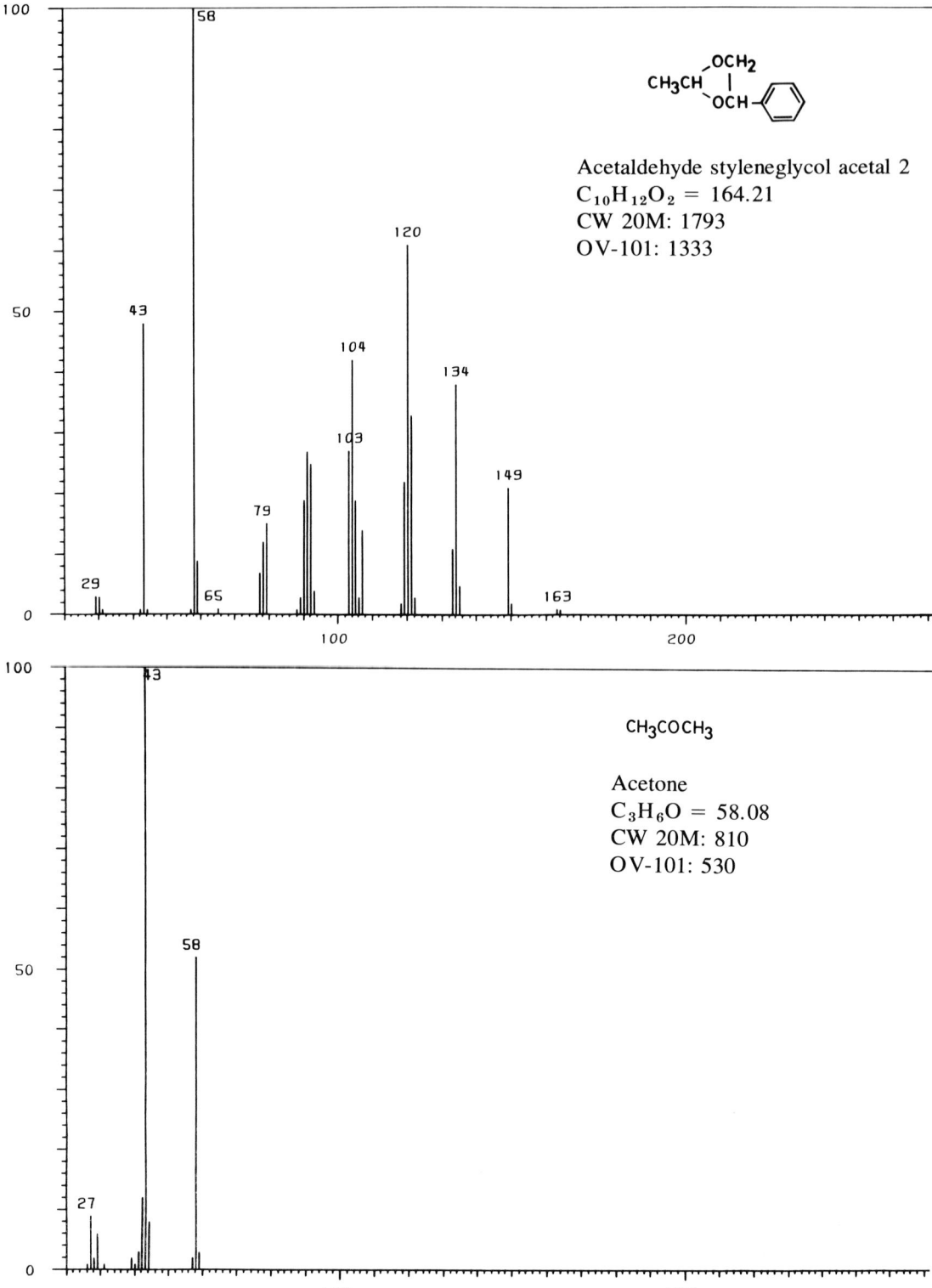

Acetaldehyde styleneglycol acetal 2
$C_{10}H_{12}O_2 = 164.21$
CW 20M: 1793
OV-101: 1333

Acetone
$C_3H_6O = 58.08$
CW 20M: 810
OV-101: 530

APPENDIX IV   MASS SPECTRA OF INDIVIDUAL COMPOUNDS   119

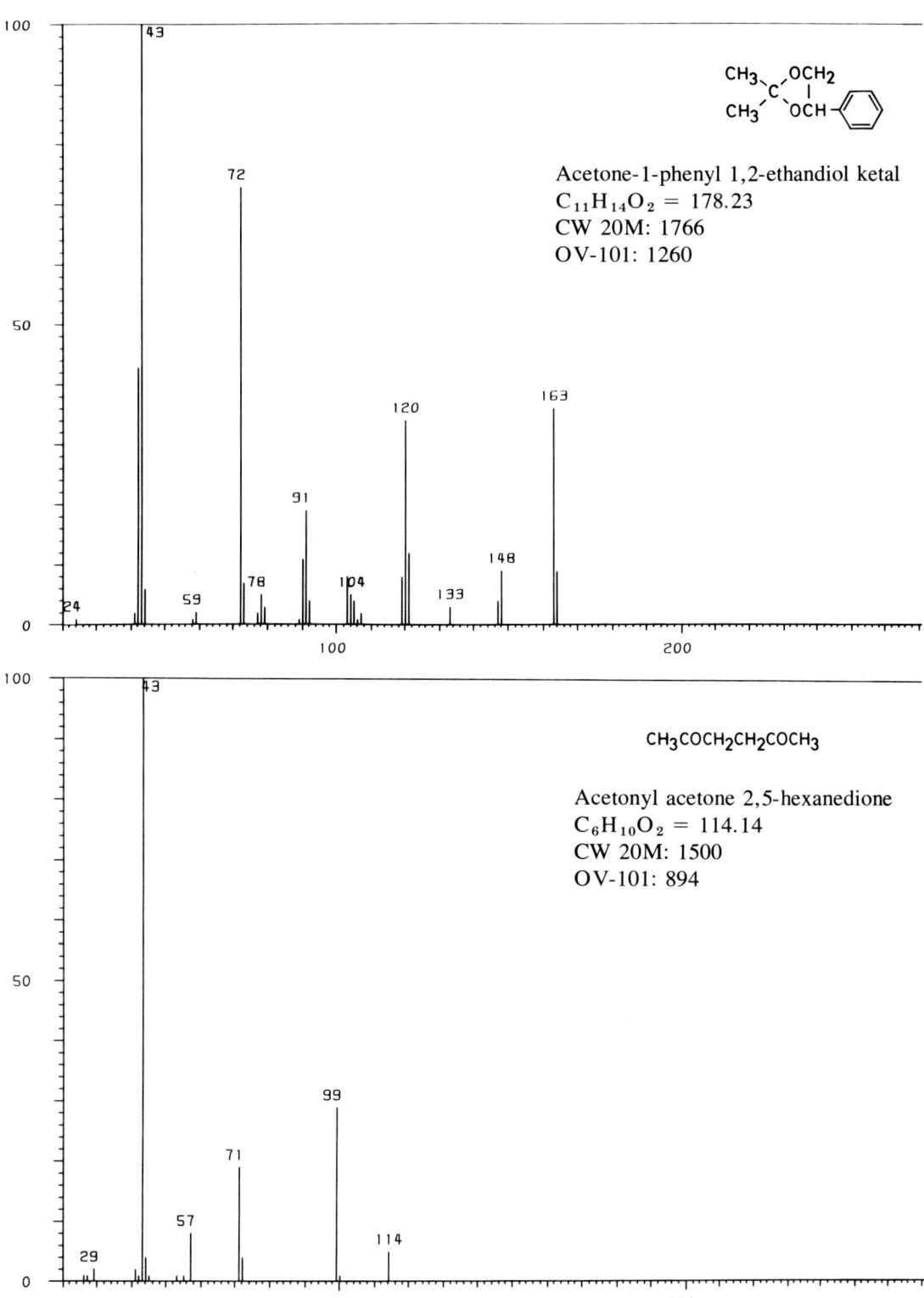

Acetone-1-phenyl 1,2-ethandiol ketal
$C_{11}H_{14}O_2 = 178.23$
CW 20M: 1766
OV-101: 1260

$CH_3COCH_2CH_2COCH_3$

Acetonyl acetone 2,5-hexanedione
$C_6H_{10}O_2 = 114.14$
CW 20M: 1500
OV-101: 894

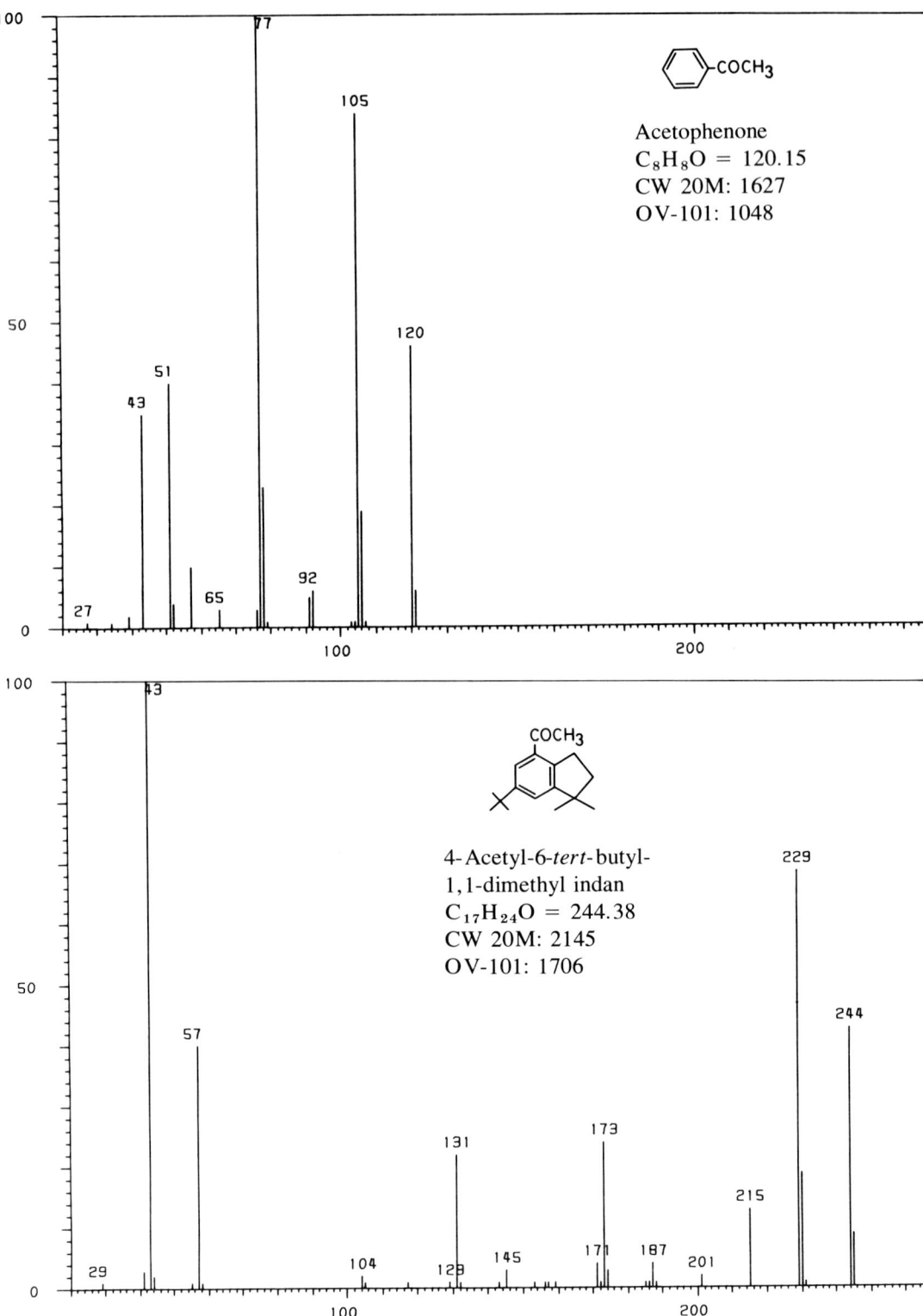

APPENDIX IV   MASS SPECTRA OF INDIVIDUAL COMPOUNDS

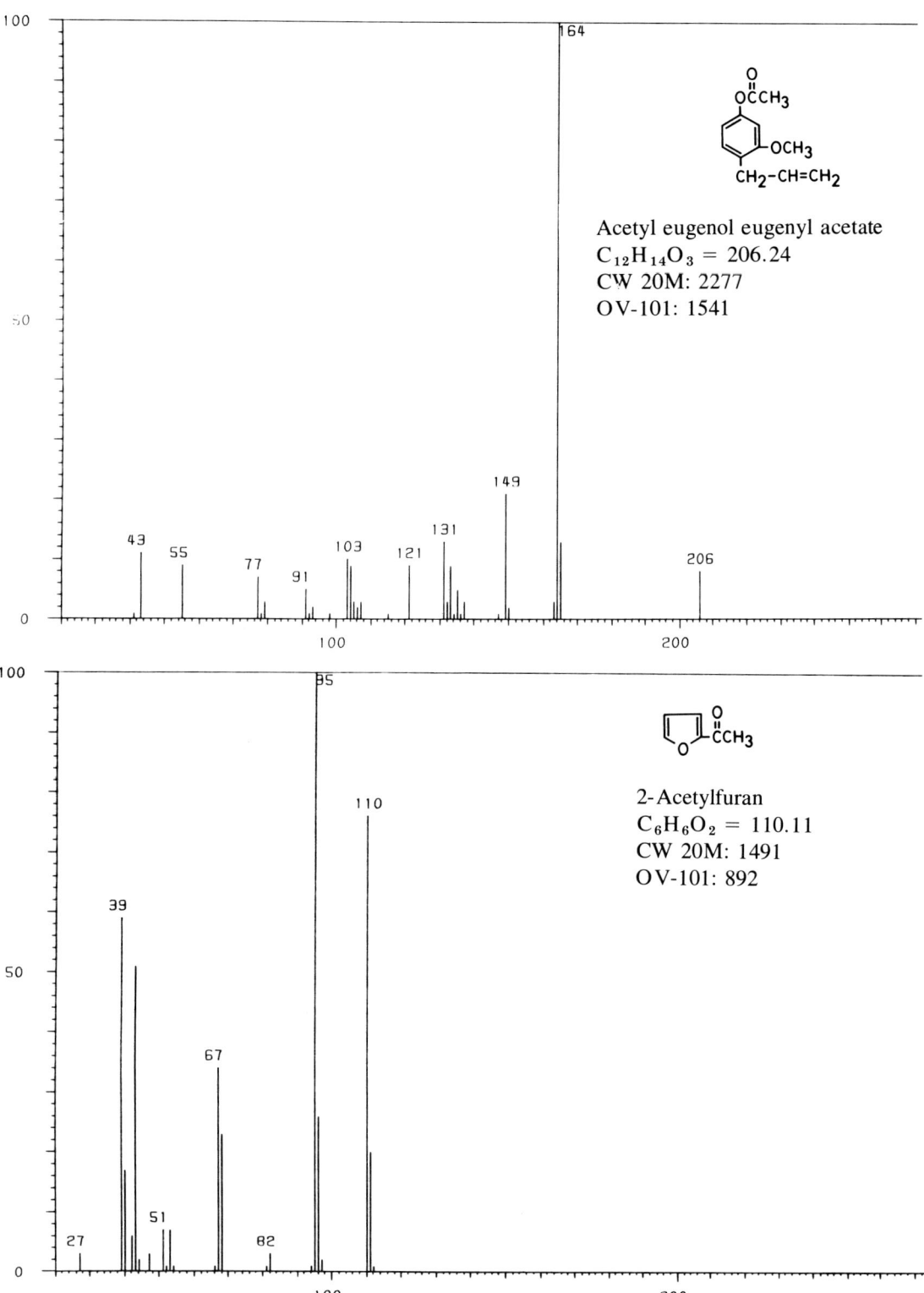

Acetyl eugenol eugenyl acetate
$C_{12}H_{14}O_3 = 206.24$
CW 20M: 2277
OV-101: 1541

2-Acetylfuran
$C_6H_6O_2 = 110.11$
CW 20M: 1491
OV-101: 892

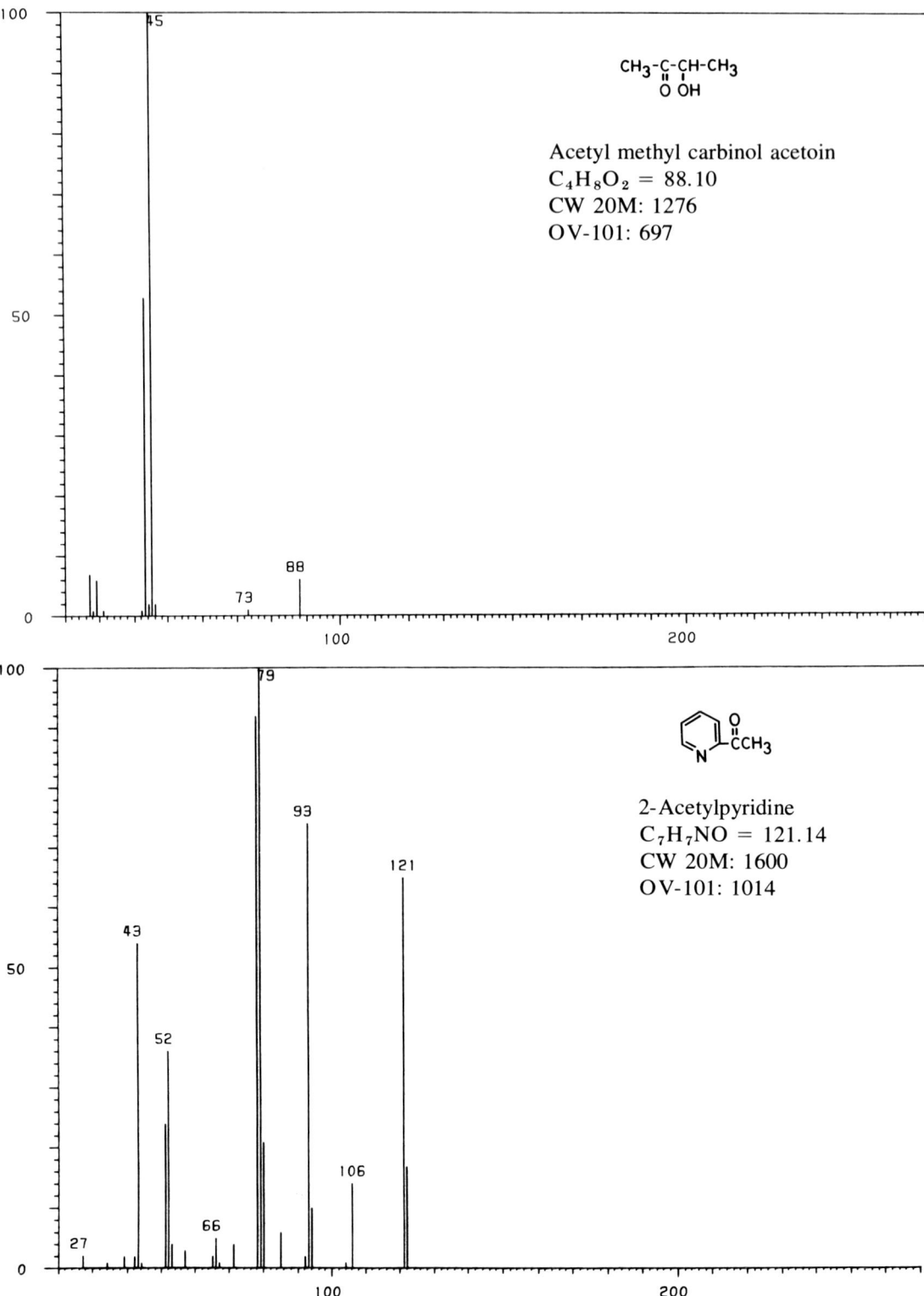

APPENDIX IV   MASS SPECTRA OF INDIVIDUAL COMPOUNDS

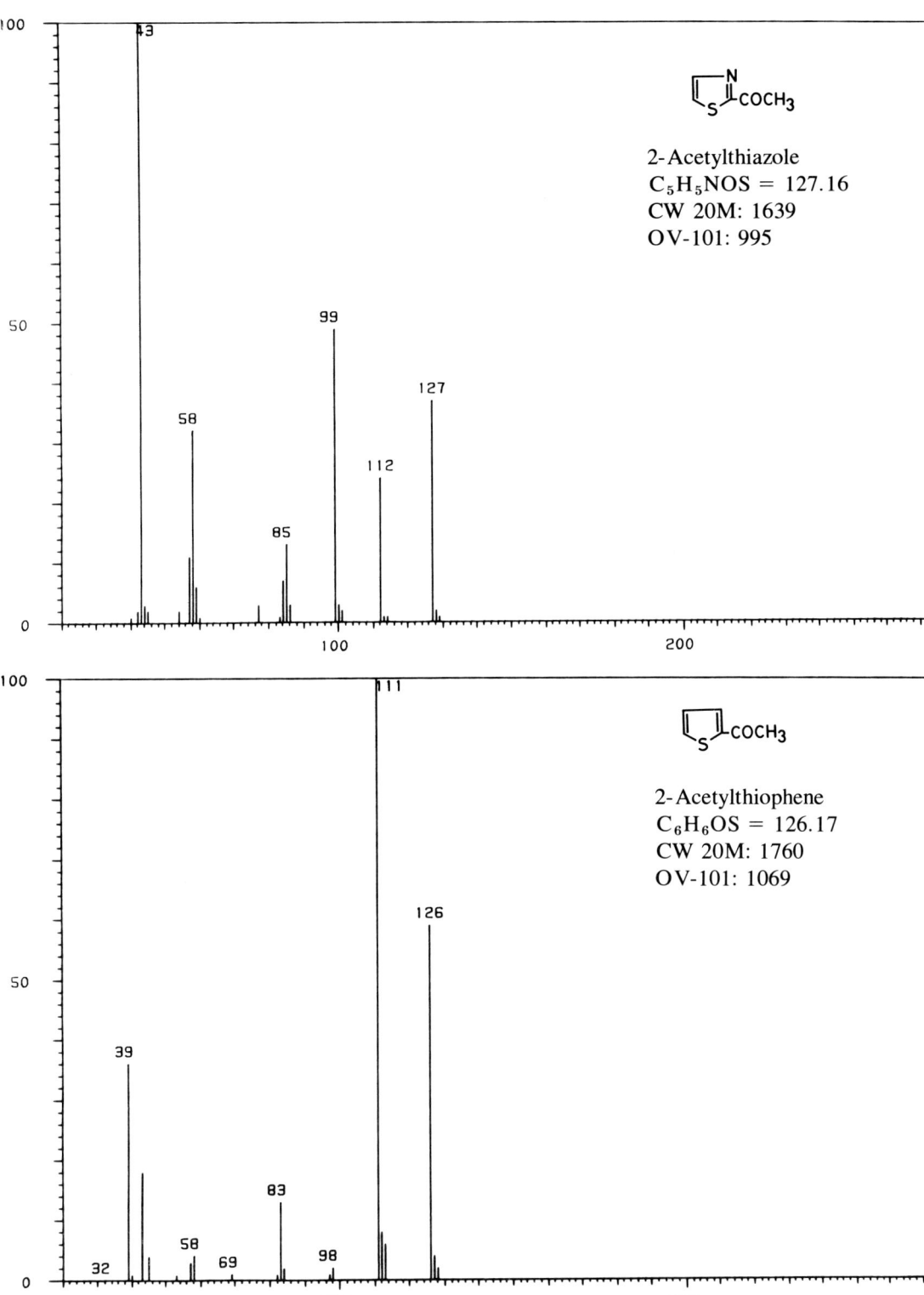

2-Acetylthiazole
$C_5H_5NOS = 127.16$
CW 20M: 1639
OV-101: 995

2-Acetylthiophene
$C_6H_6OS = 126.17$
CW 20M: 1760
OV-101: 1069

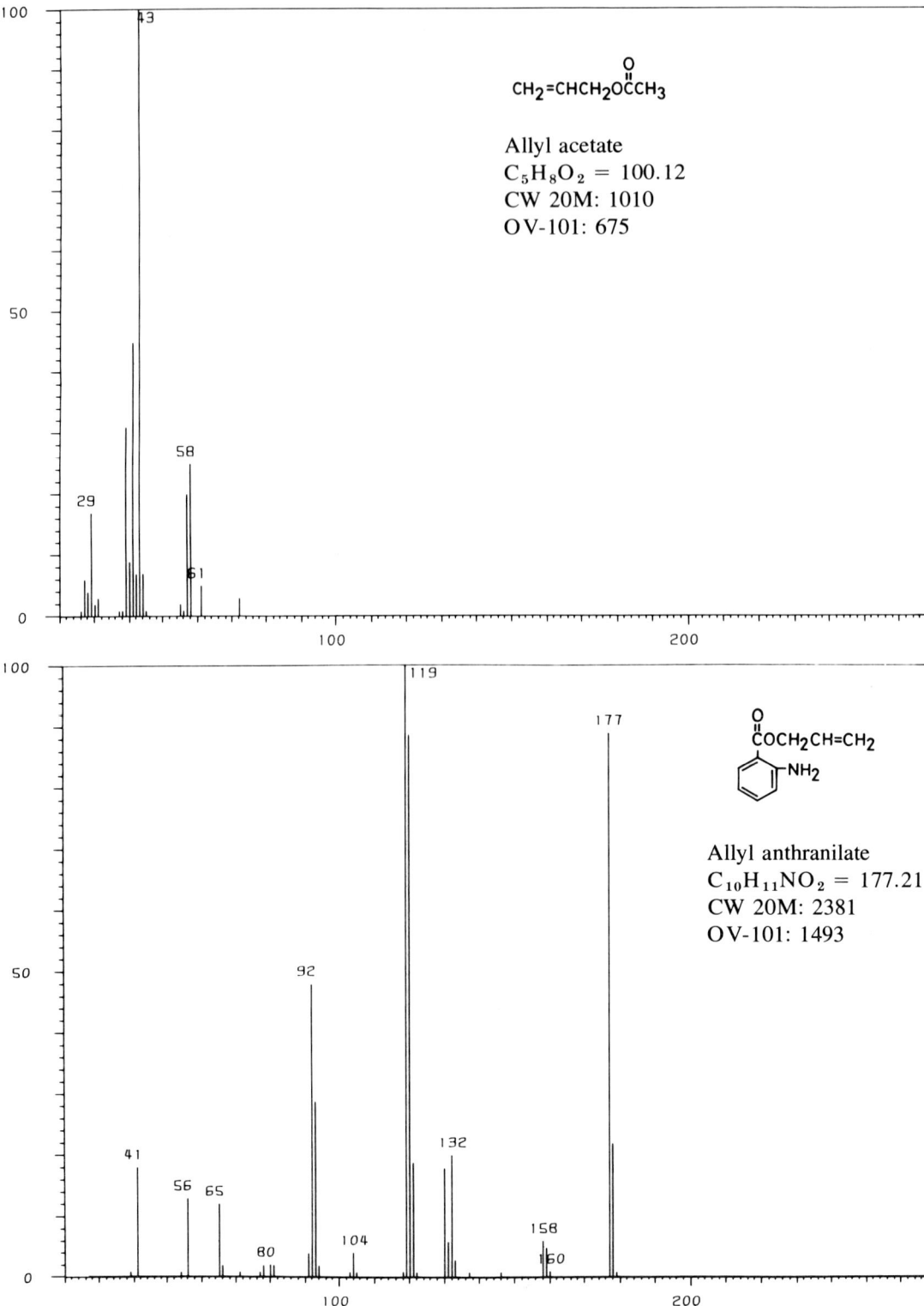

Allyl acetate
$C_5H_8O_2 = 100.12$
CW 20M: 1010
OV-101: 675

Allyl anthranilate
$C_{10}H_{11}NO_2 = 177.21$
CW 20M: 2381
OV-101: 1493

APPENDIX IV   MASS SPECTRA OF INDIVIDUAL COMPOUNDS

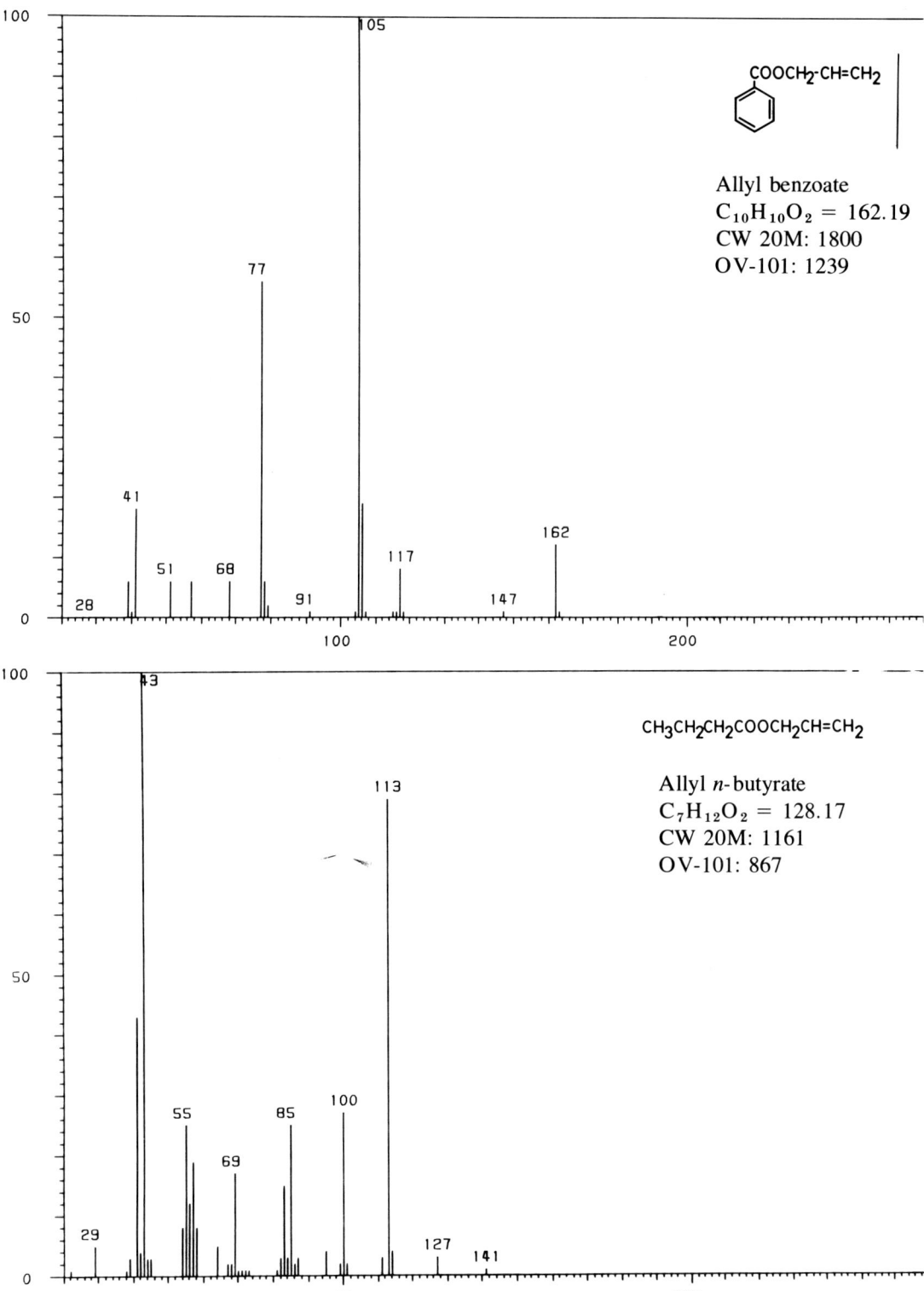

Allyl benzoate
$C_{10}H_{10}O_2 = 162.19$
CW 20M: 1800
OV-101: 1239

Allyl *n*-butyrate
$C_7H_{12}O_2 = 128.17$
CW 20M: 1161
OV-101: 867

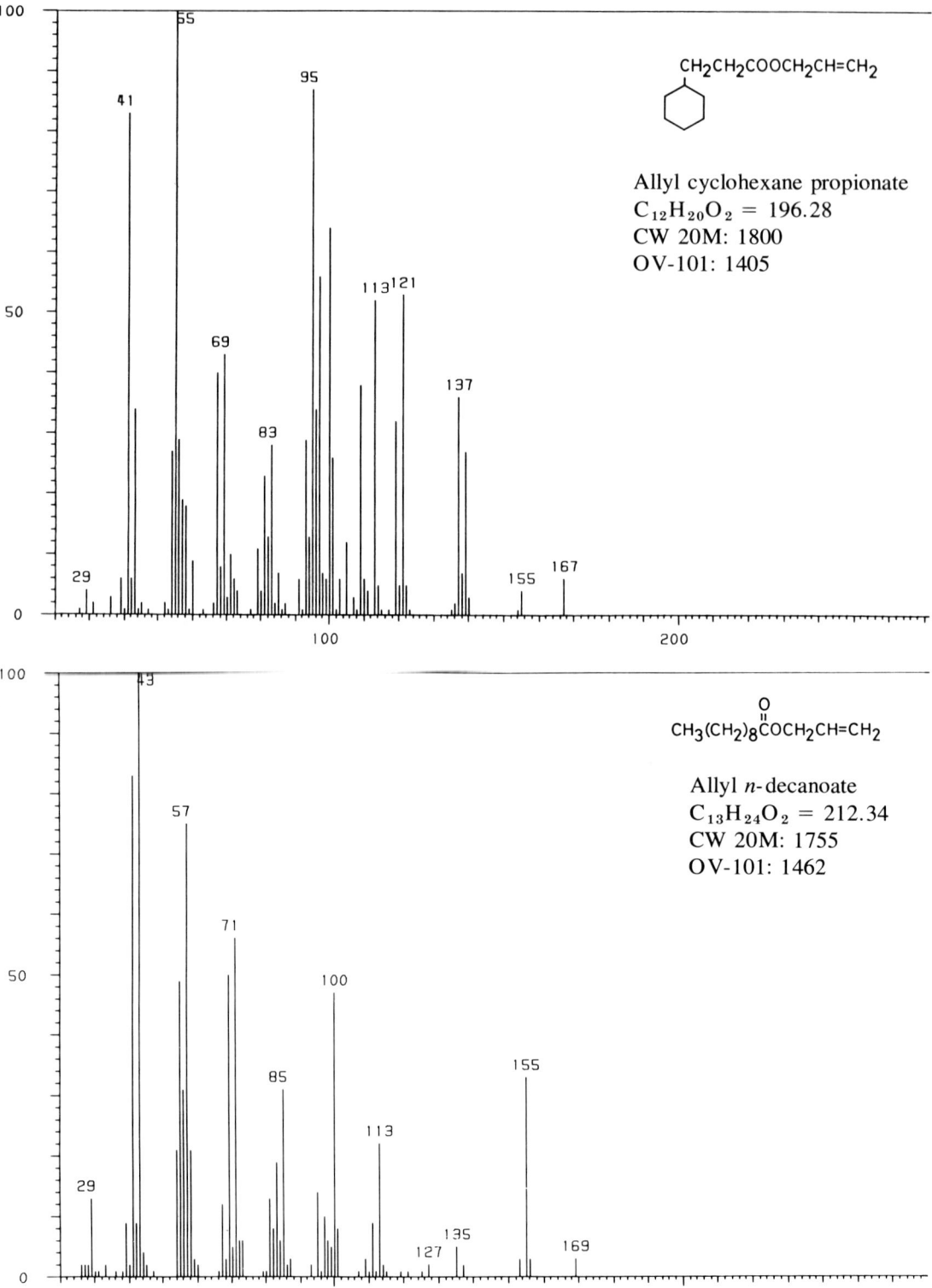

APPENDIX IV MASS SPECTRA OF INDIVIDUAL COMPOUNDS

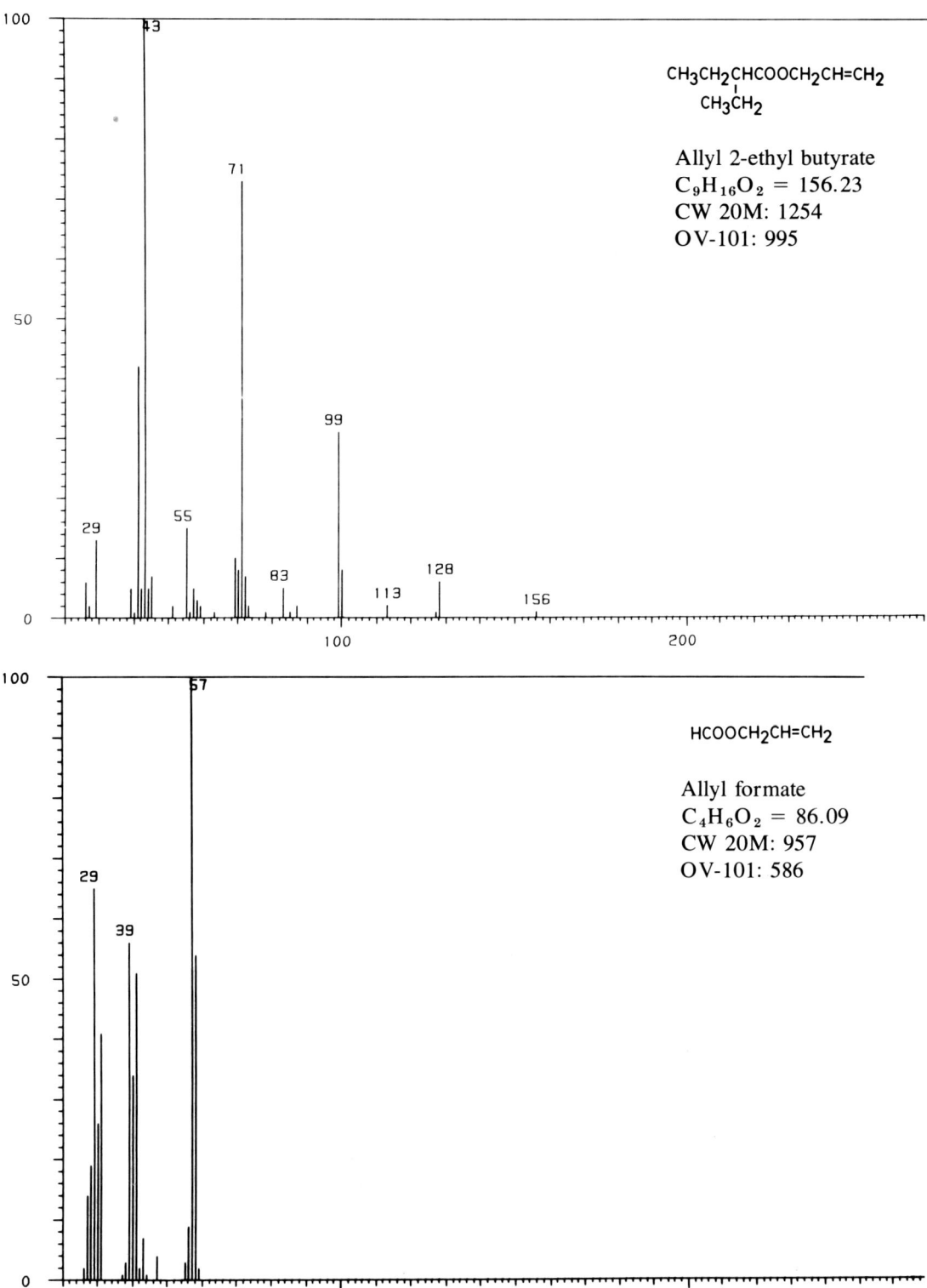

CH$_3$CH$_2$CHCOOCH$_2$CH=CH$_2$
         |
       CH$_3$CH$_2$

Allyl 2-ethyl butyrate
C$_9$H$_{16}$O$_2$ = 156.23
CW 20M: 1254
OV-101: 995

HCOOCH$_2$CH=CH$_2$

Allyl formate
C$_4$H$_6$O$_2$ = 86.09
CW 20M: 957
OV-101: 586

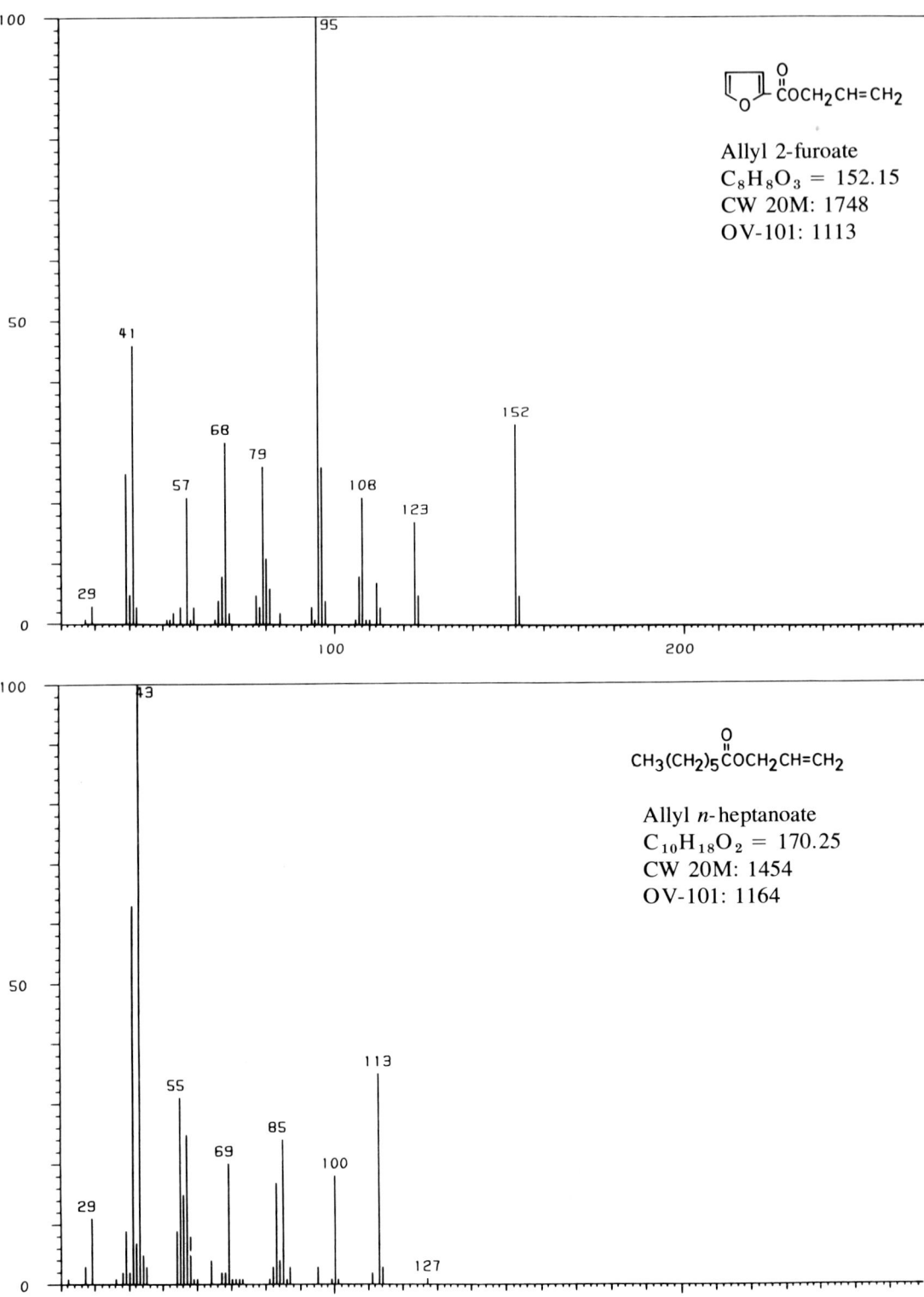

APPENDIX IV   MASS SPECTRA OF INDIVIDUAL COMPOUNDS

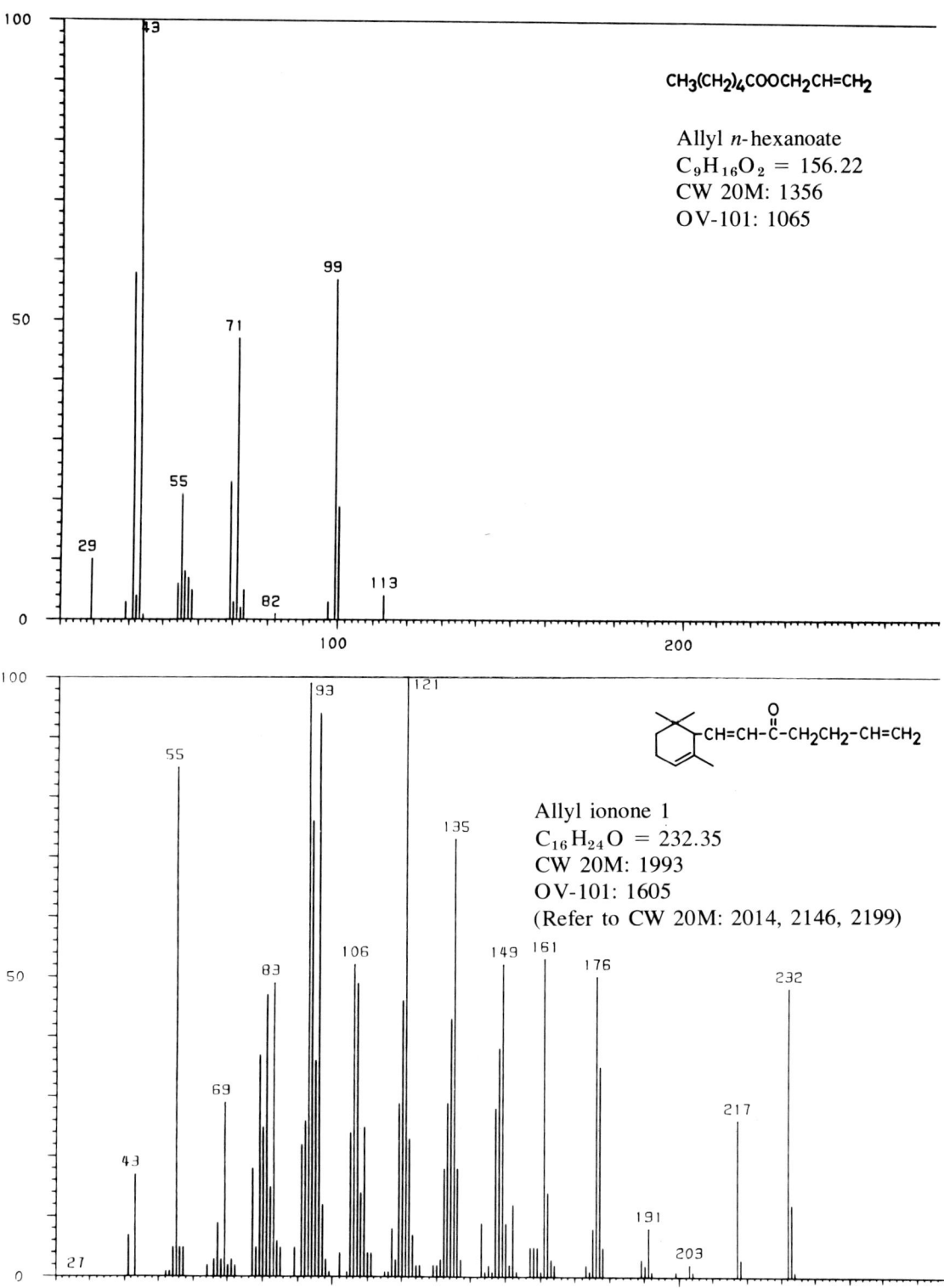

$CH_3(CH_2)_4COOCH_2CH=CH_2$

Allyl *n*-hexanoate
$C_9H_{16}O_2 = 156.22$
CW 20M: 1356
OV-101: 1065

Allyl ionone 1
$C_{16}H_{24}O = 232.35$
CW 20M: 1993
OV-101: 1605
(Refer to CW 20M: 2014, 2146, 2199)

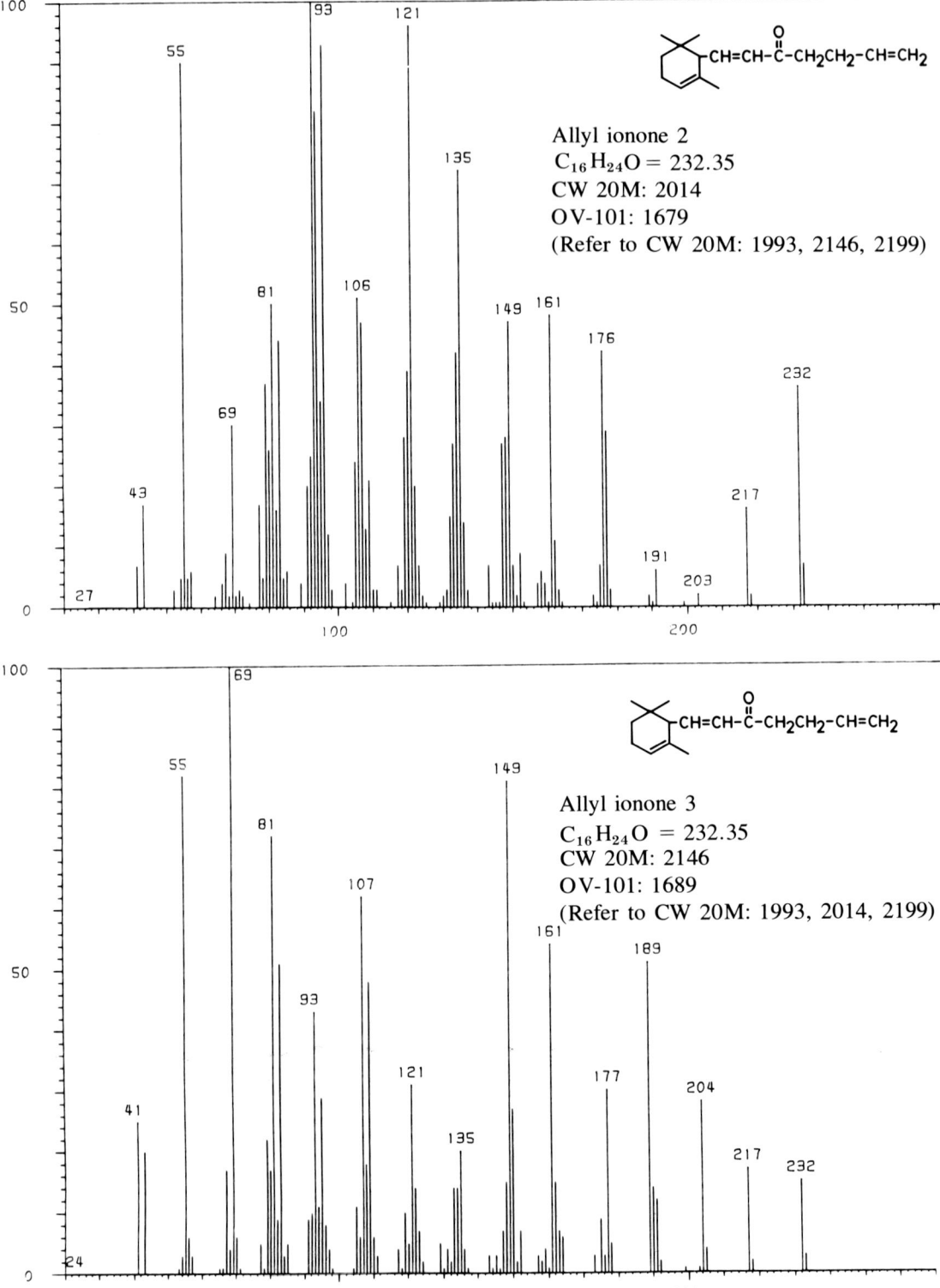

Allyl ionone 2
$C_{16}H_{24}O = 232.35$
CW 20M: 2014
OV-101: 1679
(Refer to CW 20M: 1993, 2146, 2199)

Allyl ionone 3
$C_{16}H_{24}O = 232.35$
CW 20M: 2146
OV-101: 1689
(Refer to CW 20M: 1993, 2014, 2199)

APPENDIX IV   MASS SPECTRA OF INDIVIDUAL COMPOUNDS

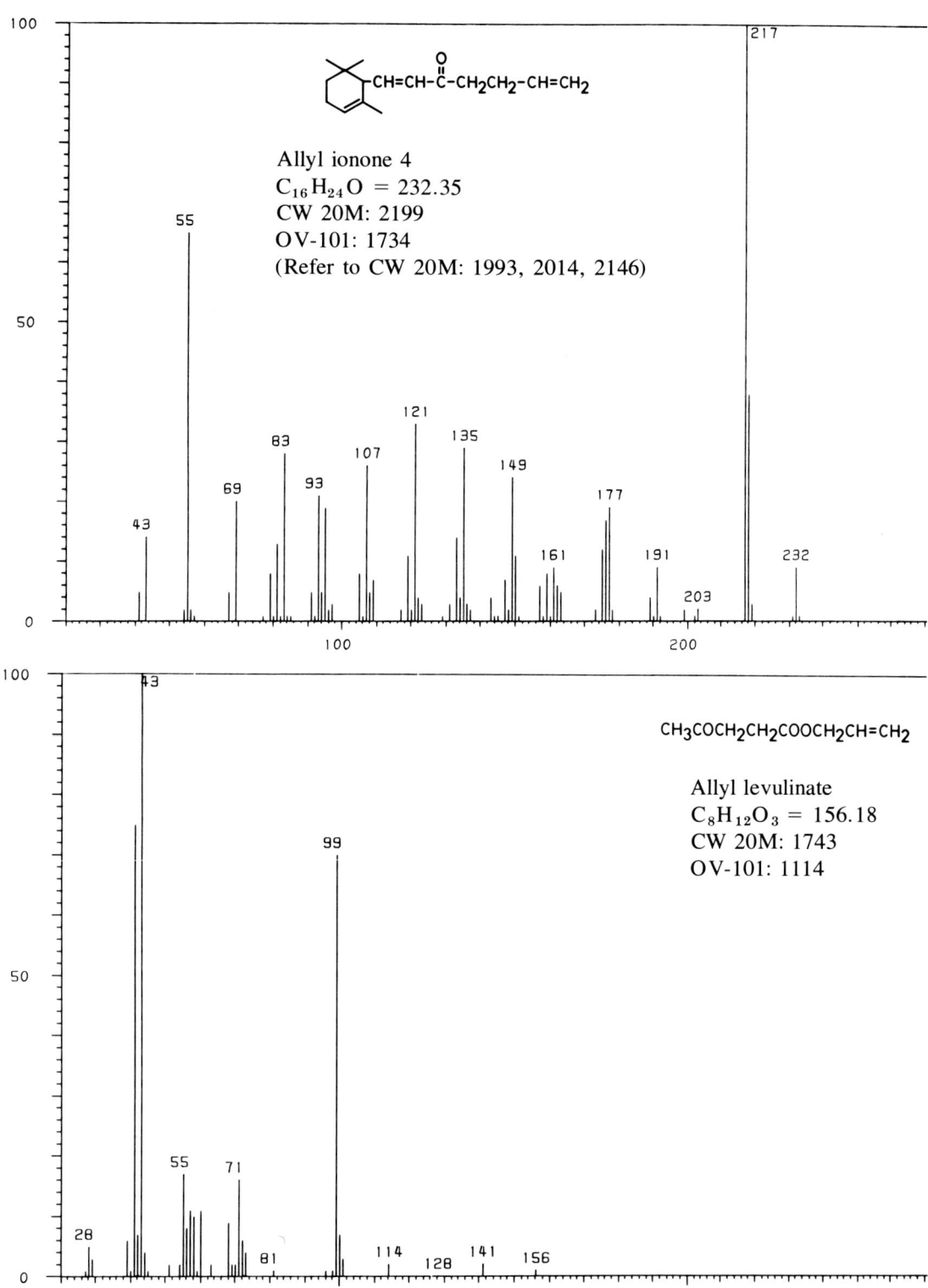

Allyl ionone 4
$C_{16}H_{24}O = 232.35$
CW 20M: 2199
OV-101: 1734
(Refer to CW 20M: 1993, 2014, 2146)

Allyl levulinate
$C_8H_{12}O_3 = 156.18$
CW 20M: 1743
OV-101: 1114

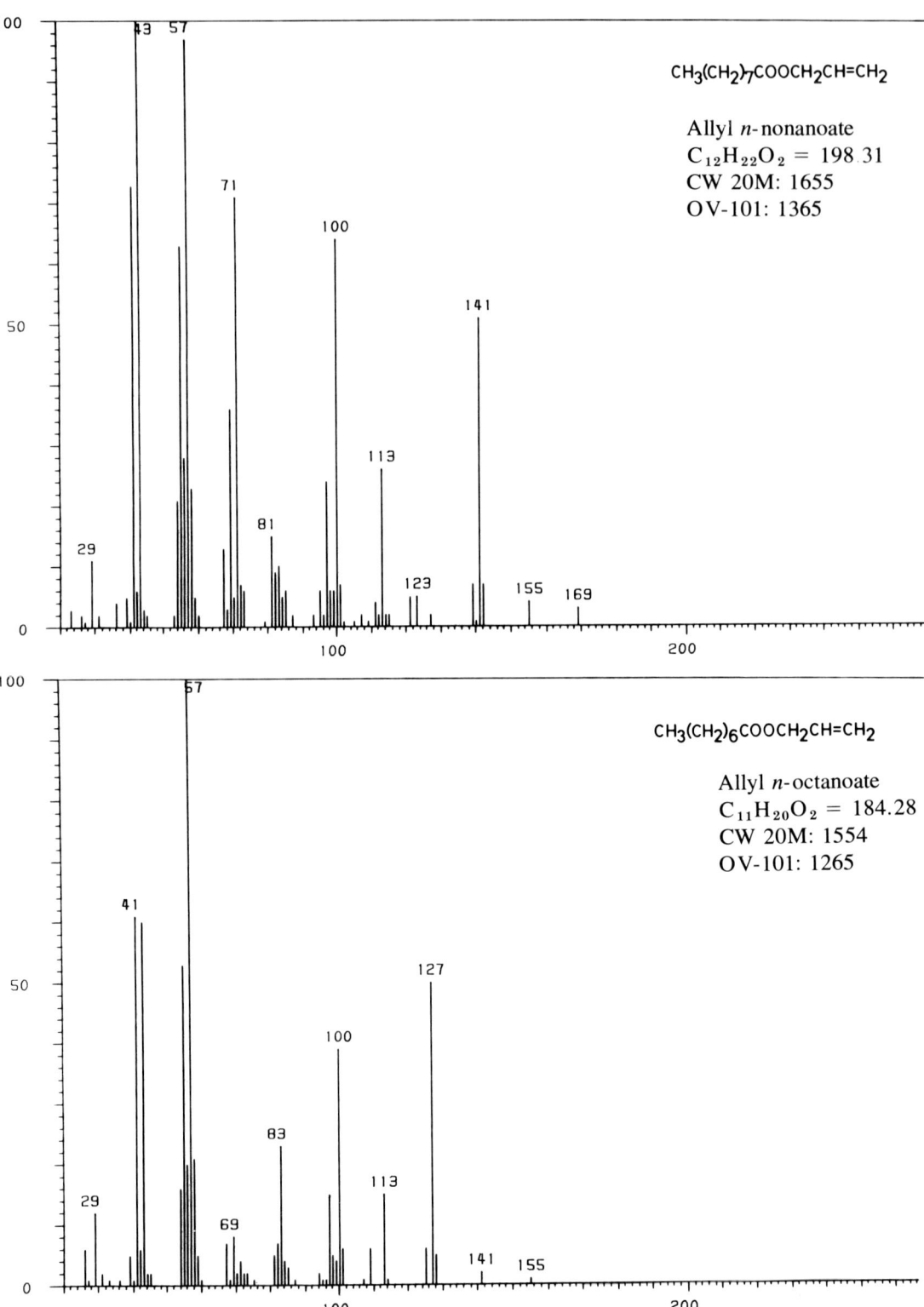

APPENDIX IV MASS SPECTRA OF INDIVIDUAL COMPOUNDS

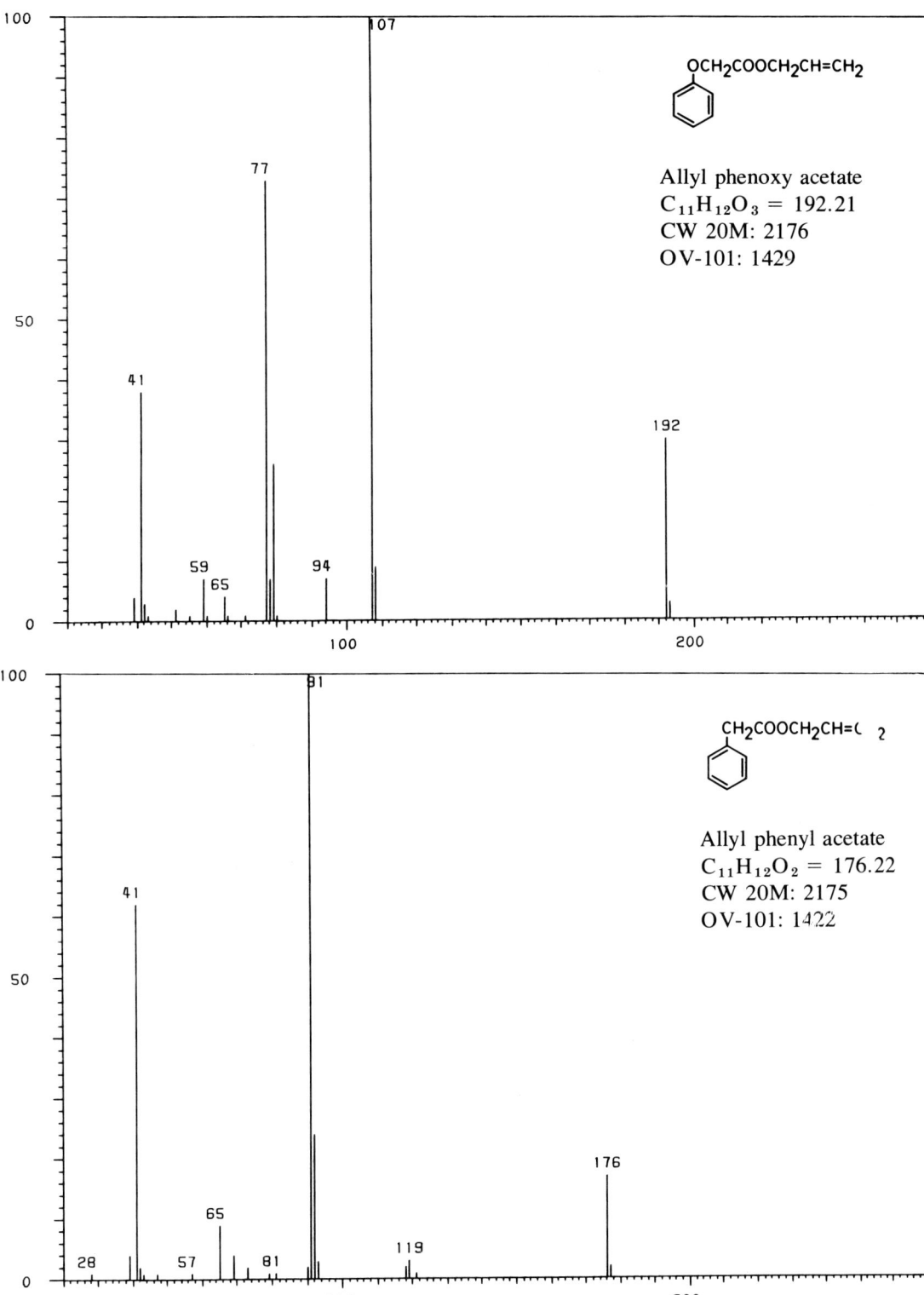

Allyl phenoxy acetate
$C_{11}H_{12}O_3 = 192.21$
CW 20M: 2176
OV-101: 1429

Allyl phenyl acetate
$C_{11}H_{12}O_2 = 176.22$
CW 20M: 2175
OV-101: 1422

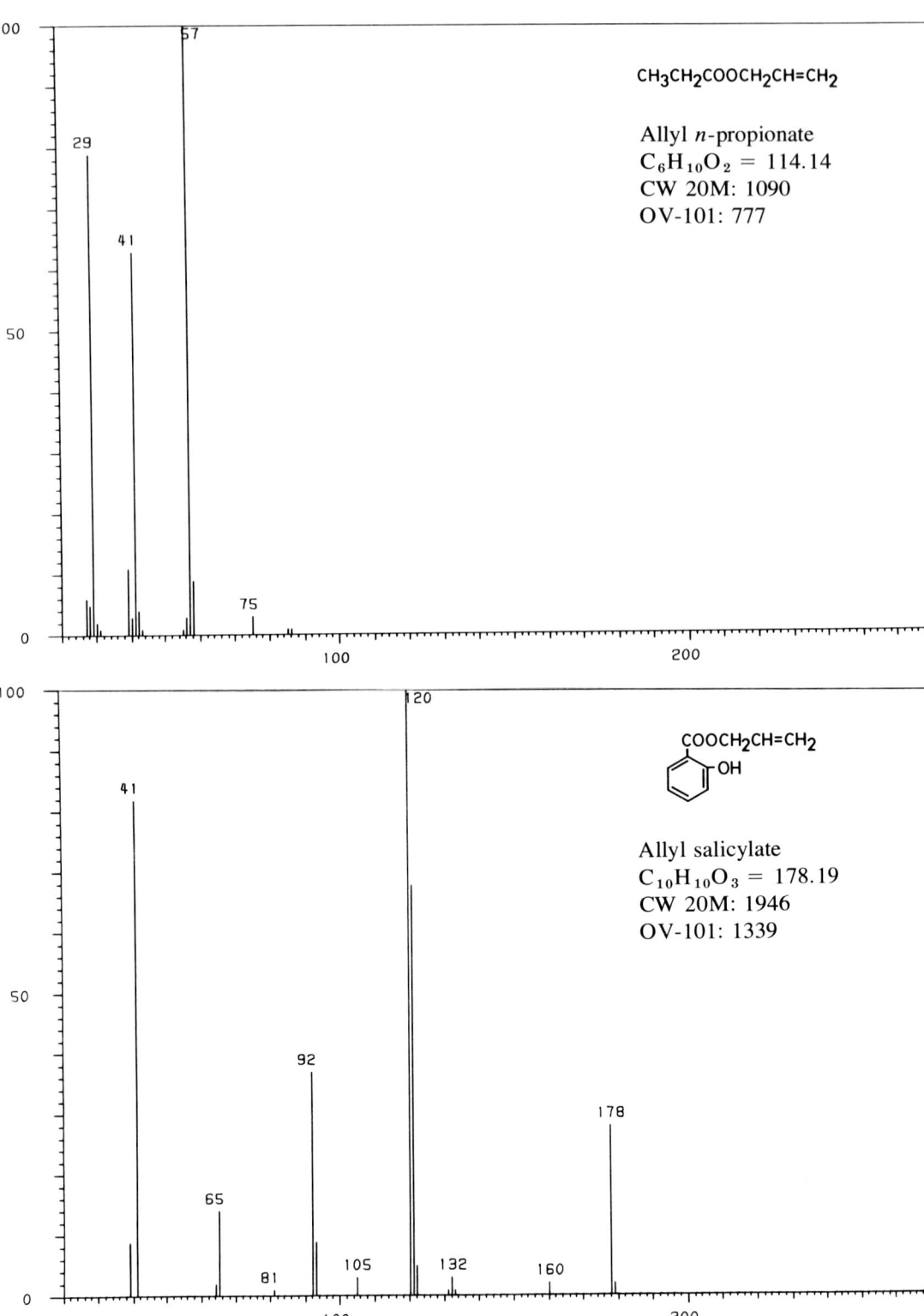

APPENDIX IV   MASS SPECTRA OF INDIVIDUAL COMPOUNDS

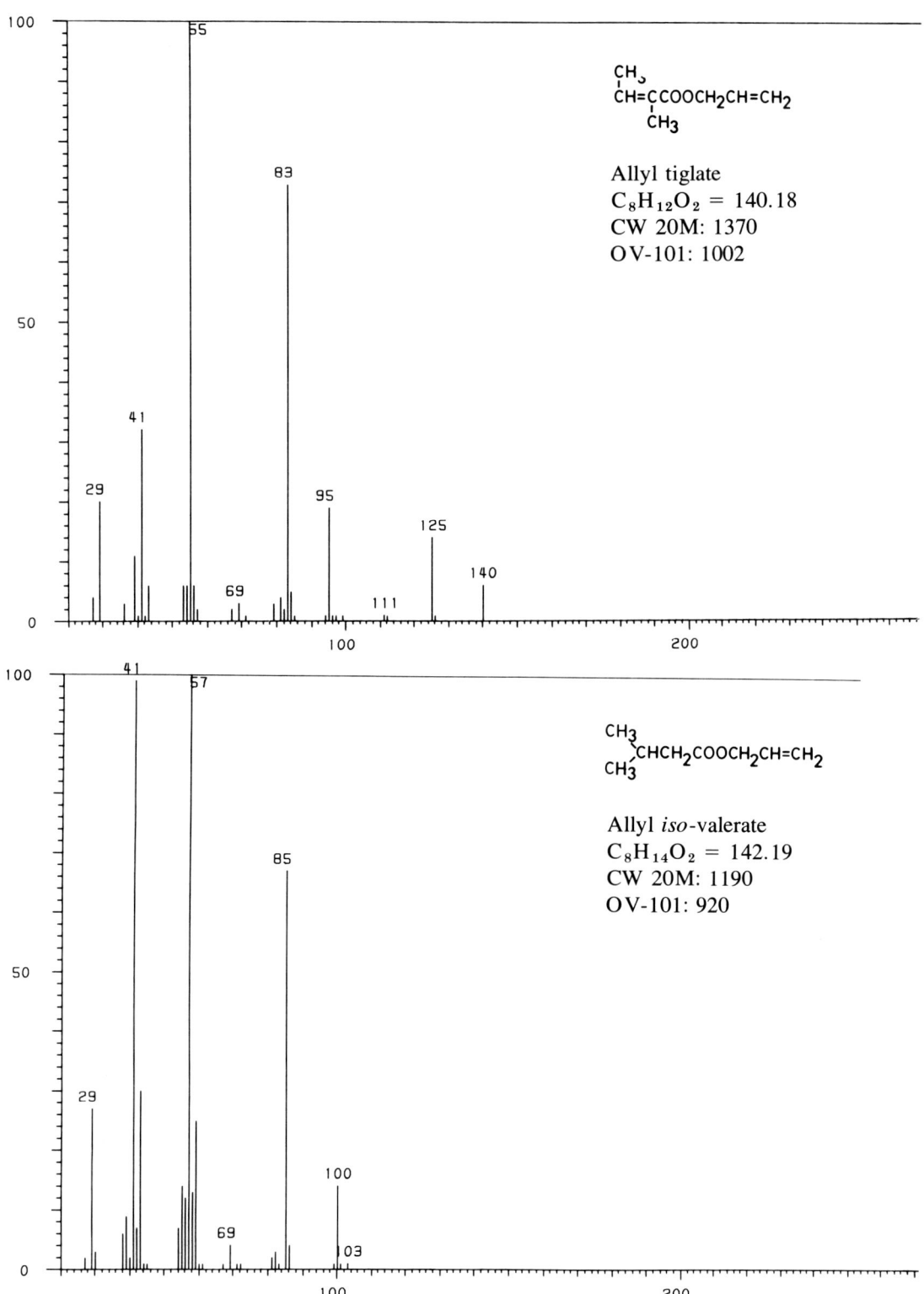

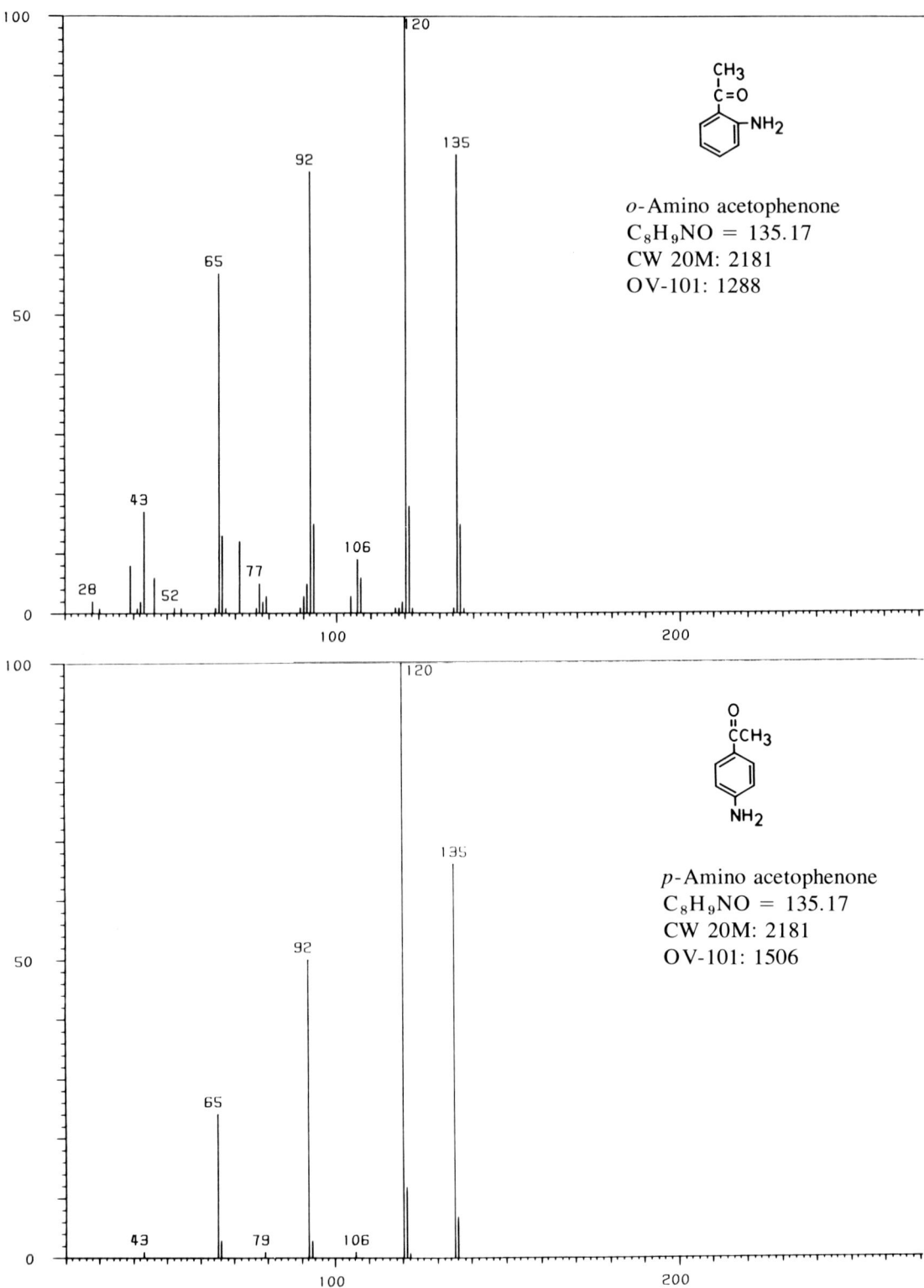

APPENDIX IV   MASS SPECTRA OF INDIVIDUAL COMPOUNDS

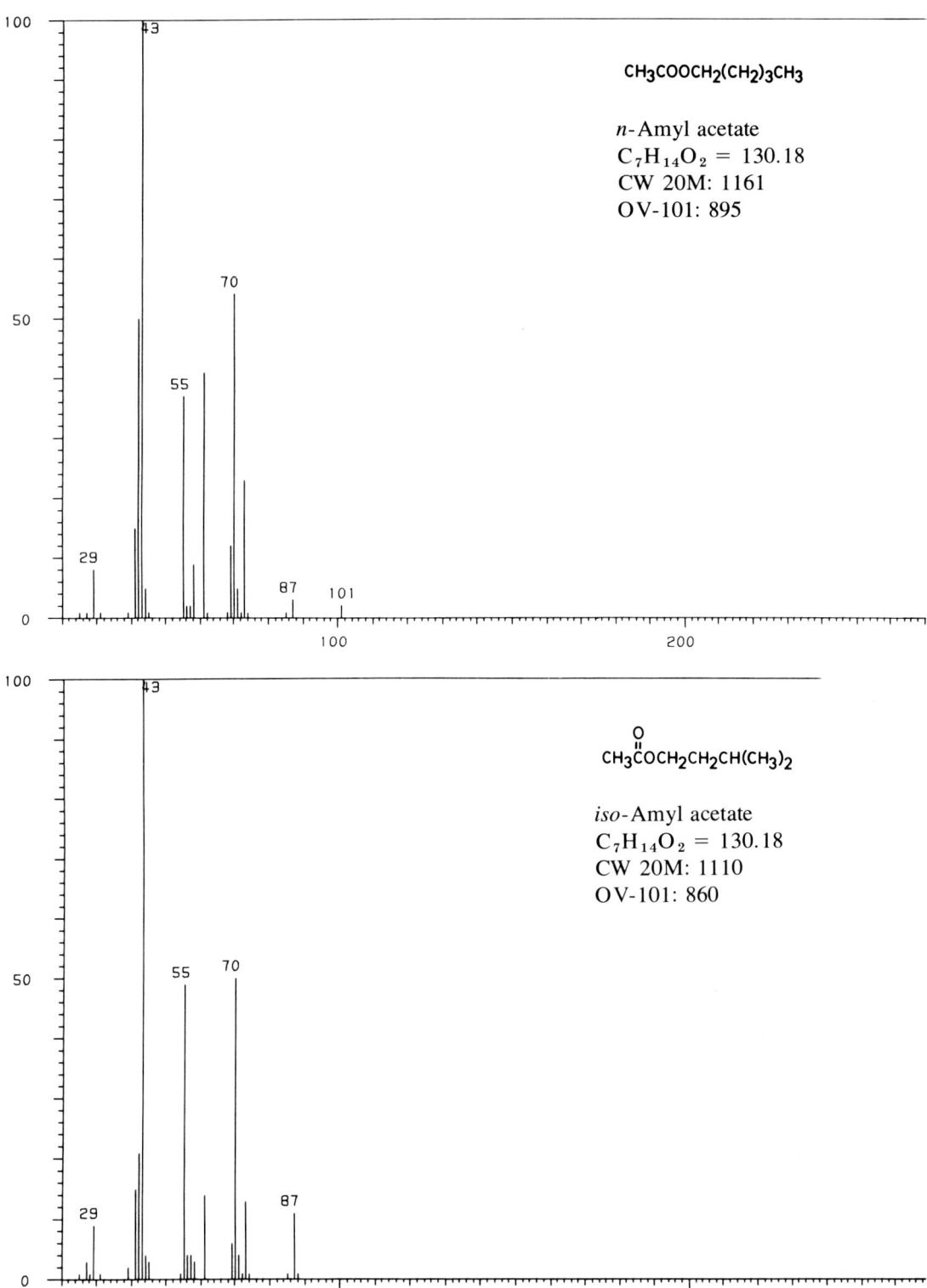

CH₃COOCH₂(CH₂)₃CH₃

n-Amyl acetate
C₇H₁₄O₂ = 130.18
CW 20M: 1161
OV-101: 895

CH₃COCH₂CH₂CH(CH₃)₂

iso-Amyl acetate
C₇H₁₄O₂ = 130.18
CW 20M: 1110
OV-101: 860

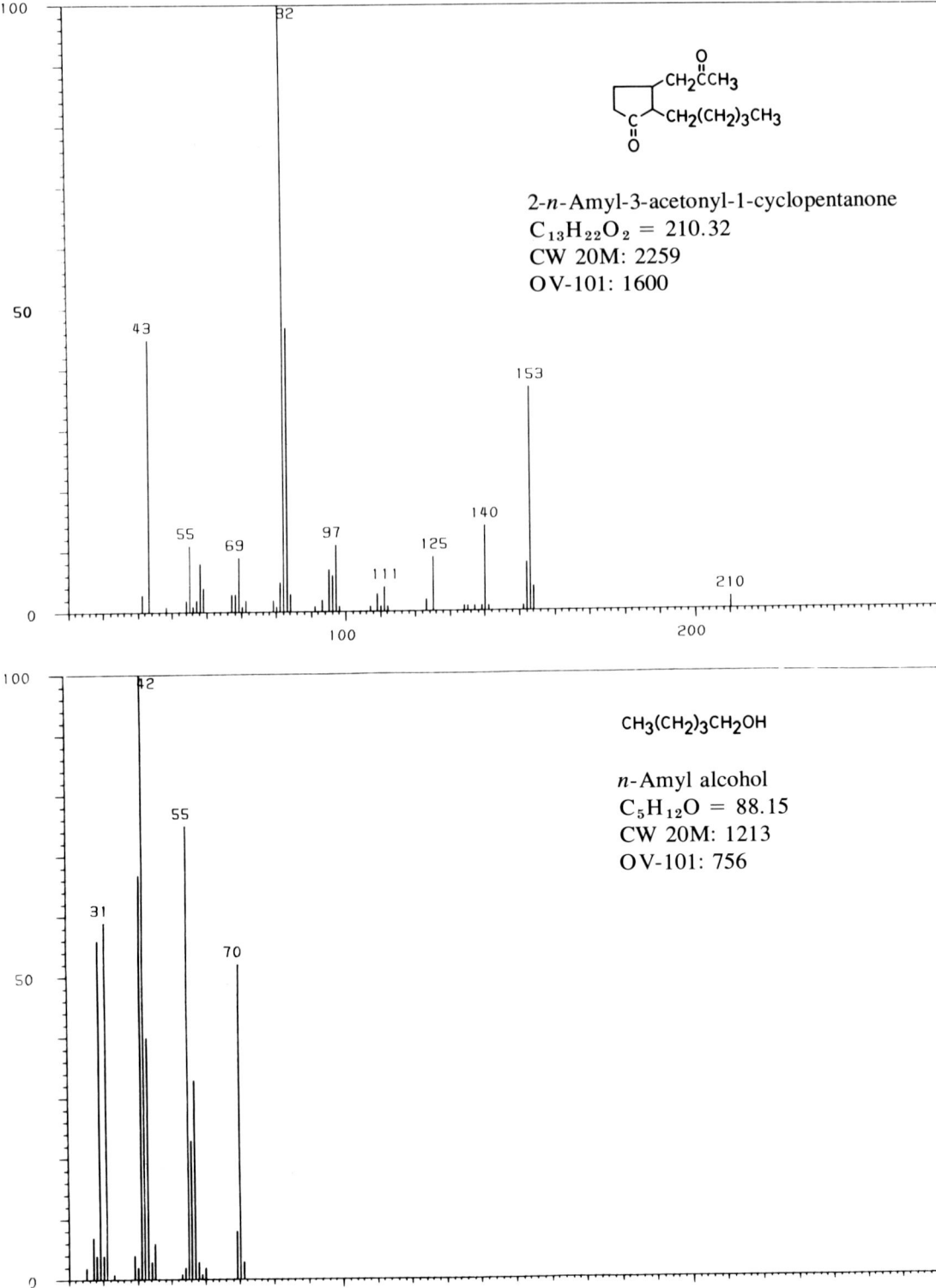

APPENDIX IV   MASS SPECTRA OF INDIVIDUAL COMPOUNDS   **139**

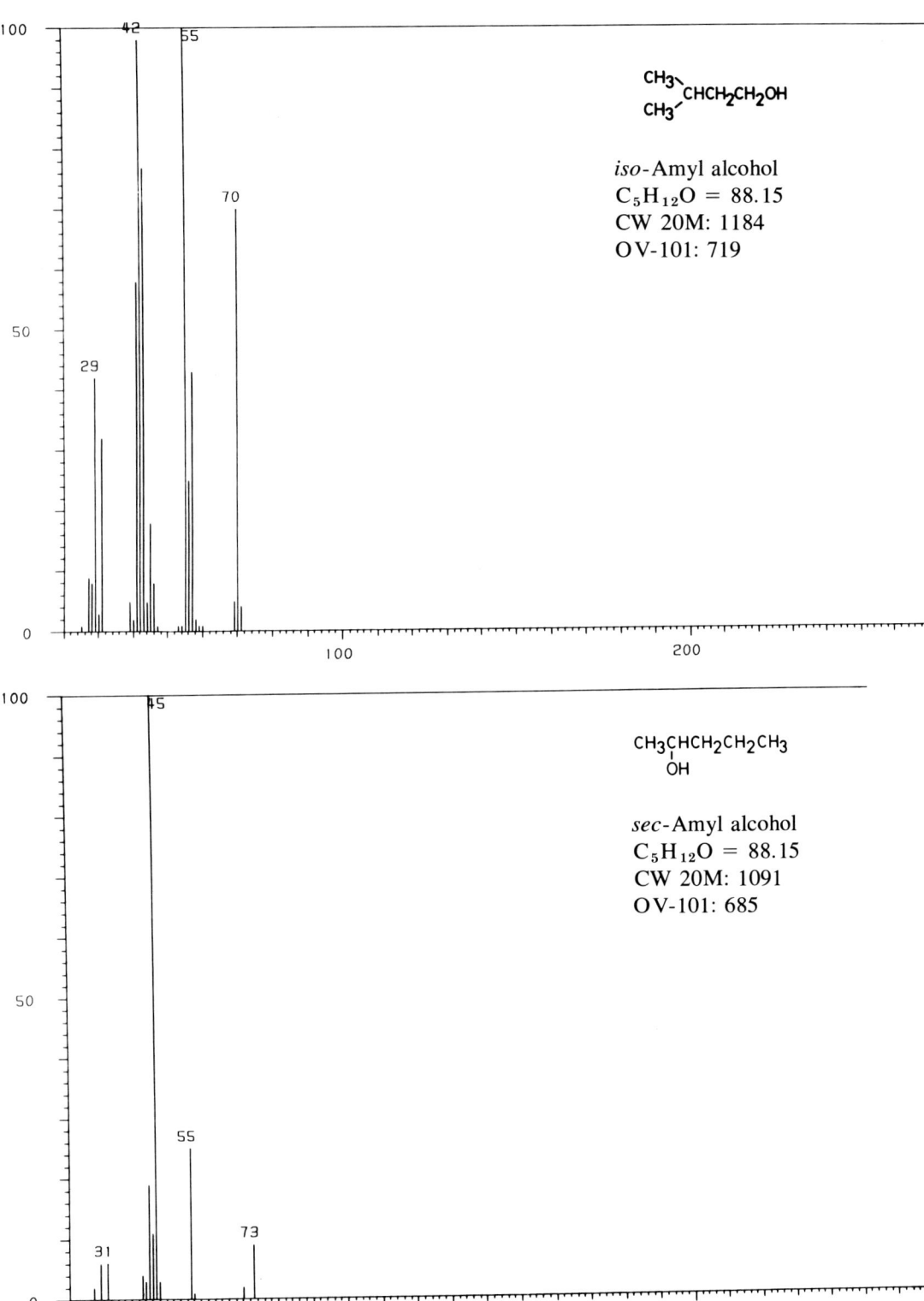

iso-Amyl alcohol
$C_5H_{12}O = 88.15$
CW 20M: 1184
OV-101: 719

sec-Amyl alcohol
$C_5H_{12}O = 88.15$
CW 20M: 1091
OV-101: 685

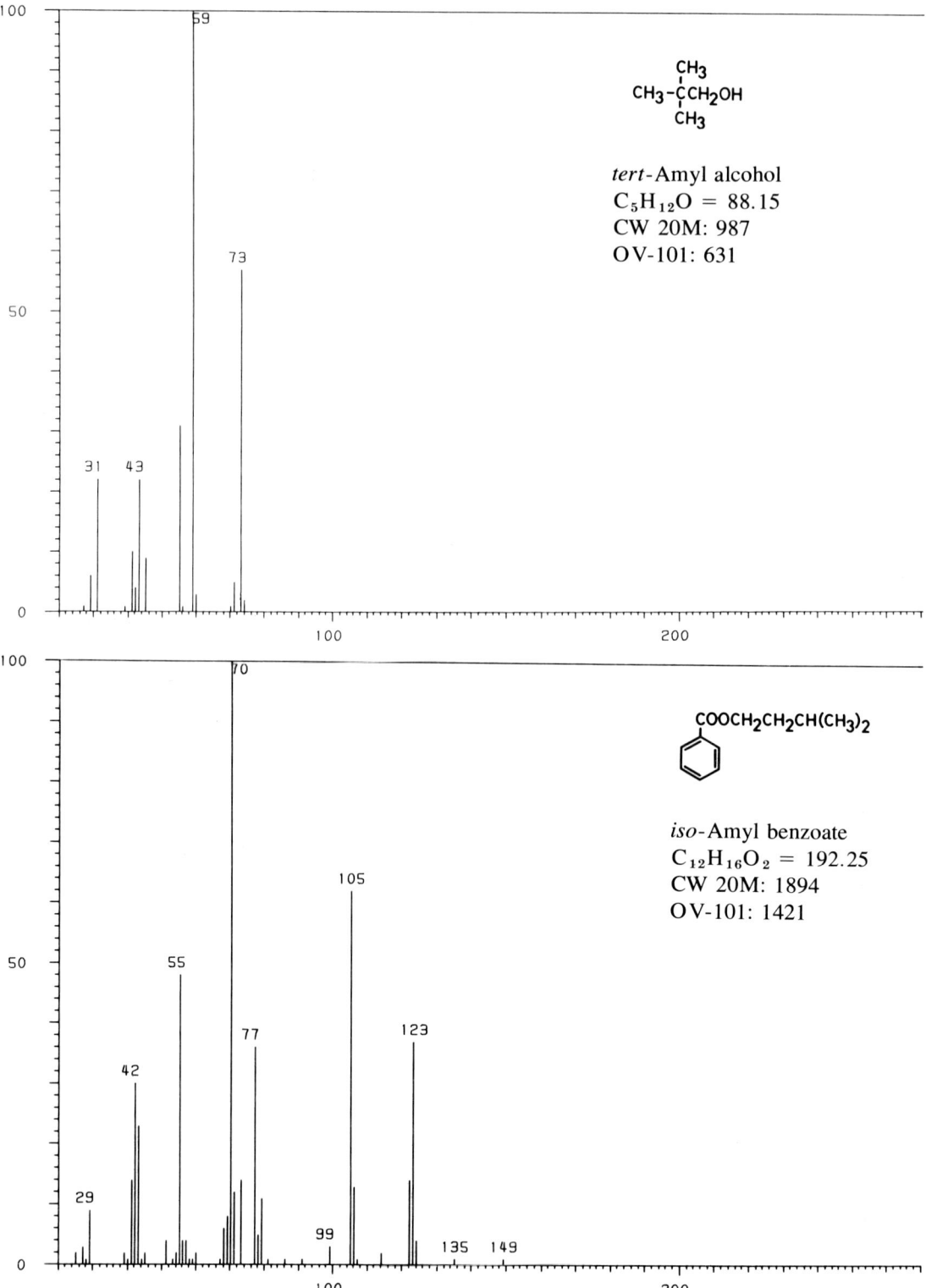

APPENDIX IV MASS SPECTRA OF INDIVIDUAL COMPOUNDS 141

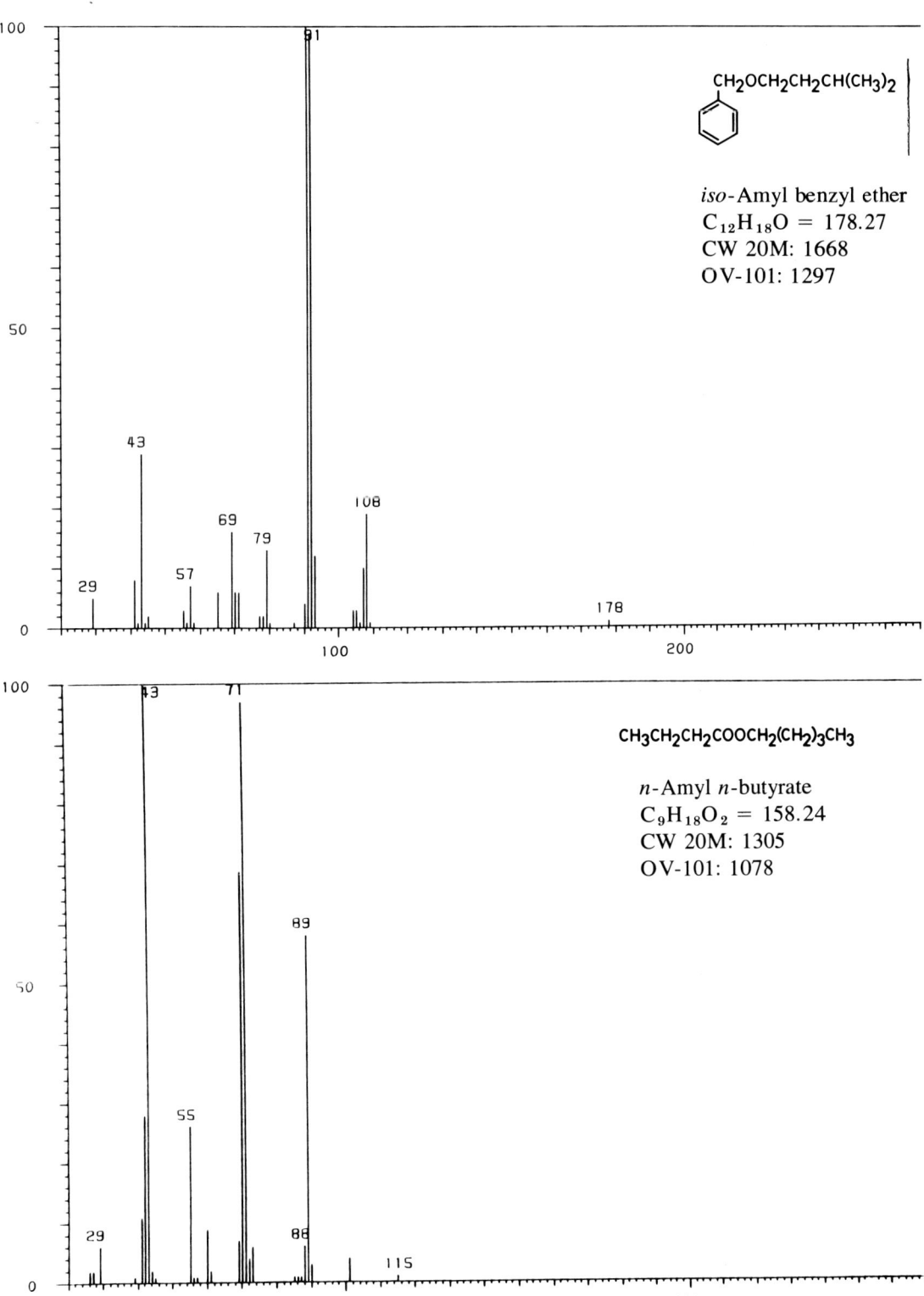

iso-Amyl benzyl ether
$C_{12}H_{18}O = 178.27$
CW 20M: 1668
OV-101: 1297

n-Amyl n-butyrate
$C_9H_{18}O_2 = 158.24$
CW 20M: 1305
OV-101: 1078

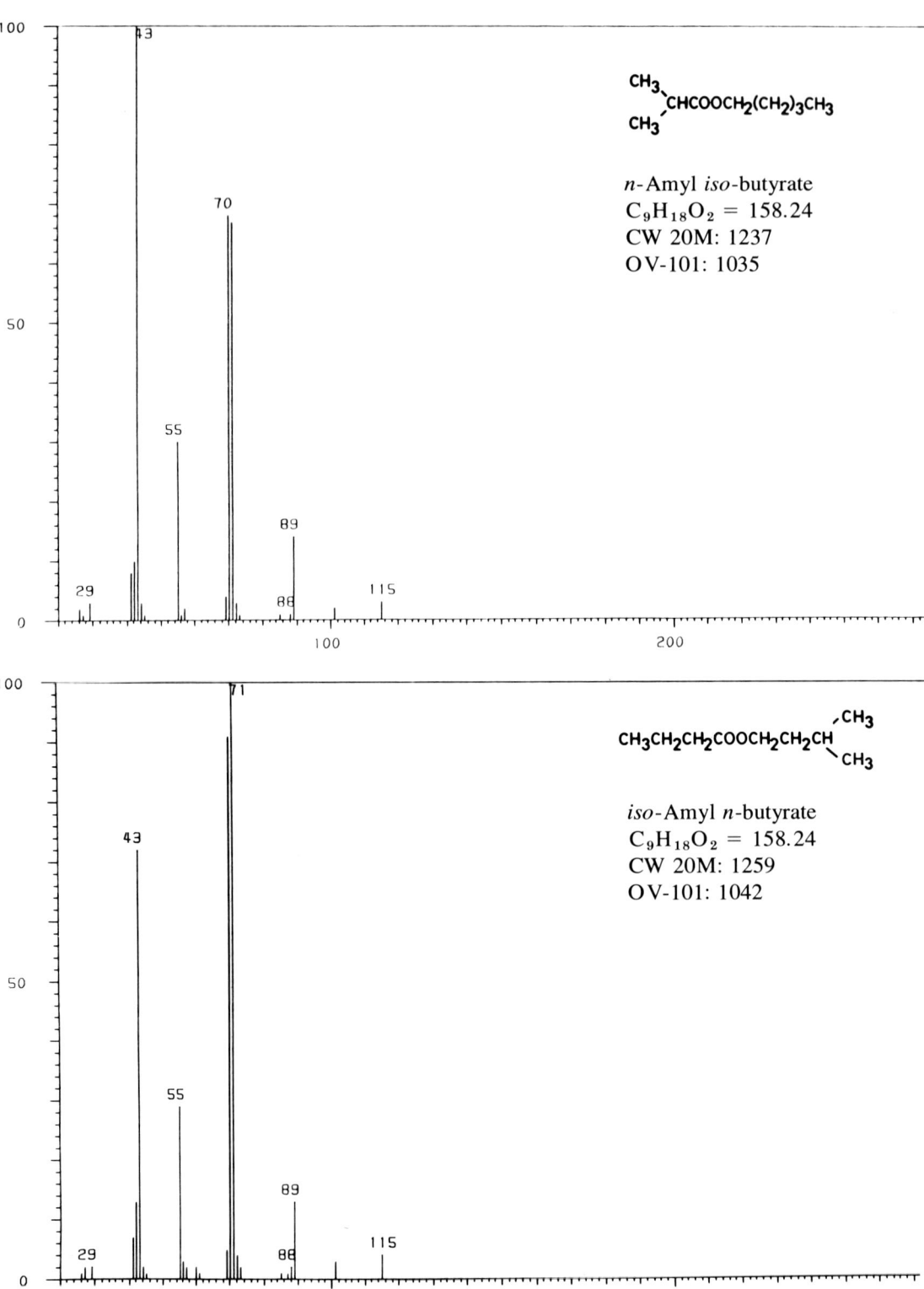

APPENDIX IV   MASS SPECTRA OF INDIVIDUAL COMPOUNDS

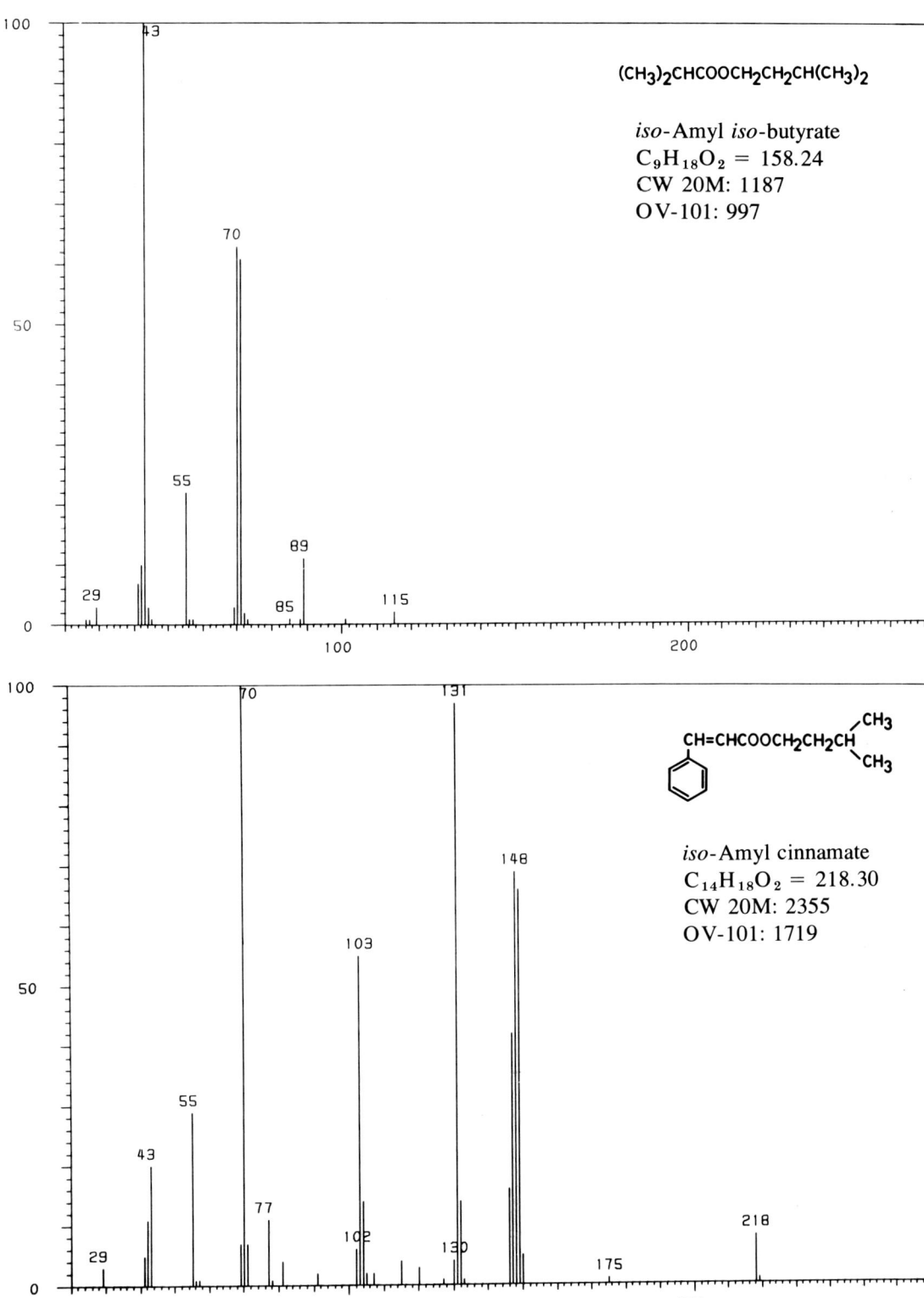

$(CH_3)_2CHCOOCH_2CH_2CH(CH_3)_2$

iso-Amyl *iso*-butyrate
$C_9H_{18}O_2 = 158.24$
CW 20M: 1187
OV-101: 997

iso-Amyl cinnamate
$C_{14}H_{18}O_2 = 218.30$
CW 20M: 2355
OV-101: 1719

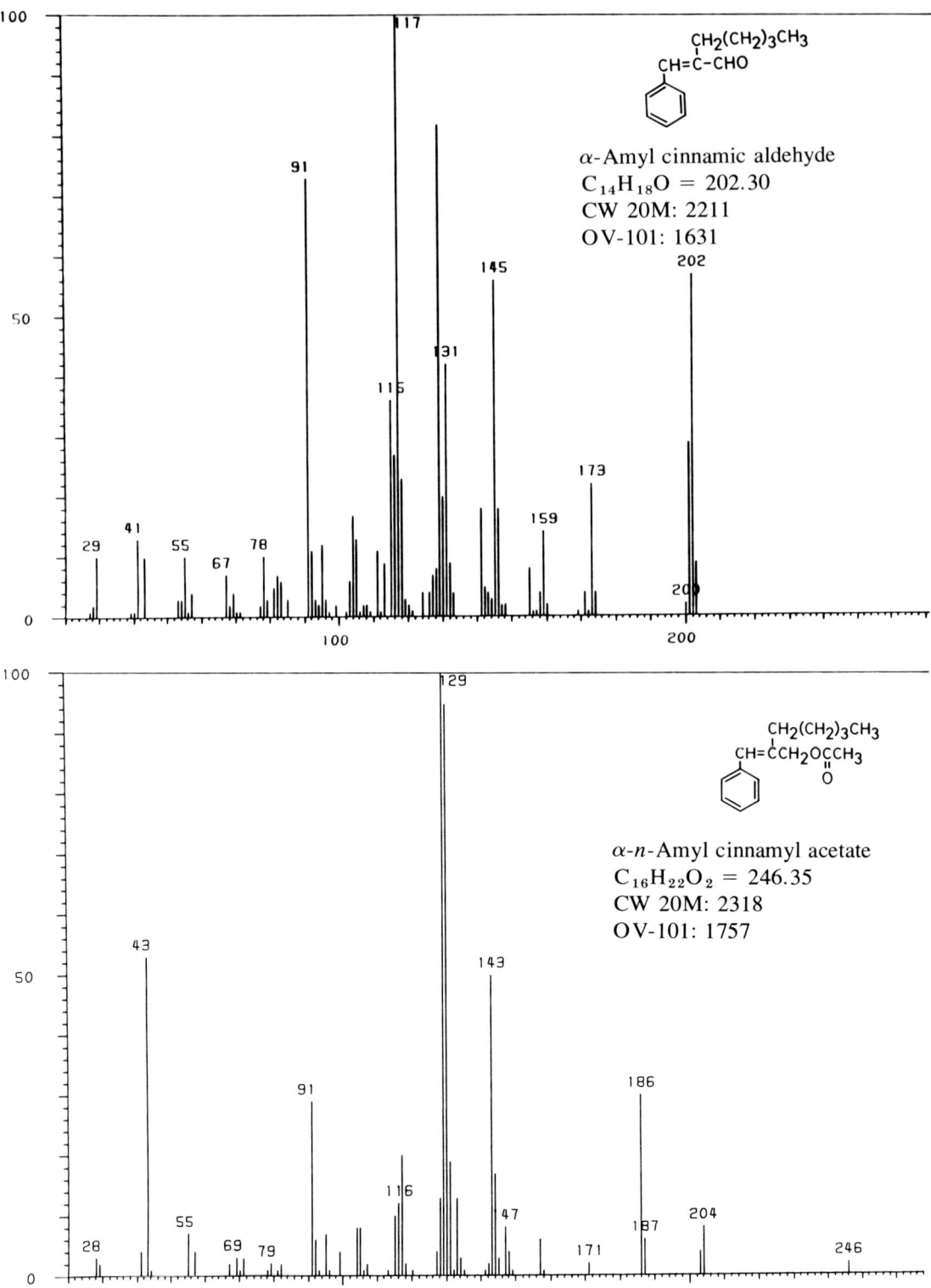

APPENDIX IV   MASS SPECTRA OF INDIVIDUAL COMPOUNDS

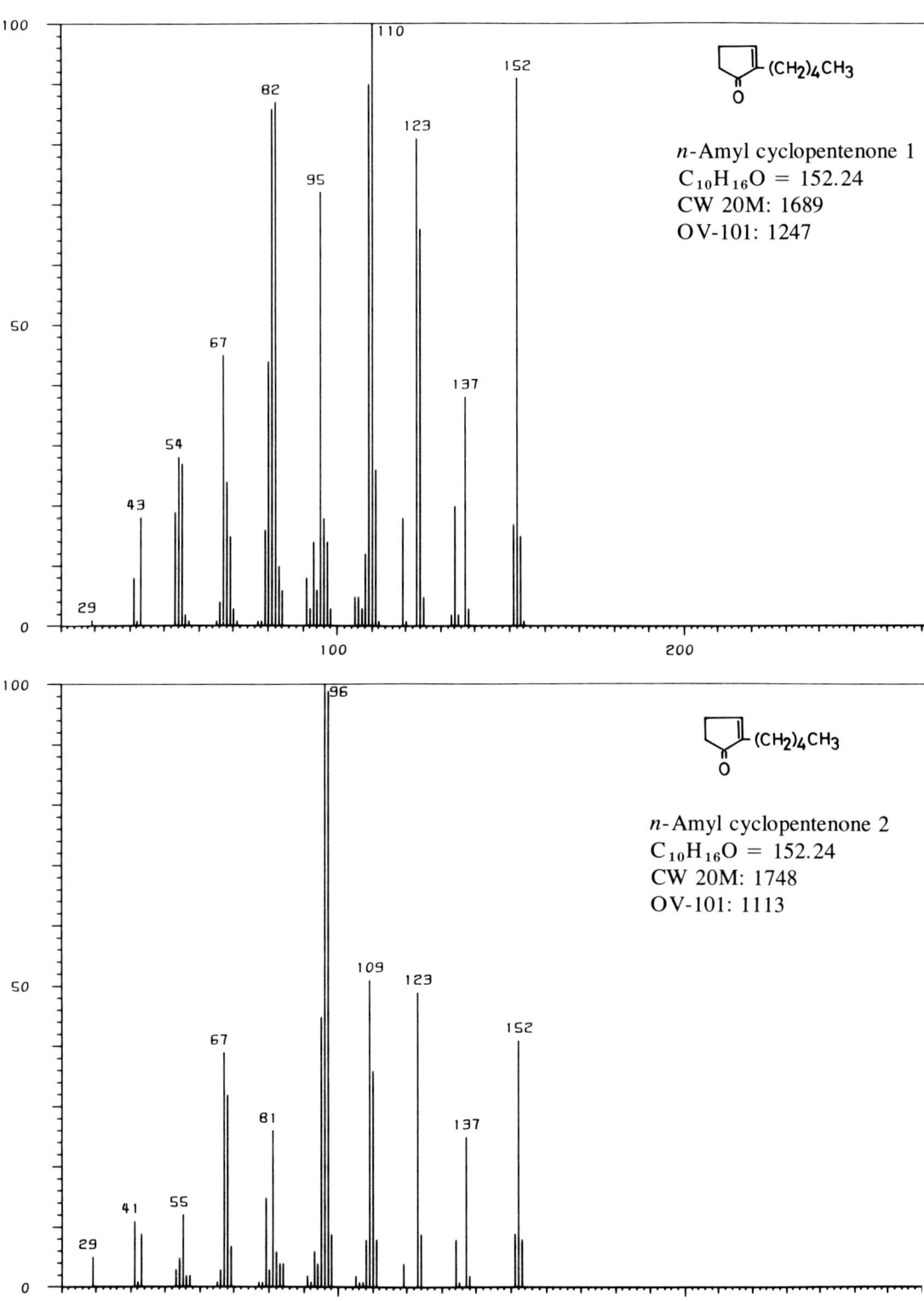

n-Amyl cyclopentenone 1
$C_{10}H_{16}O = 152.24$
CW 20M: 1689
OV-101: 1247

n-Amyl cyclopentenone 2
$C_{10}H_{16}O = 152.24$
CW 20M: 1748
OV-101: 1113

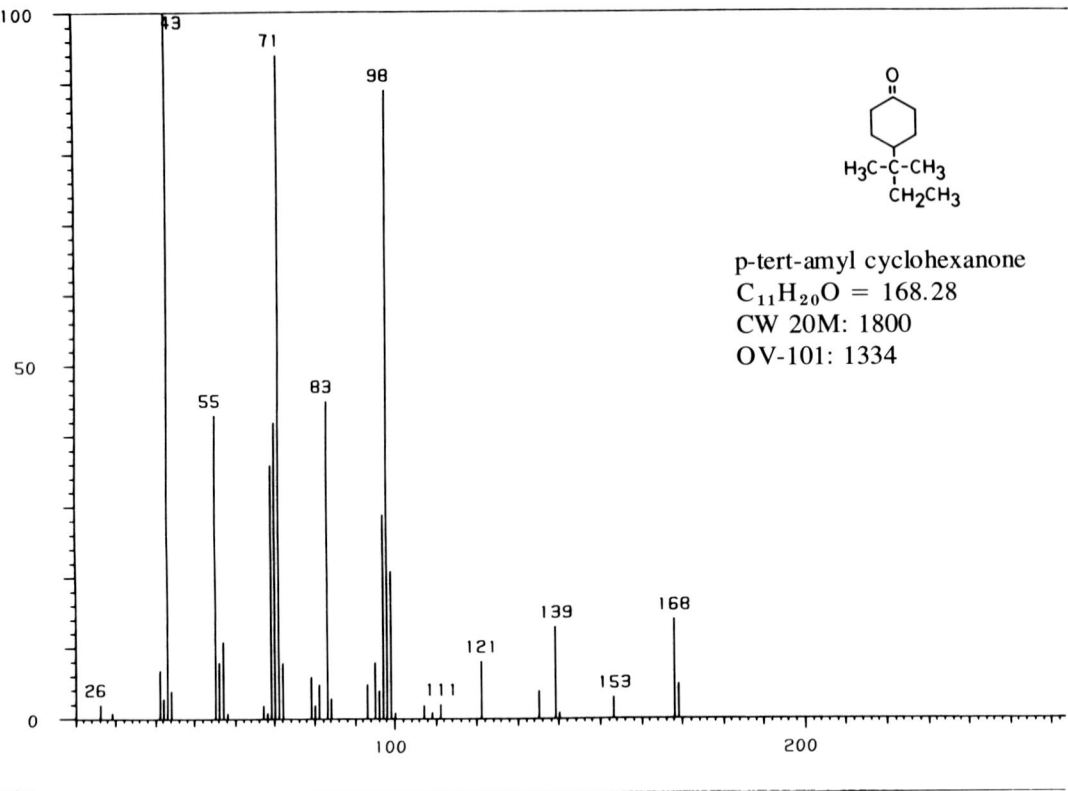

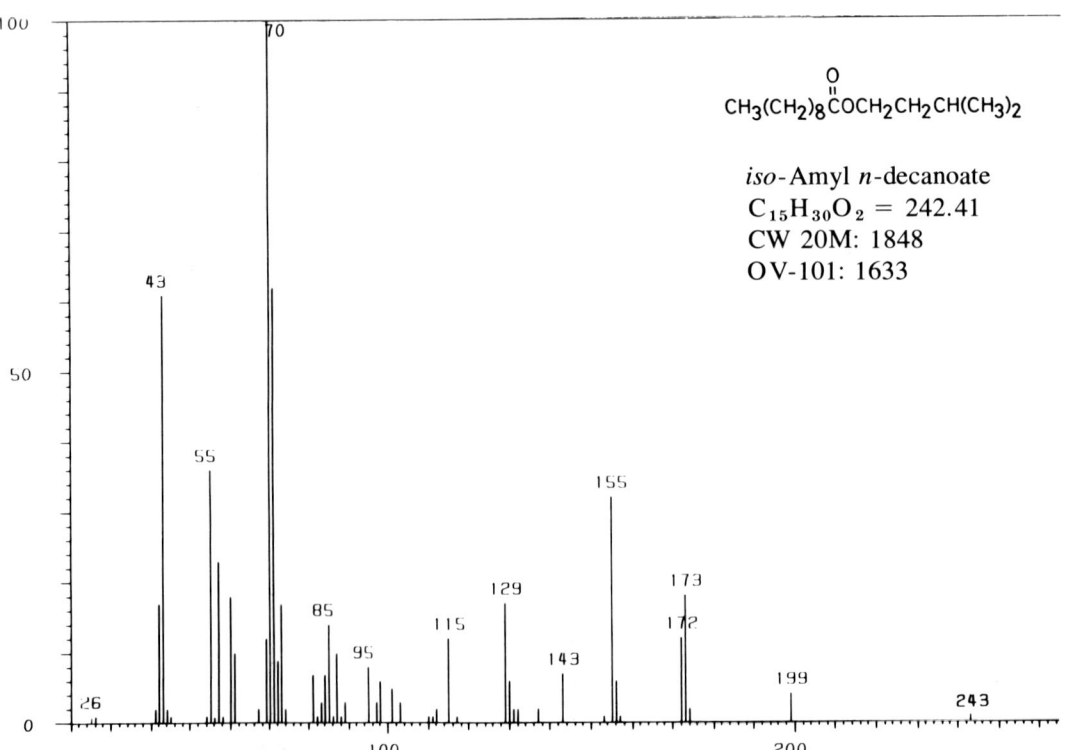

APPENDIX IV MASS SPECTRA OF INDIVIDUAL COMPOUNDS

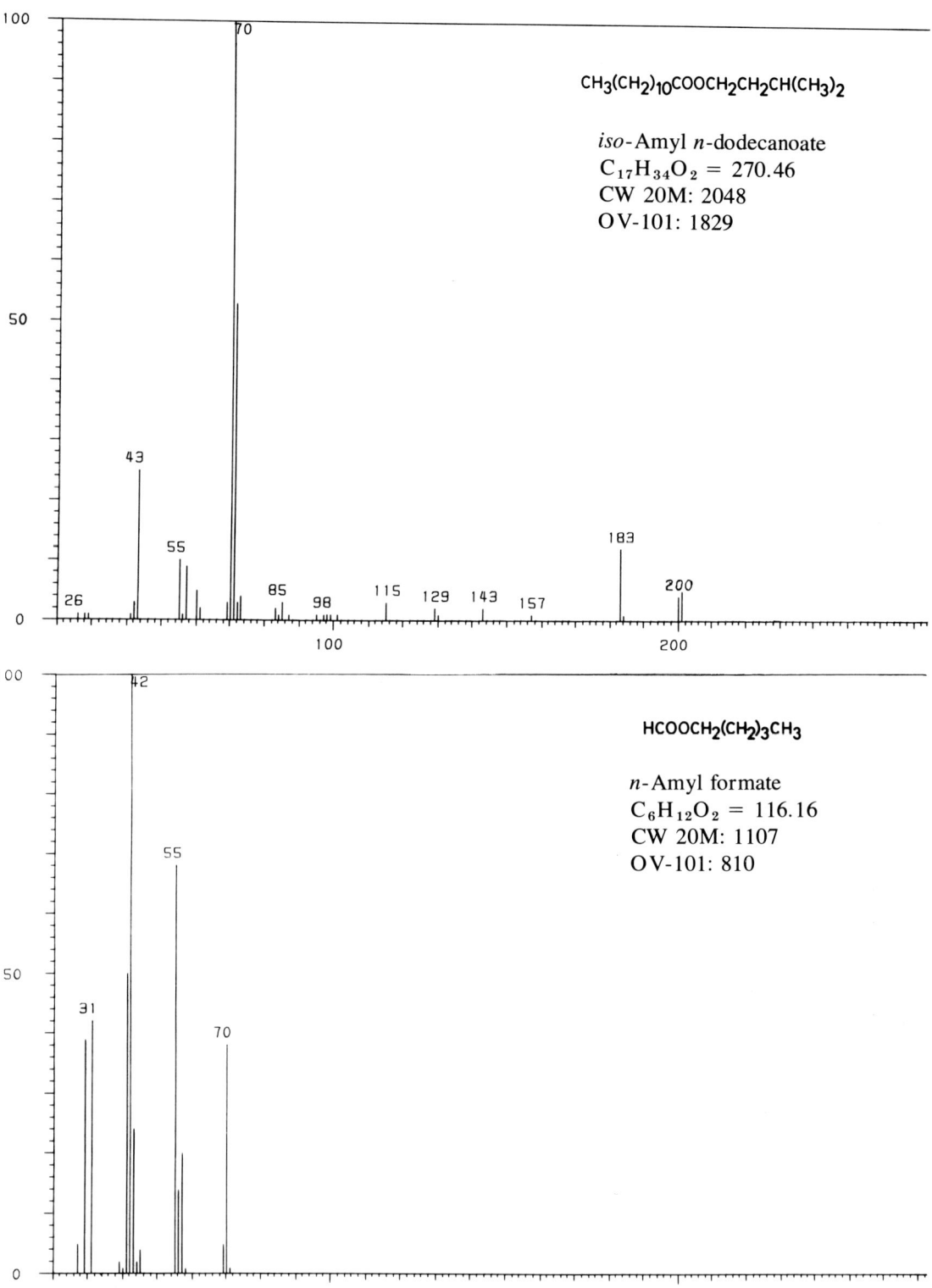

$CH_3(CH_2)_{10}COOCH_2CH_2CH(CH_3)_2$

iso-Amyl *n*-dodecanoate
$C_{17}H_{34}O_2 = 270.46$
CW 20M: 2048
OV-101: 1829

$HCOOCH_2(CH_2)_3CH_3$

n-Amyl formate
$C_6H_{12}O_2 = 116.16$
CW 20M: 1107
OV-101: 810

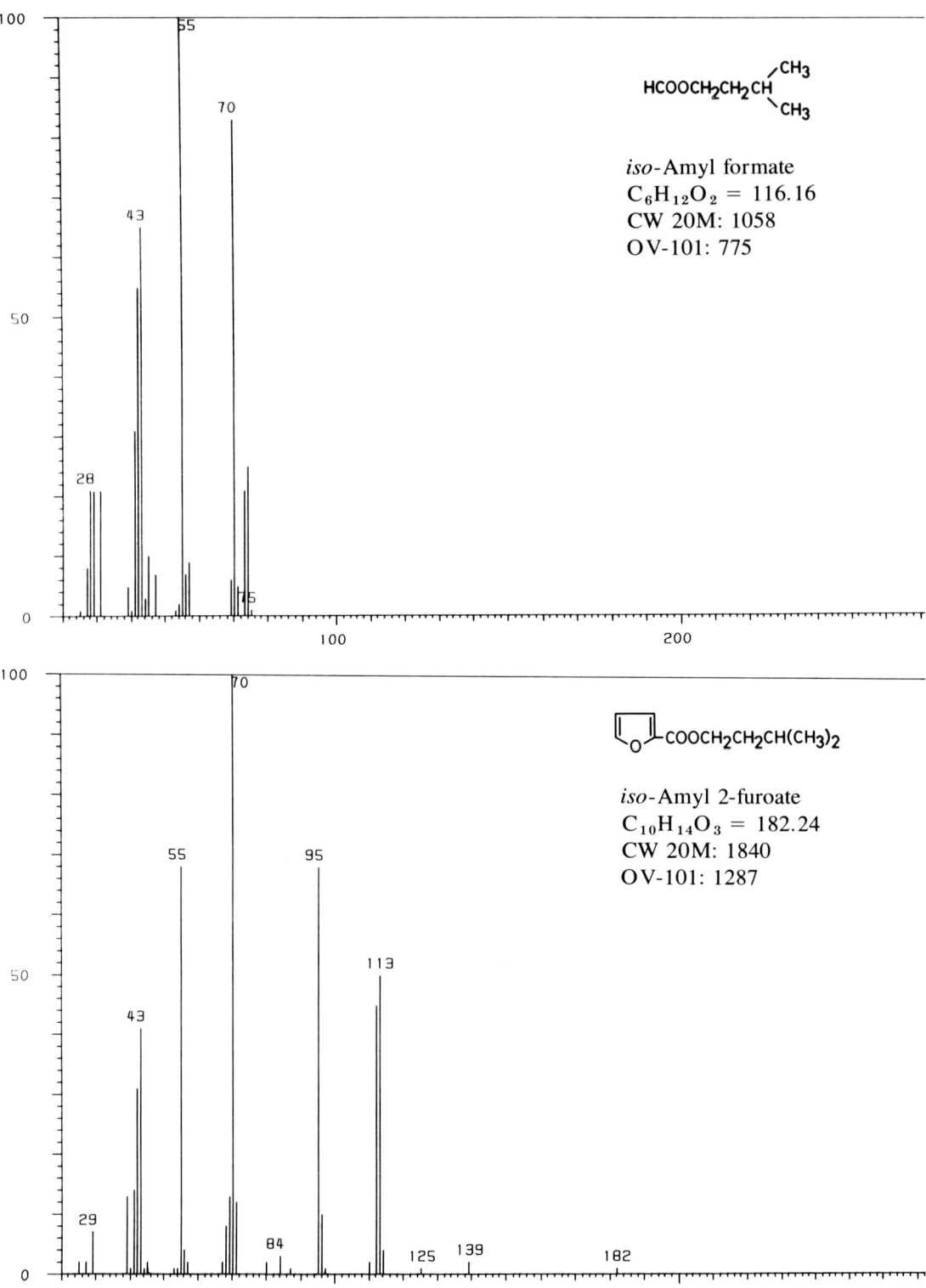

APPENDIX IV   MASS SPECTRA OF INDIVIDUAL COMPOUNDS

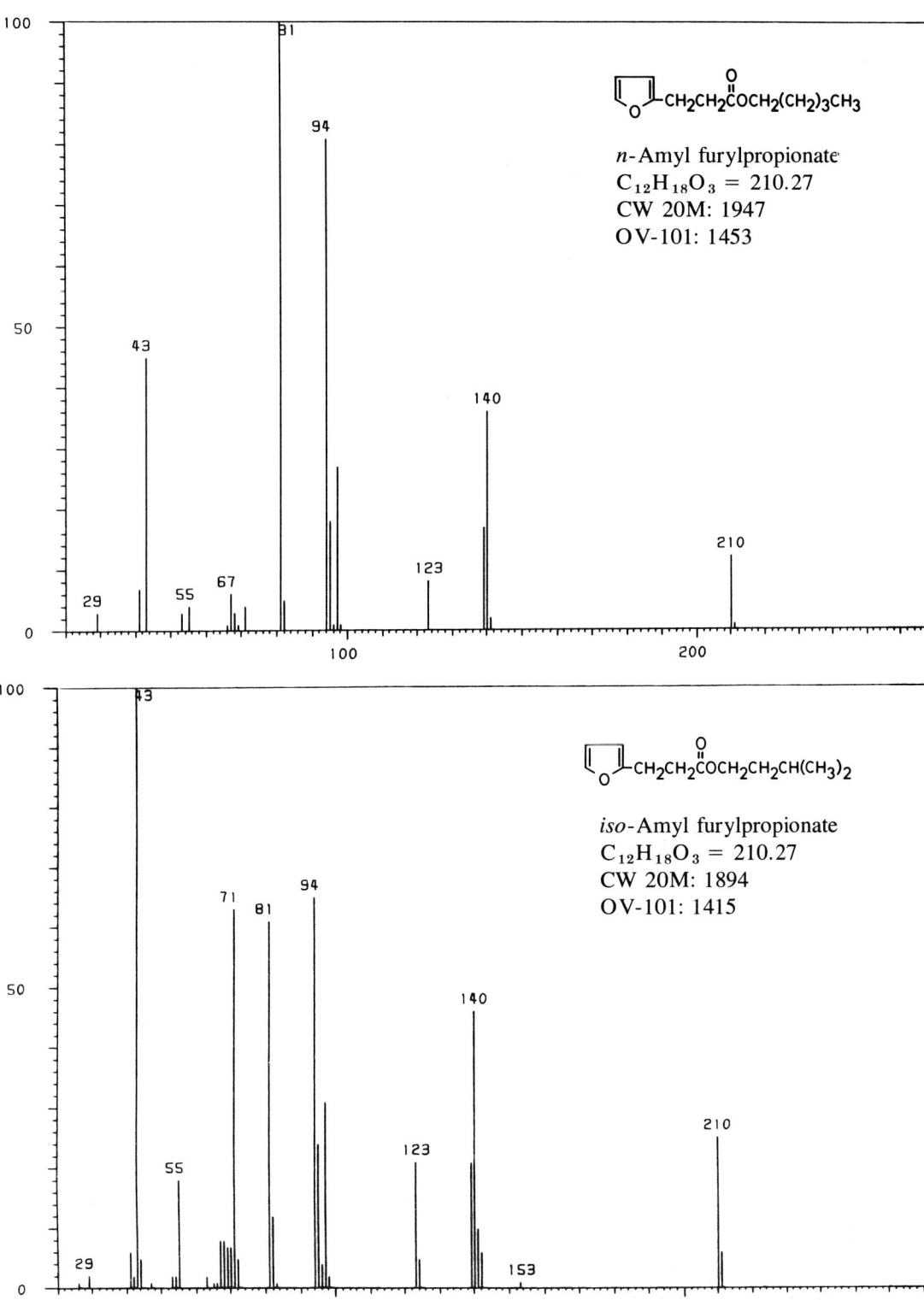

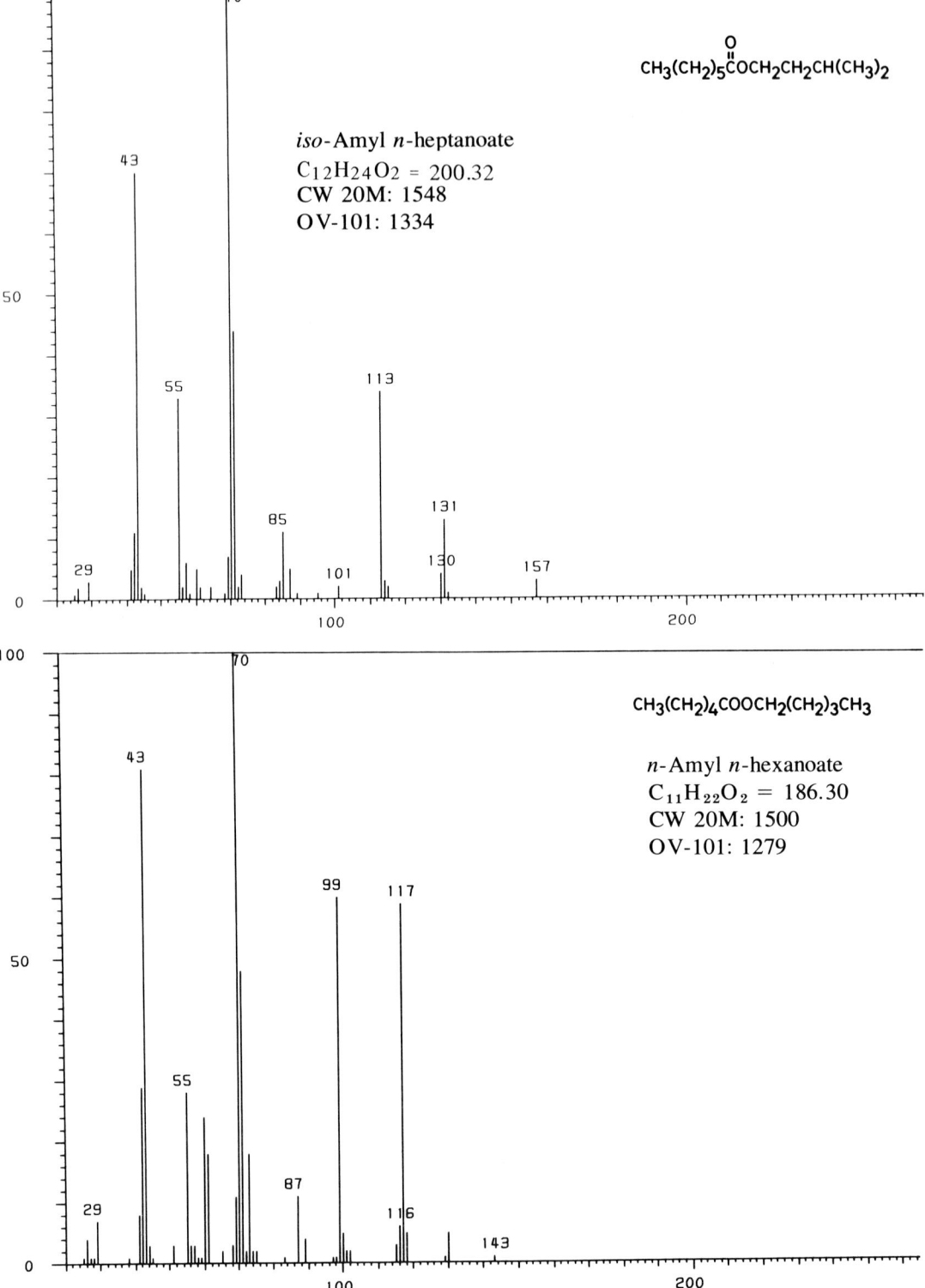

APPENDIX IV   MASS SPECTRA OF INDIVIDUAL COMPOUNDS

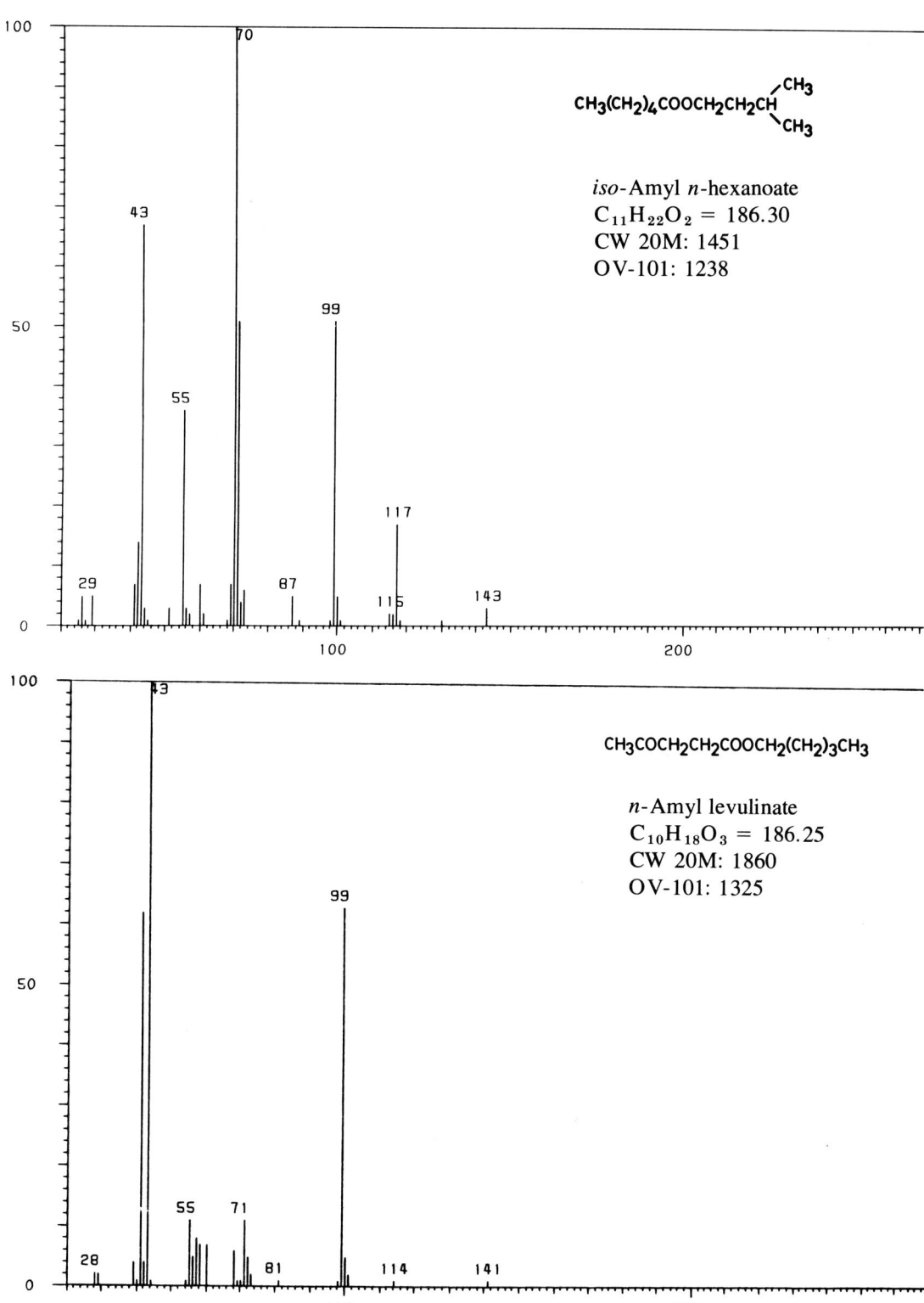

iso-Amyl n-hexanoate
$C_{11}H_{22}O_2 = 186.30$
CW 20M: 1451
OV-101: 1238

n-Amyl levulinate
$C_{10}H_{18}O_3 = 186.25$
CW 20M: 1860
OV-101: 1325

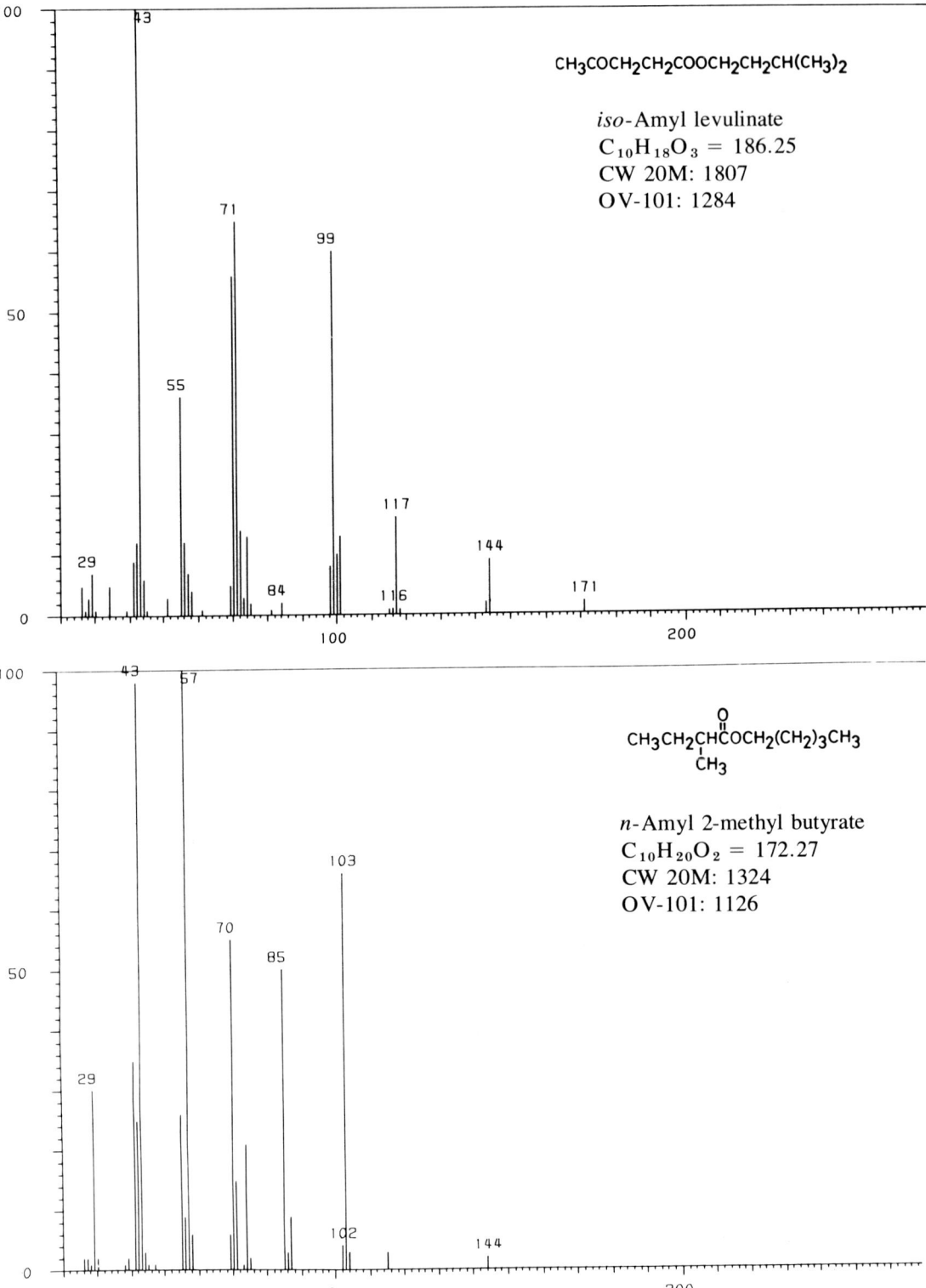

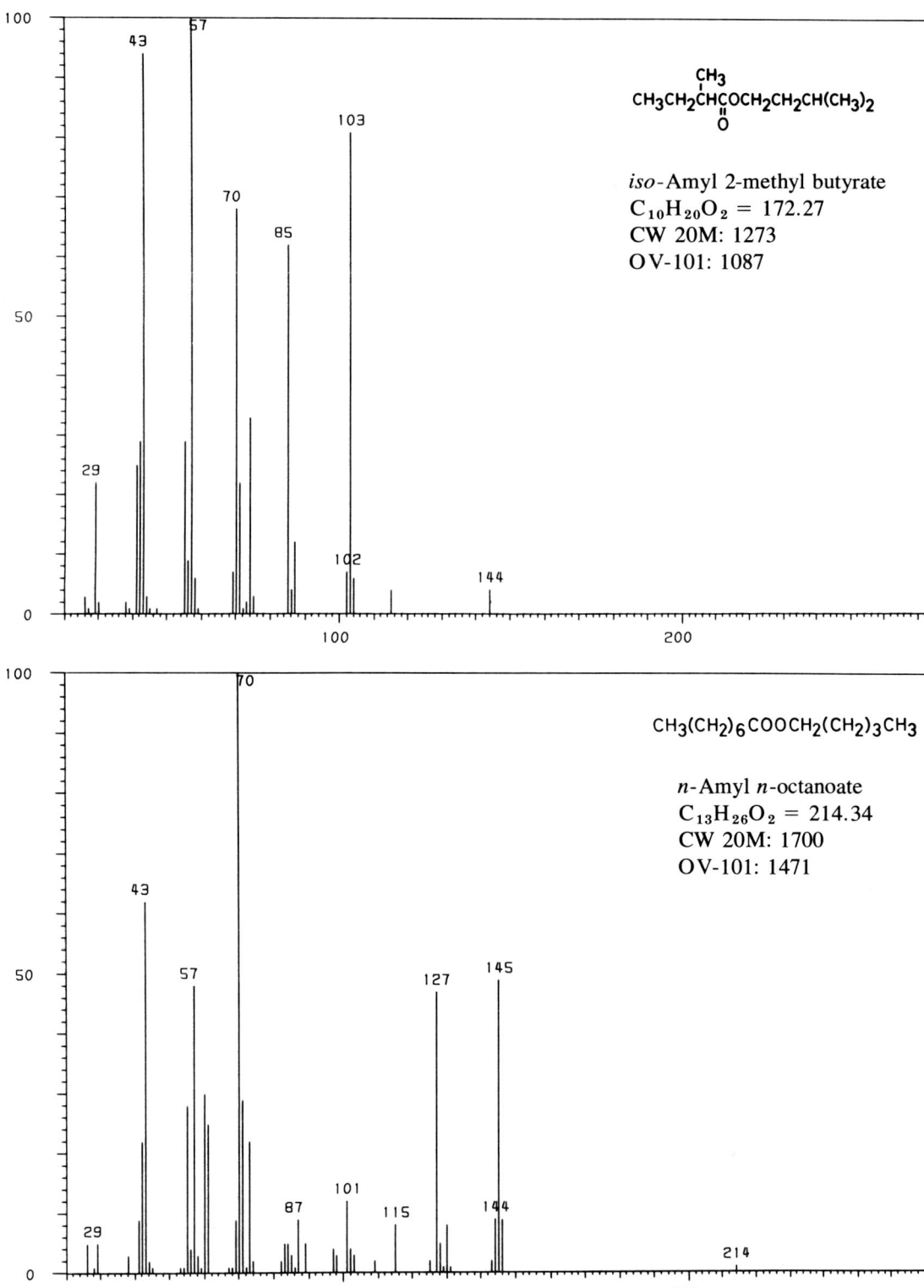

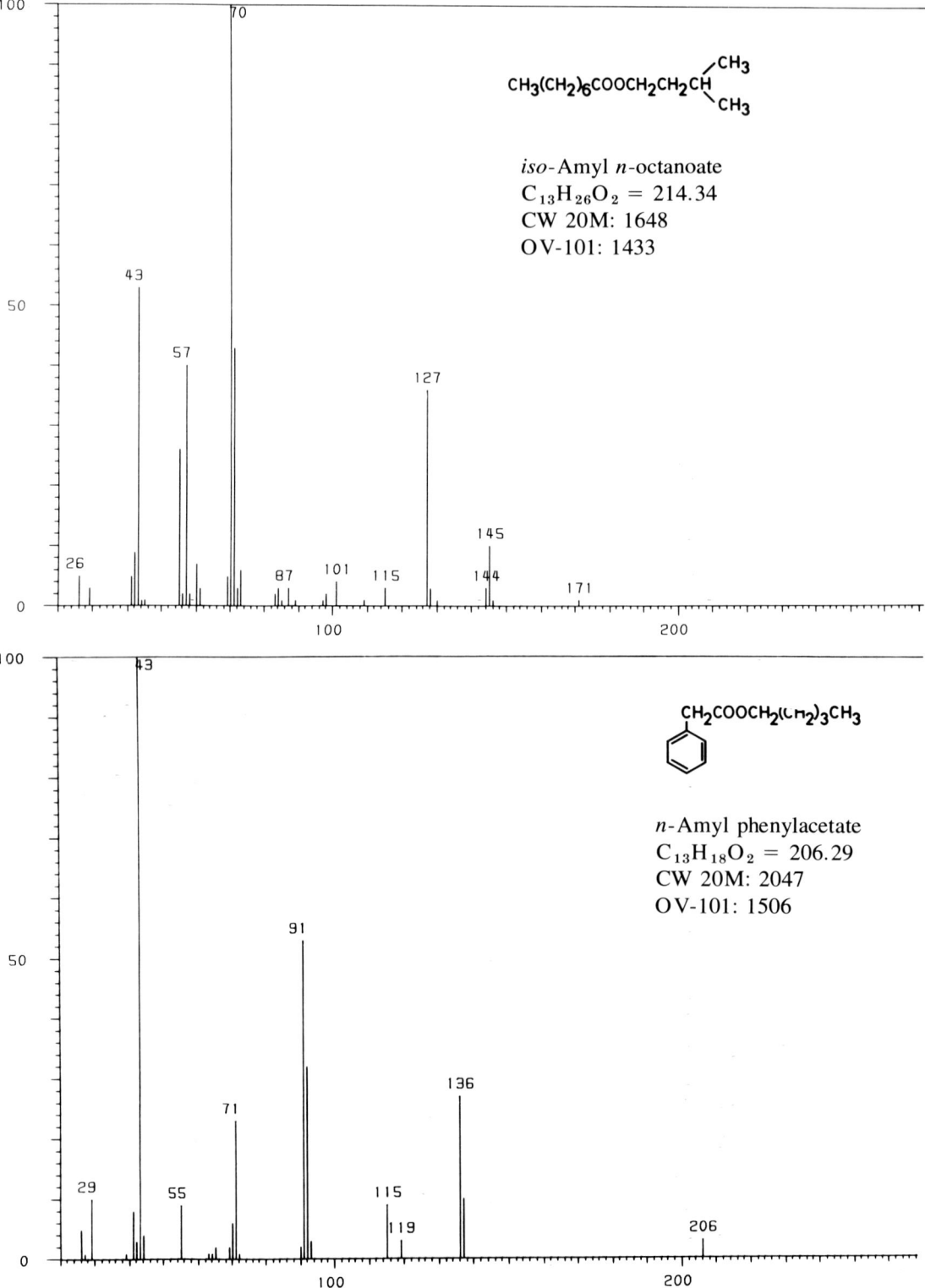

APPENDIX IV   MASS SPECTRA OF INDIVIDUAL COMPOUNDS

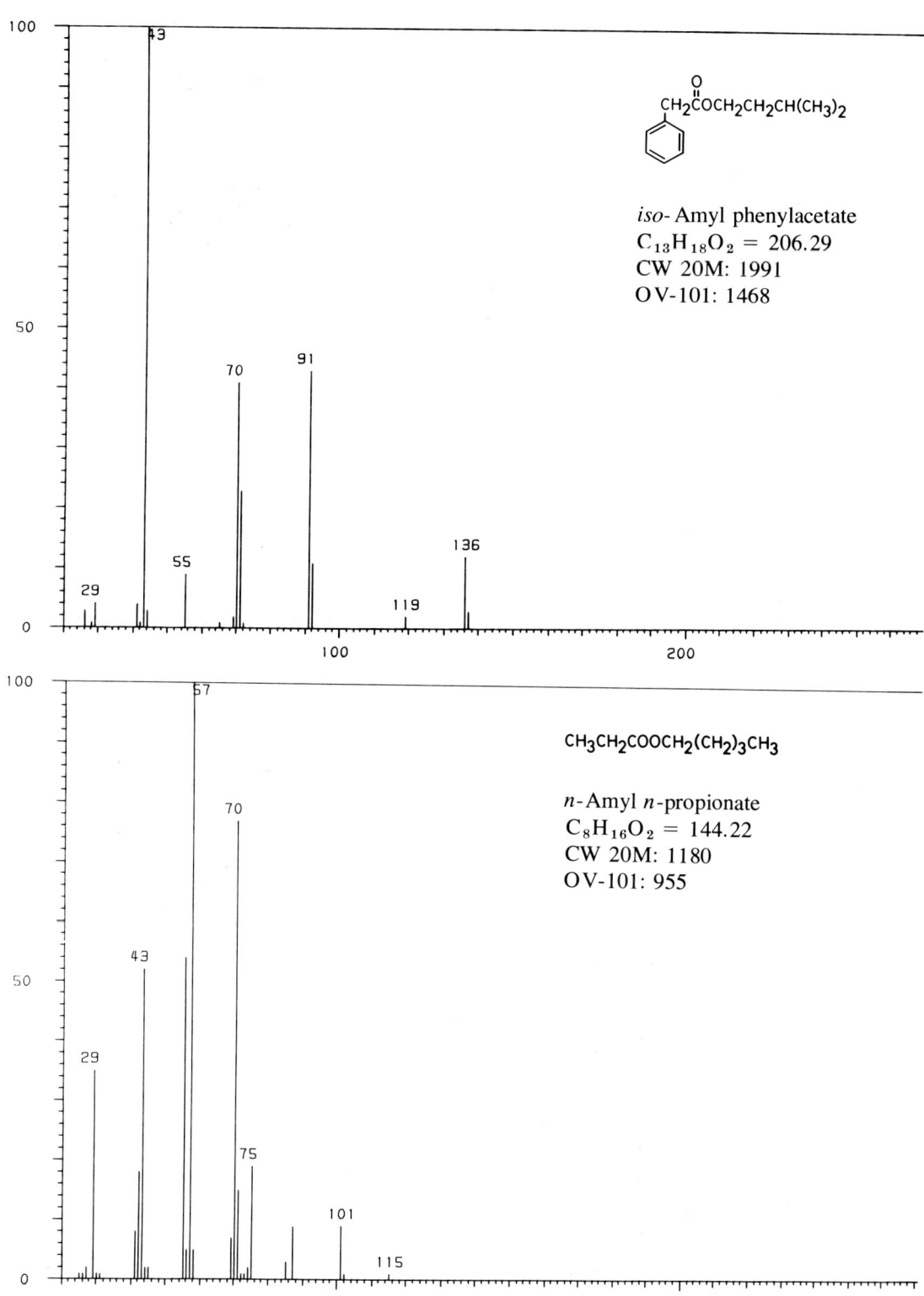

iso-Amyl phenylacetate
$C_{13}H_{18}O_2 = 206.29$
CW 20M: 1991
OV-101: 1468

n-Amyl *n*-propionate
$C_8H_{16}O_2 = 144.22$
CW 20M: 1180
OV-101: 955

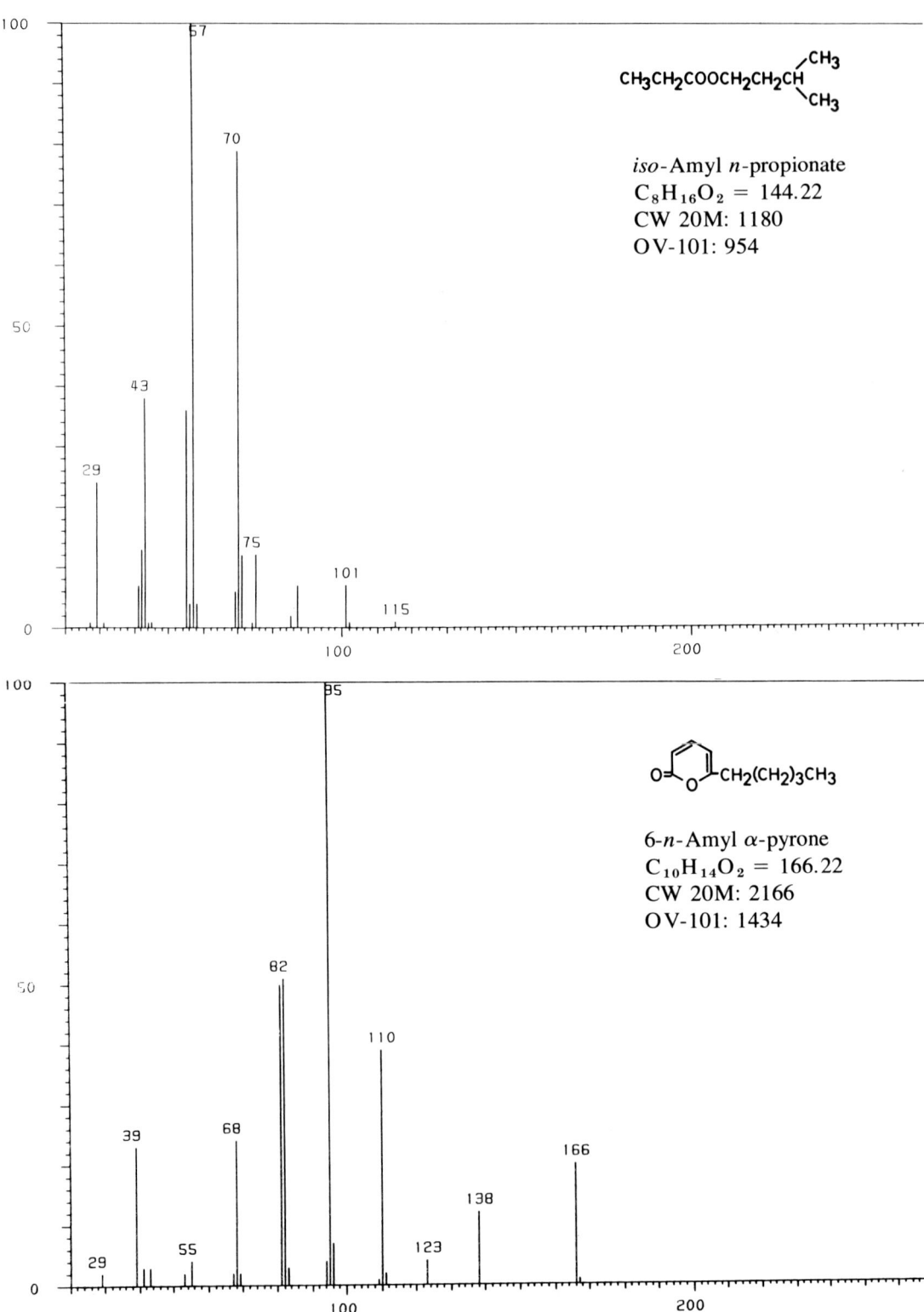

APPENDIX IV   MASS SPECTRA OF INDIVIDUAL COMPOUNDS

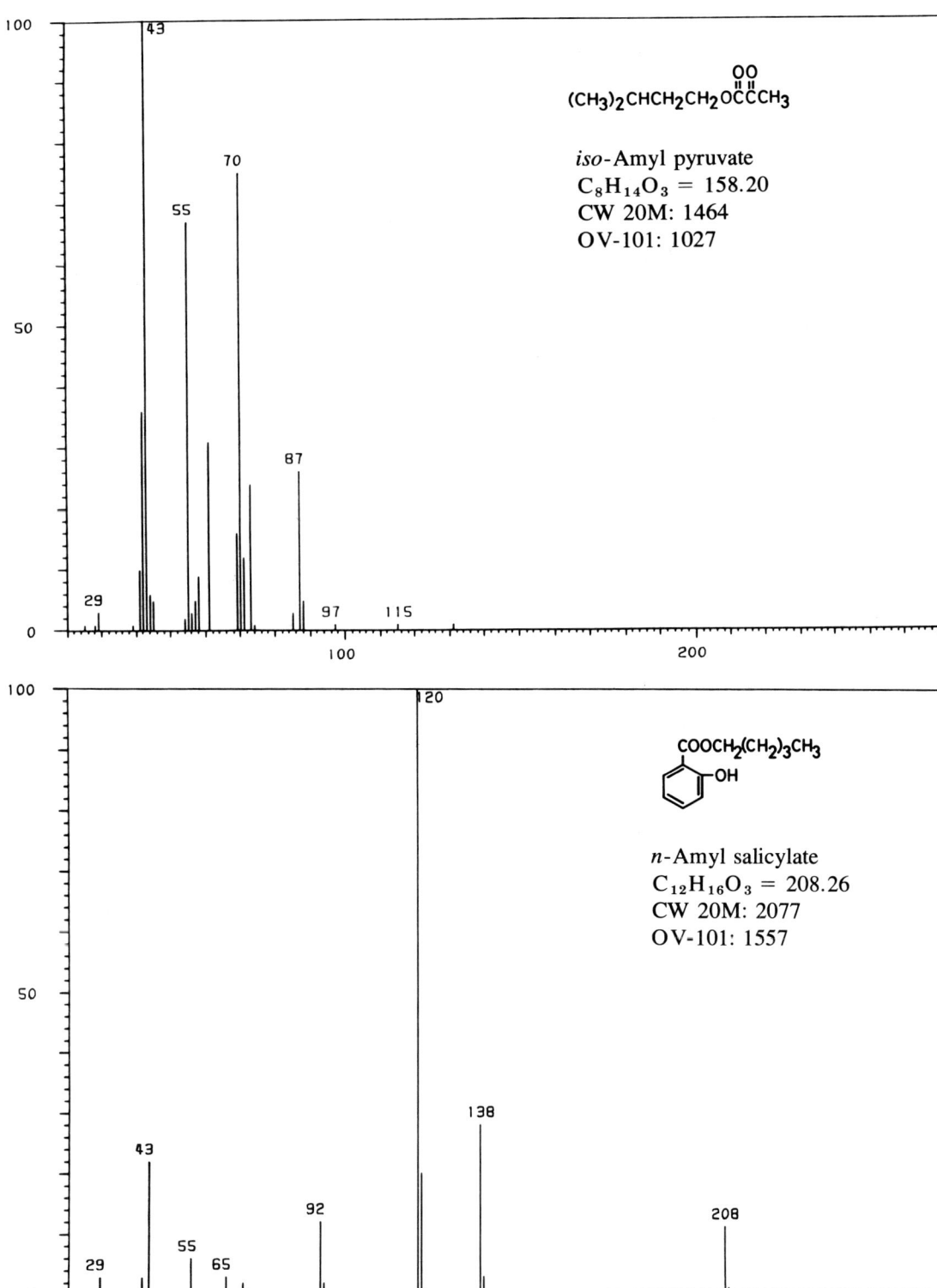

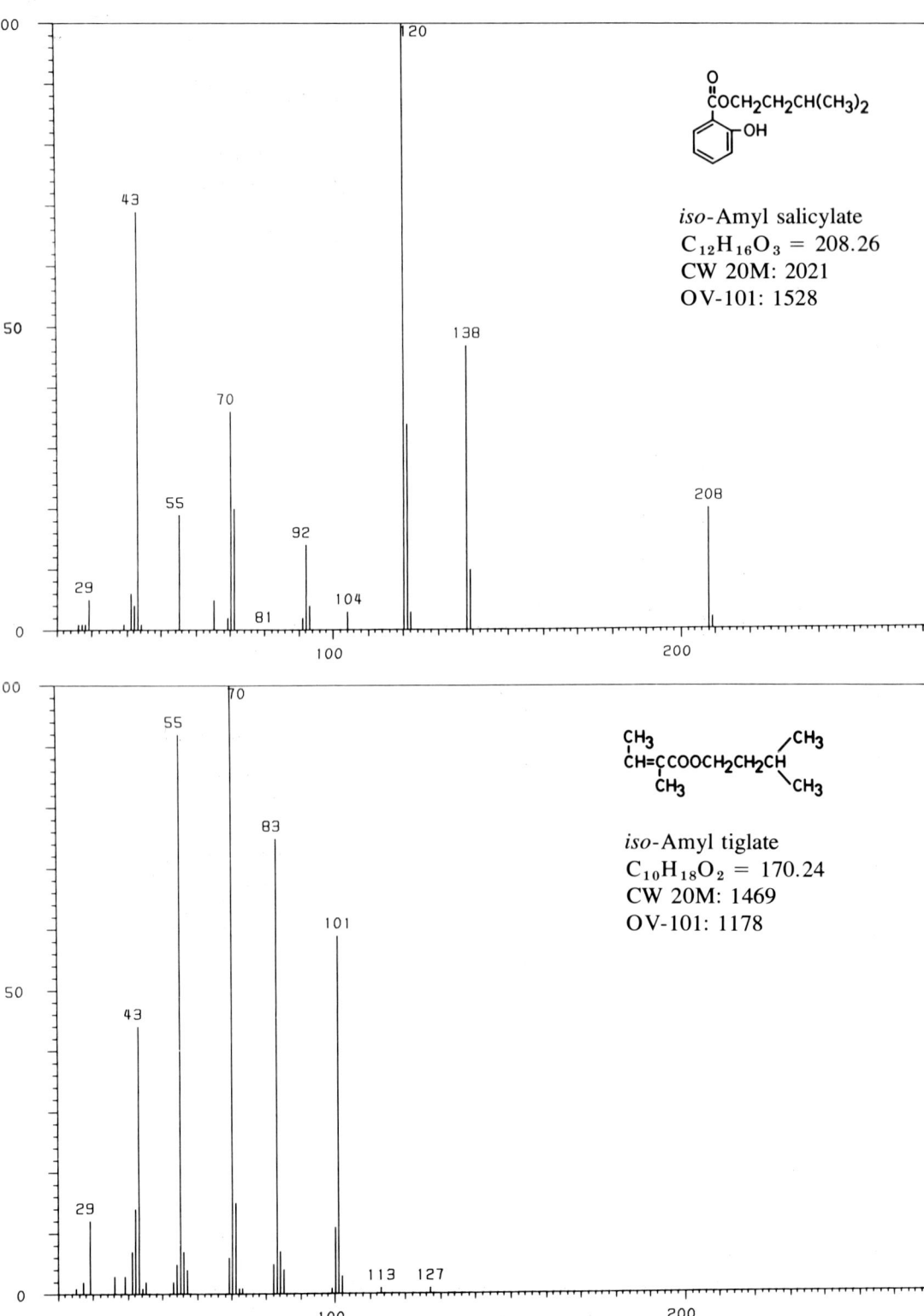

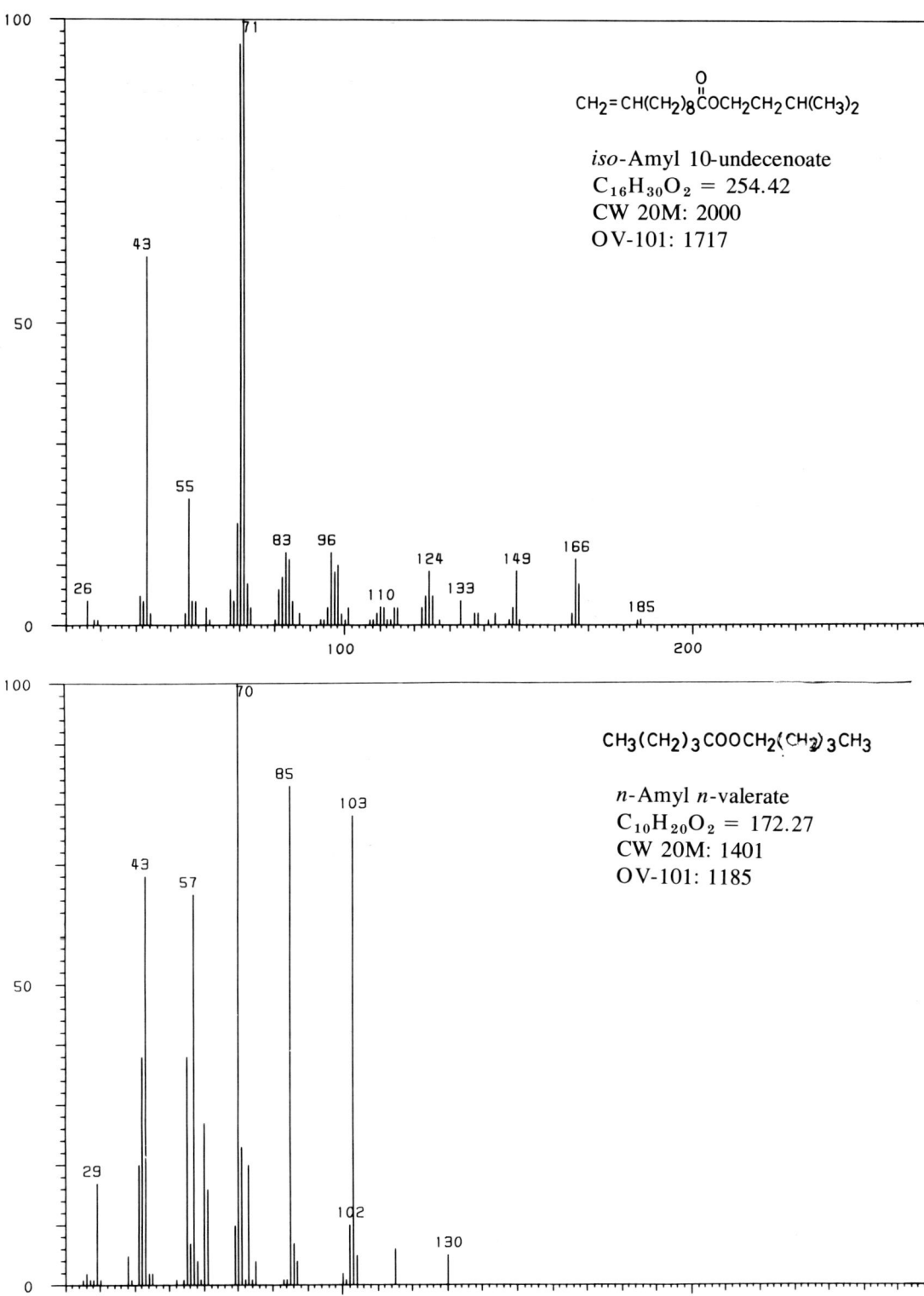

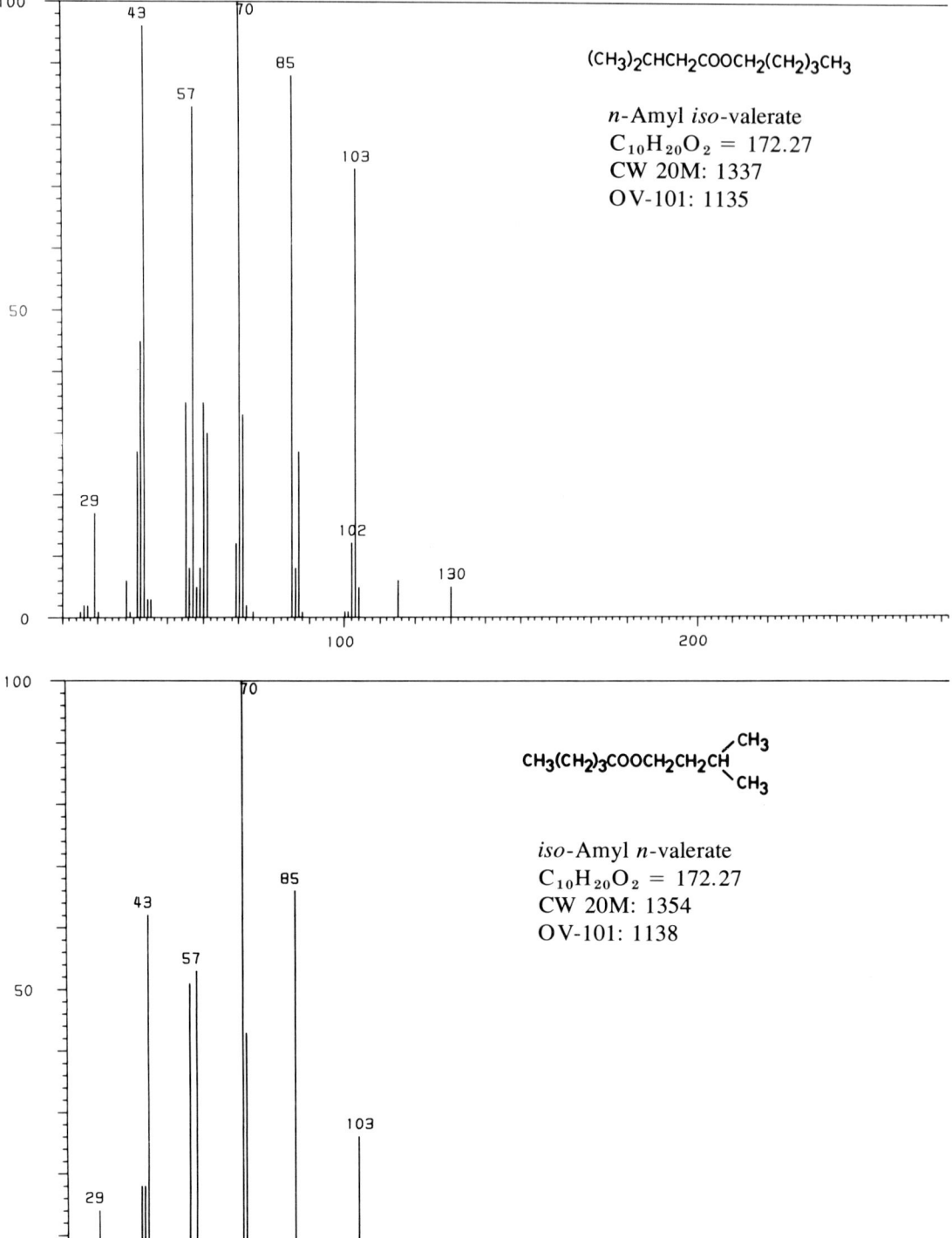

APPENDIX IV   MASS SPECTRA OF INDIVIDUAL COMPOUNDS

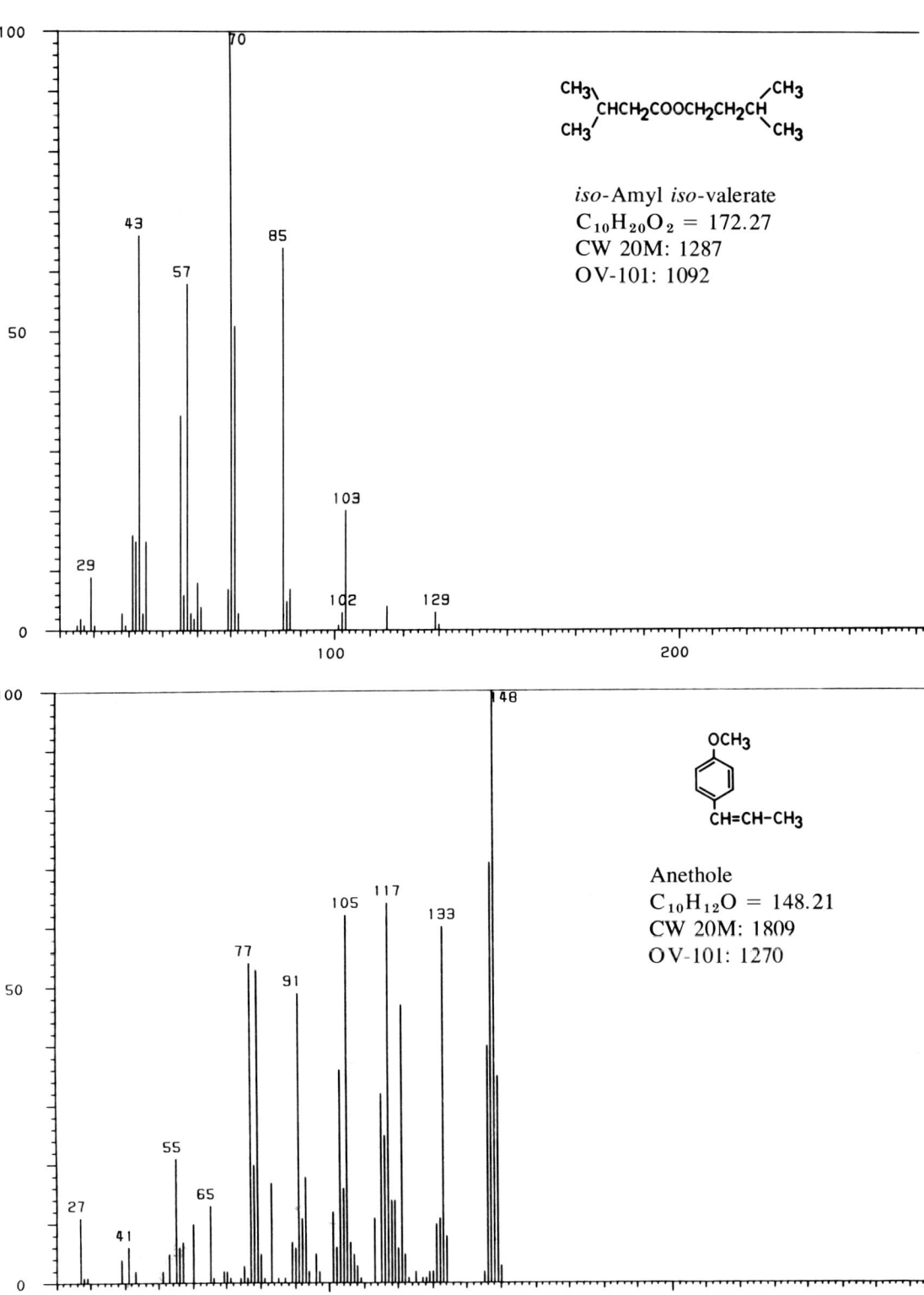

iso-Amyl iso-valerate
$C_{10}H_{20}O_2 = 172.27$
CW 20M: 1287
OV-101: 1092

Anethole
$C_{10}H_{12}O = 148.21$
CW 20M: 1809
OV-101: 1270

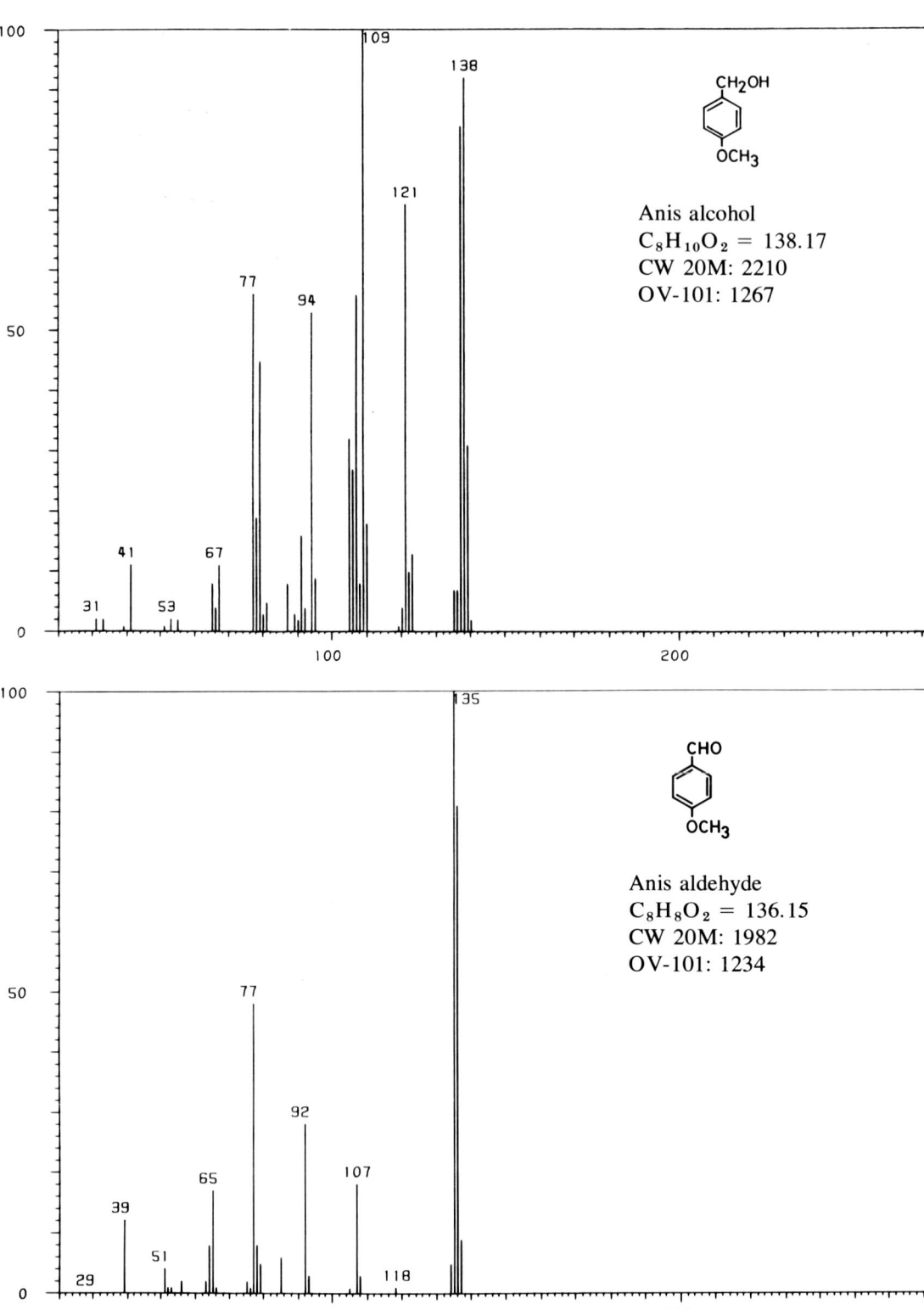

APPENDIX IV   MASS SPECTRA OF INDIVIDUAL COMPOUNDS   163

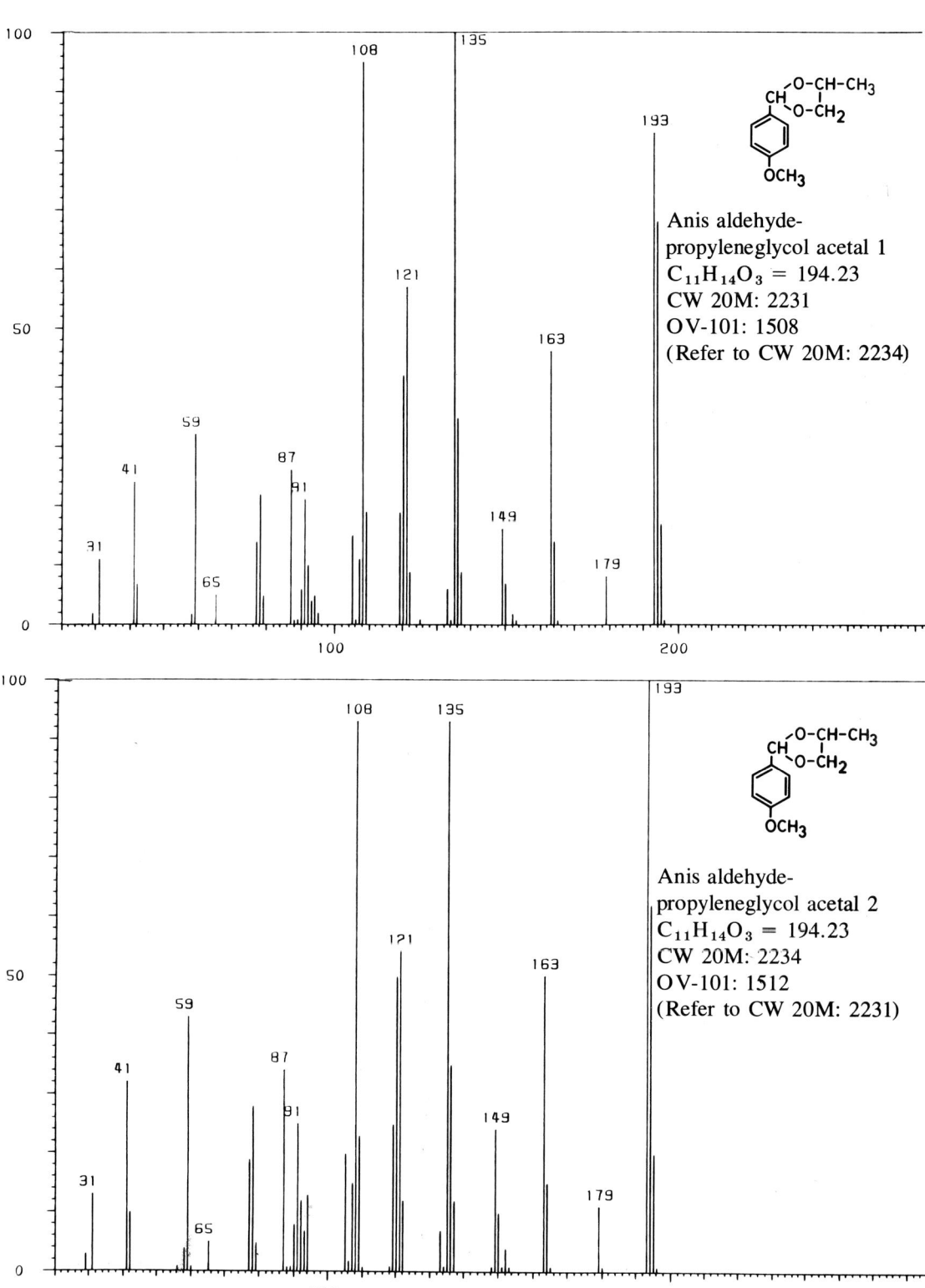

Anis aldehyde-
propyleneglycol acetal 1
$C_{11}H_{14}O_3$ = 194.23
CW 20M: 2231
OV-101: 1508
(Refer to CW 20M: 2234)

Anis aldehyde-
propyleneglycol acetal 2
$C_{11}H_{14}O_3$ = 194.23
CW 20M: 2234
OV-101: 1512
(Refer to CW 20M: 2231)

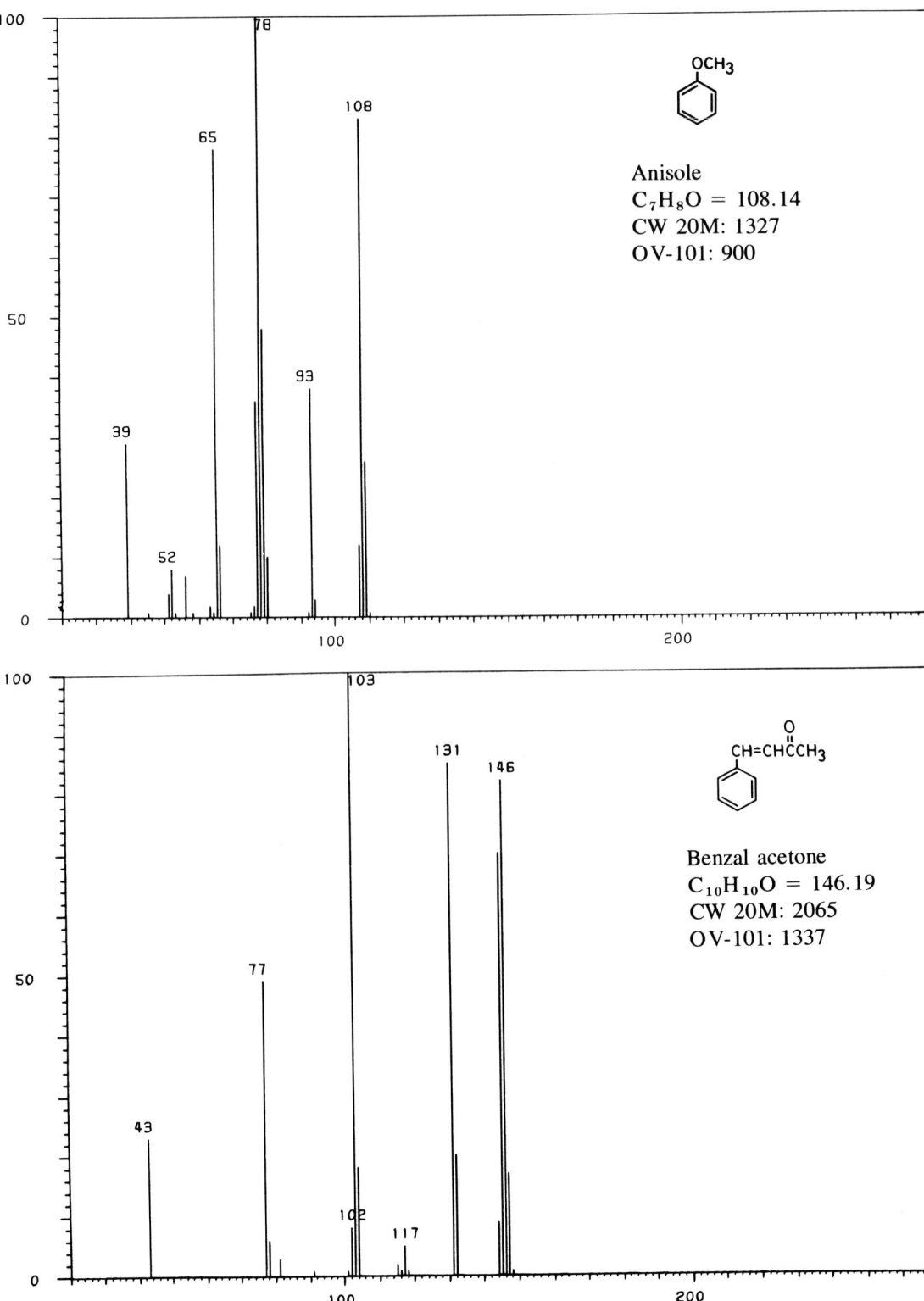

APPENDIX IV   MASS SPECTRA OF INDIVIDUAL COMPOUNDS   165

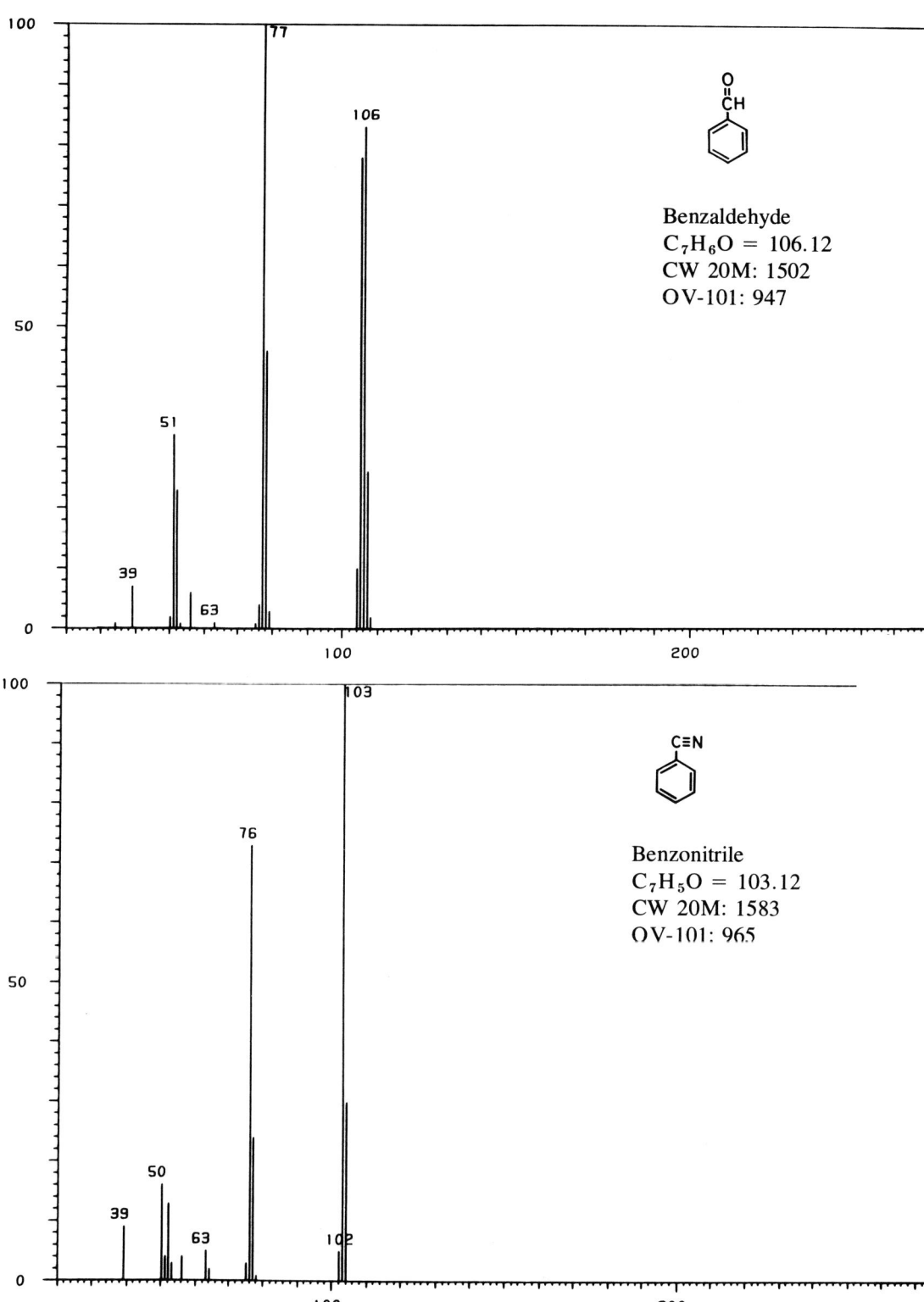

Benzaldehyde
$C_7H_6O = 106.12$
CW 20M: 1502
OV-101: 947

Benzonitrile
$C_7H_5O = 103.12$
CW 20M: 1583
OV-101: 965

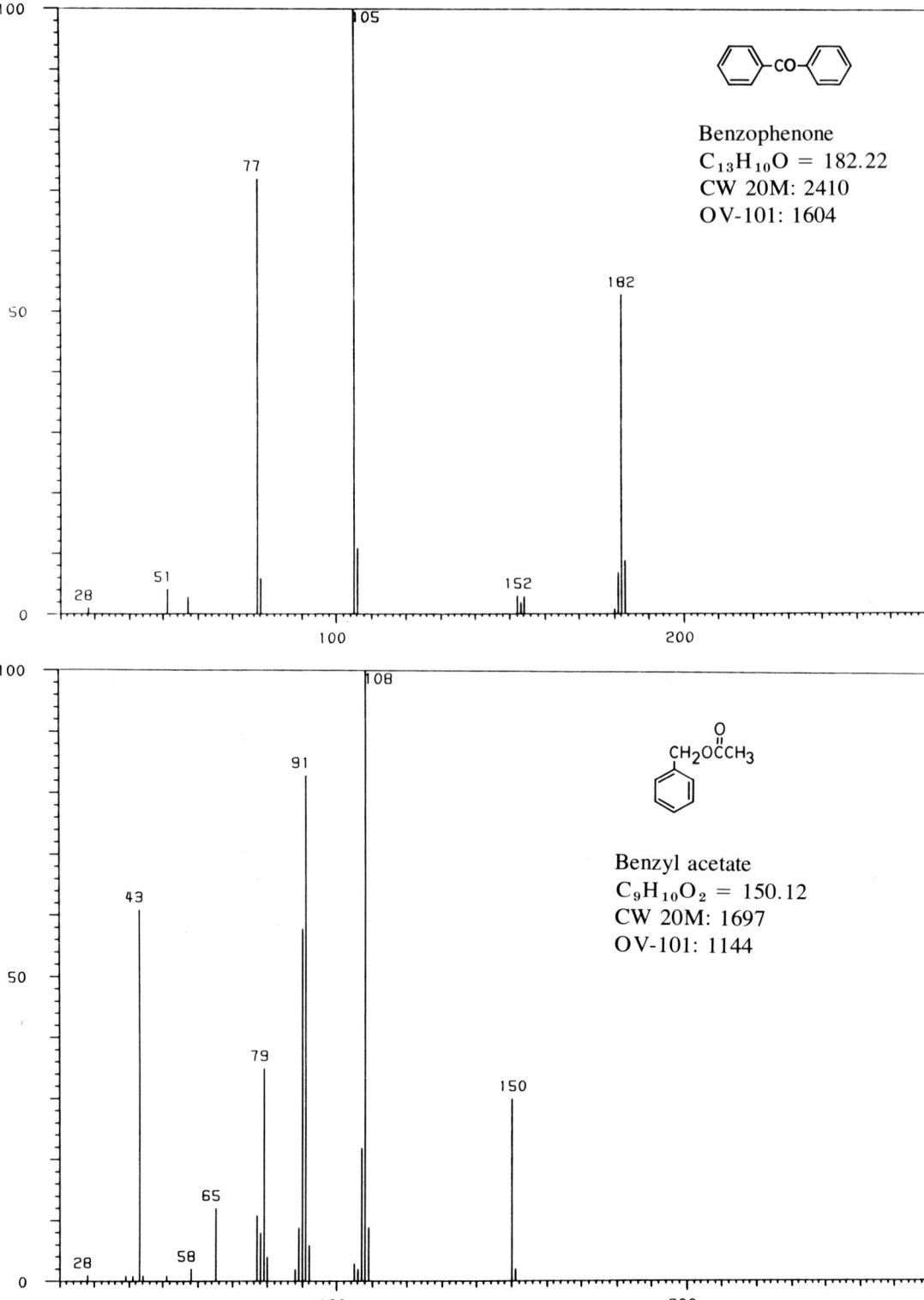

APPENDIX IV   MASS SPECTRA OF INDIVIDUAL COMPOUNDS   **167**

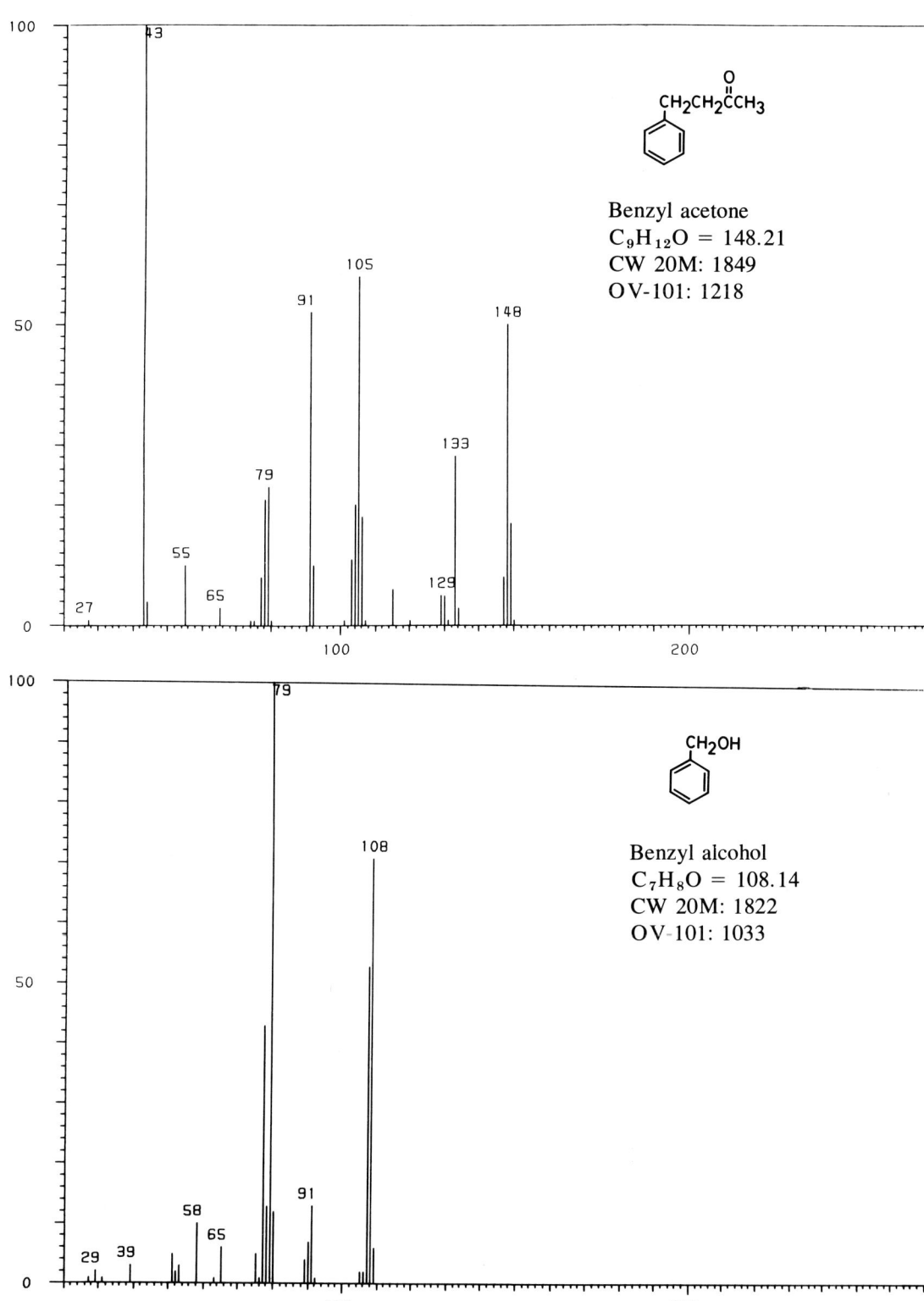

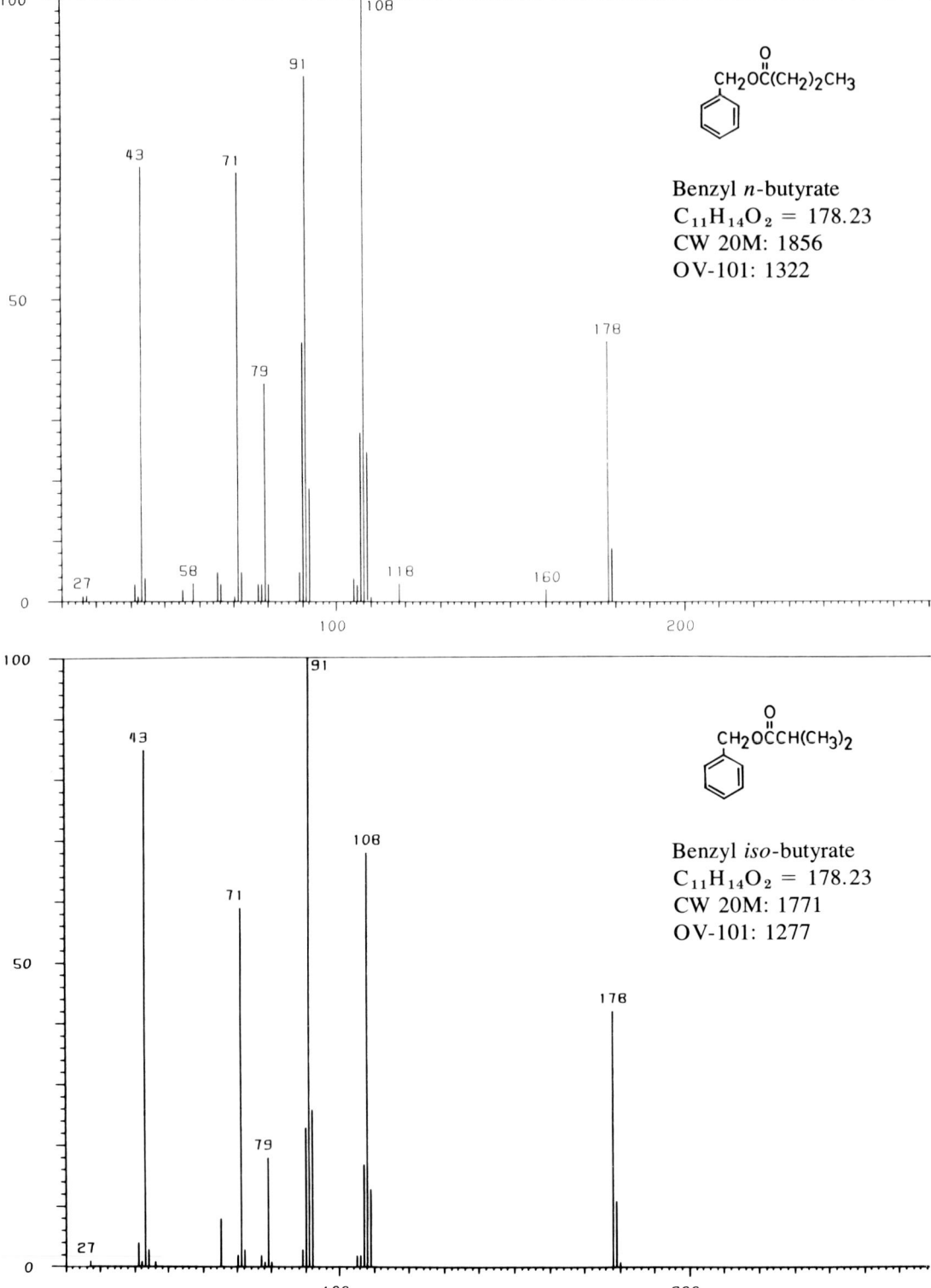

APPENDIX IV   MASS SPECTRA OF INDIVIDUAL COMPOUNDS

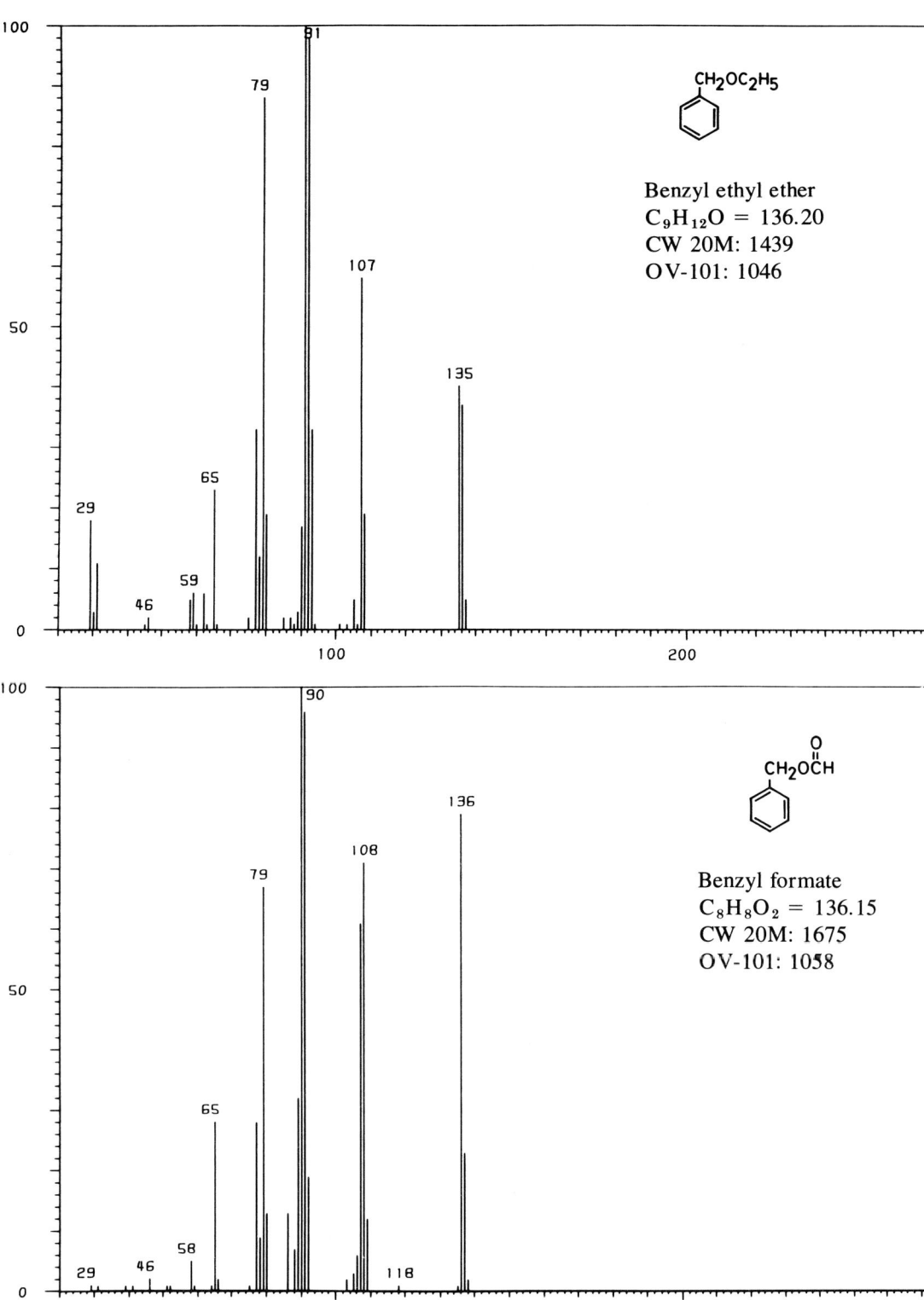

Benzyl ethyl ether
$C_9H_{12}O = 136.20$
CW 20M: 1439
OV-101: 1046

Benzyl formate
$C_8H_8O_2 = 136.15$
CW 20M: 1675
OV-101: 1058

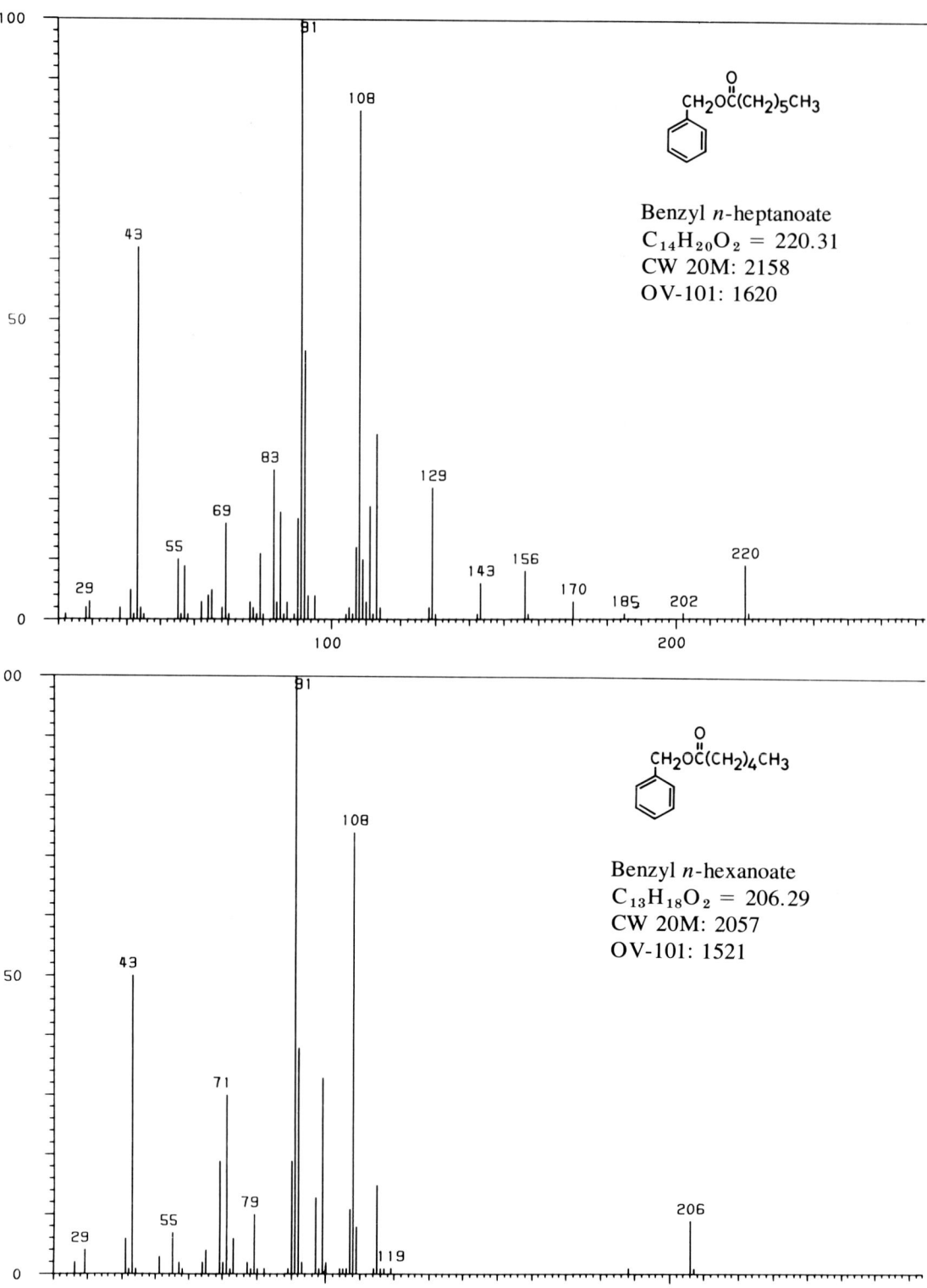

APPENDIX IV   MASS SPECTRA OF INDIVIDUAL COMPOUNDS   171

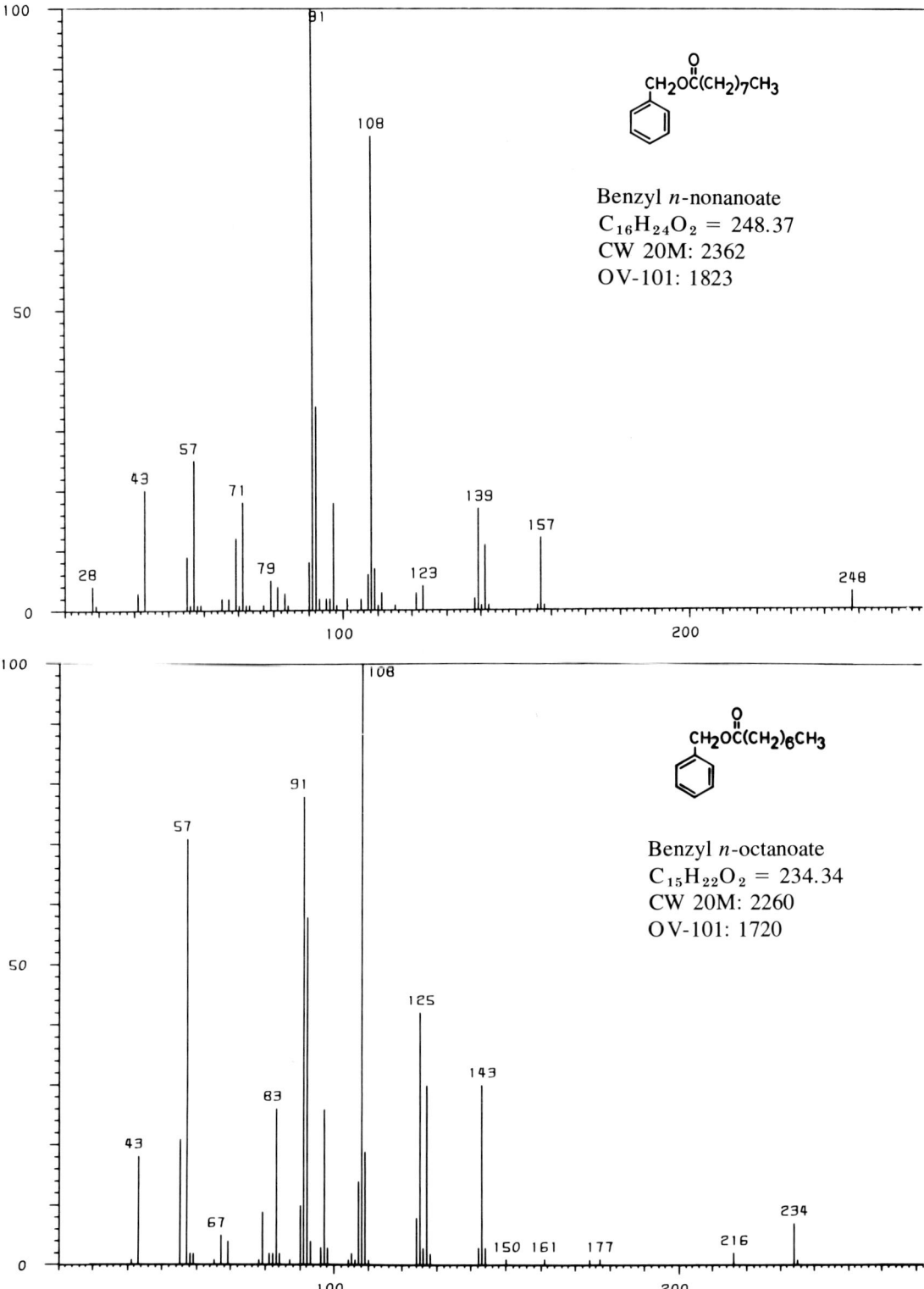

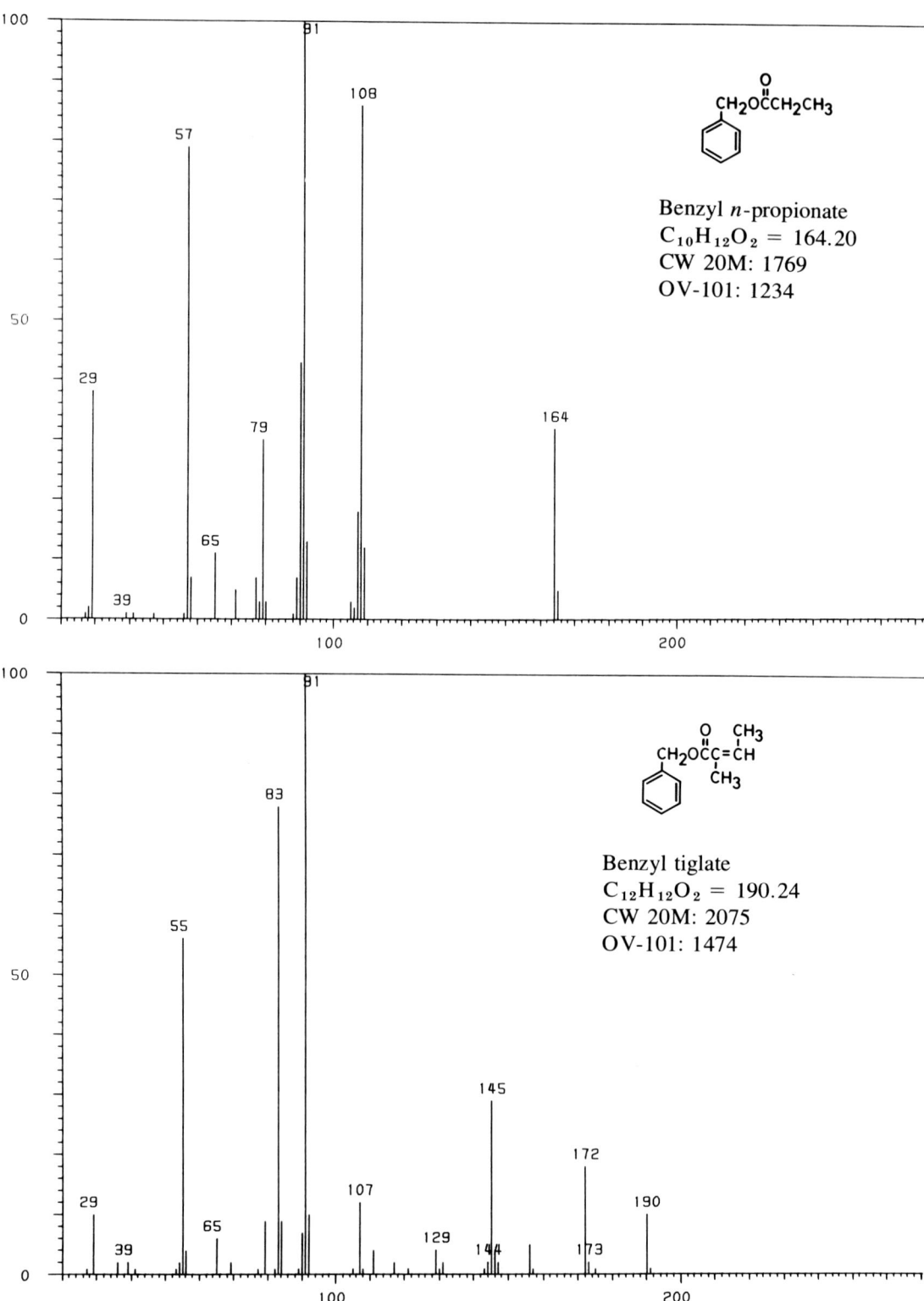

APPENDIX IV   MASS SPECTRA OF INDIVIDUAL COMPOUNDS   173

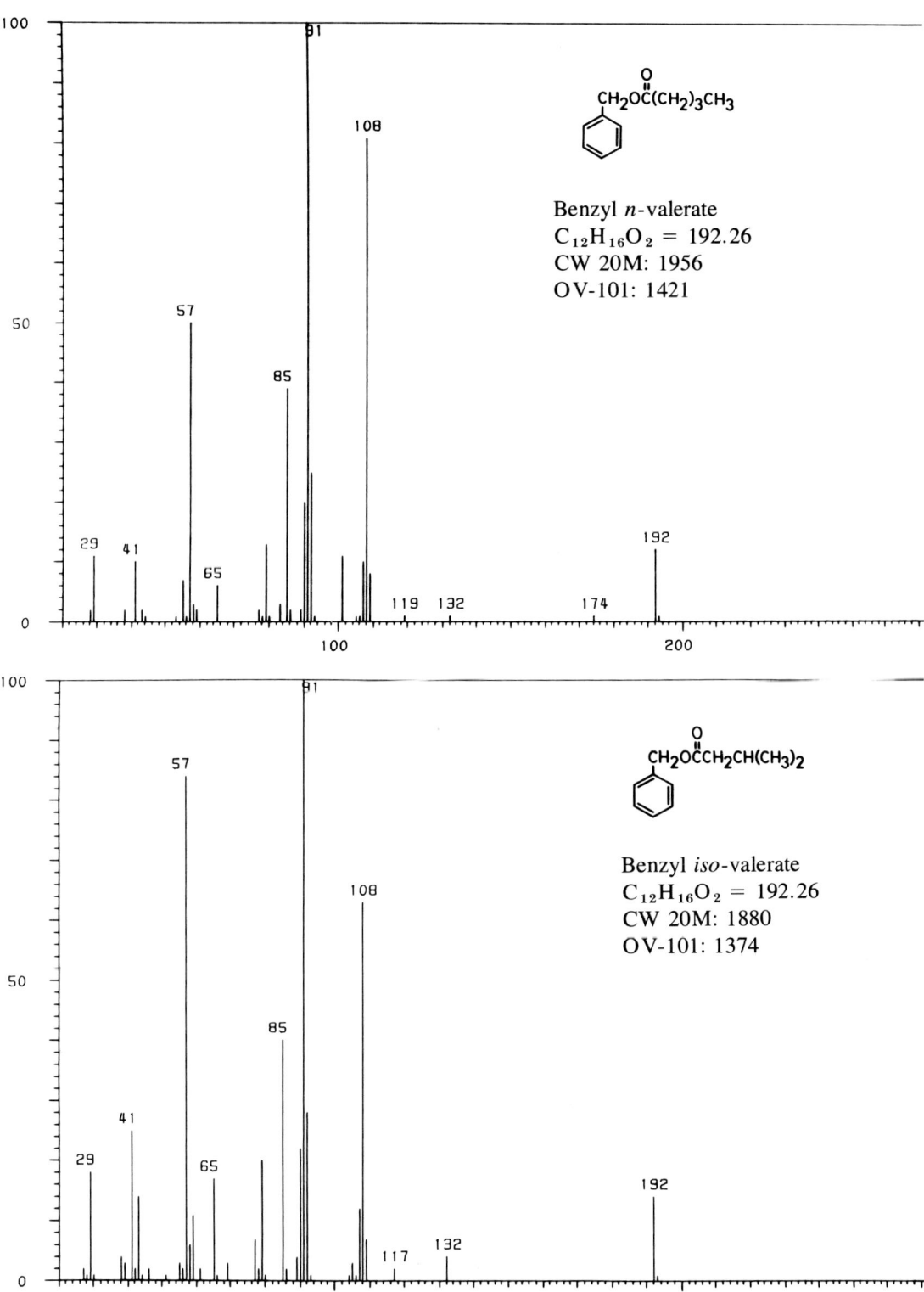

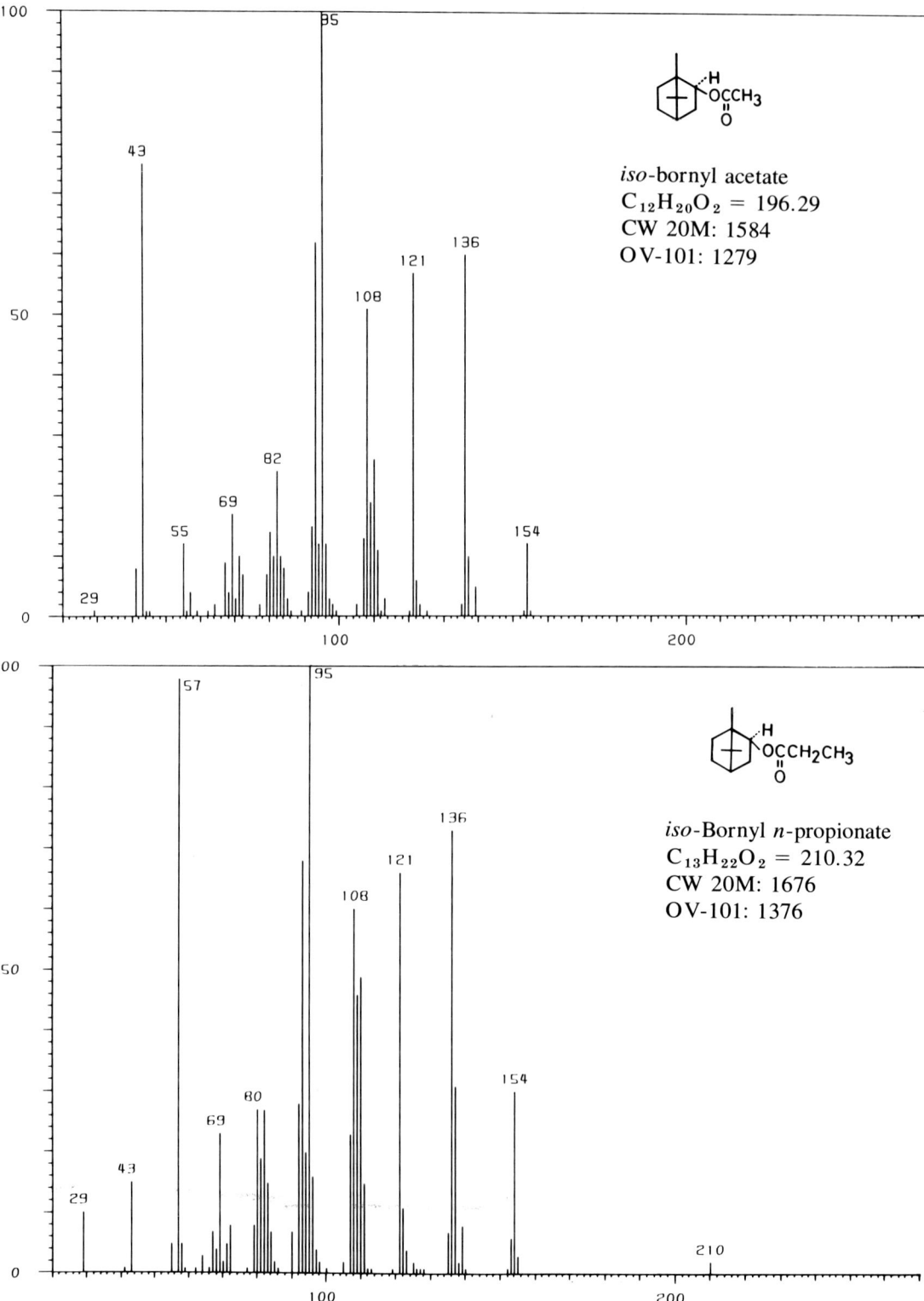

APPENDIX IV   MASS SPECTRA OF INDIVIDUAL COMPOUNDS   175

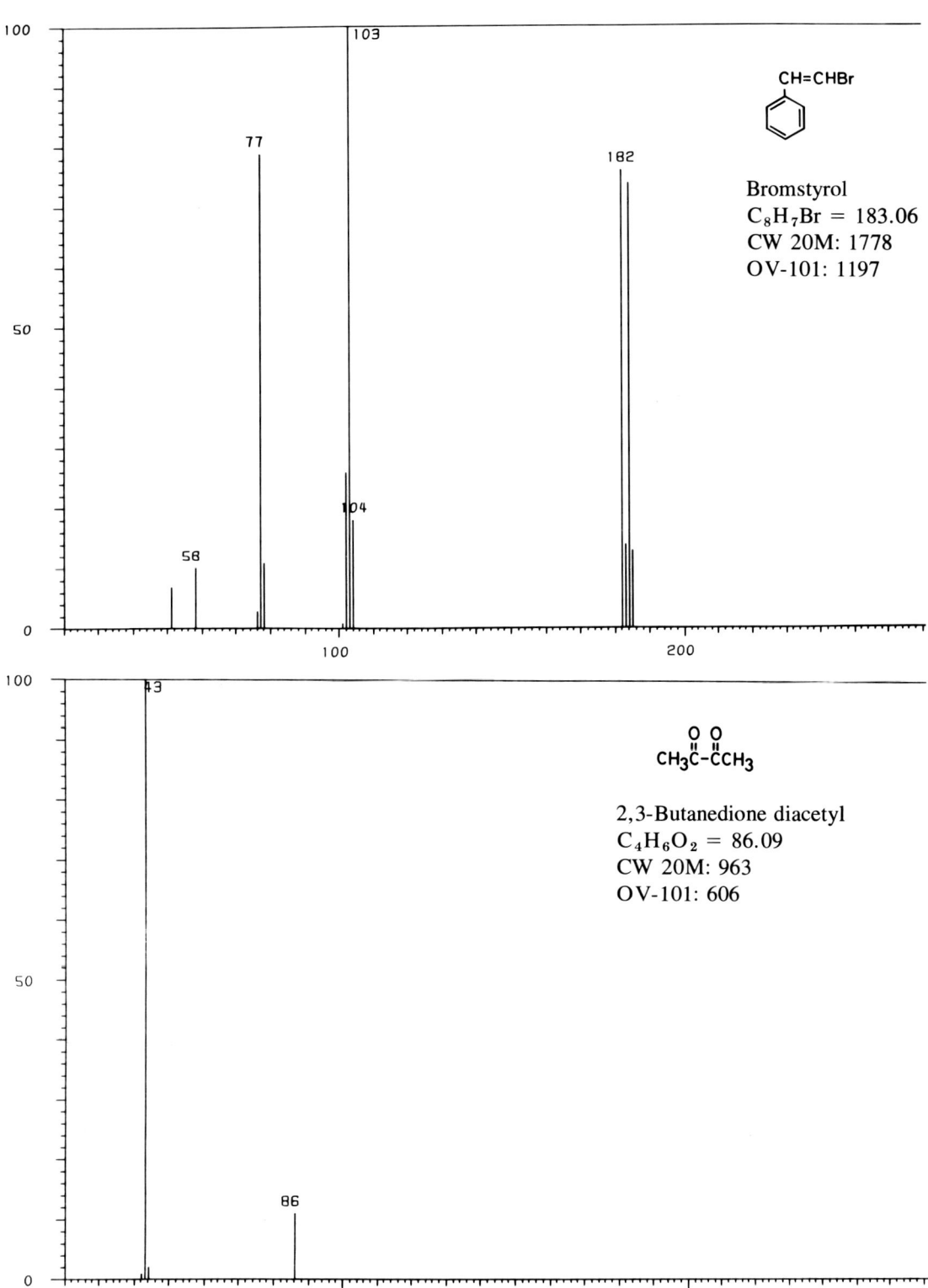

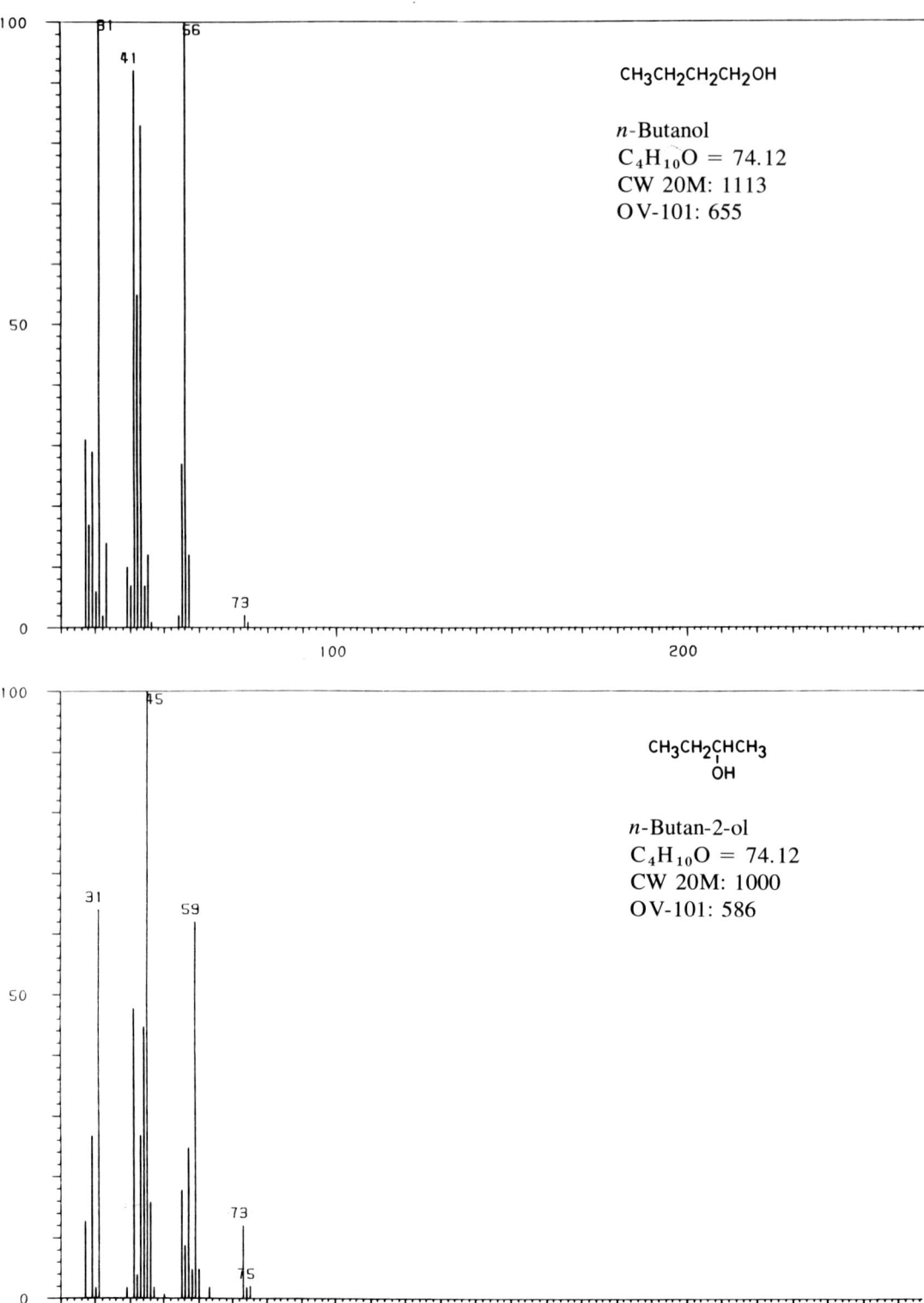

APPENDIX IV   MASS SPECTRA OF INDIVIDUAL COMPOUNDS   177

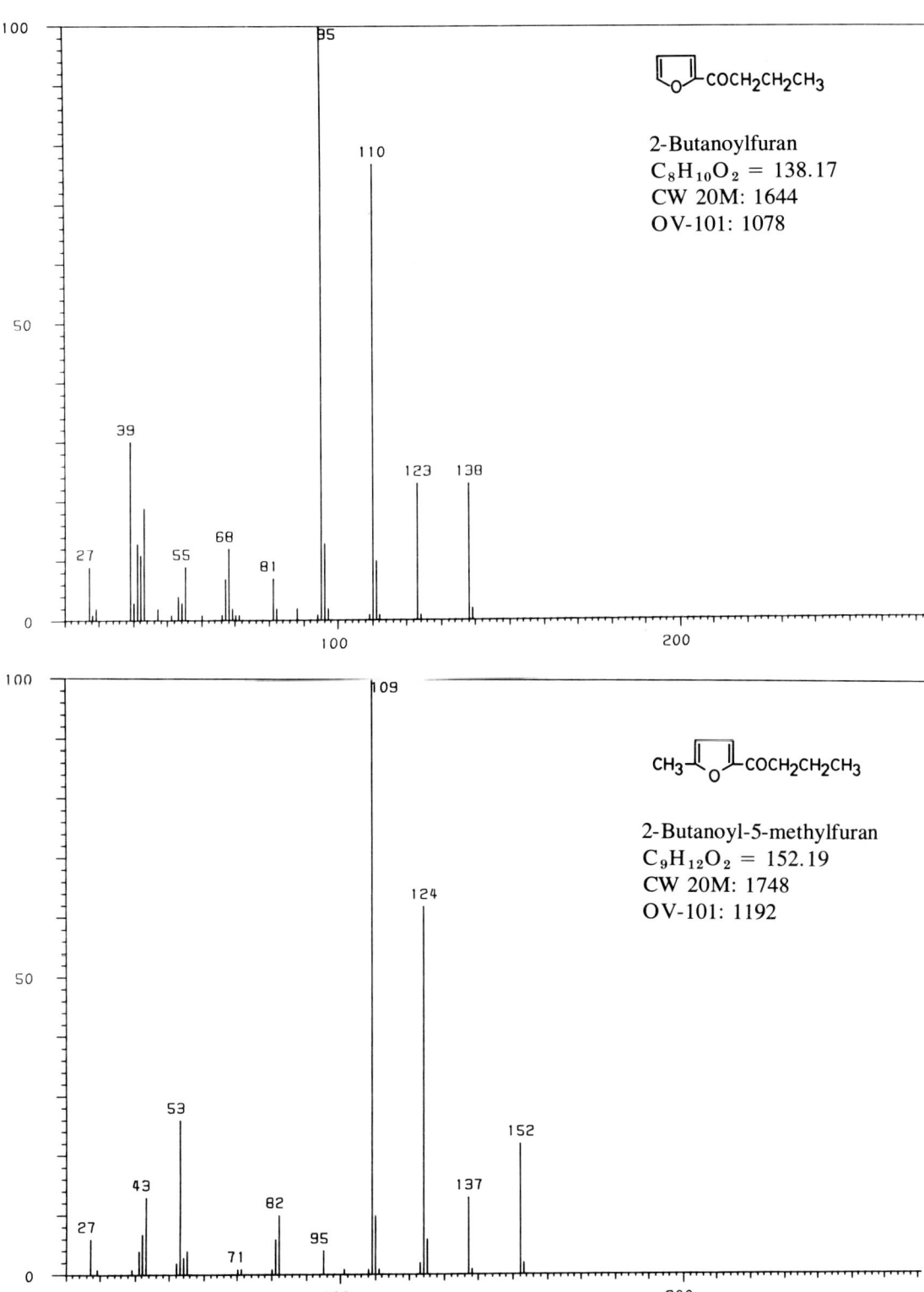

2-Butanoylfuran
$C_8H_{10}O_2 = 138.17$
CW 20M: 1644
OV-101: 1078

2-Butanoyl-5-methylfuran
$C_9H_{12}O_2 = 152.19$
CW 20M: 1748
OV-101: 1192

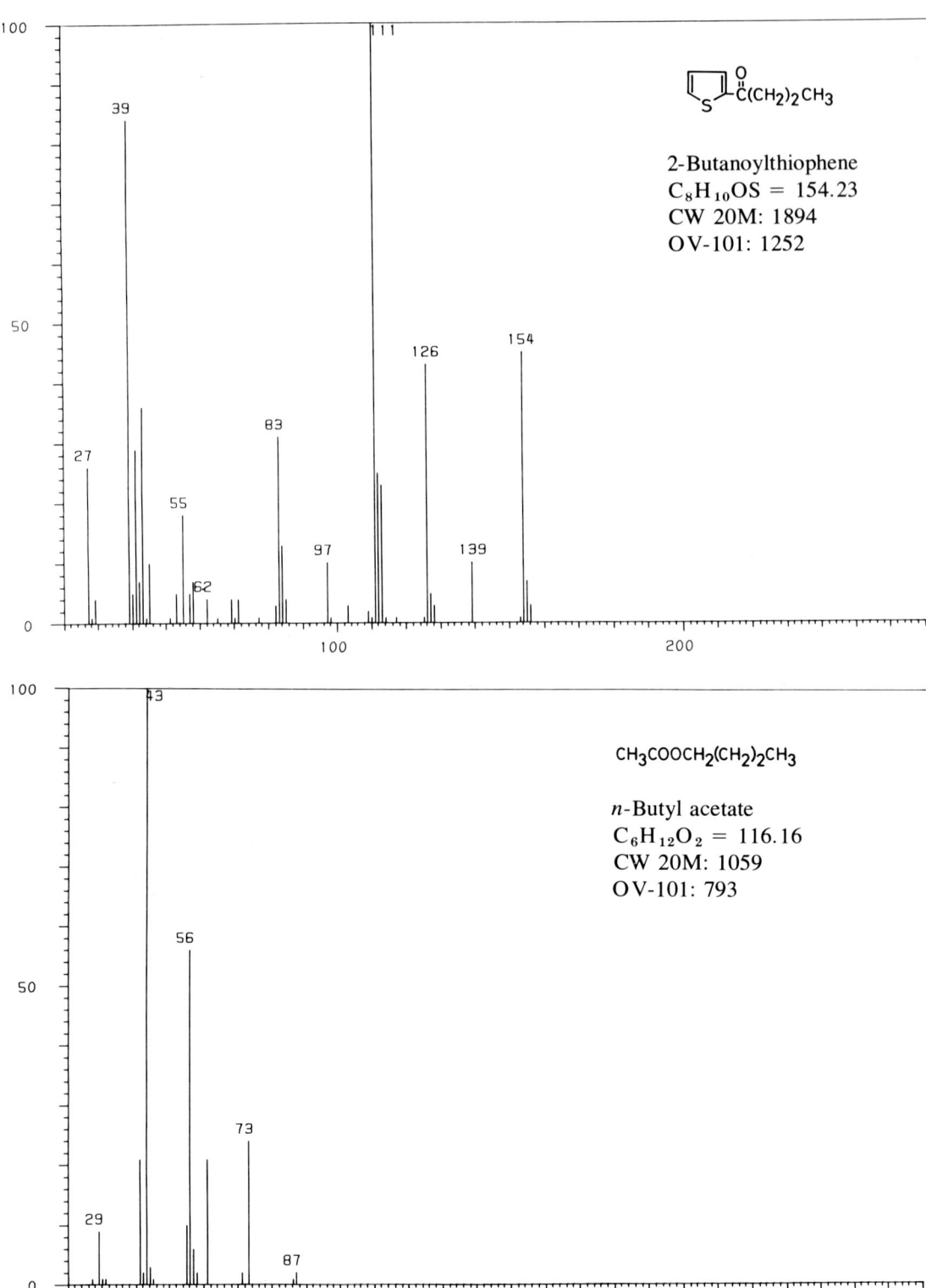

APPENDIX IV   MASS SPECTRA OF INDIVIDUAL COMPOUNDS   179

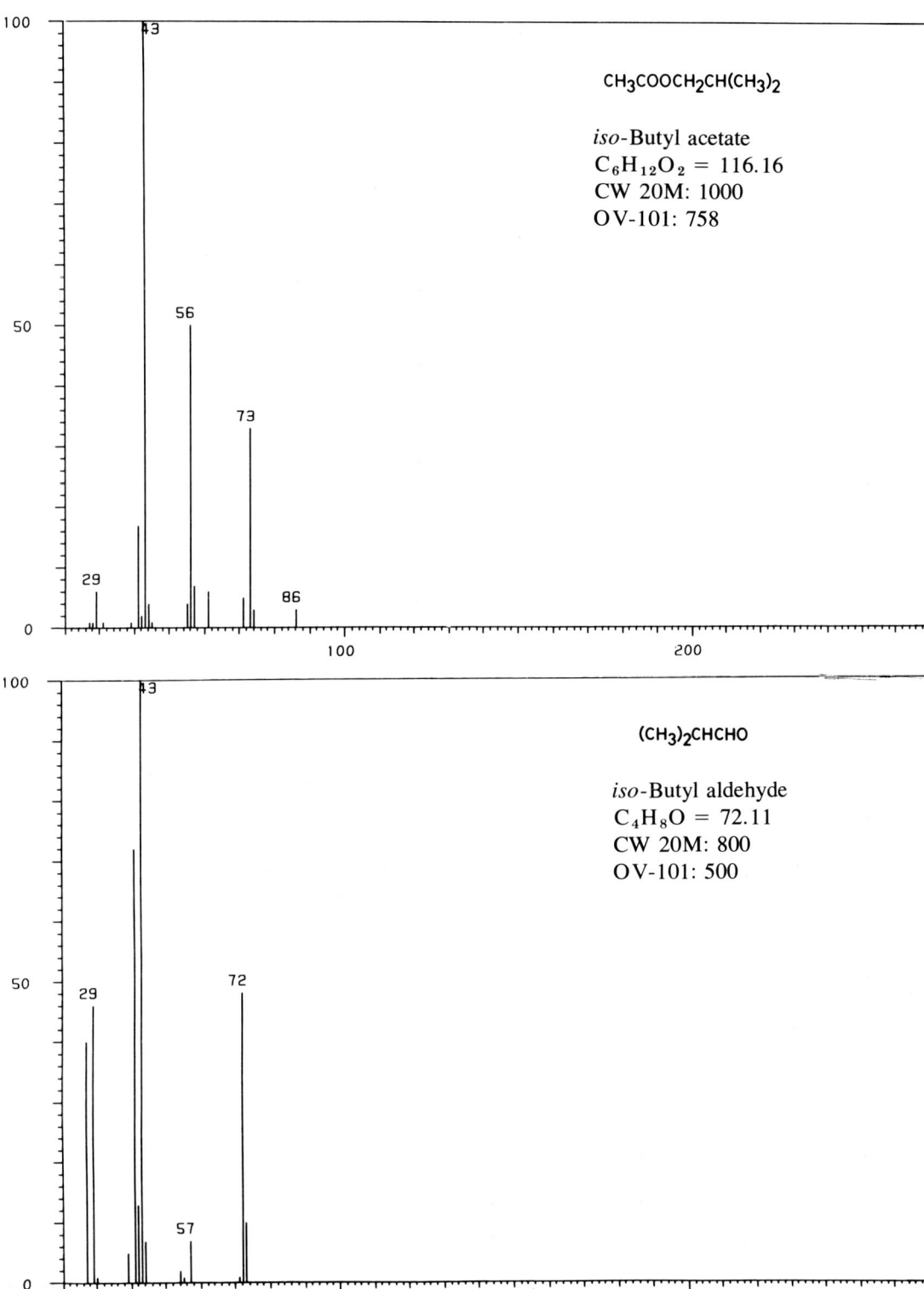

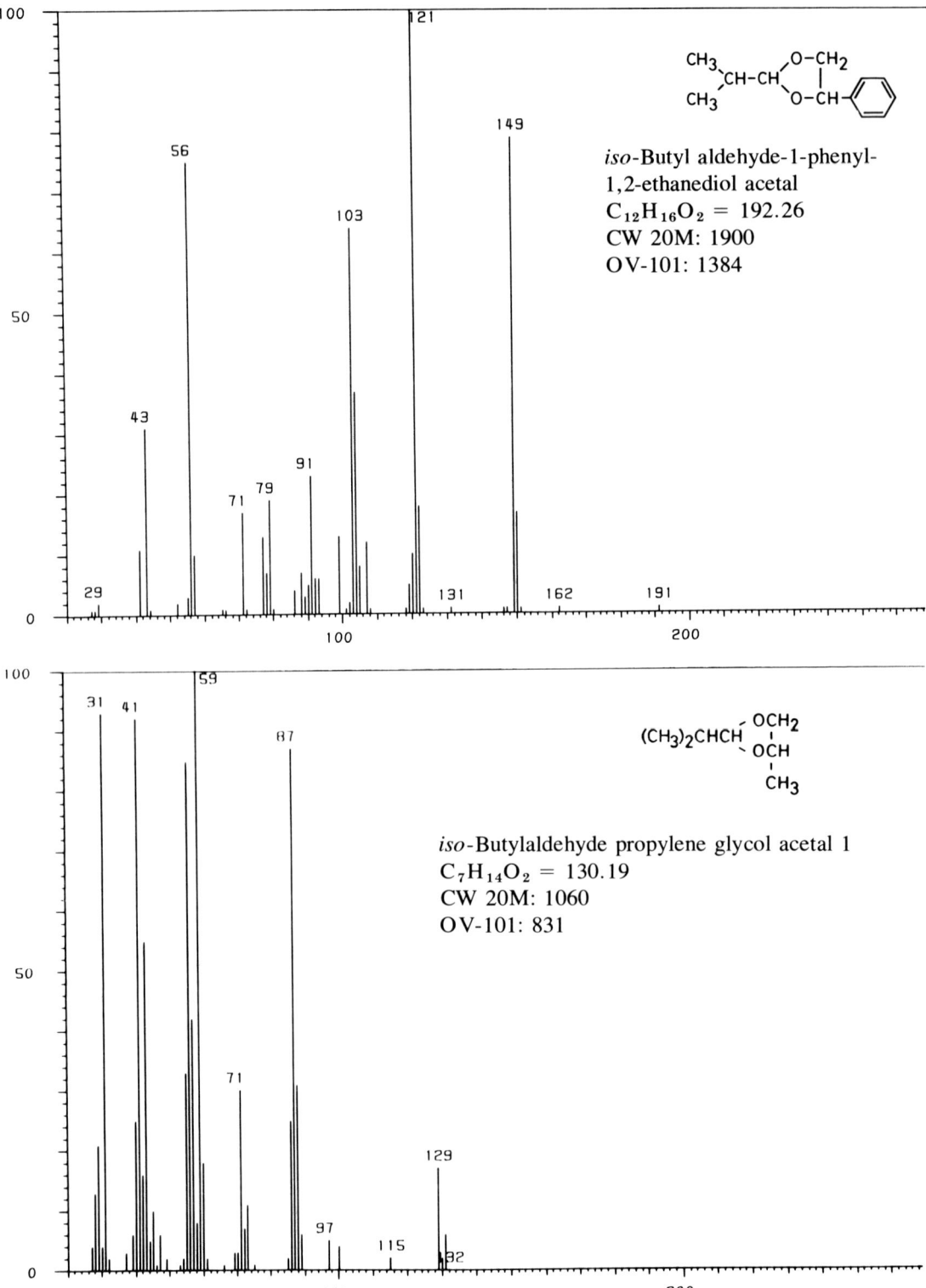

APPENDIX IV   MASS SPECTRA OF INDIVIDUAL COMPOUNDS

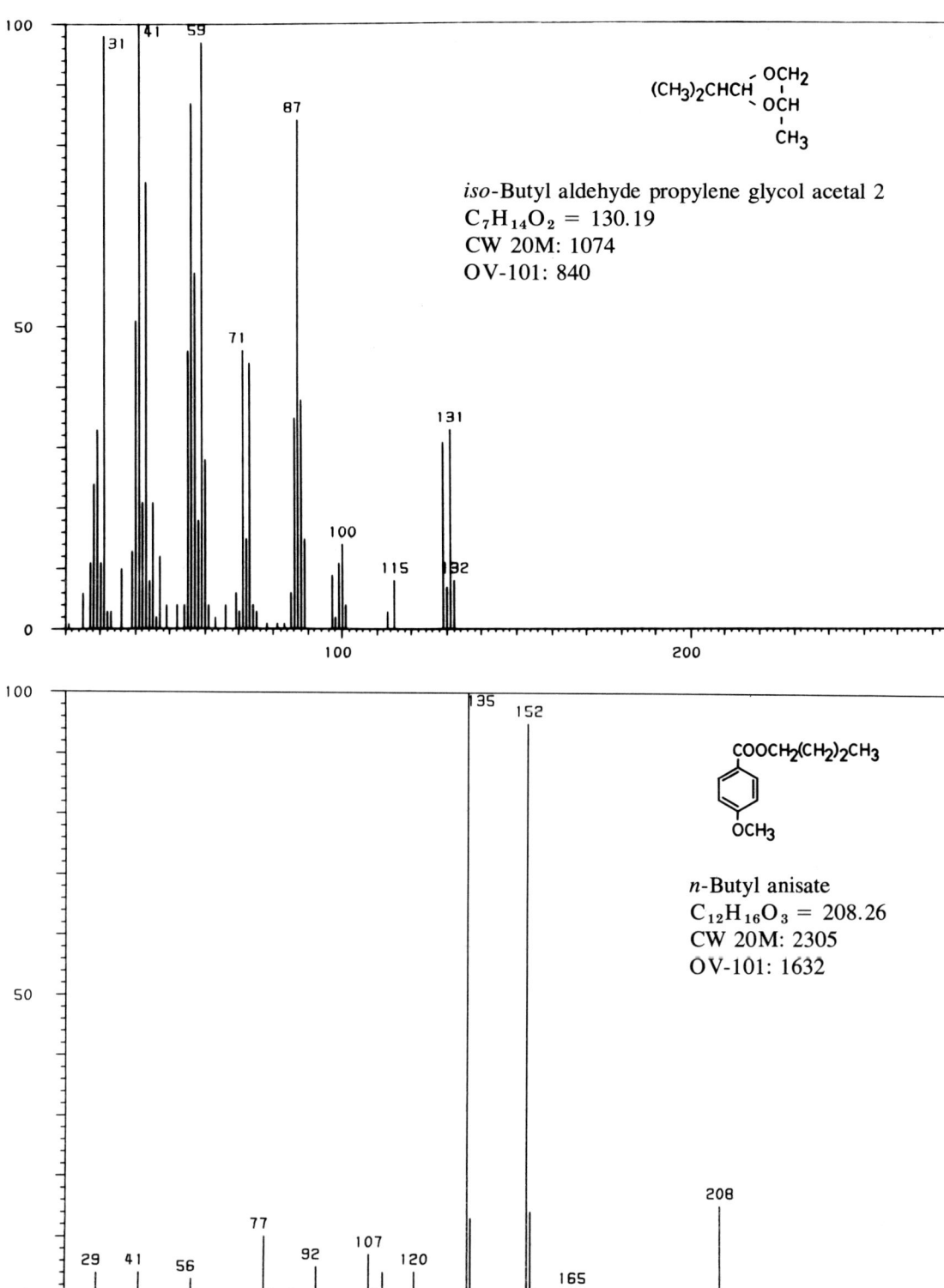

iso-Butyl aldehyde propylene glycol acetal 2
$C_7H_{14}O_2 = 130.19$
CW 20M: 1074
OV-101: 840

n-Butyl anisate
$C_{12}H_{16}O_3 = 208.26$
CW 20M: 2305
OV-101: 1632

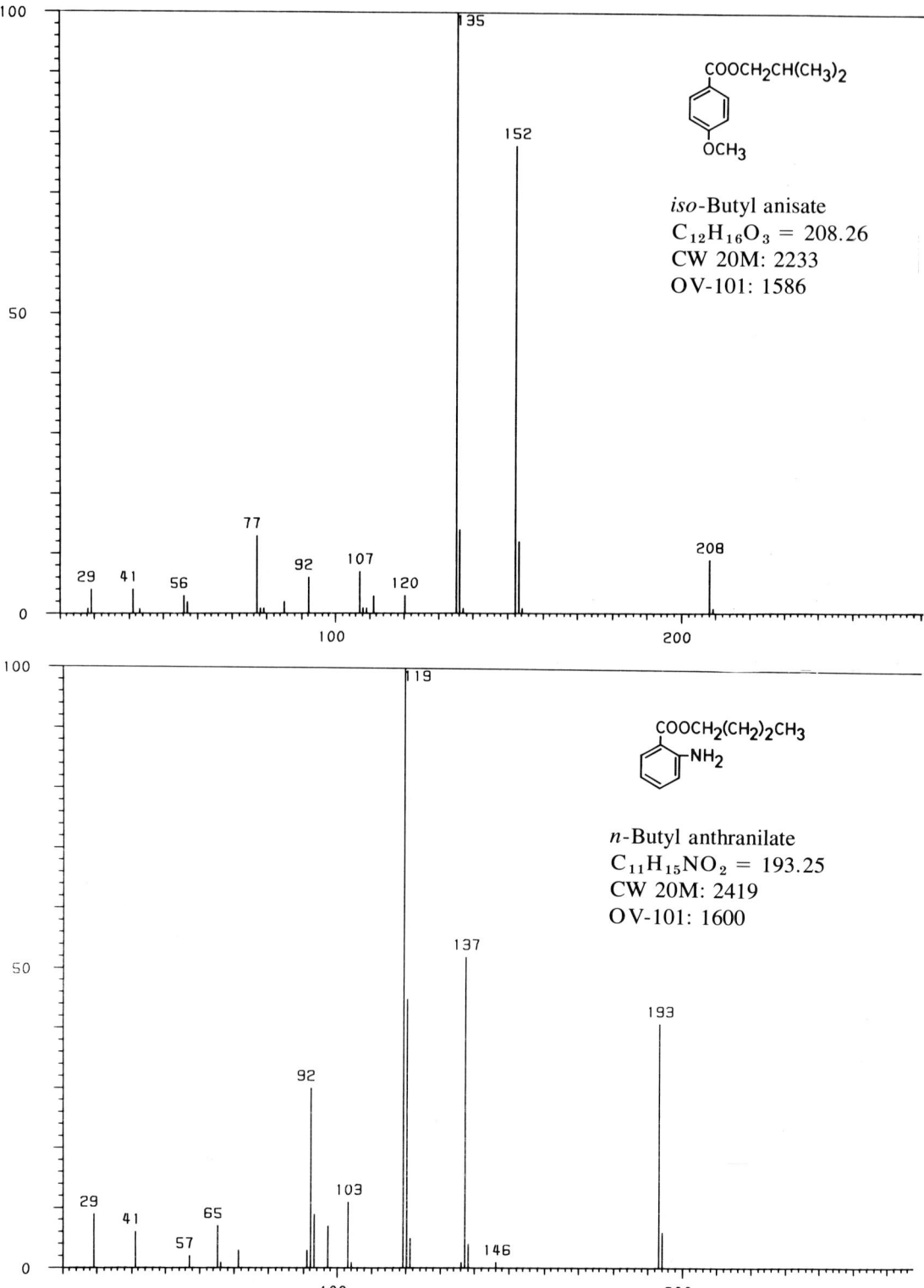

APPENDIX IV   MASS SPECTRA OF INDIVIDUAL COMPOUNDS

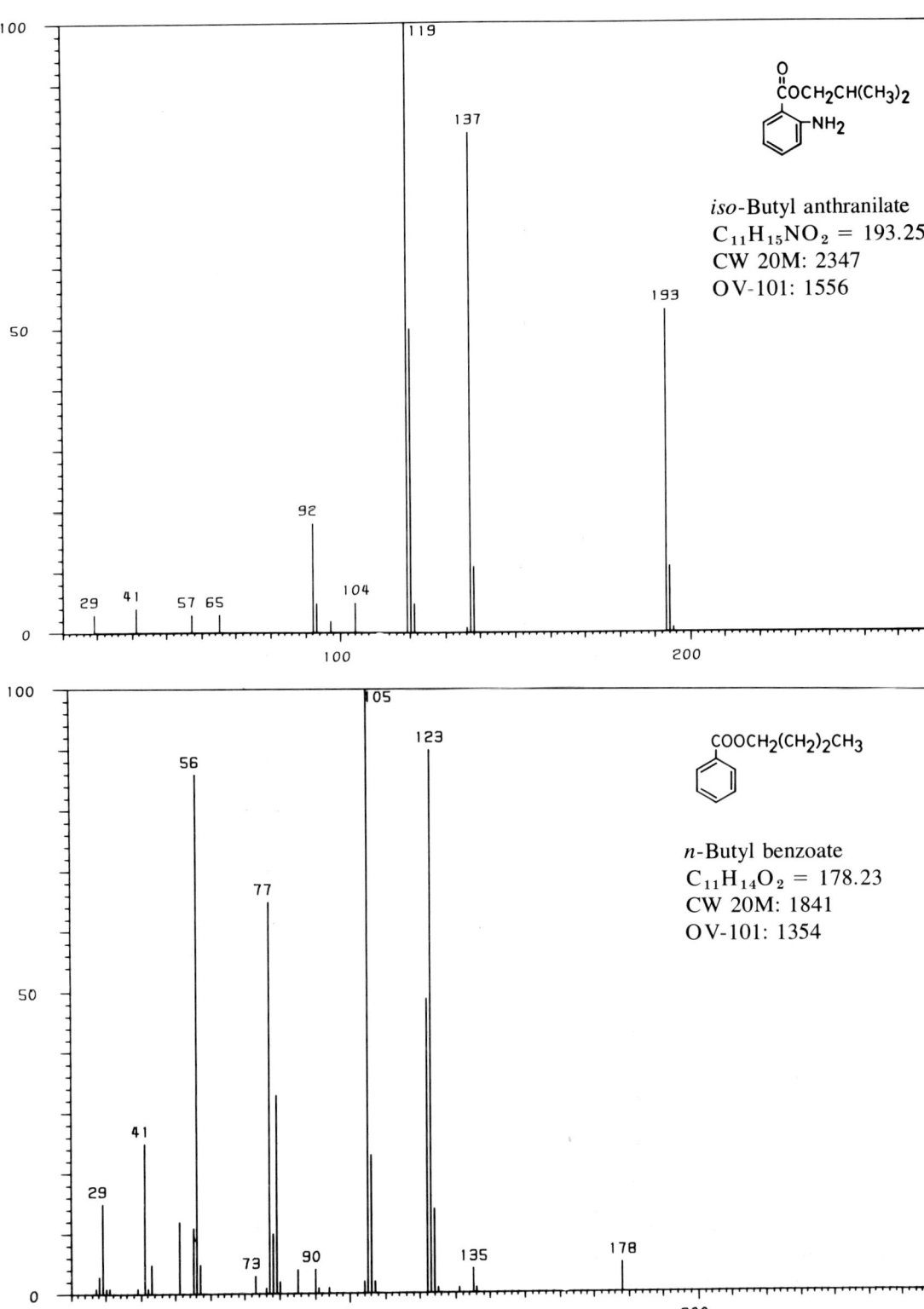

iso-Butyl anthranilate
$C_{11}H_{15}NO_2 = 193.25$
CW 20M: 2347
OV-101: 1556

n-Butyl benzoate
$C_{11}H_{14}O_2 = 178.23$
CW 20M: 1841
OV-101: 1354

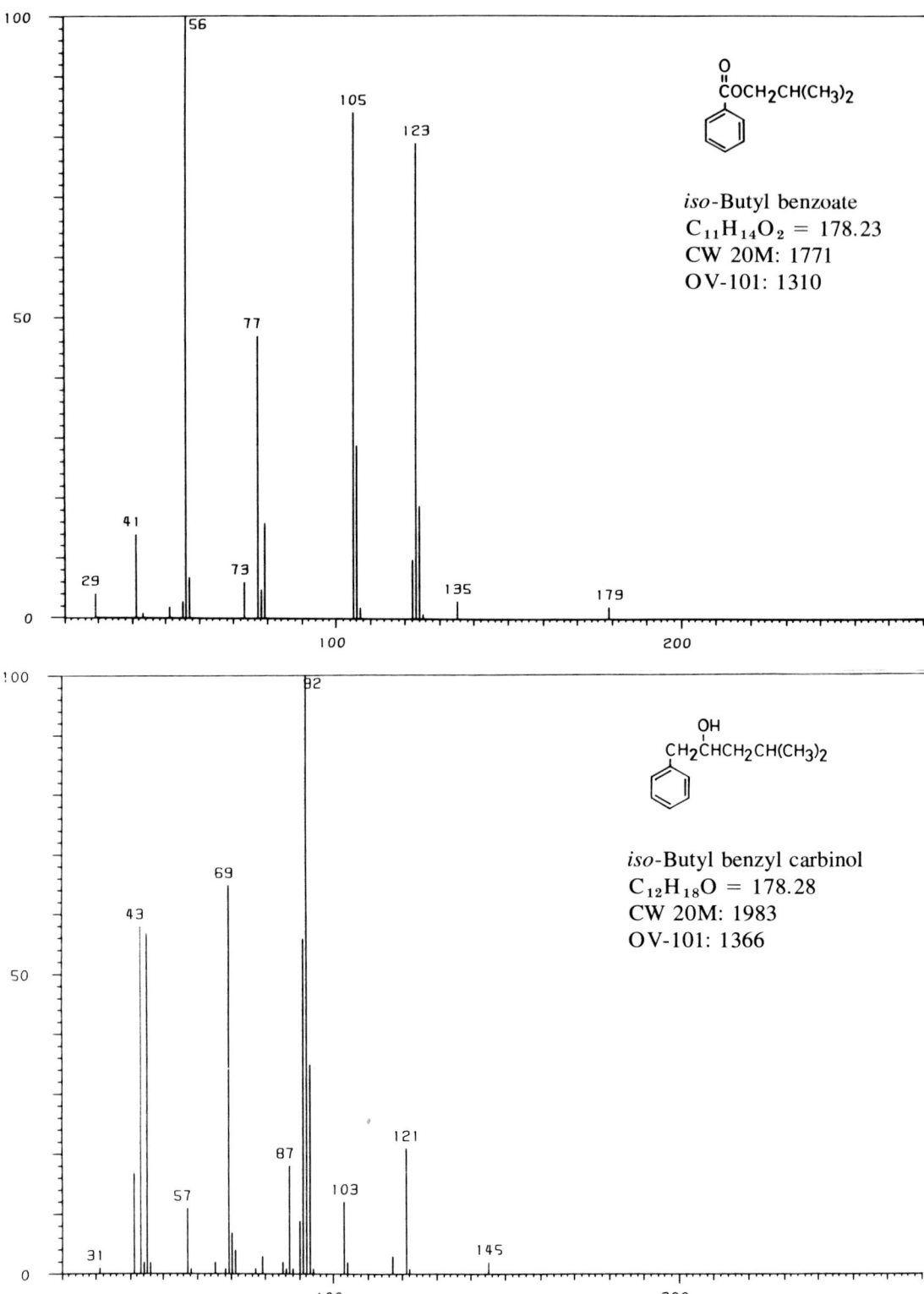

APPENDIX IV MASS SPECTRA OF INDIVIDUAL COMPOUNDS 185

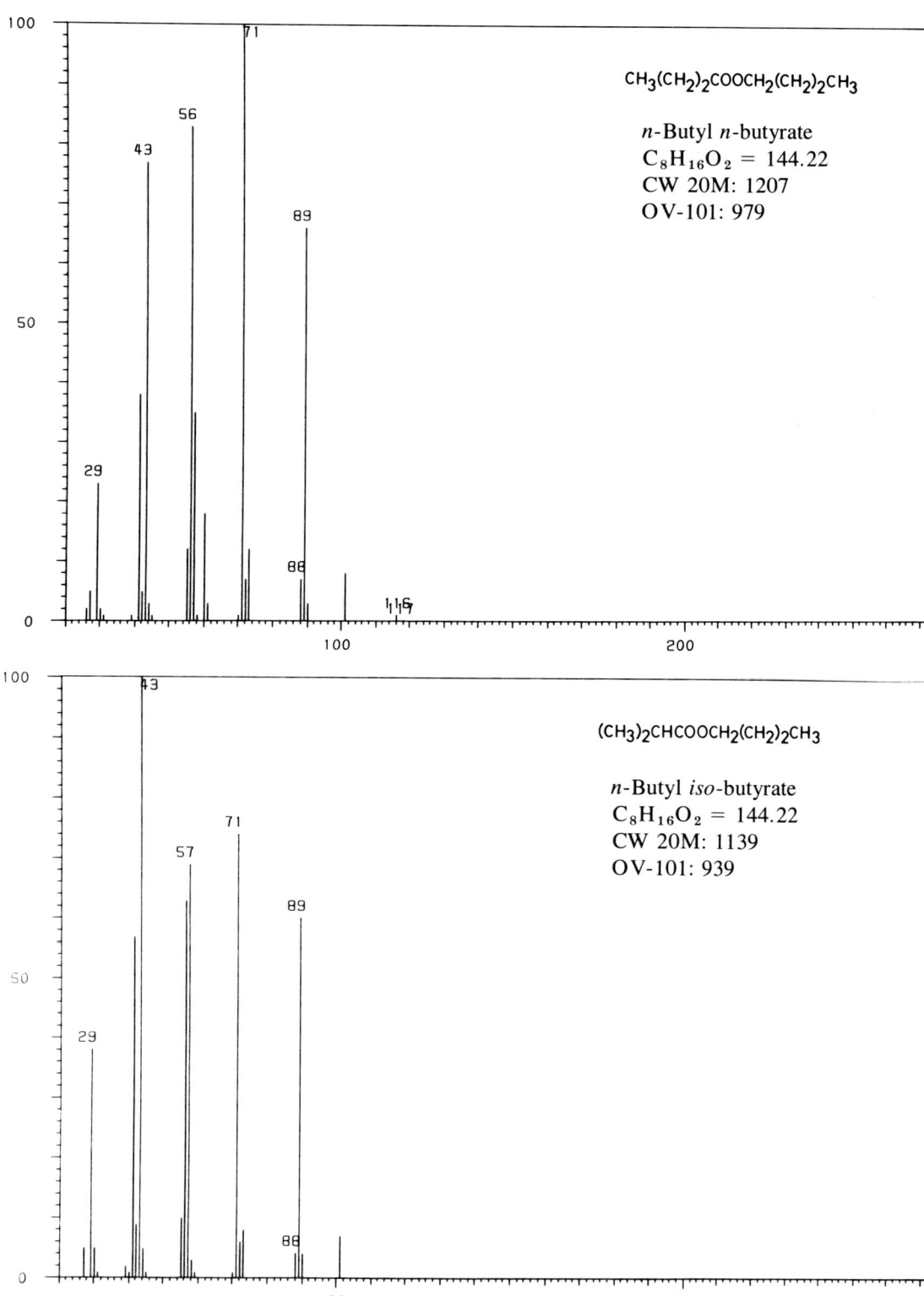

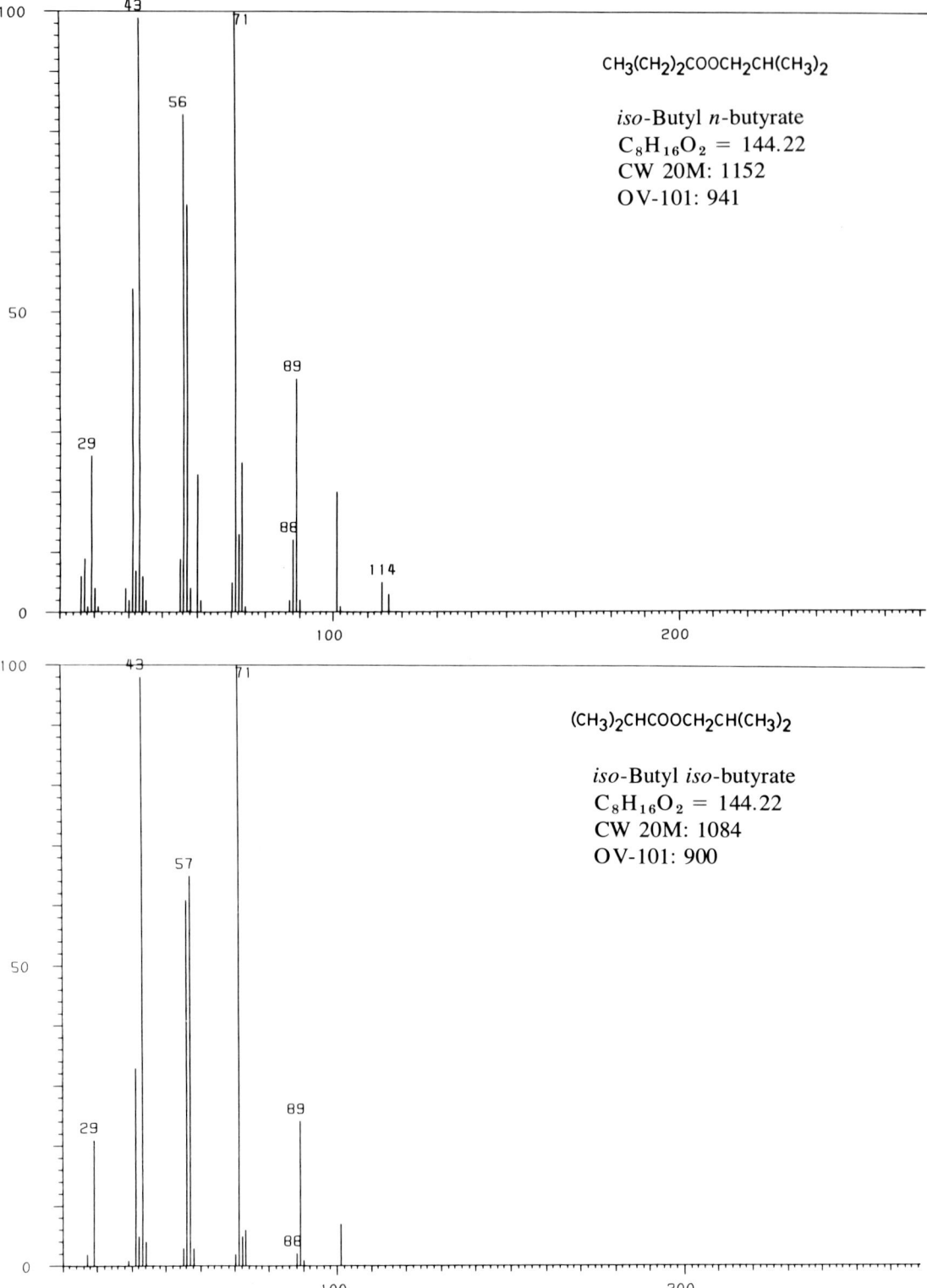

APPENDIX IV   MASS SPECTRA OF INDIVIDUAL COMPOUNDS

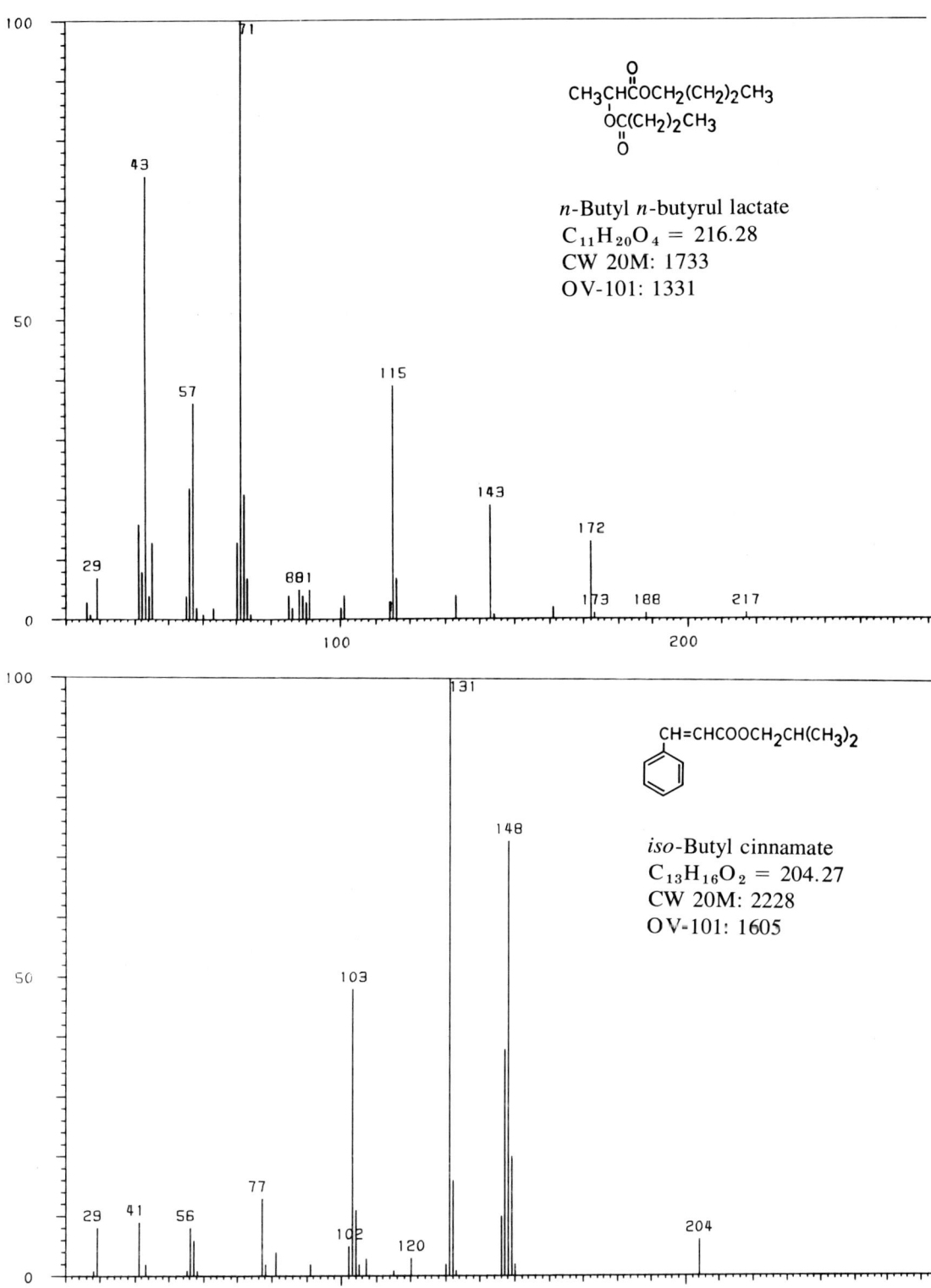

n-Butyl n-butyrul lactate
$C_{11}H_{20}O_4 = 216.28$
CW 20M: 1733
OV-101: 1331

iso-Butyl cinnamate
$C_{13}H_{16}O_2 = 204.27$
CW 20M: 2228
OV-101: 1605

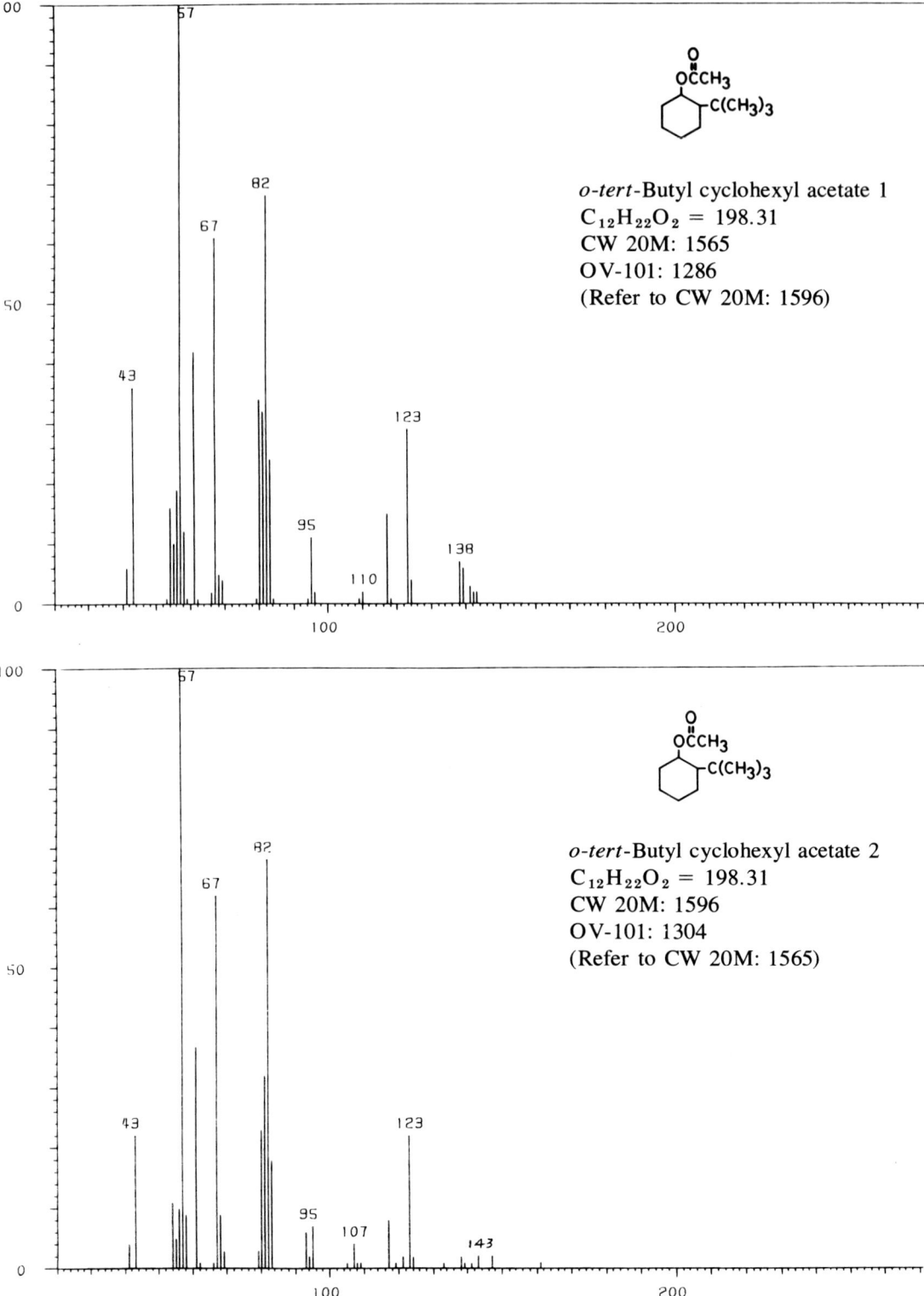

o-tert-Butyl cyclohexyl acetate 1
$C_{12}H_{22}O_2 = 198.31$
CW 20M: 1565
OV-101: 1286
(Refer to CW 20M: 1596)

o-tert-Butyl cyclohexyl acetate 2
$C_{12}H_{22}O_2 = 198.31$
CW 20M: 1596
OV-101: 1304
(Refer to CW 20M: 1565)

APPENDIX IV MASS SPECTRA OF INDIVIDUAL COMPOUNDS

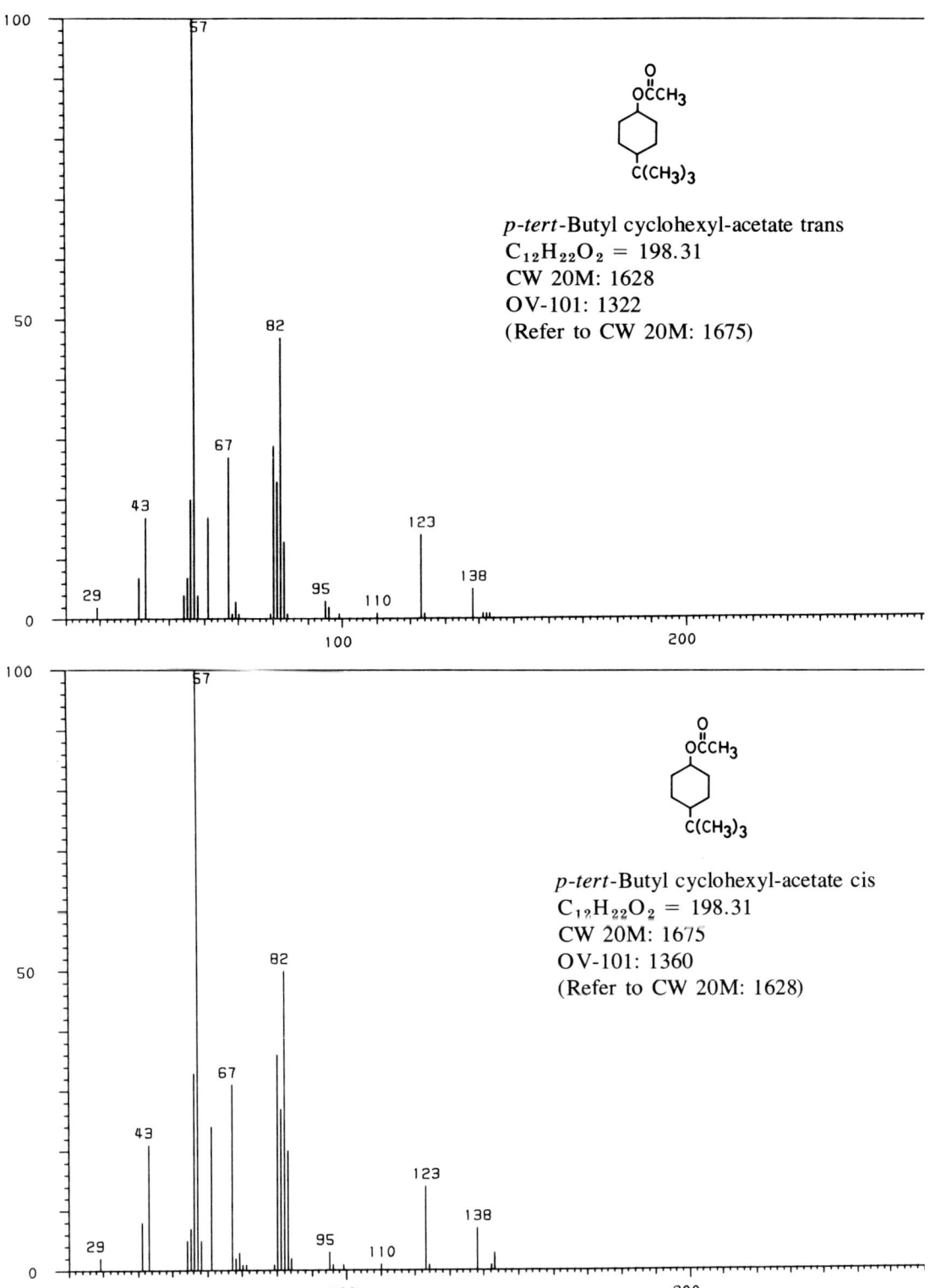

p-tert-Butyl cyclohexyl-acetate trans
$C_{12}H_{22}O_2 = 198.31$
CW 20M: 1628
OV-101: 1322
(Refer to CW 20M: 1675)

p-tert-Butyl cyclohexyl-acetate cis
$C_{12}H_{22}O_2 = 198.31$
CW 20M: 1675
OV-101: 1360
(Refer to CW 20M: 1628)

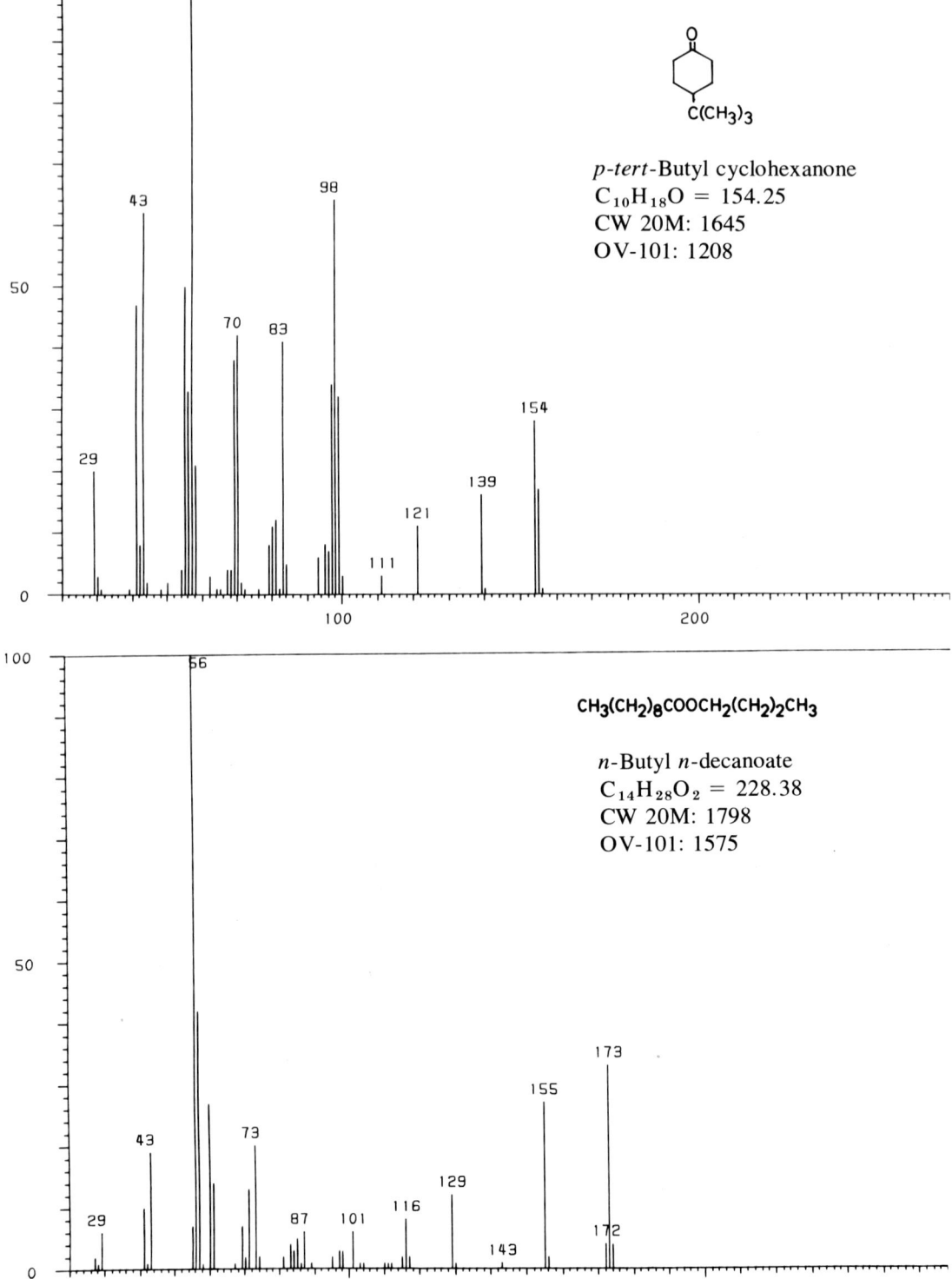

p-tert-Butyl cyclohexanone
$C_{10}H_{18}O = 154.25$
CW 20M: 1645
OV-101: 1208

$CH_3(CH_2)_8COOCH_2(CH_2)_2CH_3$

n-Butyl *n*-decanoate
$C_{14}H_{28}O_2 = 228.38$
CW 20M: 1798
OV-101: 1575

APPENDIX IV   MASS SPECTRA OF INDIVIDUAL COMPOUNDS

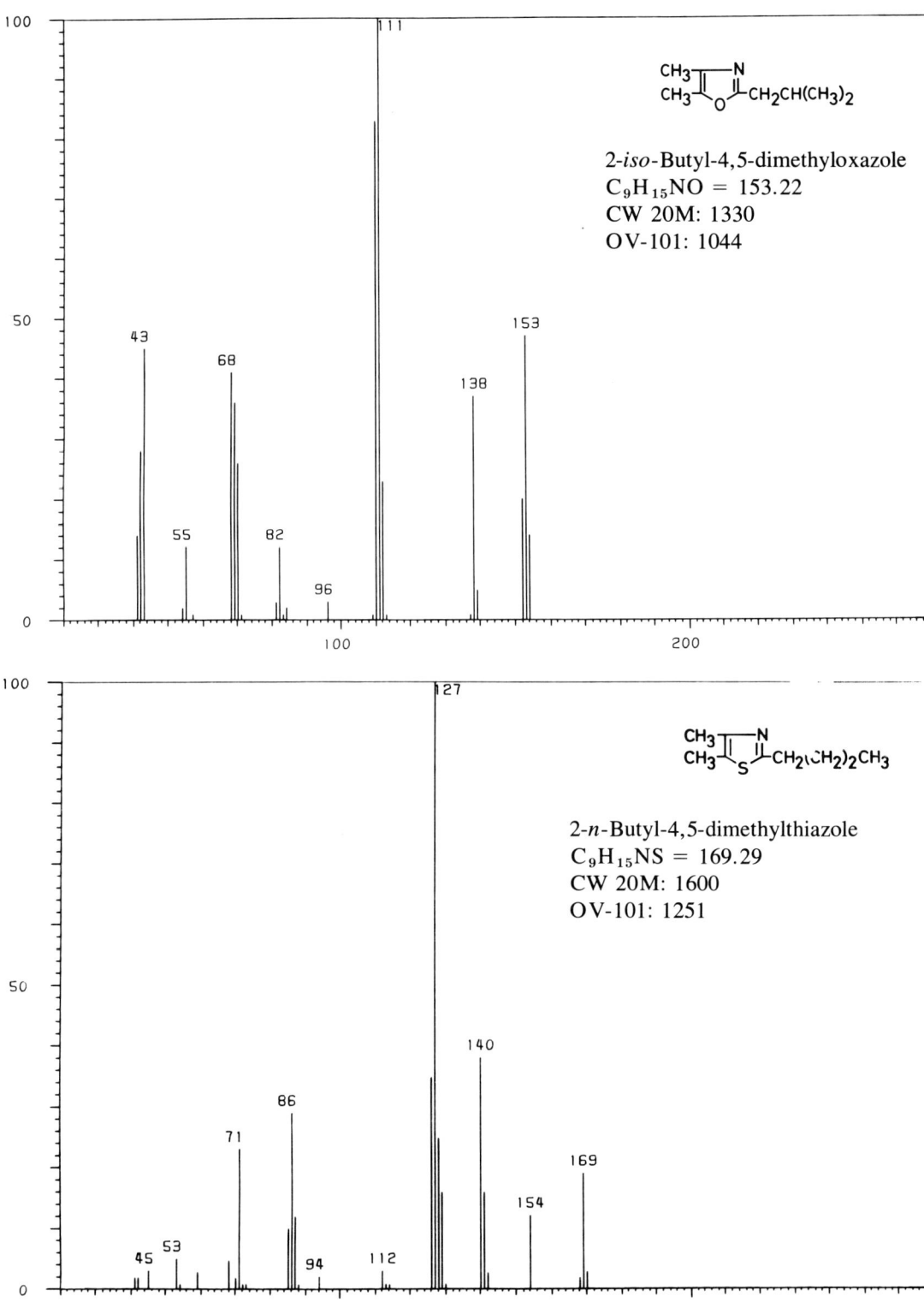

2-*iso*-Butyl-4,5-dimethyloxazole
$C_9H_{15}NO = 153.22$
CW 20M: 1330
OV-101: 1044

2-*n*-Butyl-4,5-dimethylthiazole
$C_9H_{15}NS = 169.29$
CW 20M: 1600
OV-101: 1251

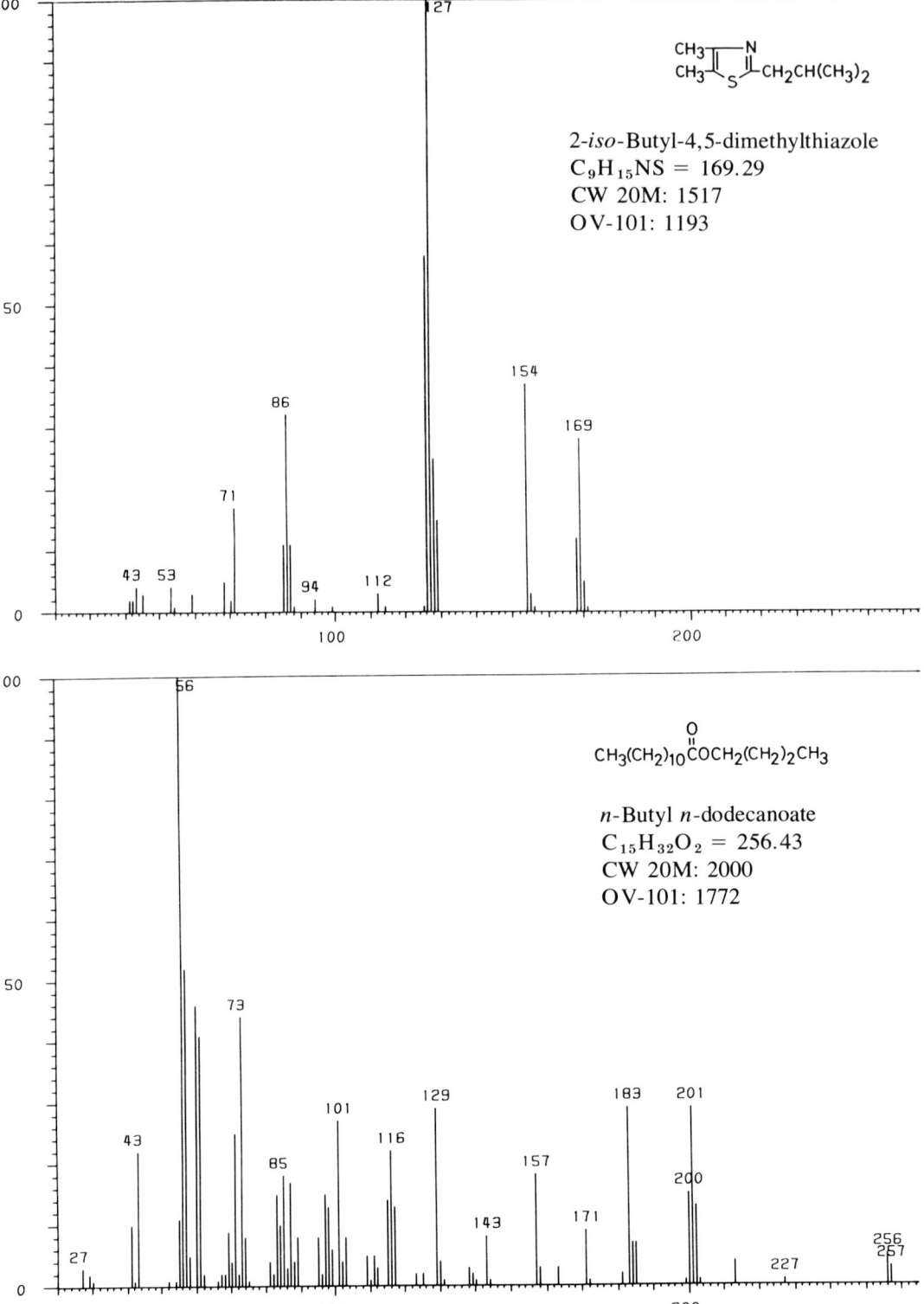

APPENDIX IV   MASS SPECTRA OF INDIVIDUAL COMPOUNDS

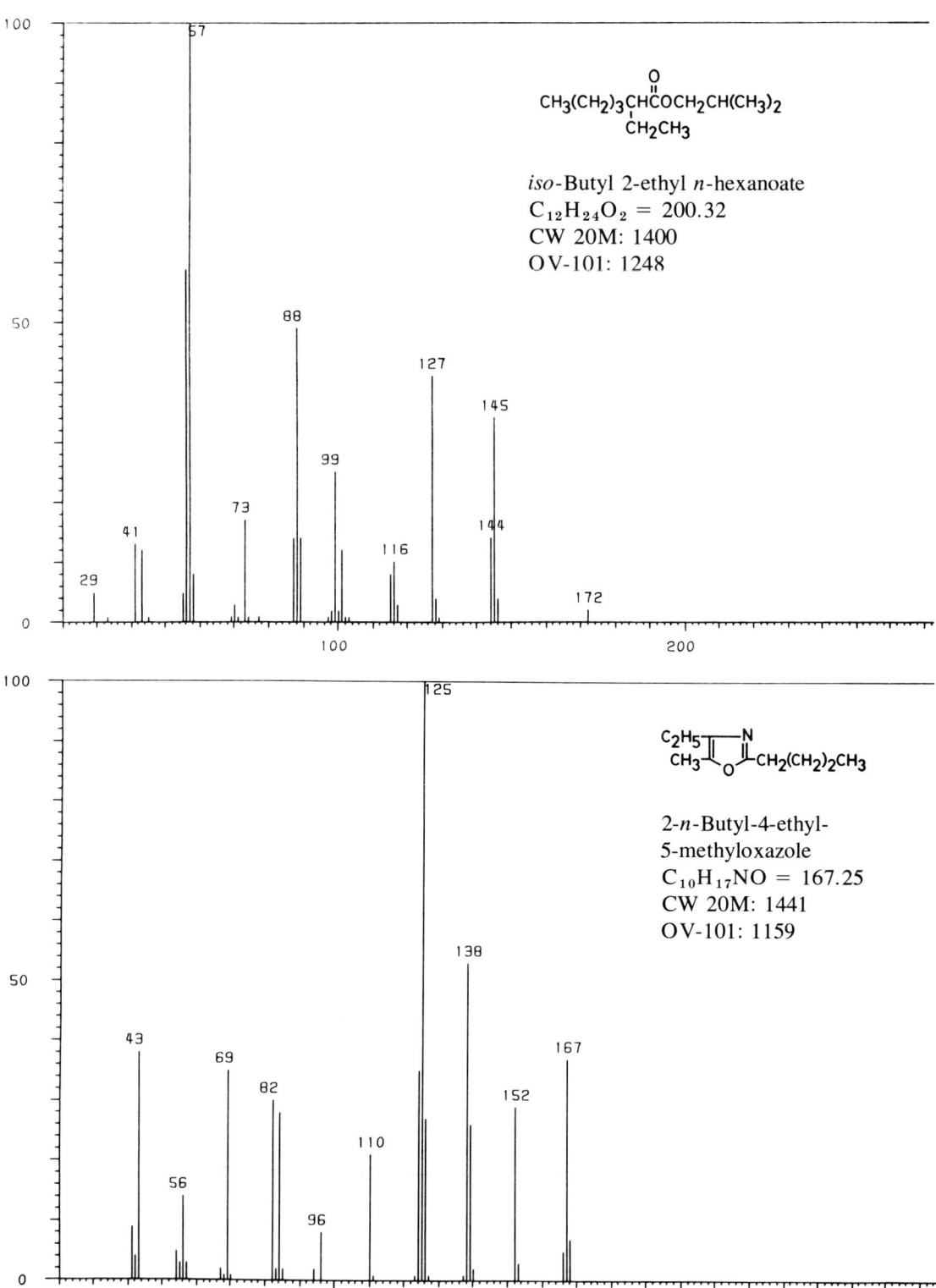

iso-Butyl 2-ethyl n-hexanoate
$C_{12}H_{24}O_2 = 200.32$
CW 20M: 1400
OV-101: 1248

2-n-Butyl-4-ethyl-5-methyloxazole
$C_{10}H_{17}NO = 167.25$
CW 20M: 1441
OV-101: 1159

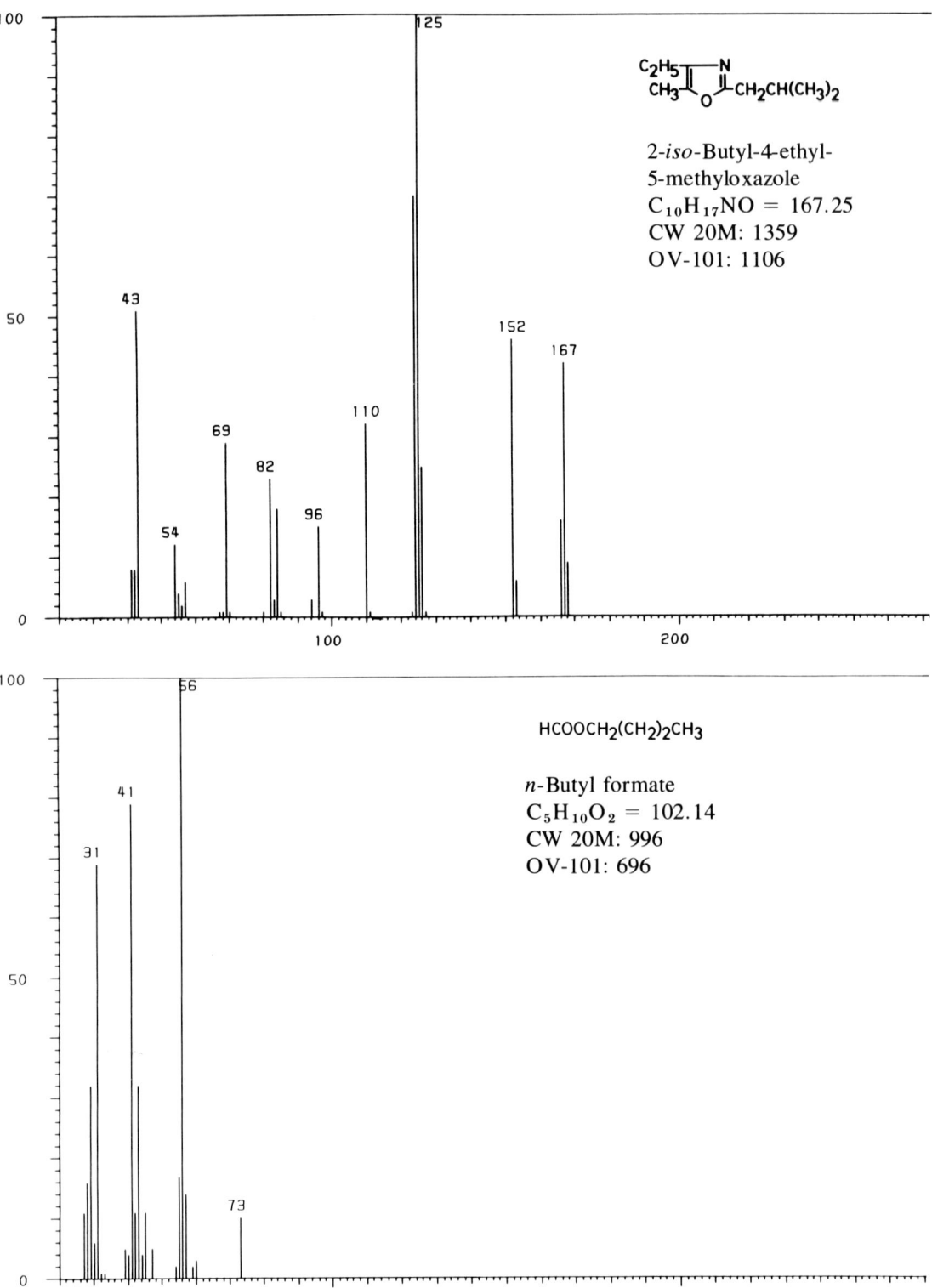

APPENDIX IV   MASS SPECTRA OF INDIVIDUAL COMPOUNDS

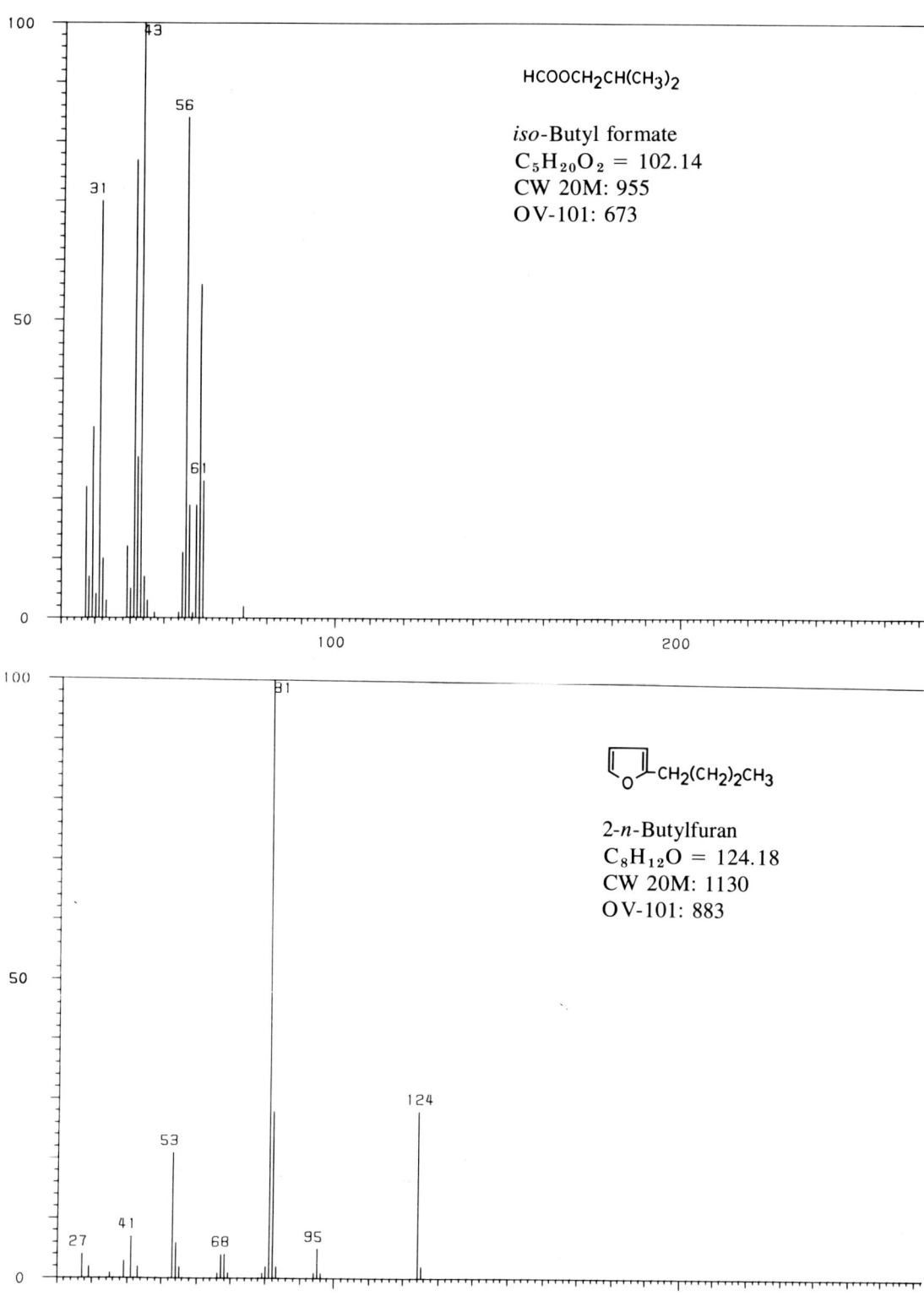

HCOOCH$_2$CH(CH$_3$)$_2$

iso-Butyl formate
C$_5$H$_{20}$O$_2$ = 102.14
CW 20M: 955
OV-101: 673

2-n-Butylfuran
C$_8$H$_{12}$O = 124.18
CW 20M: 1130
OV-101: 883

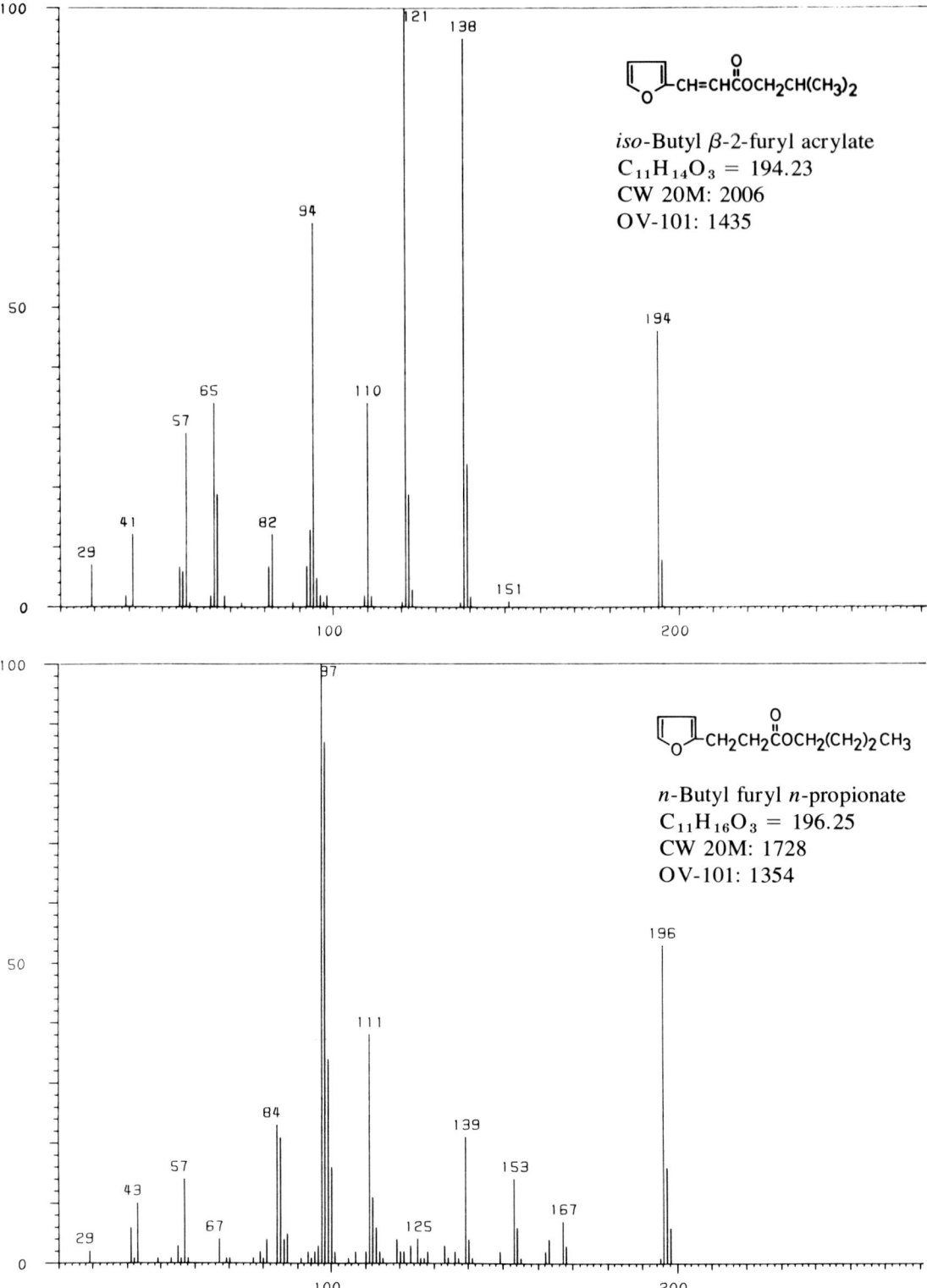

APPENDIX IV   MASS SPECTRA OF INDIVIDUAL COMPOUNDS

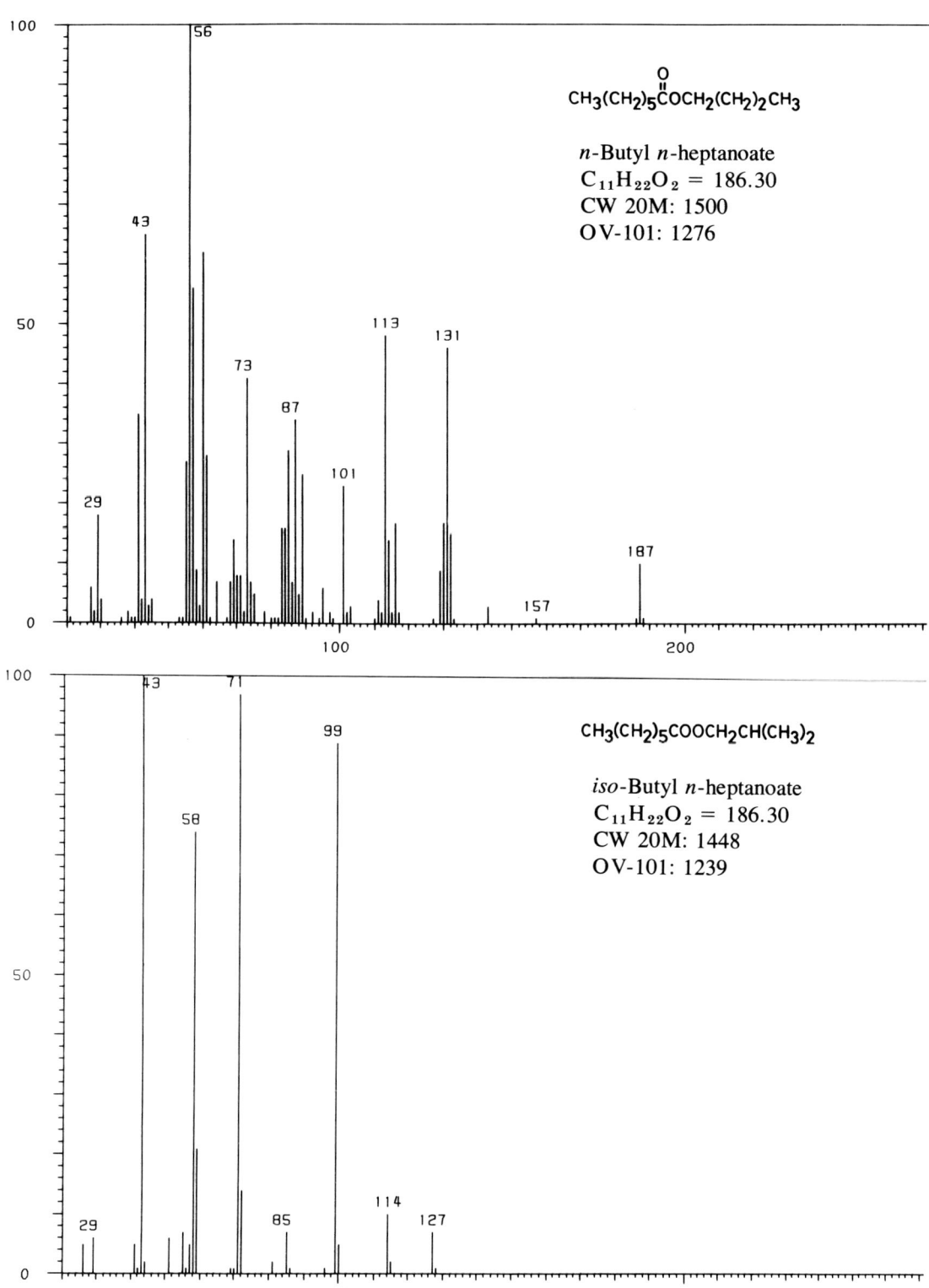

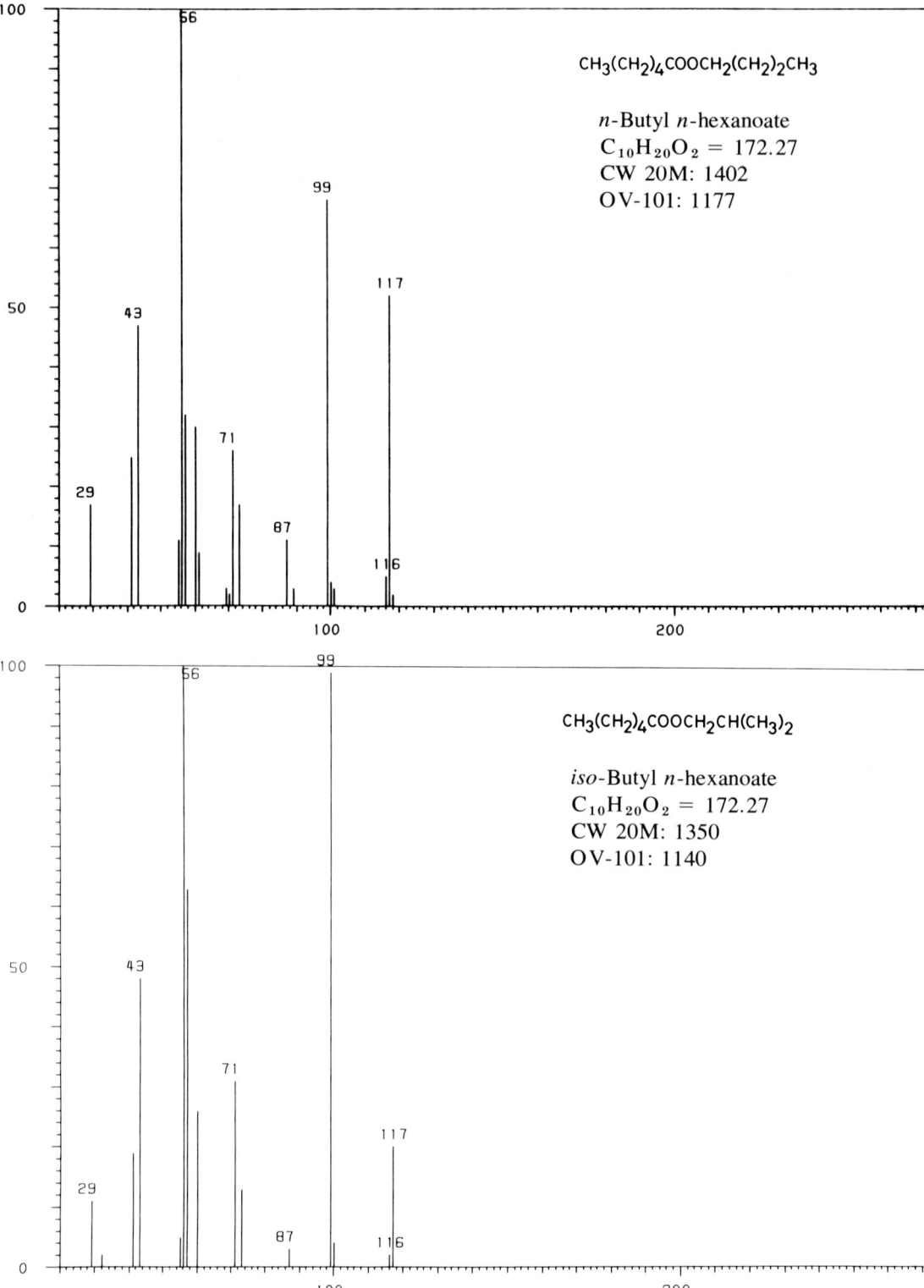

APPENDIX IV   MASS SPECTRA OF INDIVIDUAL COMPOUNDS

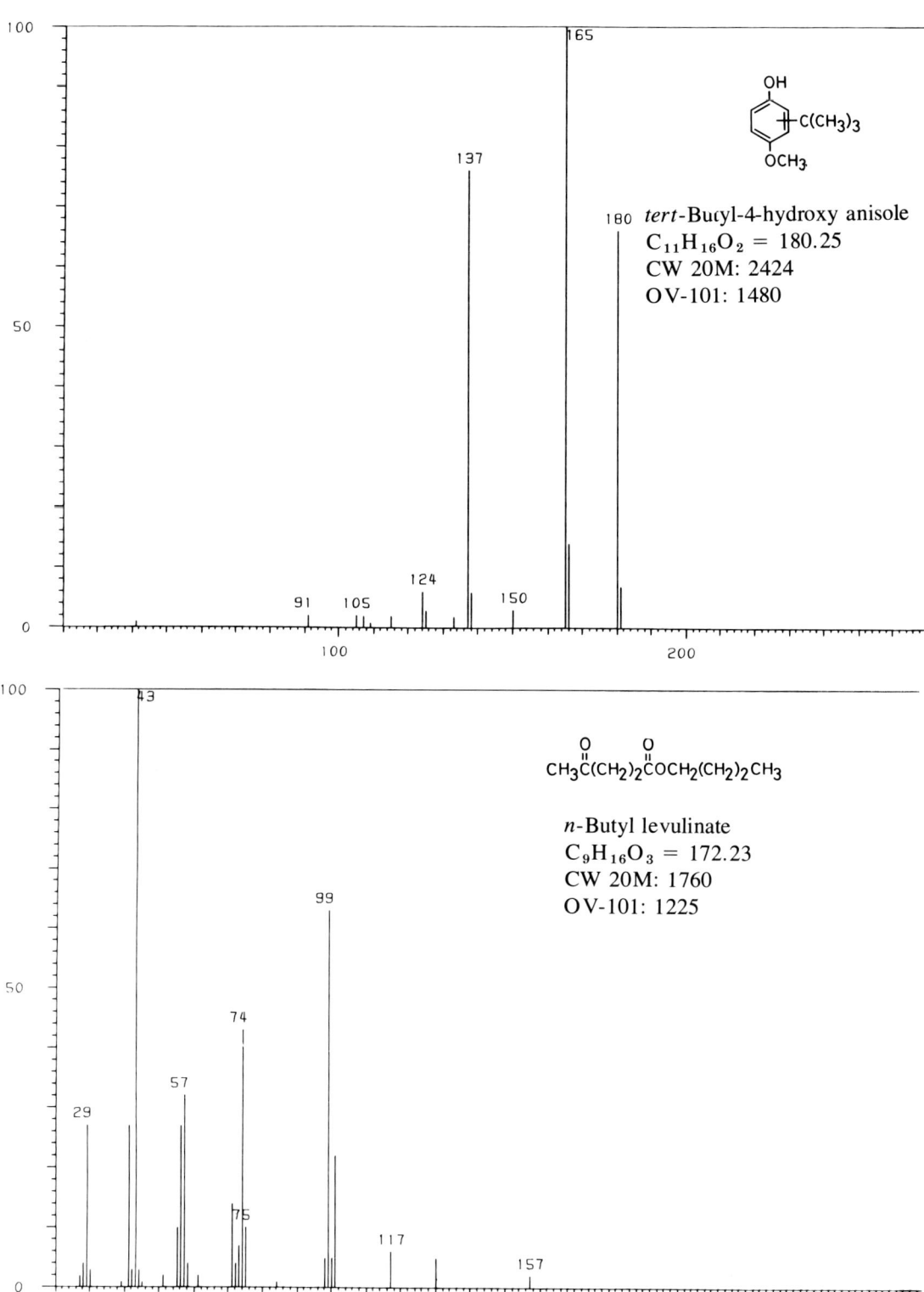

180 *tert*-Butyl-4-hydroxy anisole
$C_{11}H_{16}O_2 = 180.25$
CW 20M: 2424
OV-101: 1480

n-Butyl levulinate
$C_9H_{16}O_3 = 172.23$
CW 20M: 1760
OV-101: 1225

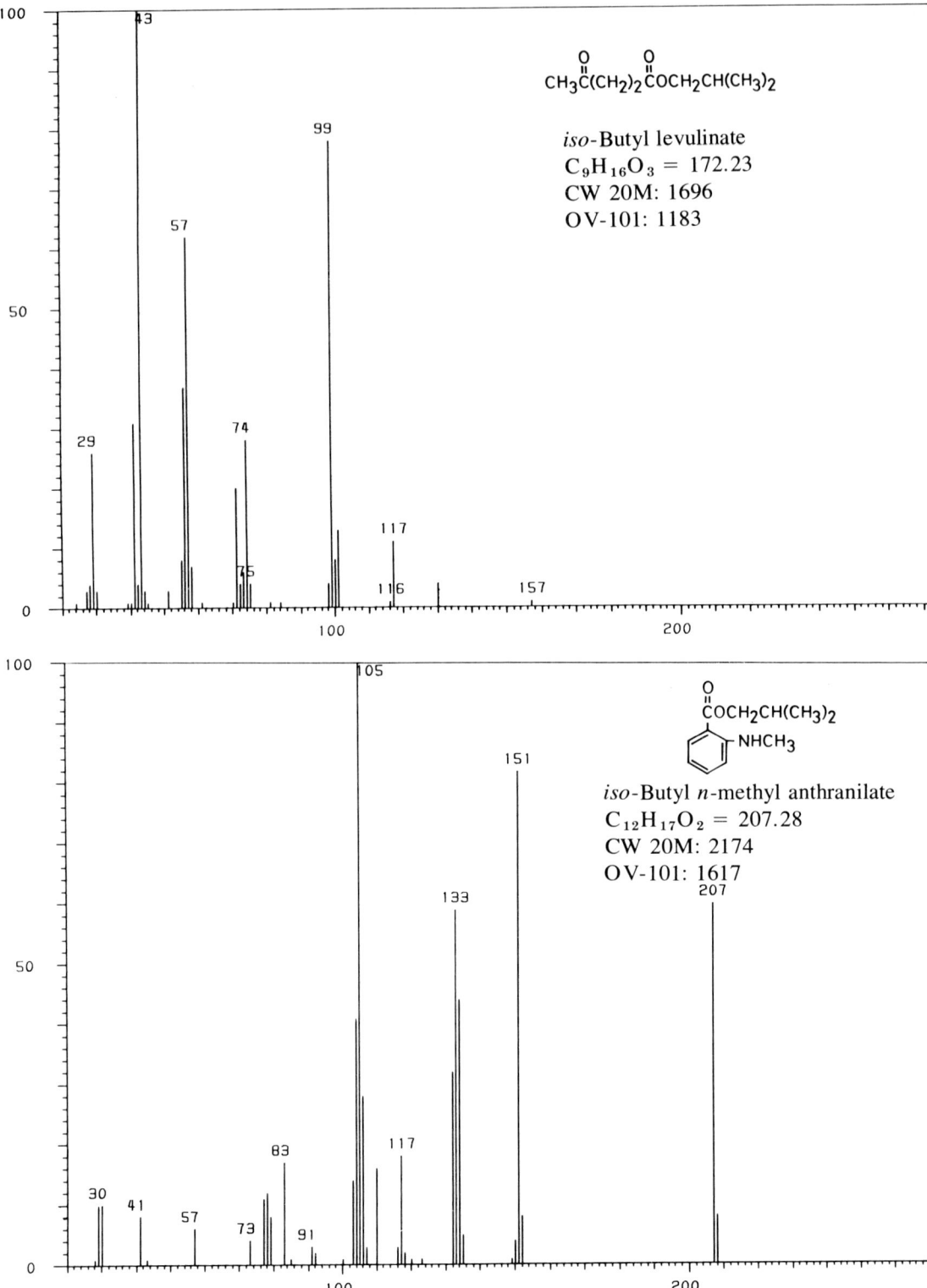

APPENDIX IV   MASS SPECTRA OF INDIVIDUAL COMPOUNDS

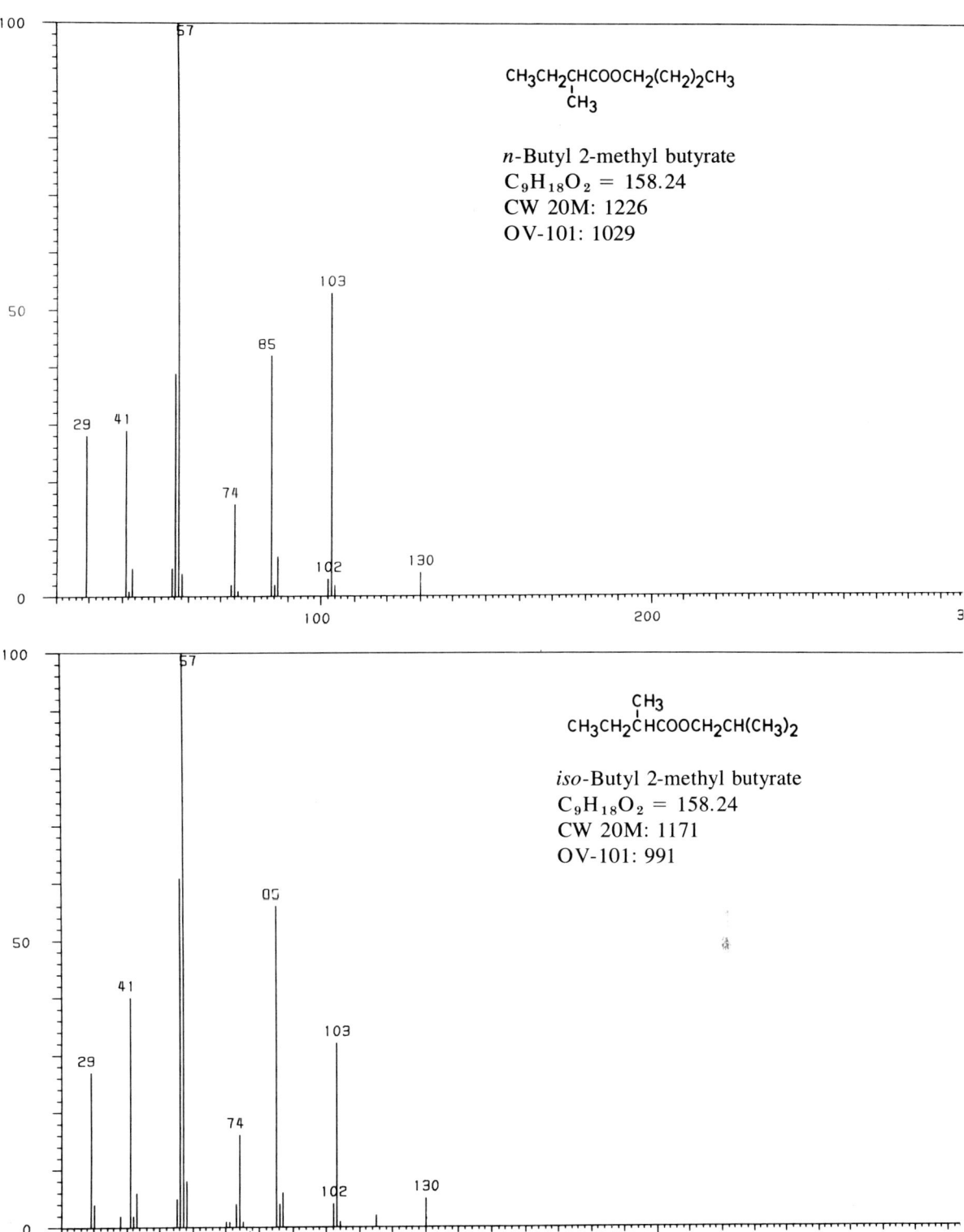

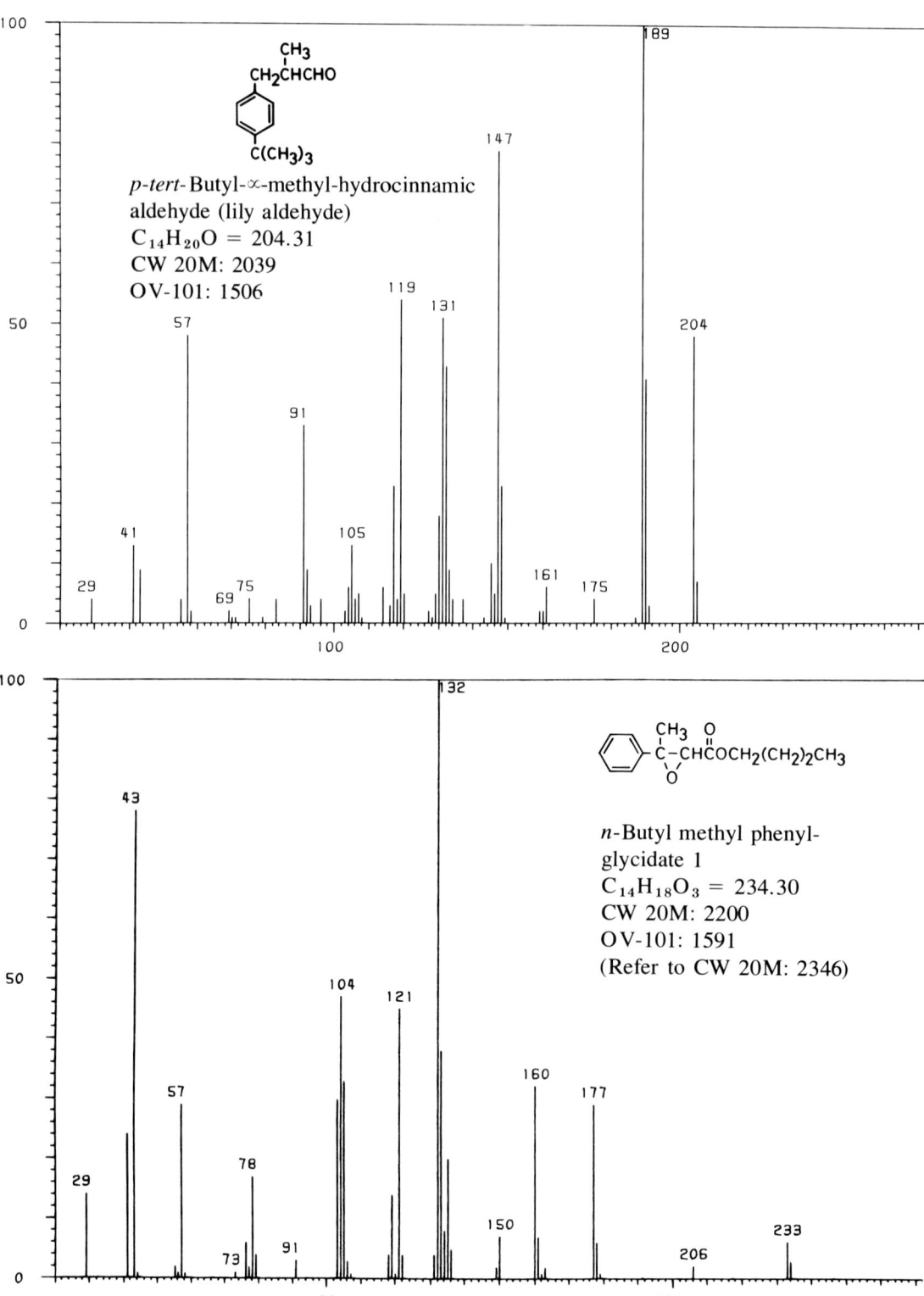

APPENDIX IV   MASS SPECTRA OF INDIVIDUAL COMPOUNDS

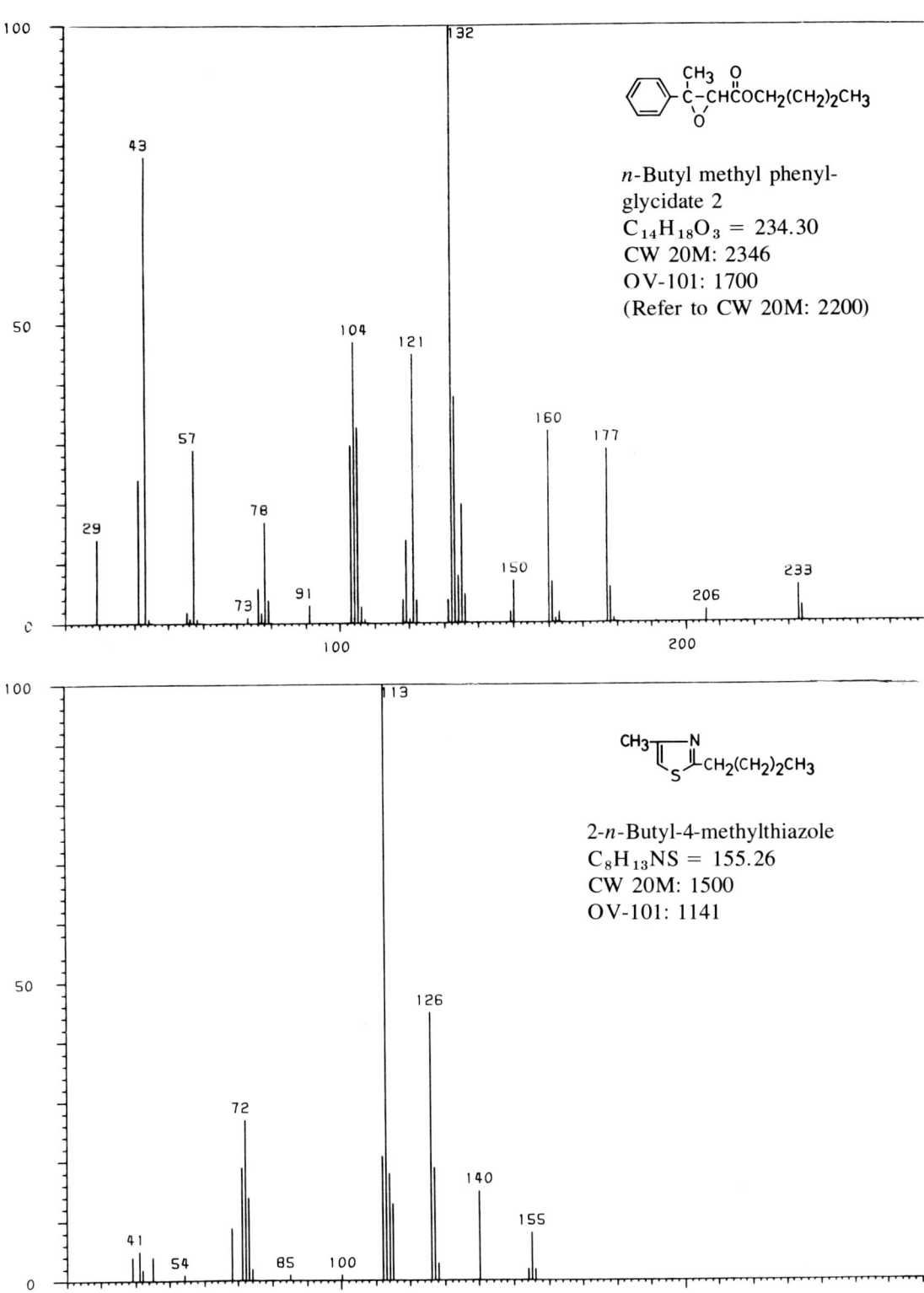

n-Butyl methyl phenyl-glycidate 2
$C_{14}H_{18}O_3 = 234.30$
CW 20M: 2346
OV-101: 1700
(Refer to CW 20M: 2200)

2-n-Butyl-4-methylthiazole
$C_8H_{13}NS = 155.26$
CW 20M: 1500
OV-101: 1141

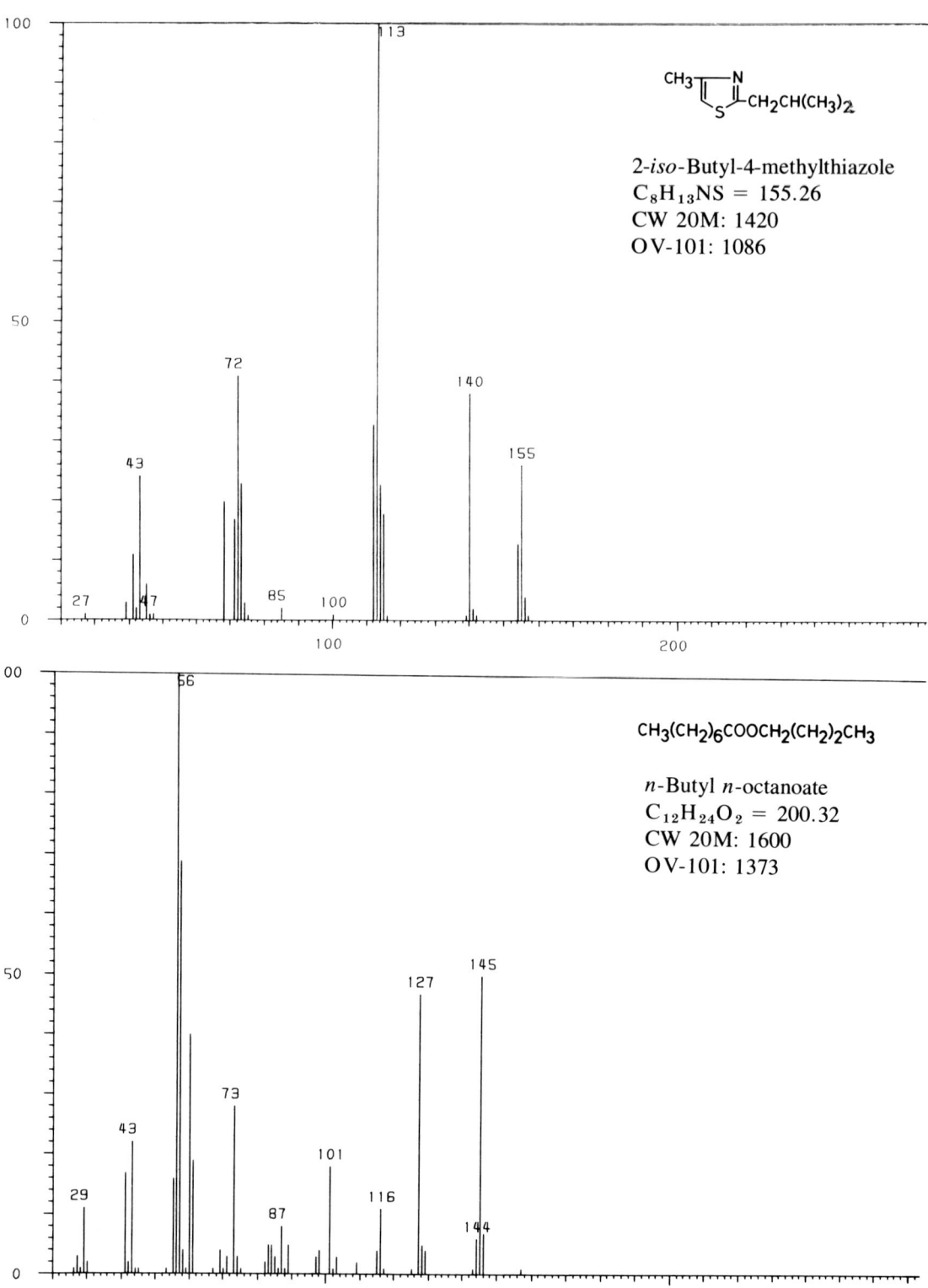

APPENDIX IV   MASS SPECTRA OF INDIVIDUAL COMPOUNDS

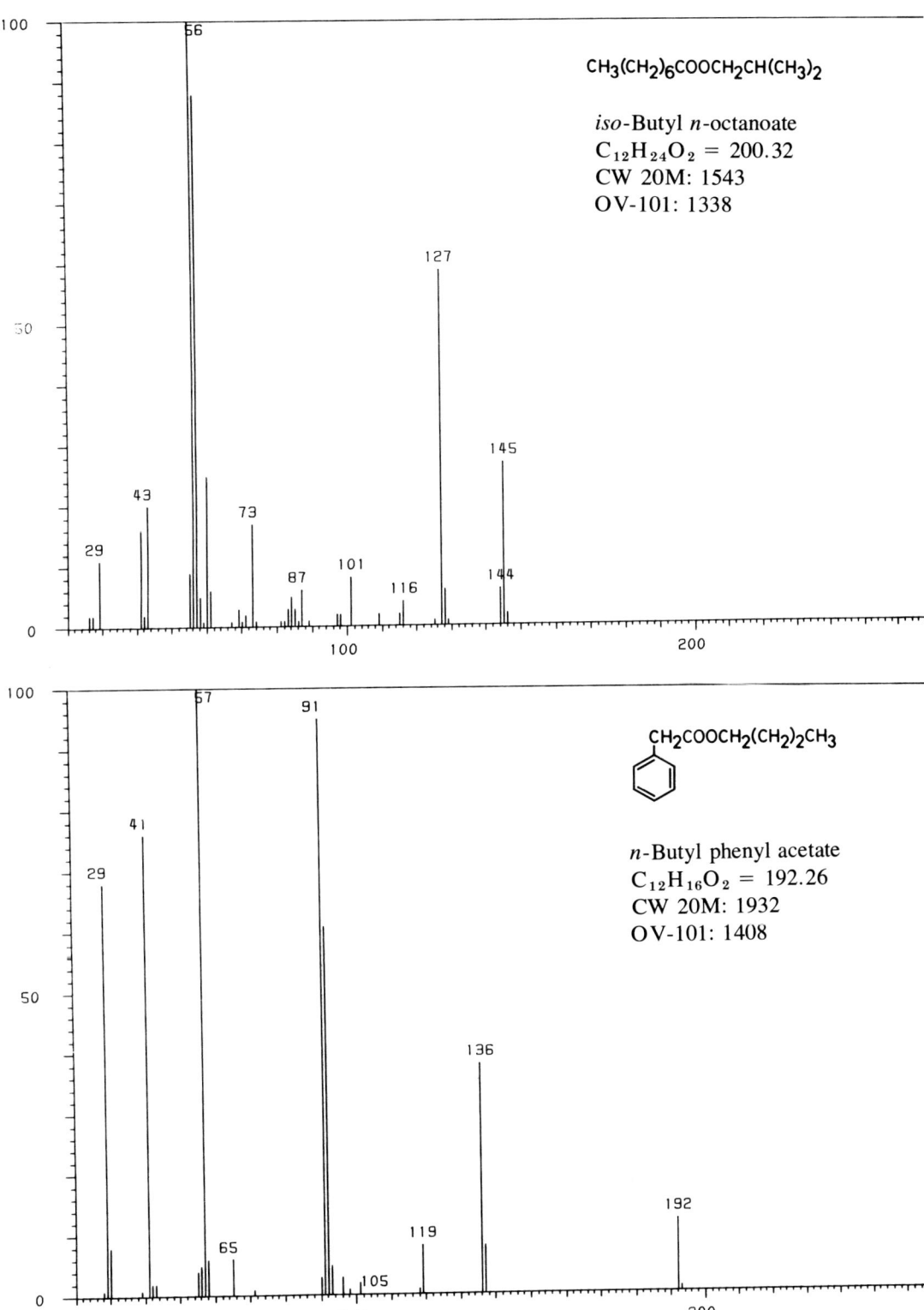

$CH_3(CH_2)_6COOCH_2CH(CH_3)_2$

iso-Butyl *n*-octanoate
$C_{12}H_{24}O_2 = 200.32$
CW 20M: 1543
OV-101: 1338

n-Butyl phenyl acetate
$C_{12}H_{16}O_2 = 192.26$
CW 20M: 1932
OV-101: 1408

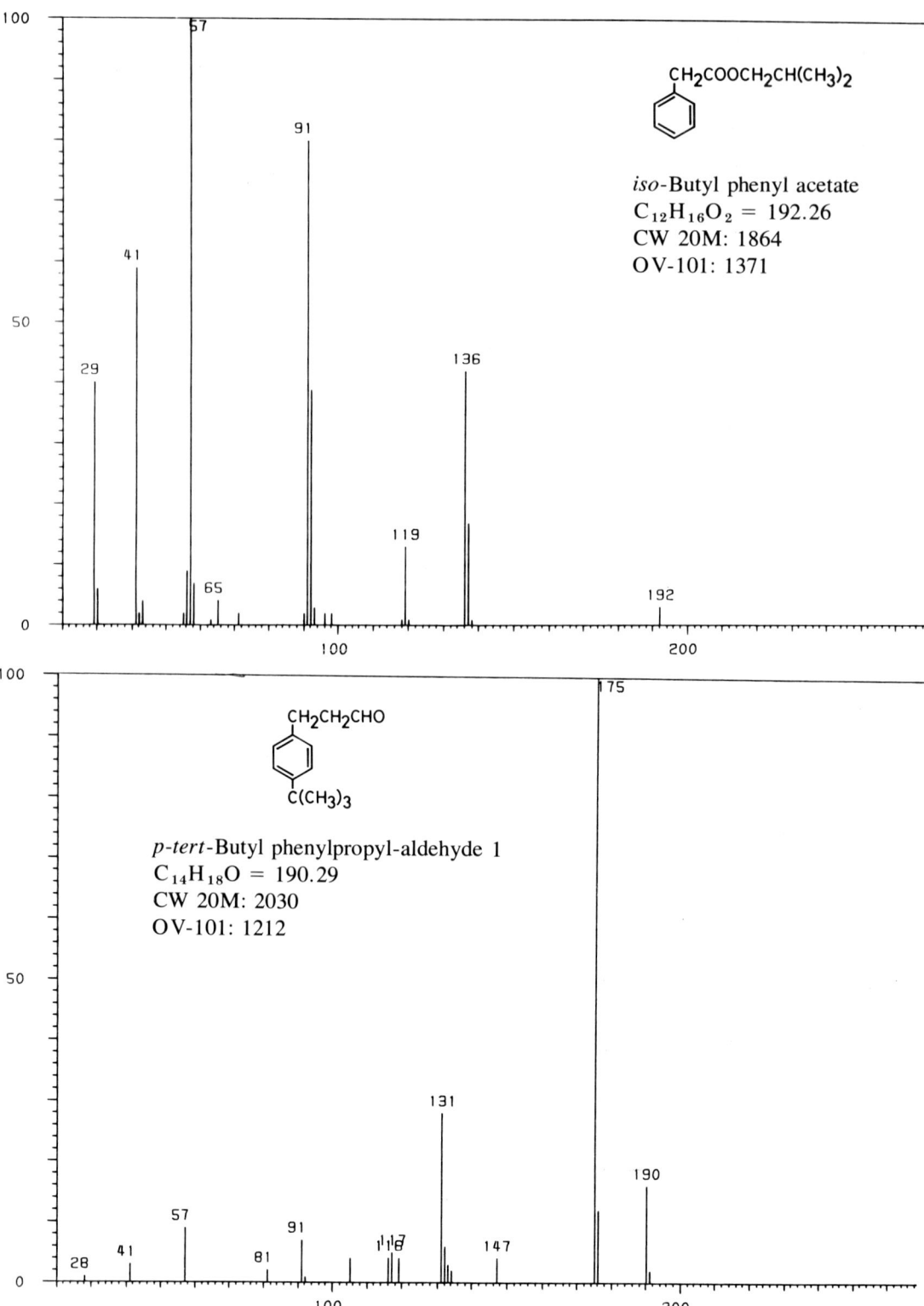

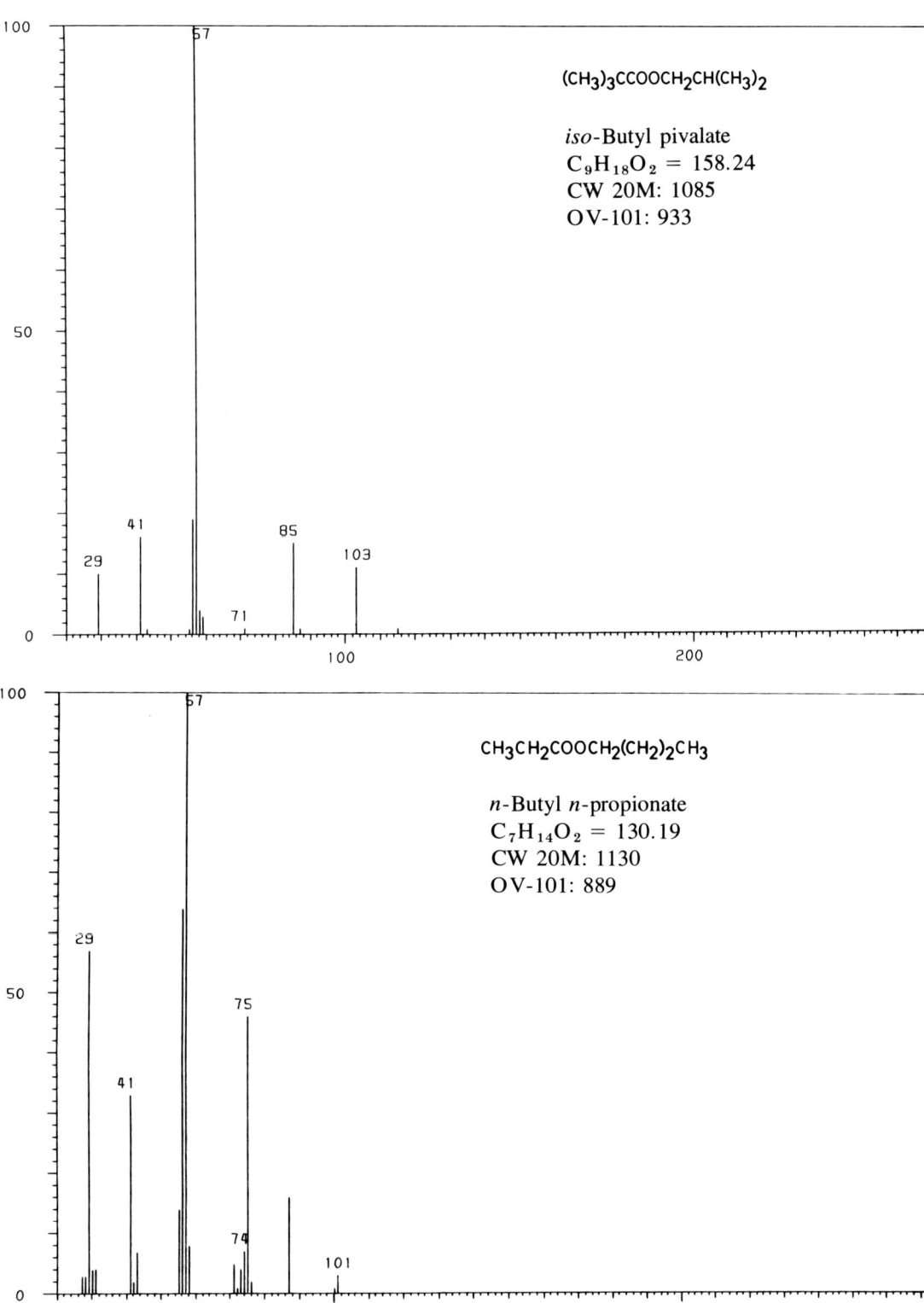

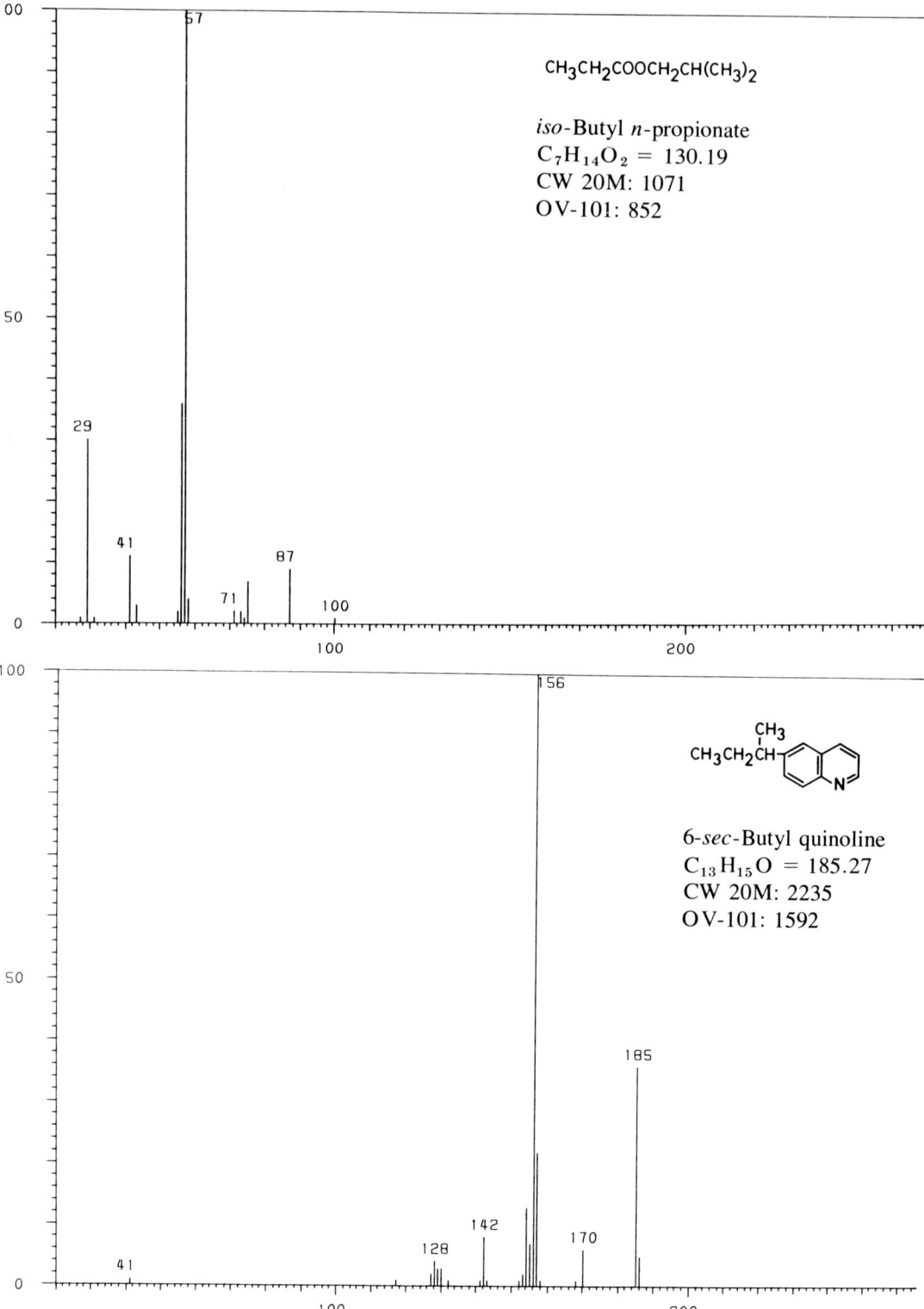

CH3CH2COOCH2CH(CH3)2

iso-Butyl *n*-propionate
$C_7H_{14}O_2 = 130.19$
CW 20M: 1071
OV-101: 852

6-*sec*-Butyl quinoline
$C_{13}H_{15}O = 185.27$
CW 20M: 2235
OV-101: 1592

APPENDIX IV   MASS SPECTRA OF INDIVIDUAL COMPOUNDS

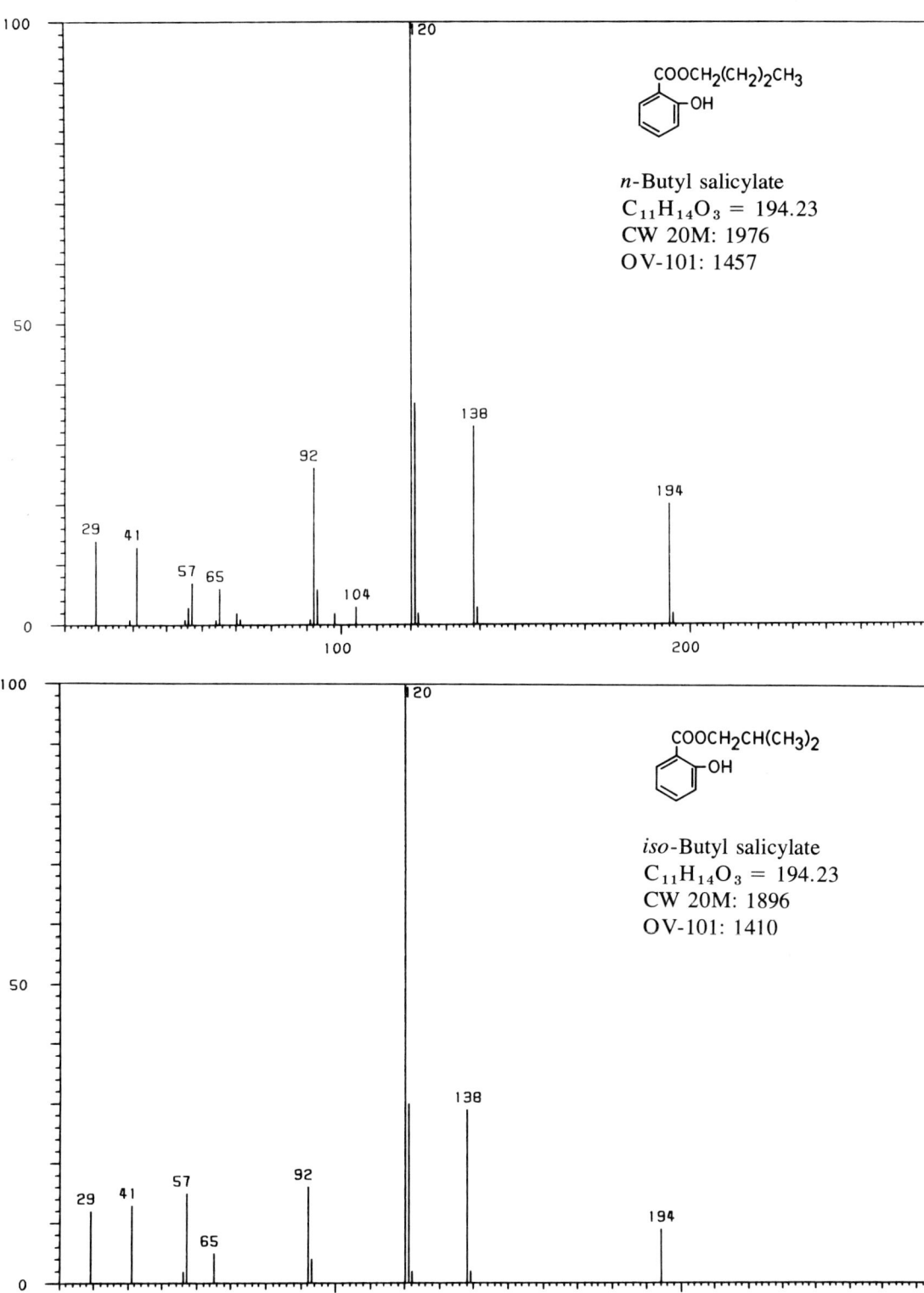

n-Butyl salicylate
$C_{11}H_{14}O_3 = 194.23$
CW 20M: 1976
OV-101: 1457

iso-Butyl salicylate
$C_{11}H_{14}O_3 = 194.23$
CW 20M: 1896
OV-101: 1410

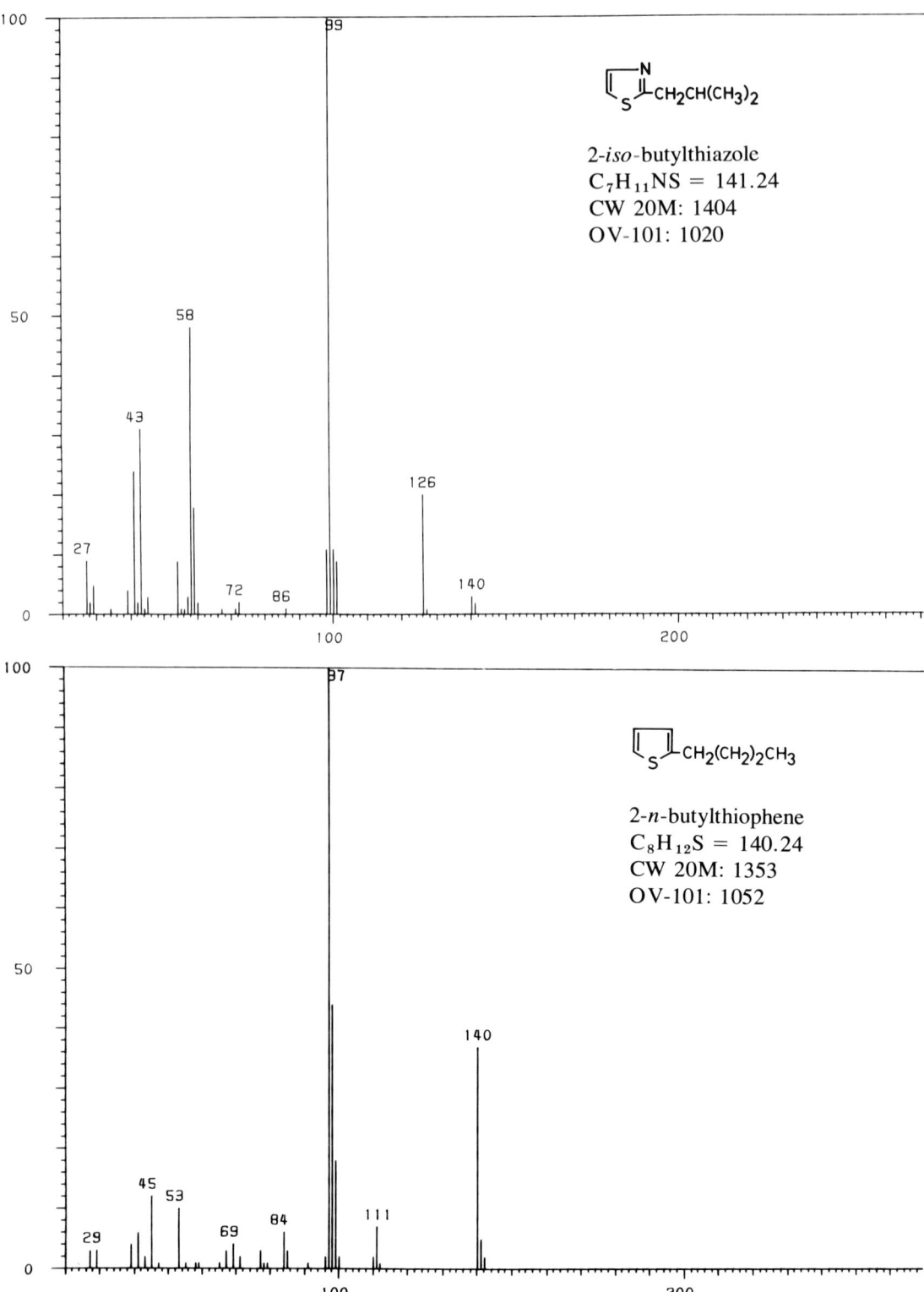

APPENDIX IV   MASS SPECTRA OF INDIVIDUAL COMPOUNDS

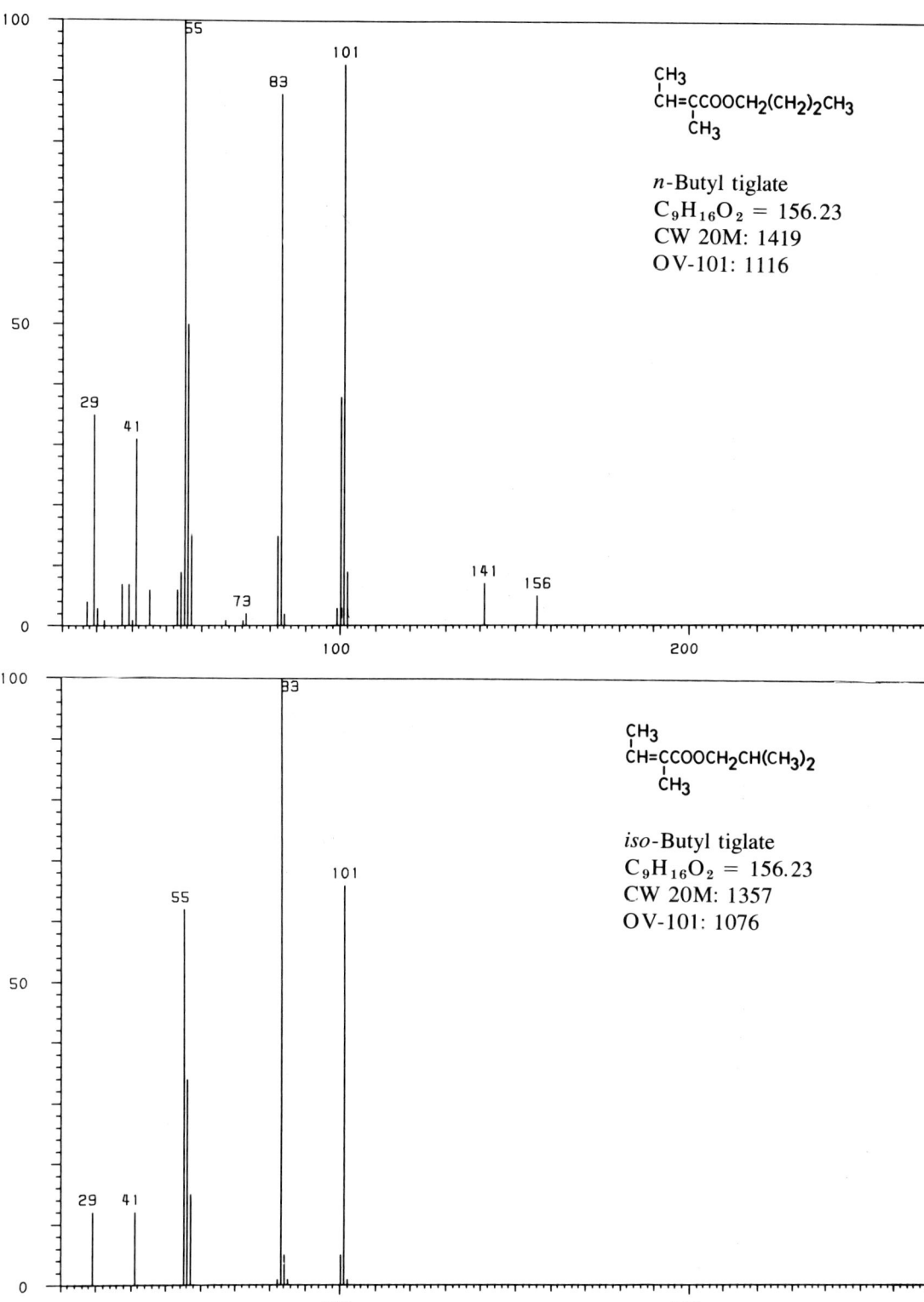

n-Butyl tiglate
$C_9H_{16}O_2 = 156.23$
CW 20M: 1419
OV-101: 1116

iso-Butyl tiglate
$C_9H_{16}O_2 = 156.23$
CW 20M: 1357
OV-101: 1076

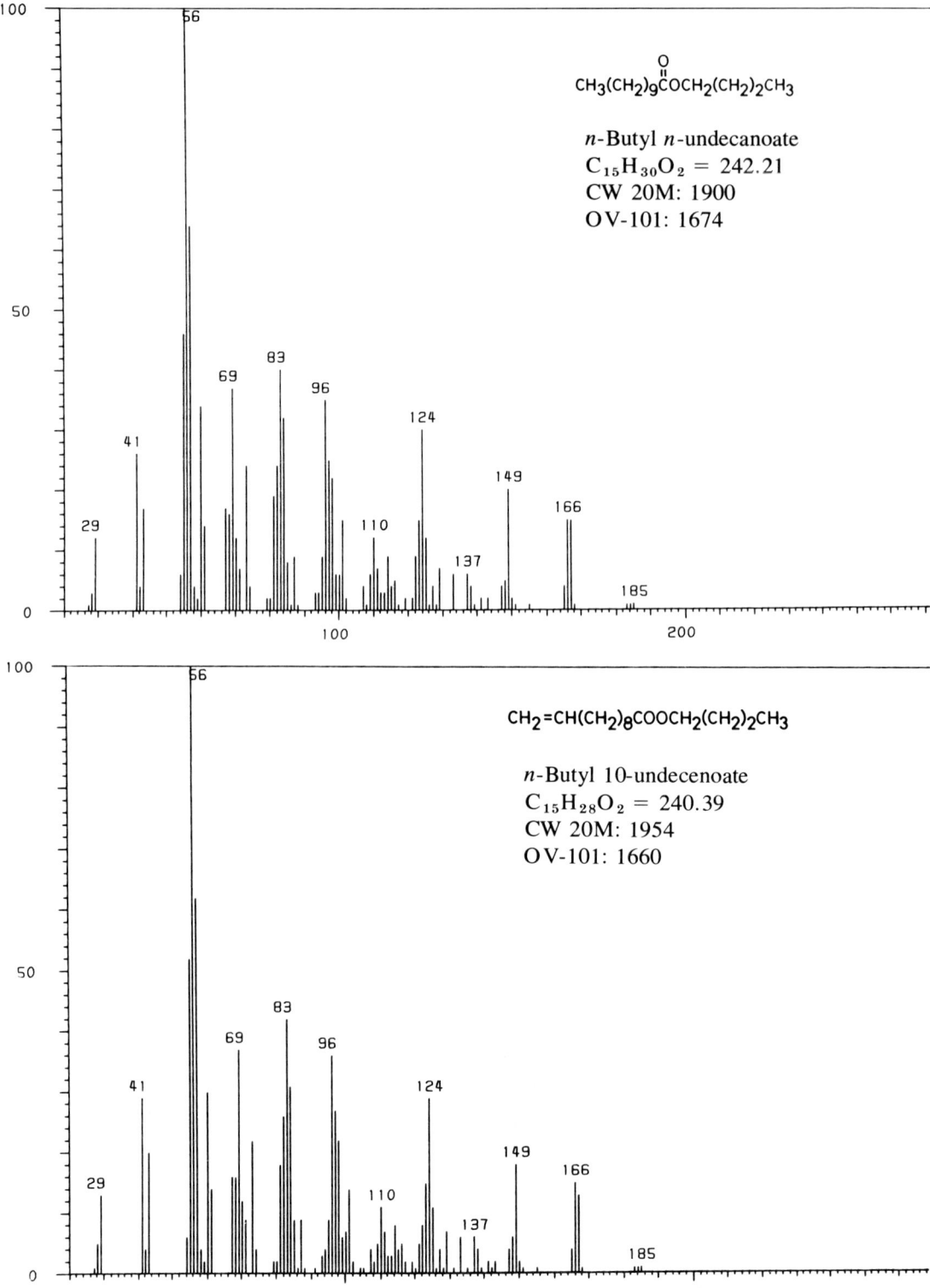

APPENDIX IV   MASS SPECTRA OF INDIVIDUAL COMPOUNDS

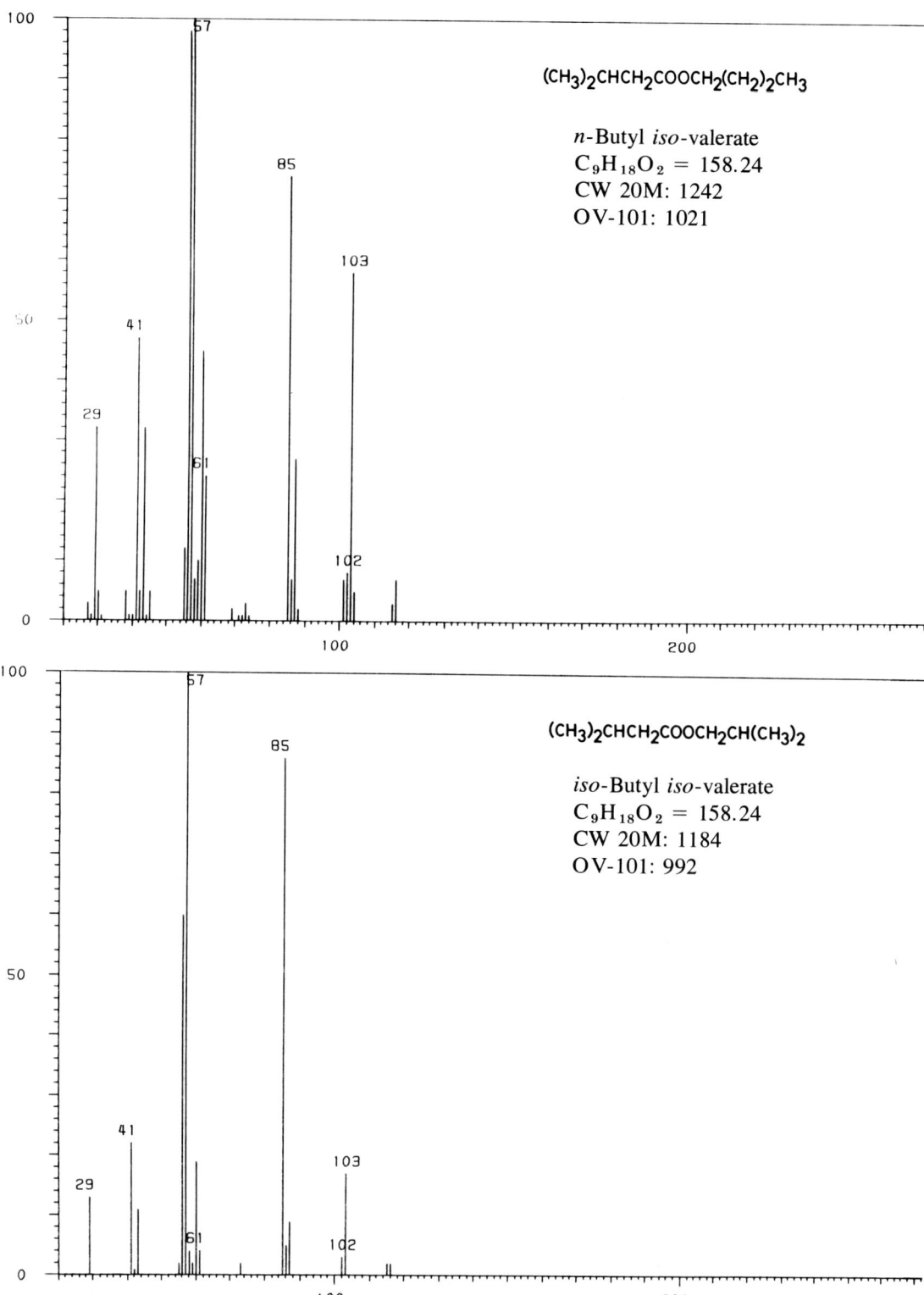

(CH$_3$)$_2$CHCH$_2$COOCH$_2$(CH$_2$)$_2$CH$_3$

n-Butyl iso-valerate
C$_9$H$_{18}$O$_2$ = 158.24
CW 20M: 1242
OV-101: 1021

(CH$_3$)$_2$CHCH$_2$COOCH$_2$CH(CH$_3$)$_2$

iso-Butyl iso-valerate
C$_9$H$_{18}$O$_2$ = 158.24
CW 20M: 1184
OV-101: 992

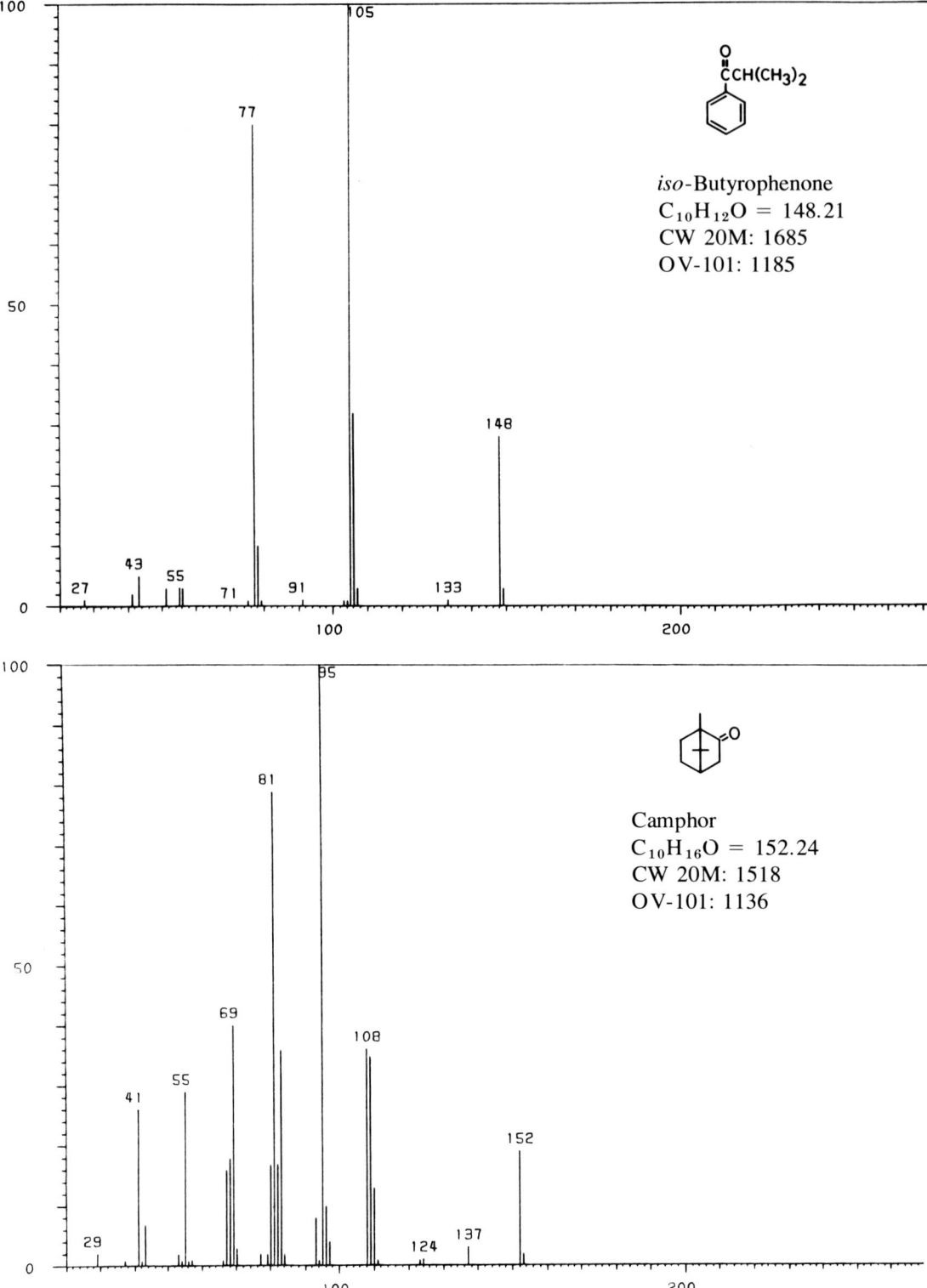

APPENDIX IV   MASS SPECTRA OF INDIVIDUAL COMPOUNDS

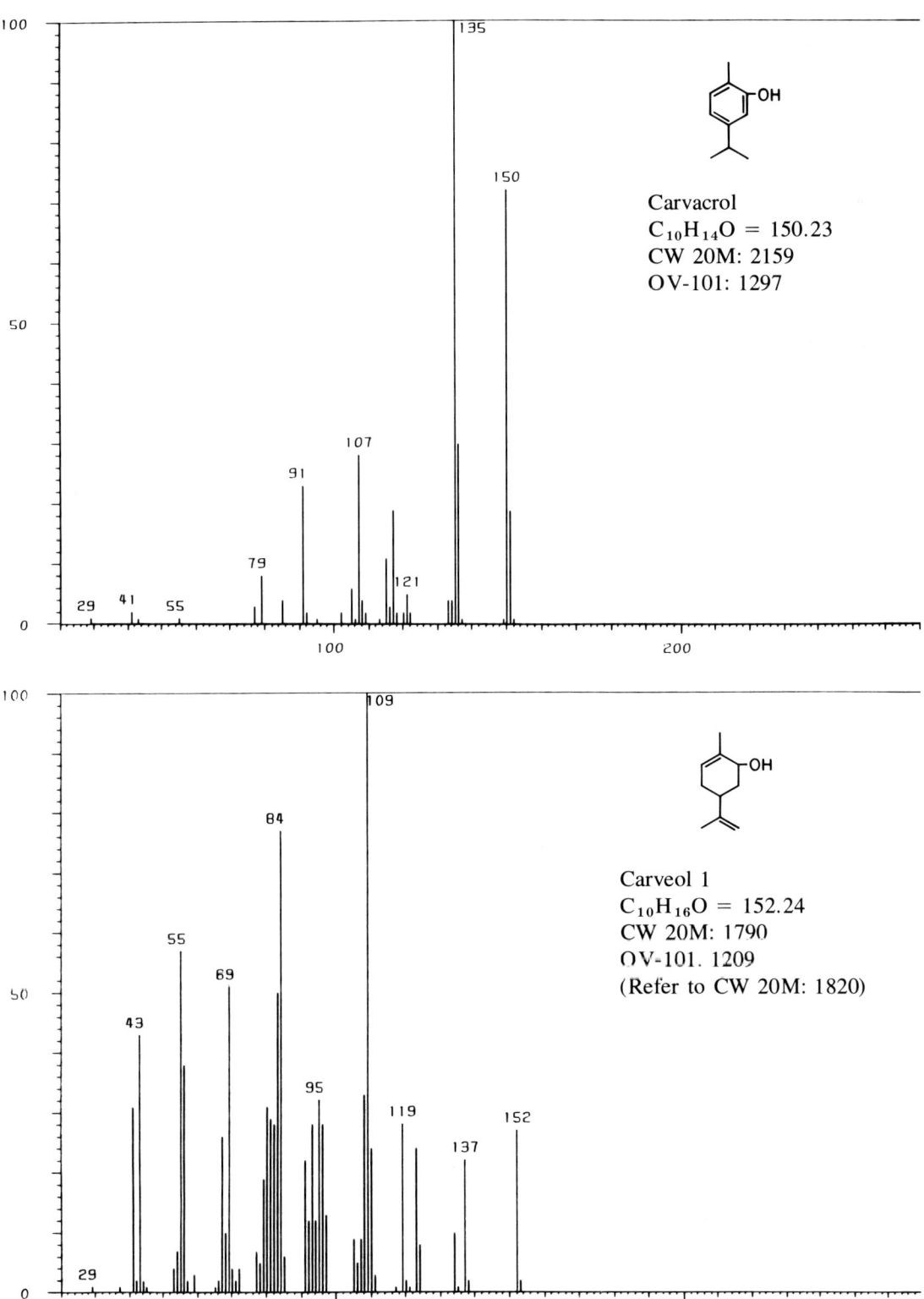

Carvacrol
$C_{10}H_{14}O = 150.23$
CW 20M: 2159
OV-101: 1297

Carveol 1
$C_{10}H_{16}O = 152.24$
CW 20M: 1790
OV-101: 1209
(Refer to CW 20M: 1820)

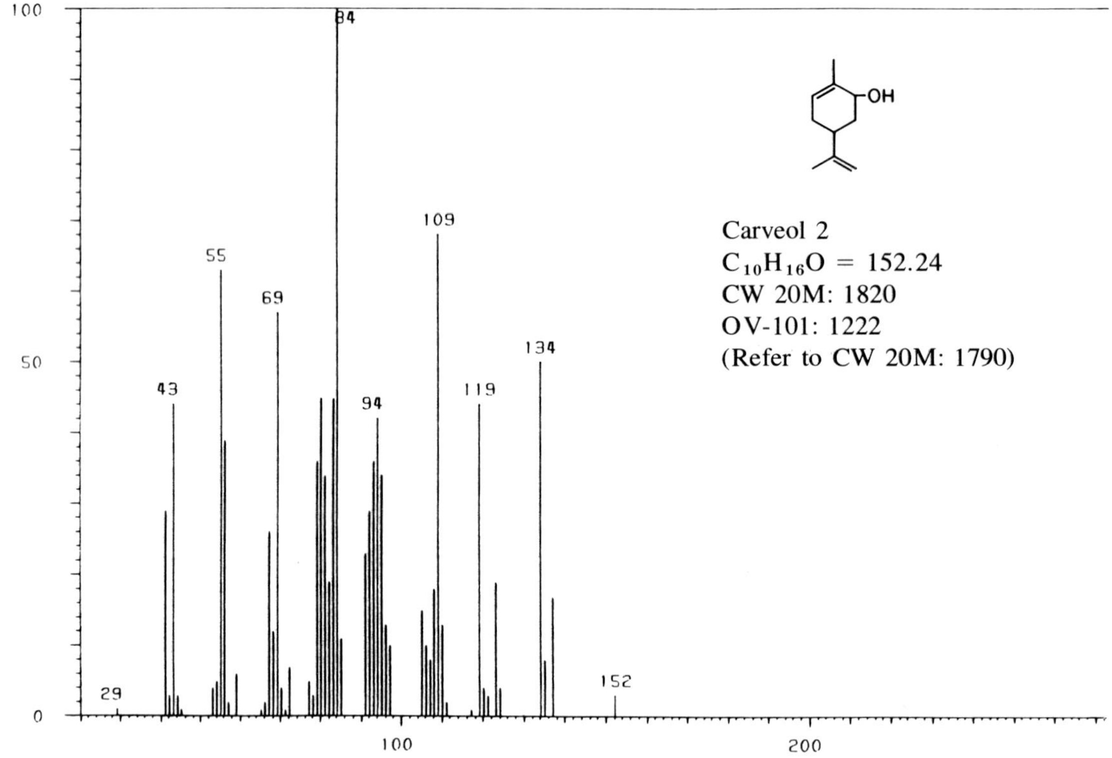

Carveol 2
$C_{10}H_{16}O = 152.24$
CW 20M: 1820
OV-101: 1222
(Refer to CW 20M: 1790)

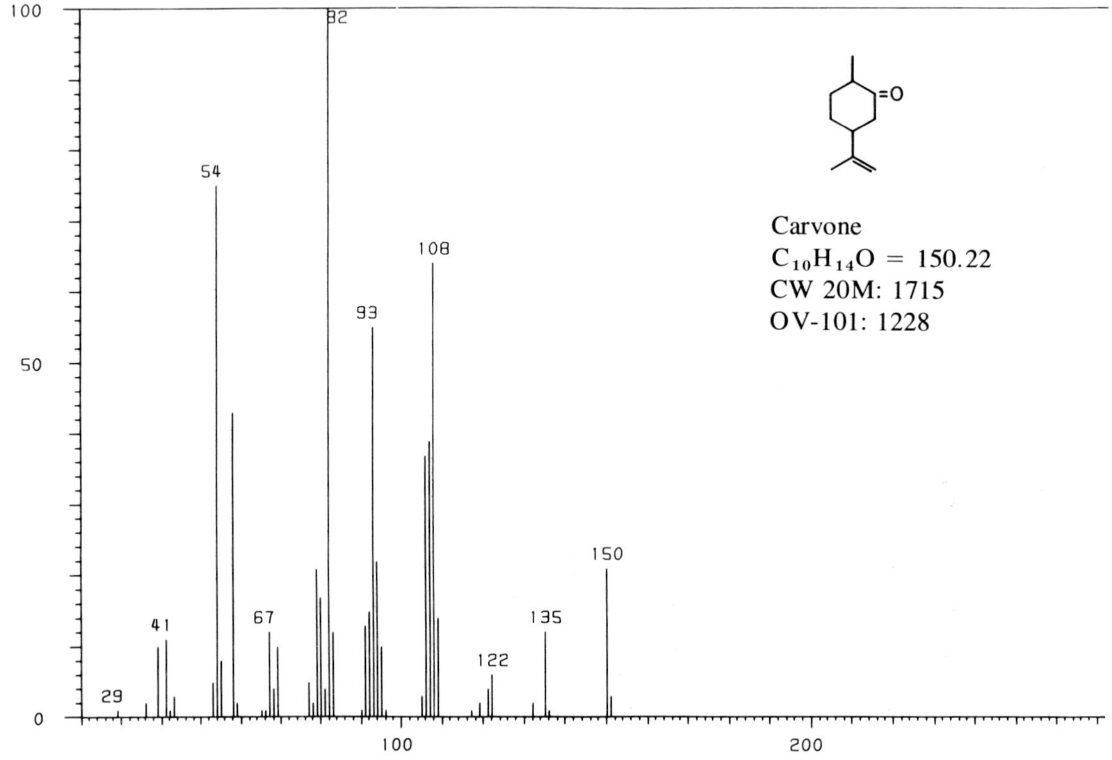

Carvone
$C_{10}H_{14}O = 150.22$
CW 20M: 1715
OV-101: 1228

APPENDIX IV   MASS SPECTRA OF INDIVIDUAL COMPOUNDS   217

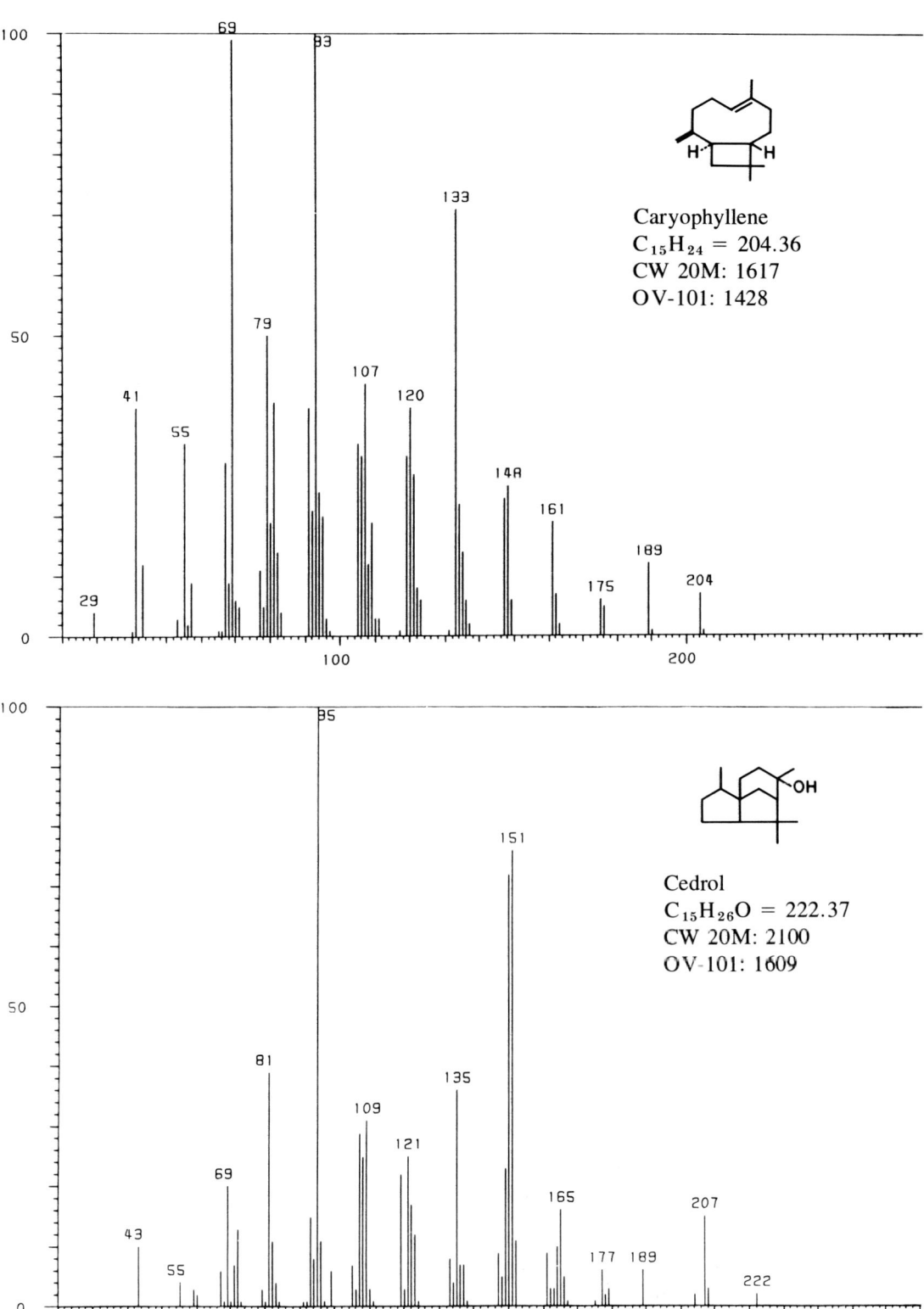

Caryophyllene
$C_{15}H_{24} = 204.36$
CW 20M: 1617
OV-101: 1428

Cedrol
$C_{15}H_{26}O = 222.37$
CW 20M: 2100
OV-101: 1609

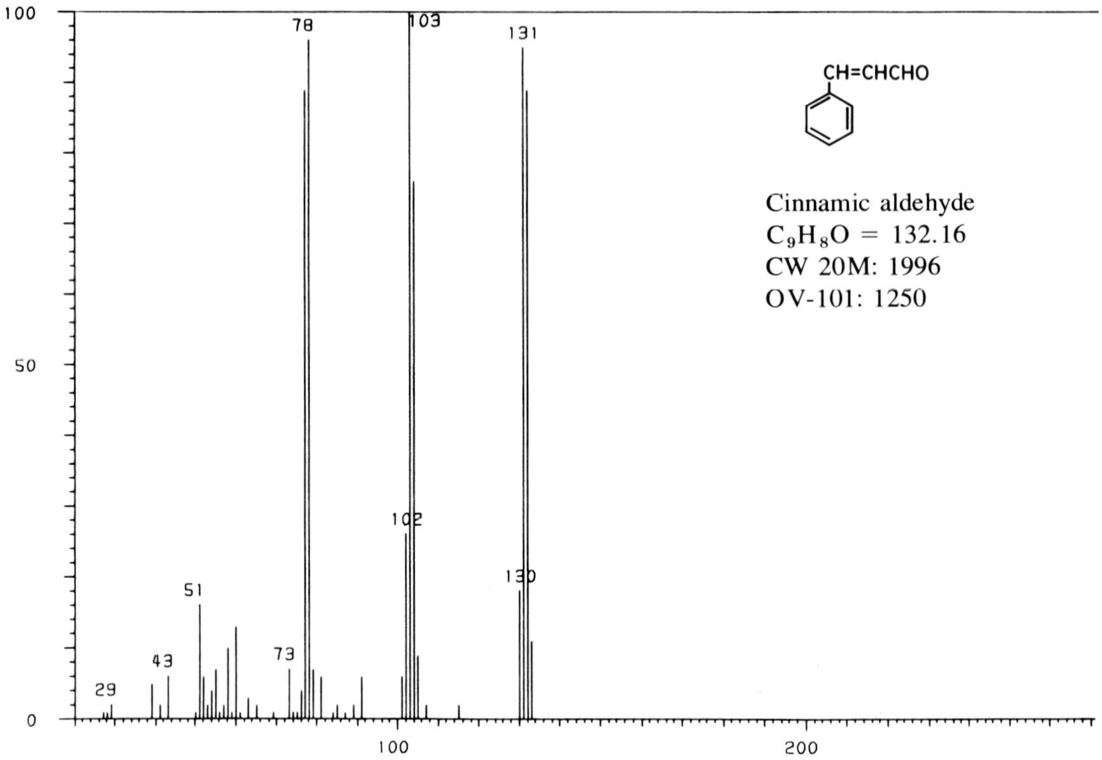

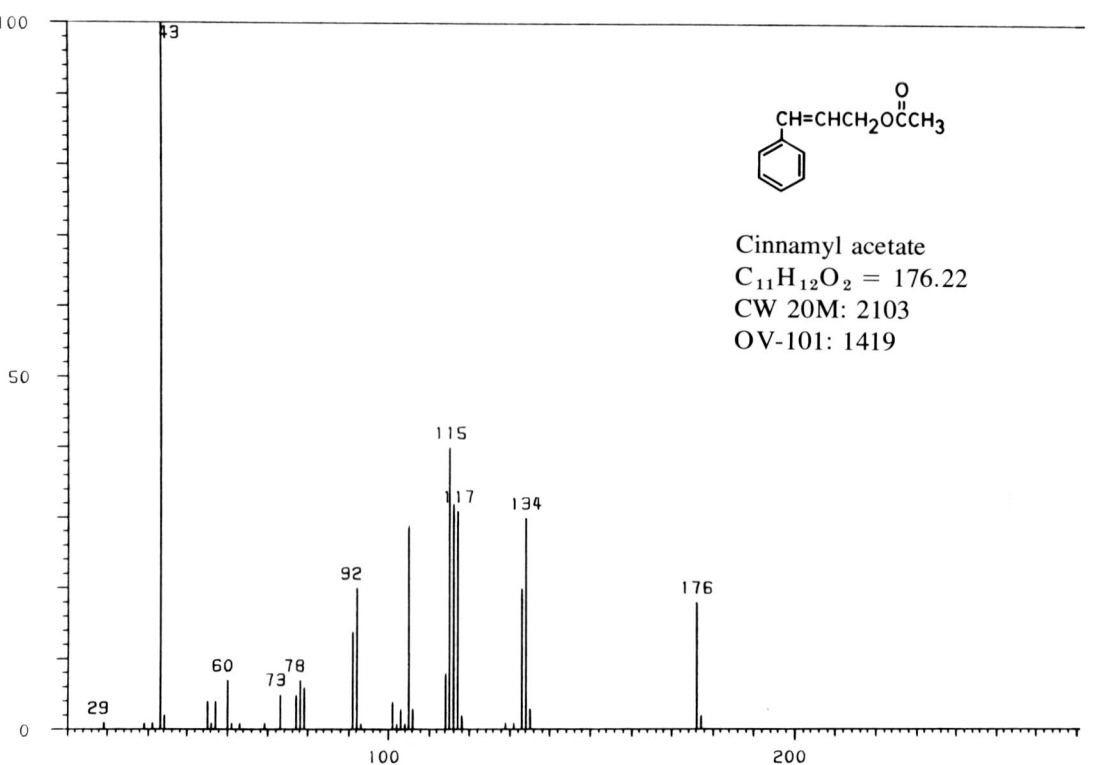

Cinnamic aldehyde
$C_9H_8O = 132.16$
CW 20M: 1996
OV-101: 1250

Cinnamyl acetate
$C_{11}H_{12}O_2 = 176.22$
CW 20M: 2103
OV-101: 1419

APPENDIX IV MASS SPECTRA OF INDIVIDUAL COMPOUNDS 219

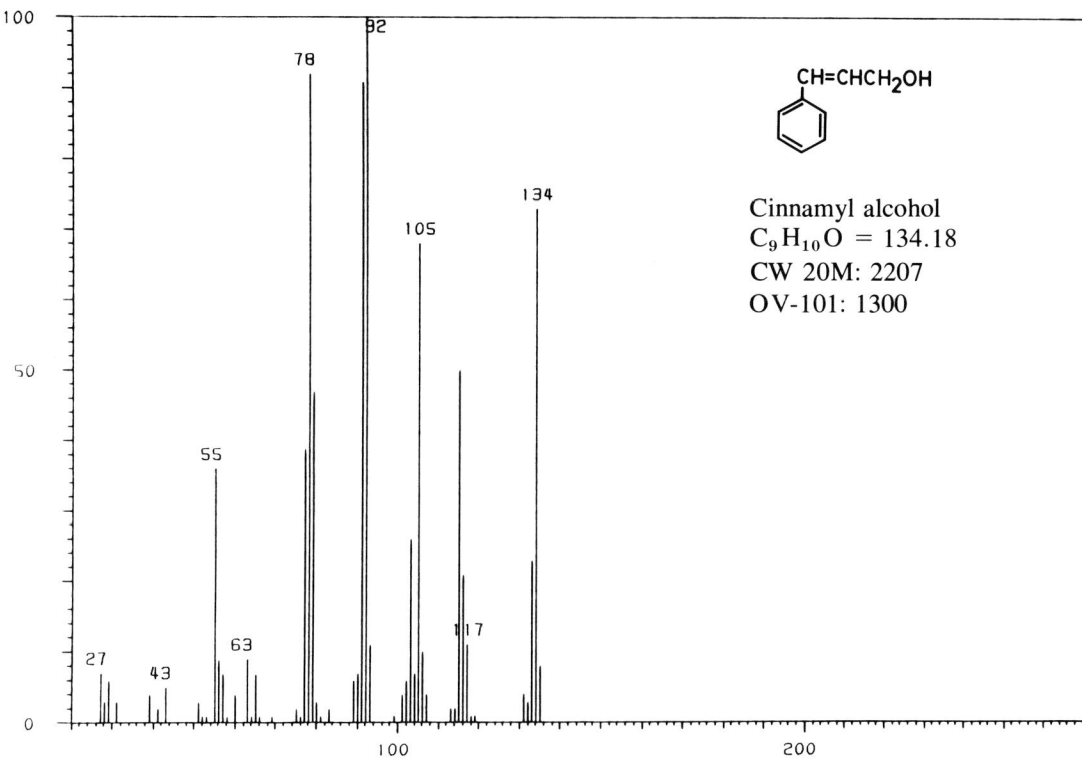

Cinnamyl alcohol
$C_9H_{10}O = 134.18$
CW 20M: 2207
OV-101: 1300

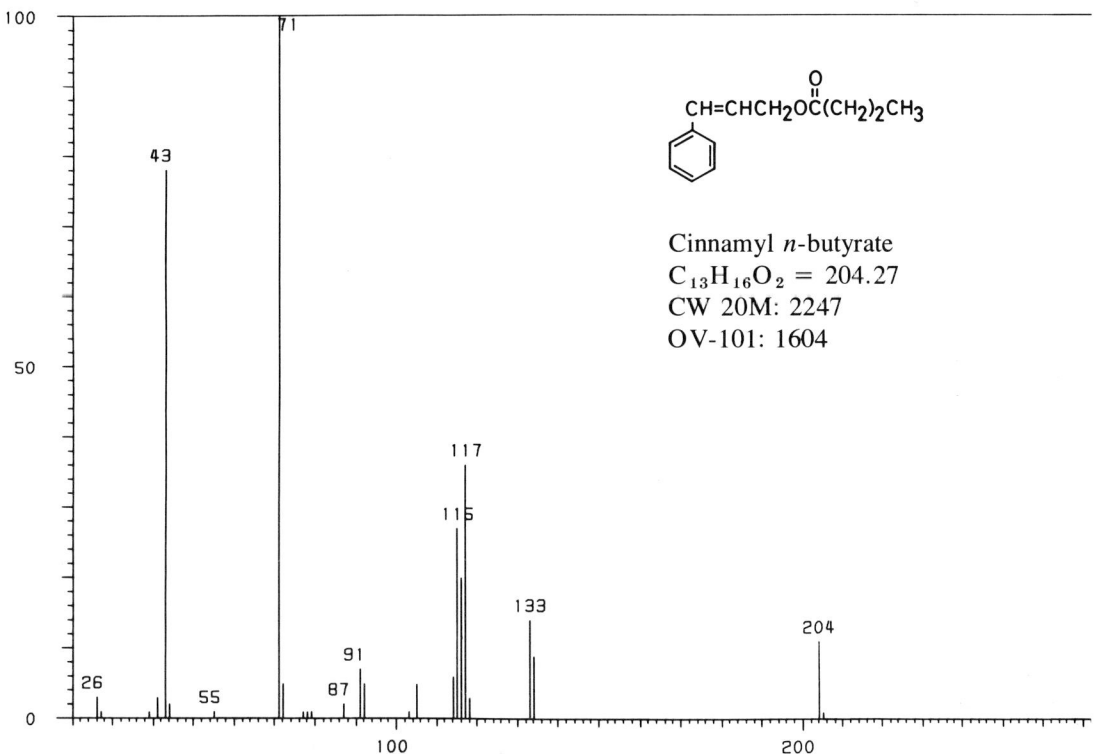

Cinnamyl n-butyrate
$C_{13}H_{16}O_2 = 204.27$
CW 20M: 2247
OV-101: 1604

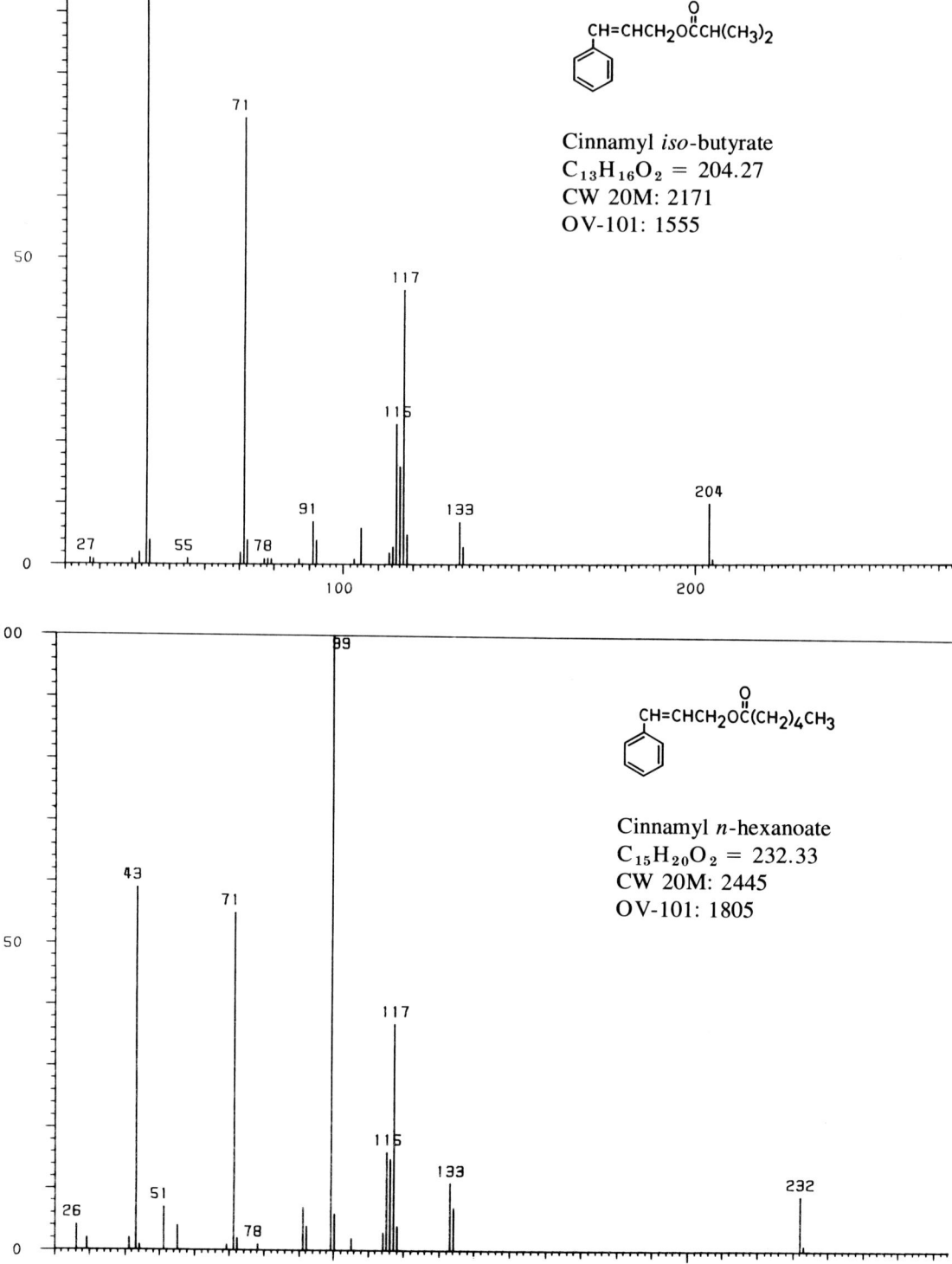

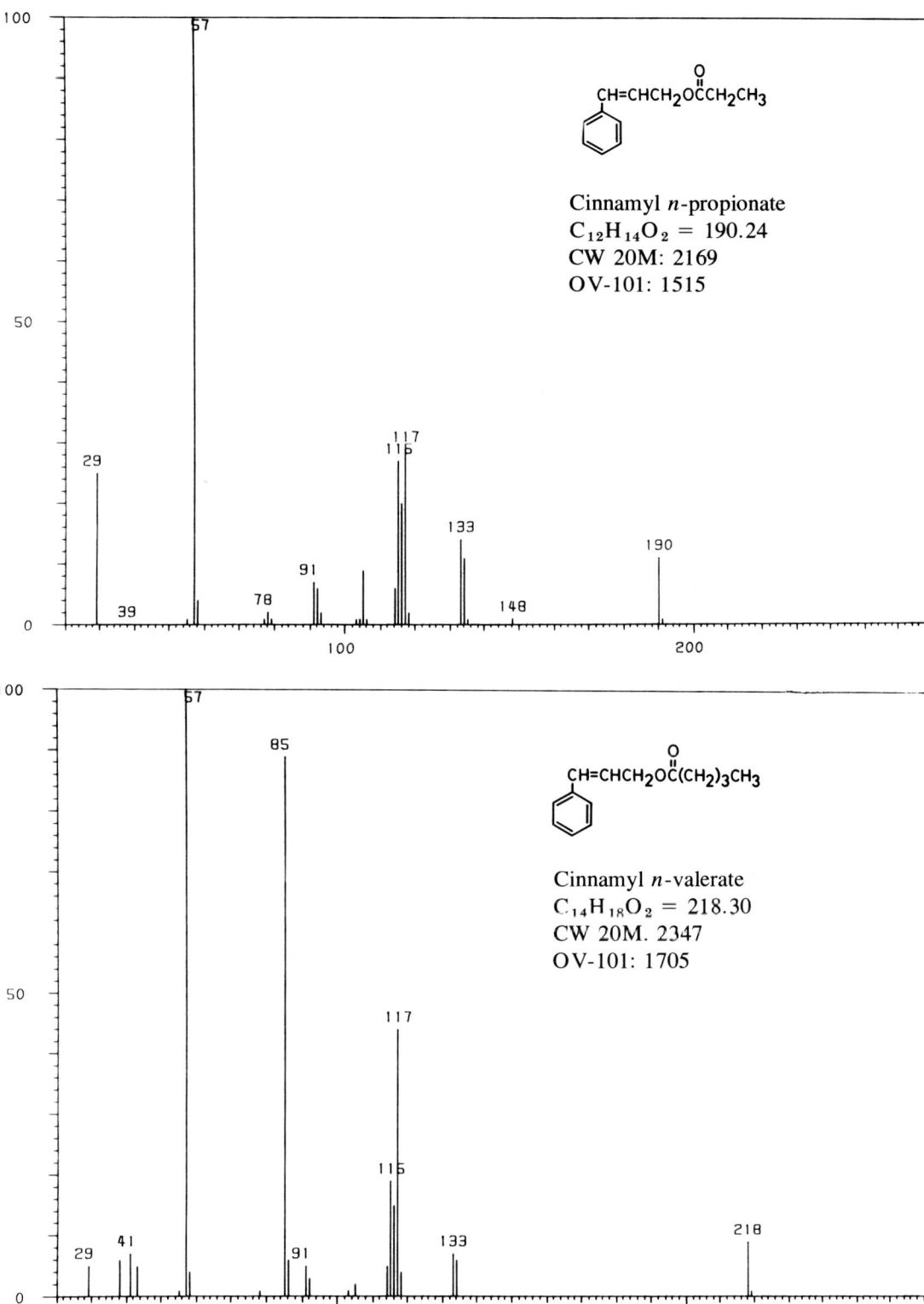

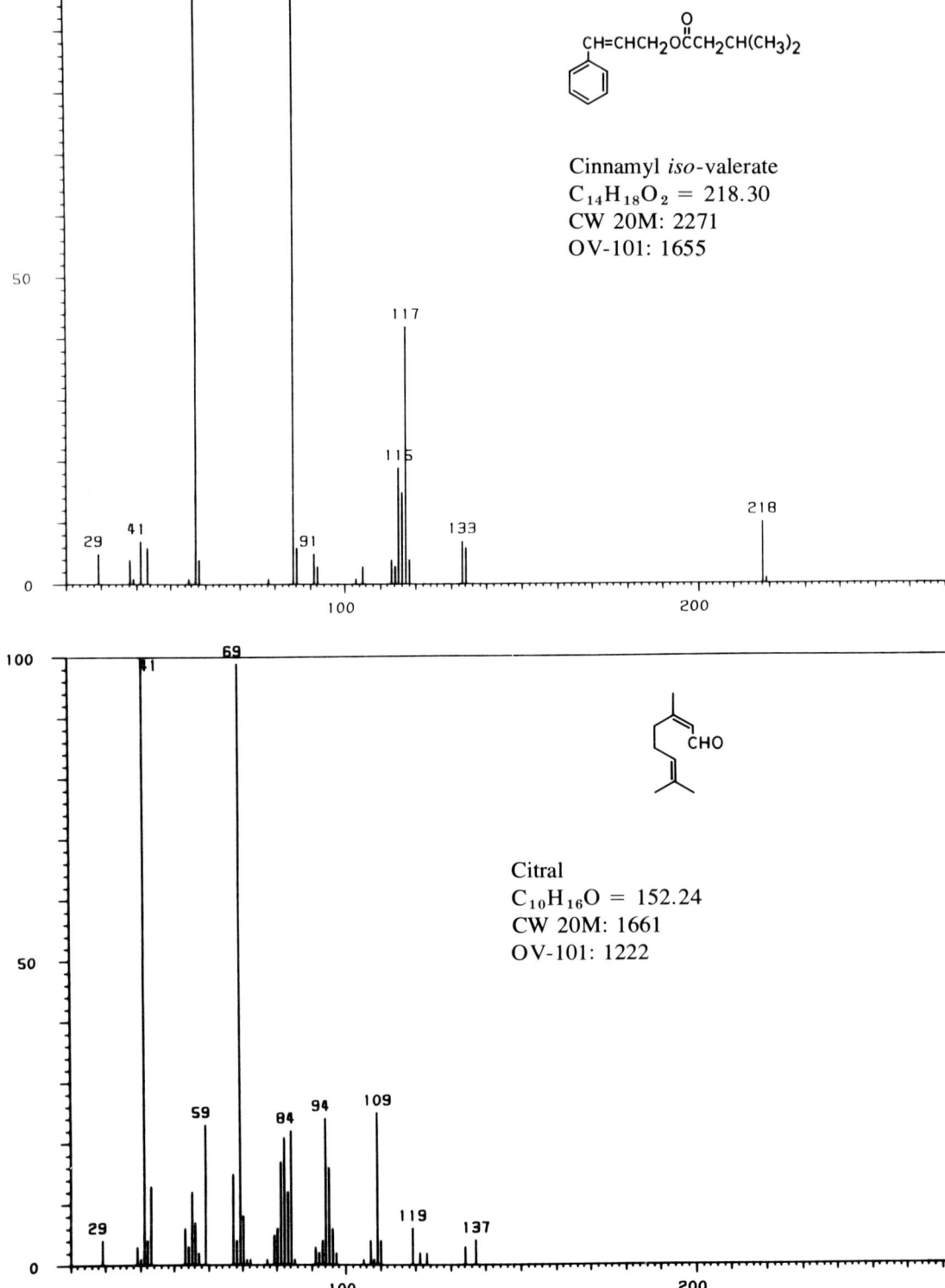

APPENDIX IV   MASS SPECTRA OF INDIVIDUAL COMPOUNDS

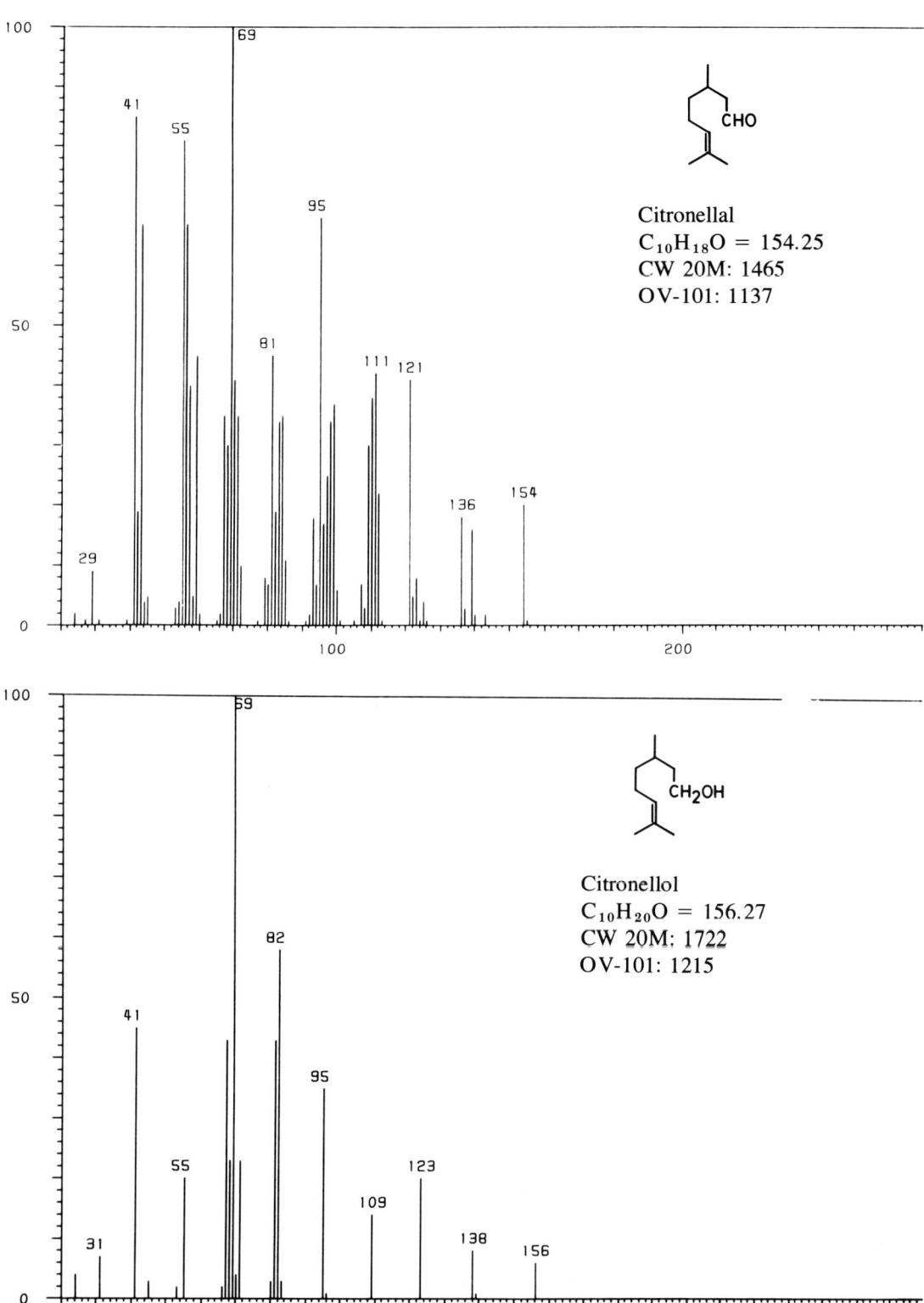

Citronellal
$C_{10}H_{18}O = 154.25$
CW 20M: 1465
OV-101: 1137

Citronellol
$C_{10}H_{20}O = 156.27$
CW 20M: 1722
OV-101: 1215

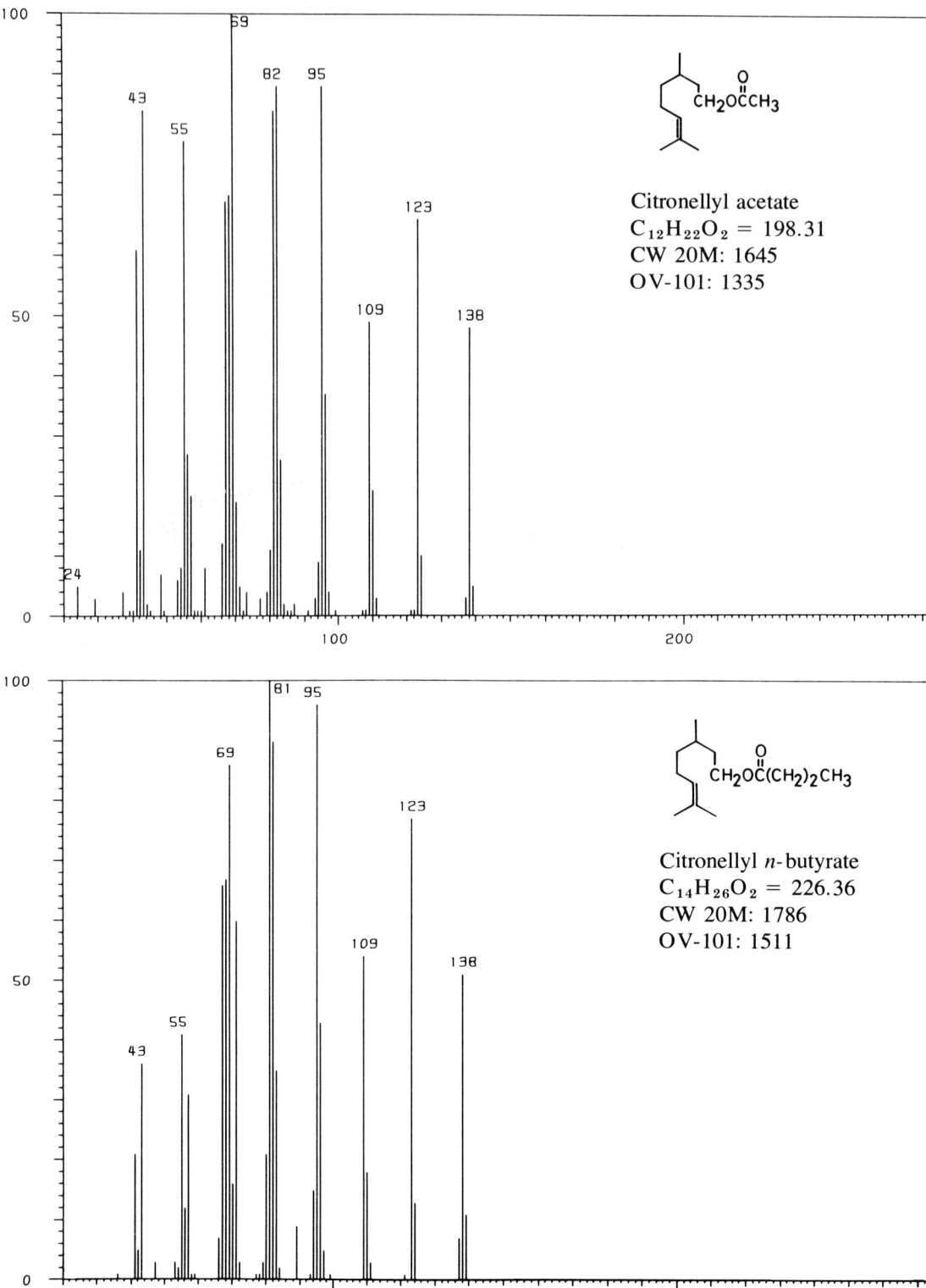

Citronellyl acetate
$C_{12}H_{22}O_2 = 198.31$
CW 20M: 1645
OV-101: 1335

Citronellyl *n*-butyrate
$C_{14}H_{26}O_2 = 226.36$
CW 20M: 1786
OV-101: 1511

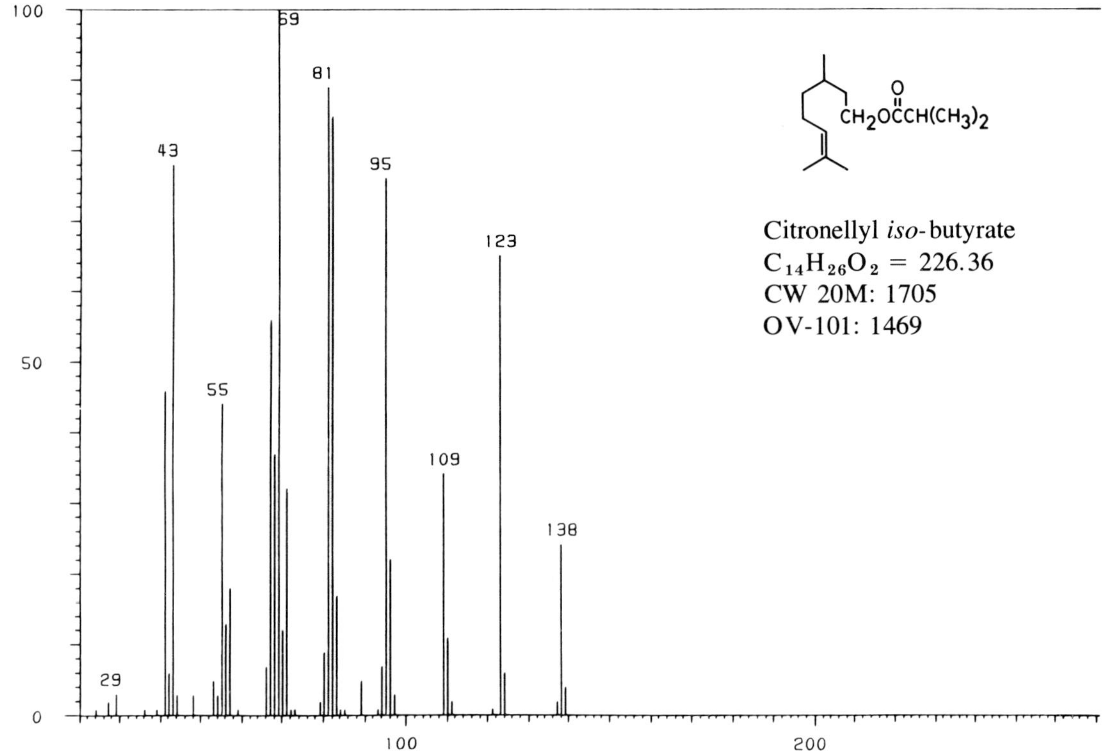

APPENDIX IV   MASS SPECTRA OF INDIVIDUAL COMPOUNDS

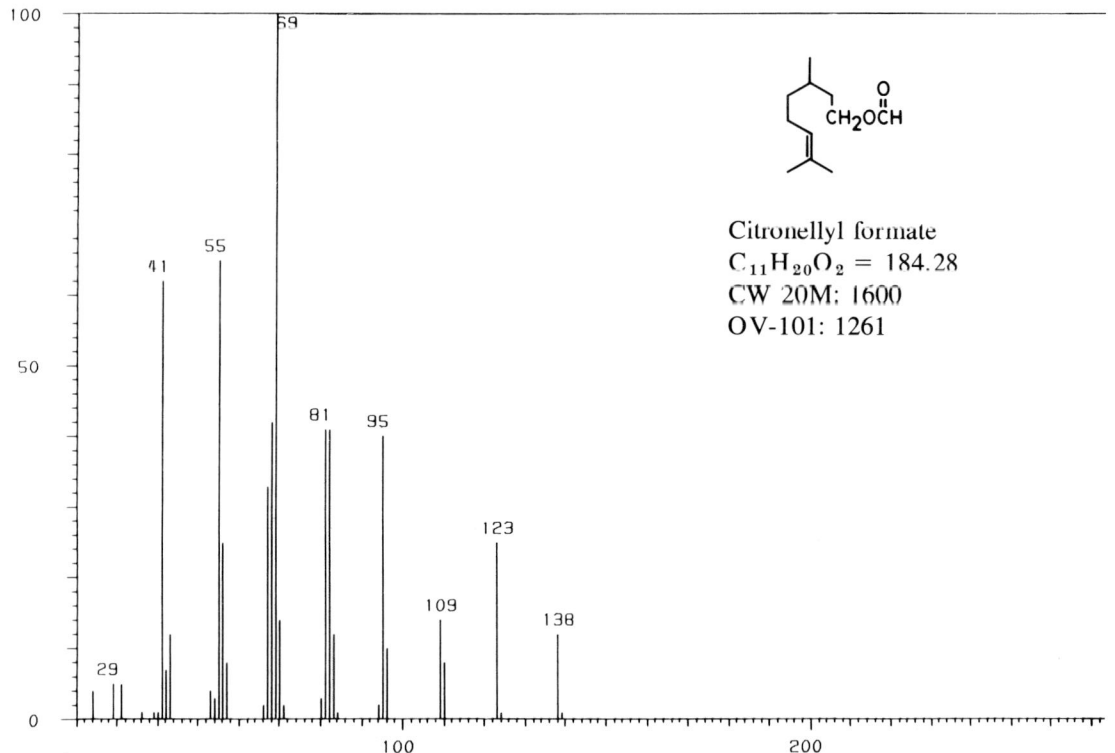

Citronellyl *iso*-butyrate
$C_{14}H_{26}O_2 = 226.36$
CW 20M: 1705
OV-101: 1469

Citronellyl formate
$C_{11}H_{20}O_2 = 184.28$
CW 20M: 1600
OV-101: 1261

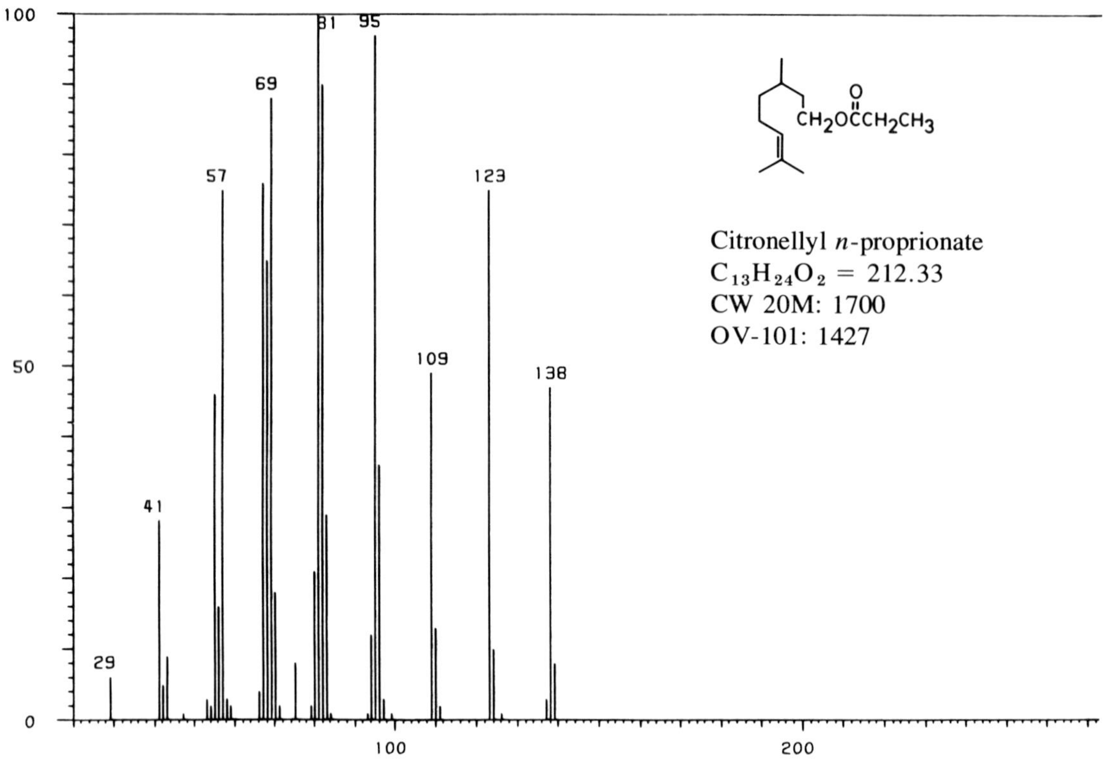

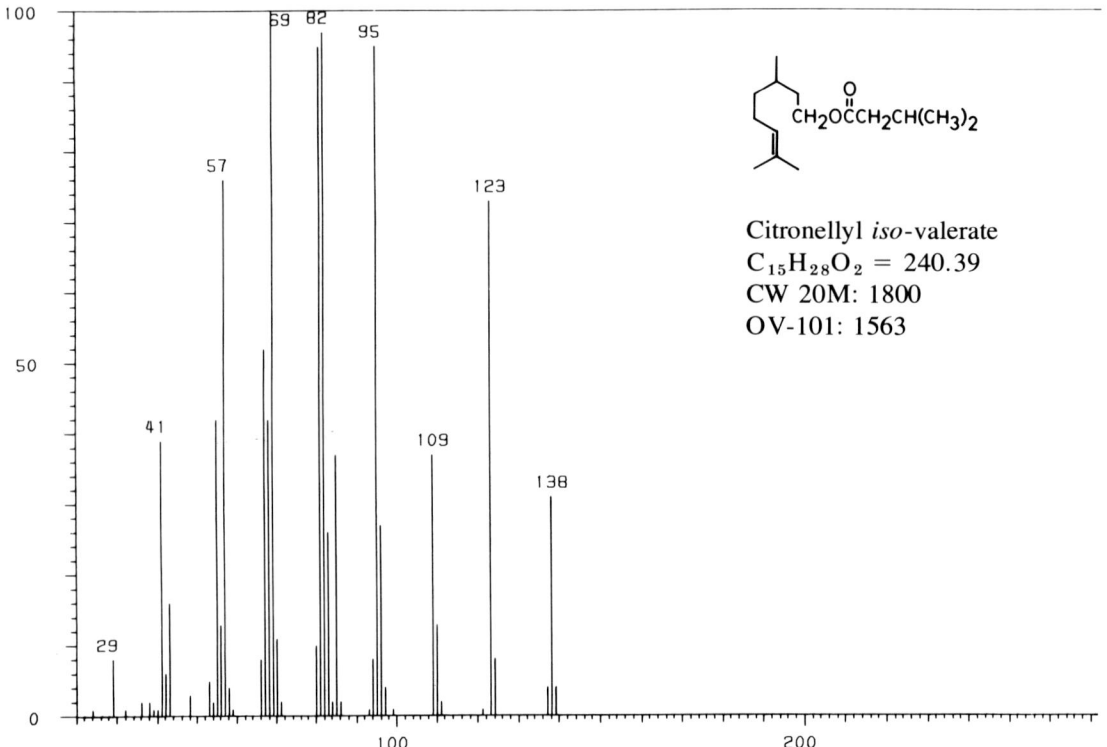

Citronellyl *n*-proprionate
$C_{13}H_{24}O_2 = 212.33$
CW 20M: 1700
OV-101: 1427

Citronellyl *iso*-valerate
$C_{15}H_{28}O_2 = 240.39$
CW 20M: 1800
OV-101: 1563

APPENDIX IV   MASS SPECTRA OF INDIVIDUAL COMPOUNDS   227

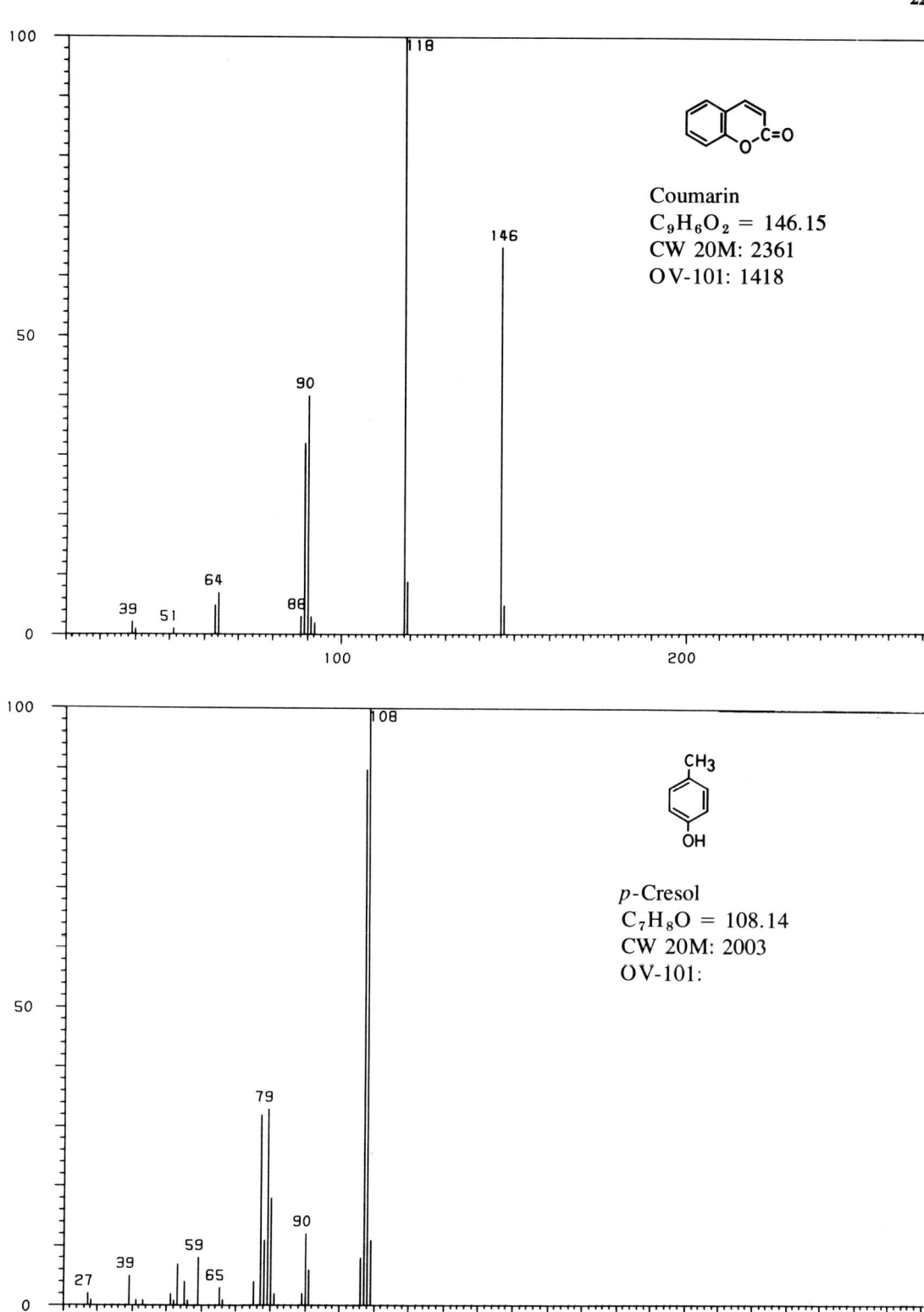

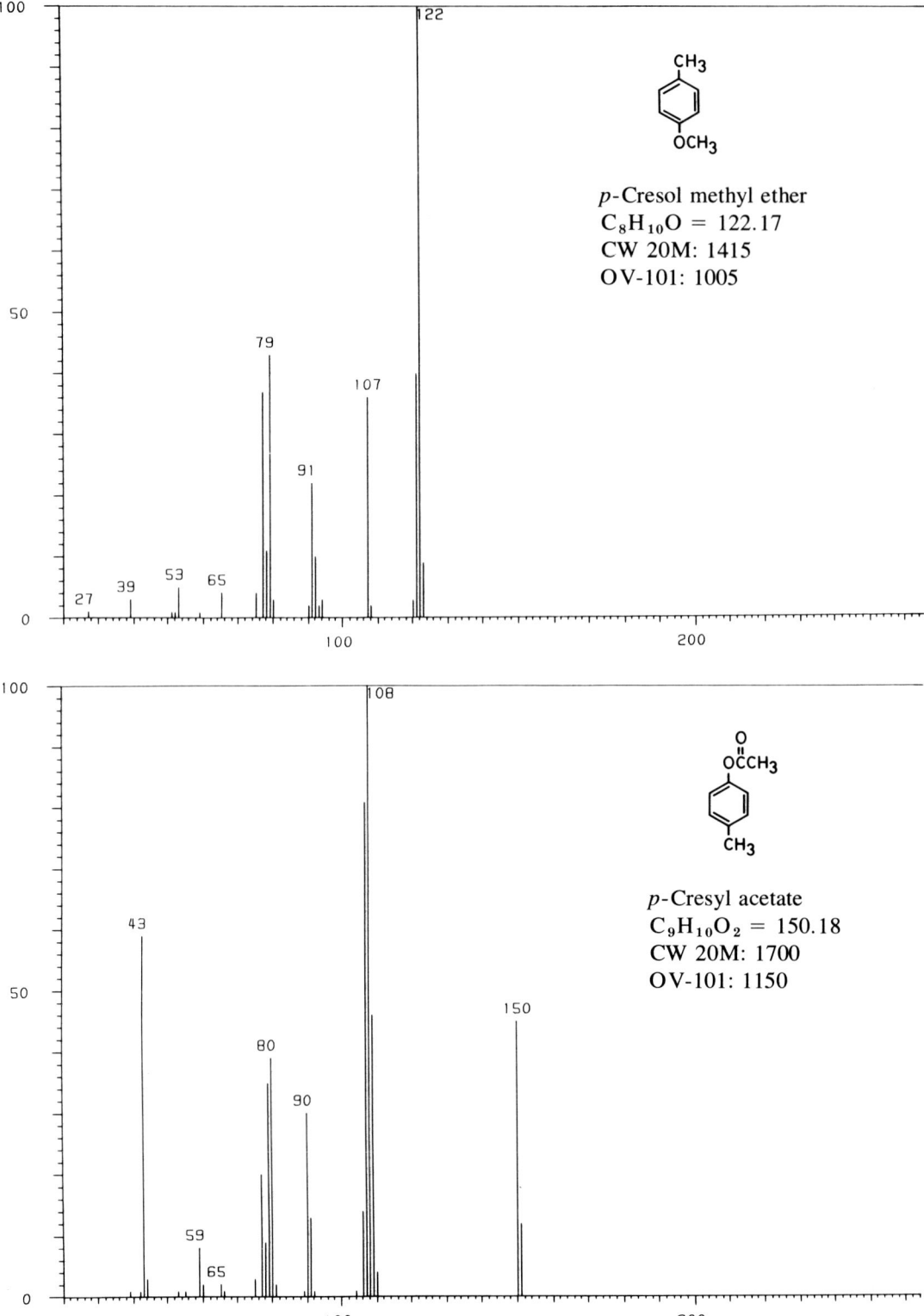

APPENDIX IV   MASS SPECTRA OF INDIVIDUAL COMPOUNDS

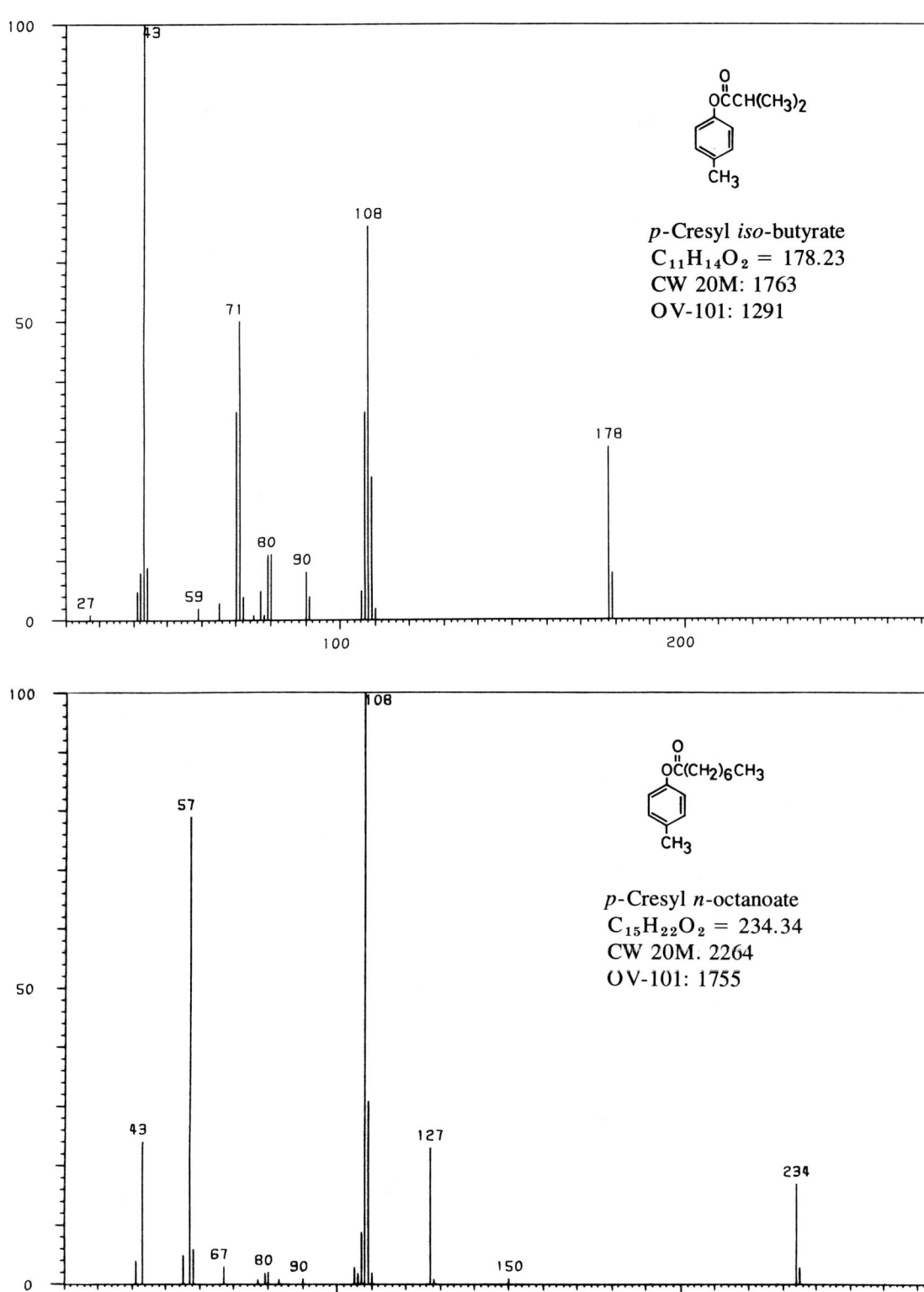

p-Cresyl *iso*-butyrate
$C_{11}H_{14}O_2 = 178.23$
CW 20M: 1763
OV-101: 1291

p-Cresyl *n*-octanoate
$C_{15}H_{22}O_2 = 234.34$
CW 20M: 2264
OV-101: 1755

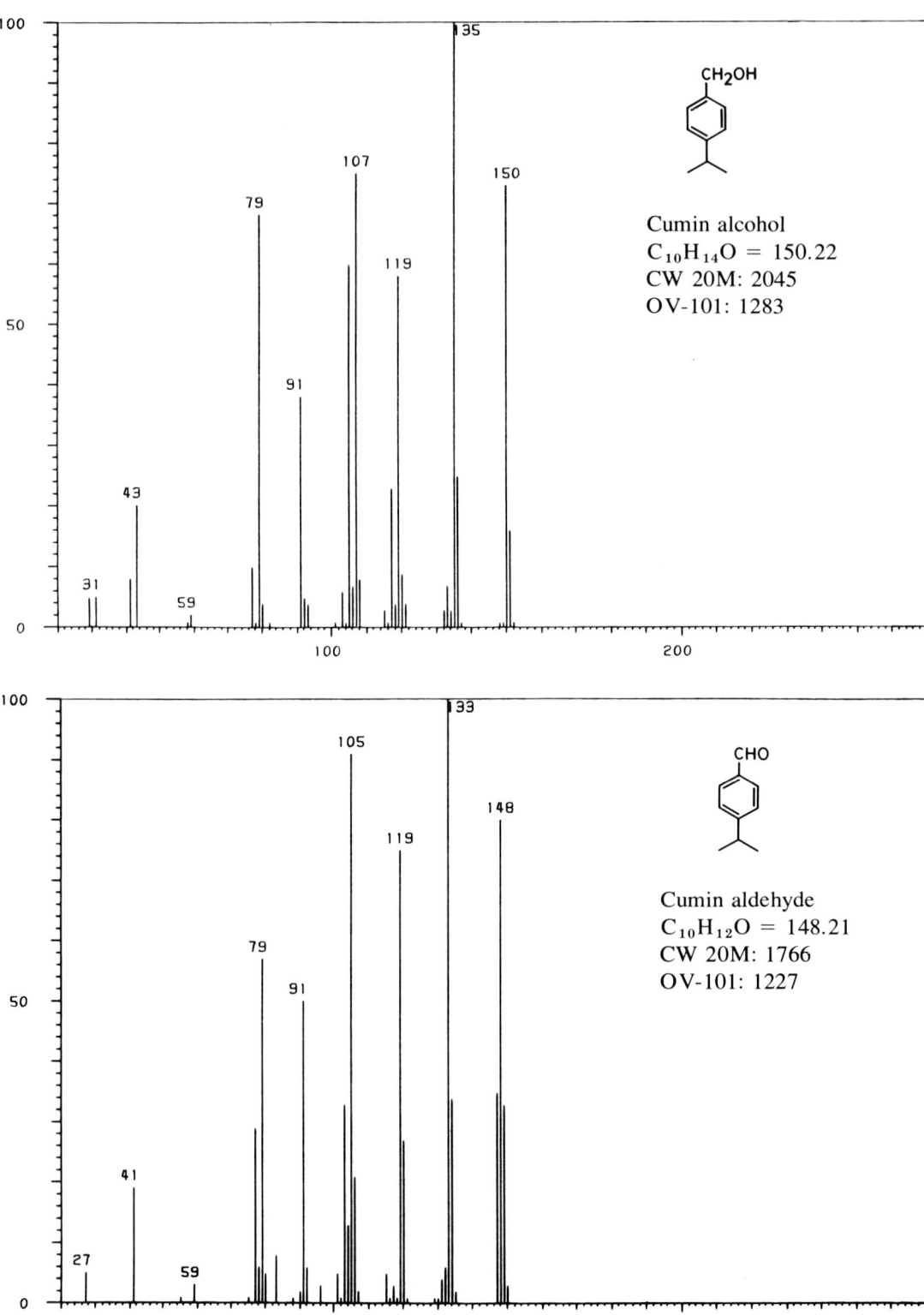

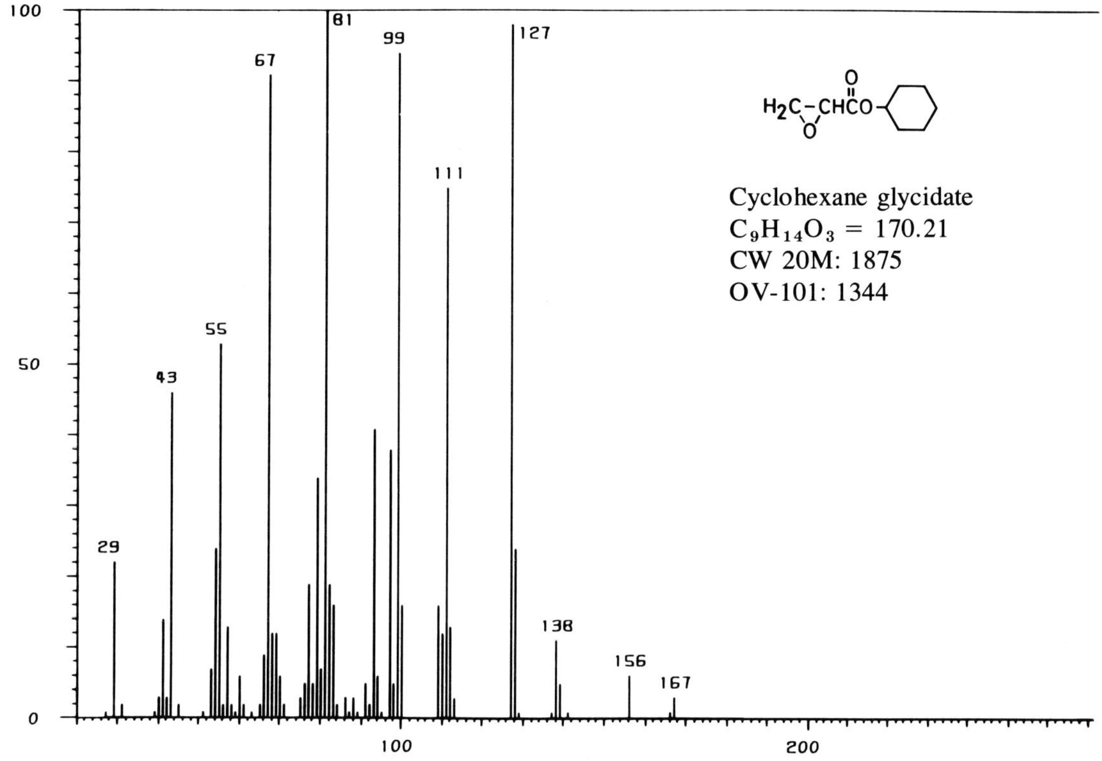

APPENDIX IV   MASS SPECTRA OF INDIVIDUAL COMPOUNDS

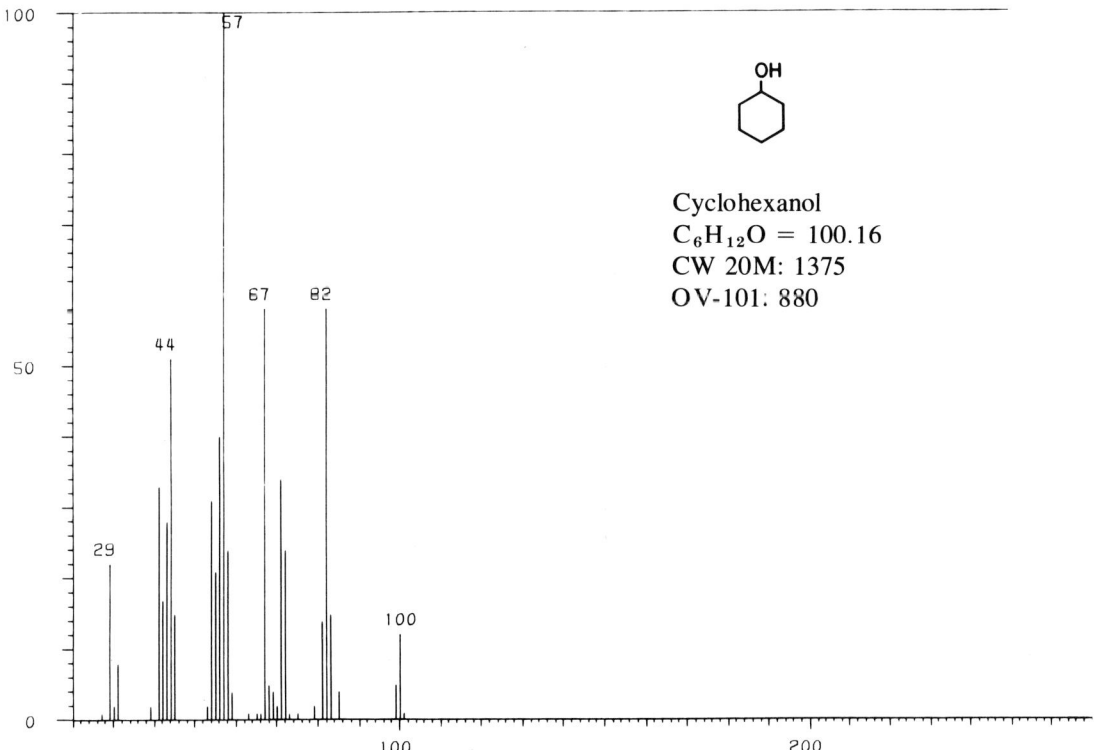

Cyclohexane glycidate
$C_9H_{14}O_3 = 170.21$
CW 20M: 1875
OV-101: 1344

Cyclohexanol
$C_6H_{12}O = 100.16$
CW 20M: 1375
OV-101: 880

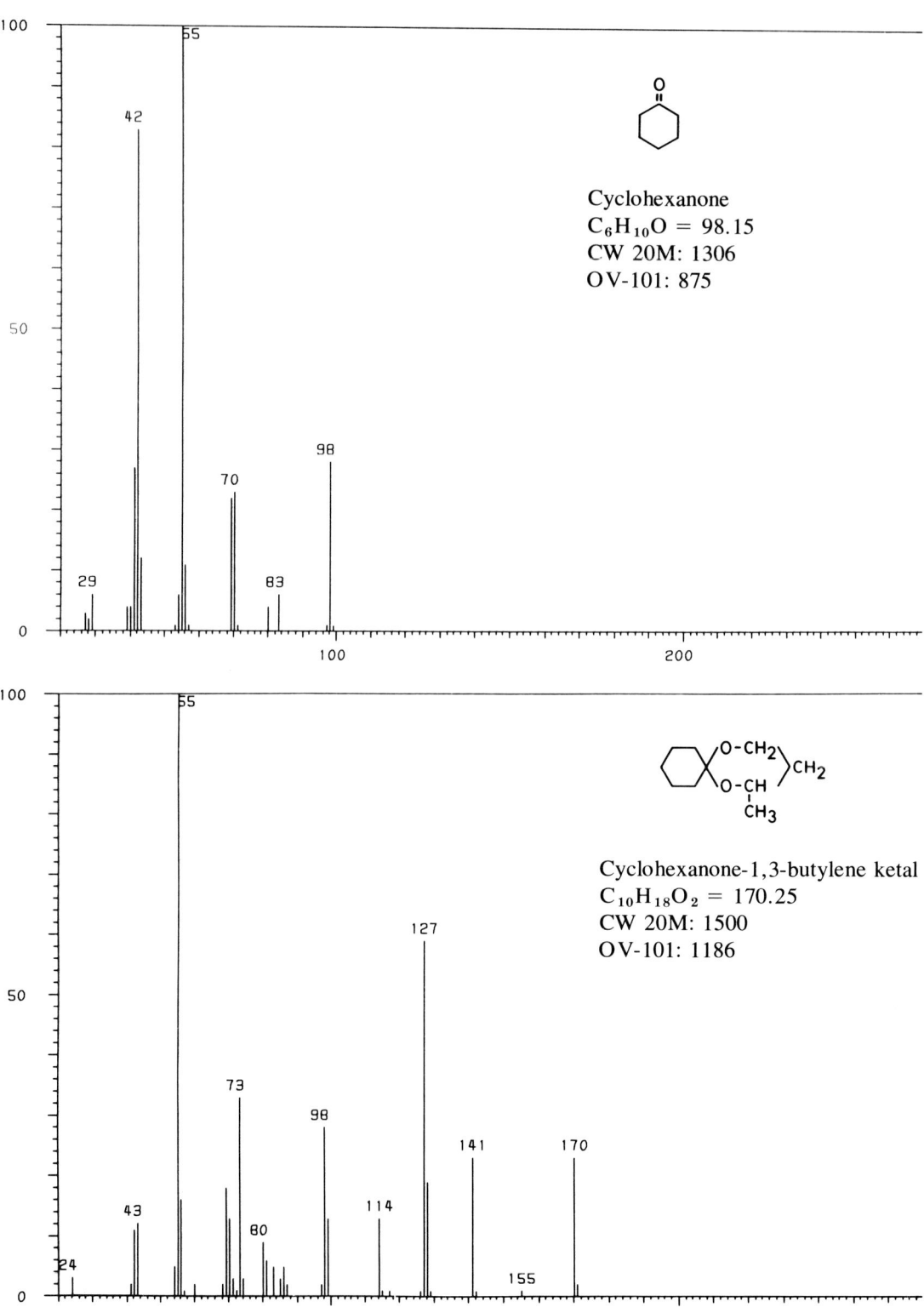

APPENDIX IV MASS SPECTRA OF INDIVIDUAL COMPOUNDS

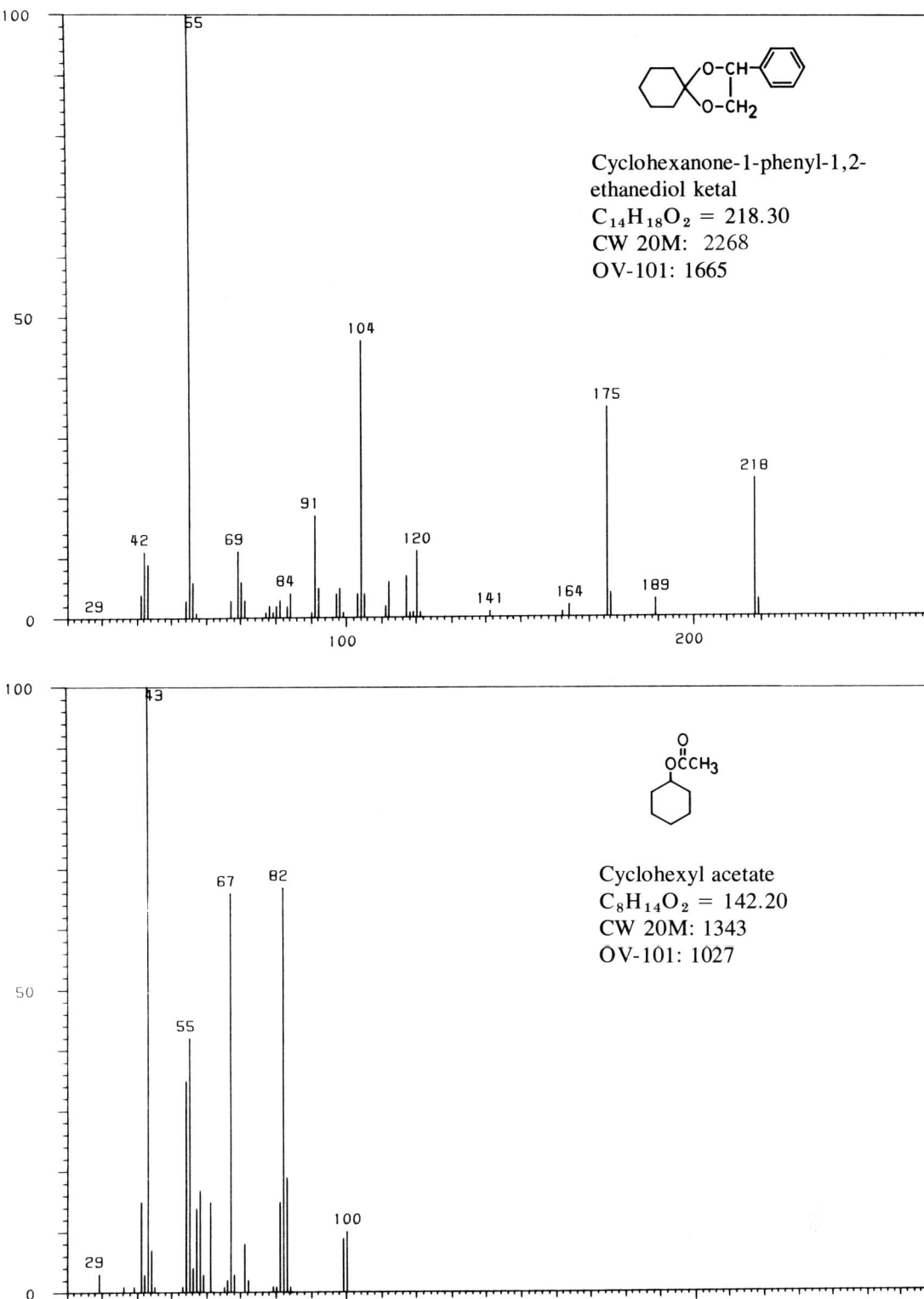

Cyclohexanone-1-phenyl-1,2-ethanediol ketal
$C_{14}H_{18}O_2 = 218.30$
CW 20M: 2268
OV-101: 1665

Cyclohexyl acetate
$C_8H_{14}O_2 = 142.20$
CW 20M: 1343
OV-101: 1027

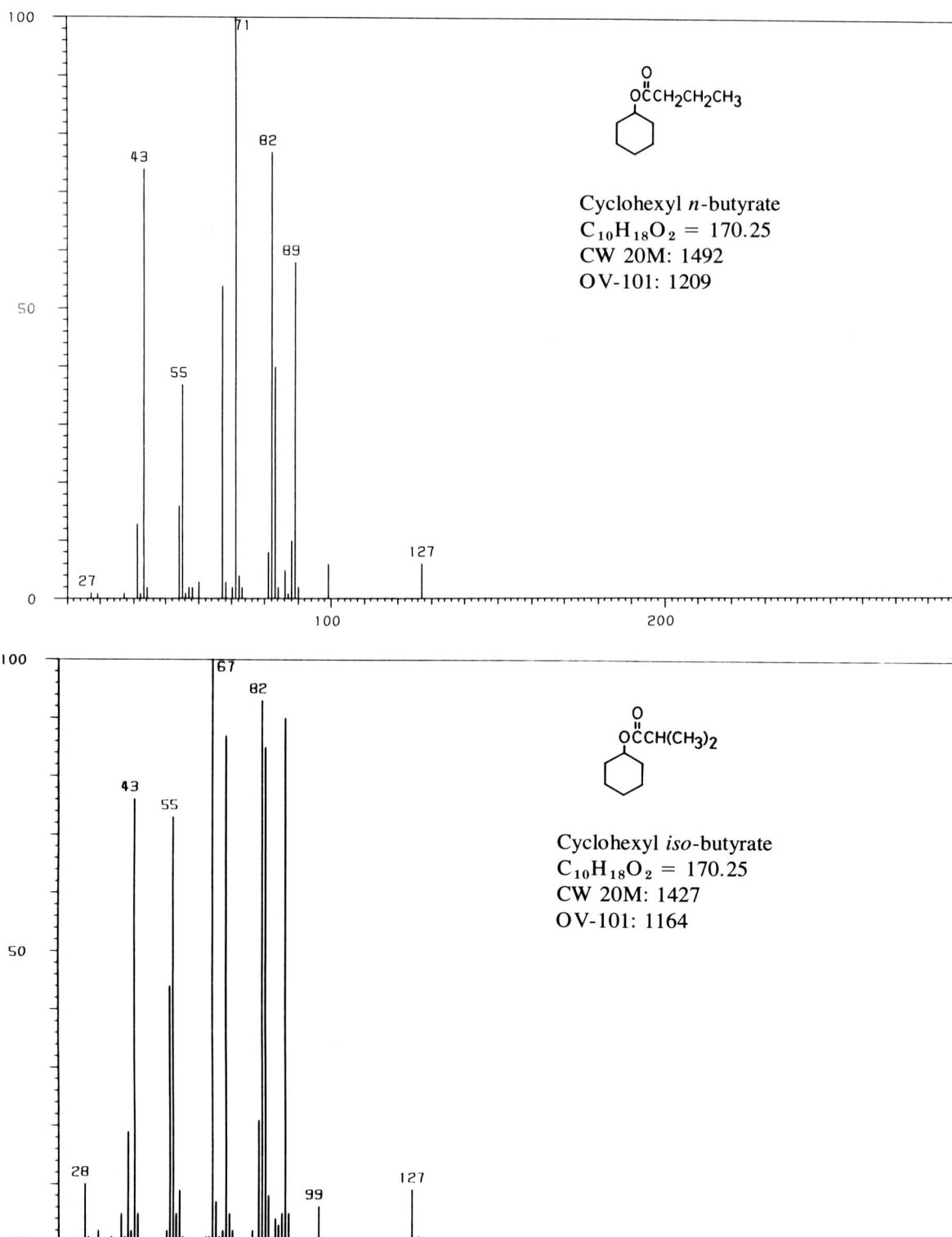

APPENDIX IV   MASS SPECTRA OF INDIVIDUAL COMPOUNDS

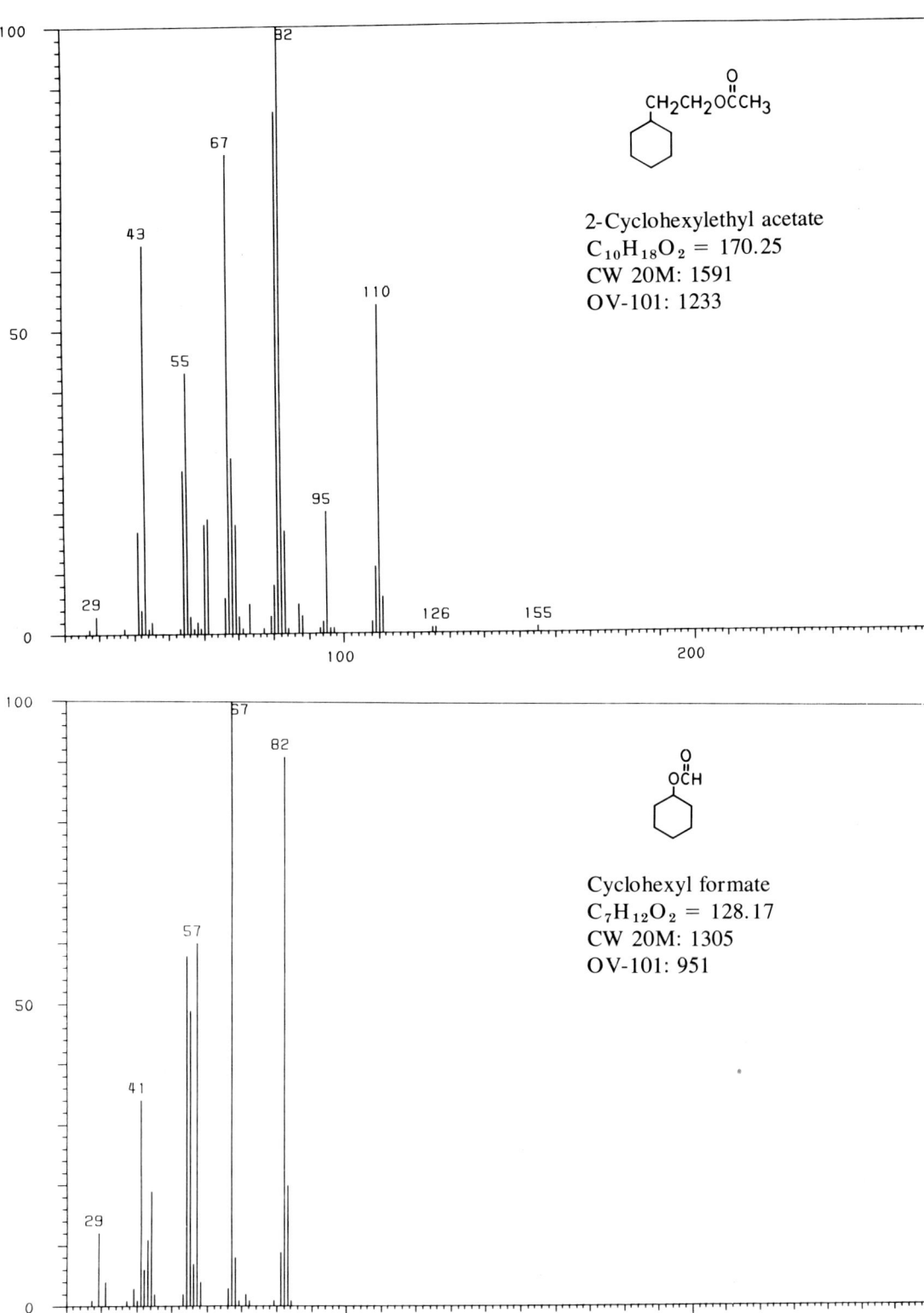

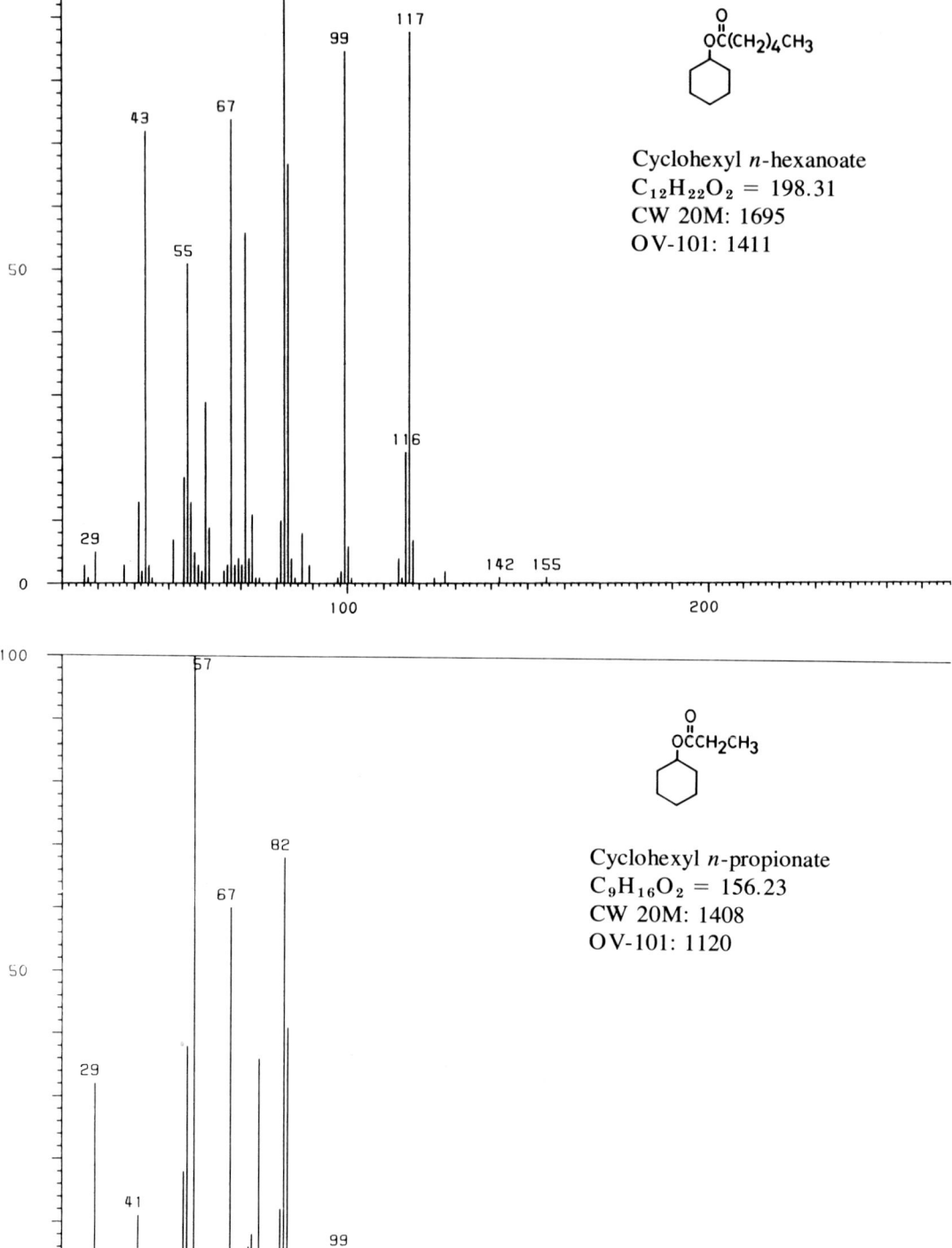

APPENDIX IV MASS SPECTRA OF INDIVIDUAL COMPOUNDS 237

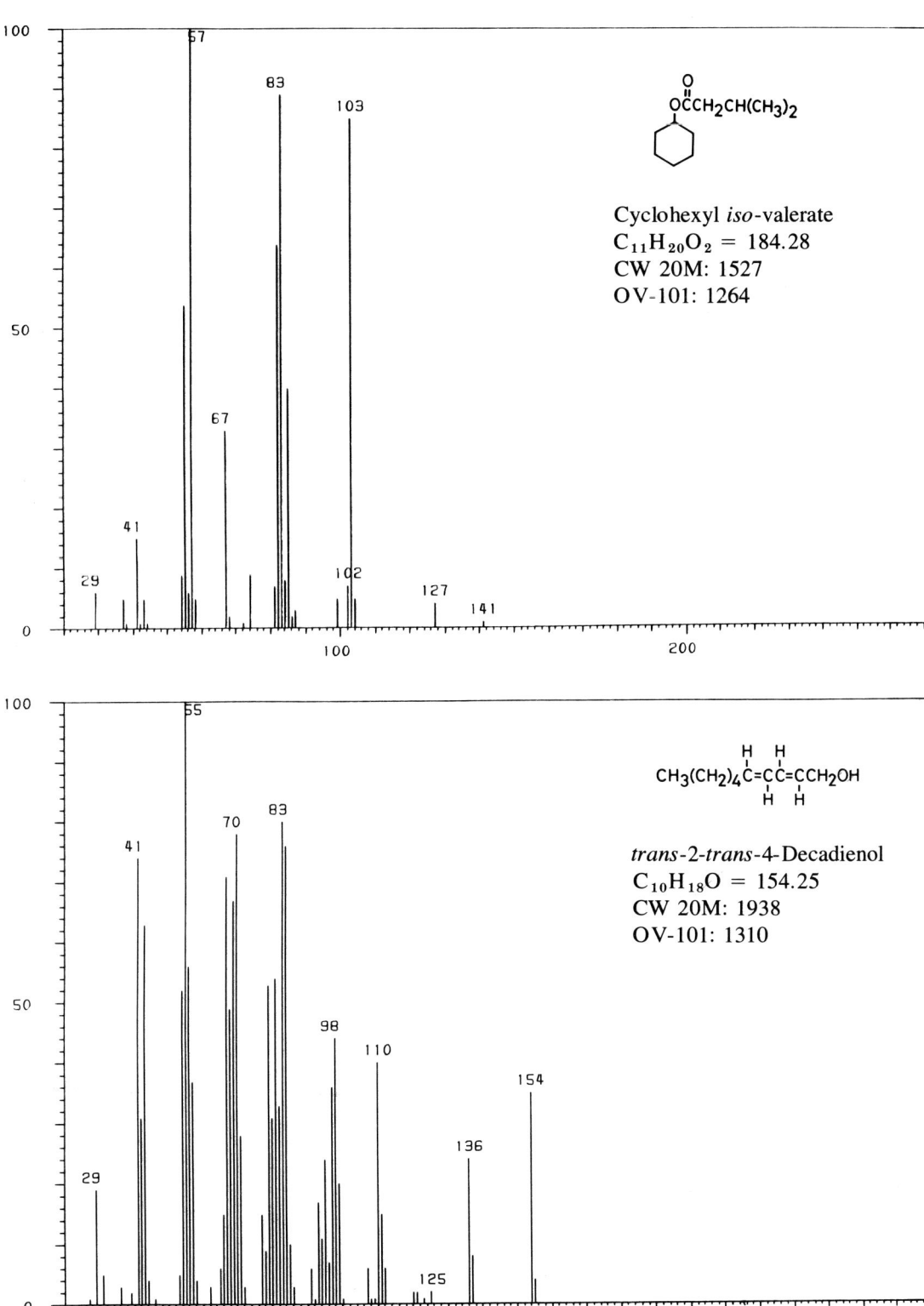

Cyclohexyl *iso*-valerate
$C_{11}H_{20}O_2 = 184.28$
CW 20M: 1527
OV-101: 1264

trans-2-*trans*-4-Decadienol
$C_{10}H_{18}O = 154.25$
CW 20M: 1938
OV-101: 1310

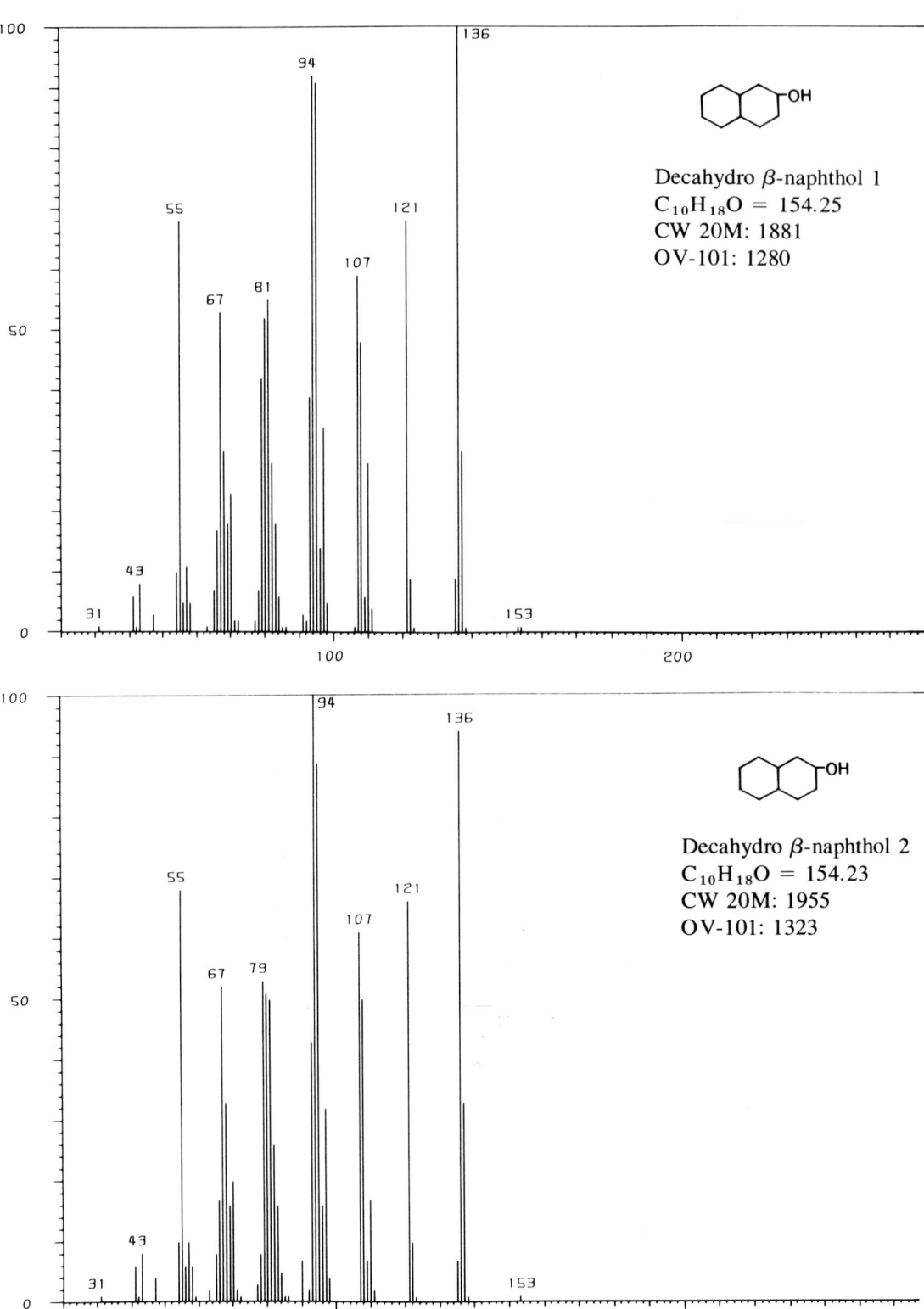

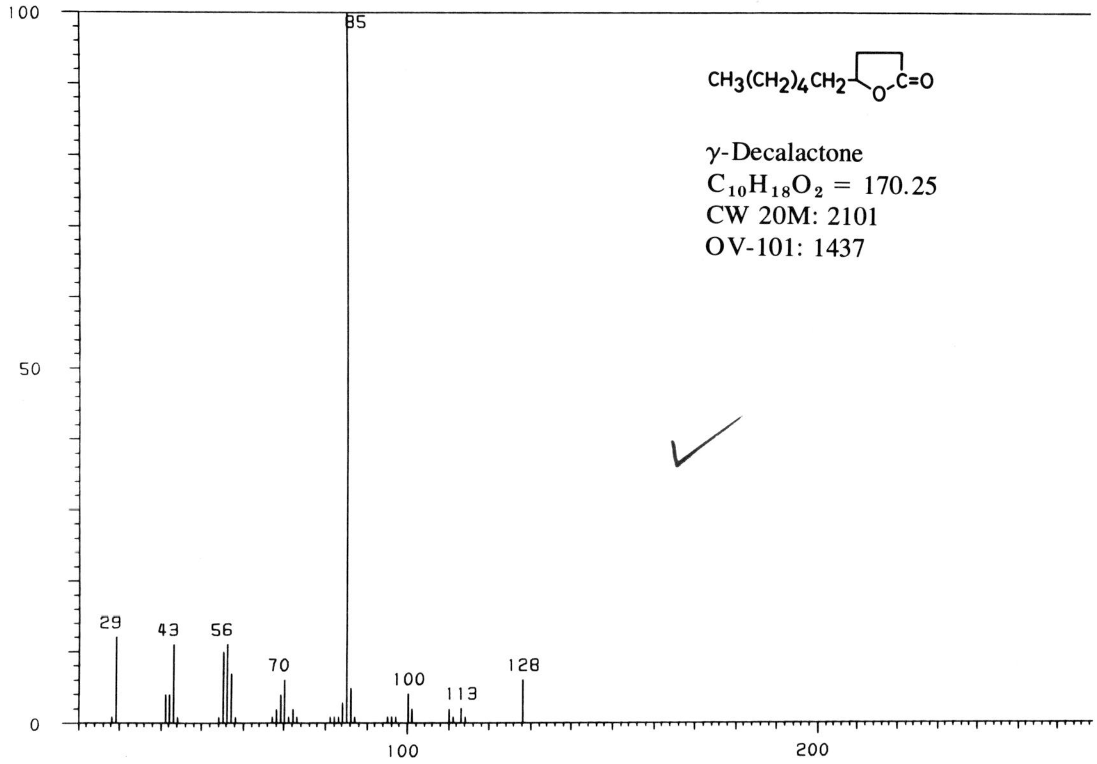

APPENDIX IV   MASS SPECTRA OF INDIVIDUAL COMPOUNDS

$CH_3(CH_2)_4CH_2\underset{O}{\underset{|}{\diagdown}}C=O$ (γ-lactone ring)

γ-Decalactone
$C_{10}H_{18}O_2 = 170.25$
CW 20M: 2101
OV-101: 1437

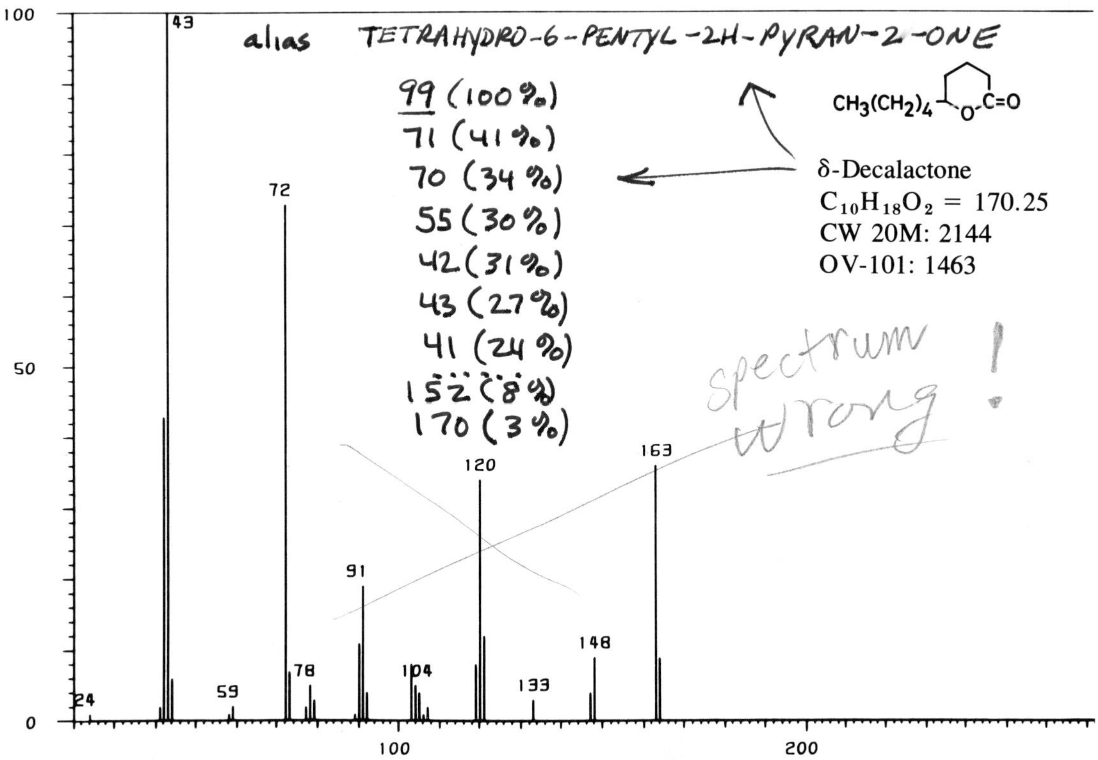

Peaks: 29, 43, 56, 70, 85, 100, 113, 128

---

alias   TETRAHYDRO-6-PENTYL-2H-PYRAN-2-ONE

99 (100%)
71 (41%)
70 (34%)
55 (30%)
42 (31%)
43 (27%)
41 (24%)
152 (8%)
170 (3%)

$CH_3(CH_2)_4\underset{O}{\underset{|}{\diagdown}}C=O$ (δ-lactone ring)

δ-Decalactone
$C_{10}H_{18}O_2 = 170.25$
CW 20M: 2144
OV-101: 1463

spectrum wrong!

Peaks: 24, 43, 59, 72, 78, 91, 104, 120, 133, 148, 163

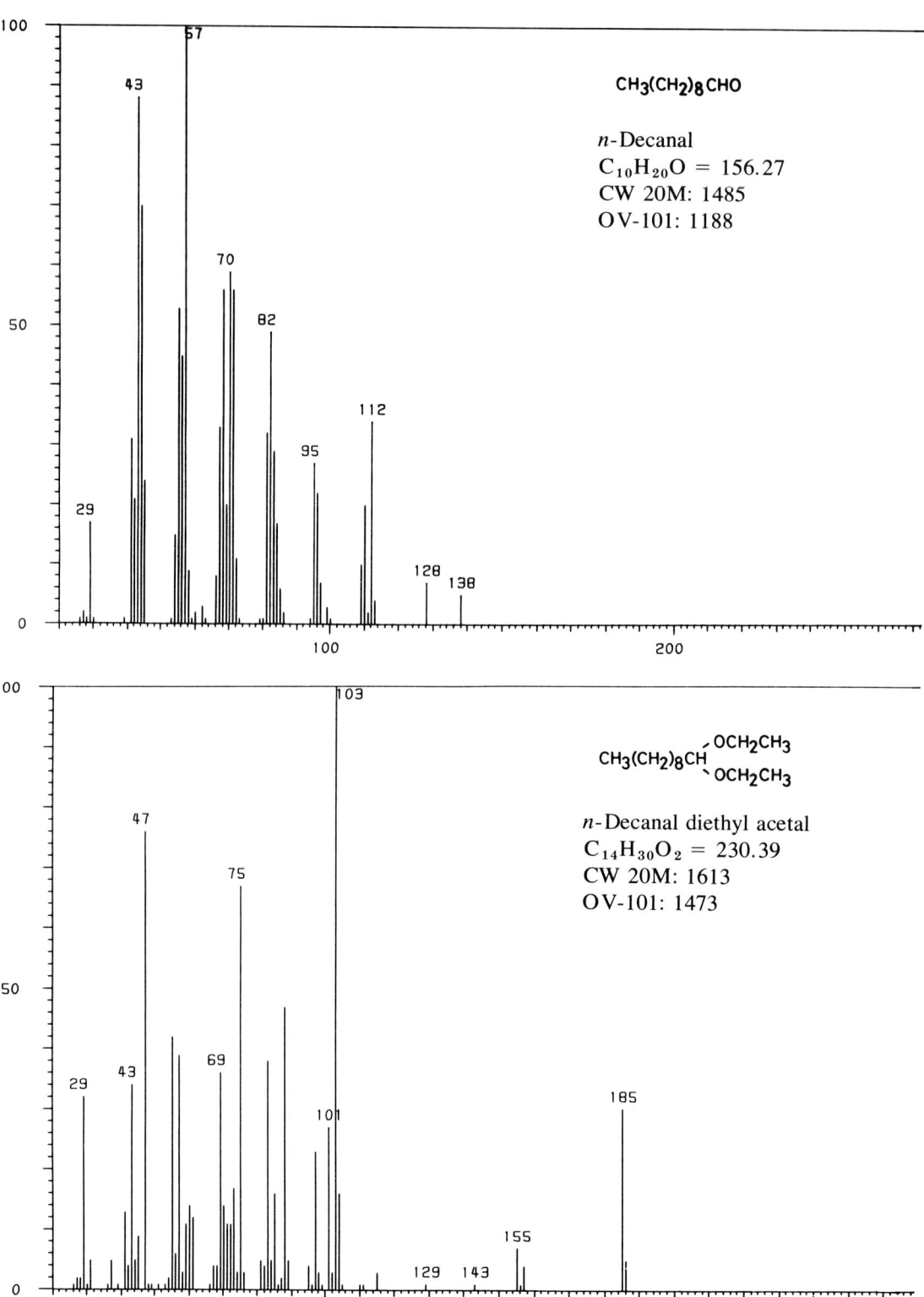

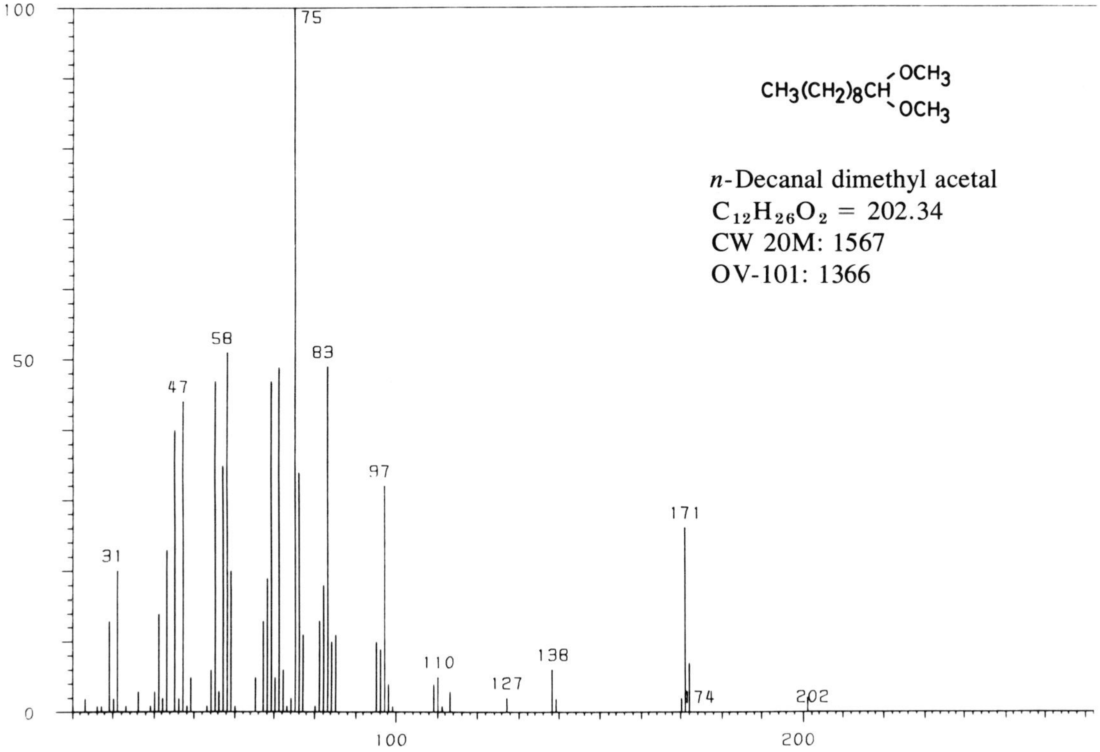

APPENDIX IV   MASS SPECTRA OF INDIVIDUAL COMPOUNDS

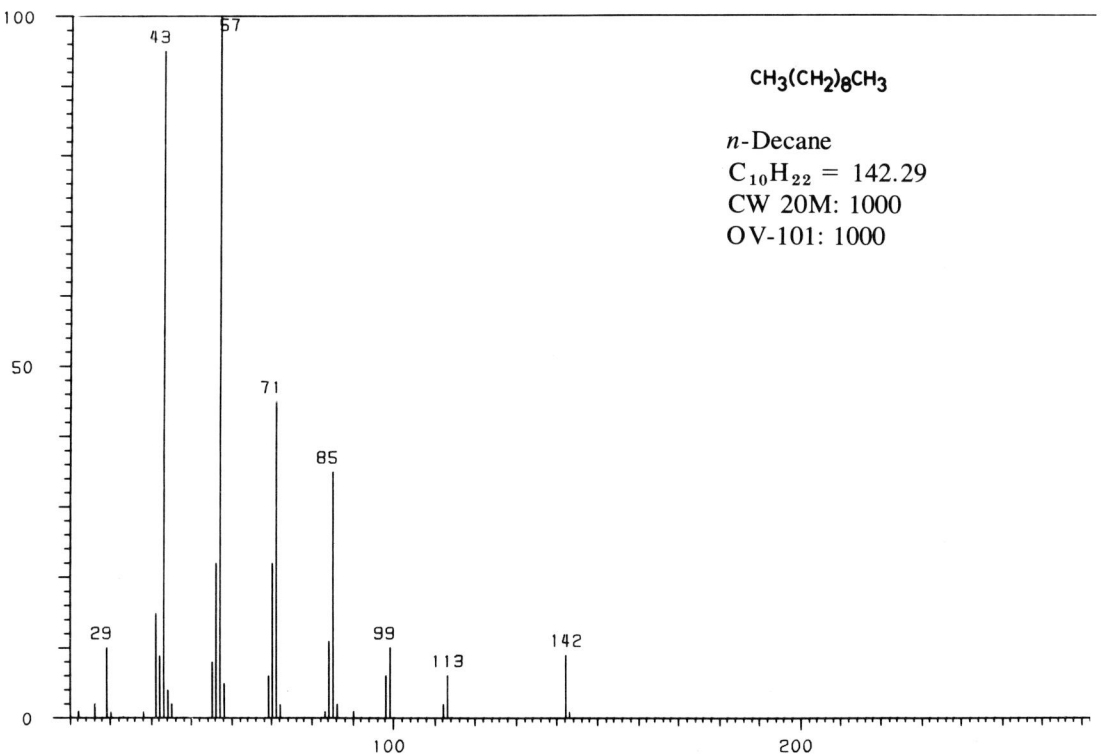

$CH_3(CH_2)_8CH \begin{smallmatrix} OCH_3 \\ OCH_3 \end{smallmatrix}$

n-Decanal dimethyl acetal
$C_{12}H_{26}O_2 = 202.34$
CW 20M: 1567
OV-101: 1366

$CH_3(CH_2)_8CH_3$

n-Decane
$C_{10}H_{22} = 142.29$
CW 20M: 1000
OV-101: 1000

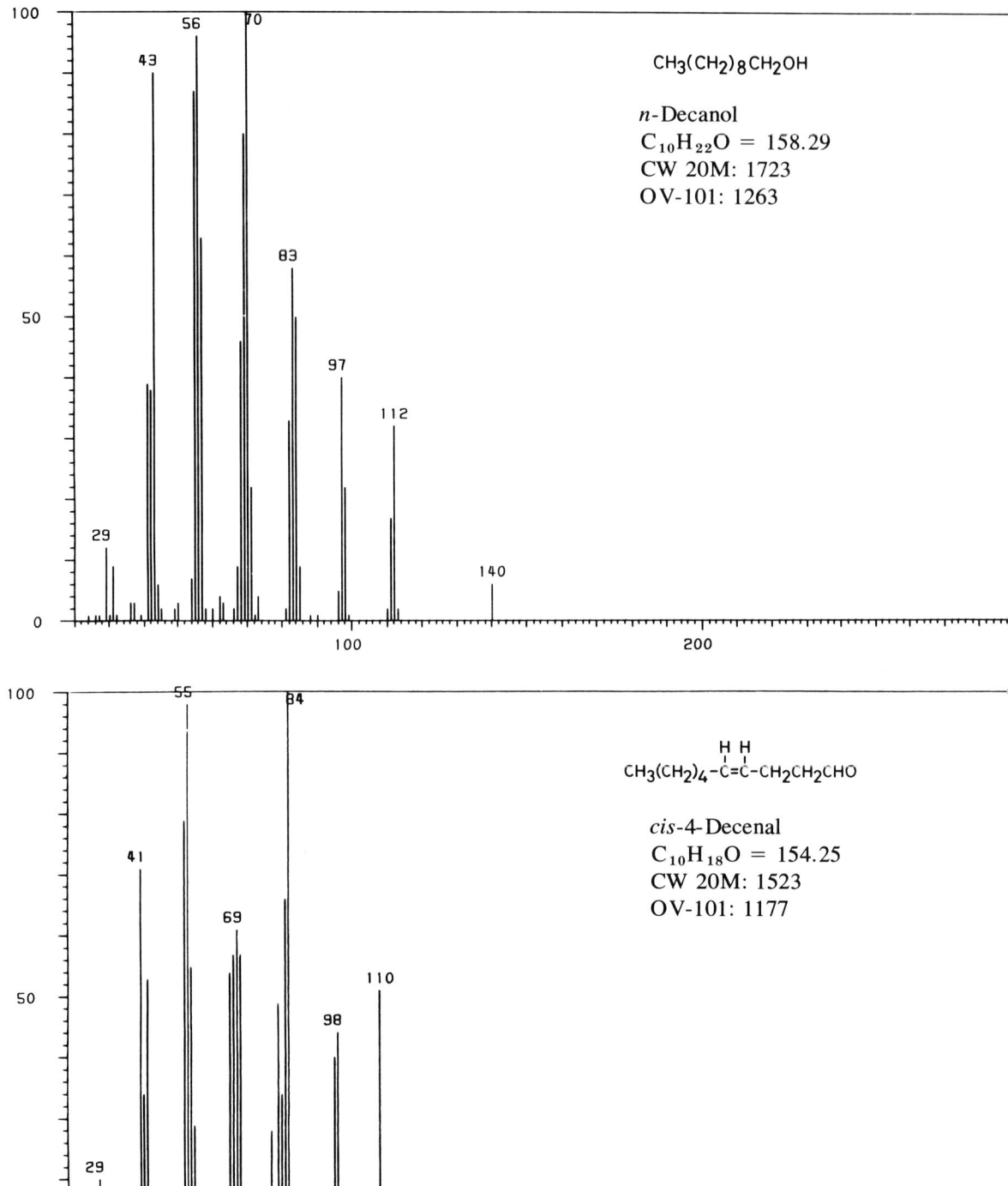

APPENDIX IV   MASS SPECTRA OF INDIVIDUAL COMPOUNDS

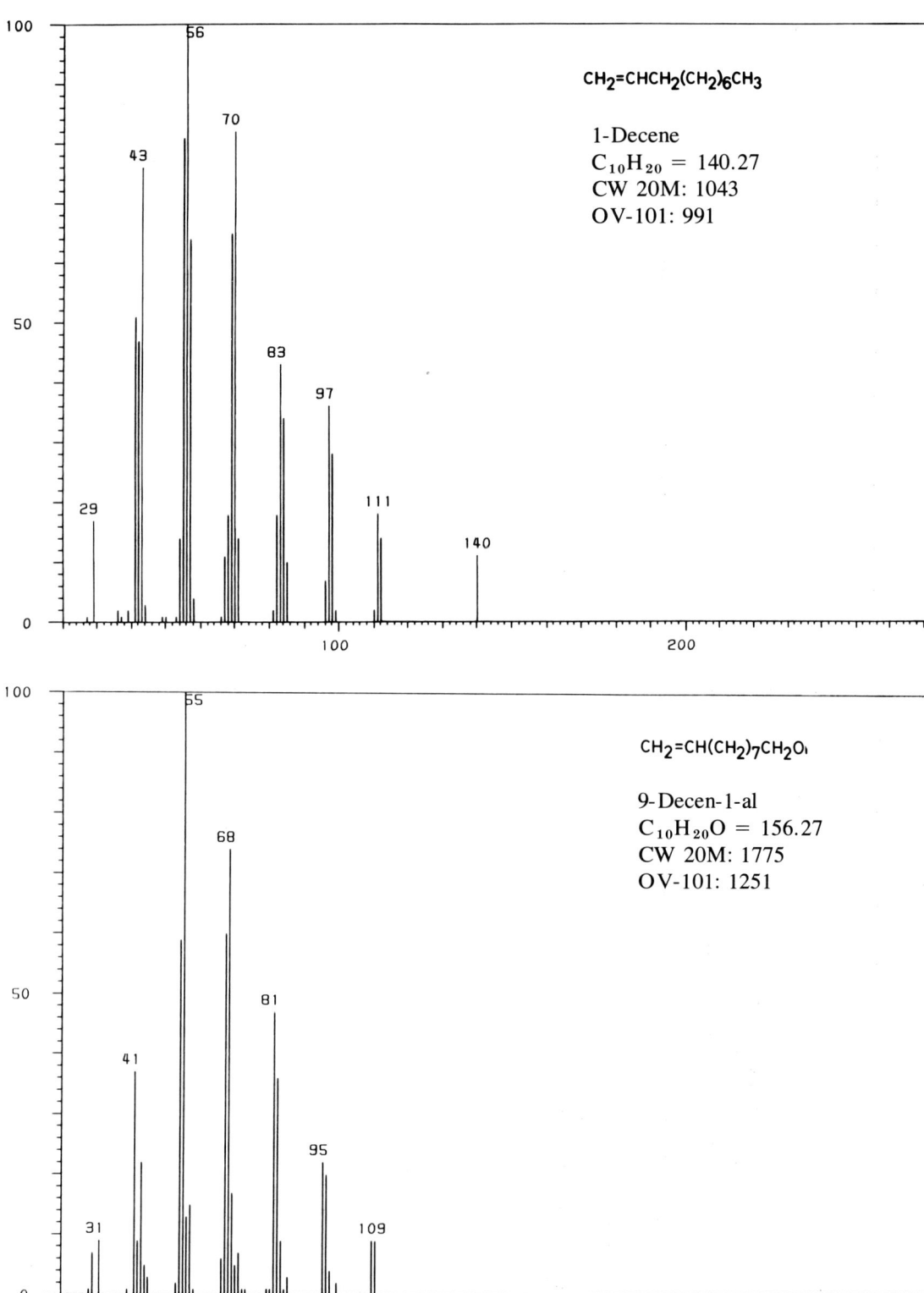

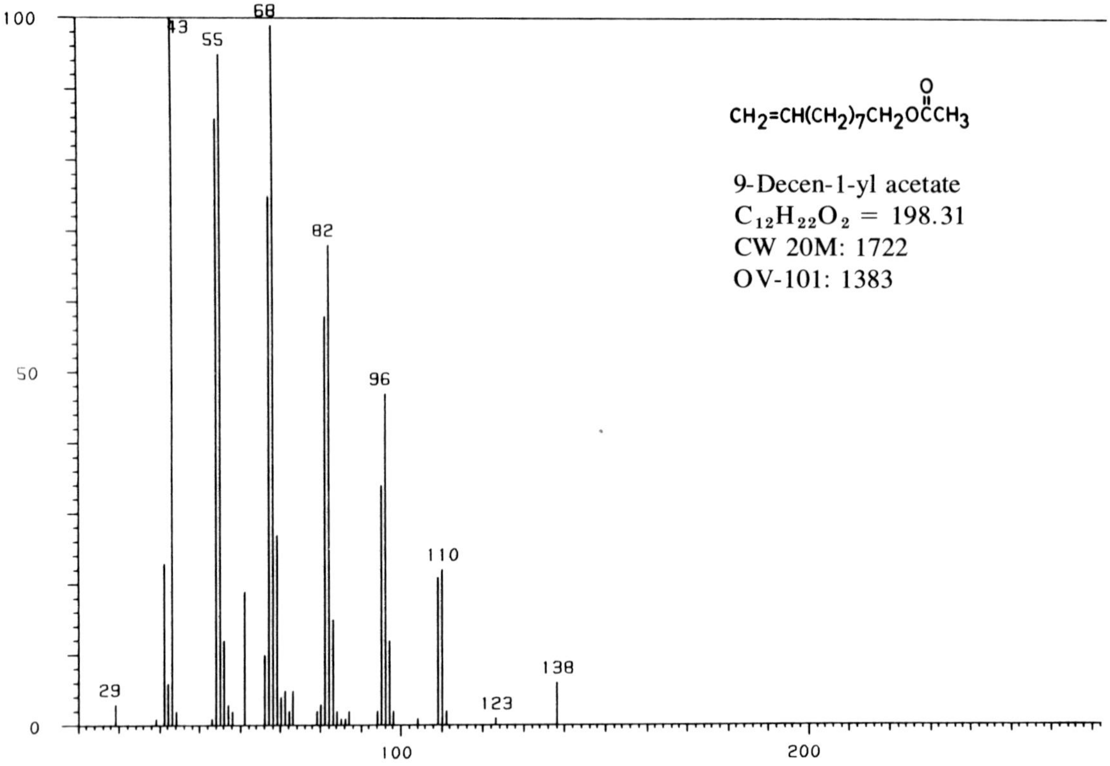

$CH_2{=}CH(CH_2)_7CH_2OC(O)CH_3$

9-Decen-1-yl acetate
$C_{12}H_{22}O_2$ = 198.31
CW 20M: 1722
OV-101: 1383

$CH_3(CH_2)_8CH_2OC(O)CH_3$

n-Decyl acetate
$C_{12}H_{24}O_2$ = 200.32
CW 20M: 1662
OV-101: 1393

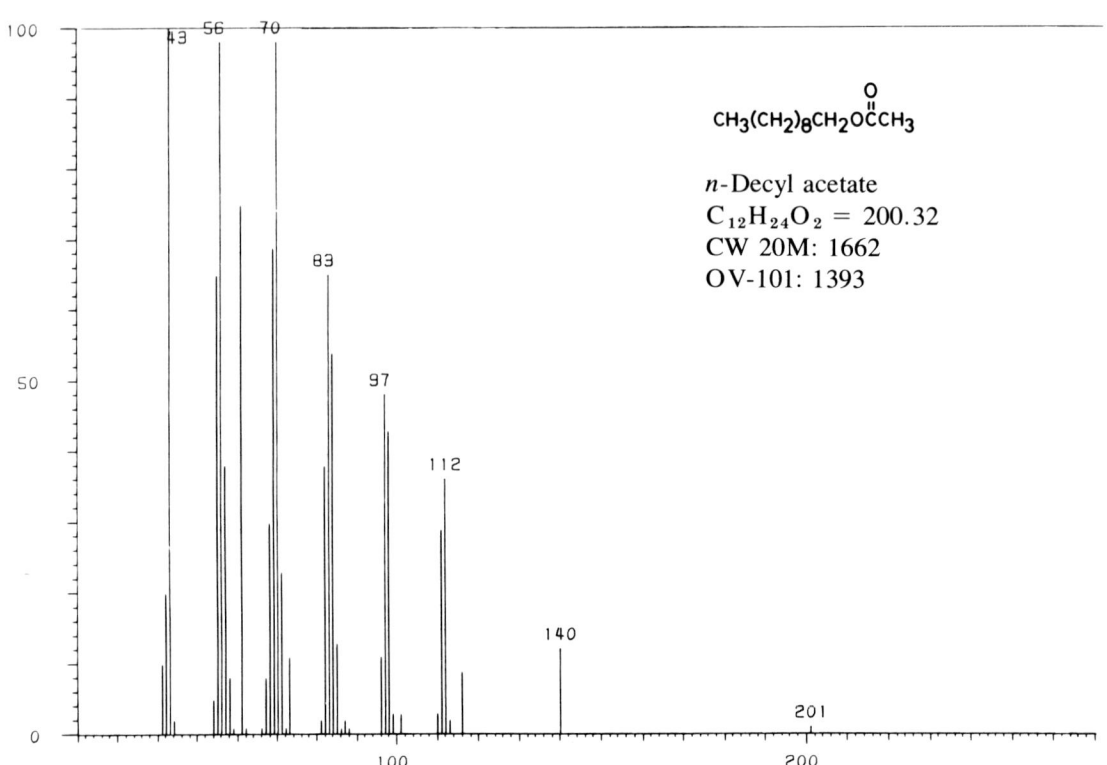

APPENDIX IV MASS SPECTRA OF INDIVIDUAL COMPOUNDS

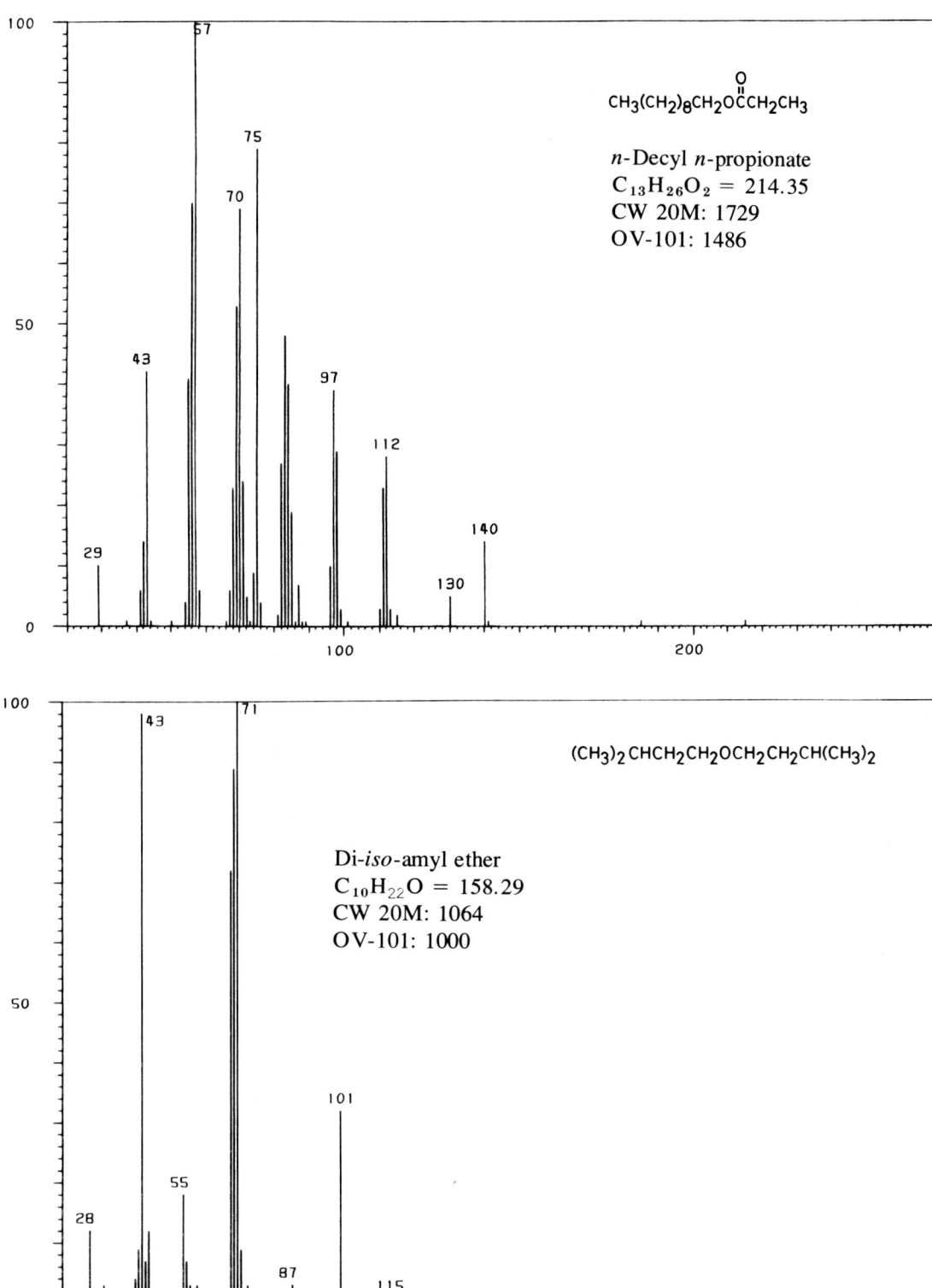

$CH_3(CH_2)_8CH_2OCCH_2CH_3$ (with C=O)

n-Decyl n-propionate
$C_{13}H_{26}O_2 = 214.35$
CW 20M: 1729
OV-101: 1486

$(CH_3)_2CHCH_2CH_2OCH_2CH_2CH(CH_3)_2$

Di-*iso*-amyl ether
$C_{10}H_{22}O = 158.29$
CW 20M: 1064
OV-101: 1000

$CH_3(CH_2)_3CH_2COCH_2(CH_2)_3CH_3$

Di-*n*-amyl ketone
$C_{11}H_{22}O$ = 170.30
CW 20M: 1528
OV-101: 1258

$CH_3(CH_2)_2CH_2OCH_2(CH_2)_2CH_3$

Di-*n*-Butyl ether
$C_8H_{18}O$ = 130.23
CW 20M: 965
OV-101: 876

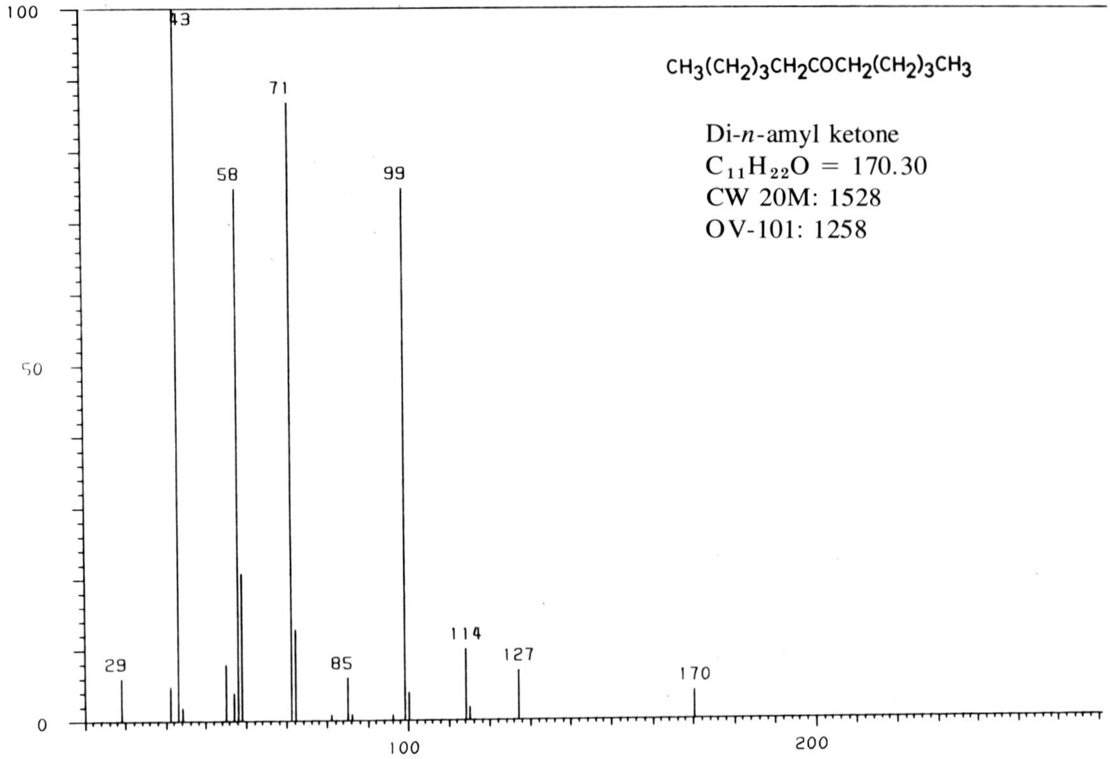

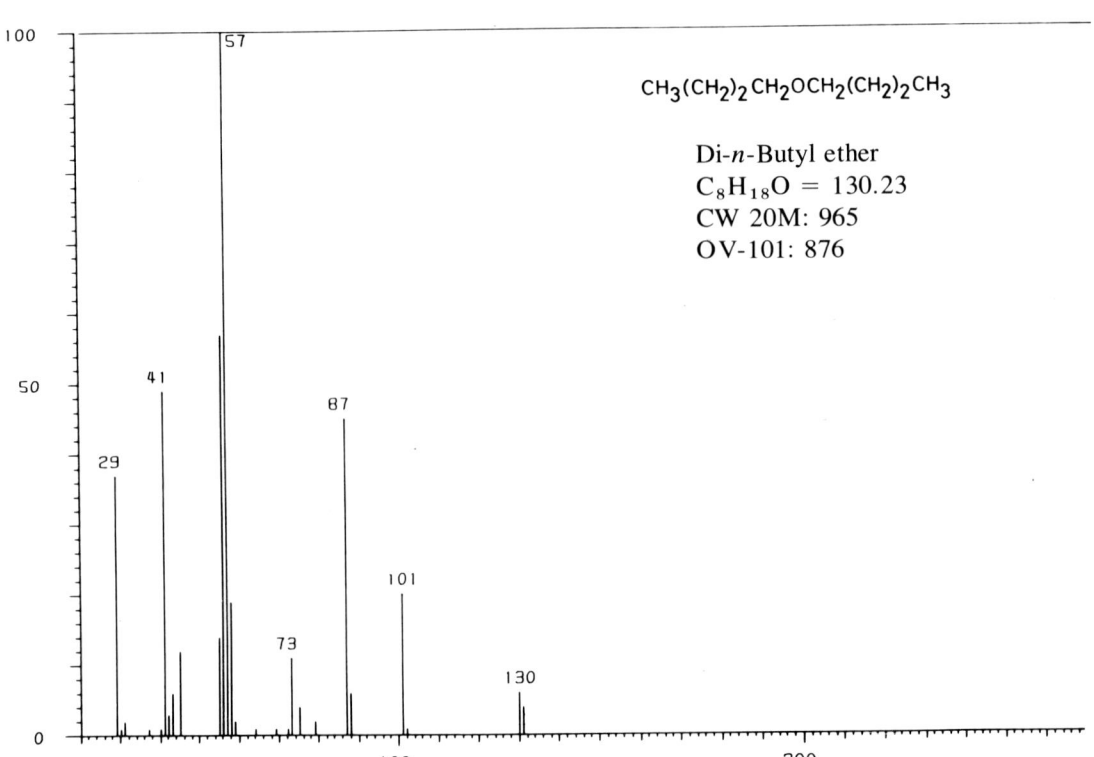

APPENDIX IV MASS SPECTRA OF INDIVIDUAL COMPOUNDS 247

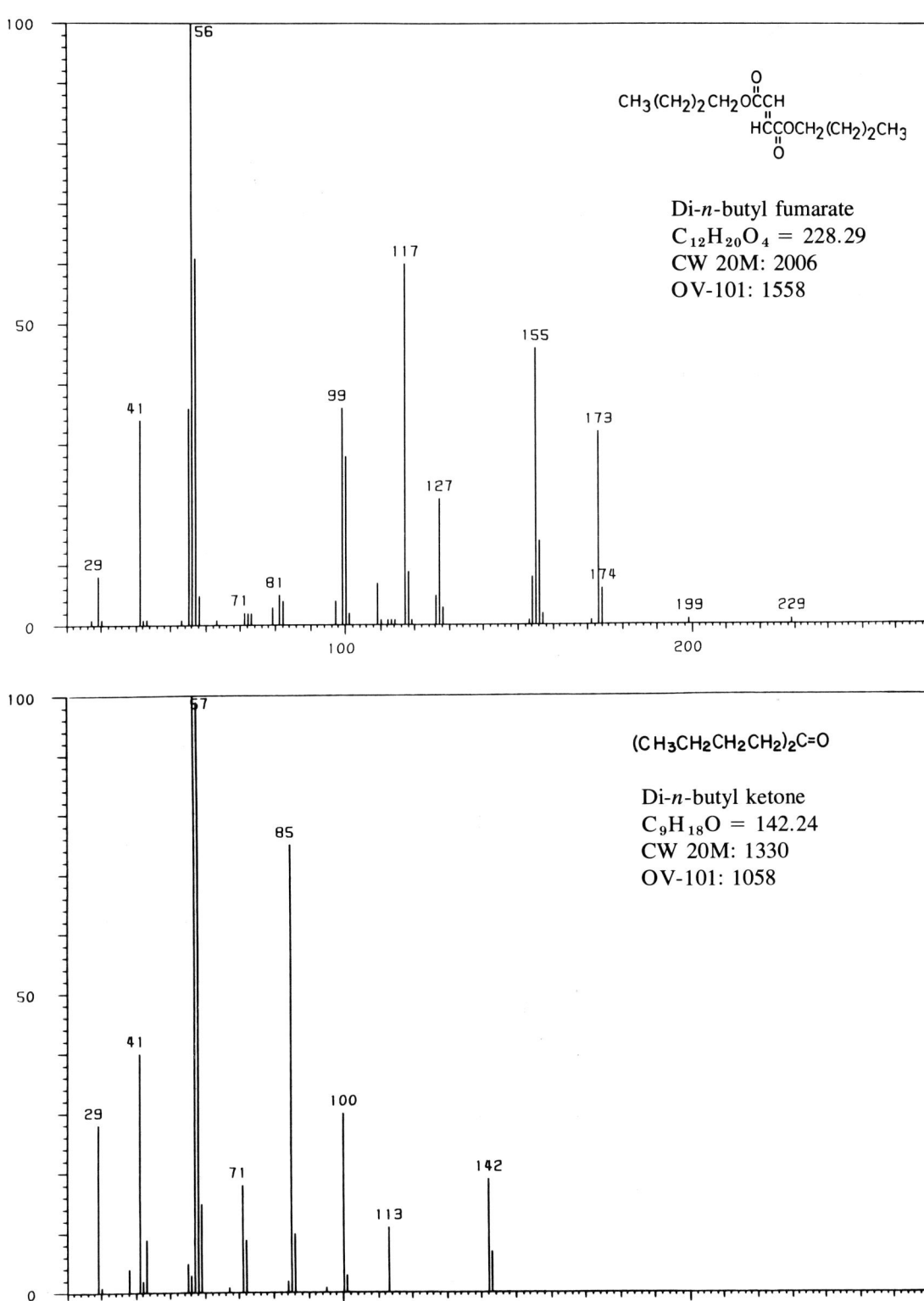

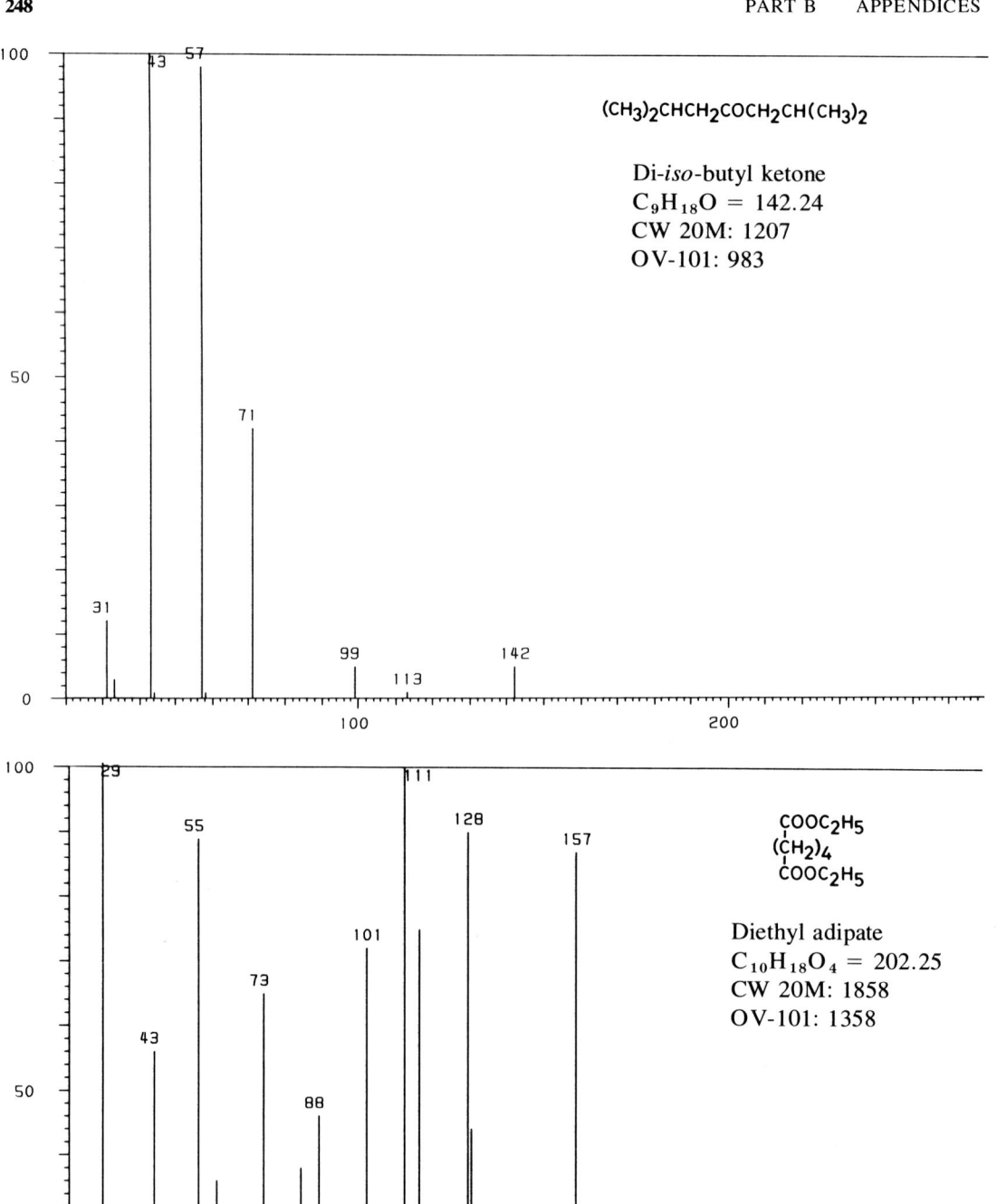

APPENDIX IV   MASS SPECTRA OF INDIVIDUAL COMPOUNDS

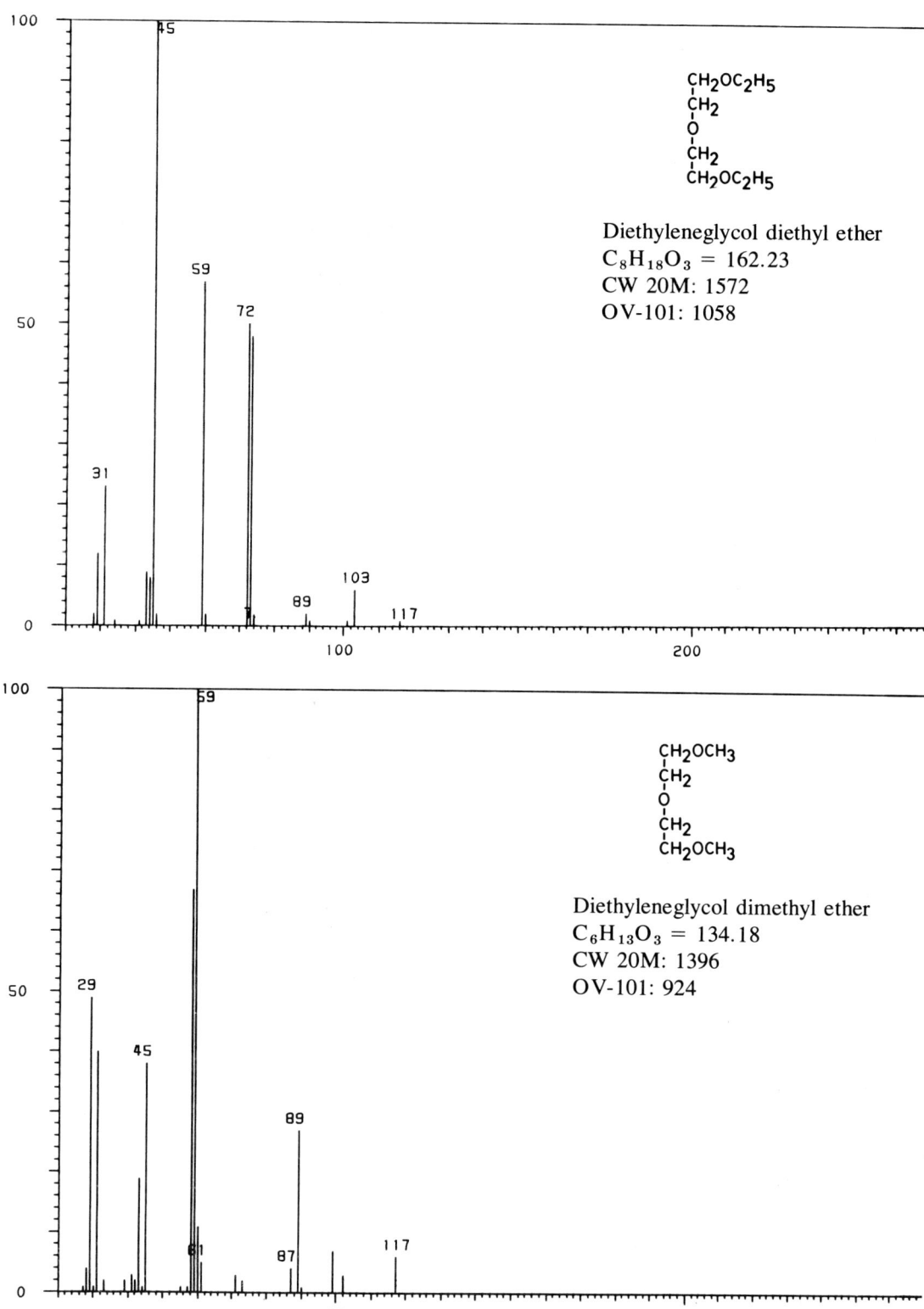

Diethyleneglycol diethyl ether
$C_8H_{18}O_3 = 162.23$
CW 20M: 1572
OV-101: 1058

Diethyleneglycol dimethyl ether
$C_6H_{13}O_3 = 134.18$
CW 20M: 1396
OV-101: 924

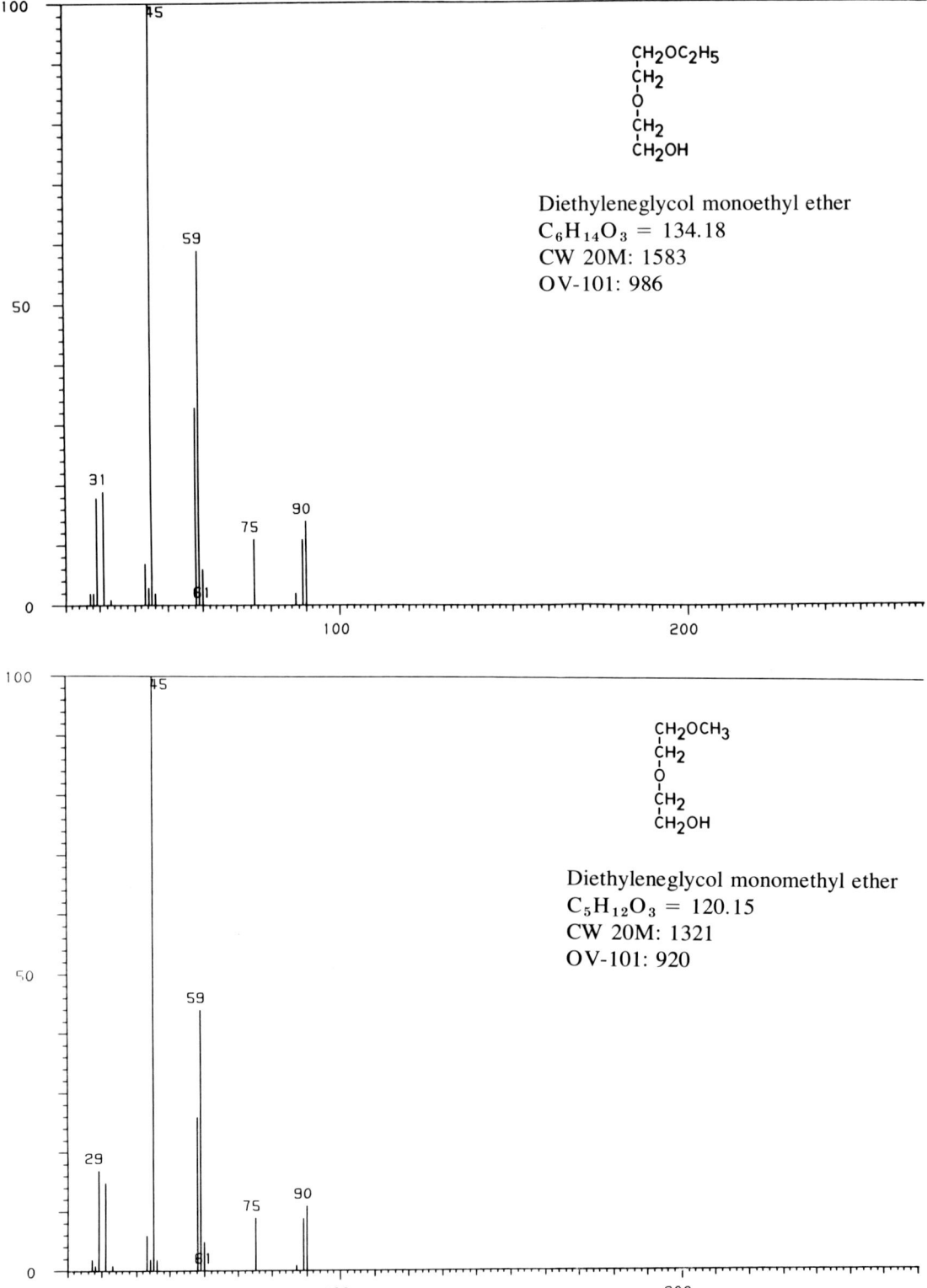

APPENDIX IV MASS SPECTRA OF INDIVIDUAL COMPOUNDS 251

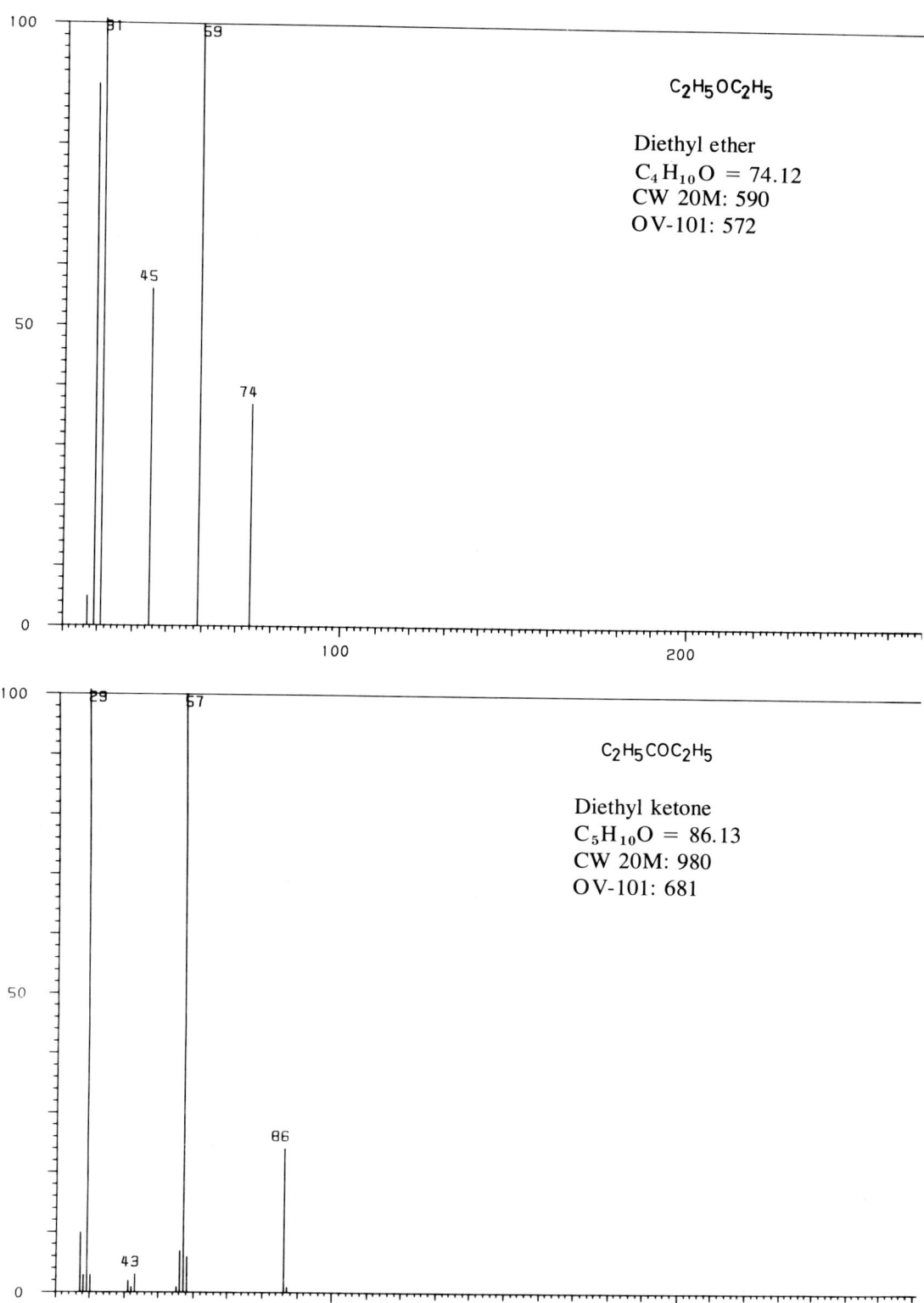

$C_2H_5OC_2H_5$

Diethyl ether
$C_4H_{10}O = 74.12$
CW 20M: 590
OV-101: 572

$C_2H_5COC_2H_5$

Diethyl ketone
$C_5H_{10}O = 86.13$
CW 20M: 980
OV-101: 681

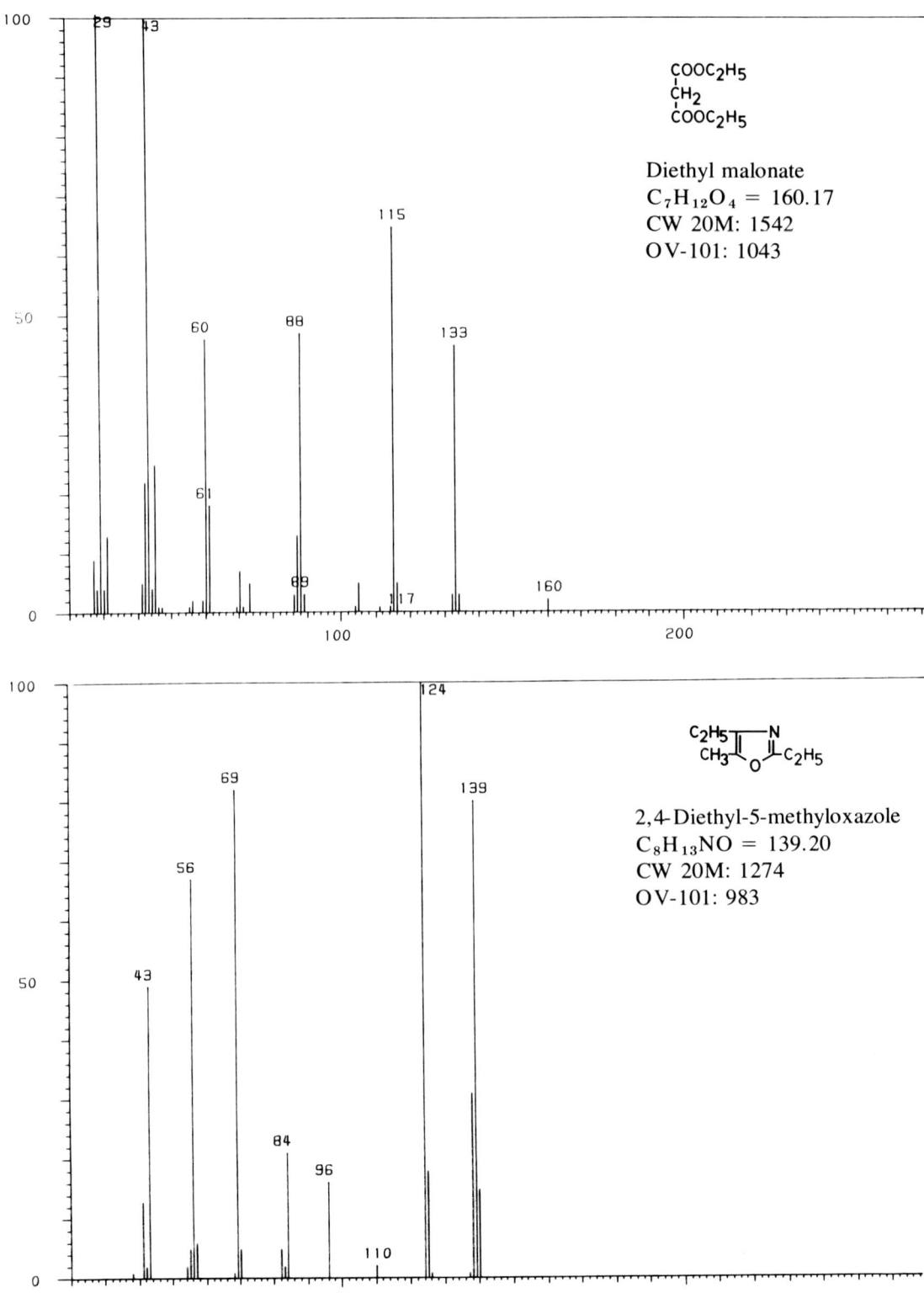

APPENDIX IV   MASS SPECTRA OF INDIVIDUAL COMPOUNDS   253

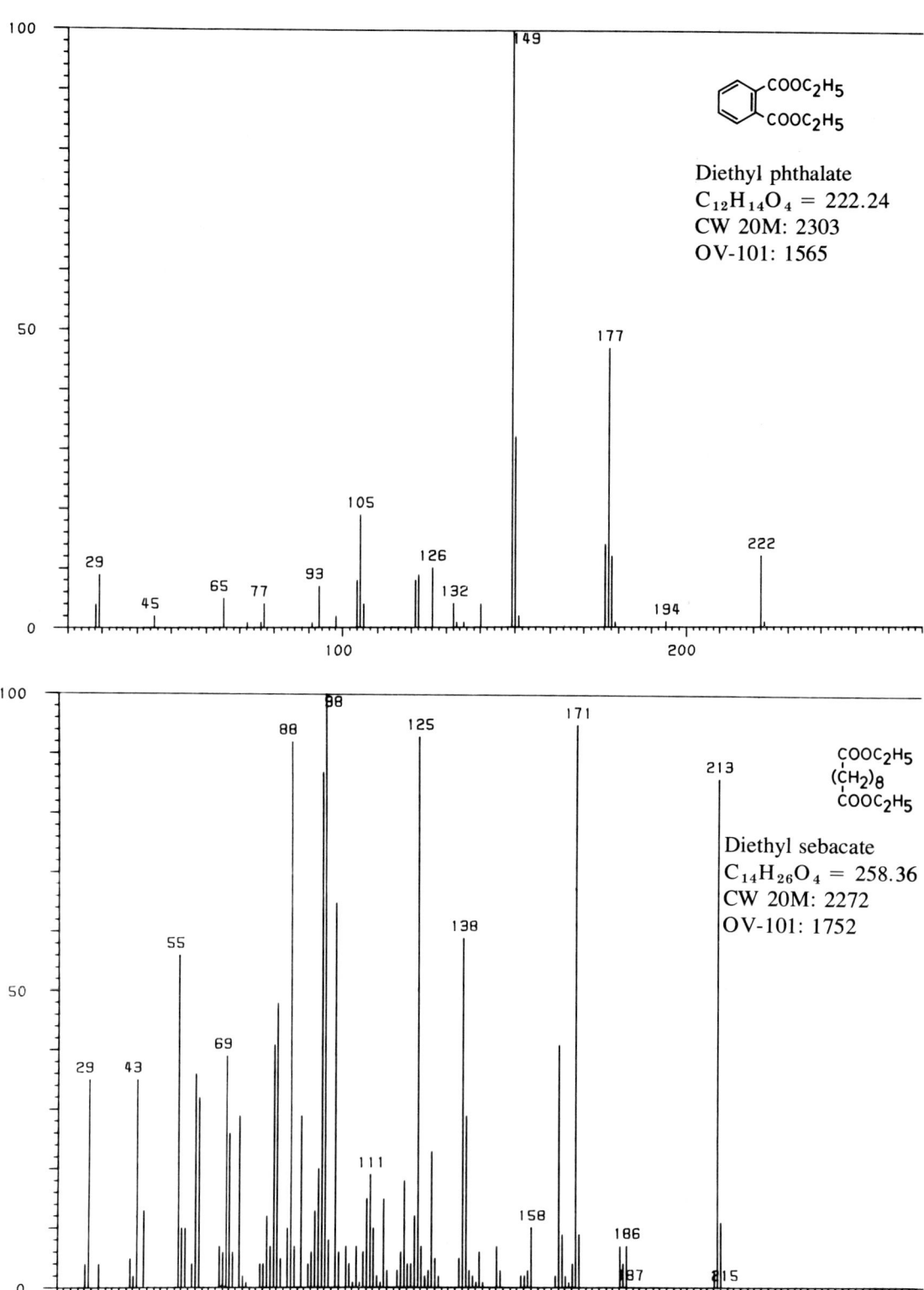

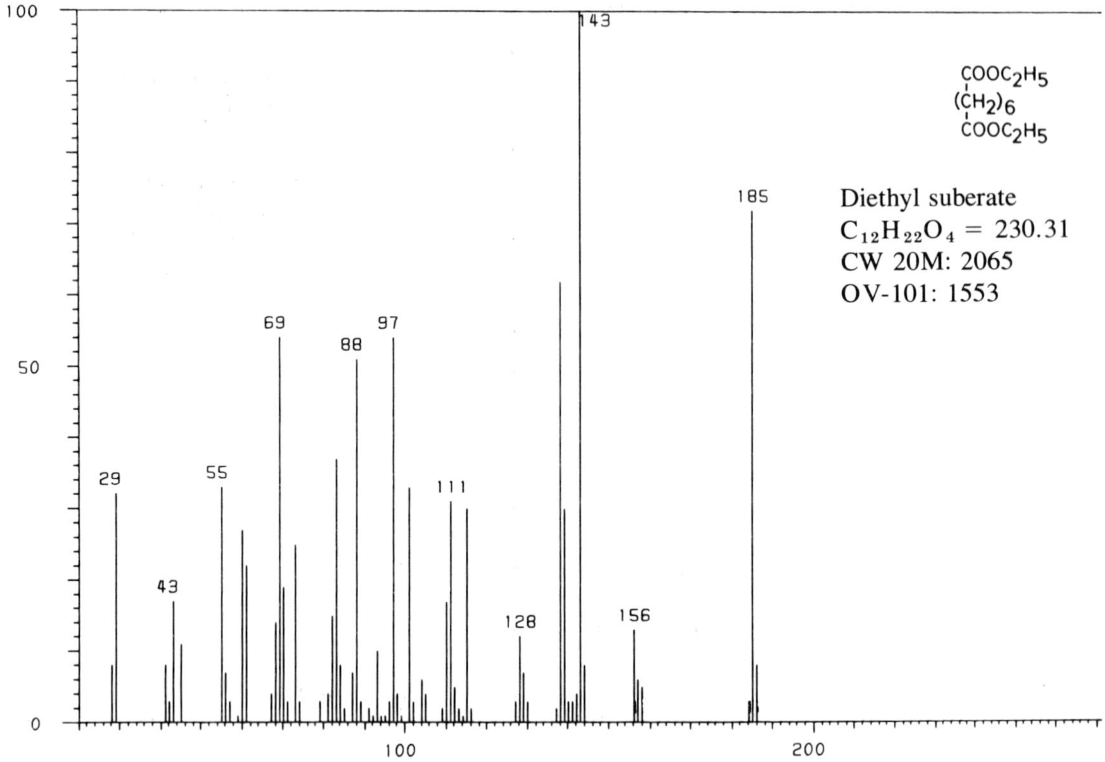

COOC₂H₅
(CH₂)₆
COOC₂H₅

Diethyl suberate
$C_{12}H_{22}O_4 = 230.31$
CW 20M: 2065
OV-101: 1553

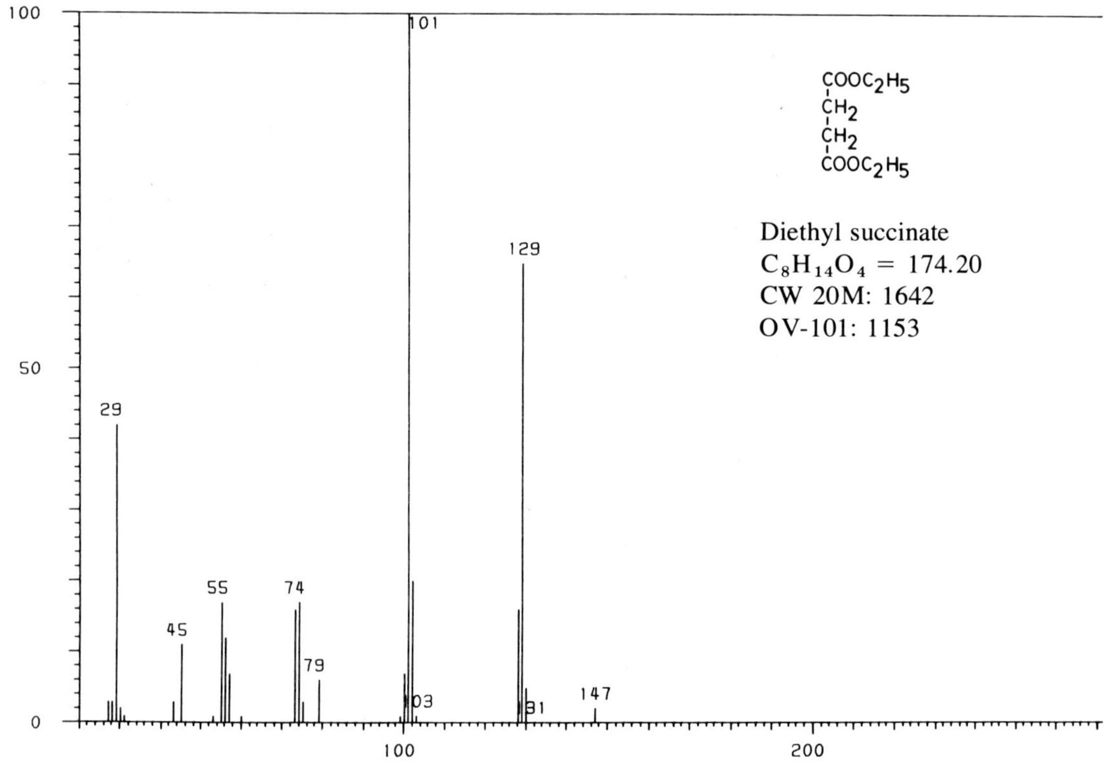

COOC₂H₅
CH₂
CH₂
COOC₂H₅

Diethyl succinate
$C_8H_{14}O_4 = 174.20$
CW 20M: 1642
OV-101: 1153

APPENDIX IV   MASS SPECTRA OF INDIVIDUAL COMPOUNDS   255

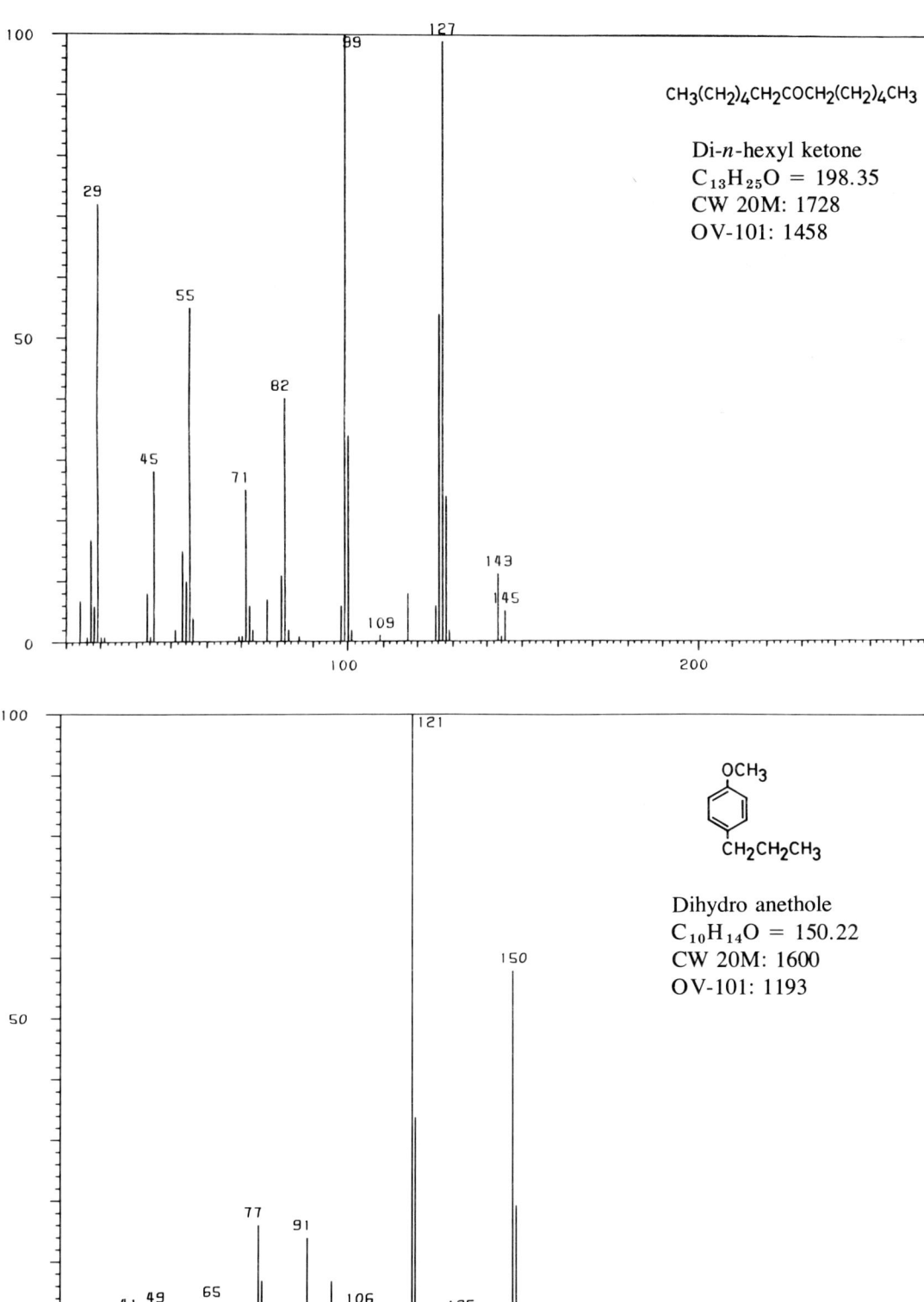

CH$_3$(CH$_2$)$_4$CH$_2$COCH$_2$(CH$_2$)$_4$CH$_3$

Di-*n*-hexyl ketone
C$_{13}$H$_{25}$O = 198.35
CW 20M: 1728
OV-101: 1458

Dihydro anethole
C$_{10}$H$_{14}$O = 150.22
CW 20M: 1600
OV-101: 1193

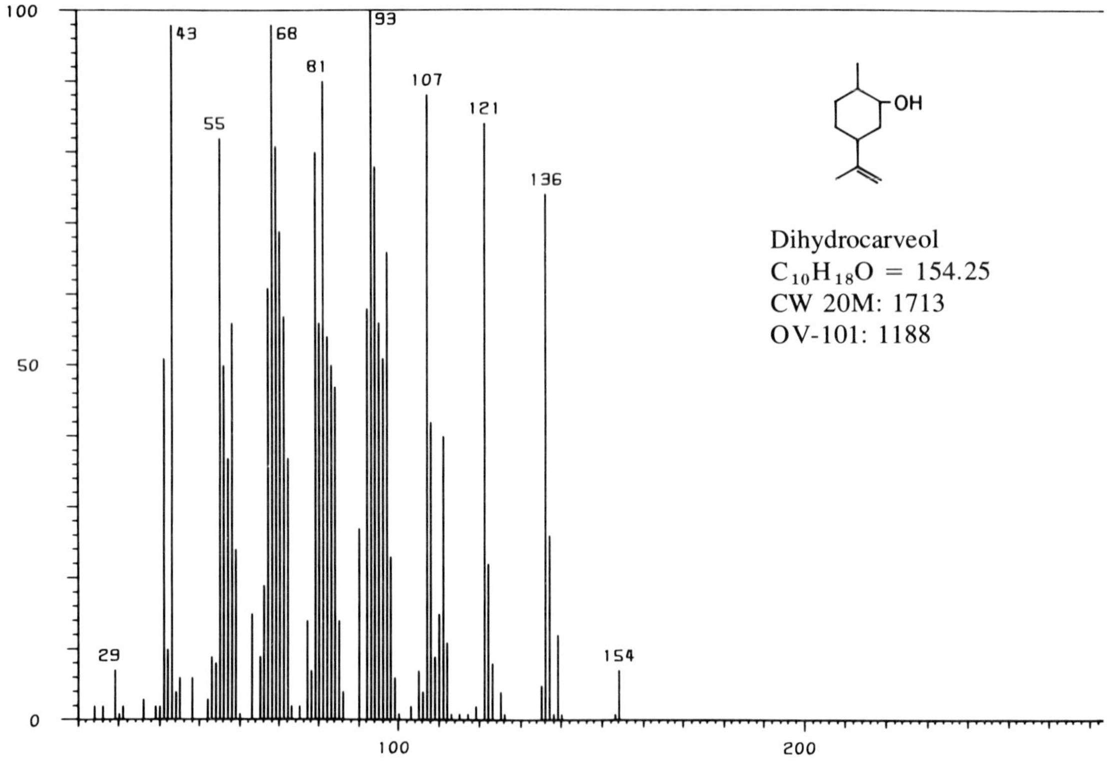

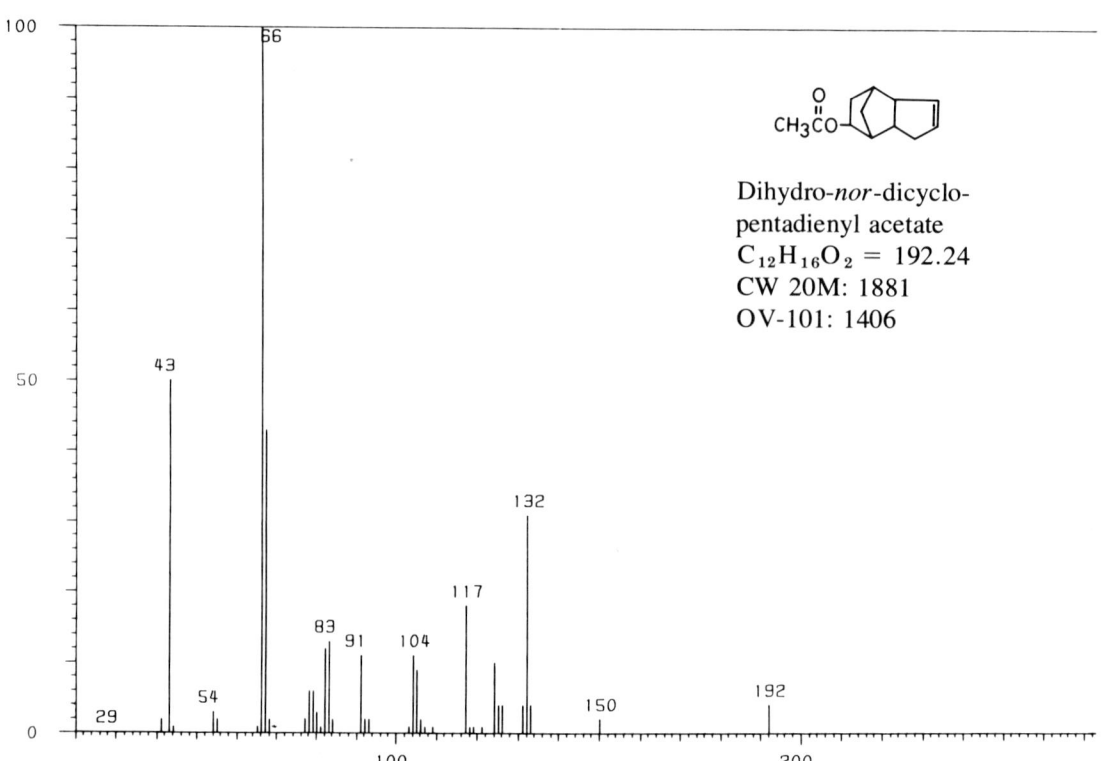

Dihydrocarveol
$C_{10}H_{18}O = 154.25$
CW 20M: 1713
OV-101: 1188

Dihydro-*nor*-dicyclo-pentadienyl acetate
$C_{12}H_{16}O_2 = 192.24$
CW 20M: 1881
OV-101: 1406

APPENDIX IV   MASS SPECTRA OF INDIVIDUAL COMPOUNDS

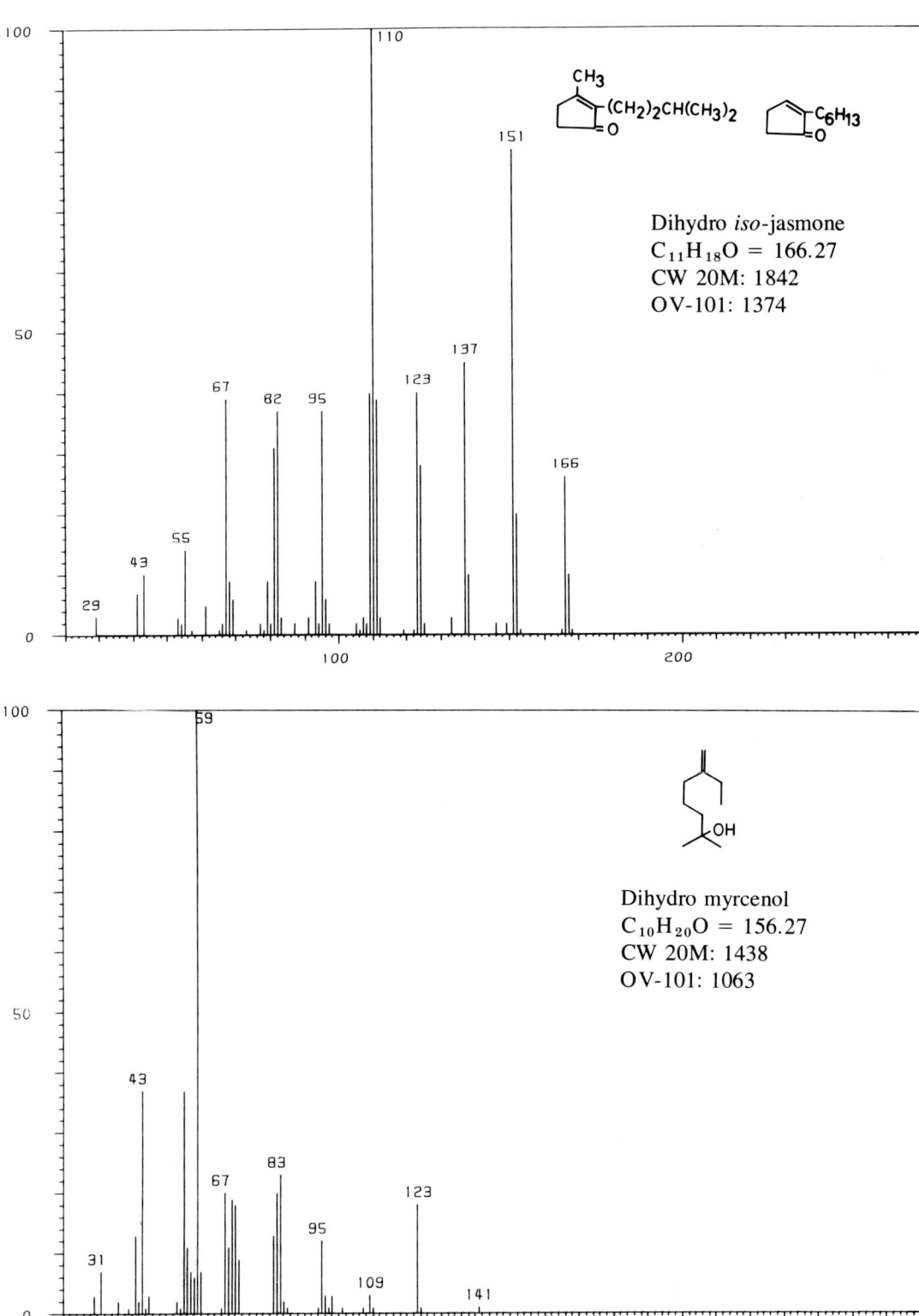

Dihydro *iso*-jasmone
$C_{11}H_{18}O = 166.27$
CW 20M: 1842
OV-101: 1374

Dihydro myrcenol
$C_{10}H_{20}O = 156.27$
CW 20M: 1438
OV-101: 1063

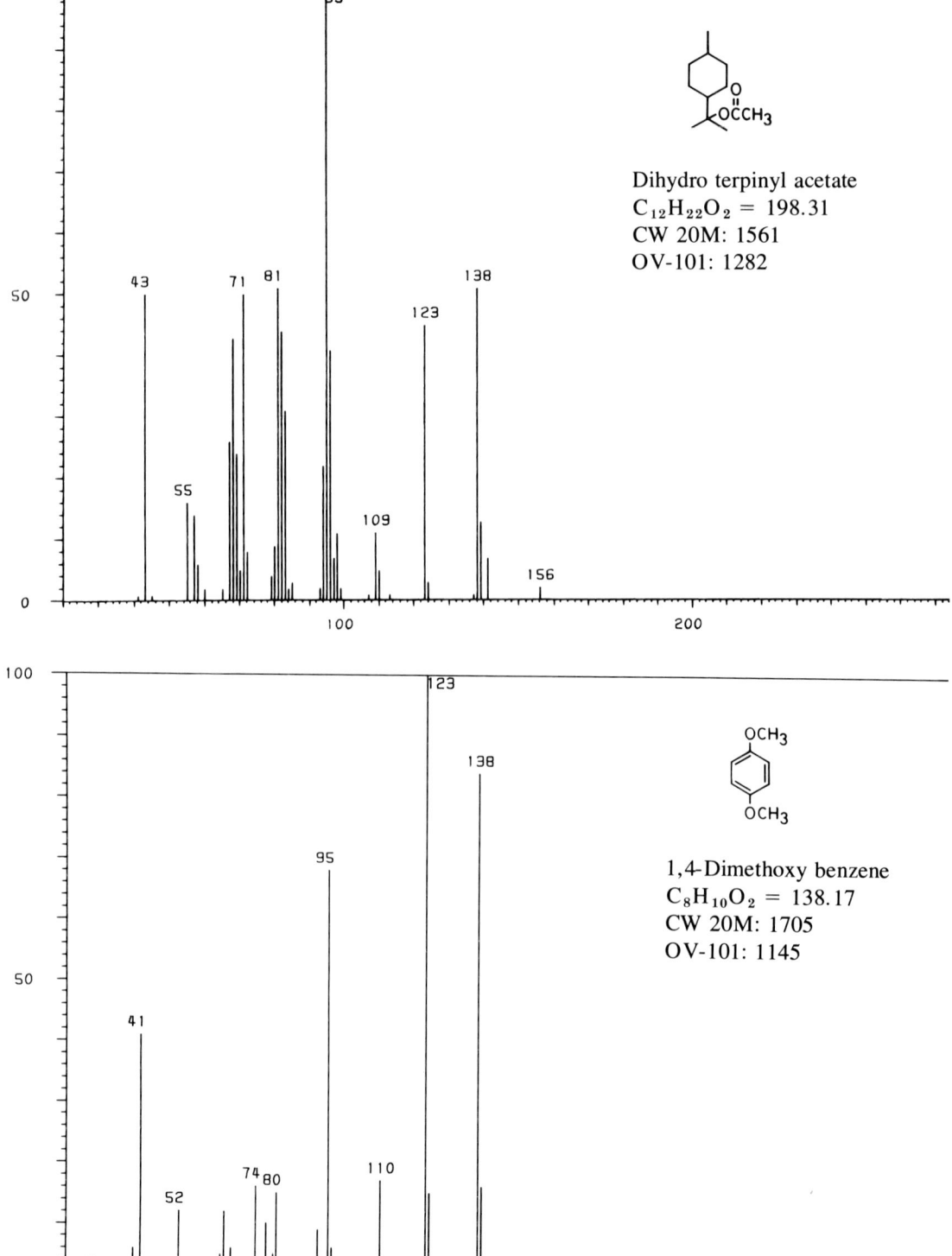

APPENDIX IV MASS SPECORA OF INDIVIDUAL COMPOUNDS

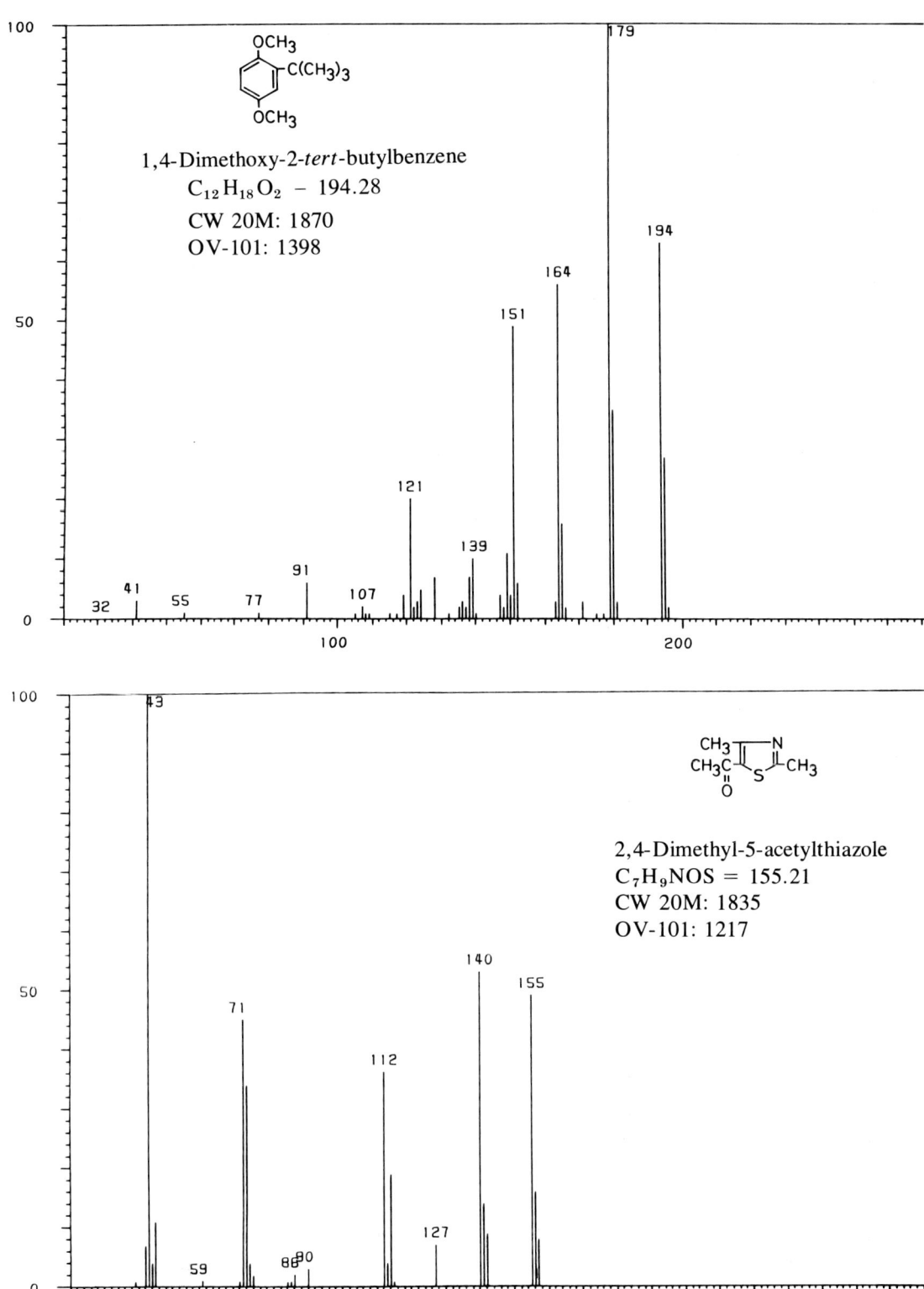

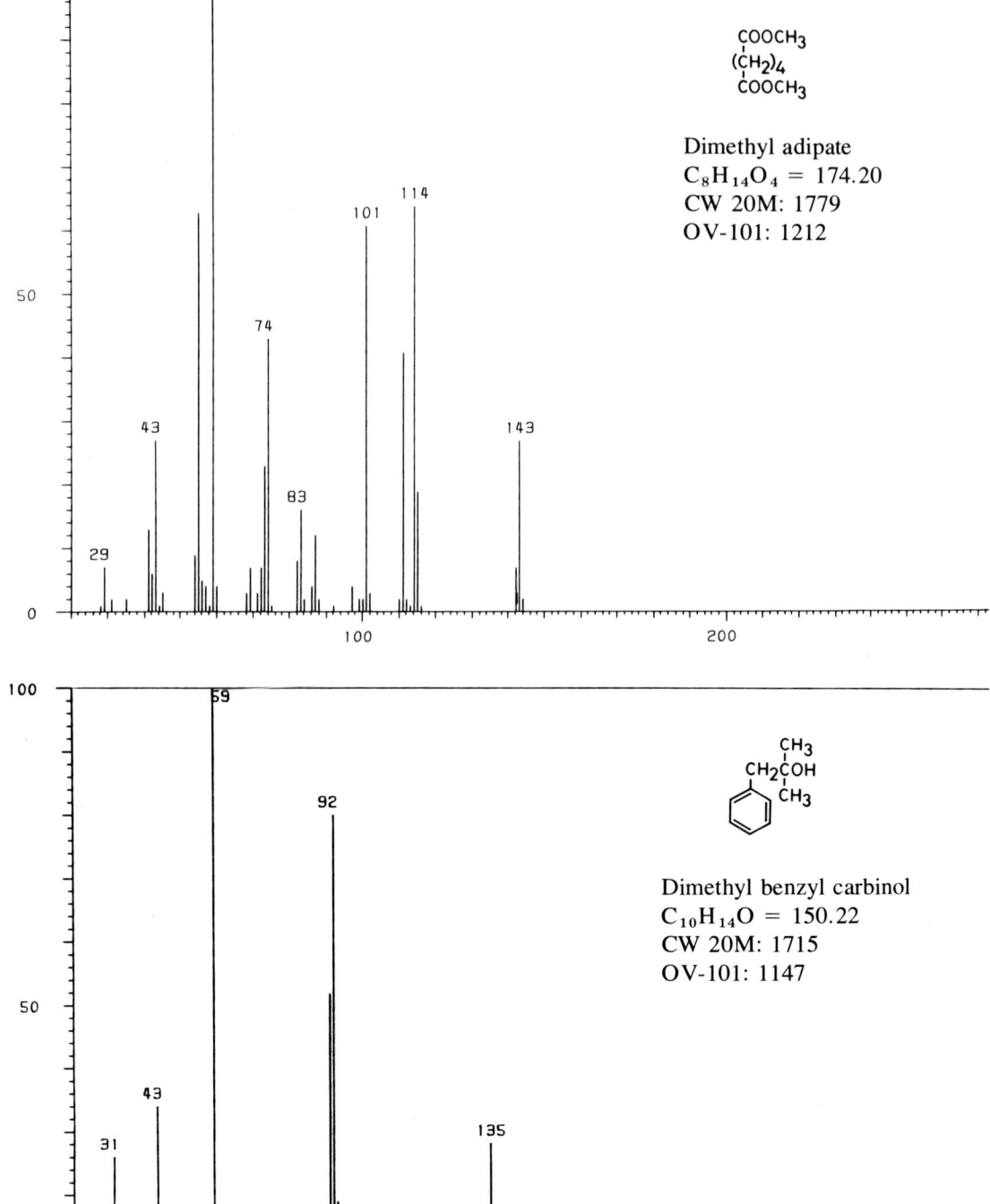

APPENDIX IV   MASS SPECTRA OF INDIVIDUAL COMPOUNDS

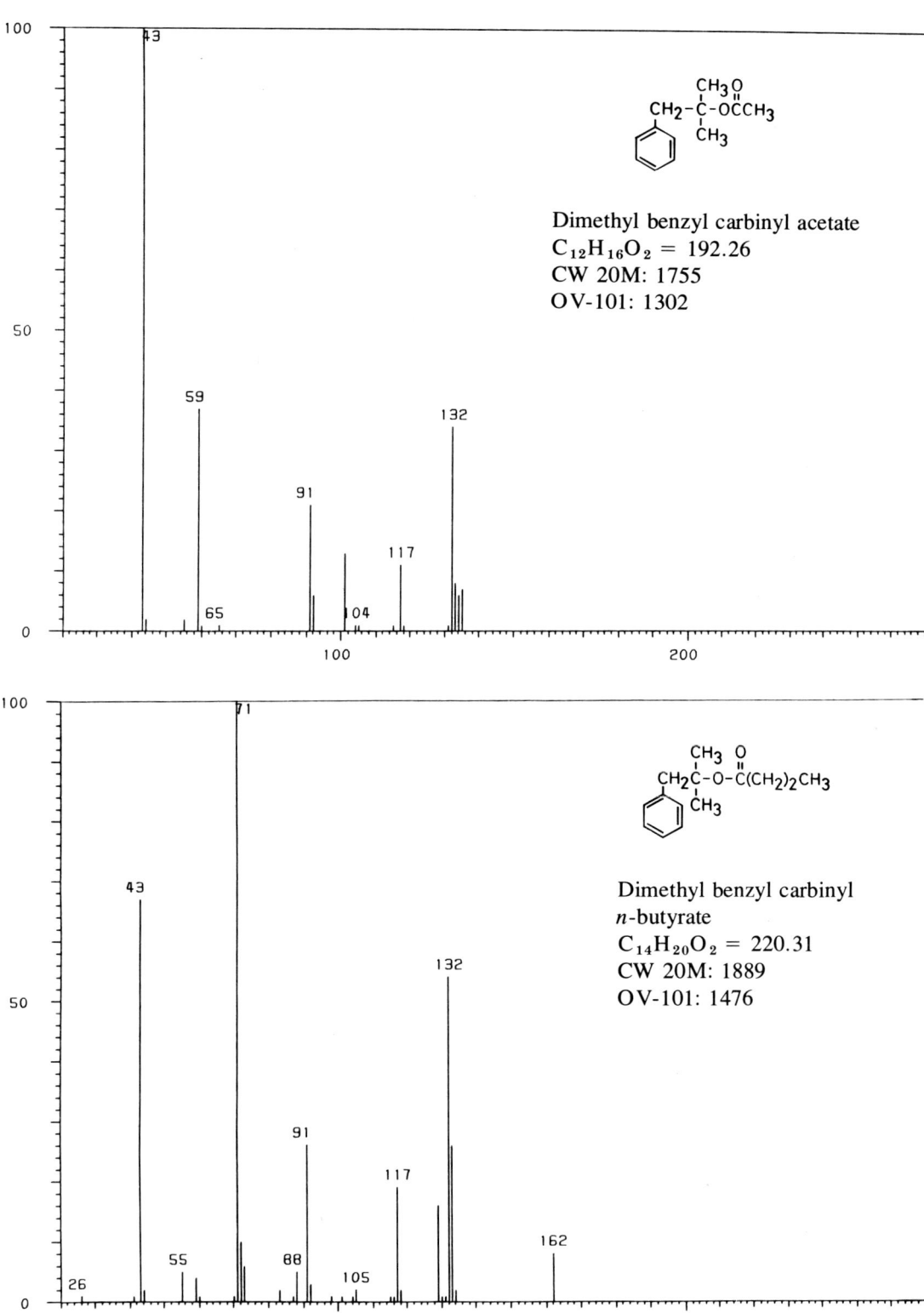

Dimethyl benzyl carbinyl acetate
$C_{12}H_{16}O_2 = 192.26$
CW 20M: 1755
OV-101: 1302

Dimethyl benzyl carbinyl
n-butyrate
$C_{14}H_{20}O_2 = 220.31$
CW 20M: 1889
OV-101: 1476

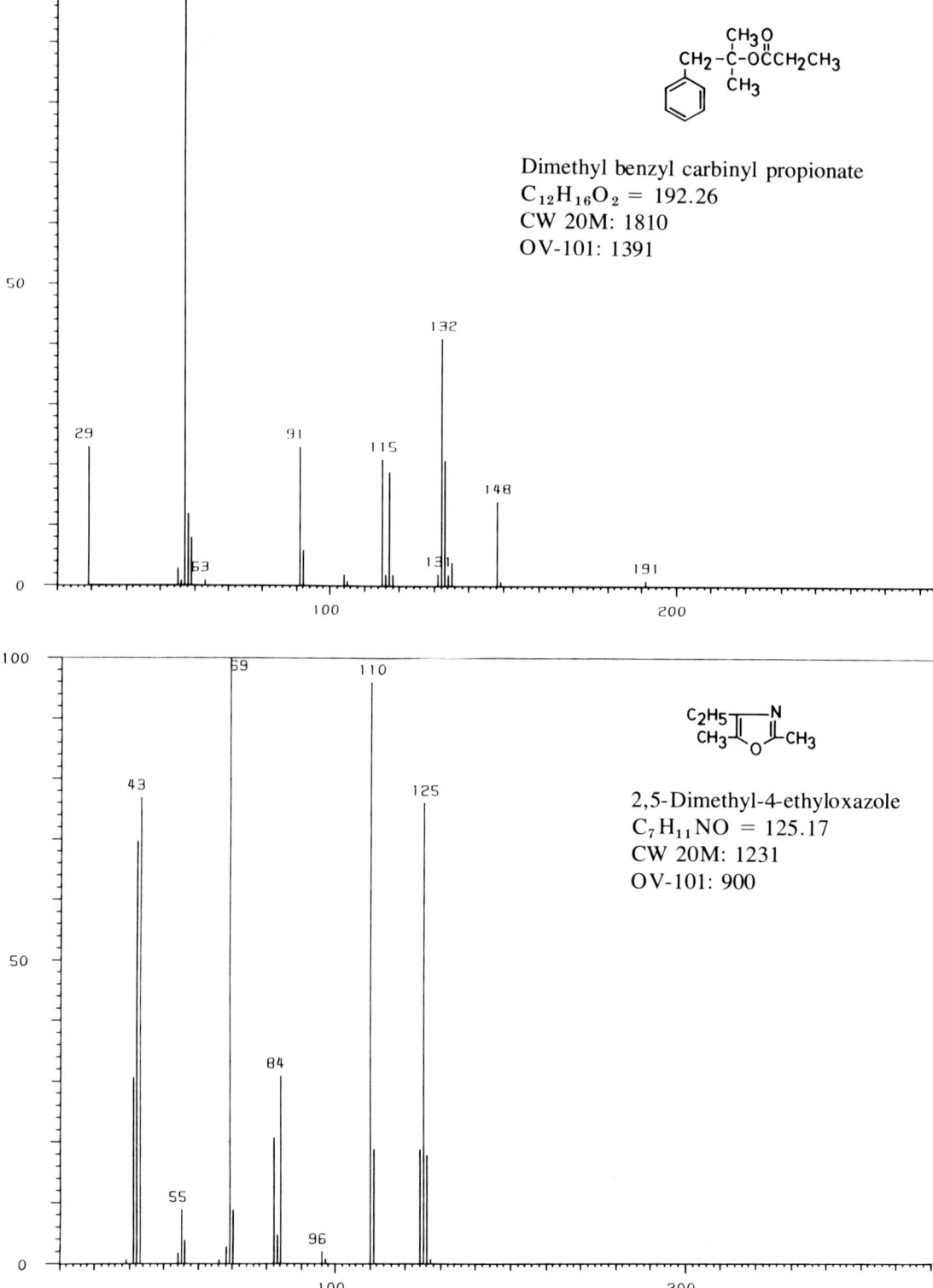

APPENDIX IV   MASS SPECTRA OF INDIVIDUAL COMPOUNDS

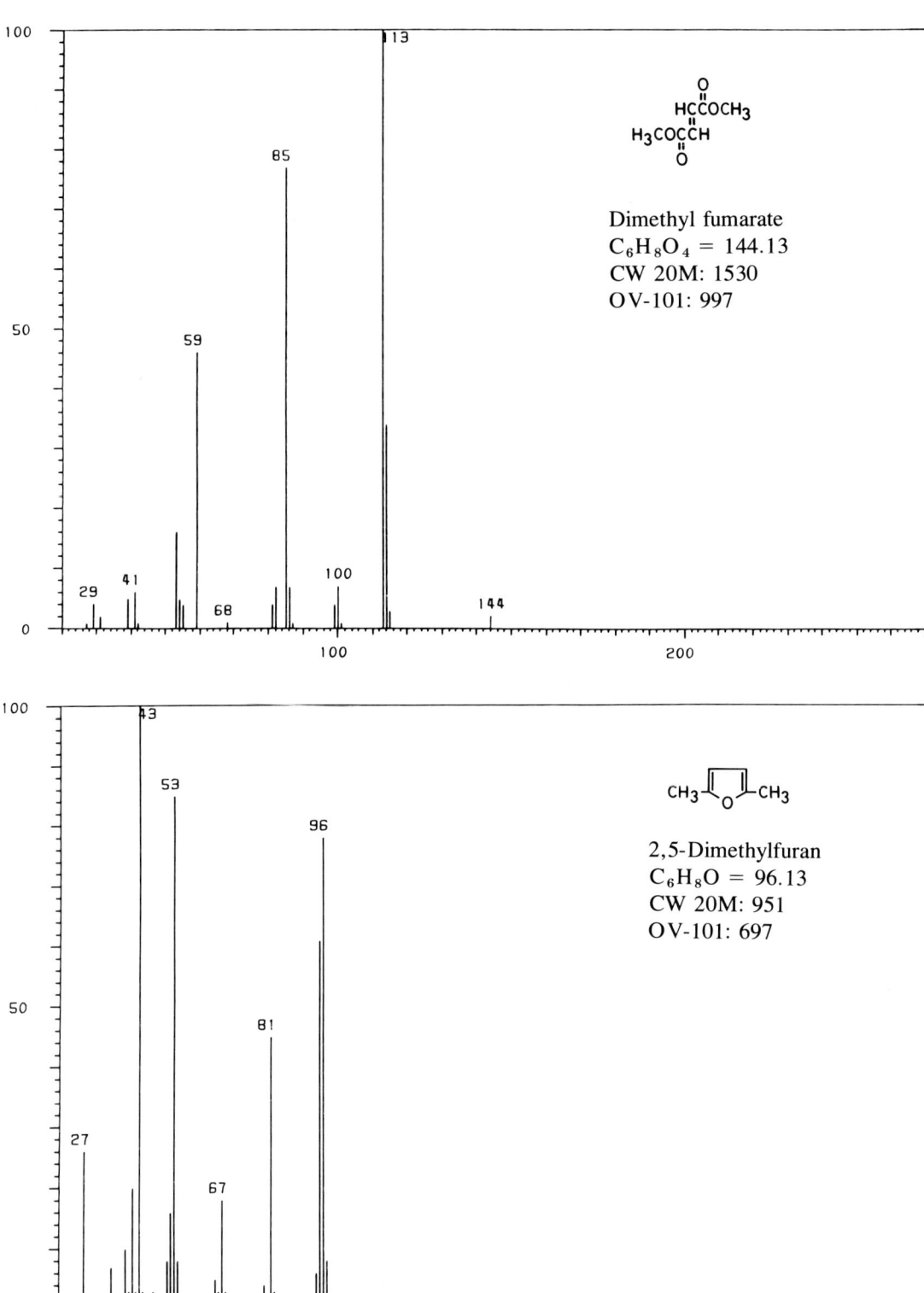

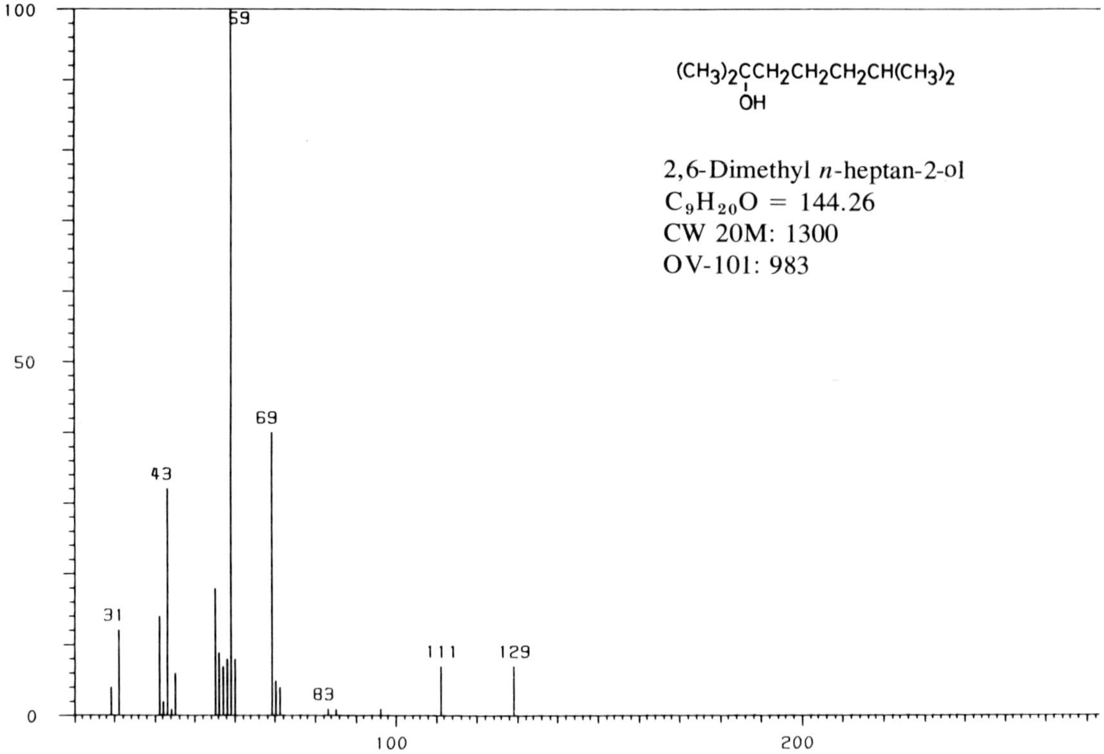
100
50
0
59
69
43
31
111
129
83
100
200
(CH3)2CCH2CH2CH2CH(CH3)2
OH

2,6-Dimethyl *n*-heptan-2-ol
$C_9H_{20}O$ = 144.26
CW 20M: 1300
OV-101: 983

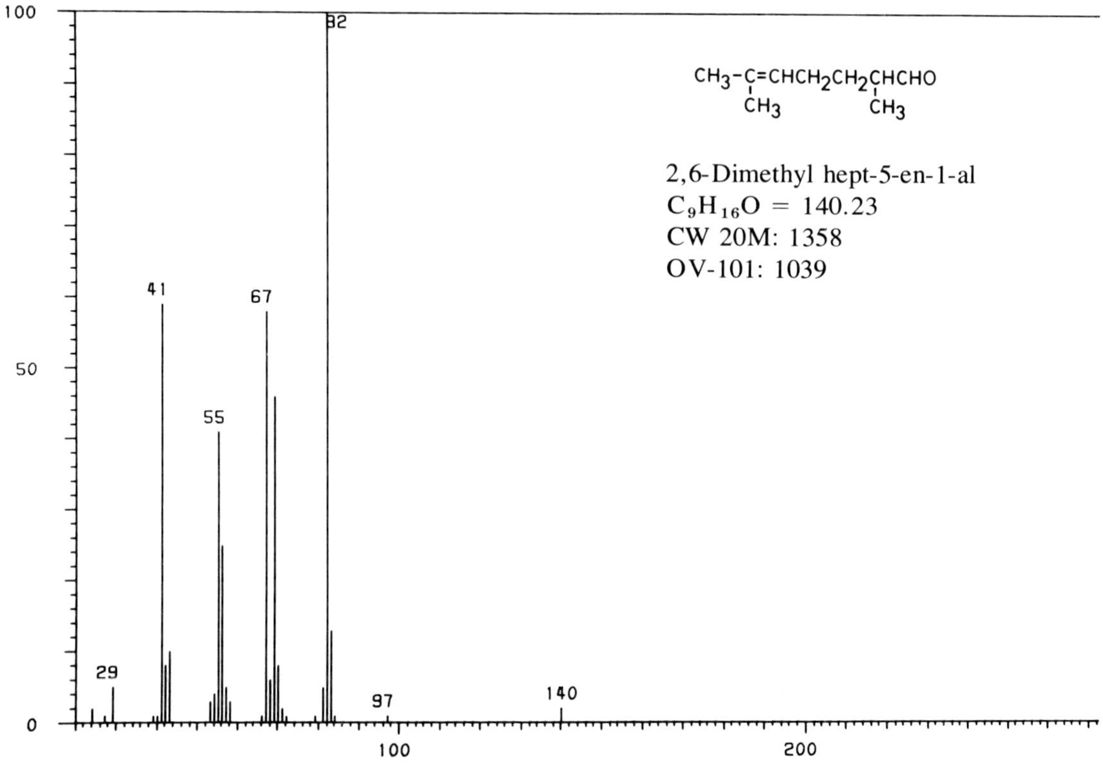
100
50
0
82
41
67
55
29
97
140
100
200
CH3-C=CHCH2CH2CHCHO
CH3 CH3

2,6-Dimethyl hept-5-en-1-al
$C_9H_{16}O$ = 140.23
CW 20M: 1358
OV-101: 1039

APPENDIX IV   MASS SPECTRA OF INDIVIDUAL COMPOUNDS

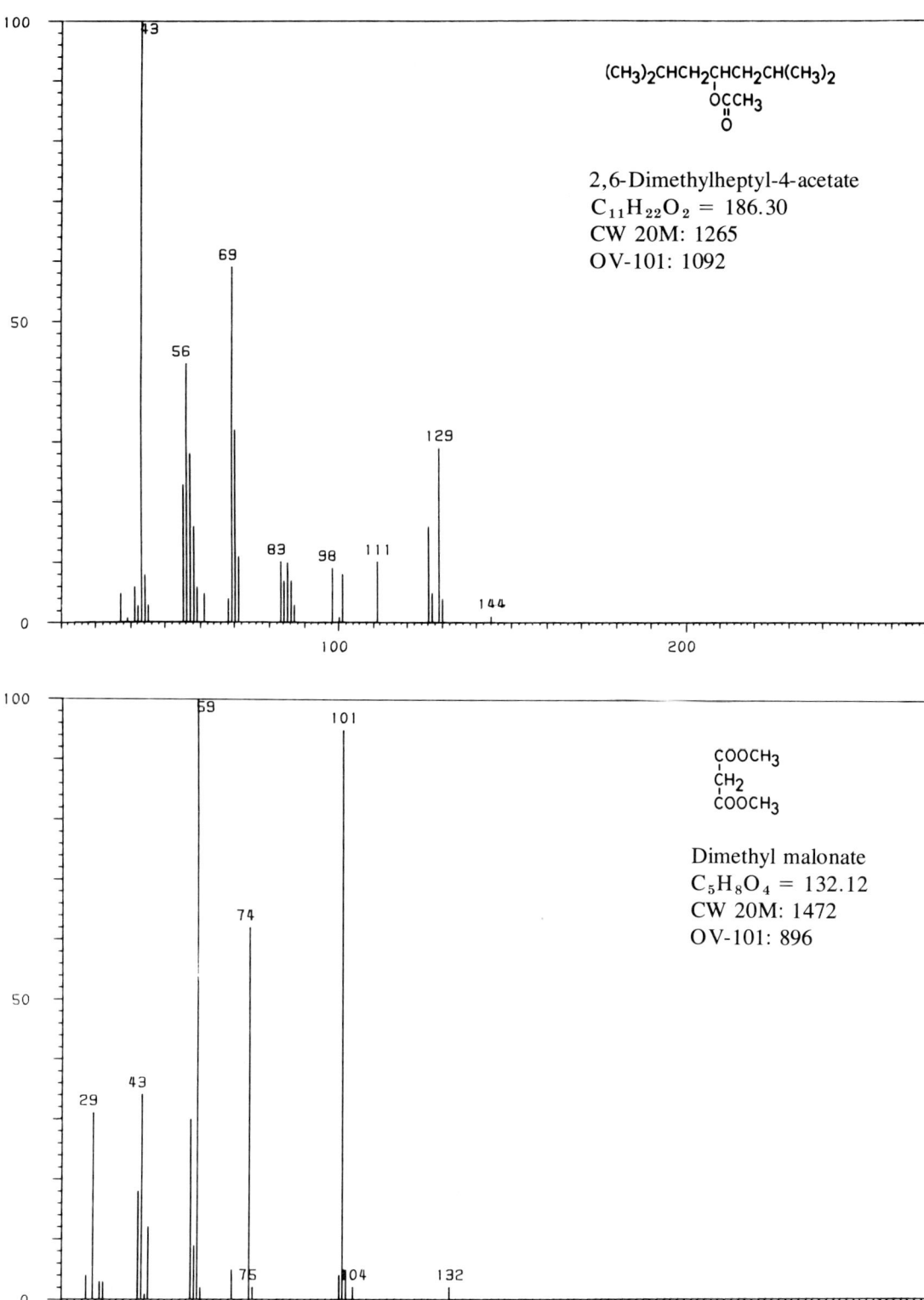

2,6-Dimethylheptyl-4-acetate
$C_{11}H_{22}O_2 = 186.30$
CW 20M: 1265
OV-101: 1092

Dimethyl malonate
$C_5H_8O_4 = 132.12$
CW 20M: 1472
OV-101: 896

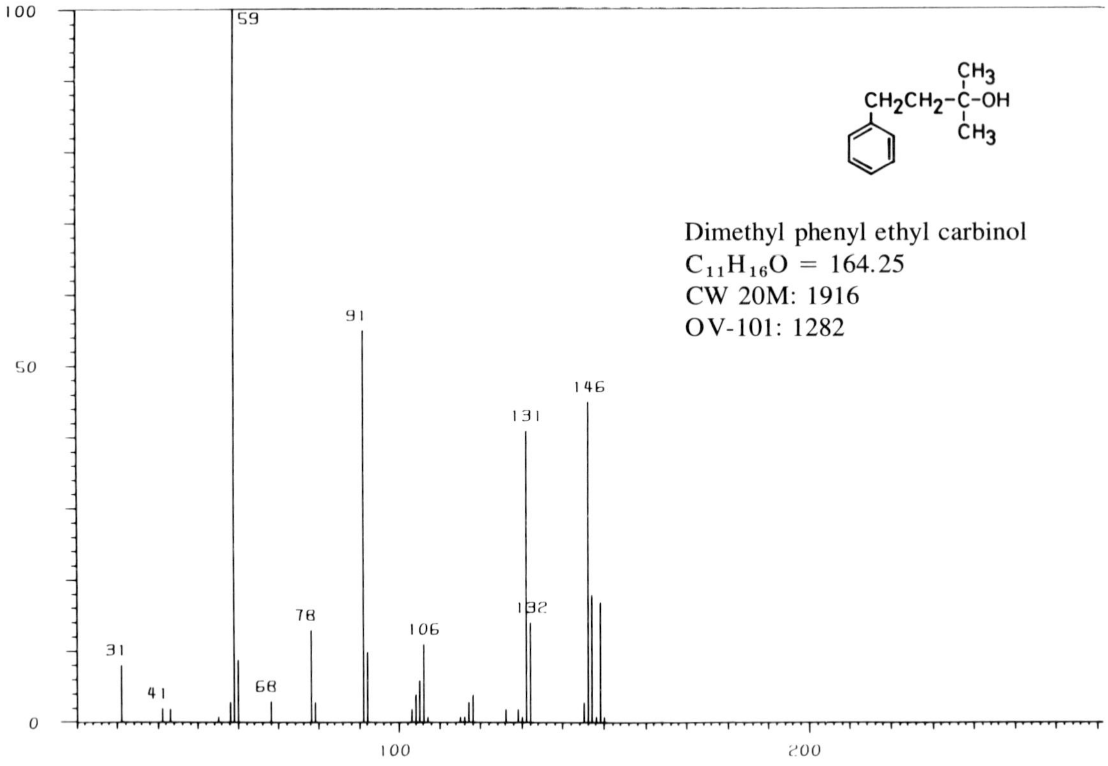

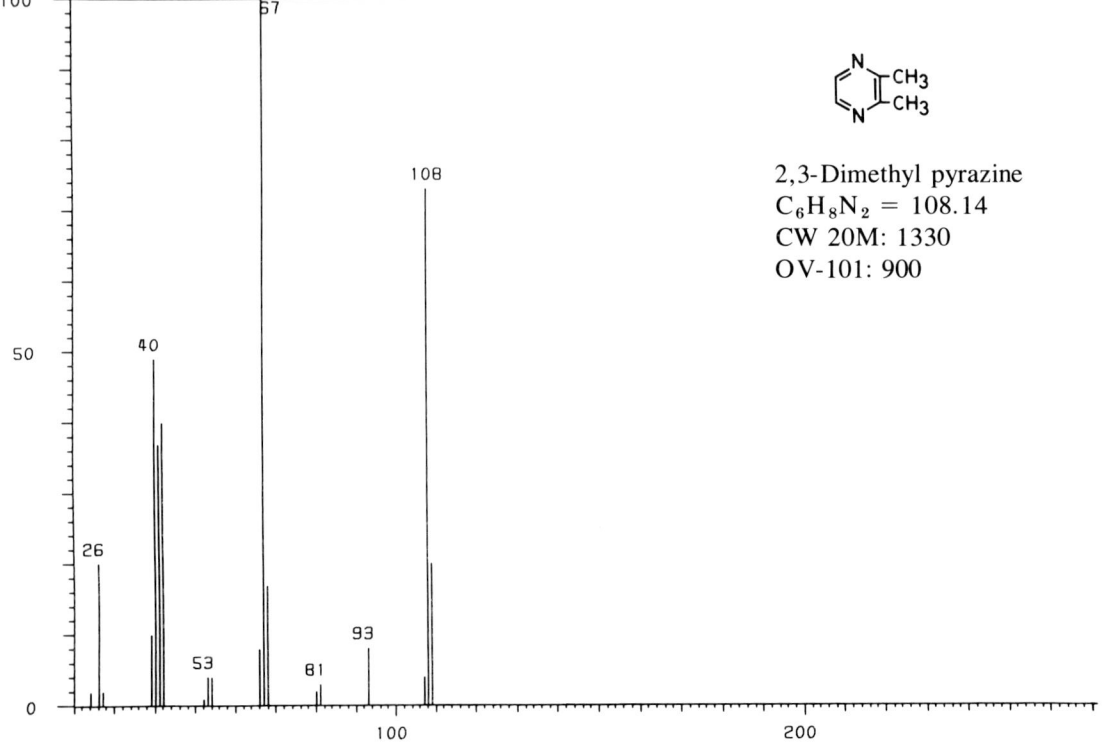

Dimethyl phenyl ethyl carbinol
$C_{11}H_{16}O = 164.25$
CW 20M: 1916
OV-101: 1282

2,3-Dimethyl pyrazine
$C_6H_8N_2 = 108.14$
CW 20M: 1330
OV-101: 900

APPENDIX IV   MASS SPECTRA OF INDIVIDUAL COMPOUNDS   267

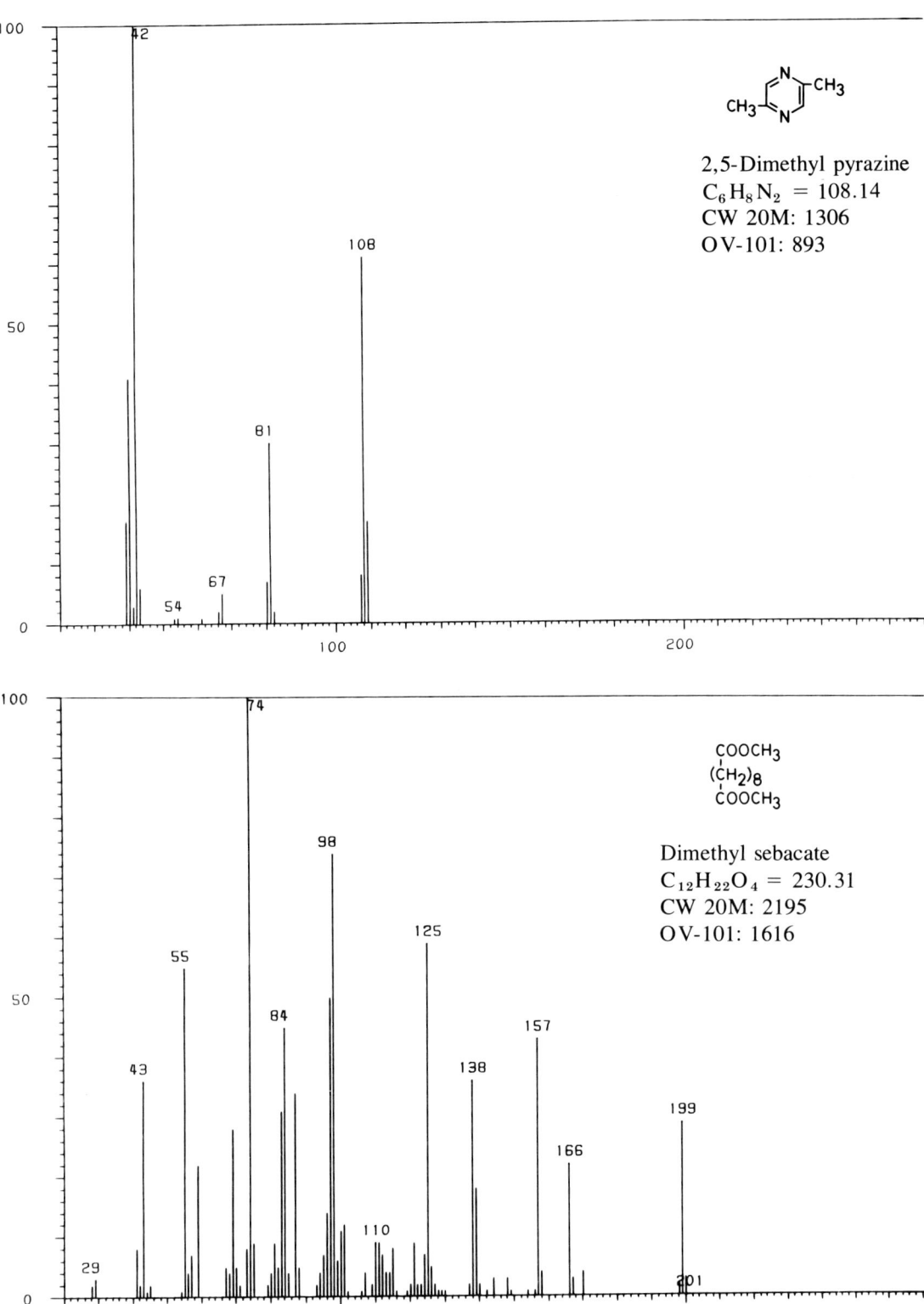

2,5-Dimethyl pyrazine
$C_6H_8N_2 = 108.14$
CW 20M: 1306
OV-101: 893

Dimethyl sebacate
$C_{12}H_{22}O_4 = 230.31$
CW 20M: 2195
OV-101: 1616

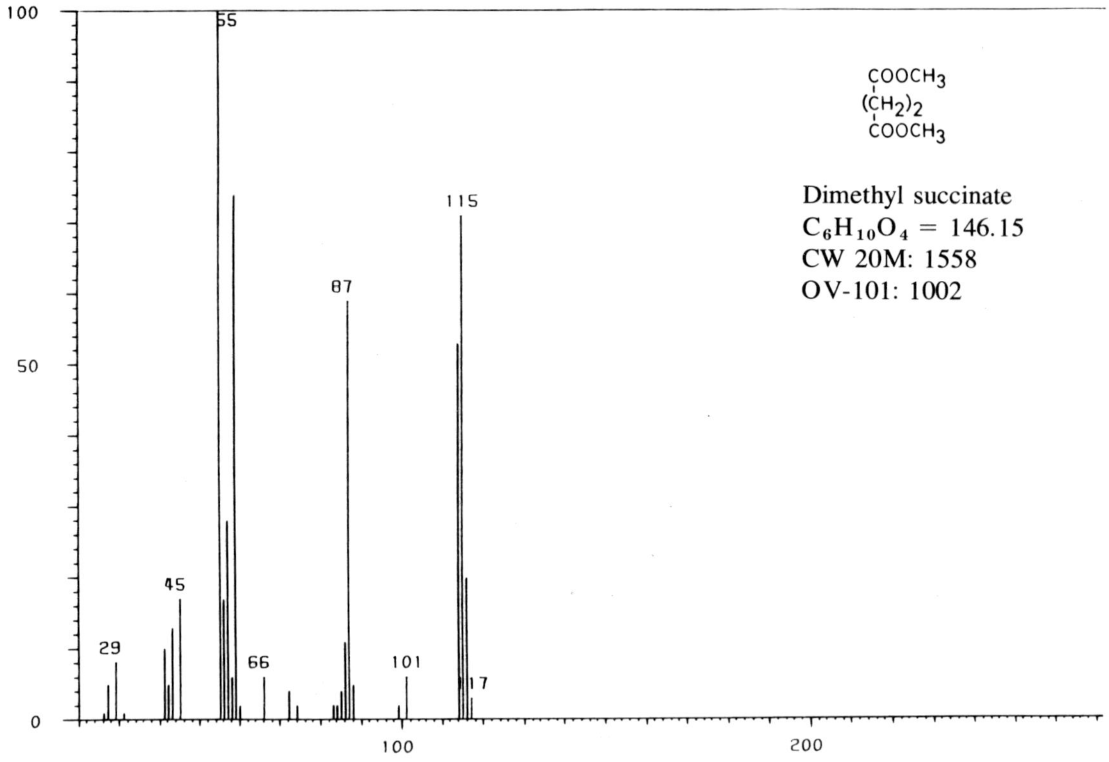

Dimethyl succinate
$C_6H_{10}O_4 = 146.15$
CW 20M: 1558
OV-101: 1002

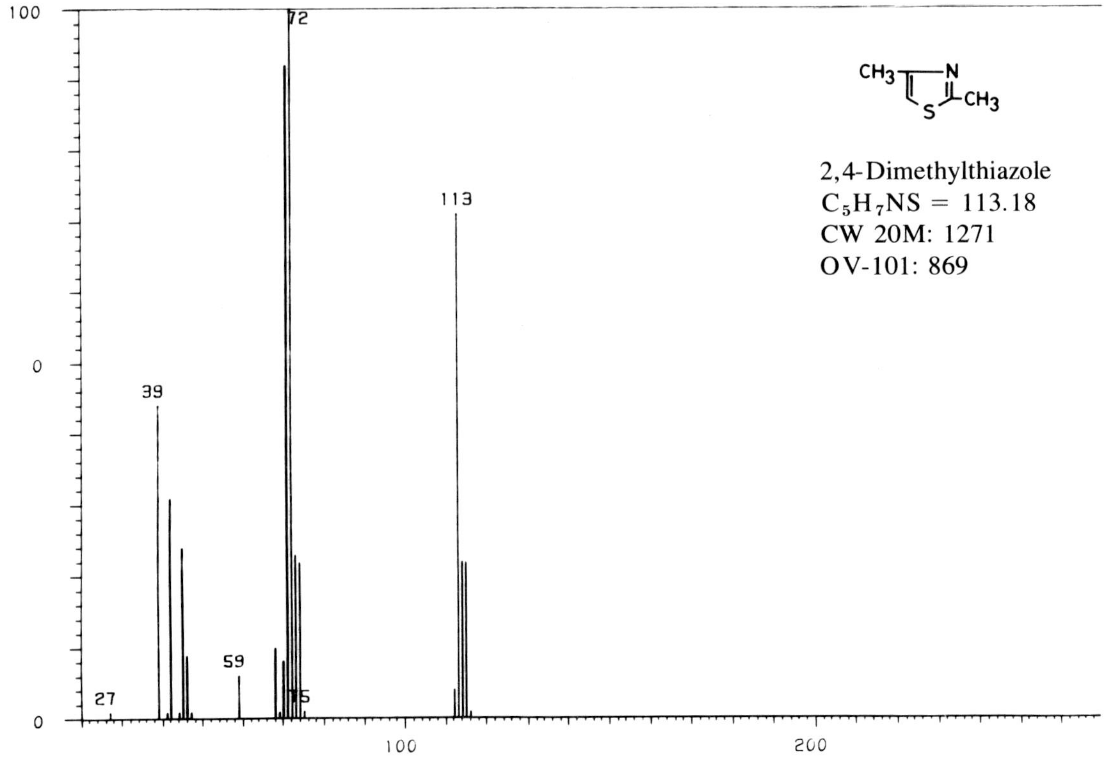

2,4-Dimethylthiazole
$C_5H_7NS = 113.18$
CW 20M: 1271
OV-101: 869

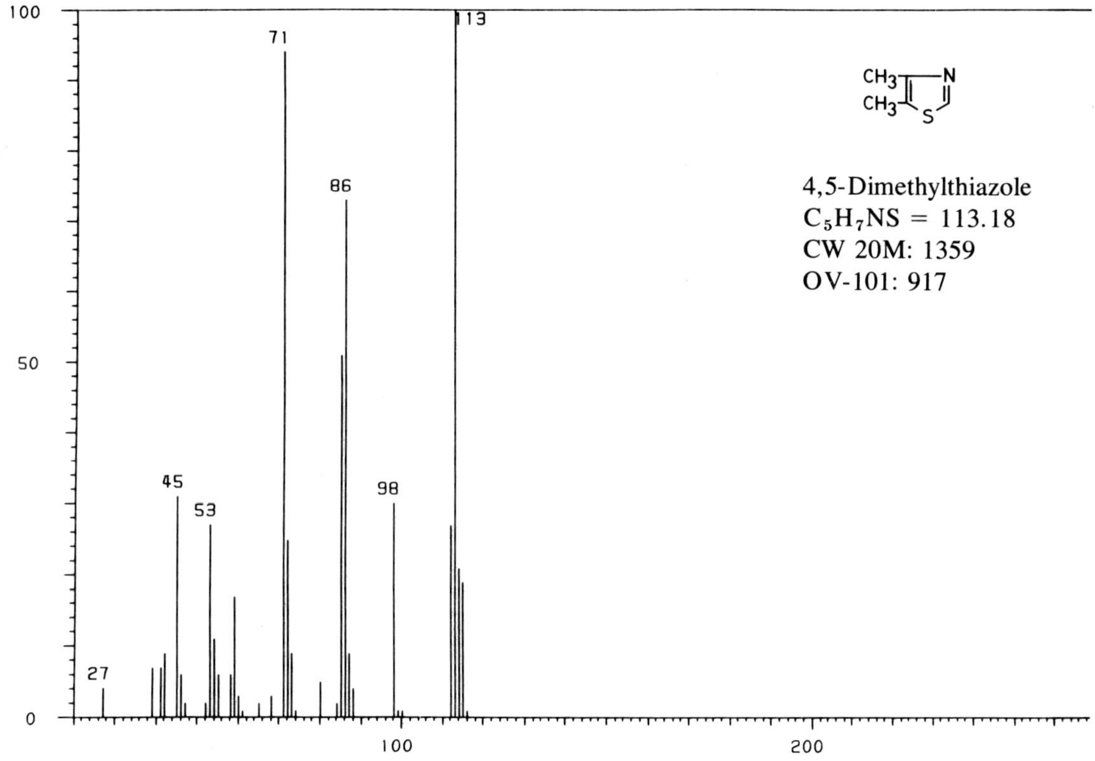

APPENDIX IV   MASS SPECTRA OF INDIVIDUAL COMPOUNDS

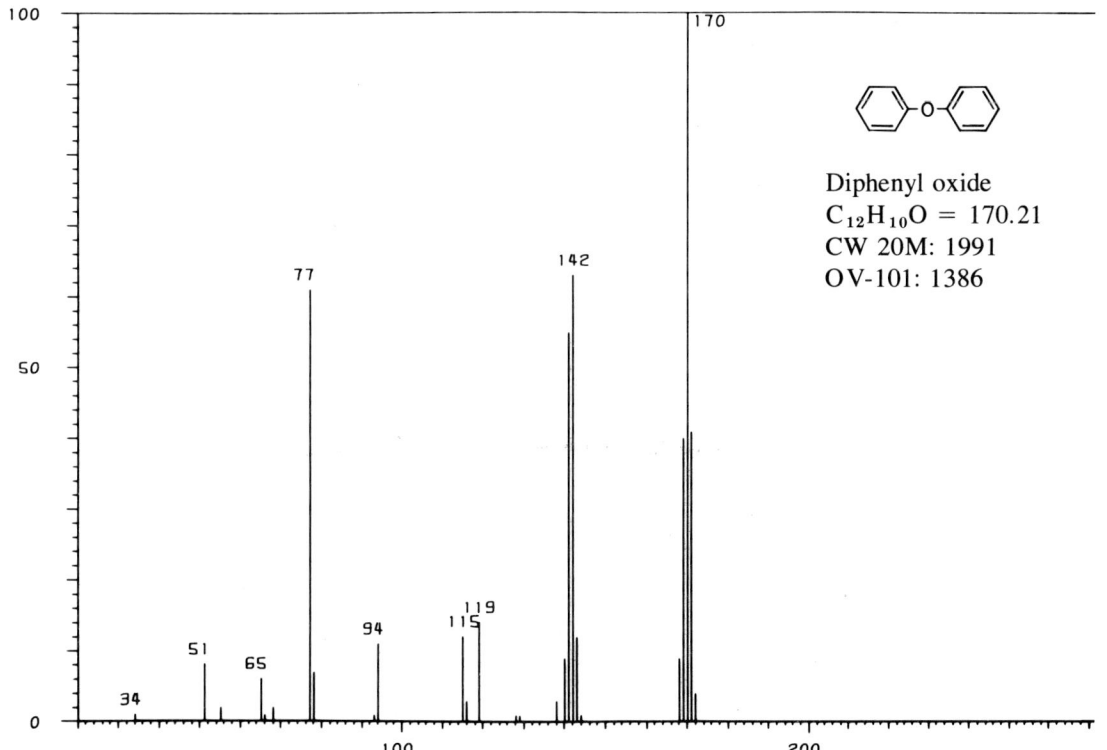

4,5-Dimethylthiazole
$C_5H_7NS = 113.18$
CW 20M: 1359
OV-101: 917

Diphenyl oxide
$C_{12}H_{10}O = 170.21$
CW 20M: 1991
OV-101: 1386

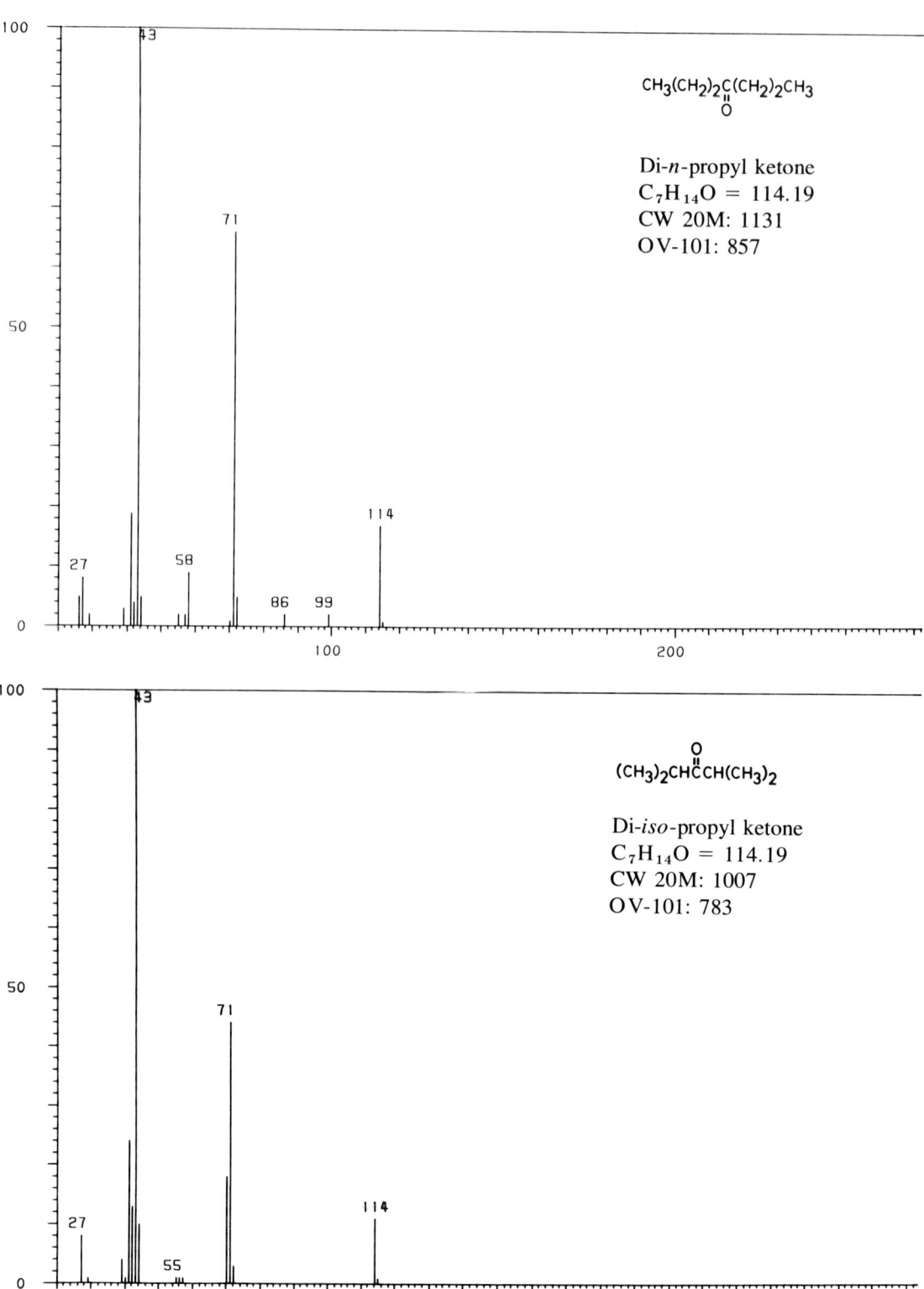

APPENDIX IV MASS SPECTRA OF INDIVIDUAL COMPOUNDS

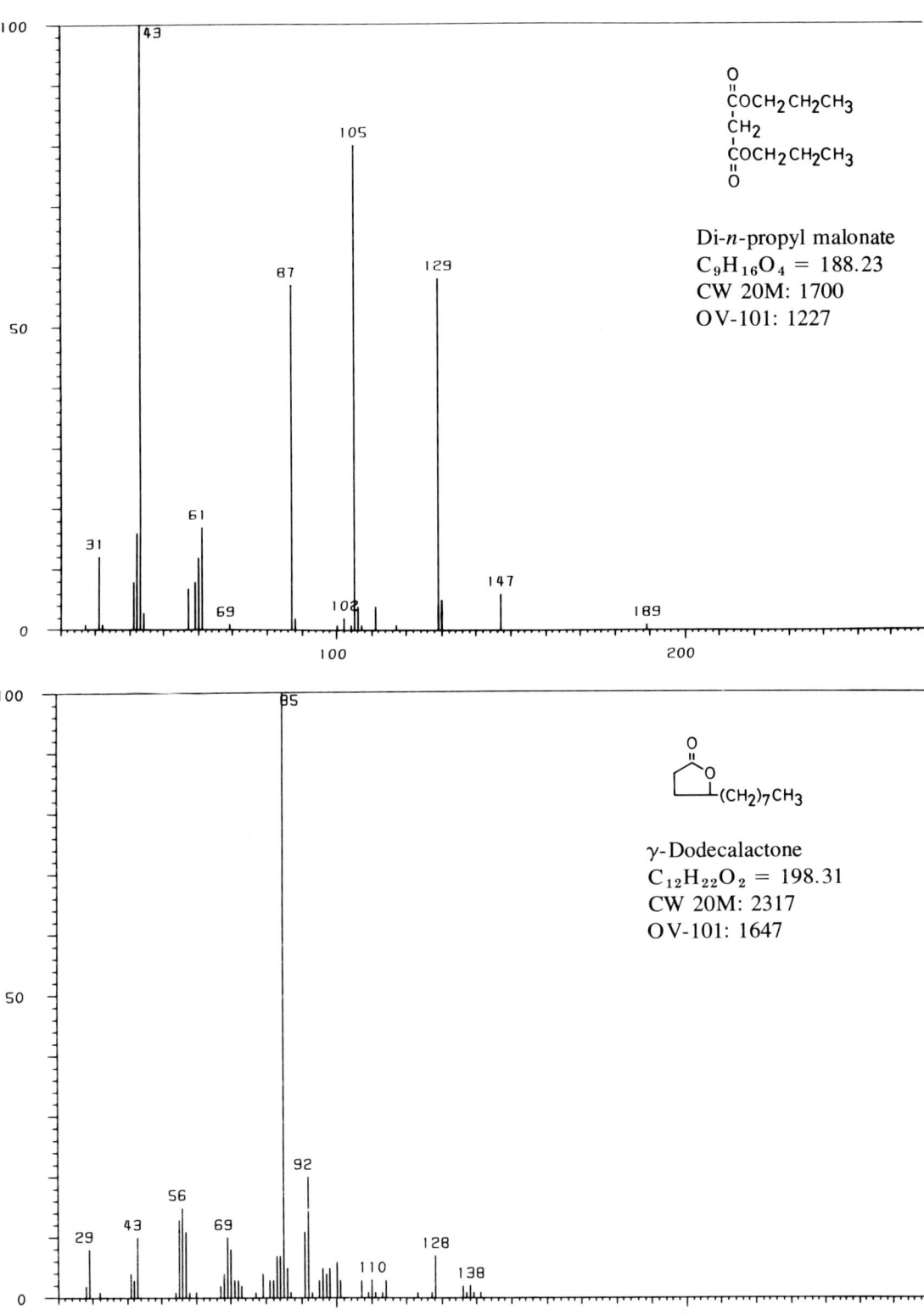

Di-n-propyl malonate
$C_9H_{16}O_4 = 188.23$
CW 20M: 1700
OV-101: 1227

γ-Dodecalactone
$C_{12}H_{22}O_2 = 198.31$
CW 20M: 2317
OV-101: 1647

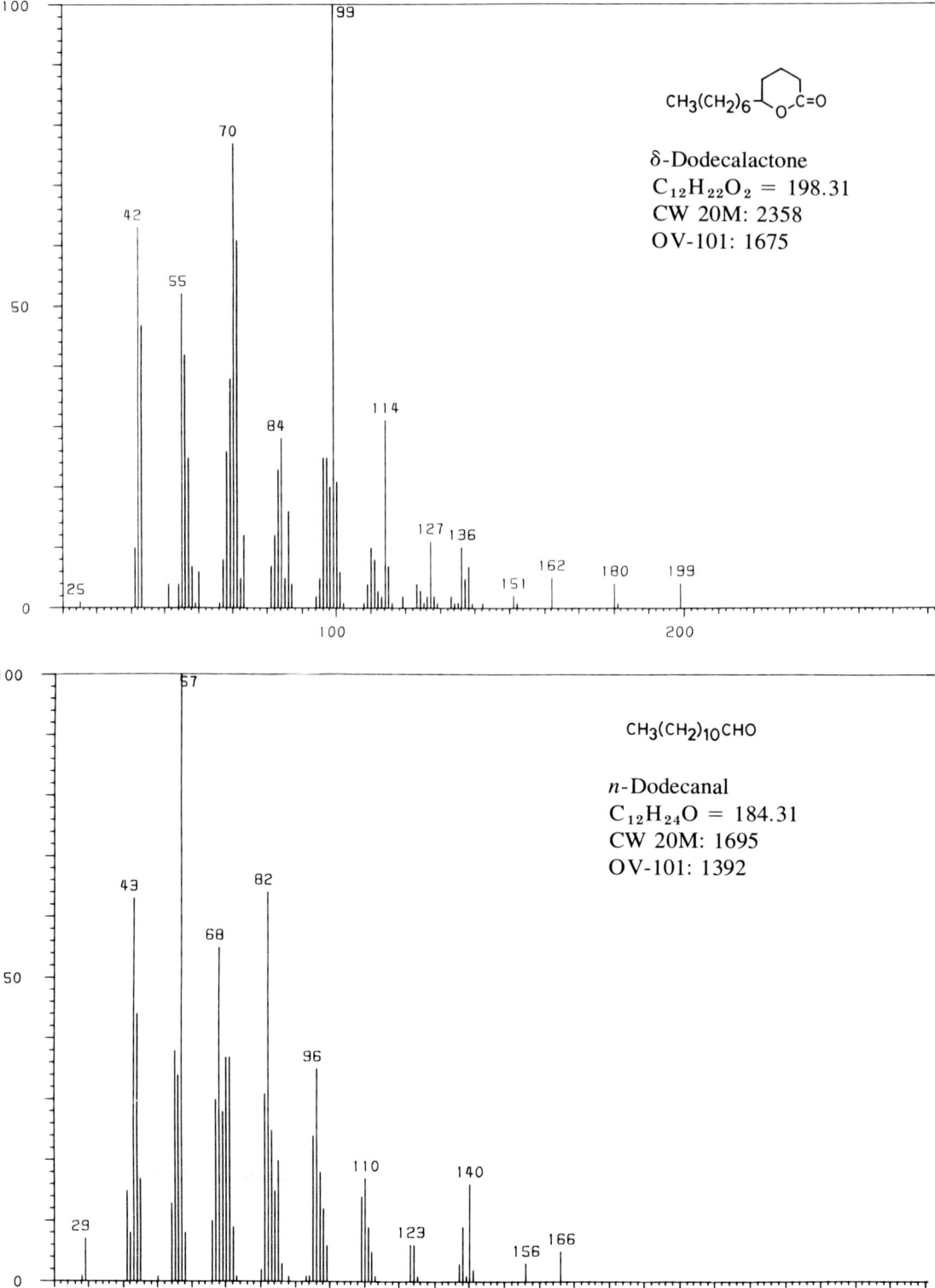

APPENDIX IV   MASS SPECTRA OF INDIVIDUAL COMPOUNDS   273

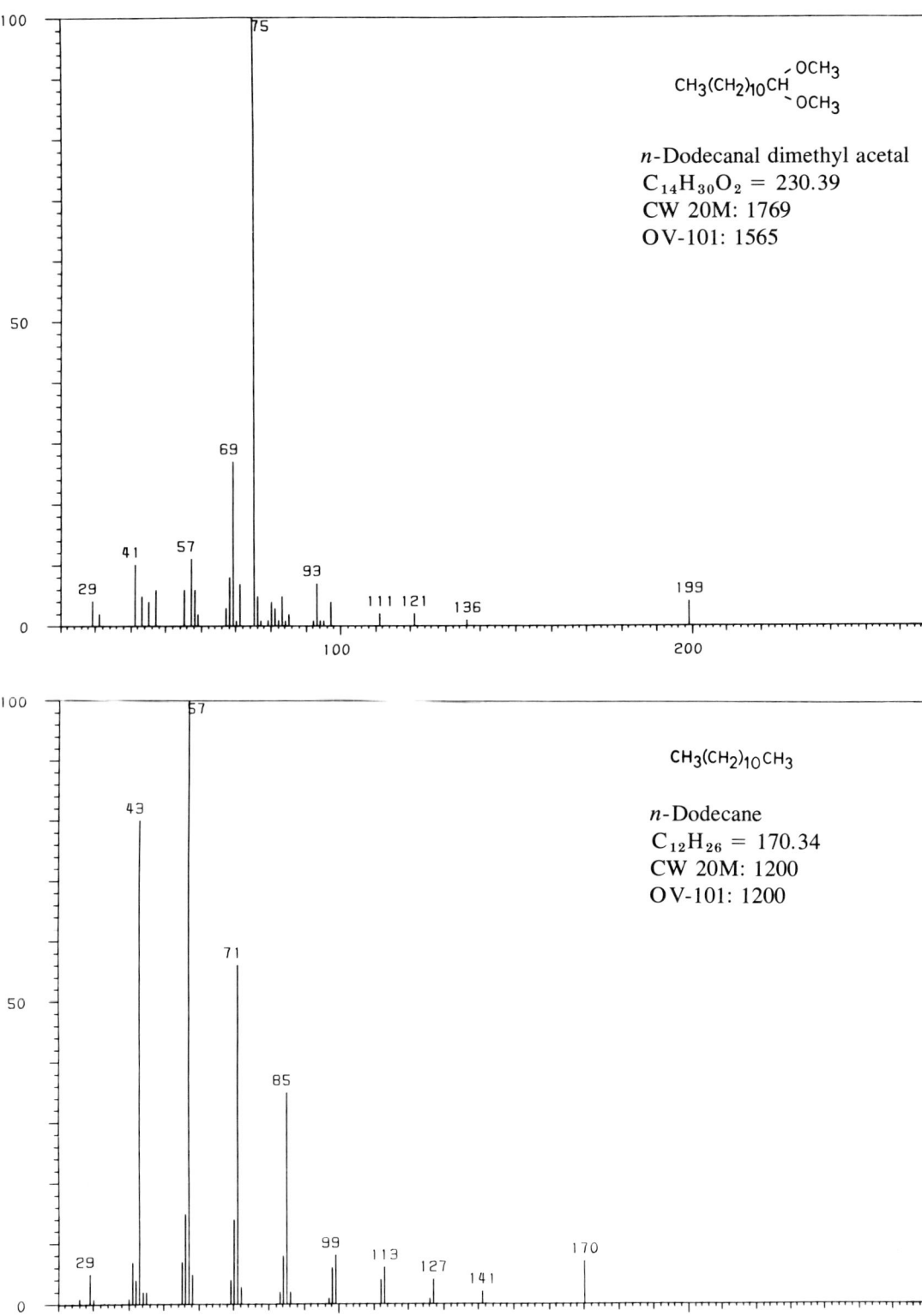

n-Dodecanol
$C_{12}H_{26}O$ = 186.33
CW 20M: 1925
OV-101: 1468

n-Eicosane
$C_{20}H_{42}$ = 282.56
CW 20M: 2000
OV-101: 2000

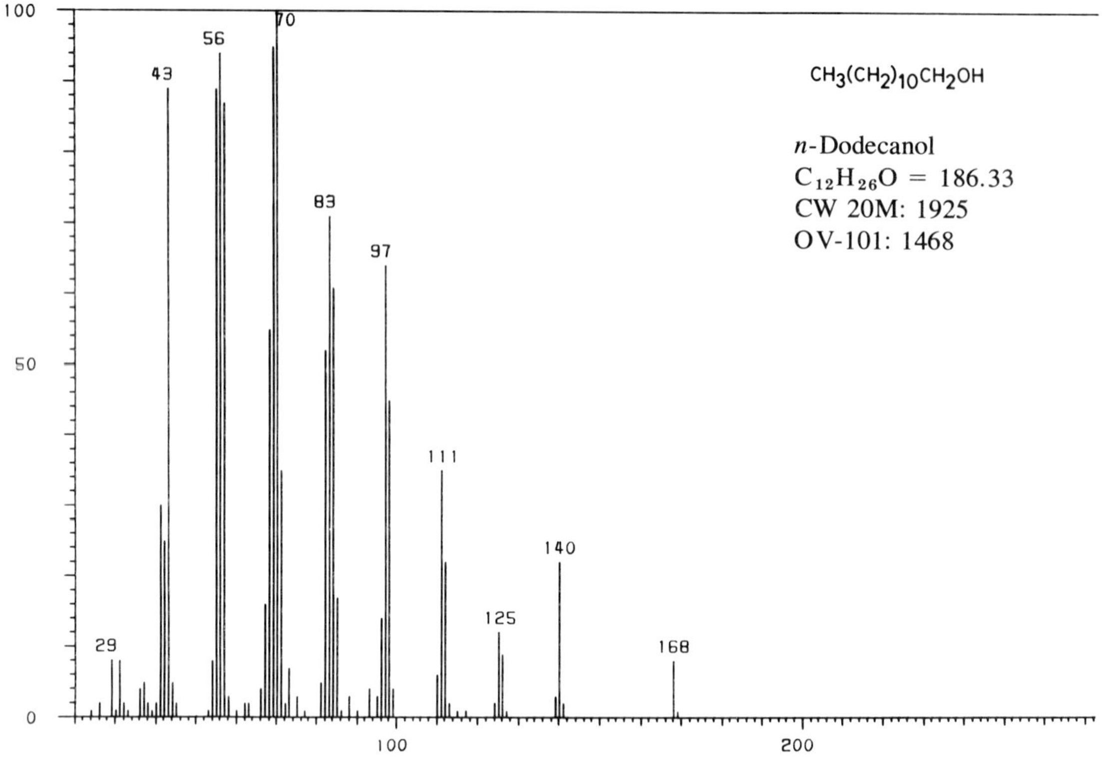

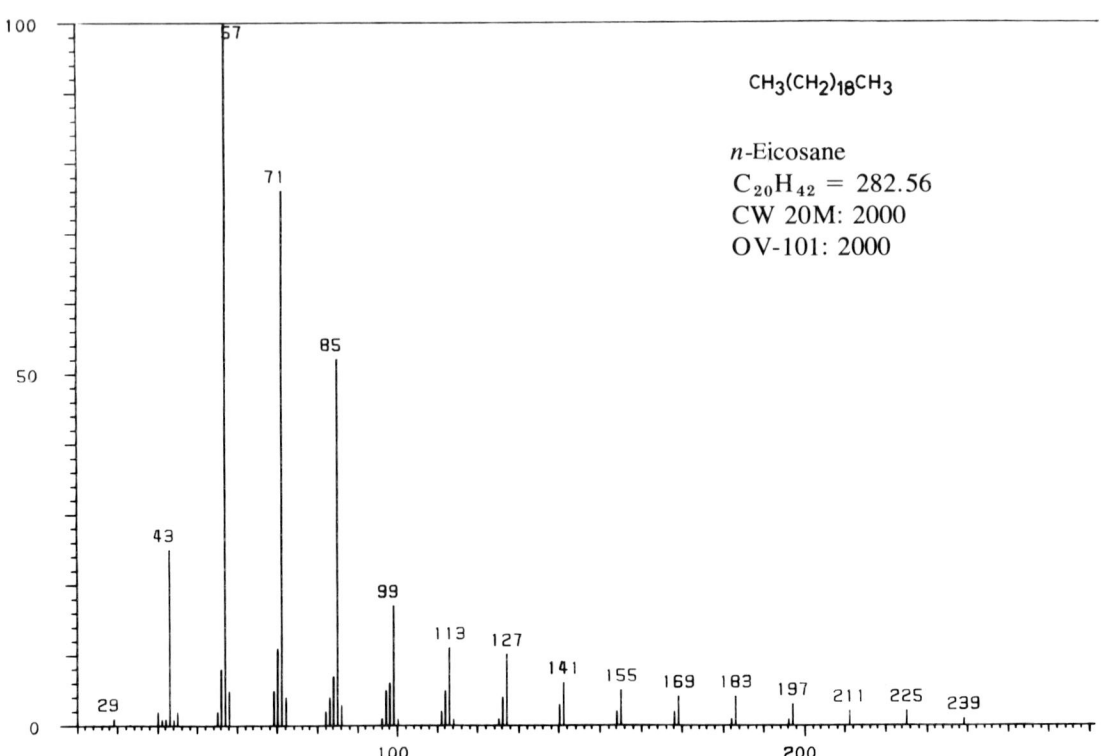

APPENDIX IV   MASS SPECTRA OF INDIVIDUAL COMPOUNDS

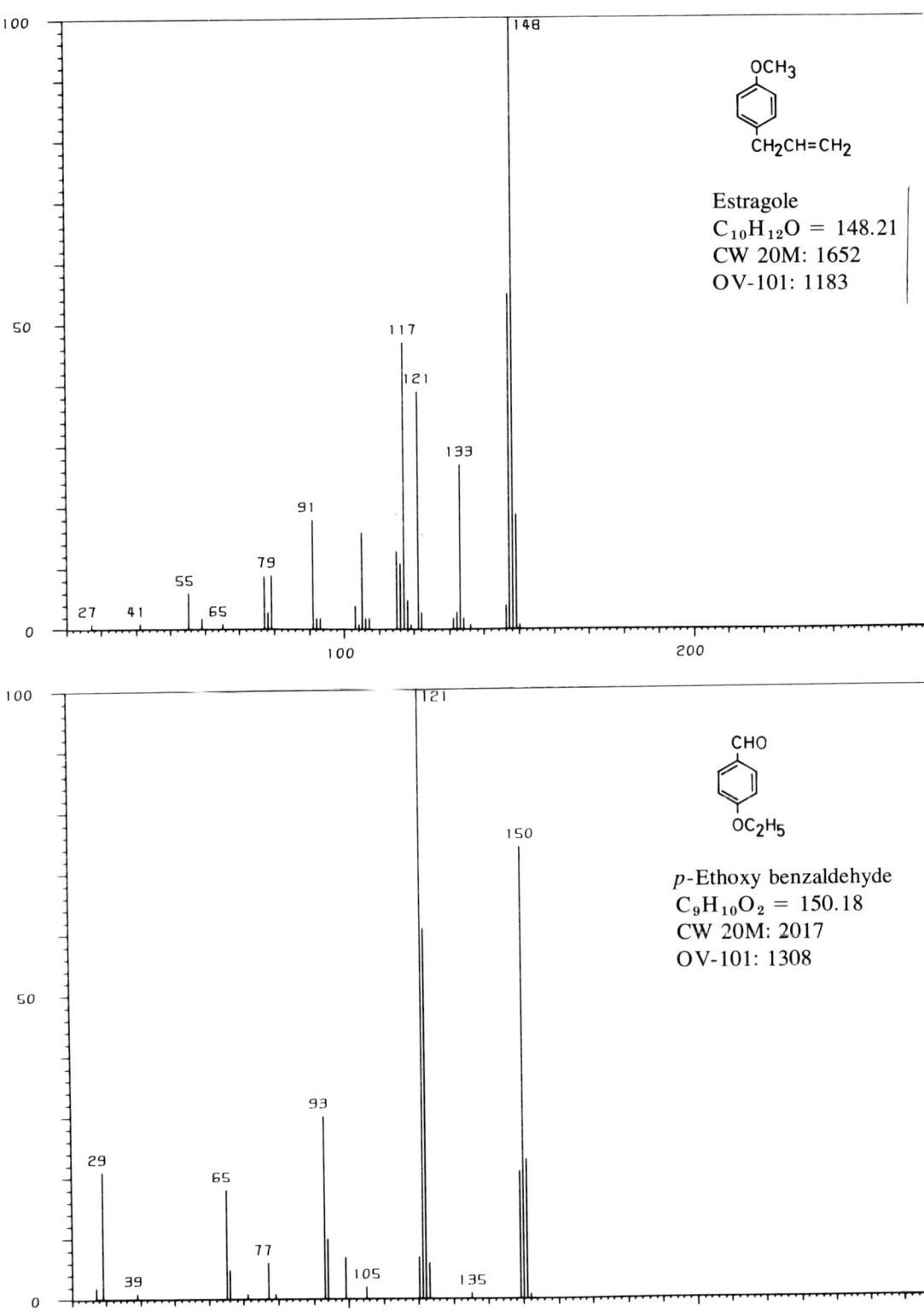

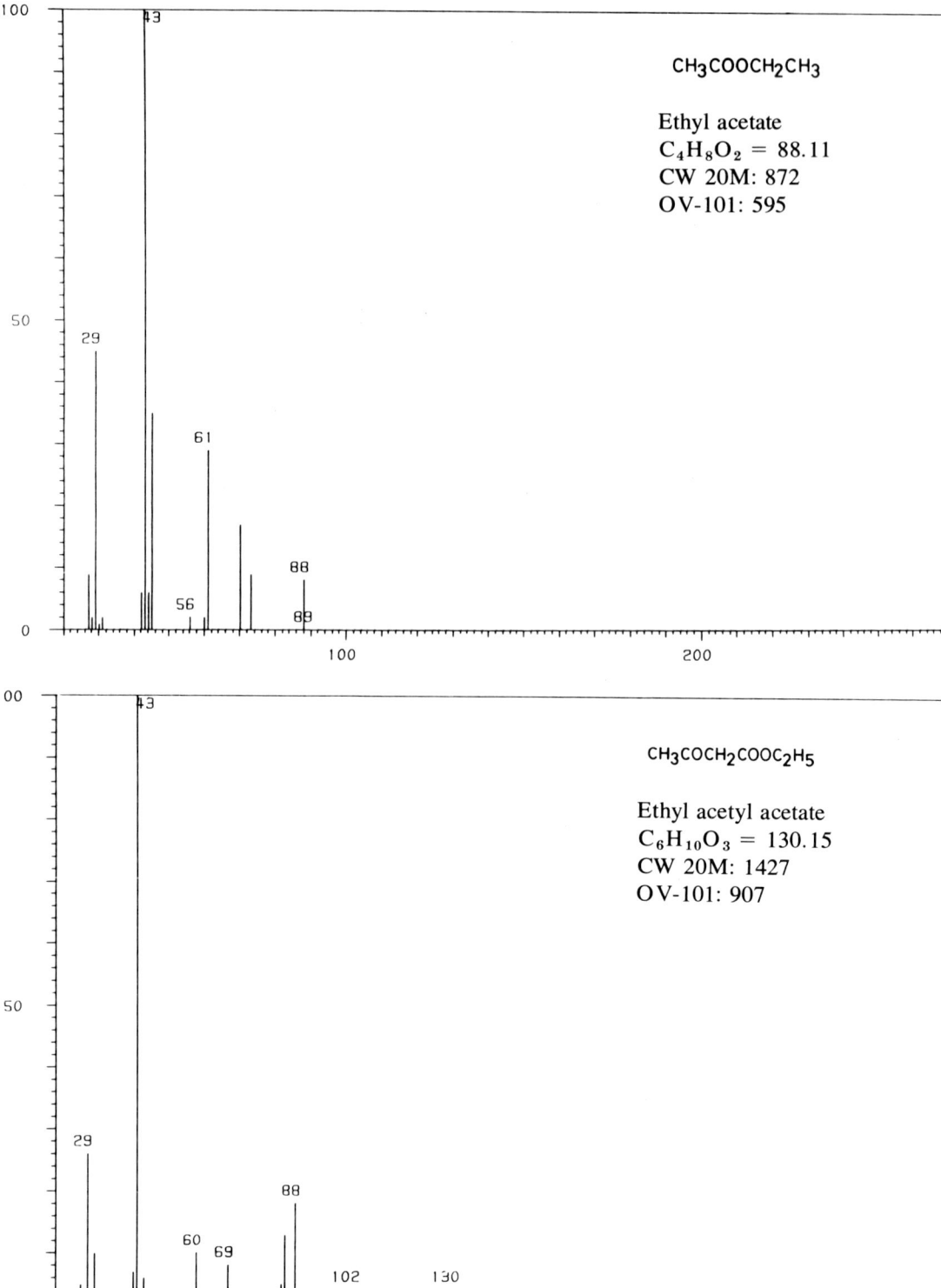

APPENDIX IV   MASS SPECTRA OF INDIVIDUAL COMPOUNDS

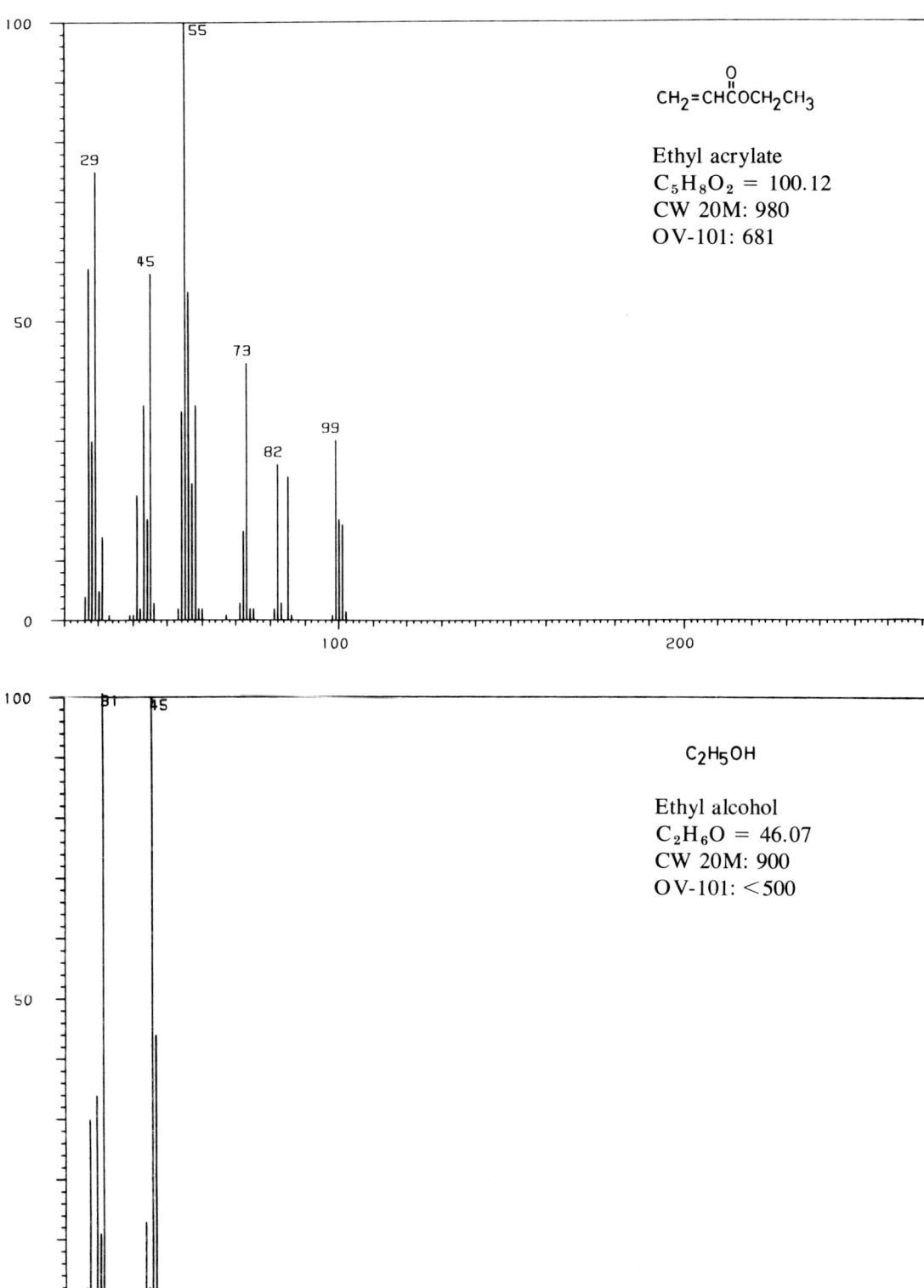

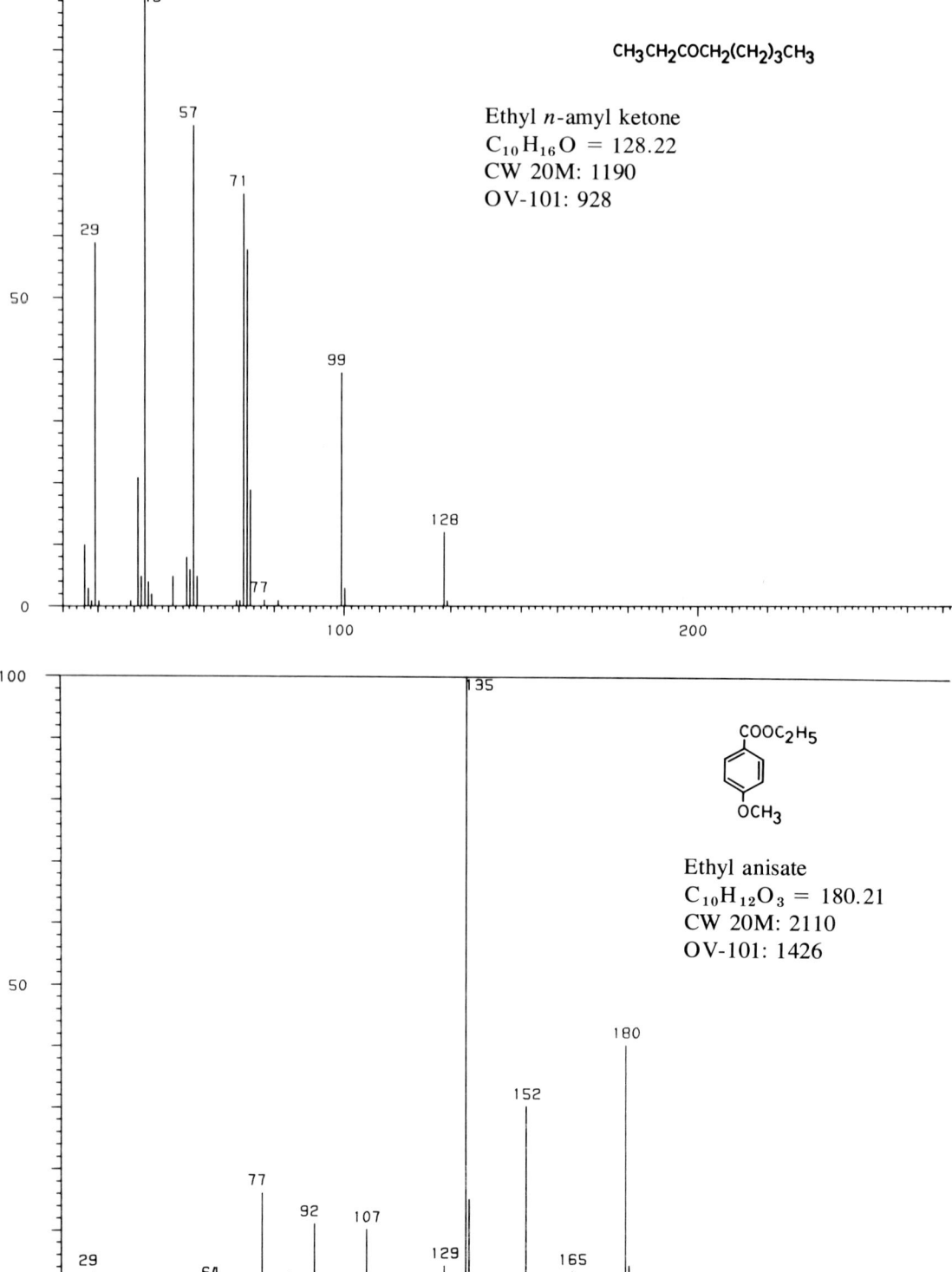

APPENDIX IV   MASS SPECTRA OF INDIVIDUAL COMPOUNDS   279

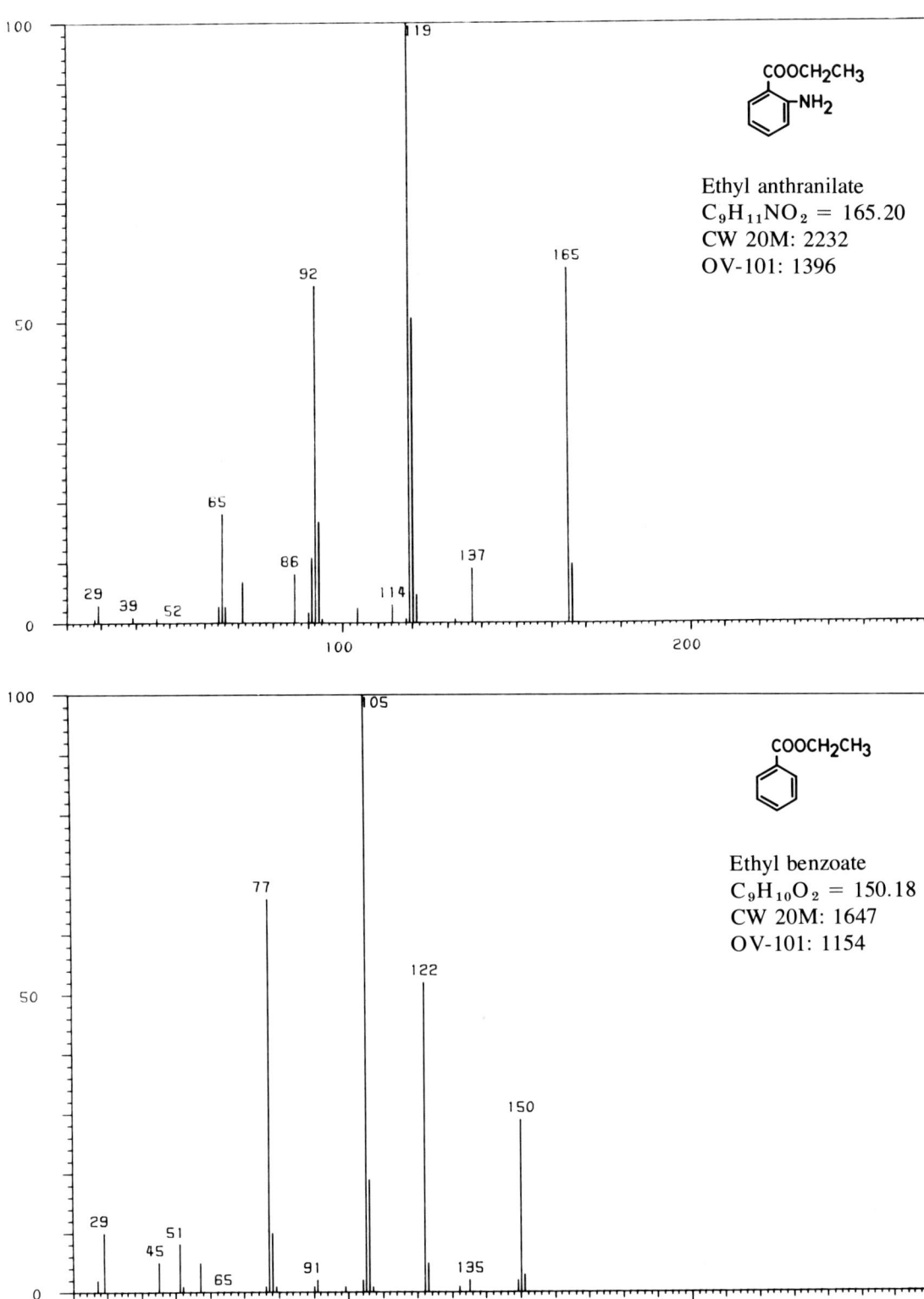

Ethyl anthranilate
$C_9H_{11}NO_2 = 165.20$
CW 20M: 2232
OV-101: 1396

Ethyl benzoate
$C_9H_{10}O_2 = 150.18$
CW 20M: 1647
OV-101: 1154

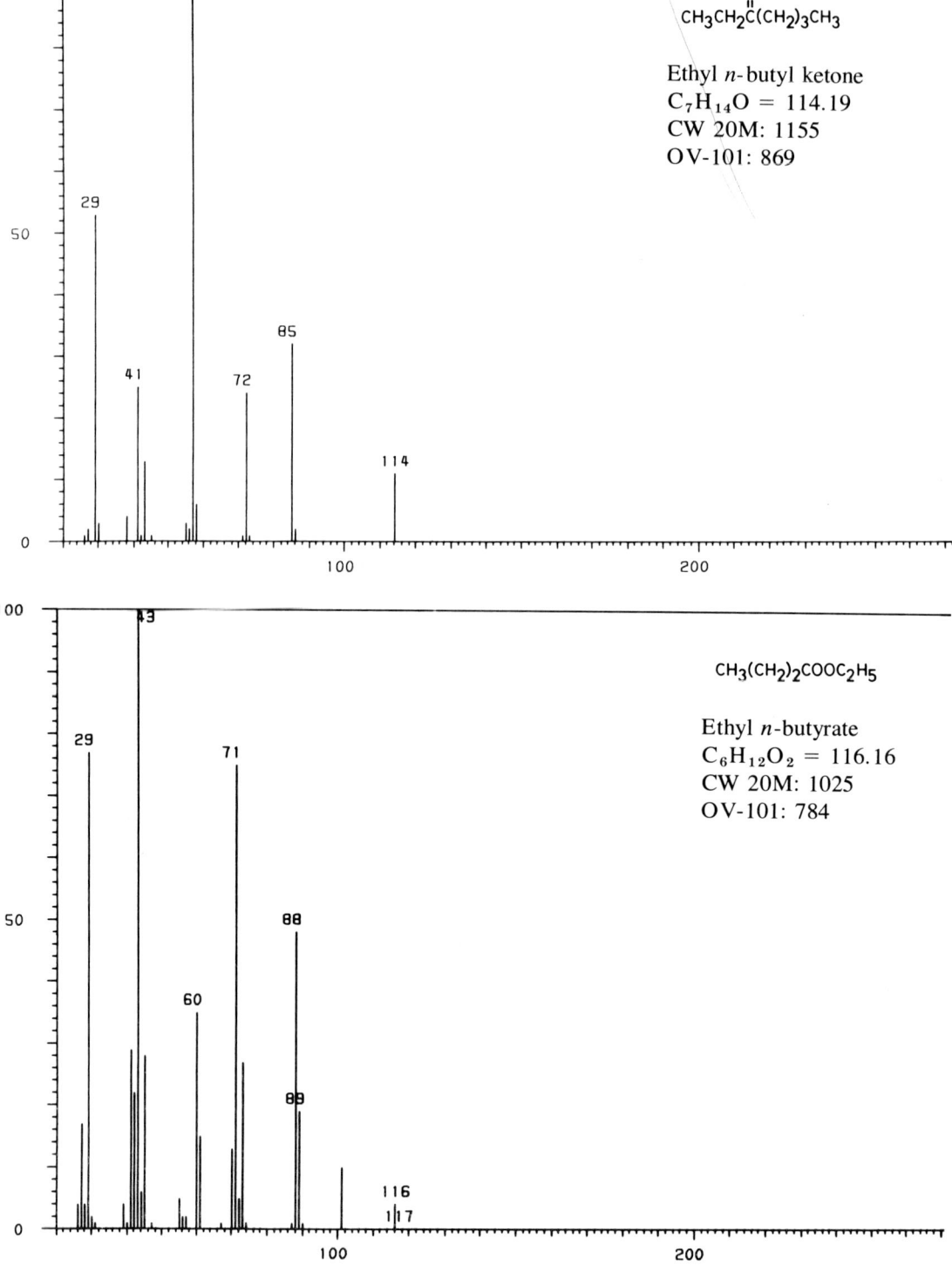

APPENDIX IV MASS SPECTRA OF INDIVIDUAL COMPOUNDS

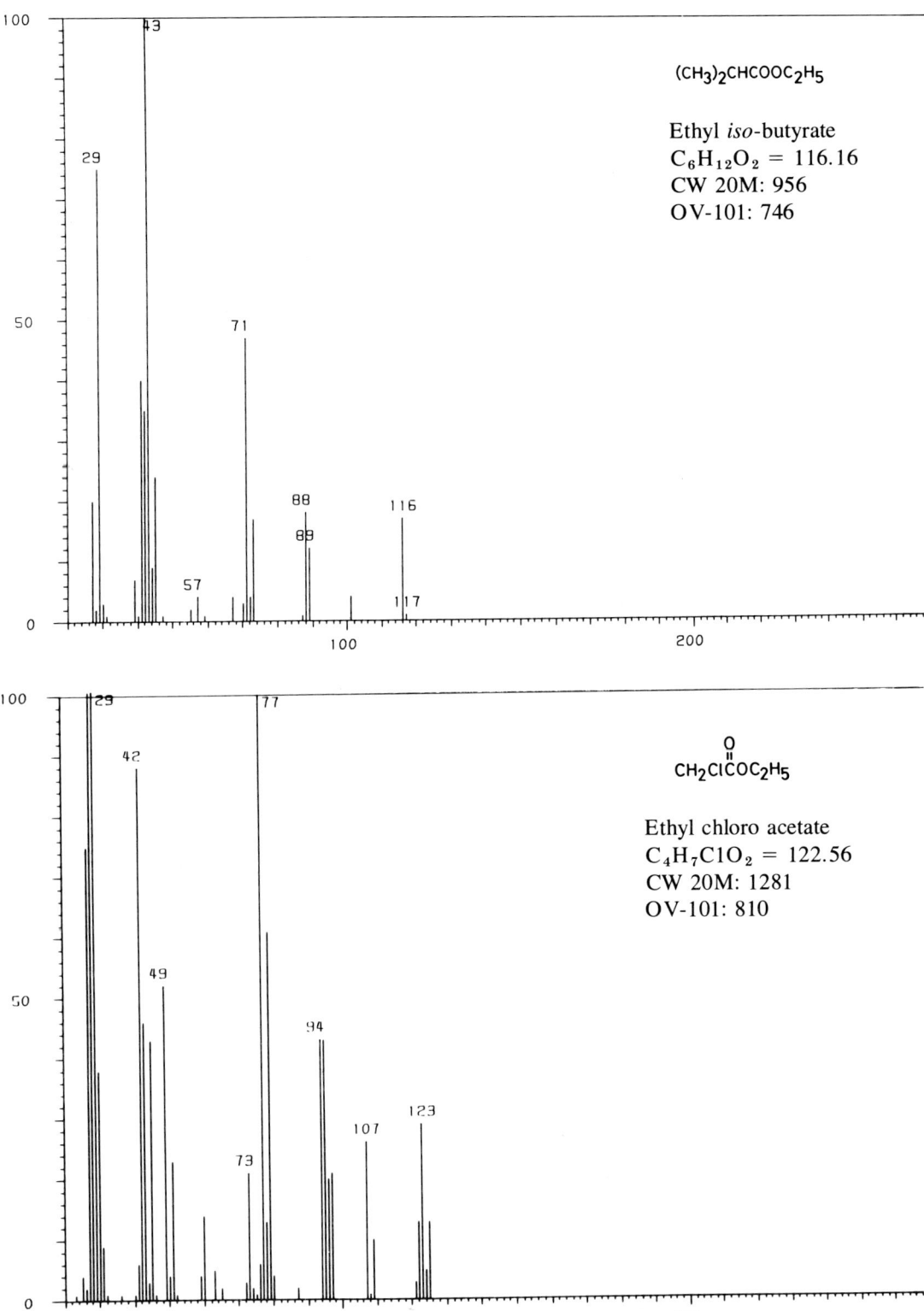

$(CH_3)_2CHCOOC_2H_5$

Ethyl *iso*-butyrate
$C_6H_{12}O_2 = 116.16$
CW 20M: 956
OV-101: 746

$CH_2ClCOOC_2H_5$

Ethyl chloro acetate
$C_4H_7ClO_2 = 122.56$
CW 20M: 1281
OV-101: 810

Ethyl cinnamate
$C_{11}H_{12}O_2$ = 176.22
CW 20M: 2095
OV-101: 1443

Ethyl crotonate
$C_6H_{10}O_2$ = 114.15
CW 20M: 1161
OV-101: 823

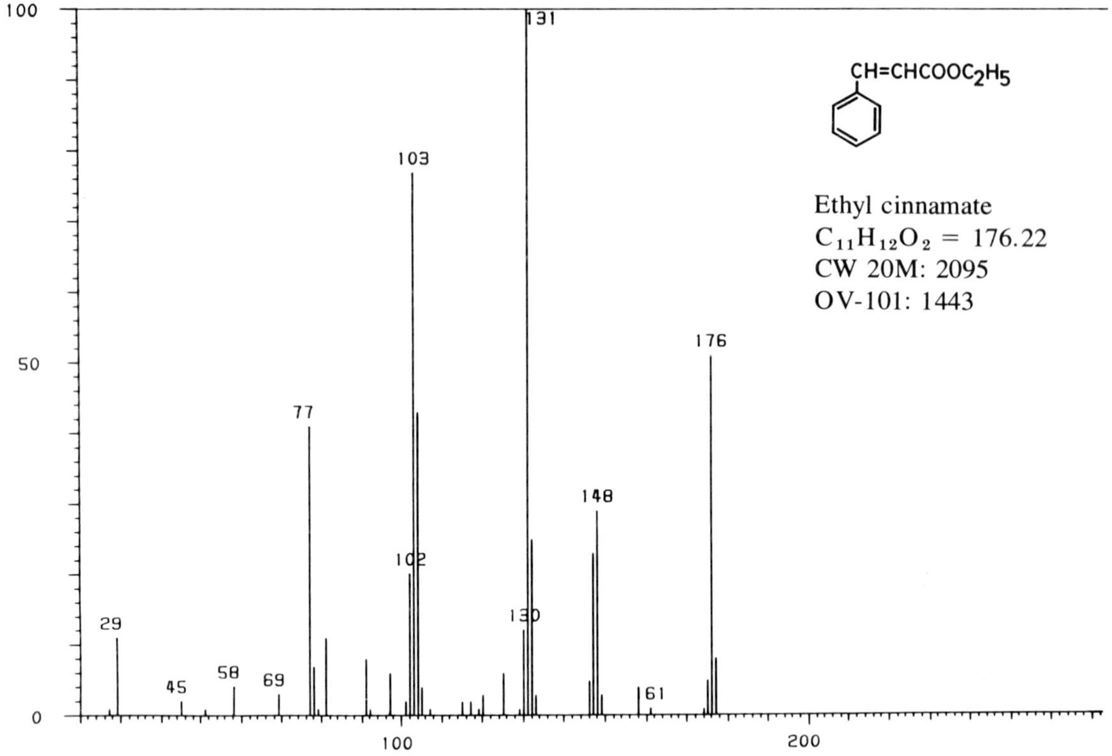

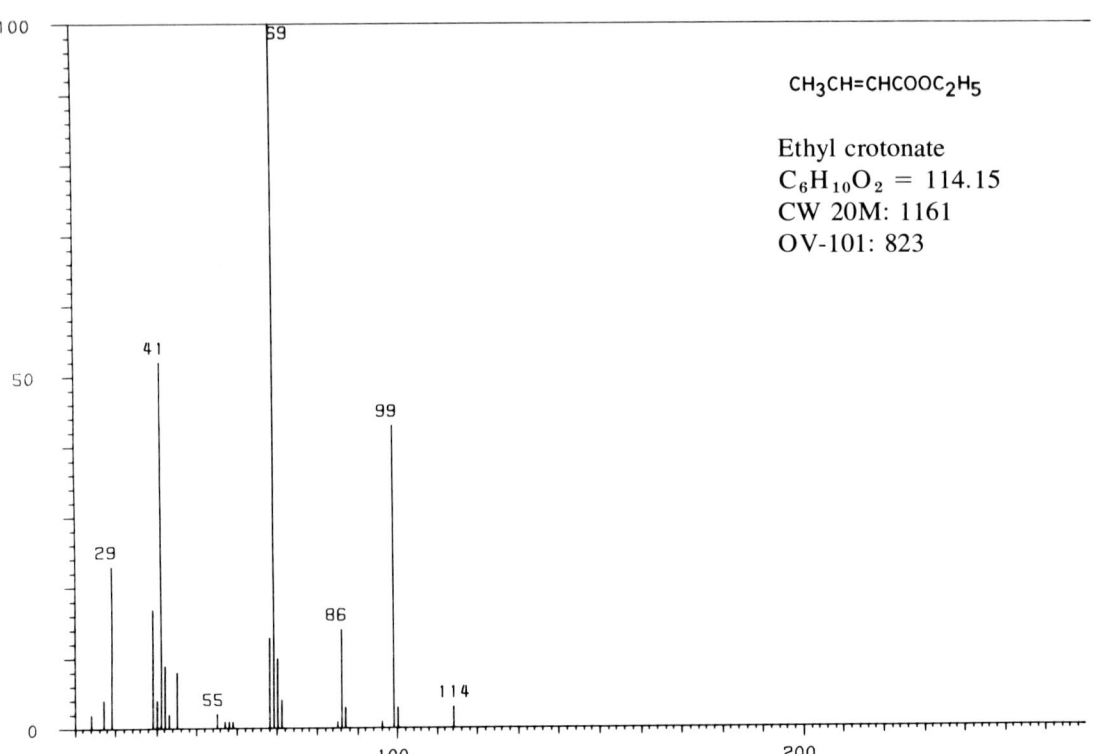

APPENDIX IV   MASS SPECTRA OF INDIVIDUAL COMPOUNDS   283

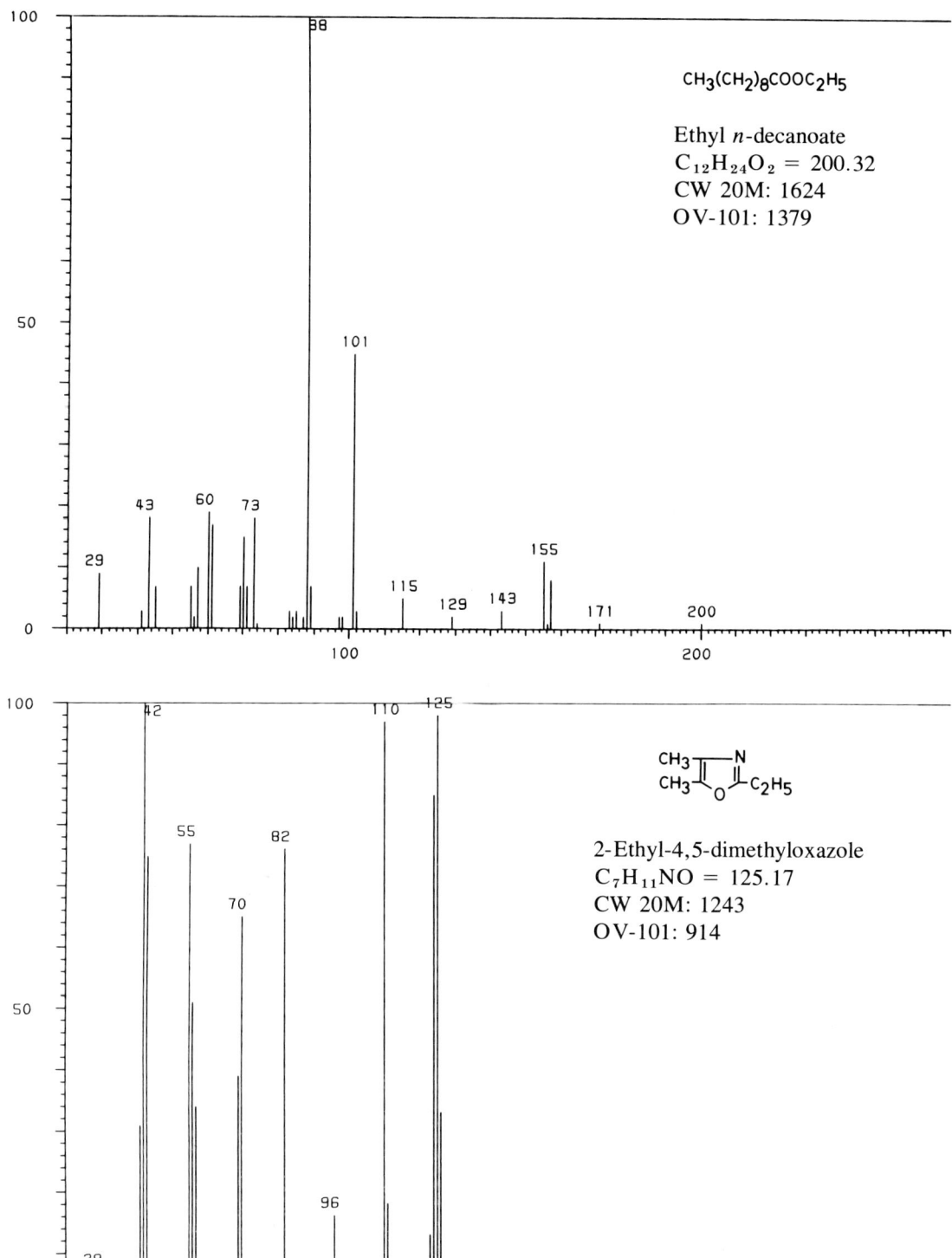

$CH_3(CH_2)_8COOC_2H_5$

Ethyl *n*-decanoate
$C_{12}H_{24}O_2 = 200.32$
CW 20M: 1624
OV-101: 1379

2-Ethyl-4,5-dimethyloxazole
$C_7H_{11}NO = 125.17$
CW 20M: 1243
OV-101: 914

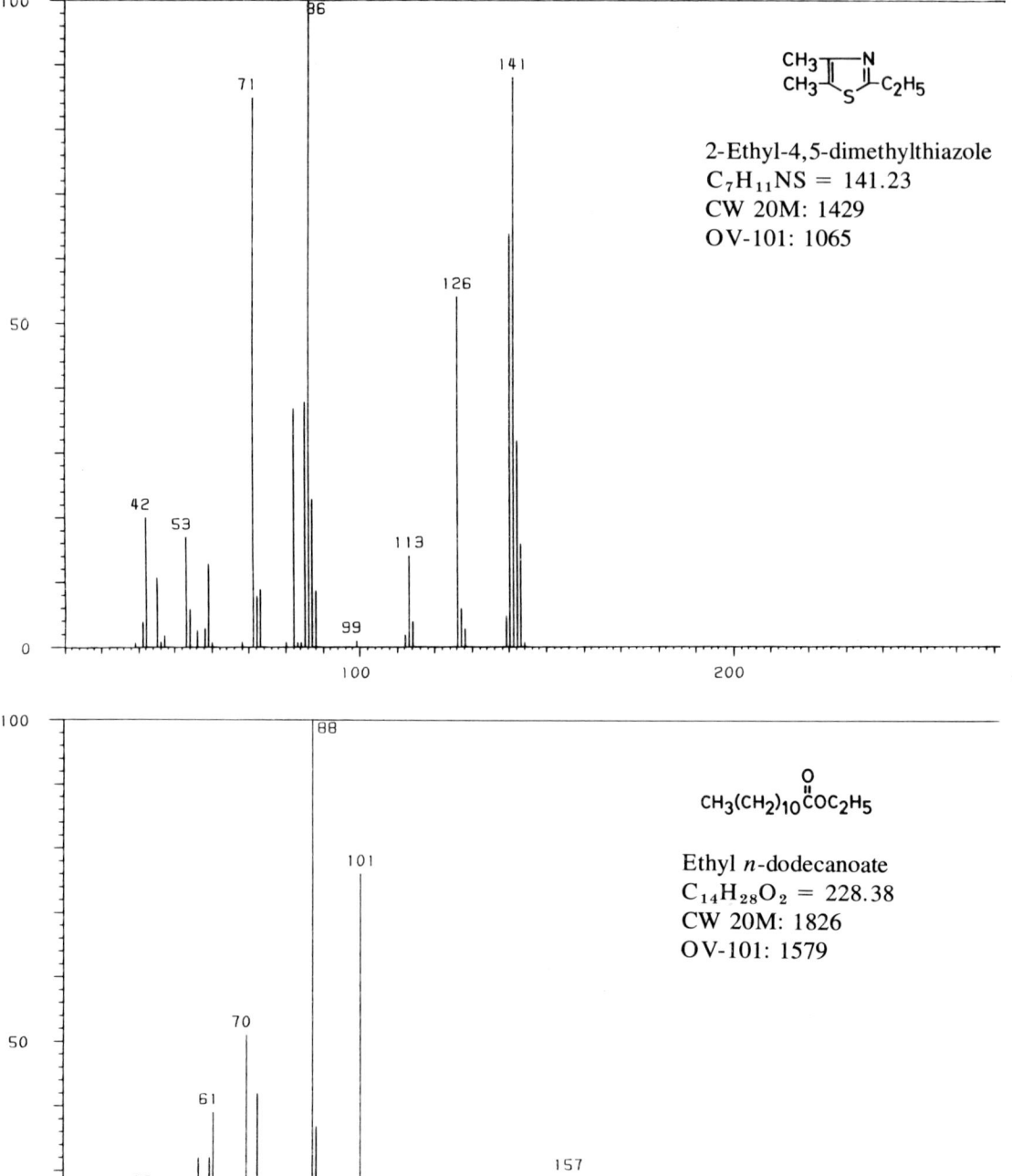

APPENDIX IV   MASS SPECTRA OF INDIVIDUAL COMPOUNDS

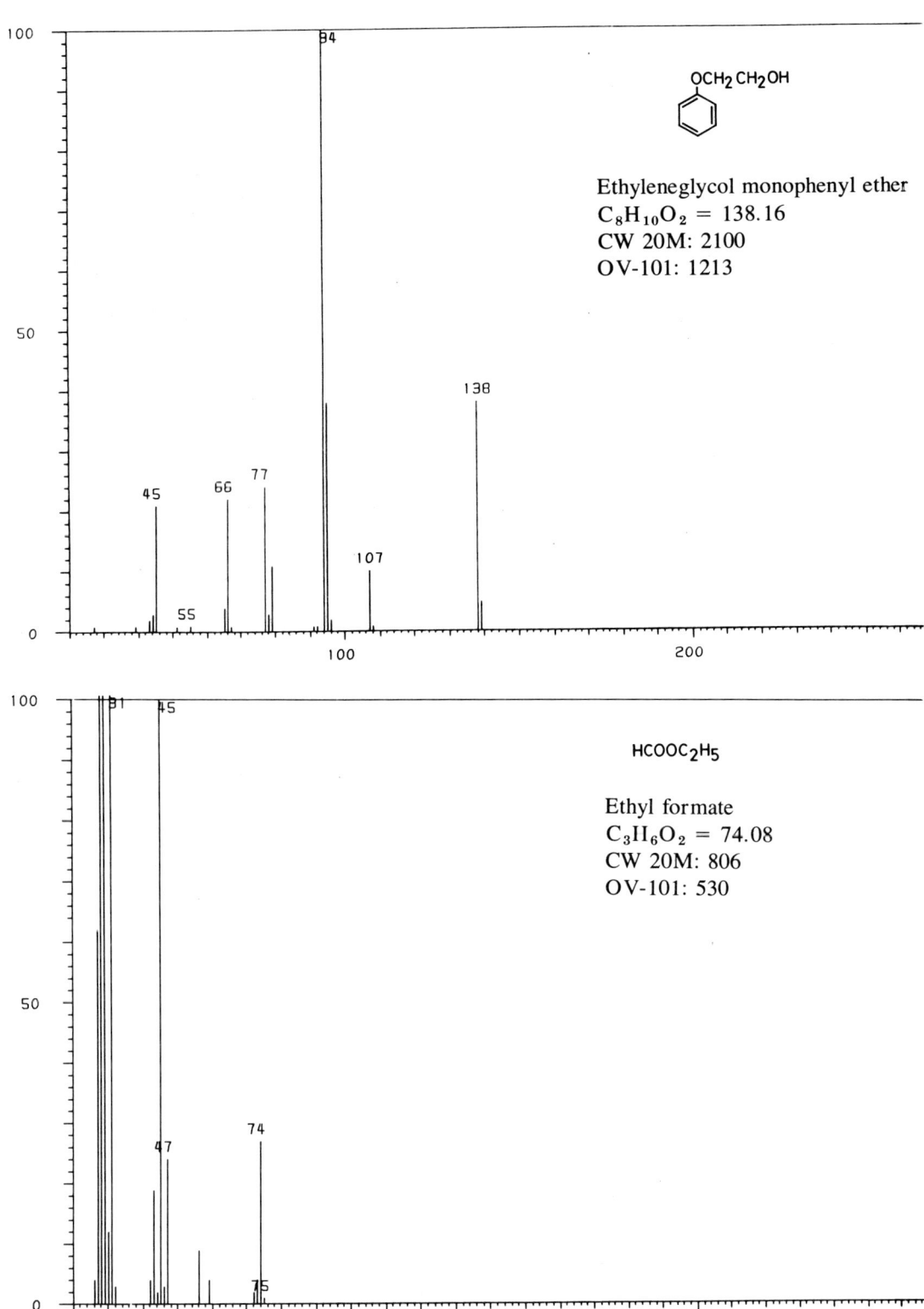

Ethyleneglycol monophenyl ether
$C_8H_{10}O_2 = 138.16$
CW 20M: 2100
OV-101: 1213

Ethyl formate
$C_3H_6O_2 = 74.08$
CW 20M: 806
OV-101: 530

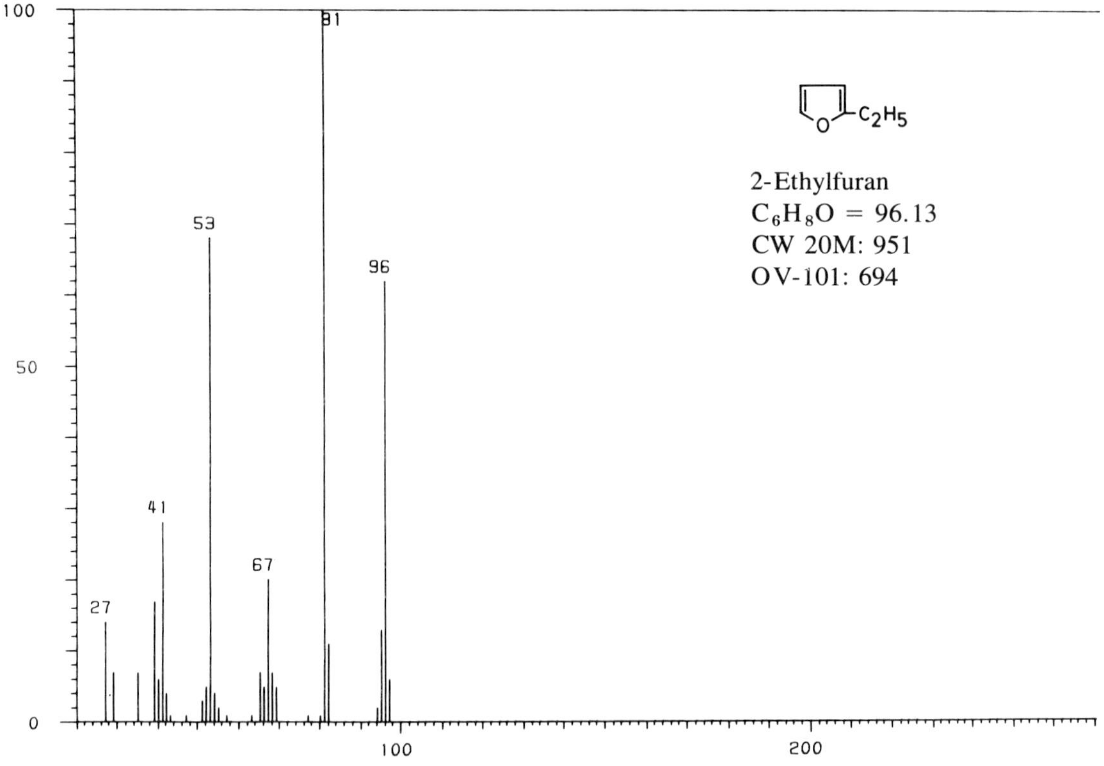

2-Ethylfuran
$C_6H_8O = 96.13$
CW 20M: 951
OV-101: 694

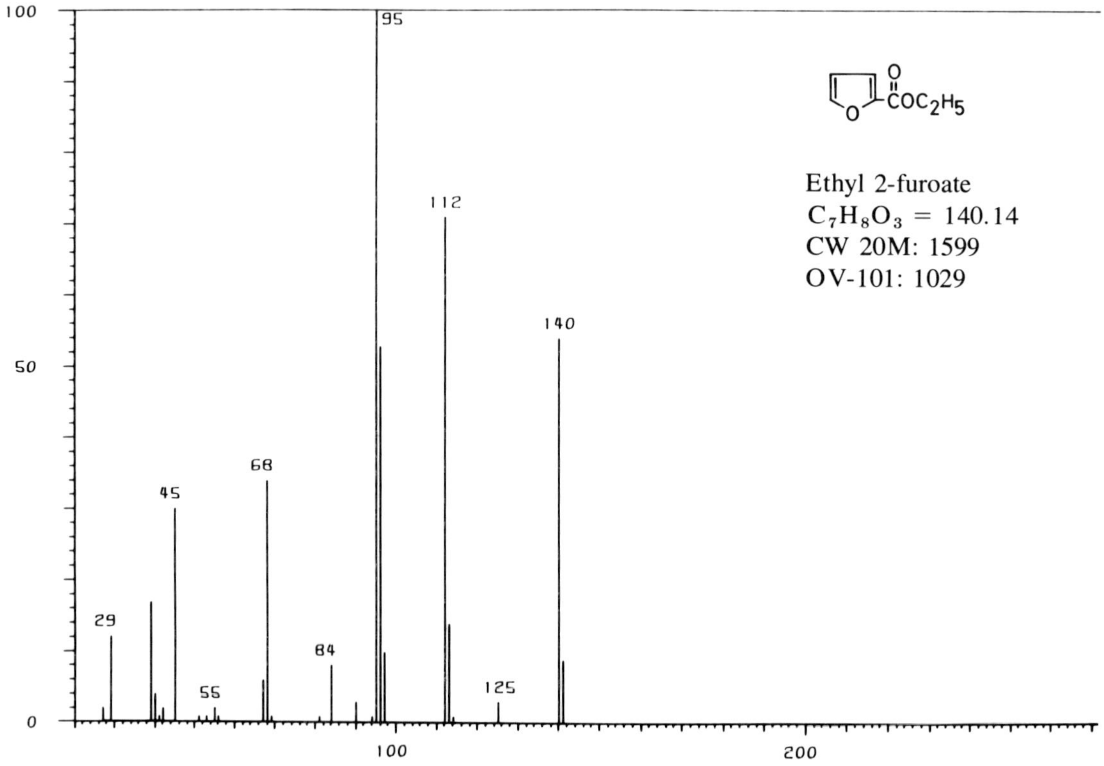

Ethyl 2-furoate
$C_7H_8O_3 = 140.14$
CW 20M: 1599
OV-101: 1029

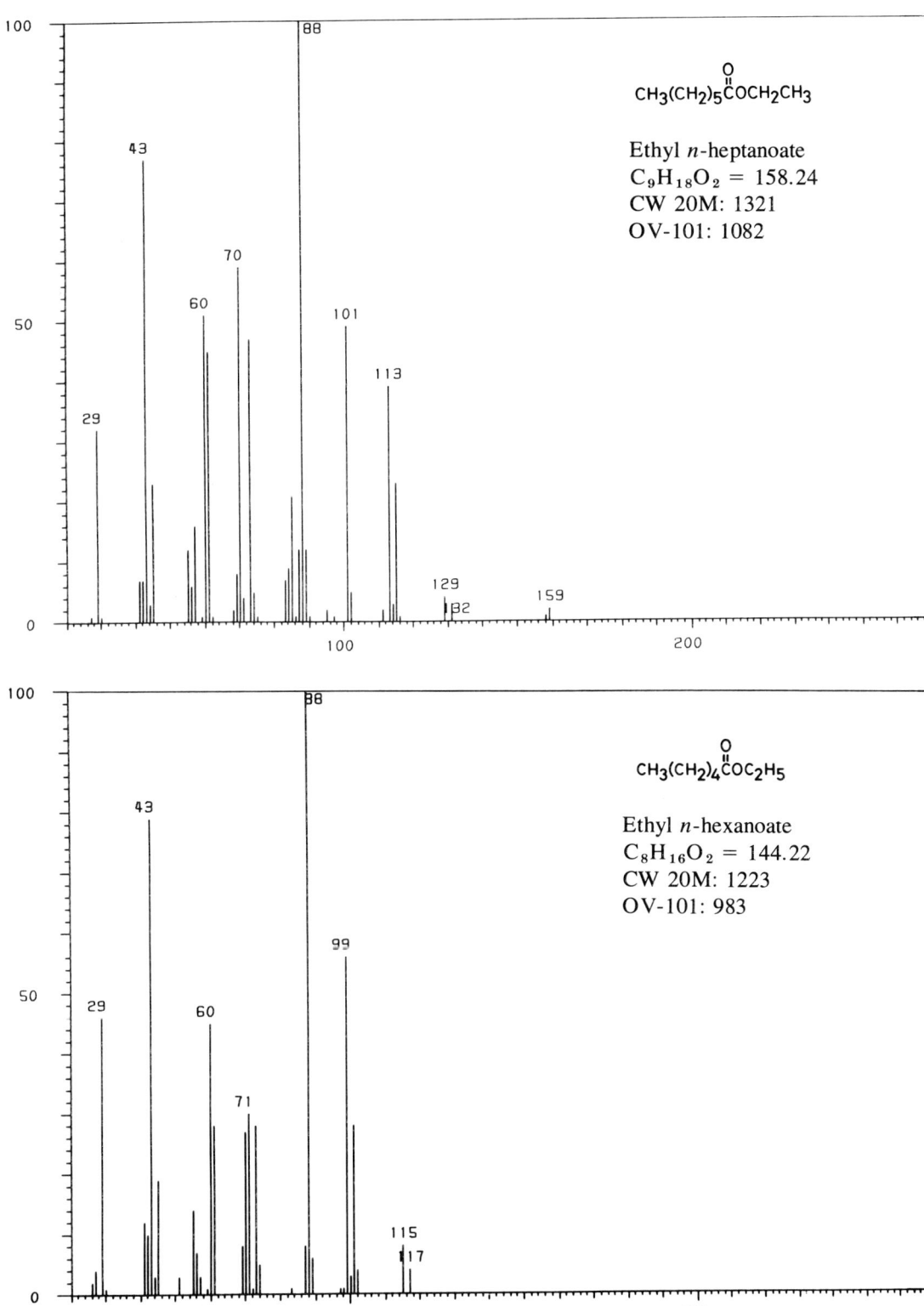

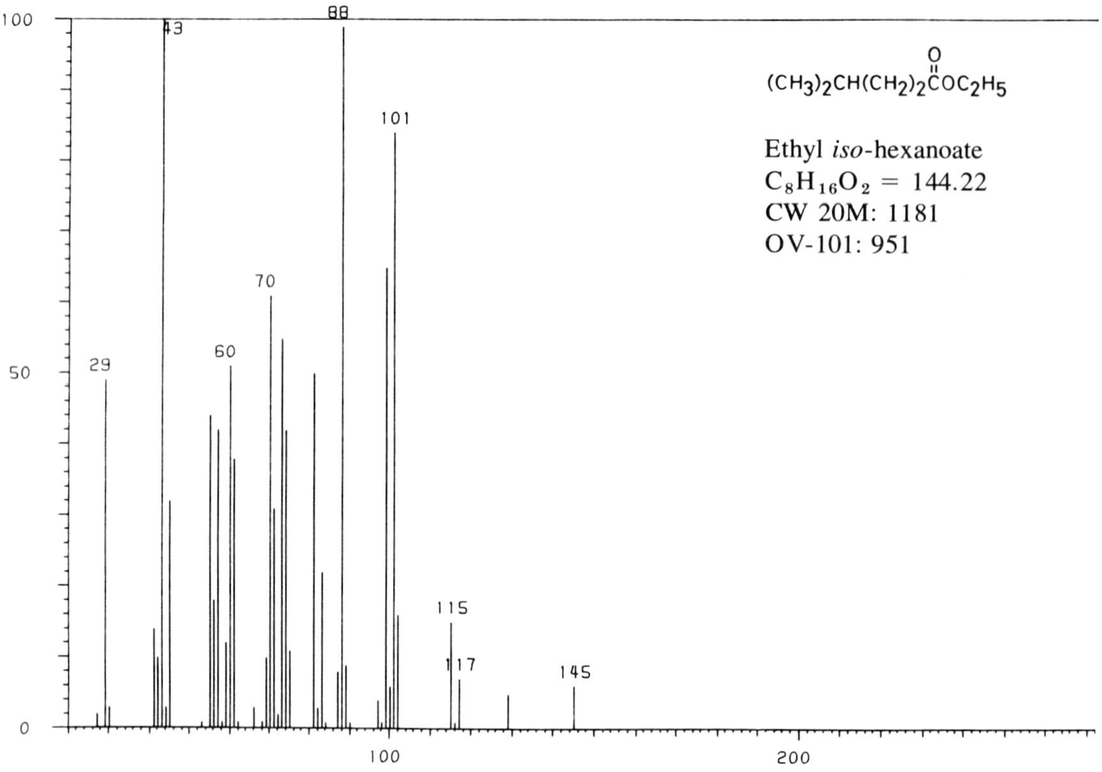

$(CH_3)_2CH(CH_2)_2\overset{O}{\underset{\|}{C}}OC_2H_5$

Ethyl *iso*-hexanoate
$C_8H_{16}O_2 = 144.22$
CW 20M: 1181
OV-101: 951

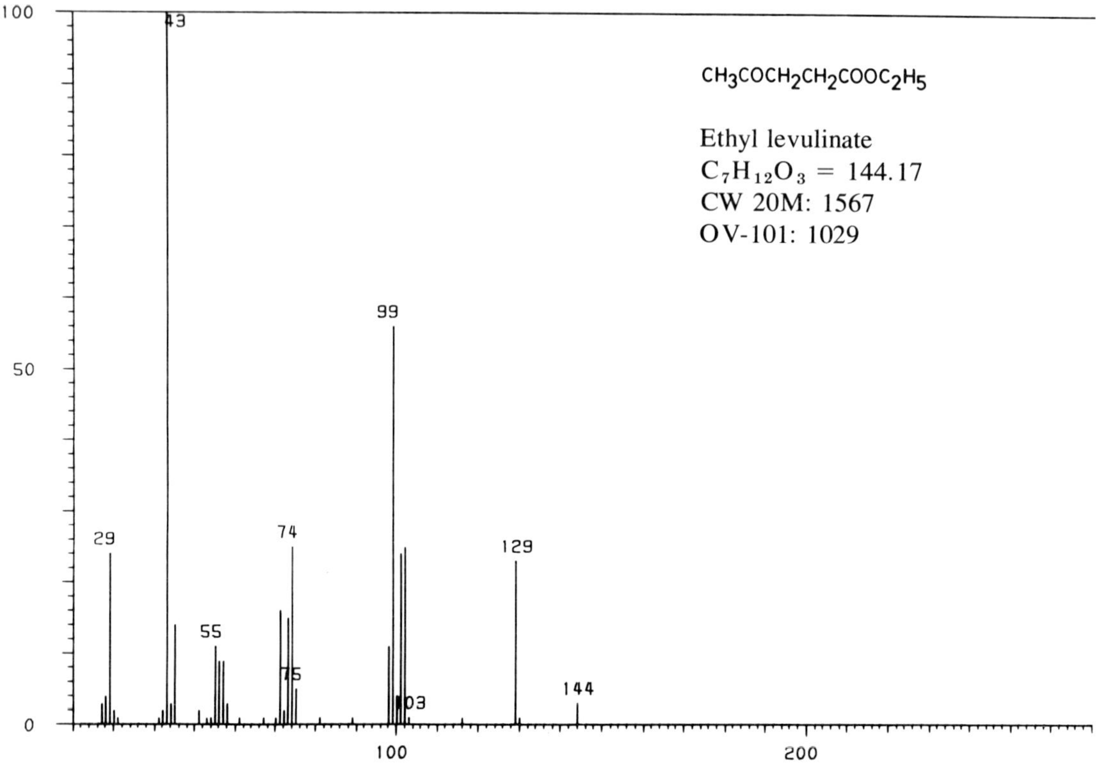

$CH_3COCH_2CH_2COOC_2H_5$

Ethyl levulinate
$C_7H_{12}O_3 = 144.17$
CW 20M: 1567
OV-101: 1029

APPENDIX IV   MASS SPECTRA OF INDIVIDUAL COMPOUNDS   289

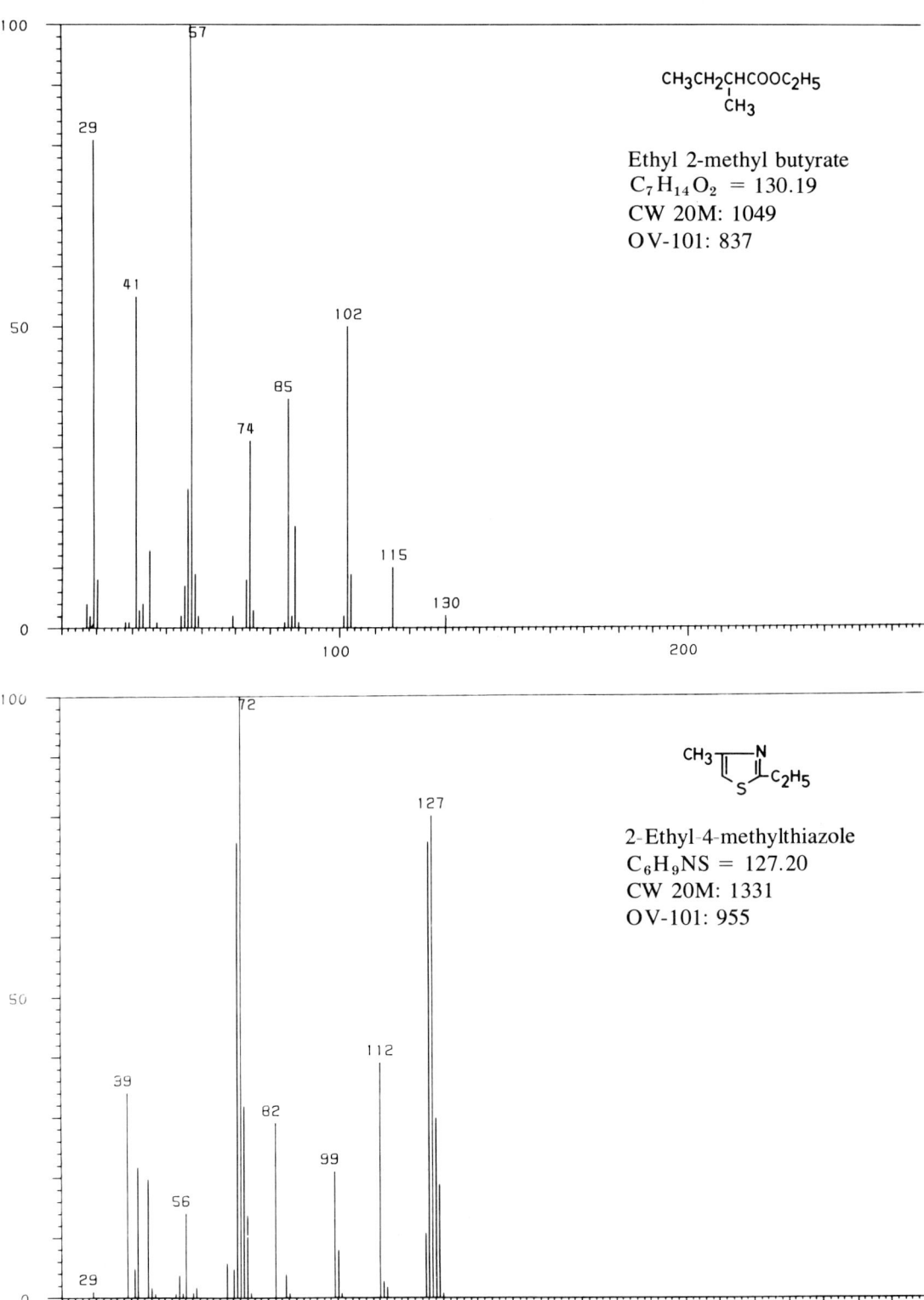

CH$_3$CH$_2$CHCOOC$_2$H$_5$
       |
      CH$_3$

Ethyl 2-methyl butyrate
C$_7$H$_{14}$O$_2$ = 130.19
CW 20M: 1049
OV-101: 837

2-Ethyl-4-methylthiazole
C$_6$H$_9$NS = 127.20
CW 20M: 1331
OV-101: 955

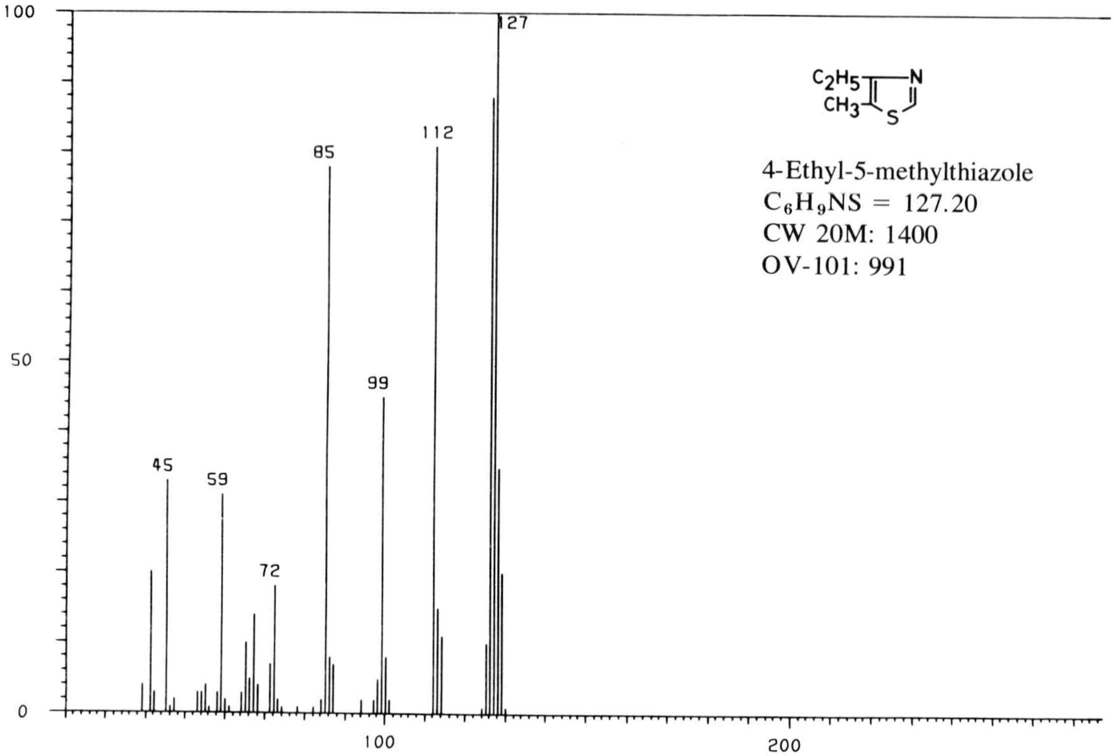

4-Ethyl-5-methylthiazole

$C_6H_9NS = 127.20$
CW 20M: 1400
OV-101: 991

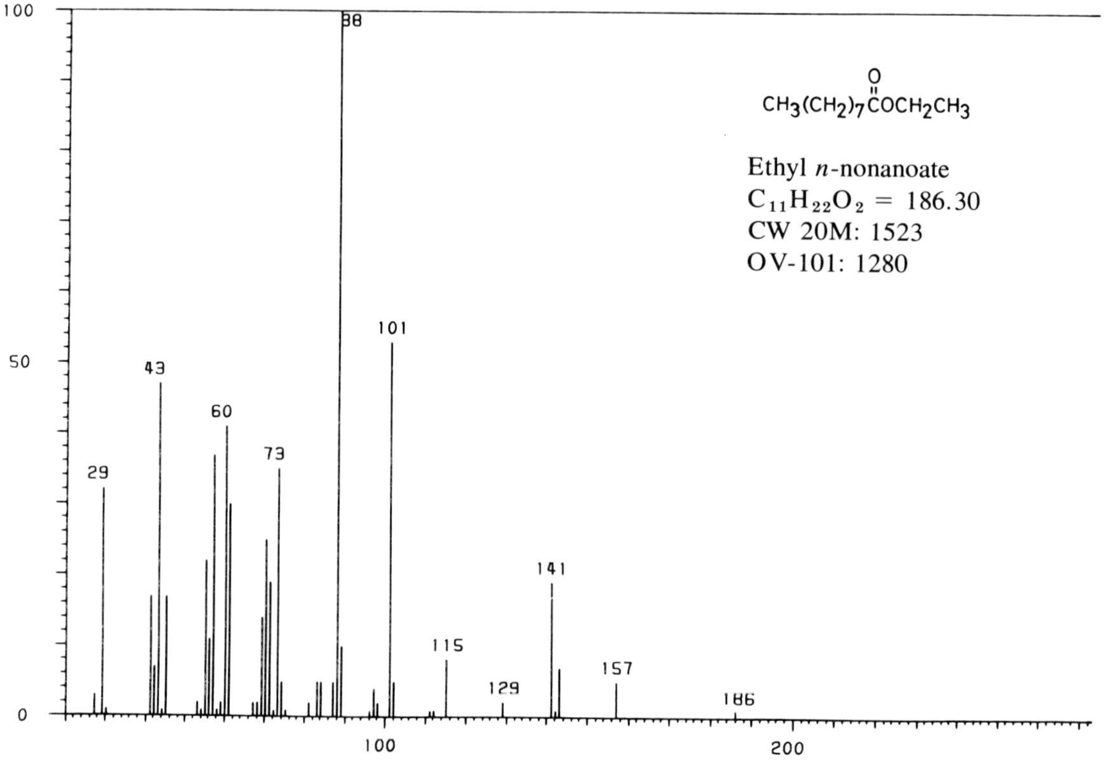

Peaks: 45, 59, 72, 85, 99, 112, 127

Ethyl n-nonanoate

$CH_3(CH_2)_7COOCH_2CH_3$

$C_{11}H_{22}O_2 = 186.30$
CW 20M: 1523
OV-101: 1280

Peaks: 29, 43, 60, 73, 88, 101, 115, 129, 141, 157, 186

APPENDIX IV MASS SPECTRA OF INDIVIDUAL COMPOUNDS

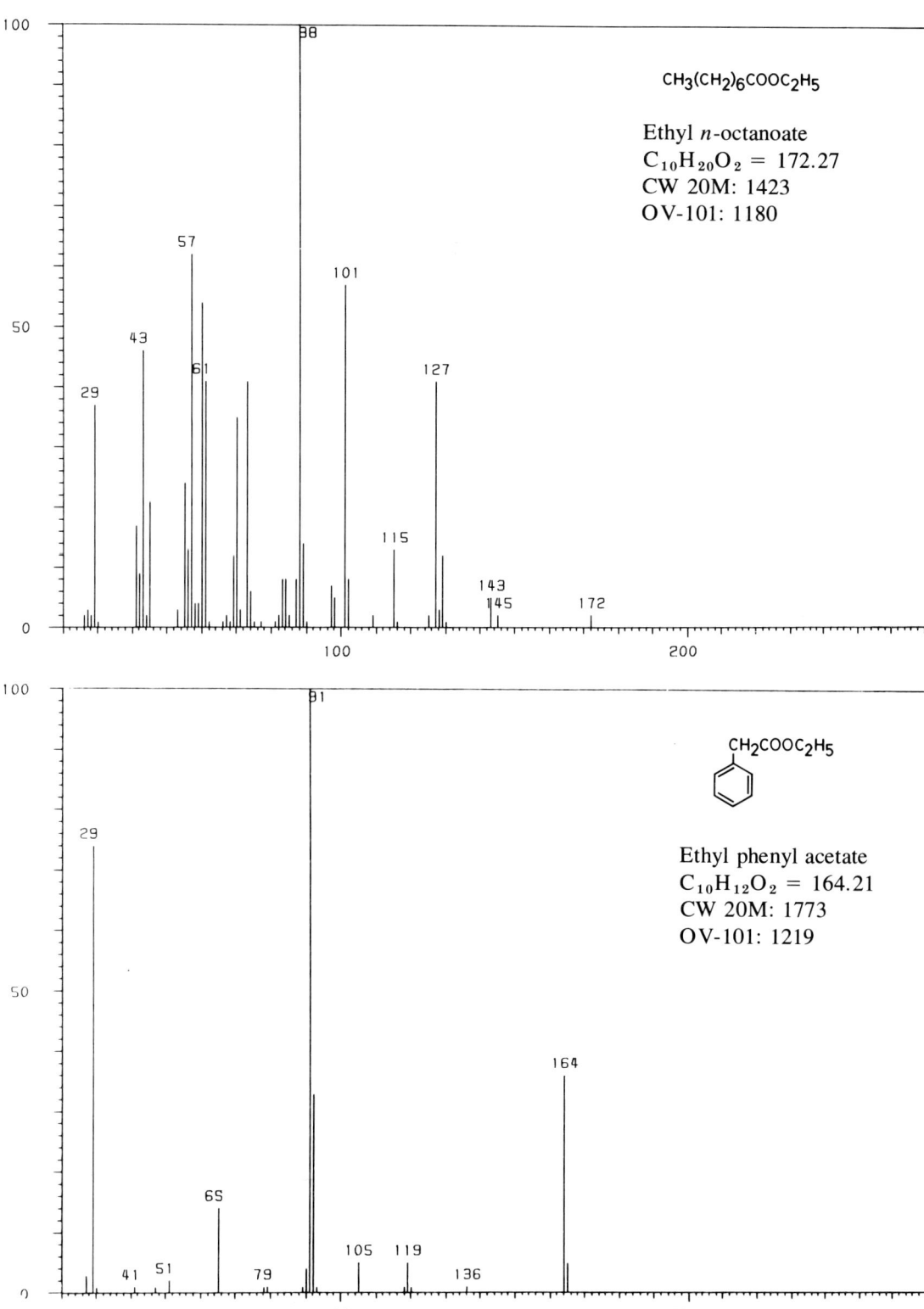

$CH_3(CH_2)_6COOC_2H_5$

Ethyl n-octanoate
$C_{10}H_{20}O_2 = 172.27$
CW 20M: 1423
OV-101: 1180

Ethyl phenyl acetate
$C_{10}H_{12}O_2 = 164.21$
CW 20M: 1773
OV-101: 1219

$(CH_3)_3CCOOC_2H_5$

Ethyl pivalate
$C_7H_{14}O_2$ = 130.19
CW 20M: 947
OV-101: 776

$CH_3CH_2COOC_2H_5$

Ethyl *n*-propionate
$C_5H_{10}O_2$ = 102.14
CW 20M: 944
OV-101: 691

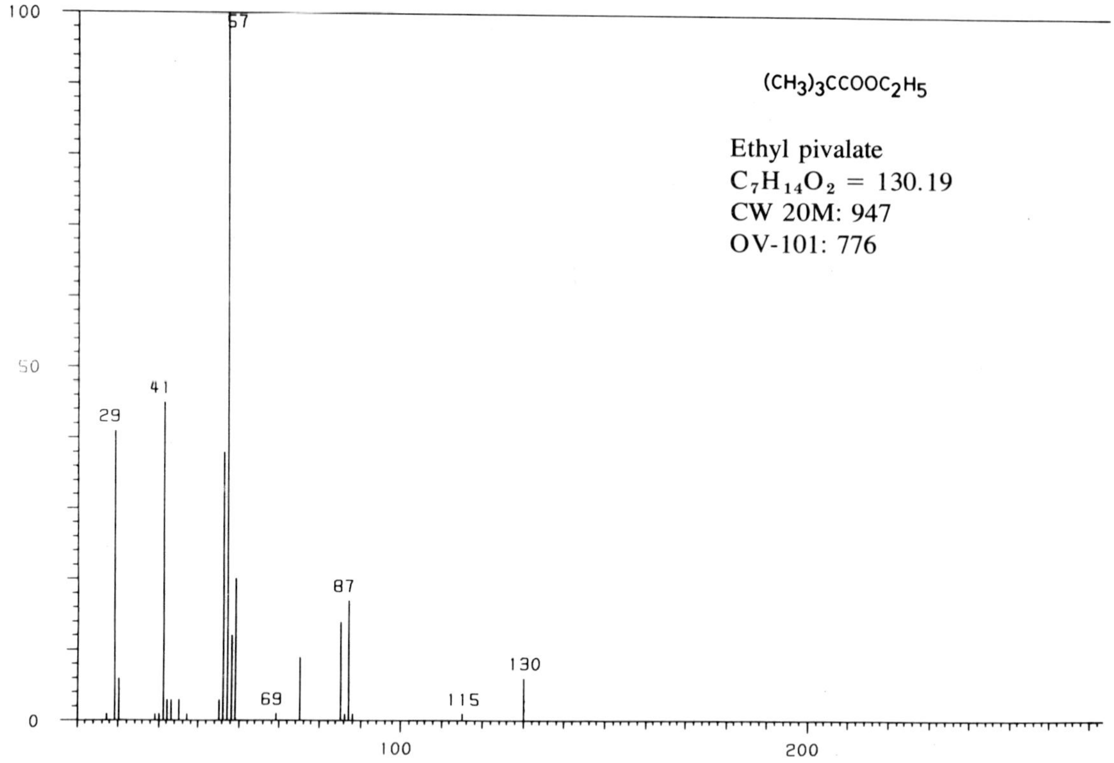

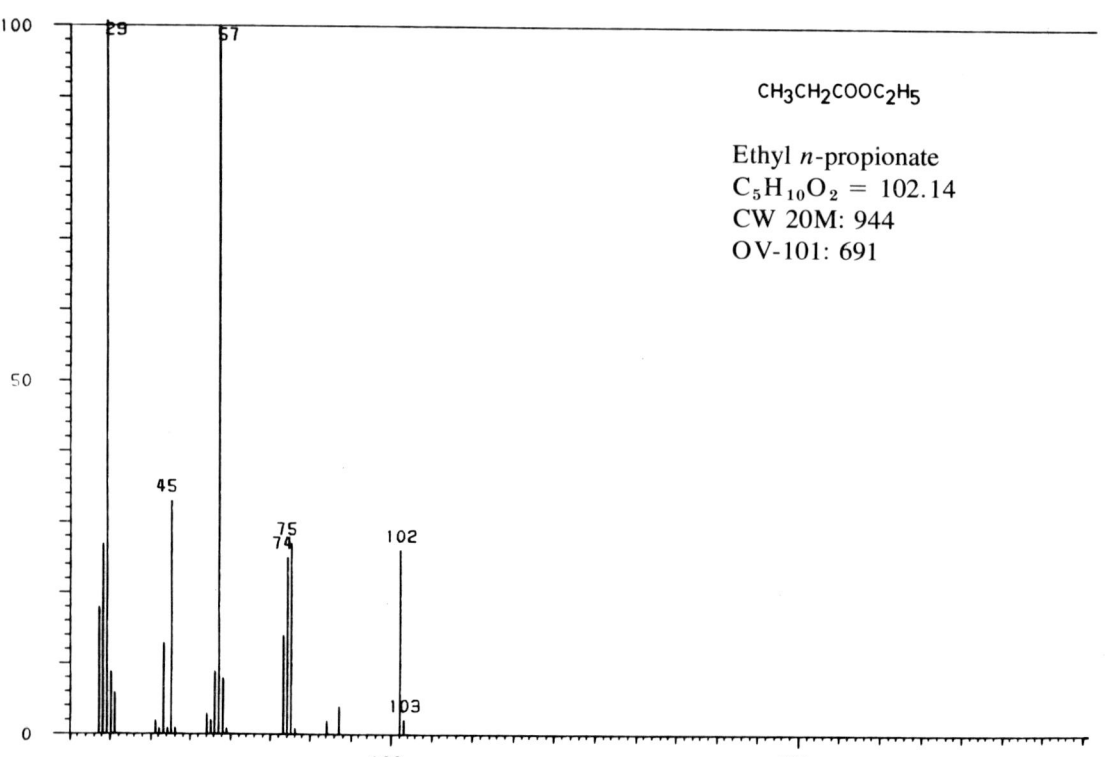

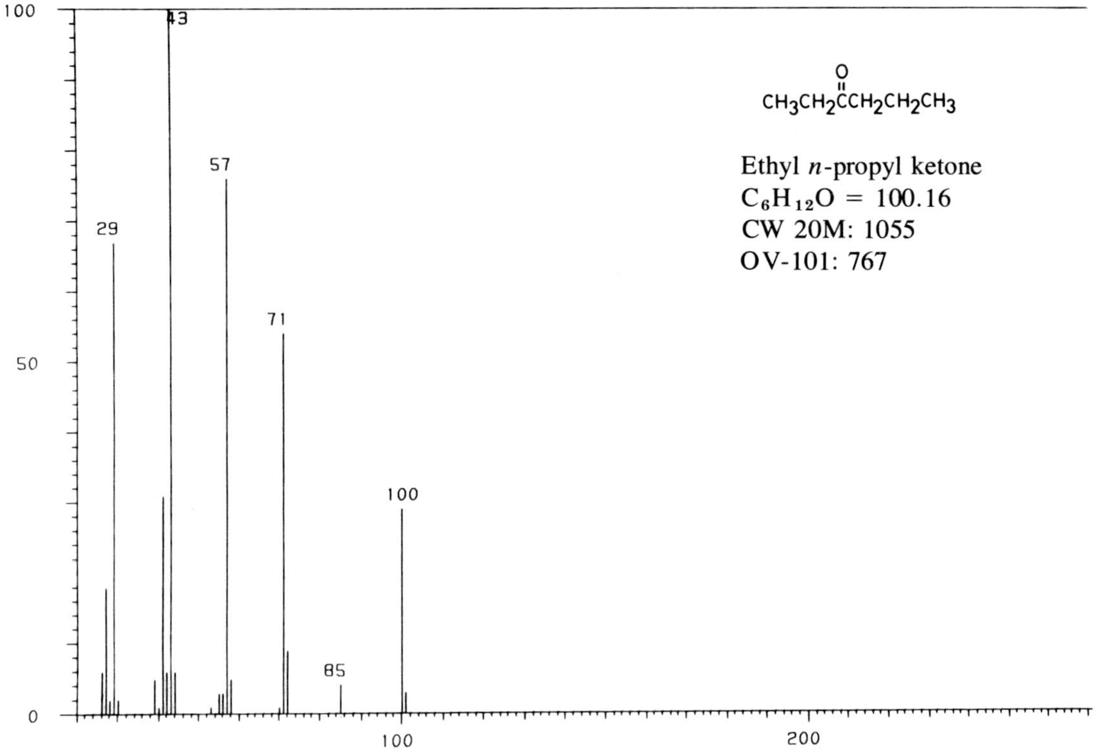

APPENDIX IV   MASS SPECTRA OF INDIVIDUAL COMPOUNDS

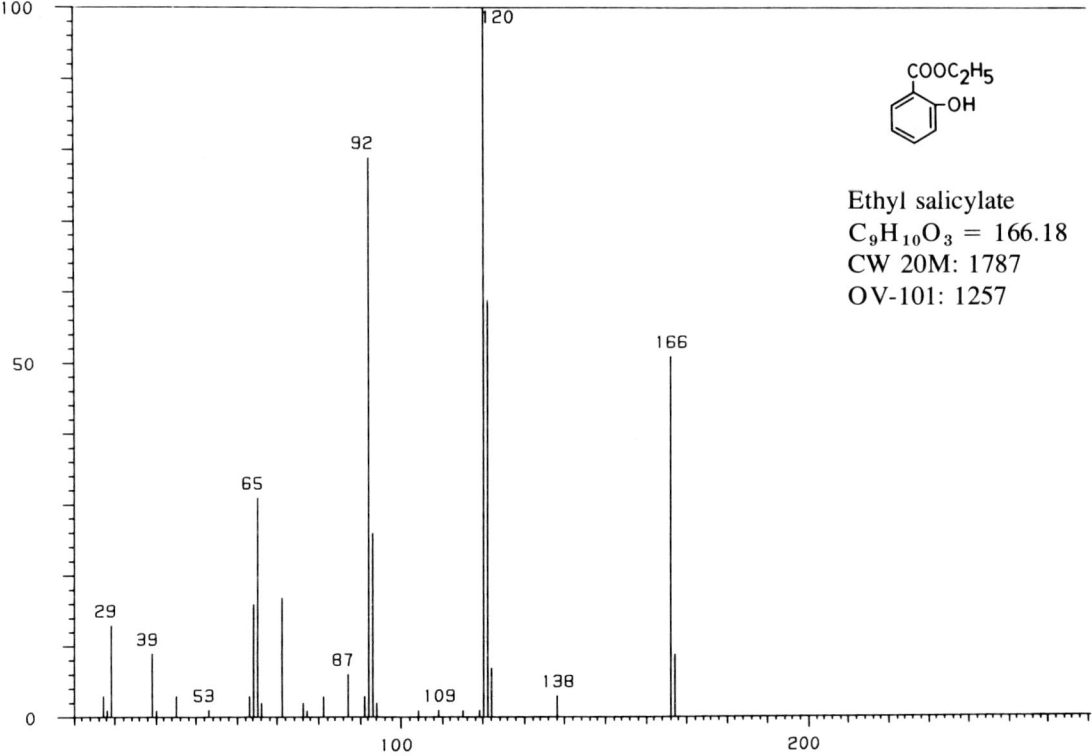

Ethyl *n*-propyl ketone
$C_6H_{12}O = 100.16$
CW 20M: 1055
OV-101: 767

Ethyl salicylate
$C_9H_{10}O_3 = 166.18$
CW 20M: 1787
OV-101: 1257

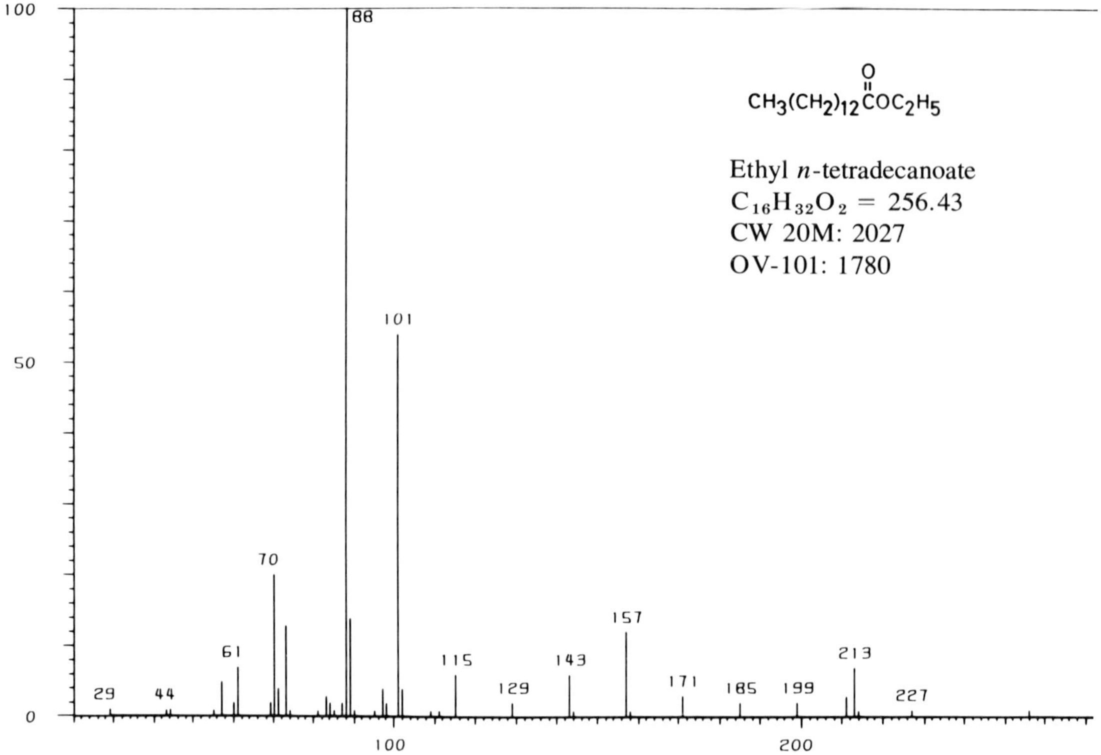

Ethyl n-tetradecanoate

CH₃(CH₂)₁₂COC₂H₅ (with C=O)

Ethyl *n*-tetradecanoate
$C_{16}H_{32}O_2 = 256.43$
CW 20M: 2027
OV-101: 1780

2-Ethylthiazole

$C_5H_7NS = 113.18$
CW 20M: 1300
OV-101: 879

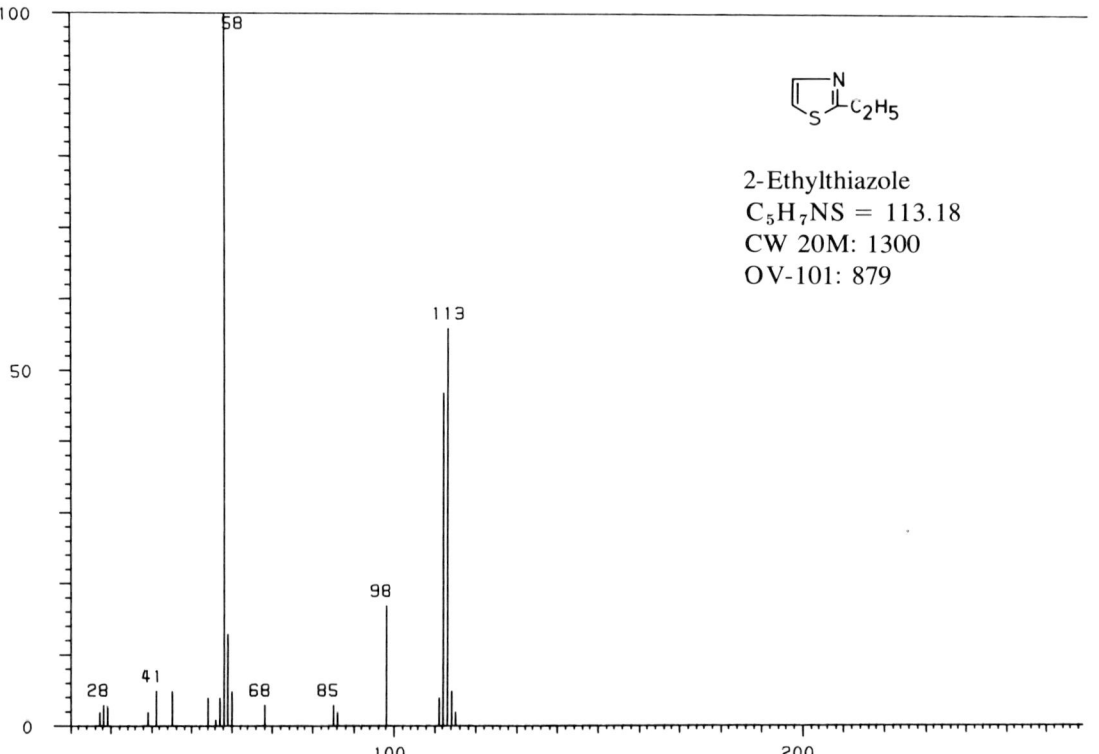

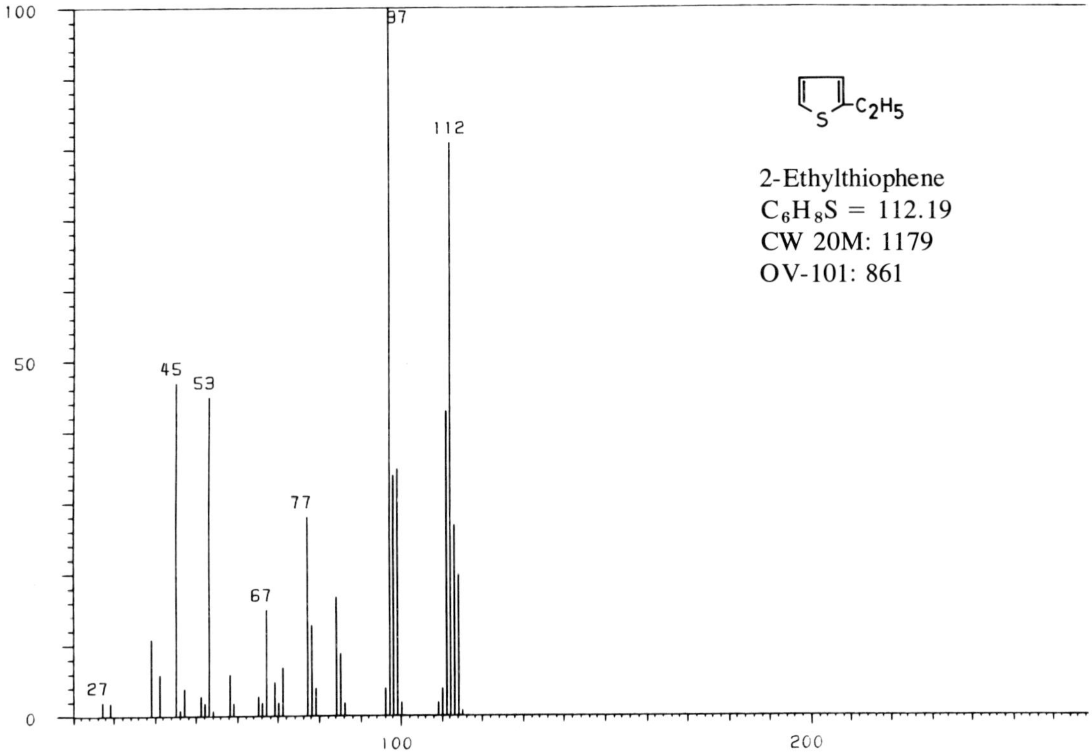

APPENDIX IV   MASS SPECTRA OF INDIVIDUAL COMPOUNDS

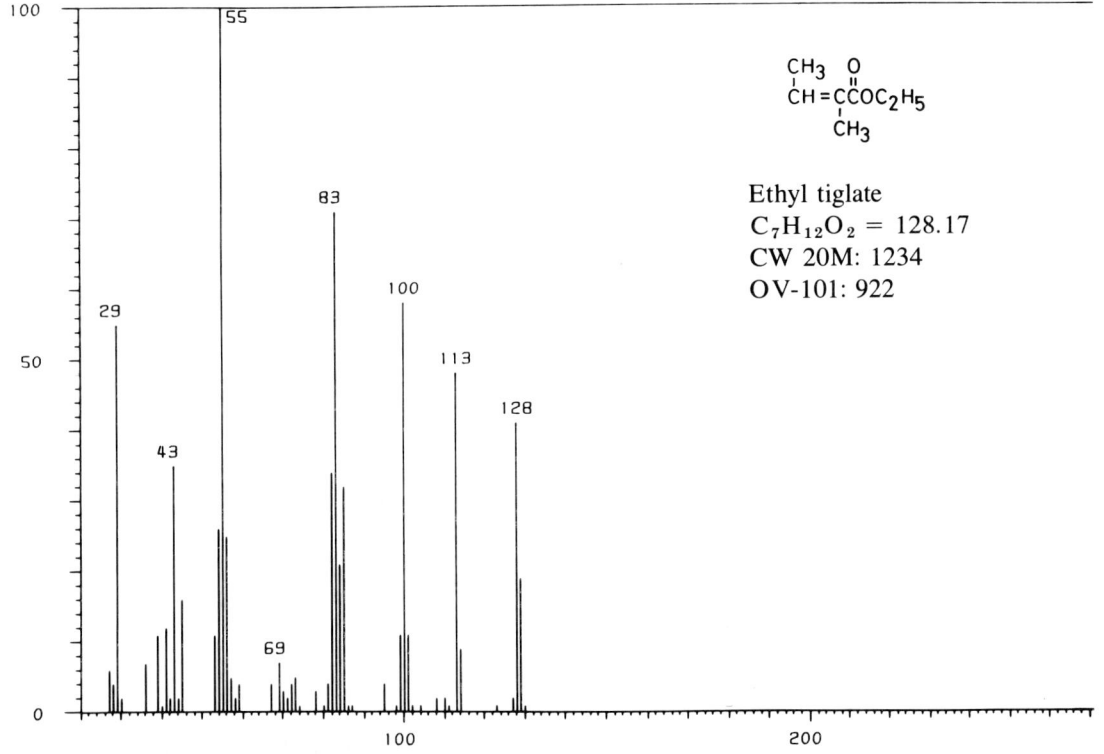

2-Ethylthiophene
$C_6H_8S = 112.19$
CW 20M: 1179
OV-101: 861

Ethyl tiglate
$C_7H_{12}O_2 = 128.17$
CW 20M: 1234
OV-101: 922

$CH_3(CH_2)_9\overset{O}{\overset{\|}{C}}OC_2H_5$

Ethyl *n*-undecanoate
$C_{13}H_{26}O_2$ = 214.35
CW 20M: 1725
OV-101: 1479

$CH_2{=}CH(CH_2)_8\overset{O}{\overset{\|}{C}}OCH_2CH_3$

Ethyl 10-undecenoate
$C_{13}H_{24}O_2$ = 212.34
CW 20M: 1775
OV-101: 1469

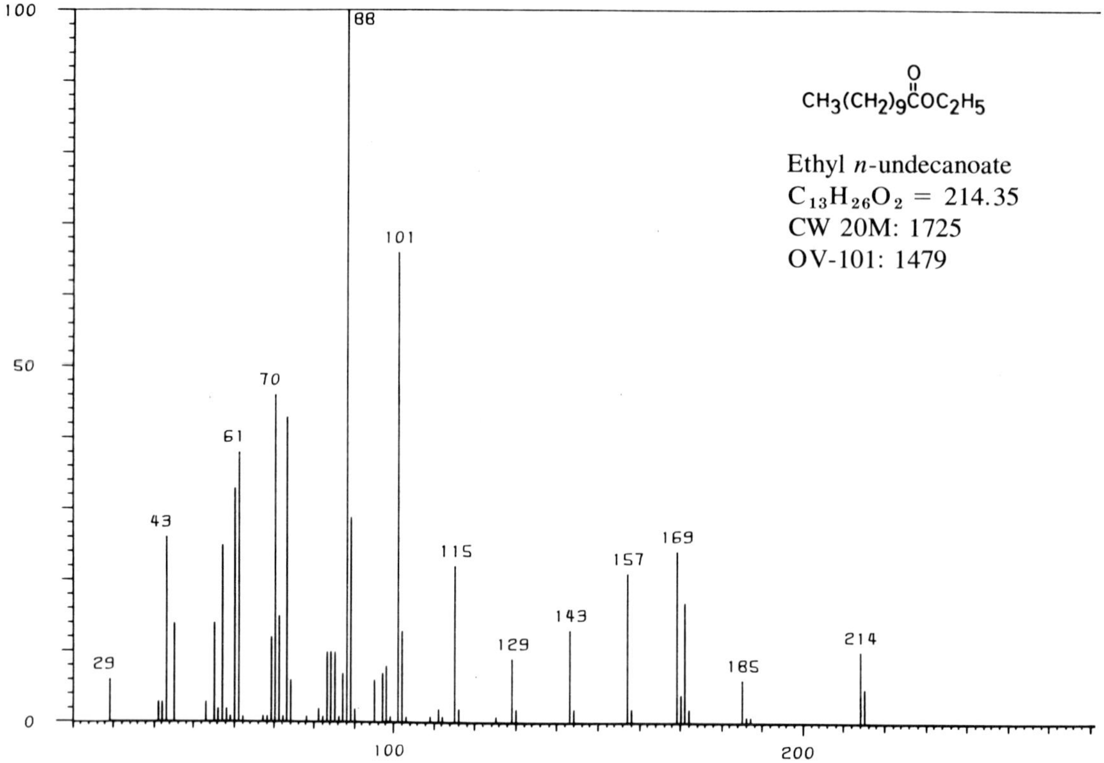

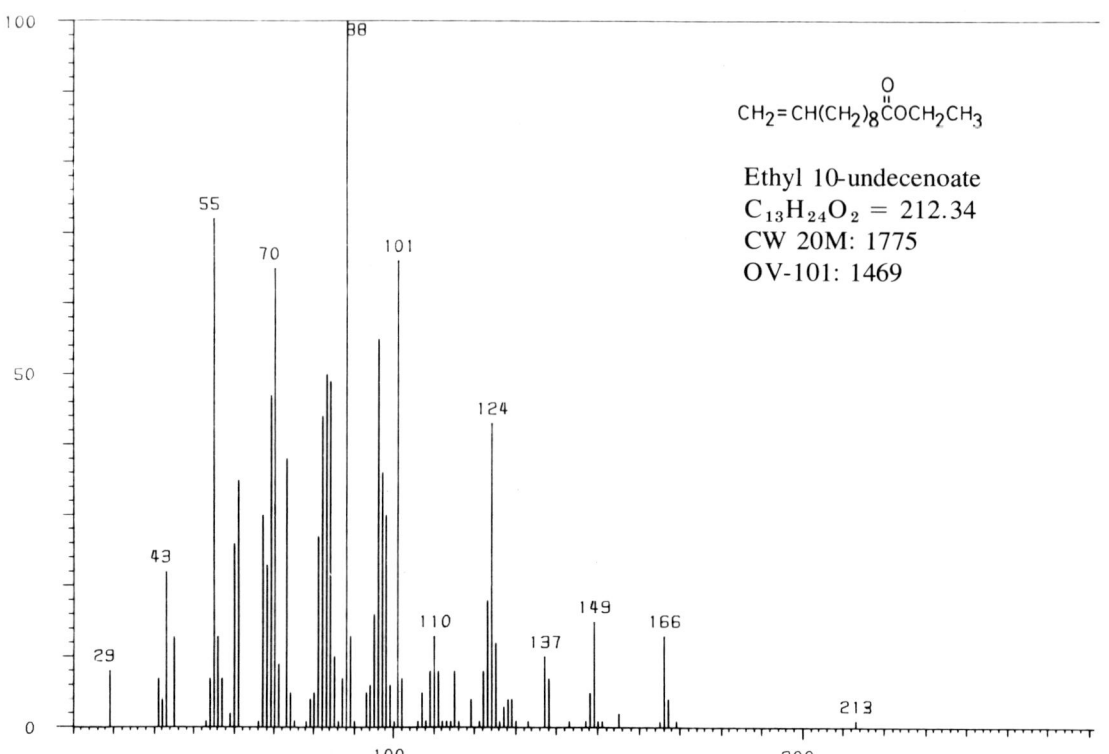

APPENDIX IV   MASS SPECTRA OF INDIVIDUAL COMPOUNDS   297

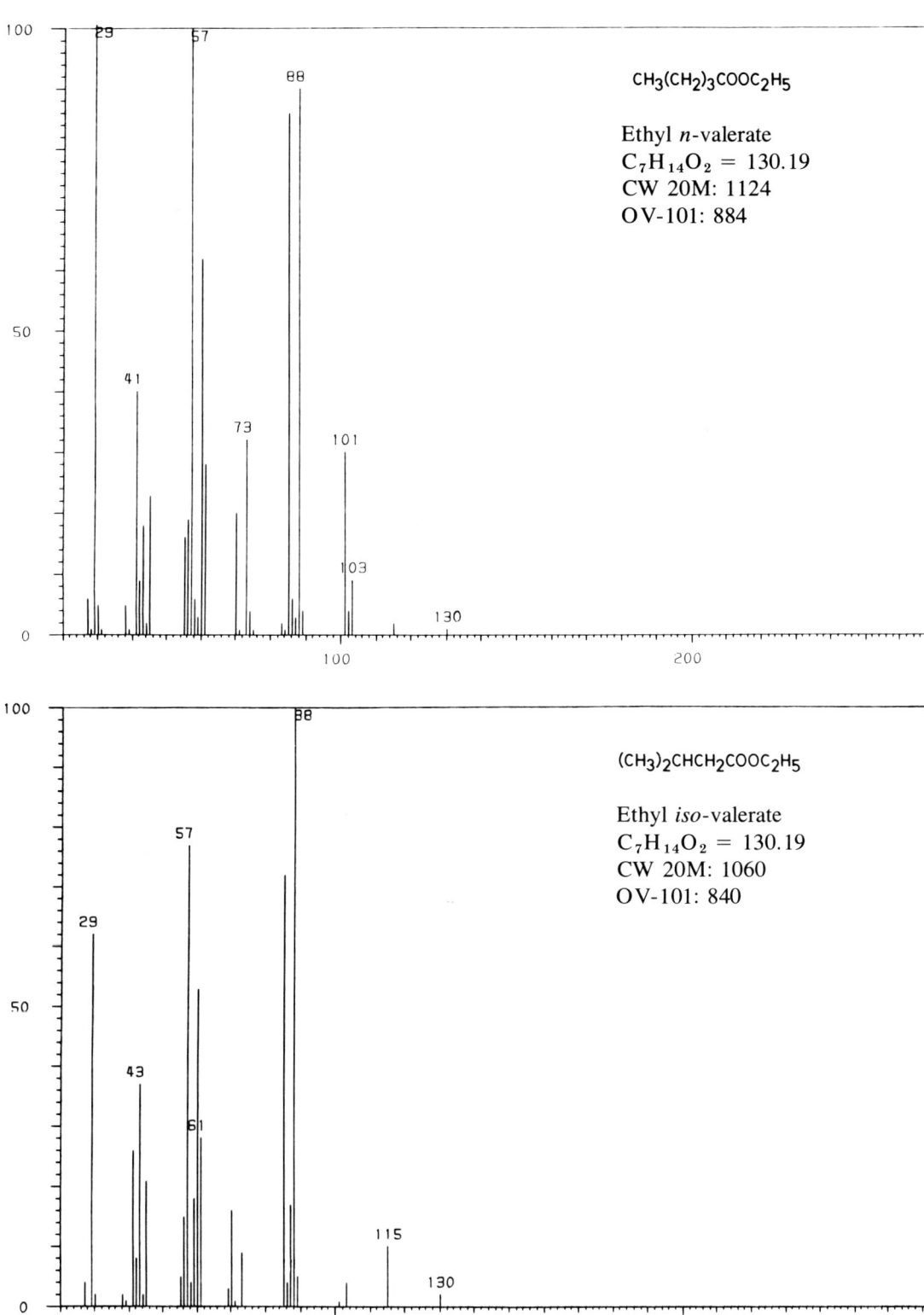

CH$_3$(CH$_2$)$_3$COOC$_2$H$_5$

Ethyl *n*-valerate
C$_7$H$_{14}$O$_2$ = 130.19
CW 20M: 1124
OV-101: 884

(CH$_3$)$_2$CHCH$_2$COOC$_2$H$_5$

Ethyl *iso*-valerate
C$_7$H$_{14}$O$_2$ = 130.19
CW 20M: 1060
OV-101: 840

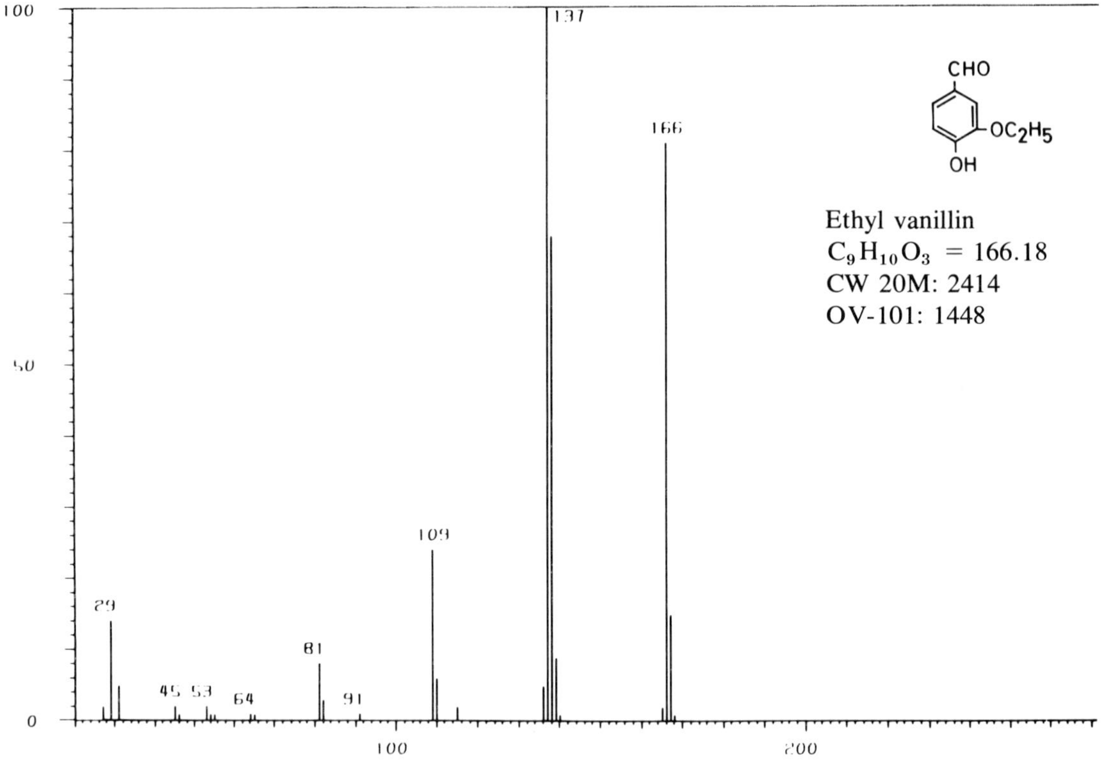

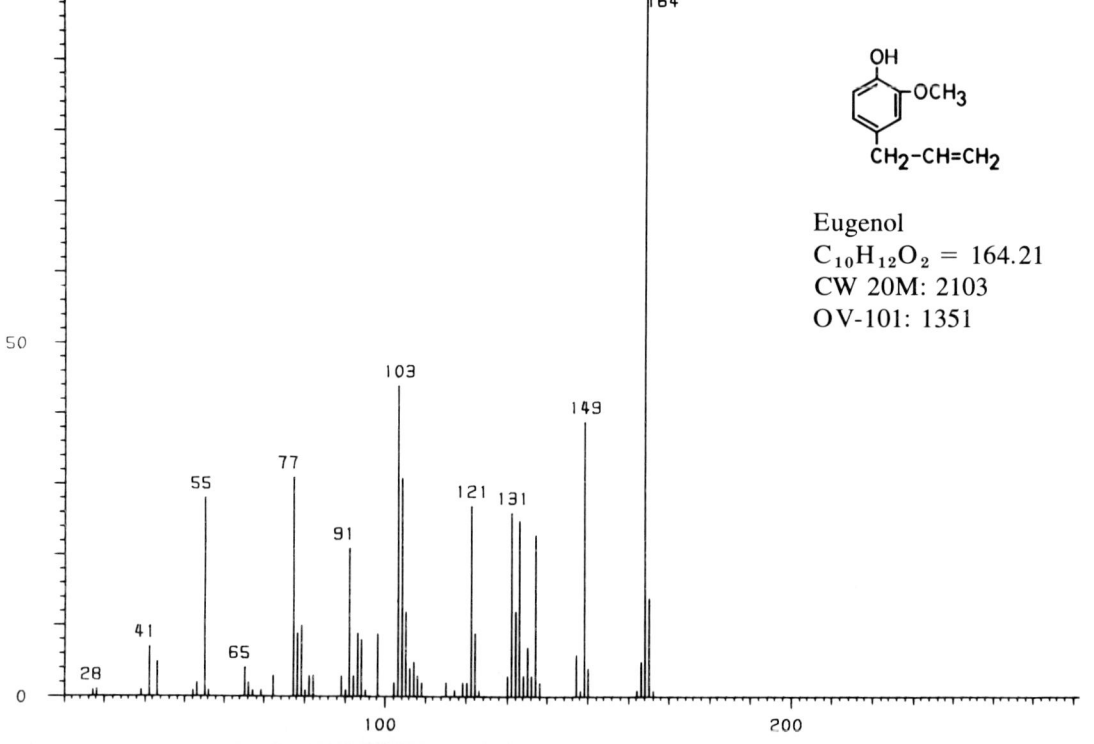

Ethyl vanillin
$C_9H_{10}O_3 = 166.18$
CW 20M: 2414
OV-101: 1448

Eugenol
$C_{10}H_{12}O_2 = 164.21$
CW 20M: 2103
OV-101: 1351

APPENDIX IV   MASS SPECTRA OF INDIVIDUAL COMPOUNDS

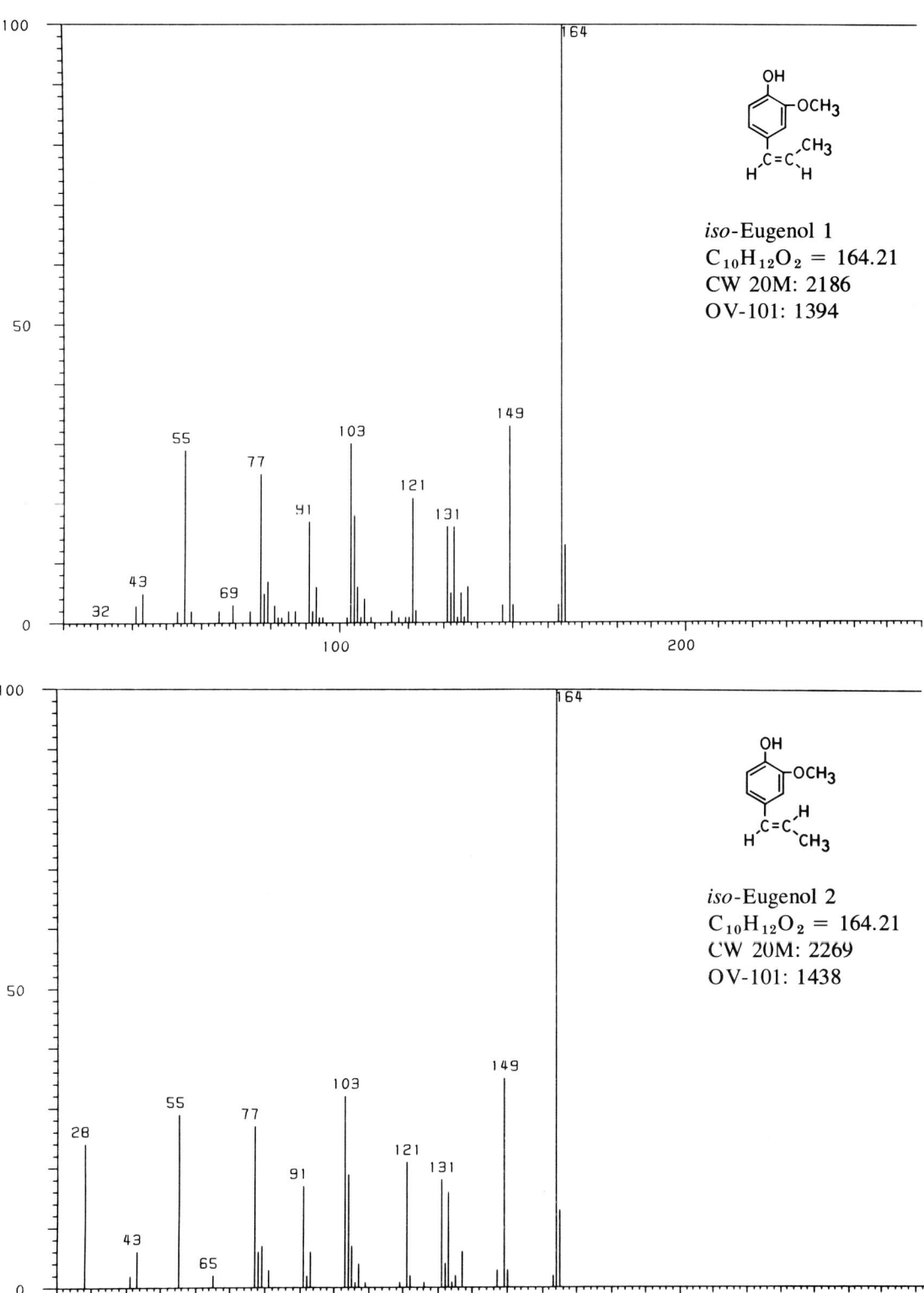

iso-Eugenol 1
$C_{10}H_{12}O_2 = 164.21$
CW 20M: 2186
OV-101: 1394

iso-Eugenol 2
$C_{10}H_{12}O_2 = 164.21$
CW 20M: 2269
OV-101: 1438

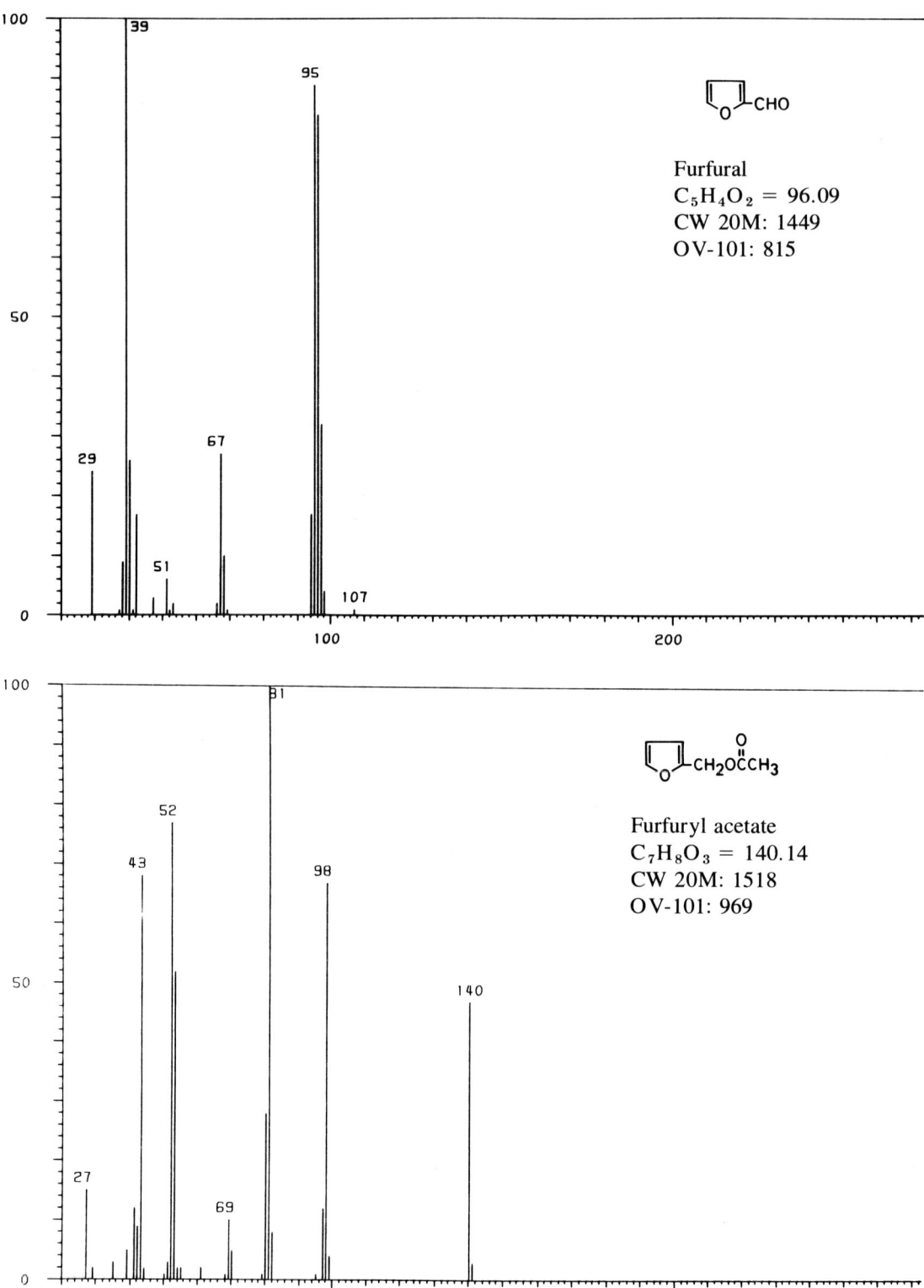

APPENDIX IV   MASS SPECTRA OF INDIVIDUAL COMPOUNDS   **301**

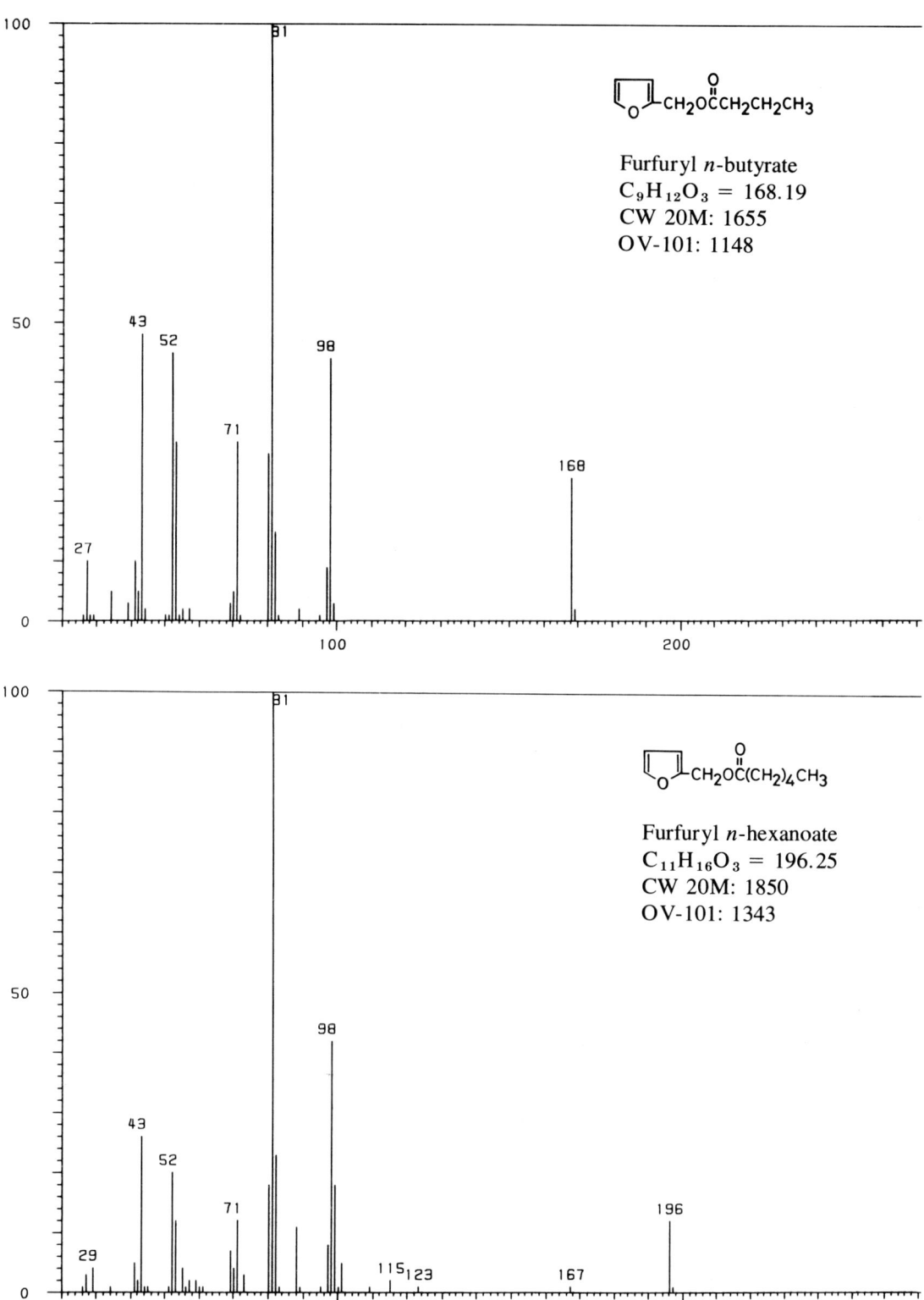

Furfuryl n-butyrate
$C_9H_{12}O_3 = 168.19$
CW 20M: 1655
OV-101: 1148

Furfuryl n-hexanoate
$C_{11}H_{16}O_3 = 196.25$
CW 20M: 1850
OV-101: 1343

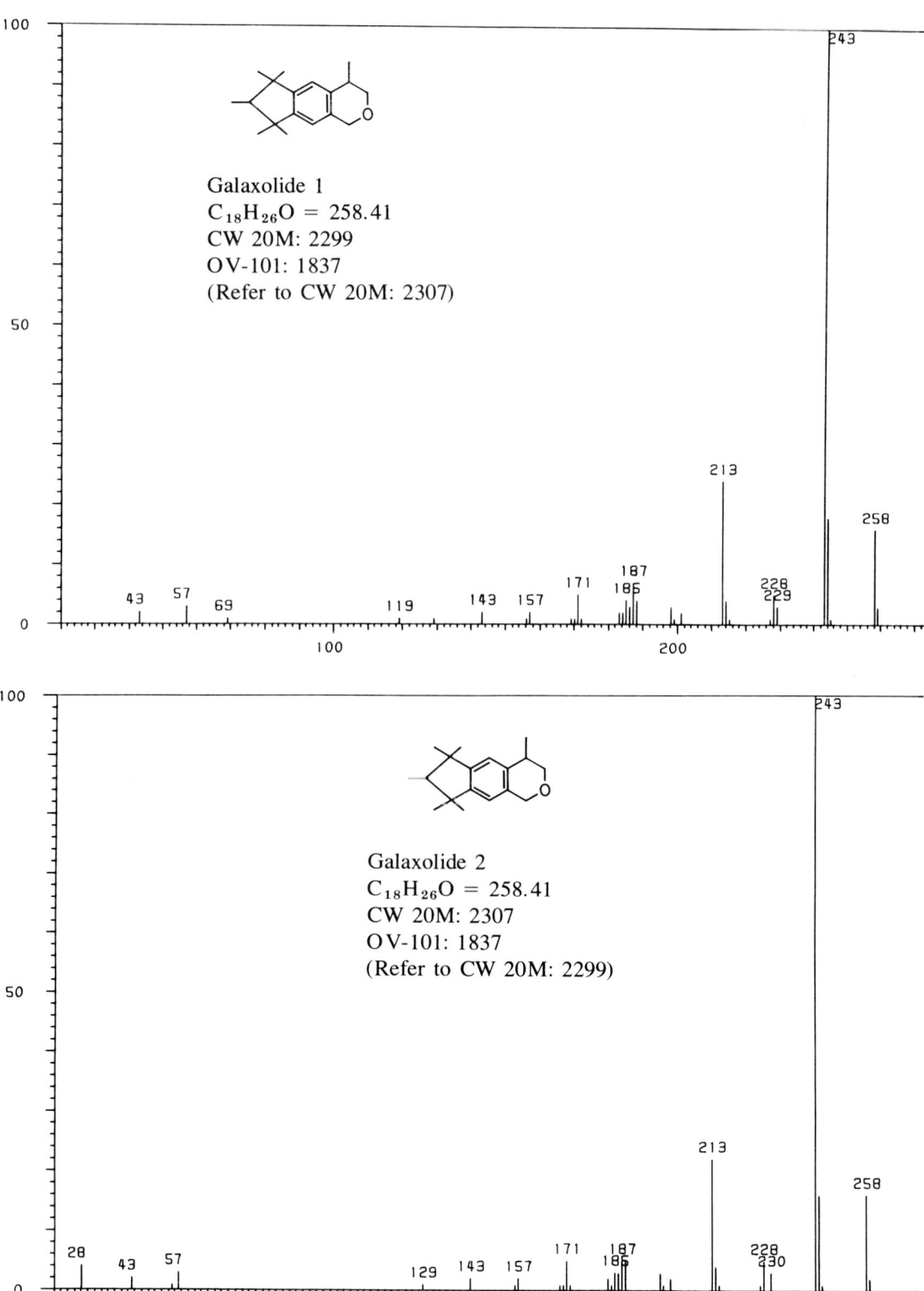

APPENDIX IV   MASS SPECTRA OF INDIVIDUAL COMPOUNDS

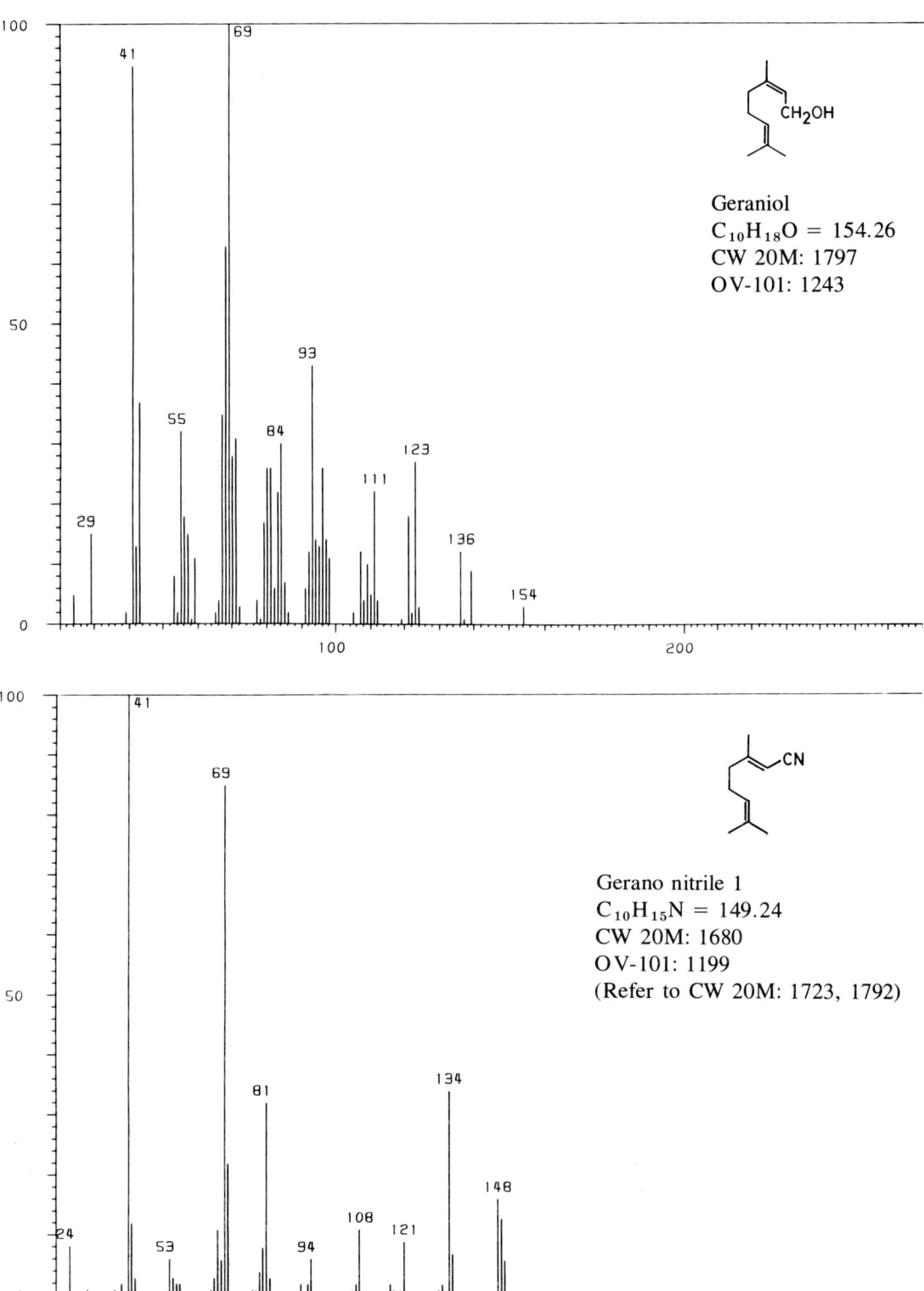

Geraniol
$C_{10}H_{18}O = 154.26$
CW 20M: 1797
OV-101: 1243

Gerano nitrile 1
$C_{10}H_{15}N = 149.24$
CW 20M: 1680
OV-101: 1199
(Refer to CW 20M: 1723, 1792)

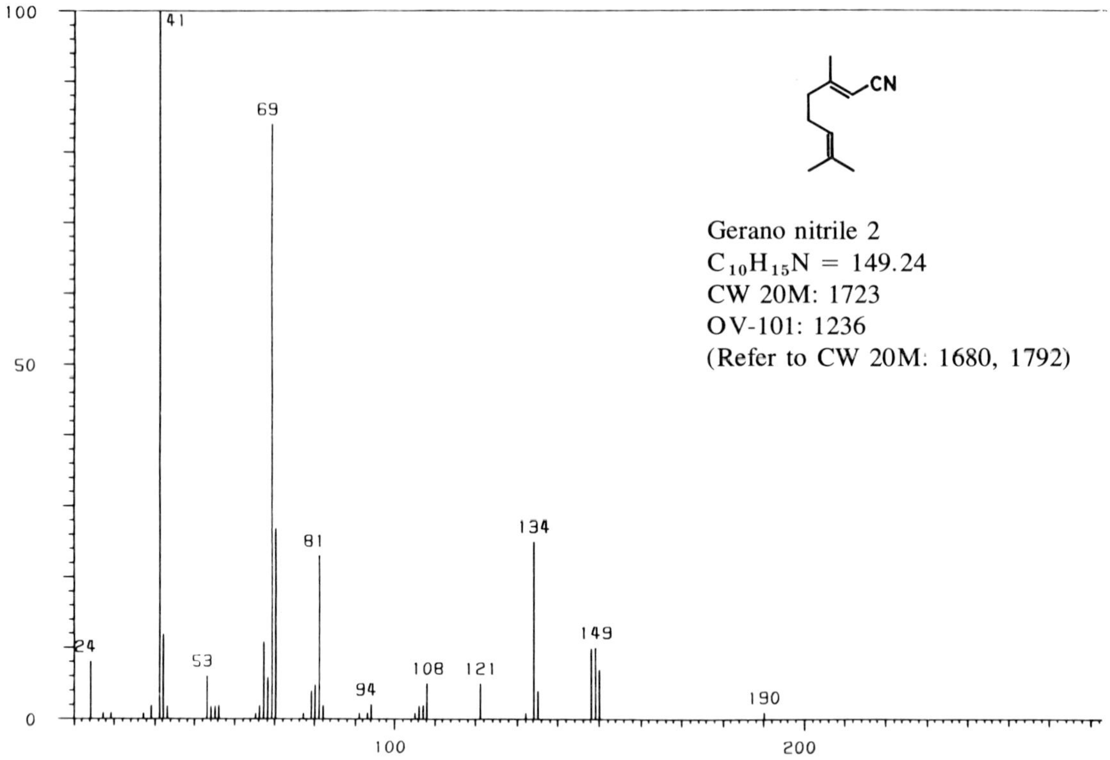

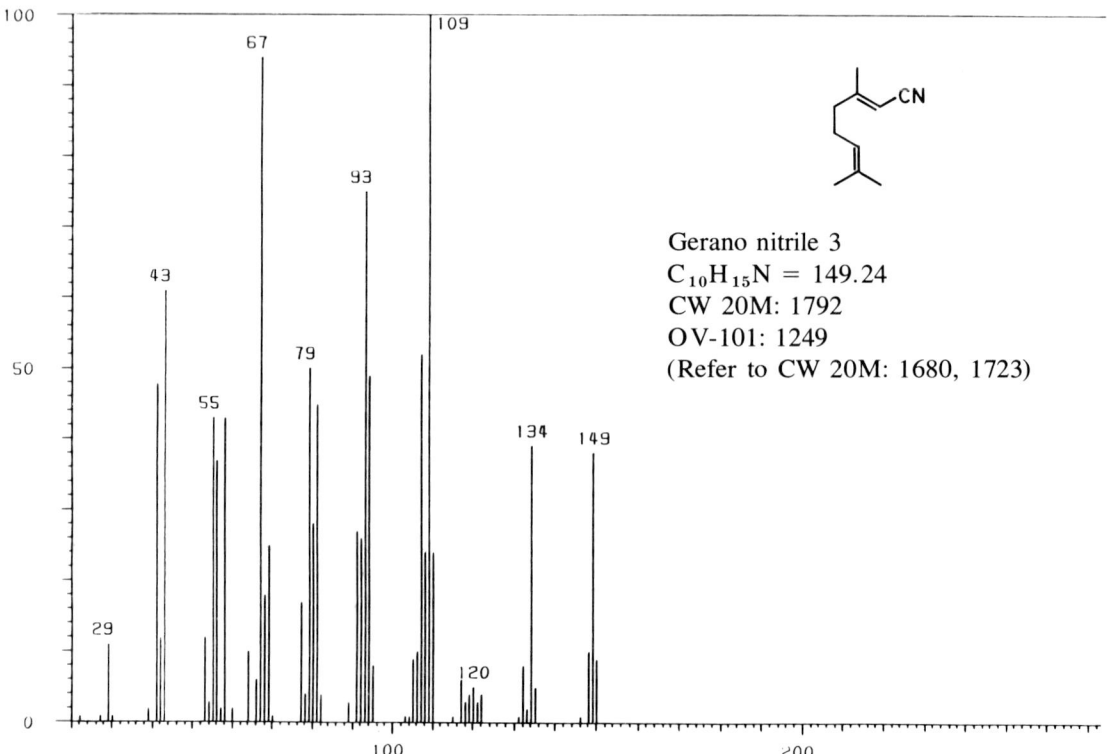

Gerano nitrile 2
$C_{10}H_{15}N = 149.24$
CW 20M: 1723
OV-101: 1236
(Refer to CW 20M: 1680, 1792)

Gerano nitrile 3
$C_{10}H_{15}N = 149.24$
CW 20M: 1792
OV-101: 1249
(Refer to CW 20M: 1680, 1723)

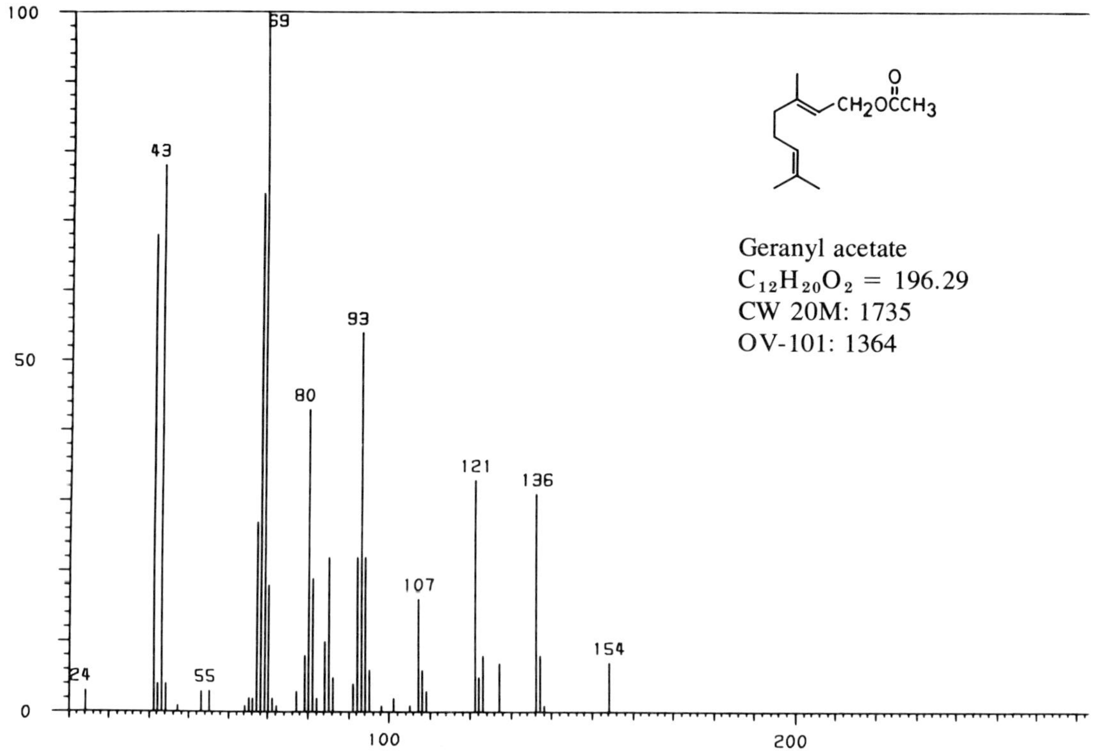

APPENDIX IV   MASS SPECTRA OF INDIVIDUAL COMPOUNDS   305

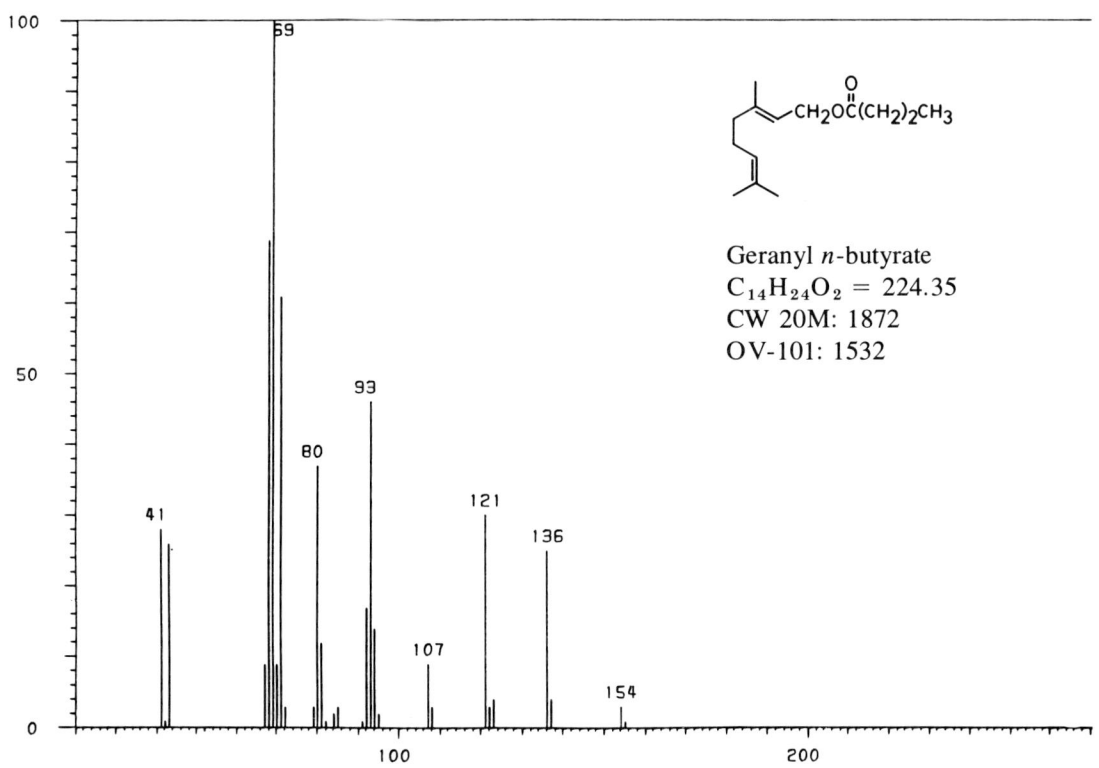

Geranyl acetate
$C_{12}H_{20}O_2 = 196.29$
CW 20M: 1735
OV-101: 1364

Geranyl n-butyrate
$C_{14}H_{24}O_2 = 224.35$
CW 20M: 1872
OV-101: 1532

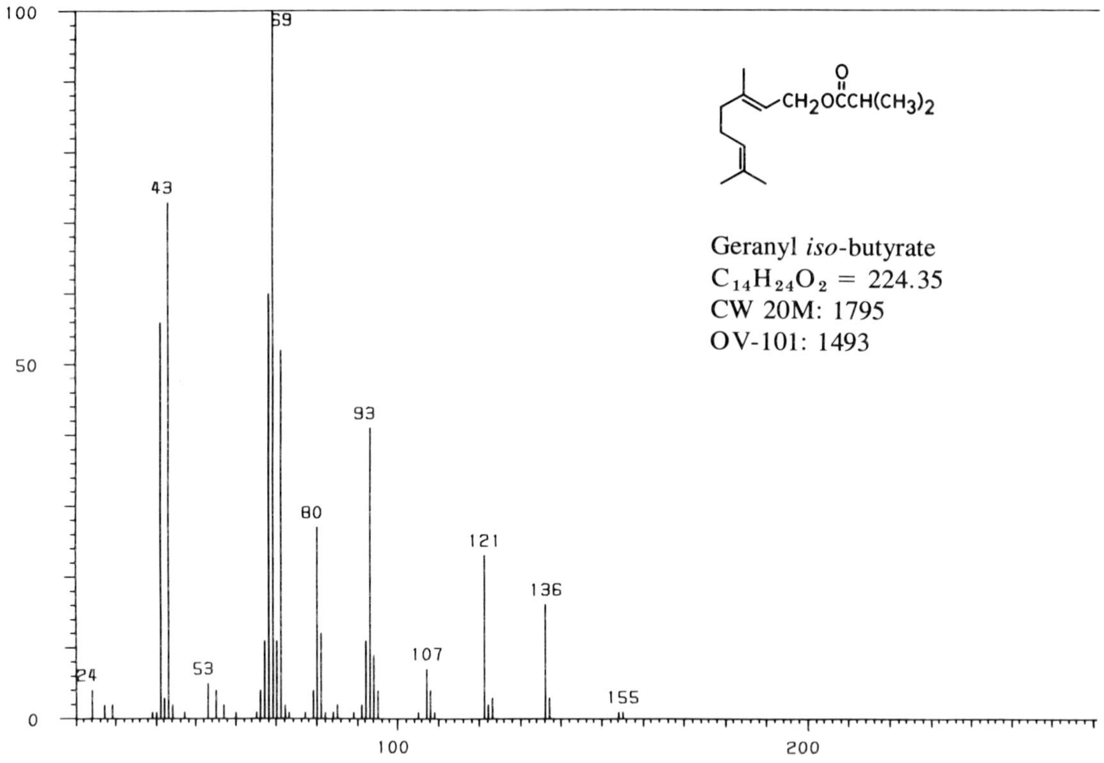

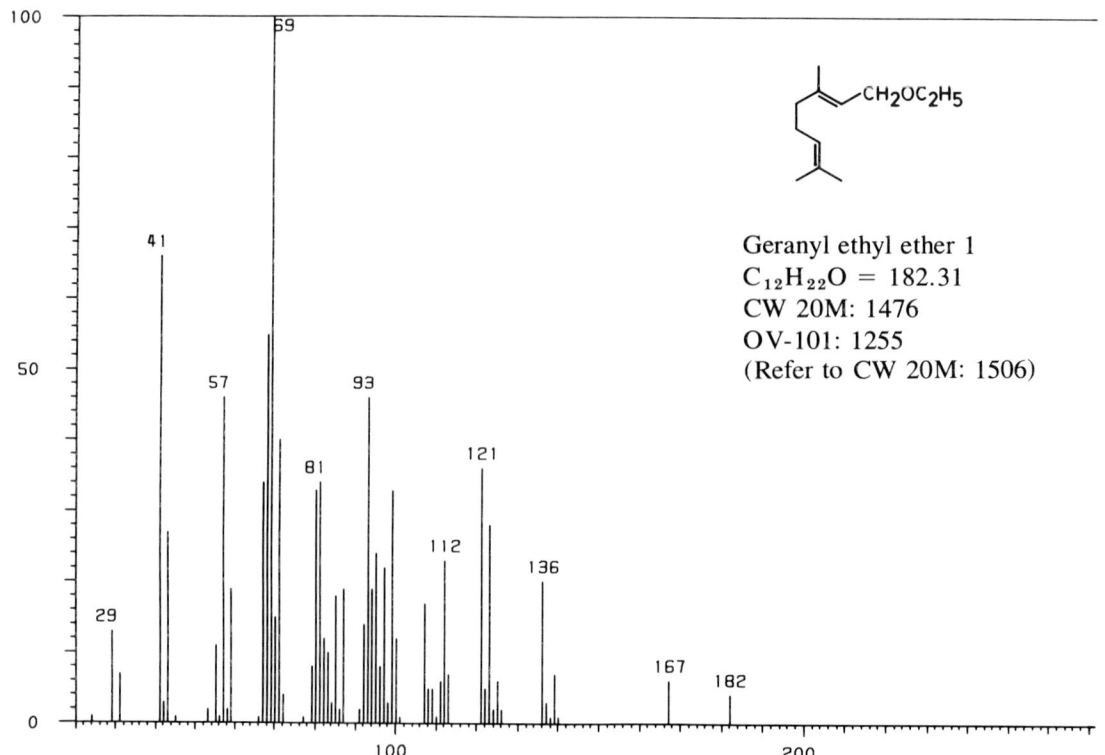

Geranyl *iso*-butyrate
$C_{14}H_{24}O_2 = 224.35$
CW 20M: 1795
OV-101: 1493

Geranyl ethyl ether 1
$C_{12}H_{22}O = 182.31$
CW 20M: 1476
OV-101: 1255
(Refer to CW 20M: 1506)

APPENDIX IV   MASS SPECTRA OF INDIVIDUAL COMPOUNDS

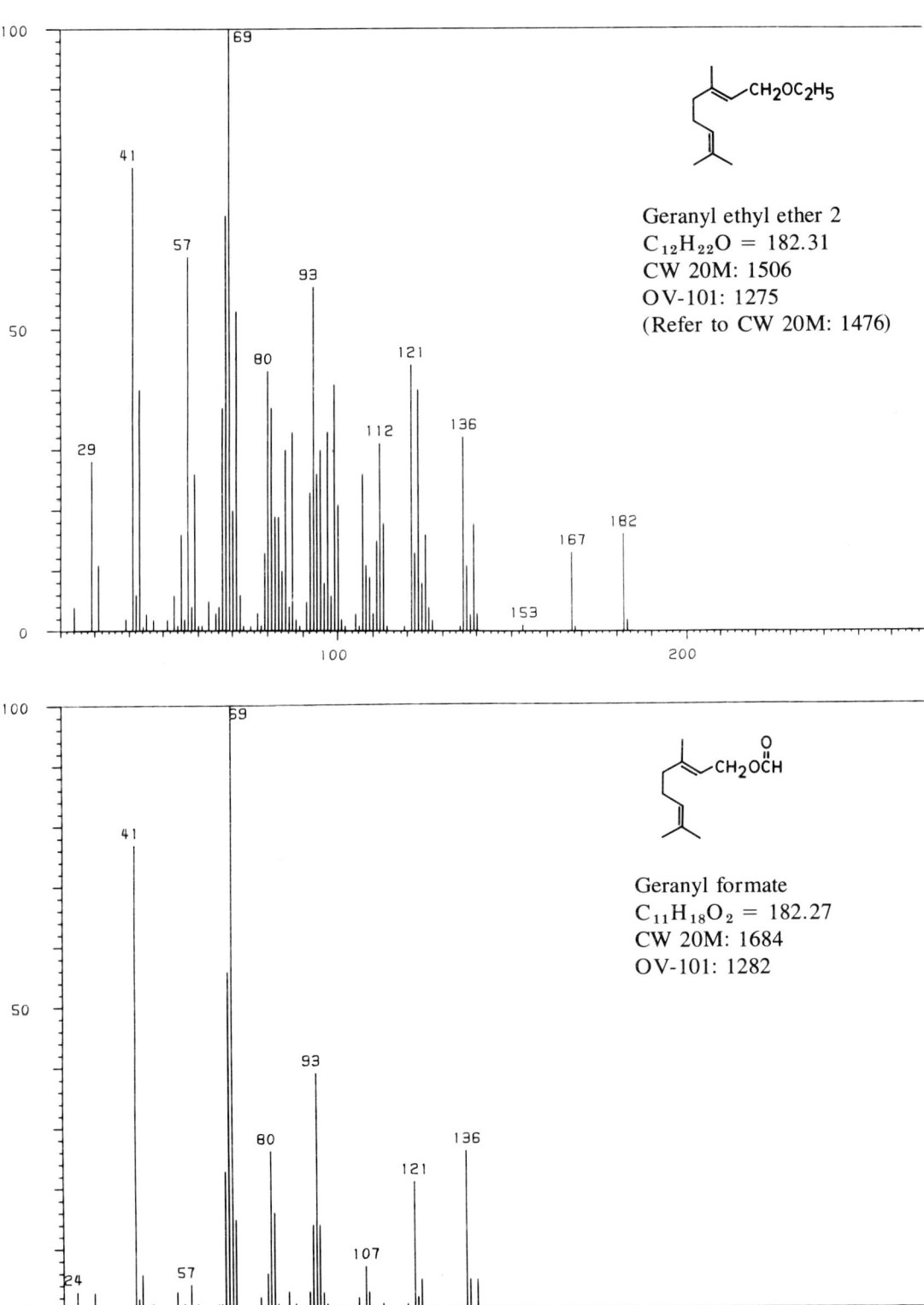

Geranyl ethyl ether 2
$C_{12}H_{22}O = 182.31$
CW 20M: 1506
OV-101: 1275
(Refer to CW 20M: 1476)

Geranyl formate
$C_{11}H_{18}O_2 = 182.27$
CW 20M: 1684
OV-101: 1282

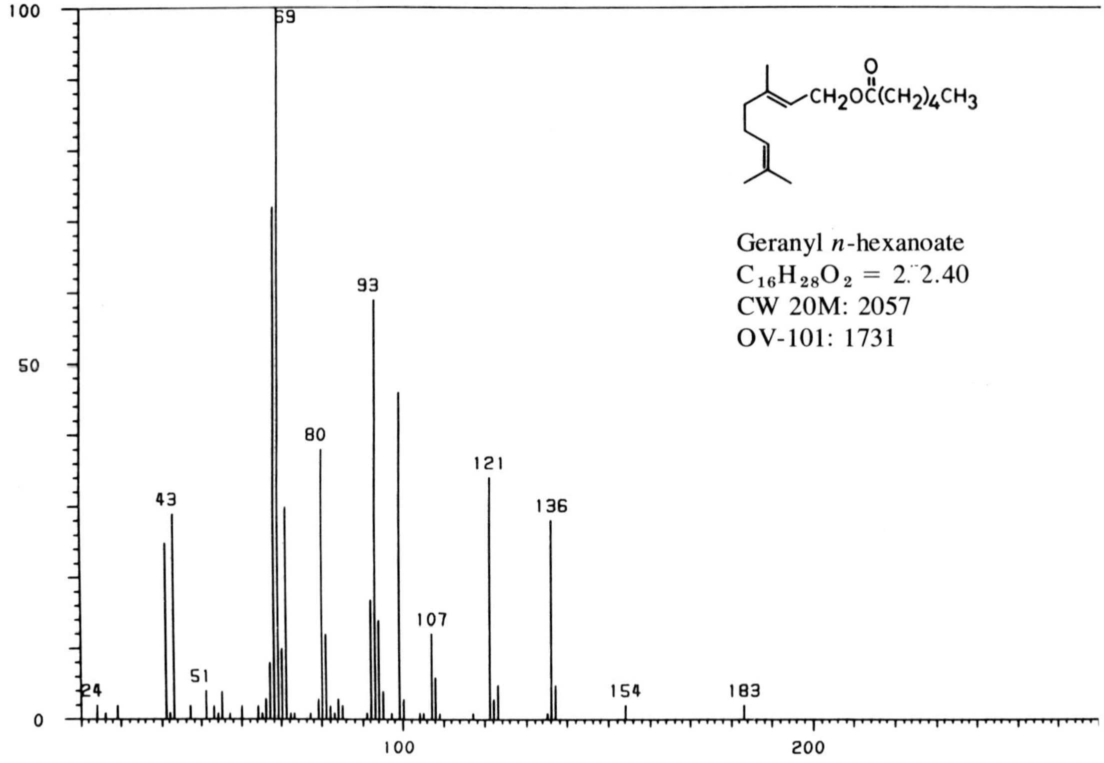

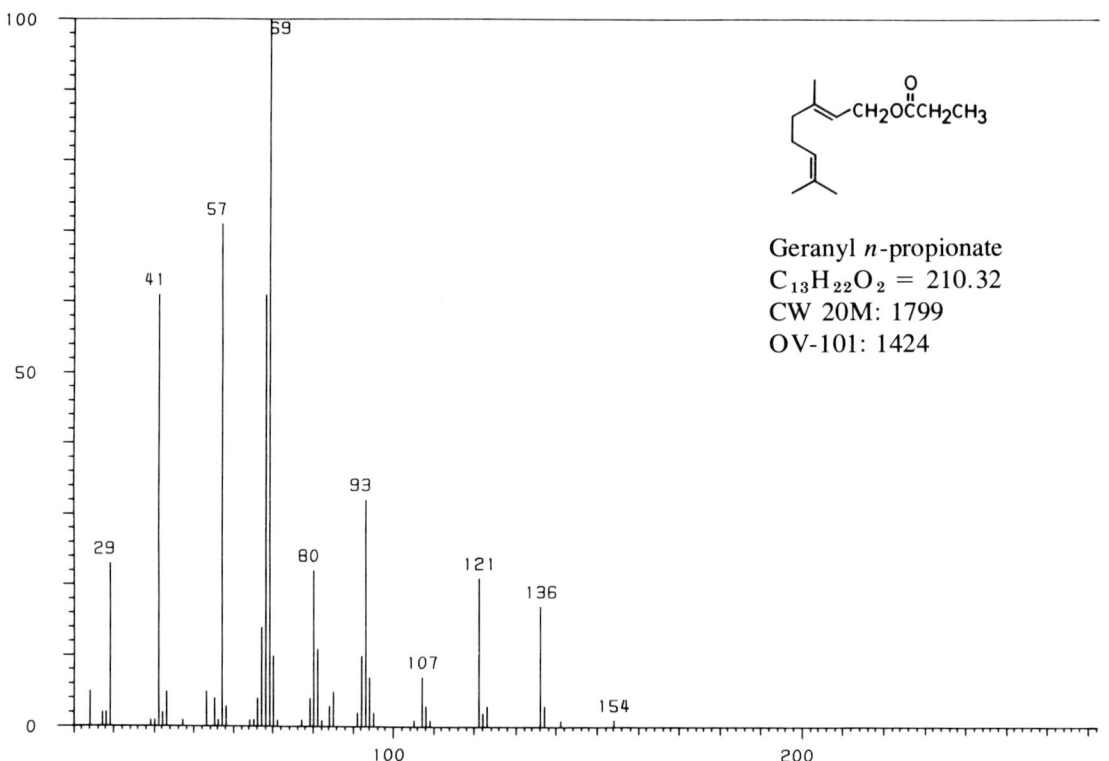

Geranyl *n*-hexanoate
$C_{16}H_{28}O_2 = 2\,2.40$
CW 20M: 2057
OV-101: 1731

Geranyl *n*-propionate
$C_{13}H_{22}O_2 = 210.32$
CW 20M: 1799
OV-101: 1424

APPENDIX IV   MASS SPECTRA OF INDIVIDUAL COMPOUNDS

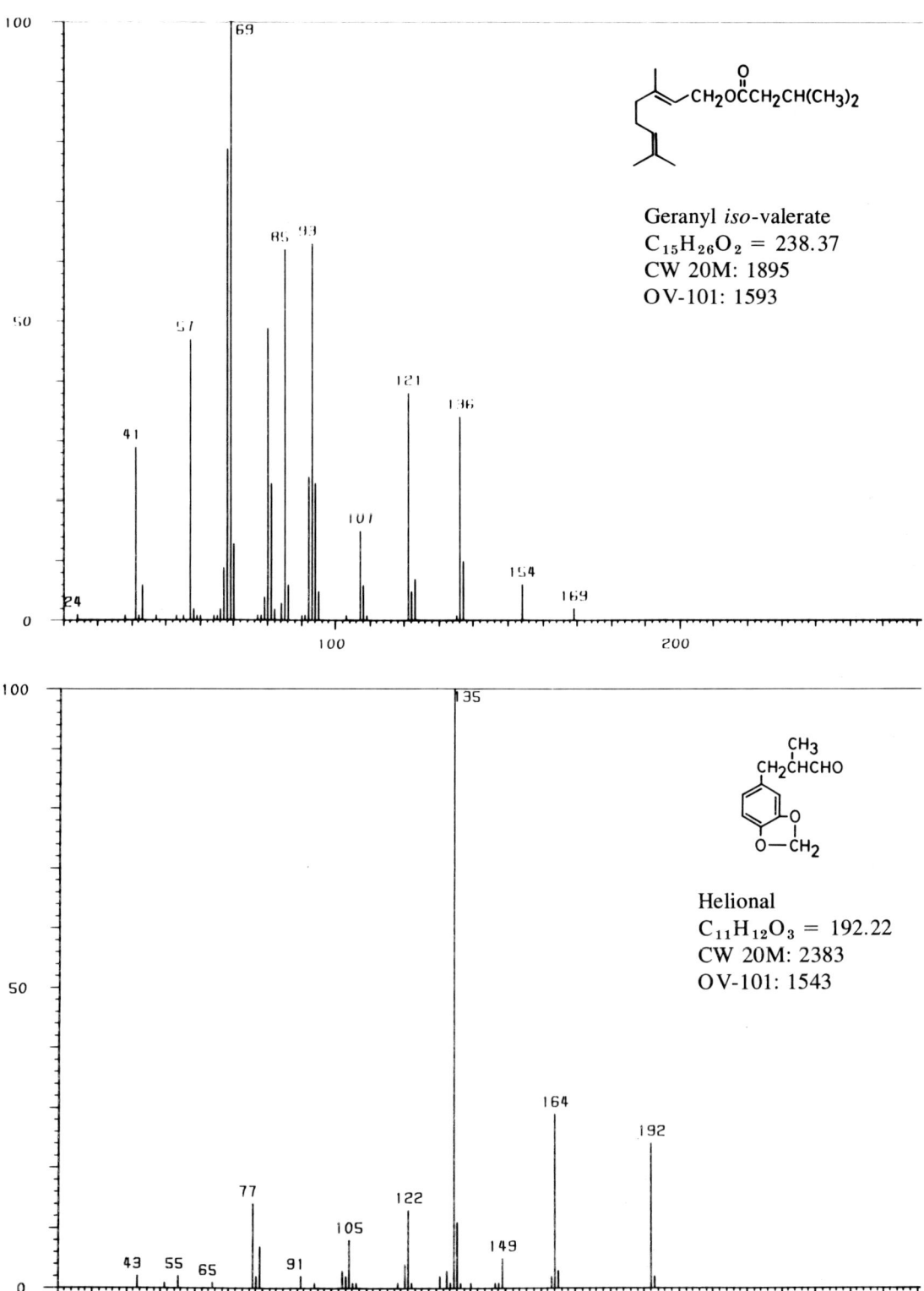

Geranyl *iso*-valerate
$C_{15}H_{26}O_2 = 238.37$
CW 20M: 1895
OV-101: 1593

Helional
$C_{11}H_{12}O_3 = 192.22$
CW 20M: 2383
OV-101: 1543

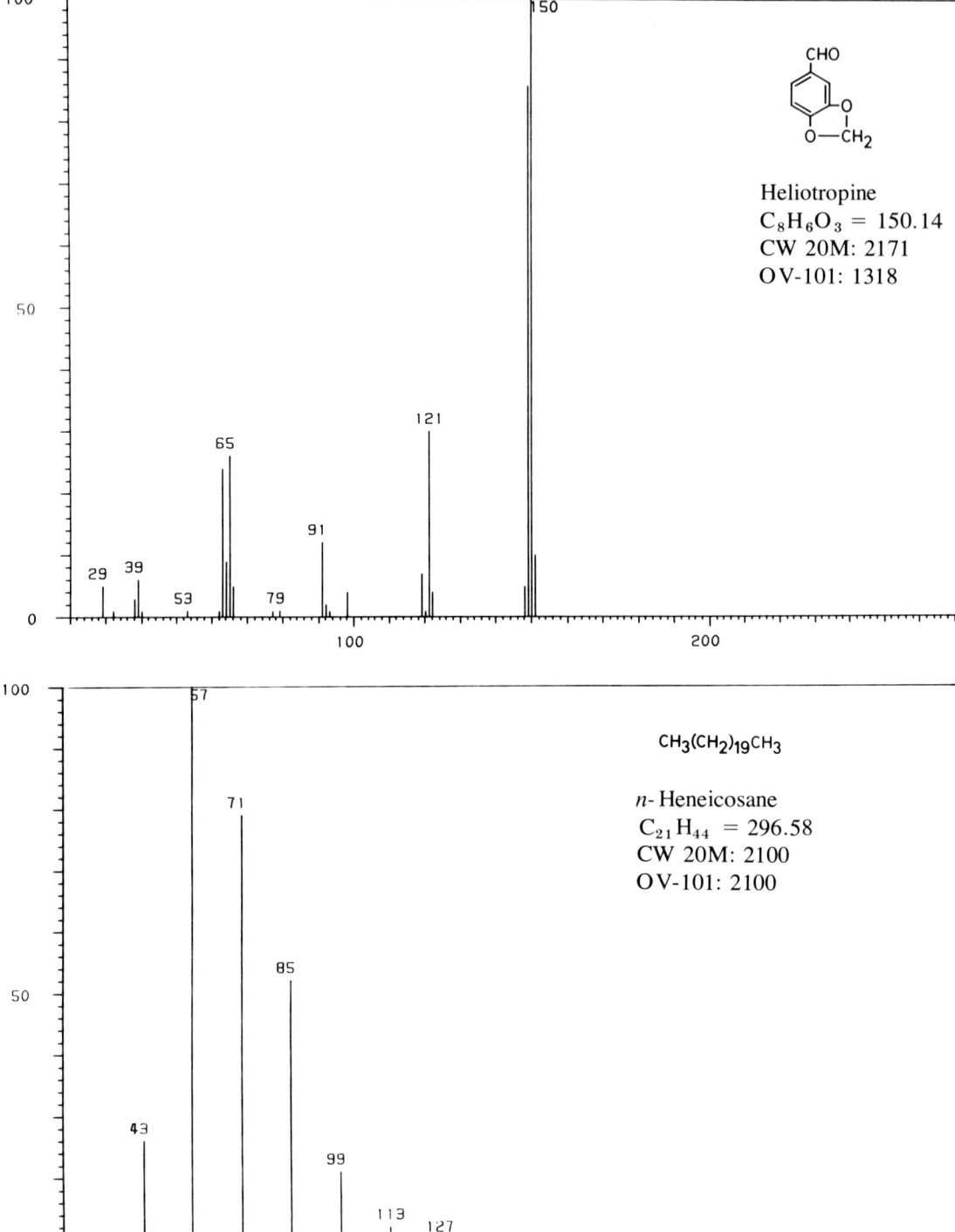

APPENDIX IV   MASS SPECTRA OF INDIVIDUAL COMPOUNDS   311

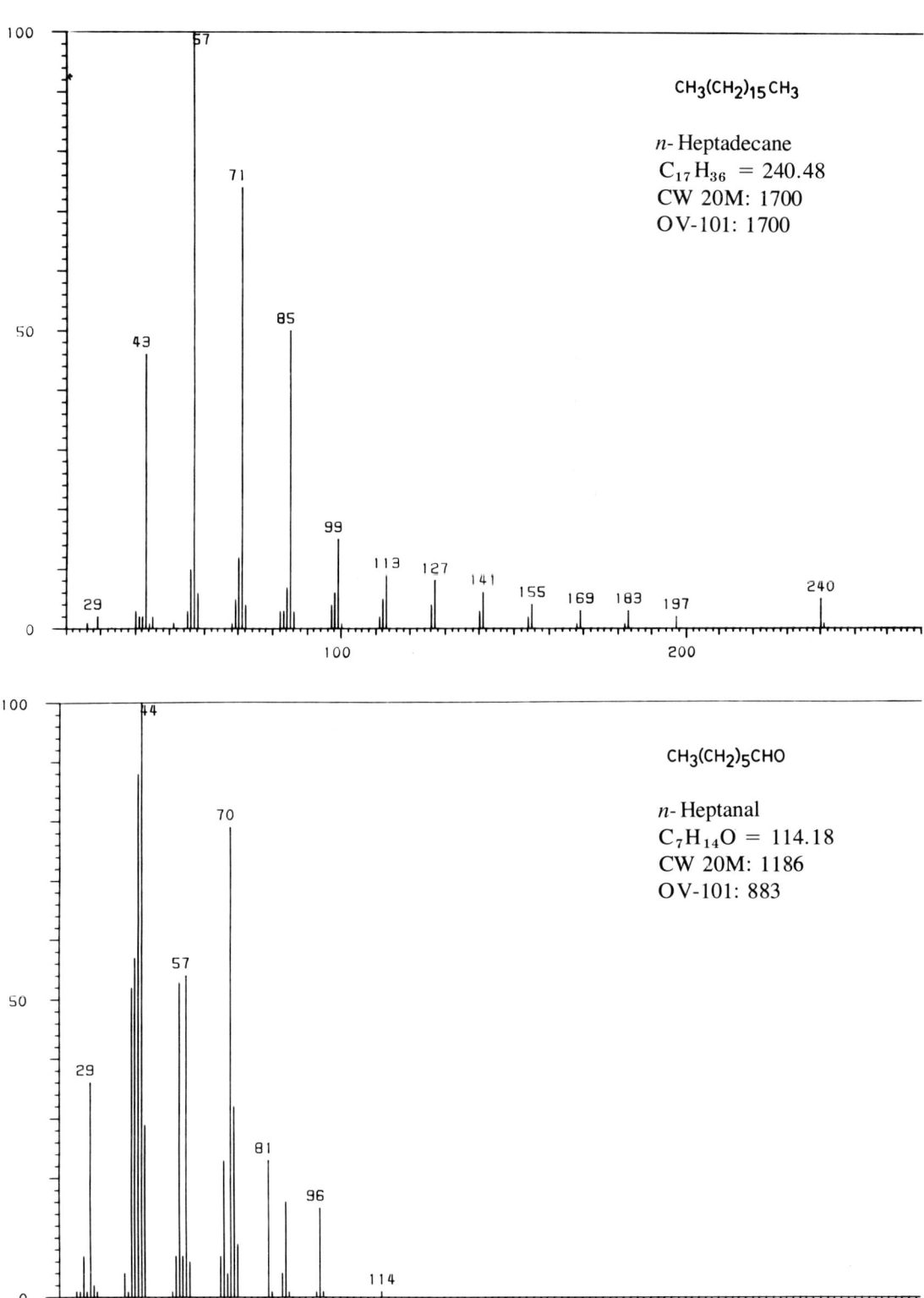

$CH_3(CH_2)_{15}CH_3$

n- Heptadecane
$C_{17}H_{36} = 240.48$
CW 20M: 1700
OV-101: 1700

$CH_3(CH_2)_5CHO$

n- Heptanal
$C_7H_{14}O = 114.18$
CW 20M: 1186
OV-101: 883

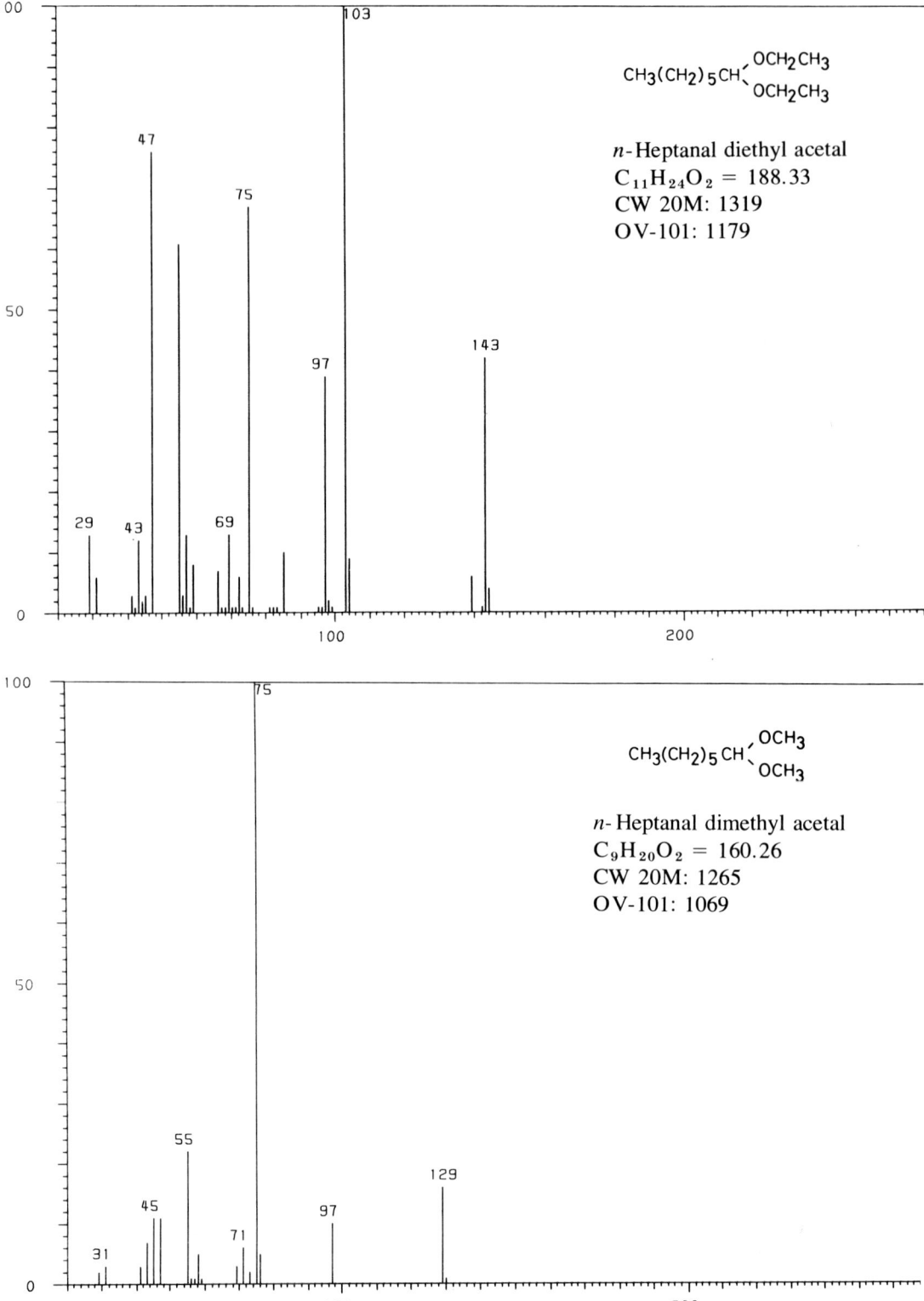

APPENDIX IV   MASS SPECTRA OF INDIVIDUAL COMPOUNDS   313

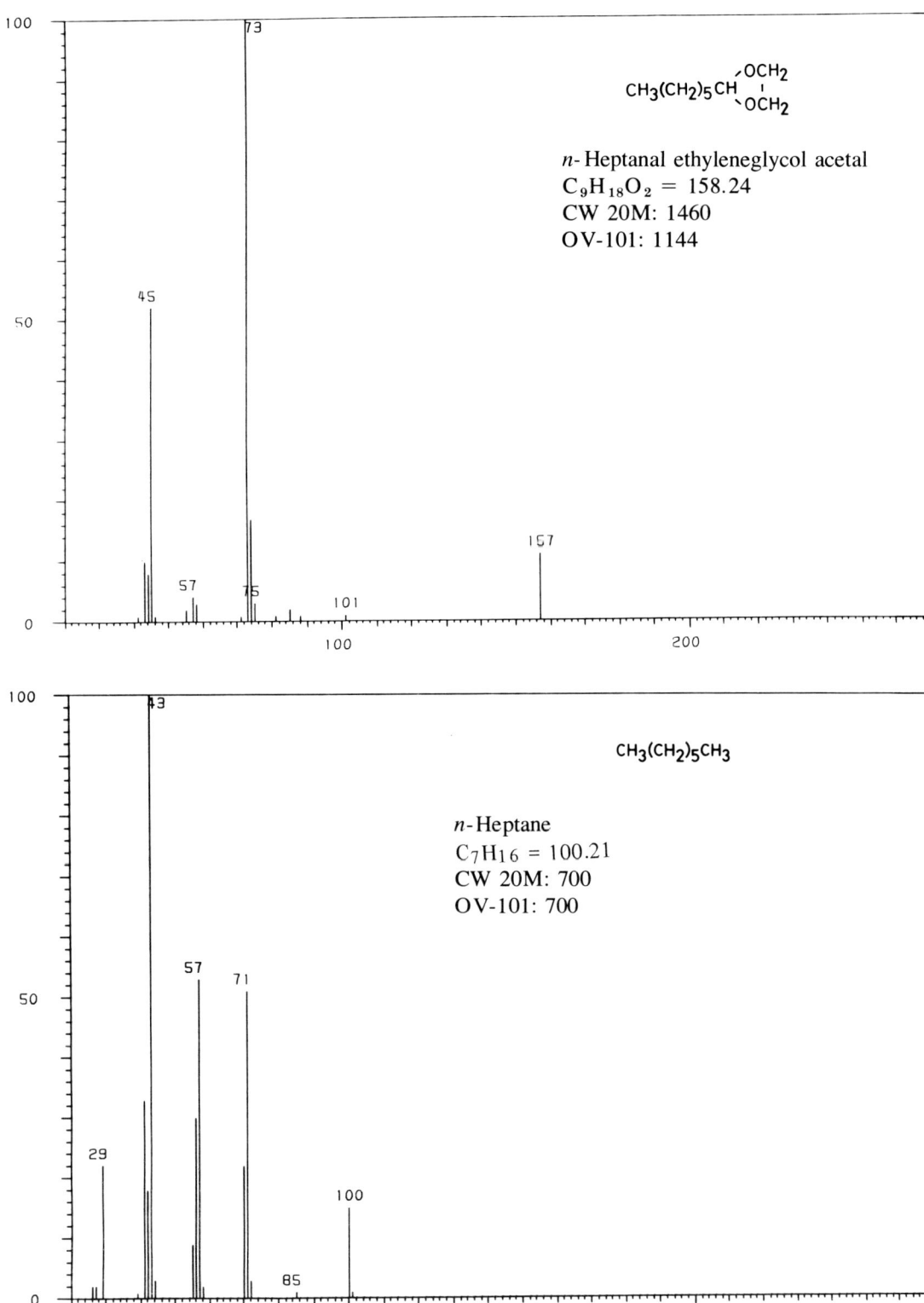

n-Heptanal ethyleneglycol acetal
$C_9H_{18}O_2 = 158.24$
CW 20M: 1460
OV-101: 1144

n-Heptane
$C_7H_{16} = 100.21$
CW 20M: 700
OV-101: 700

CH3-CO-CO-CH2(CH2)2CH3

2,3-Heptanedione
$C_7H_{12}O_2$ = 128.17
CW 20M: 1138
OV-101: 816

CH3(CH2)5CH2OH

n-Heptanol
$C_7H_{16}O$ = 116.21
CW 20M: 1419
OV-101: 957

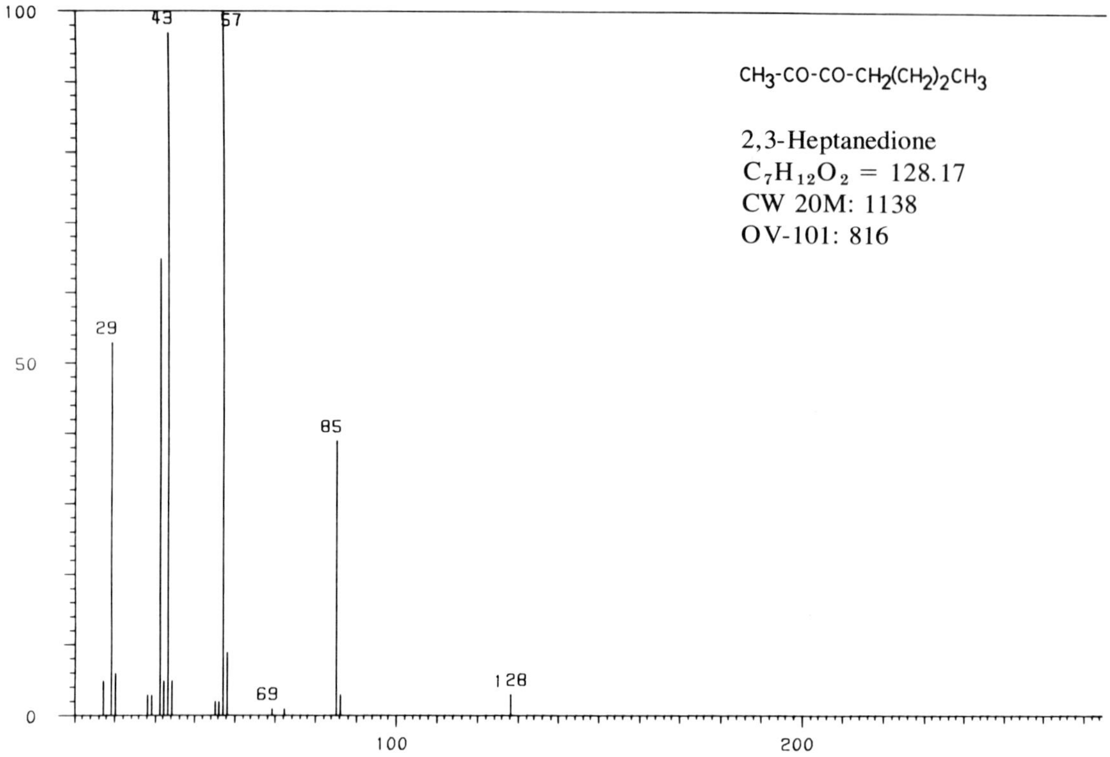

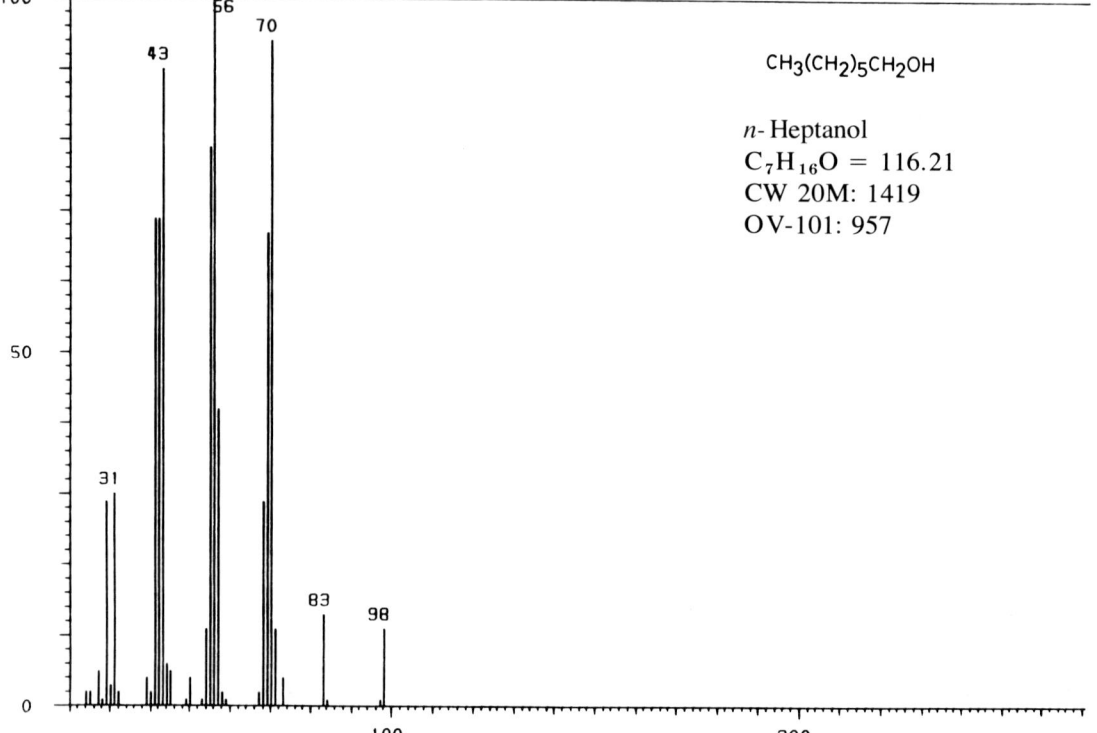

APPENDIX IV   MASS SPECTRA OF INDIVIDUAL COMPOUNDS

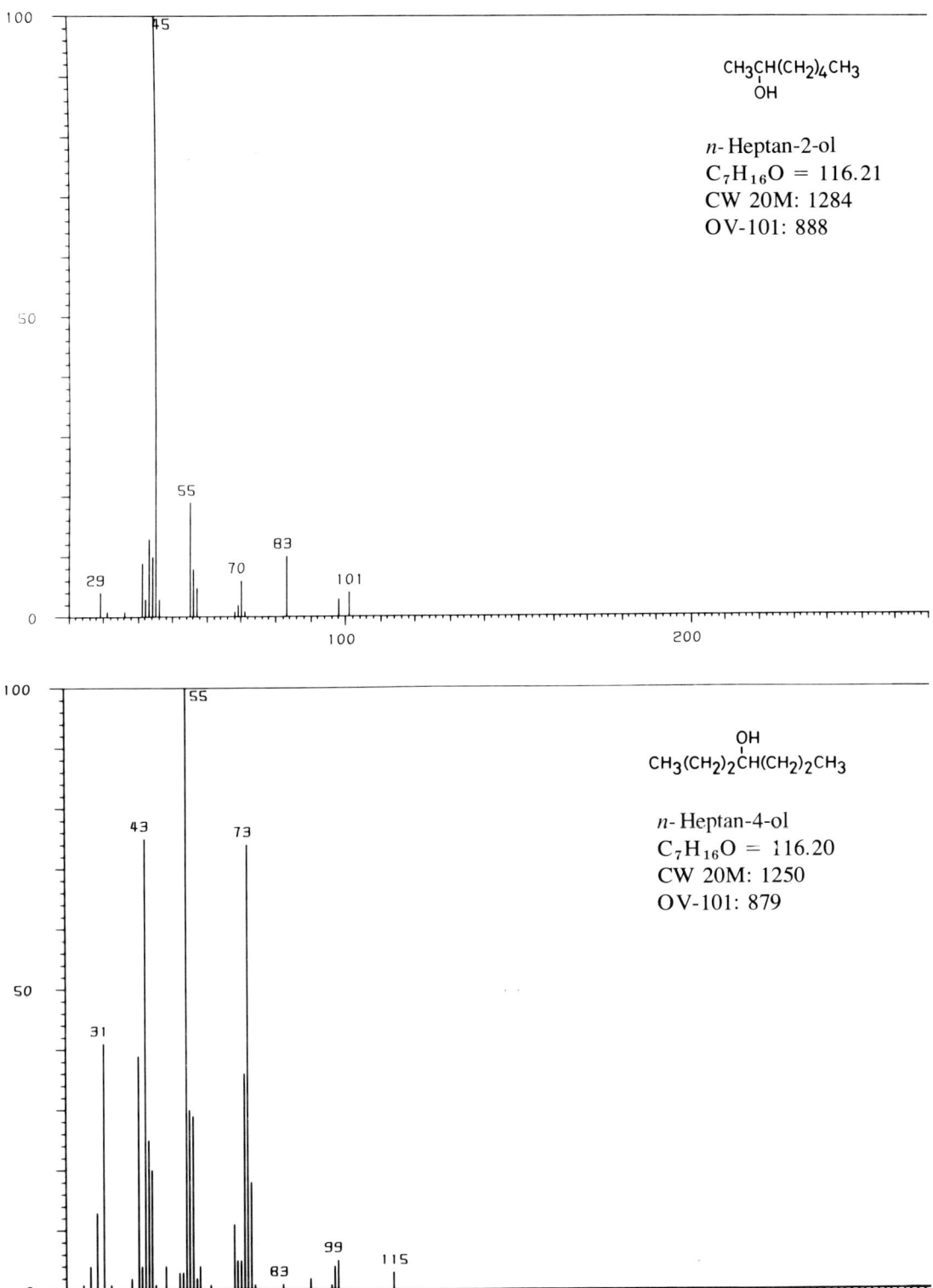

CH$_3$CH(CH$_2$)$_4$CH$_3$
   |
   OH

n-Heptan-2-ol
C$_7$H$_{16}$O = 116.21
CW 20M: 1284
OV-101: 888

       OH
       |
CH$_3$(CH$_2$)$_2$CH(CH$_2$)$_2$CH$_3$

n-Heptan-4-ol
C$_7$H$_{16}$O = 116.20
CW 20M: 1250
OV-101: 879

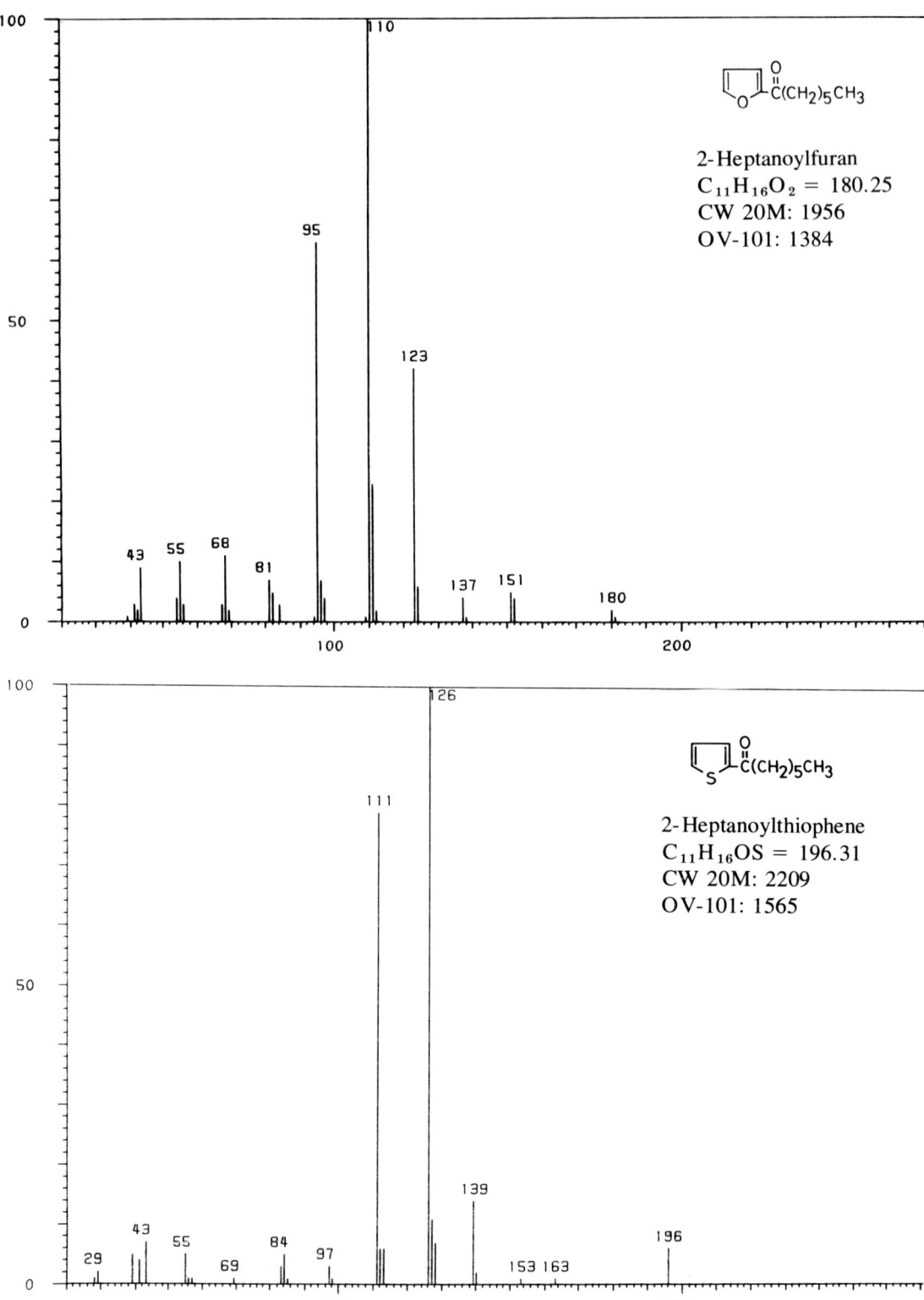

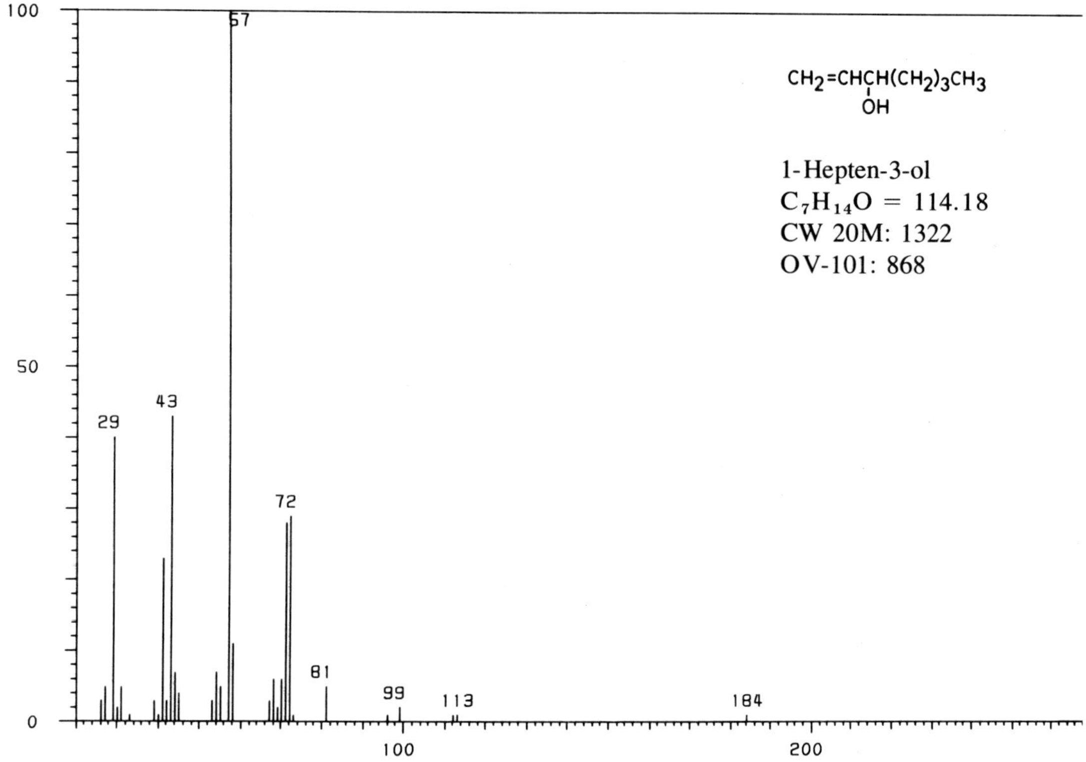

APPENDIX IV MASS SPECTRA OF INDIVIDUAL COMPOUNDS

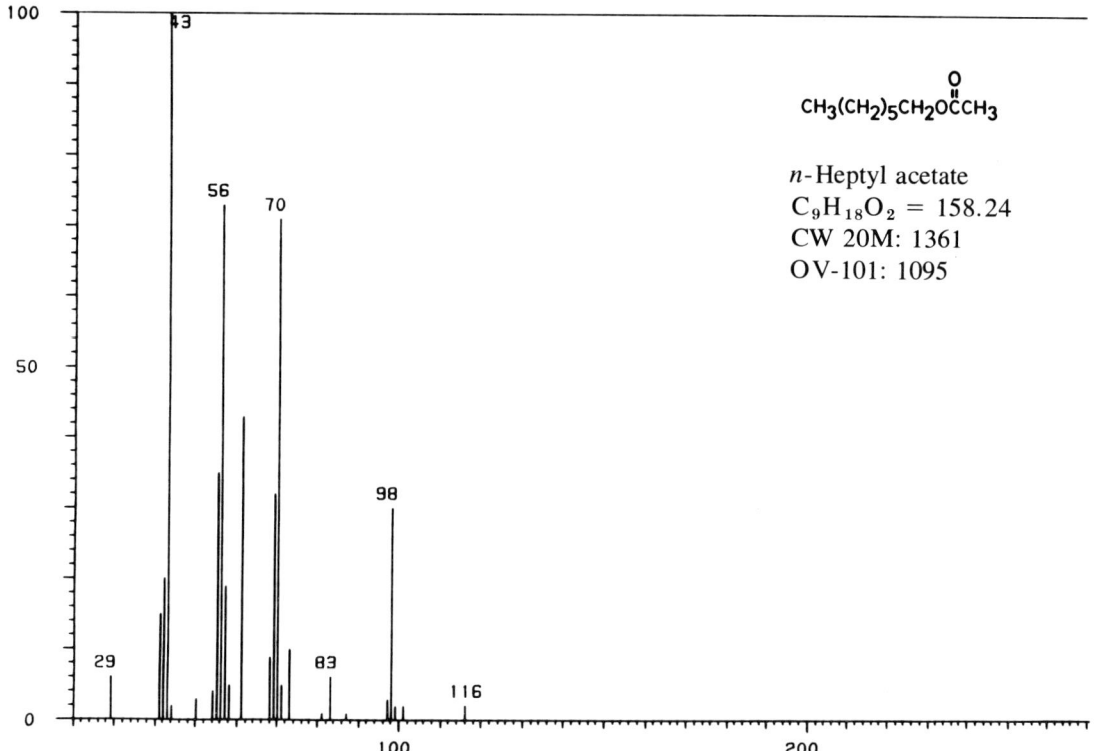

1-Hepten-3-ol
$C_7H_{14}O = 114.18$
CW 20M: 1322
OV-101: 868

n-Heptyl acetate
$C_9H_{18}O_2 = 158.24$
CW 20M: 1361
OV-101: 1095

$CH_3CH_2CH_2C(=O)OCH_2(CH_2)_5CH_3$

n-Heptyl *n*-butyrate
$C_{11}H_{22}O_2$ = 186.30
CW 20M: 1503
OV-101: 1275

$(CH_3)_2CHC(=O)OCH_2(CH_2)_5CH_3$

n-Heptyl *iso*-butyrate
$C_{11}H_{22}O_2$ = 186.30
CW 20M: 1433
OV-101: 1233

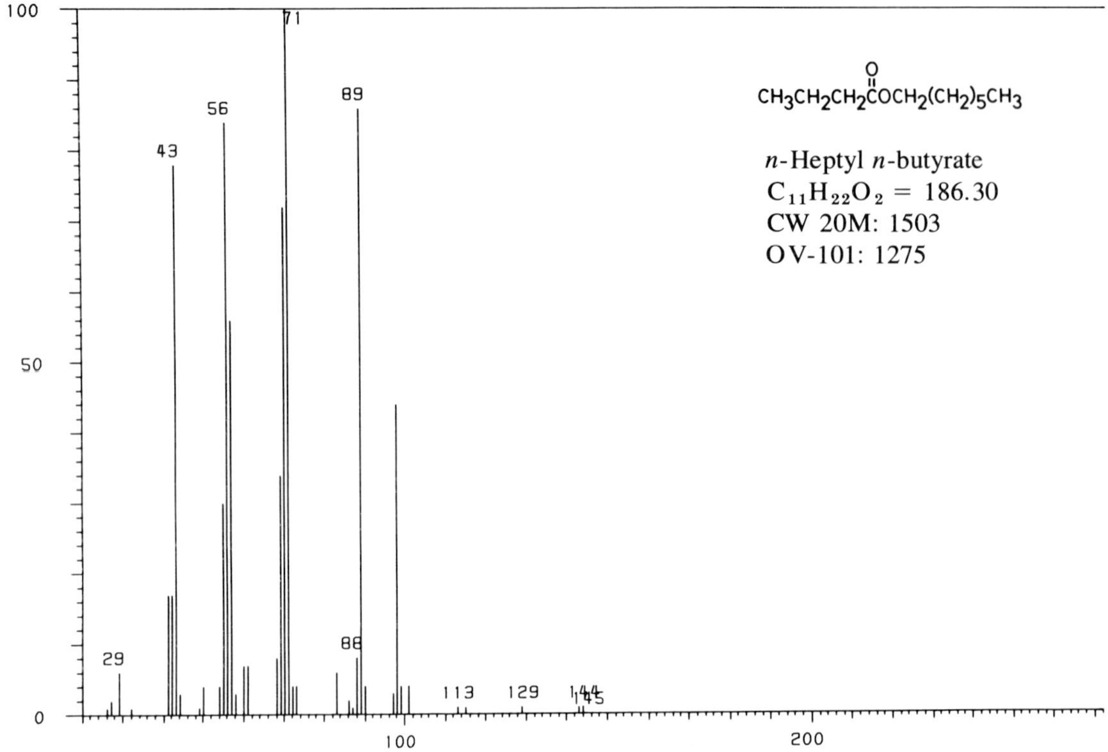

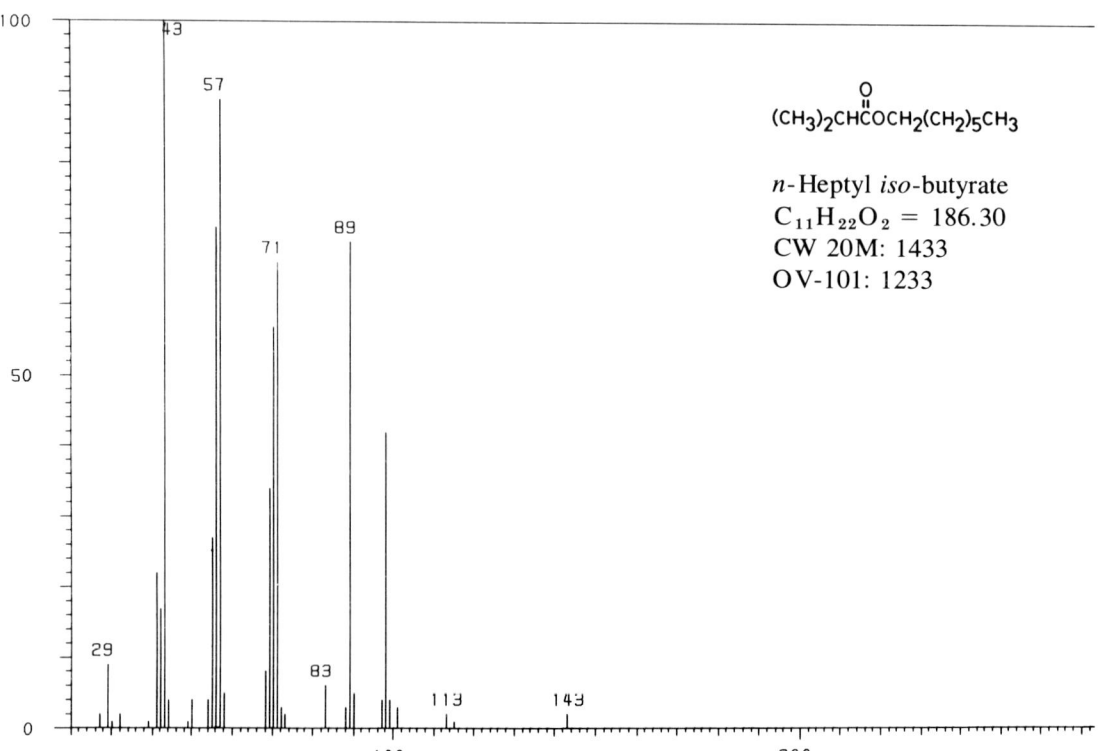

APPENDIX IV   MASS SPECTRA OF INDIVIDUAL COMPOUNDS

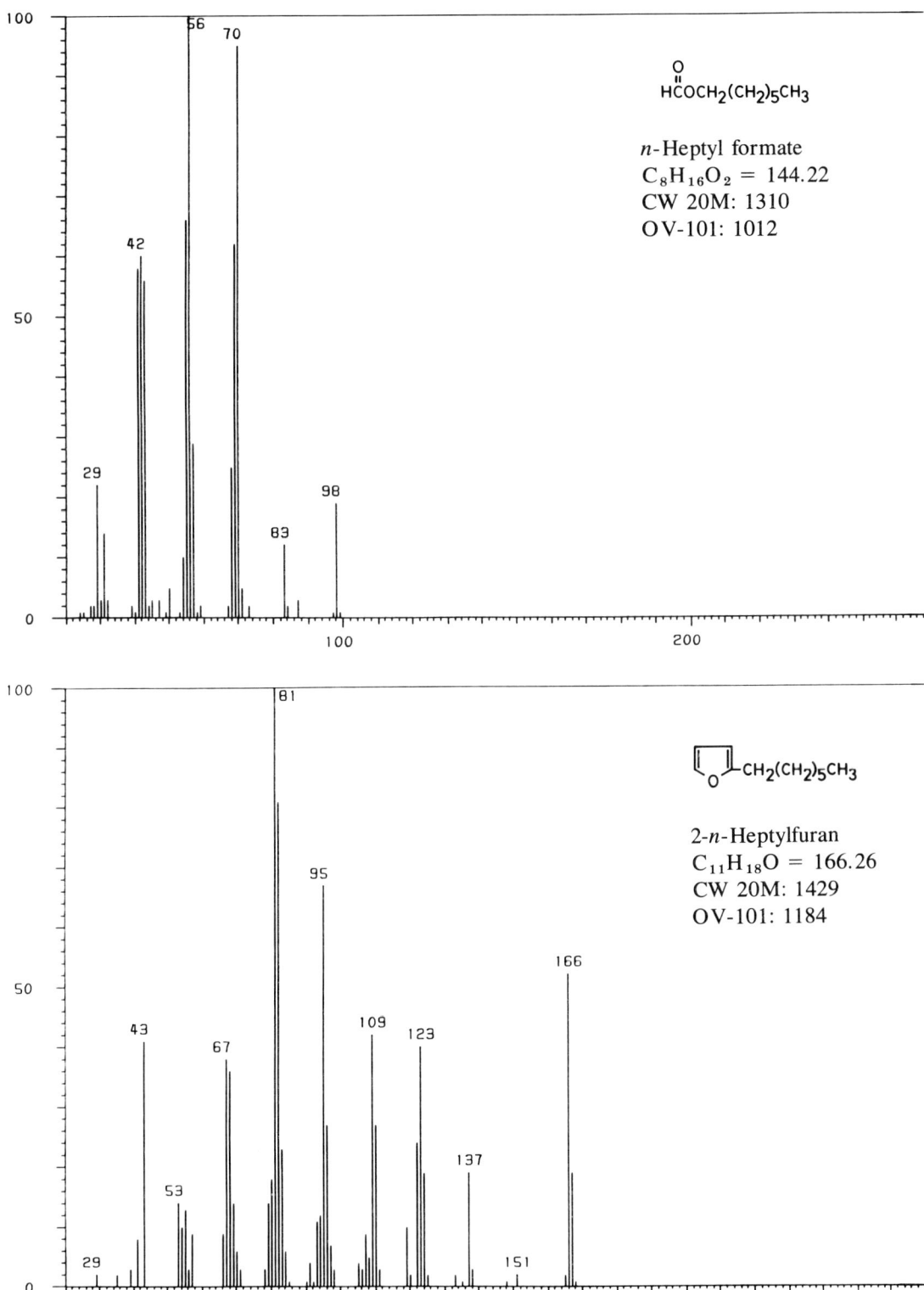

n-Heptyl formate
$C_8H_{16}O_2 = 144.22$
CW 20M: 1310
OV-101: 1012

2-n-Heptylfuran
$C_{11}H_{18}O = 166.26$
CW 20M: 1429
OV-101: 1184

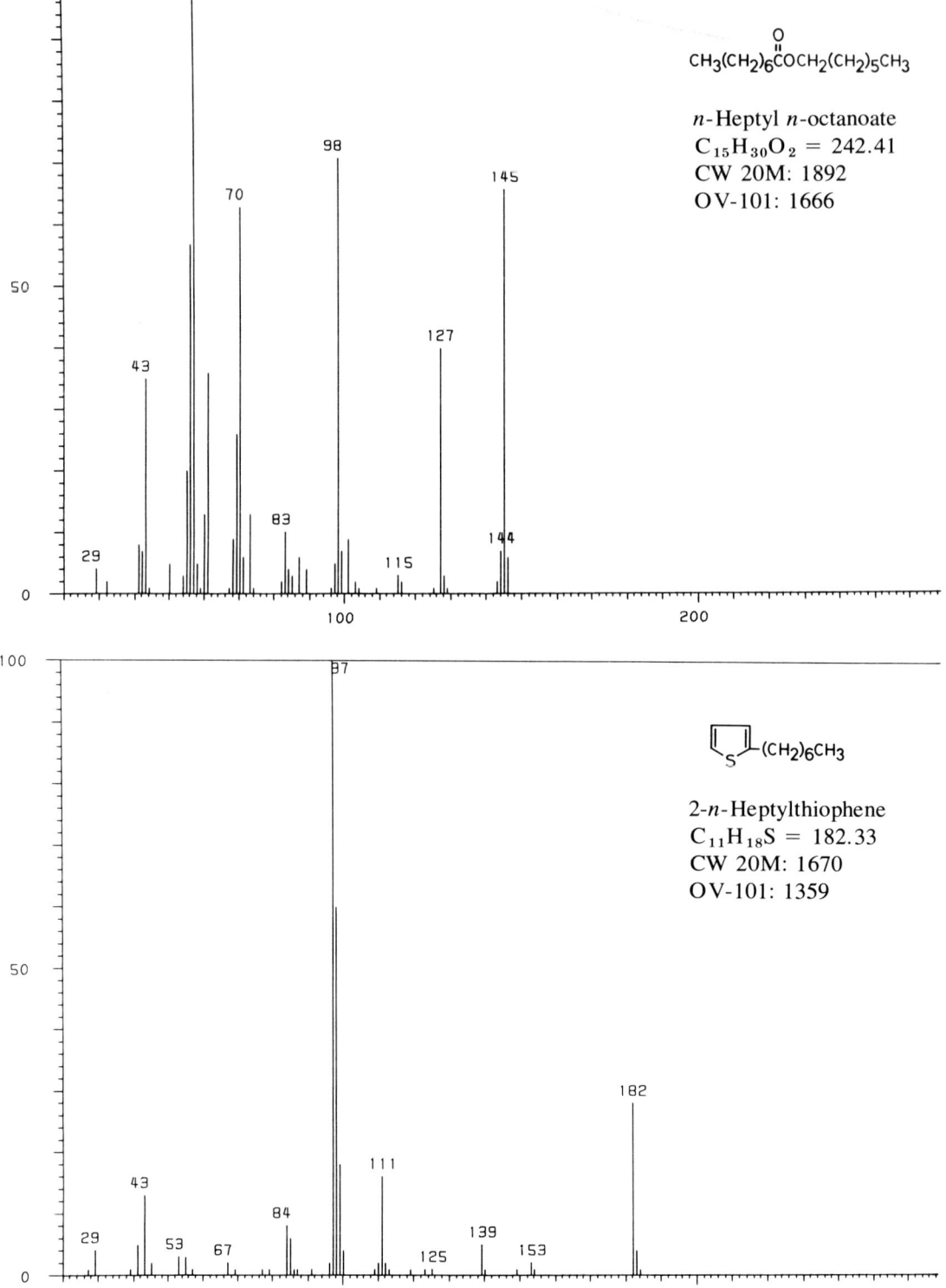

APPENDIX IV   MASS SPECTRA OF INDIVIDUAL COMPOUNDS

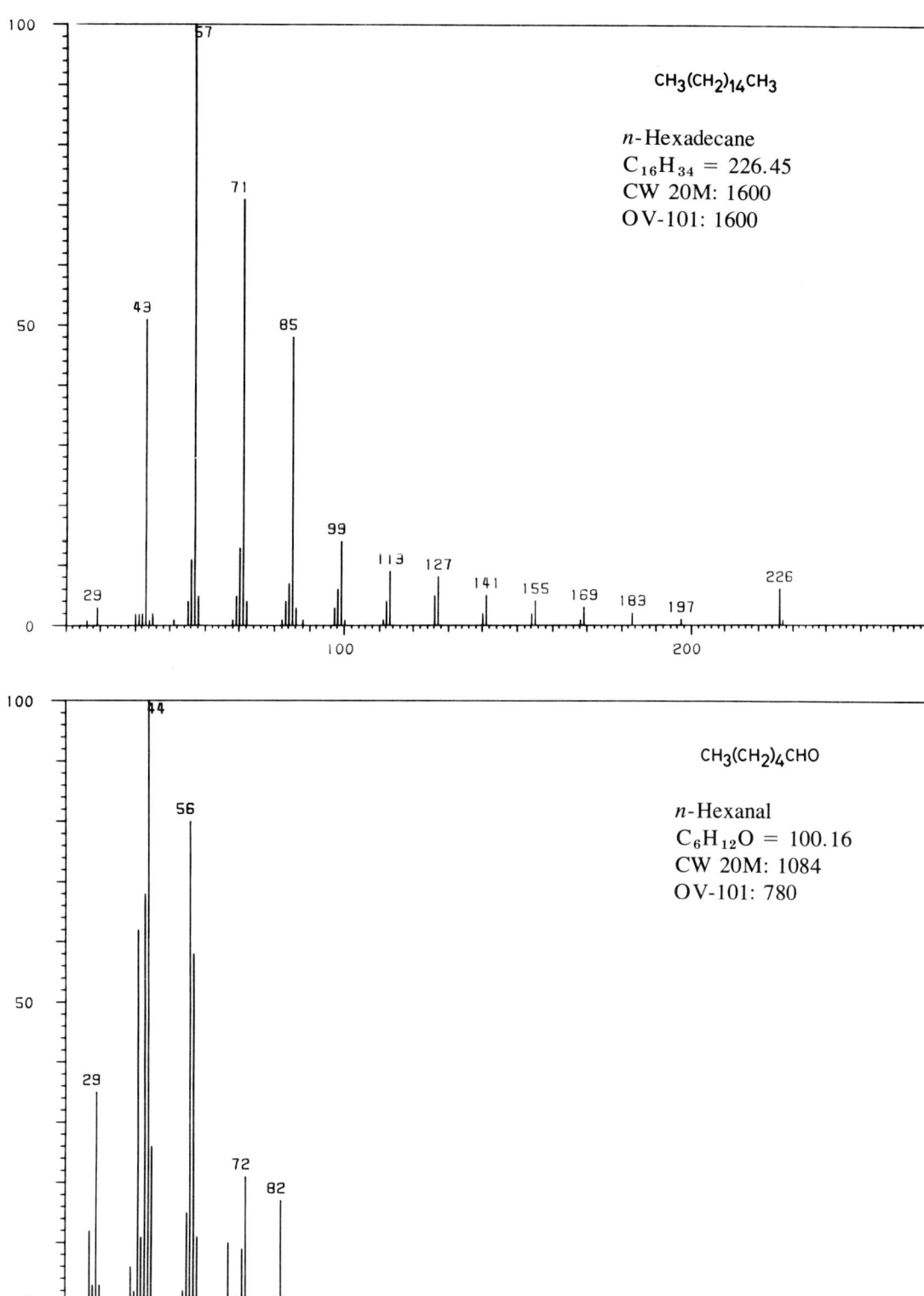

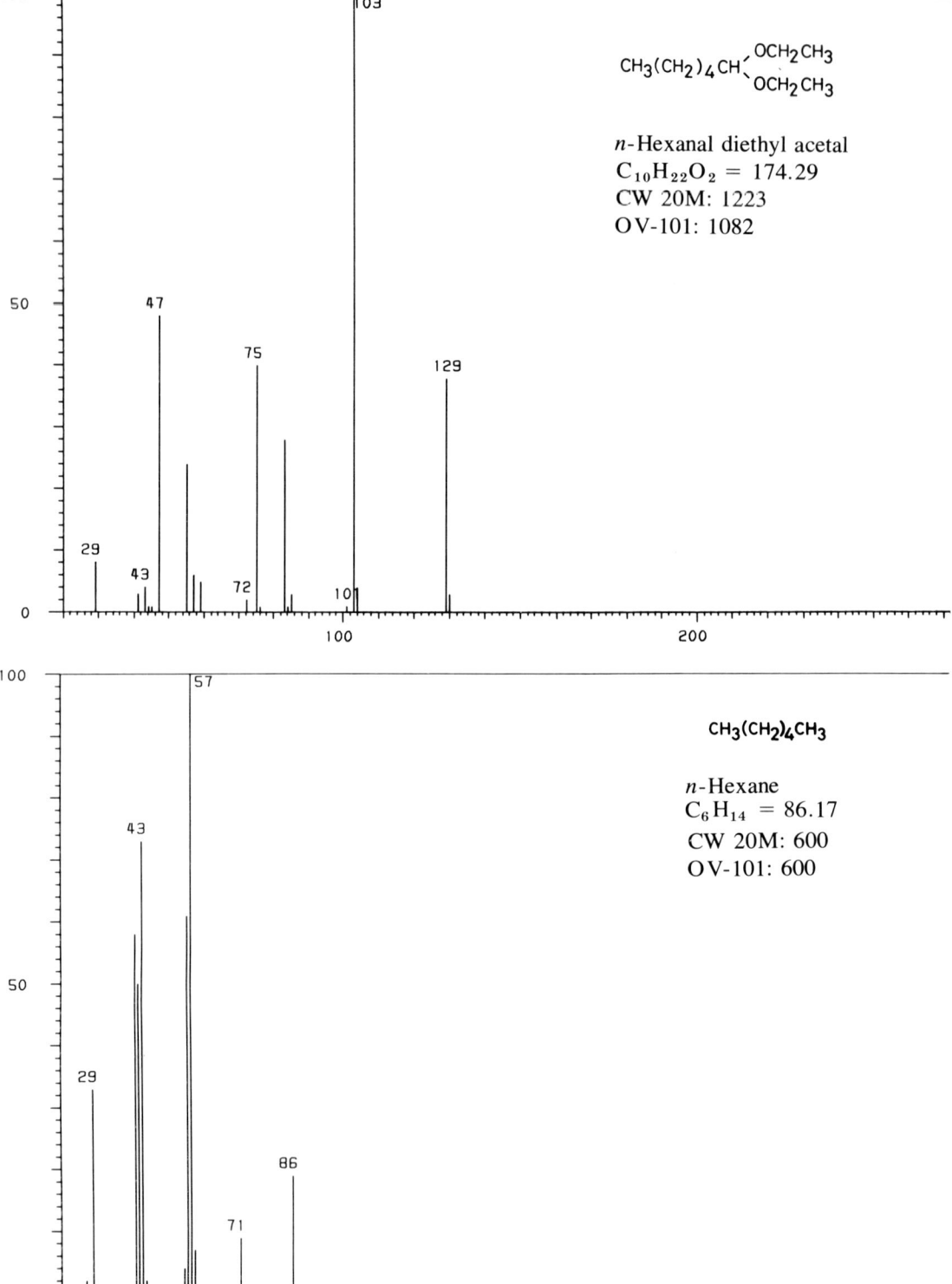

APPENDIX IV   MASS SPECTRA OF INDIVIDUAL COMPOUNDS   323

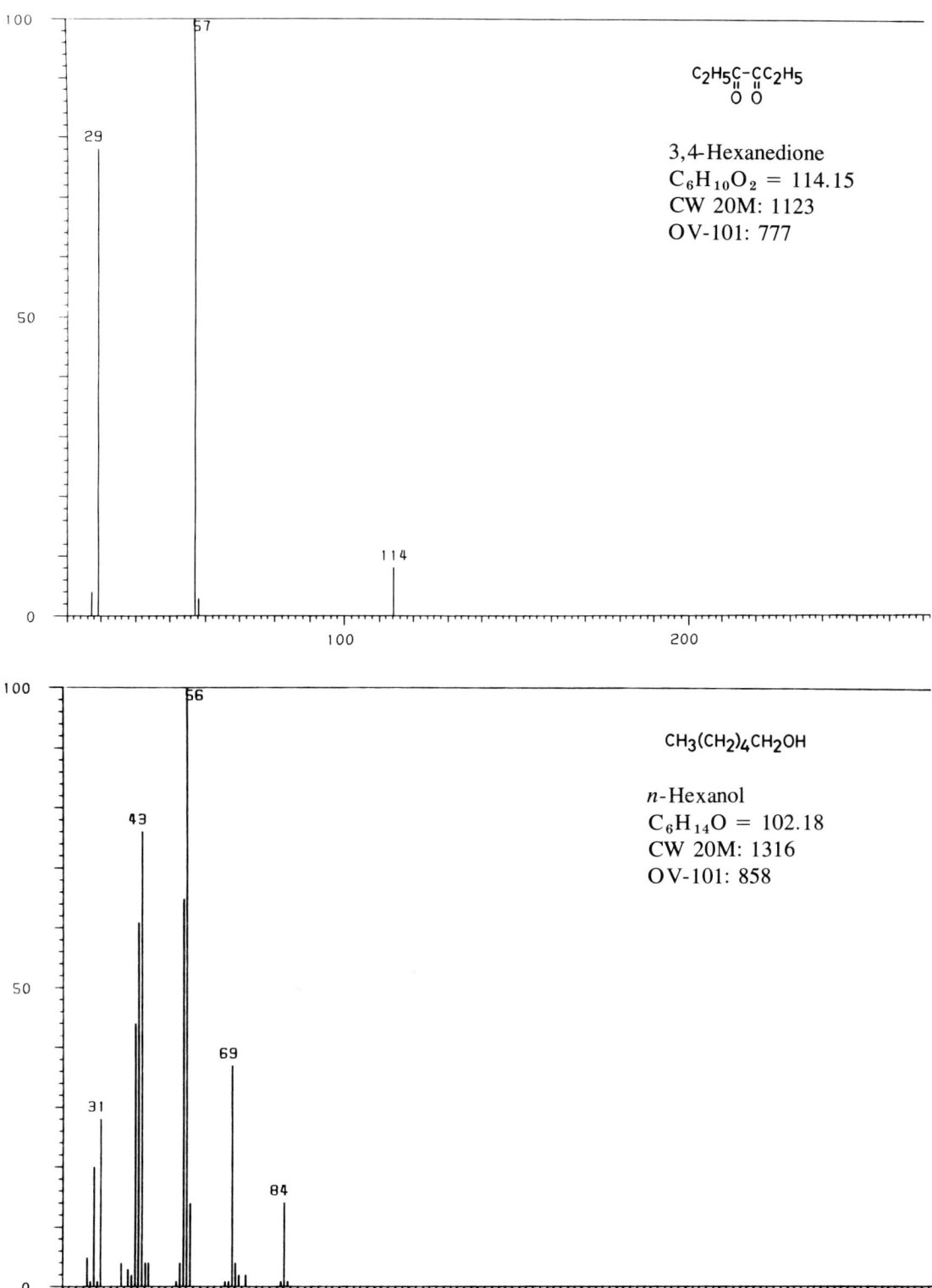

$$C_2H_5\underset{O}{C}-\underset{O}{C}C_2H_5$$

3,4-Hexanedione
$C_6H_{10}O_2 = 114.15$
CW 20M: 1123
OV-101: 777

$CH_3(CH_2)_4CH_2OH$

n-Hexanol
$C_6H_{14}O = 102.18$
CW 20M: 1316
OV-101: 858

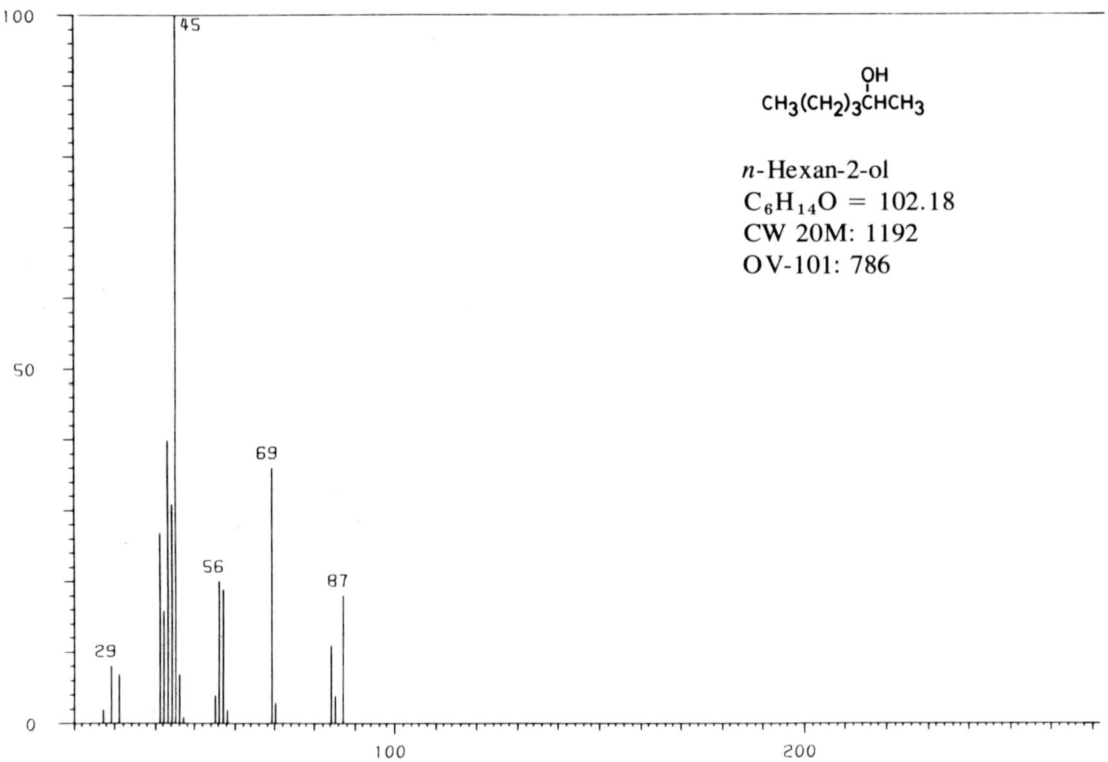

n-Hexan-2-ol
CH$_3$(CH$_2$)$_3$CHCH$_3$ (OH)
C$_6$H$_{14}$O = 102.18
CW 20M: 1192
OV-101: 786

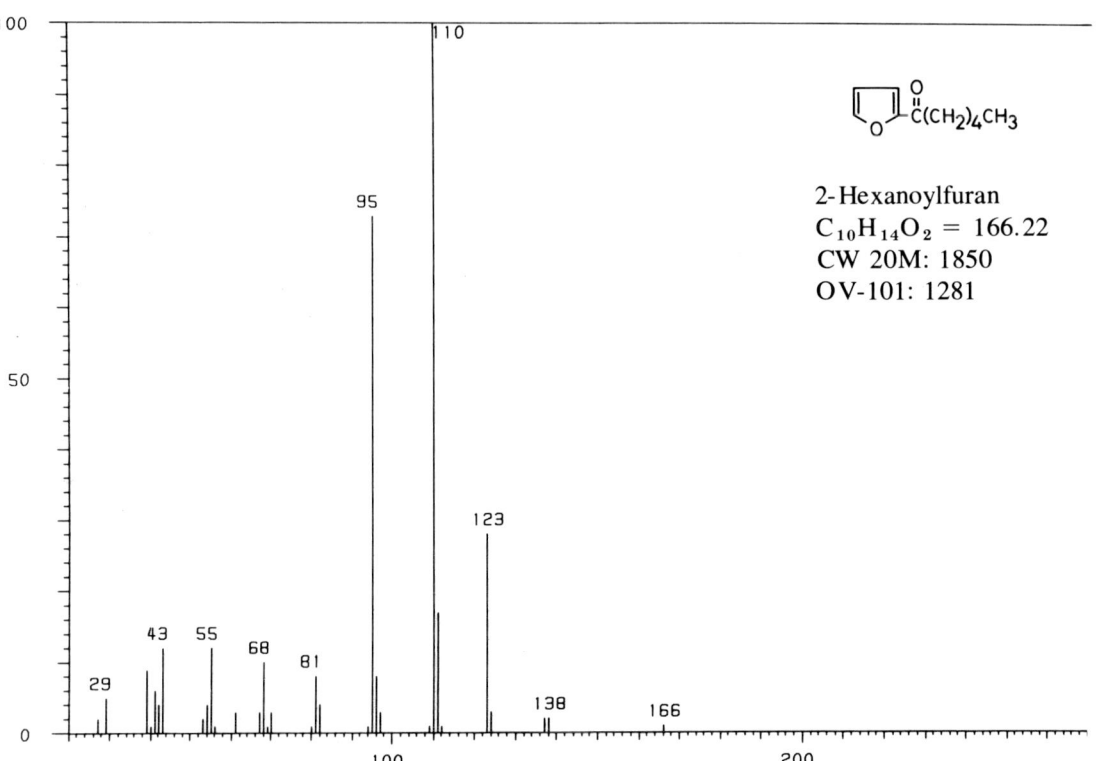

2-Hexanoylfuran
C$_{10}$H$_{14}$O$_2$ = 166.22
CW 20M: 1850
OV-101: 1281

APPENDIX IV   MASS SPECTRA OF INDIVIDUAL COMPOUNDS

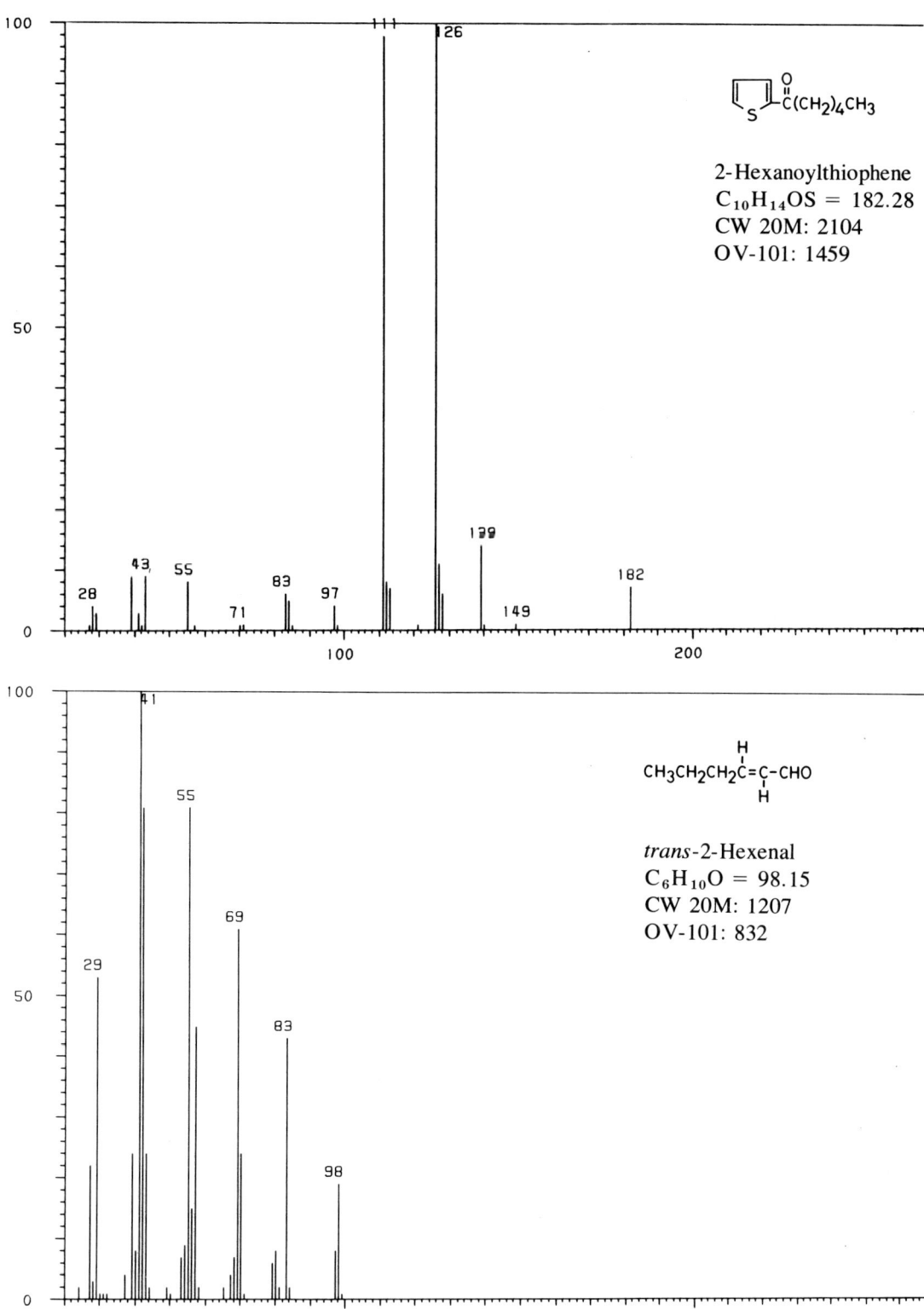

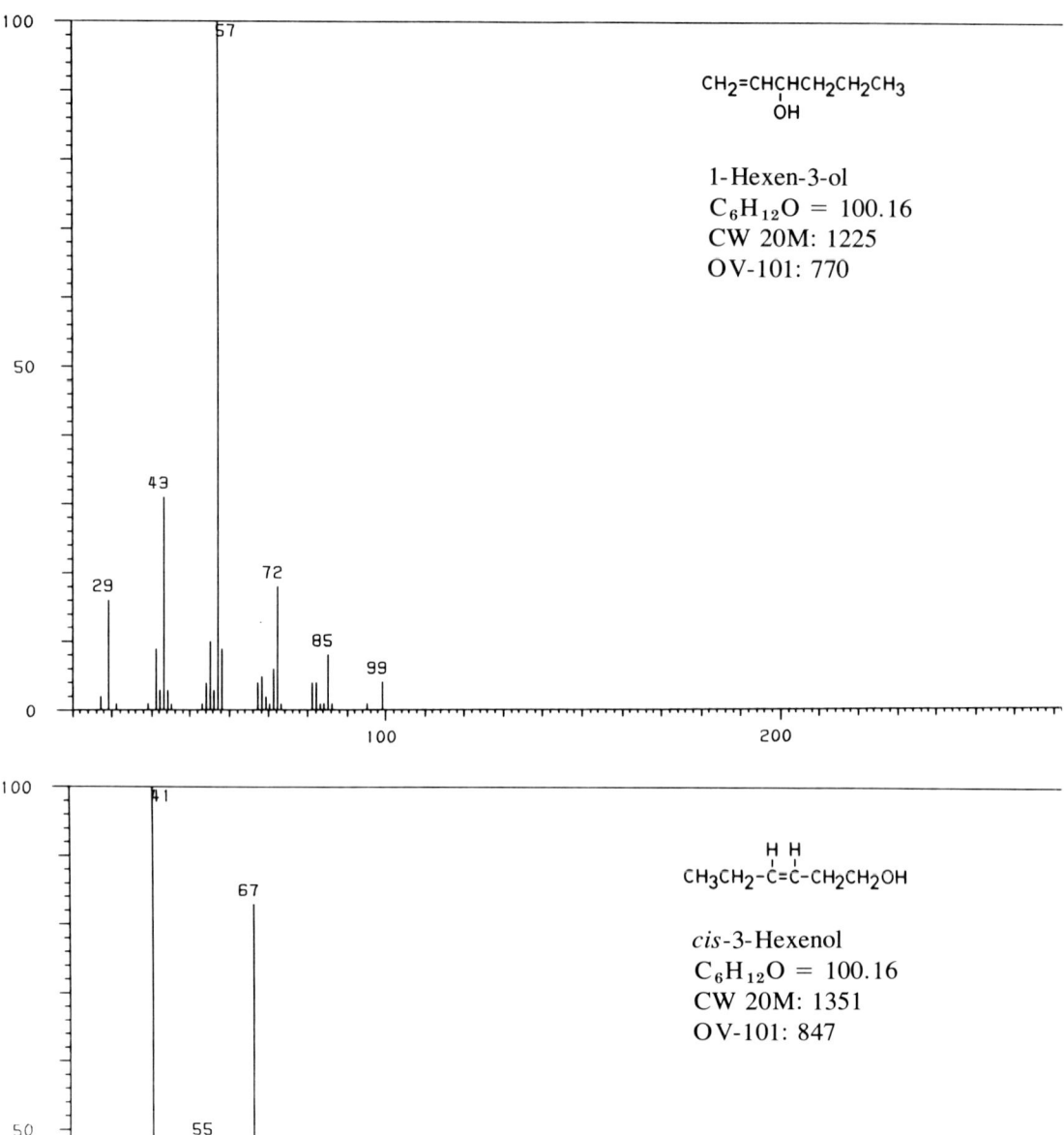

APPENDIX IV   MASS SPECTRA OF INDIVIDUAL COMPOUNDS   327

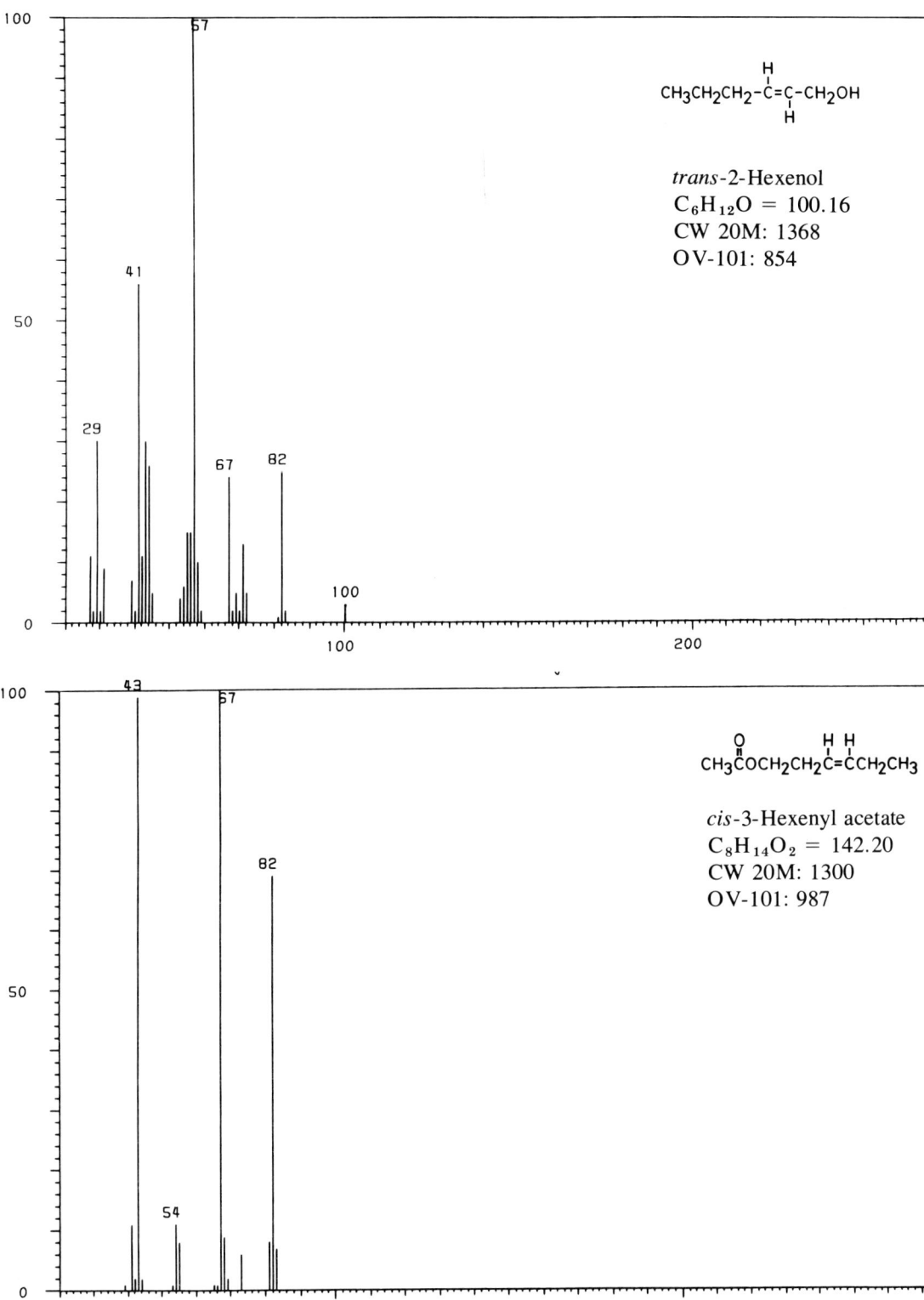

trans-2-Hexenol
$C_6H_{12}O = 100.16$
CW 20M: 1368
OV-101: 854

cis-3-Hexenyl acetate
$C_8H_{14}O_2 = 142.20$
CW 20M: 1300
OV-101: 987

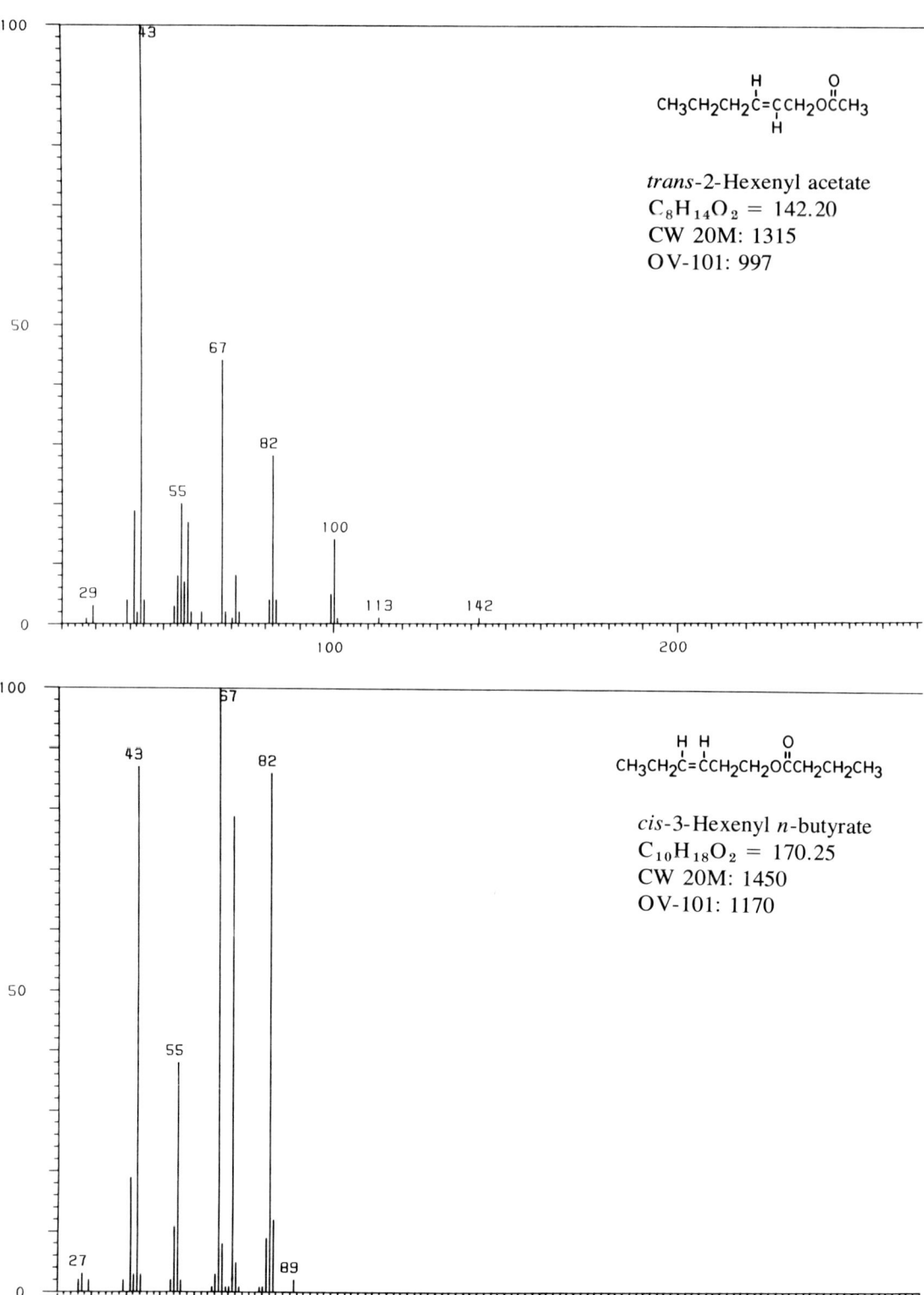

APPENDIX IV   MASS SPECTRA OF INDIVIDUAL COMPOUNDS

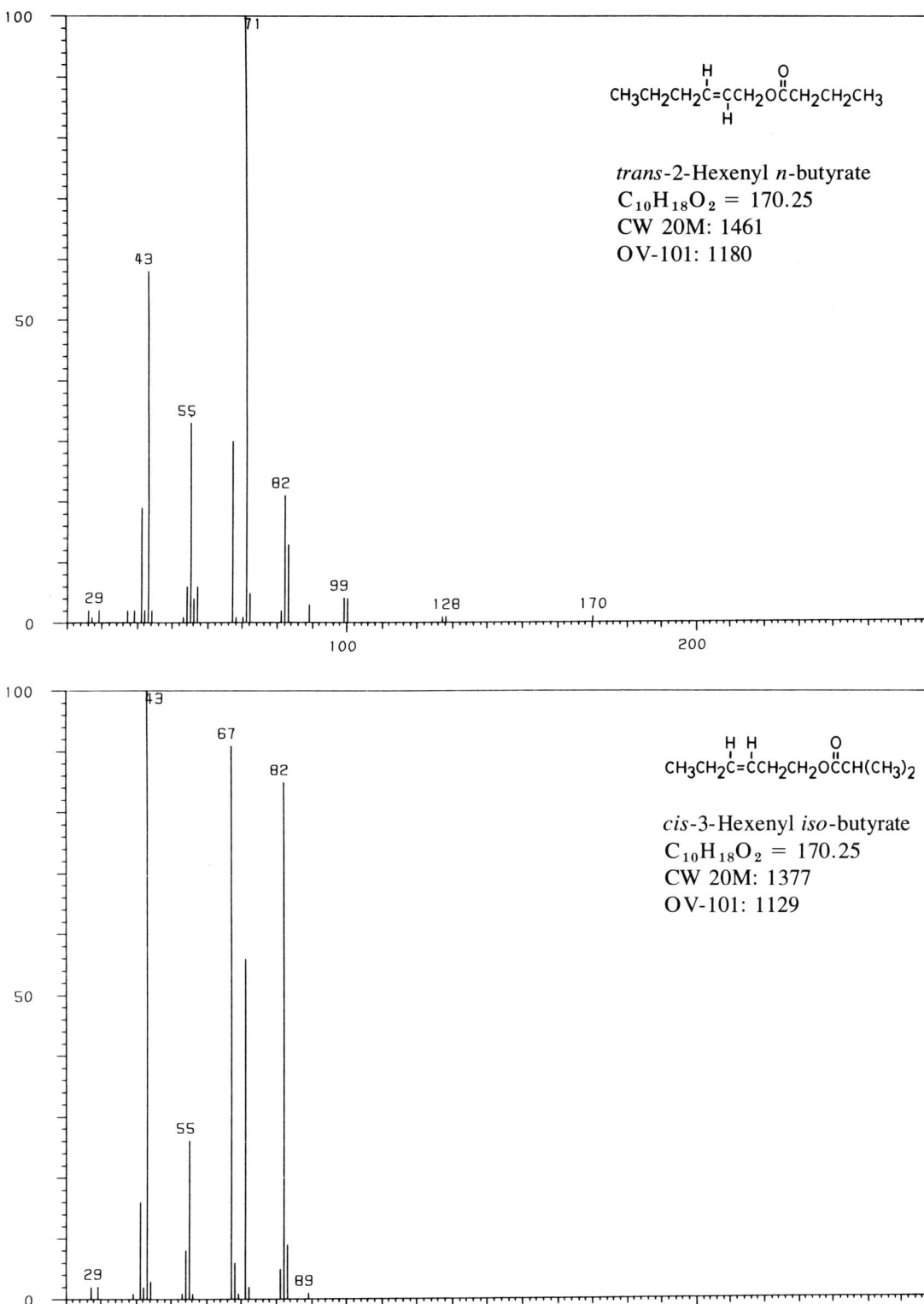

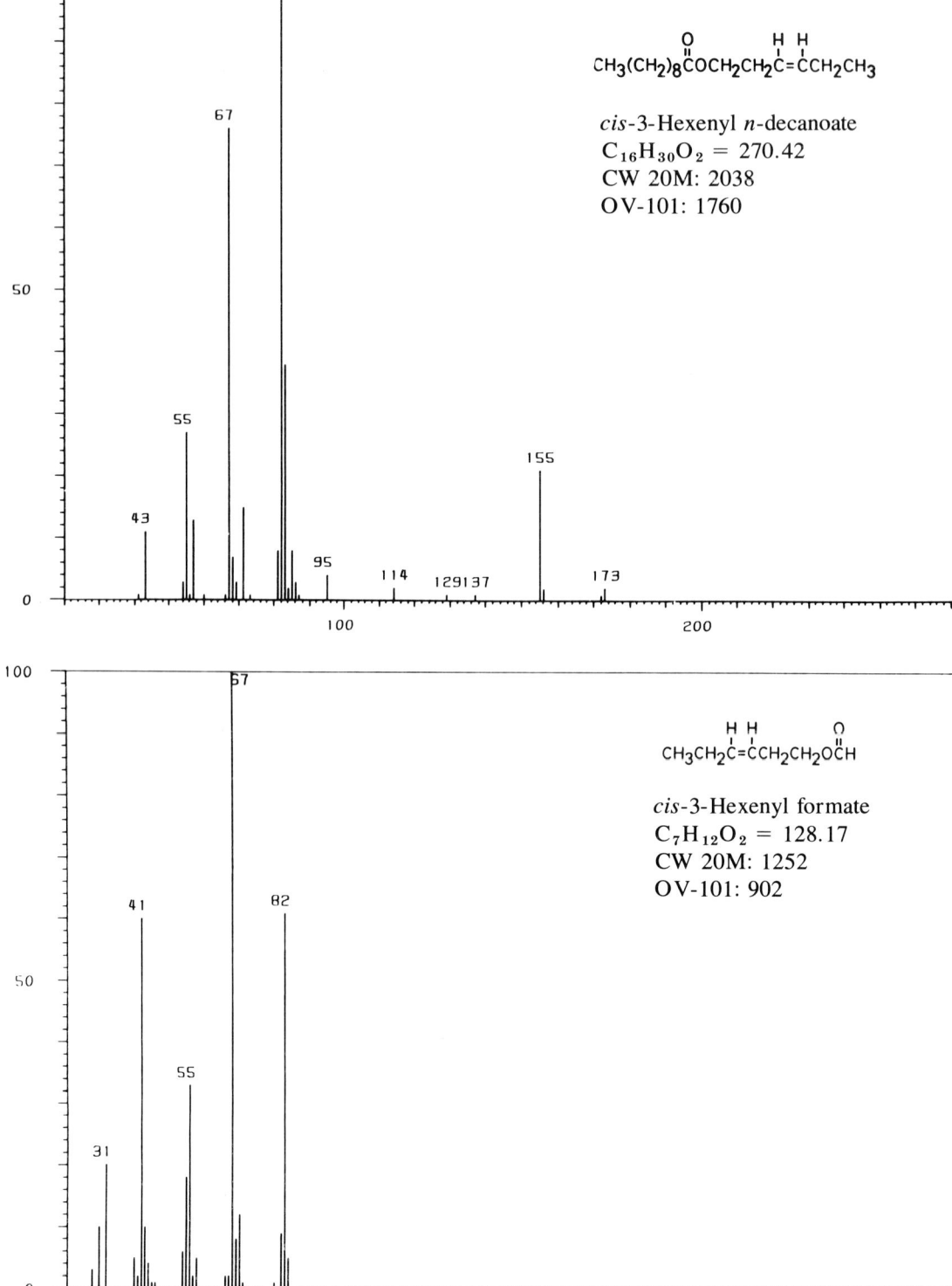

APPENDIX IV   MASS SPECTRA OF INDIVIDUAL COMPOUNDS   331

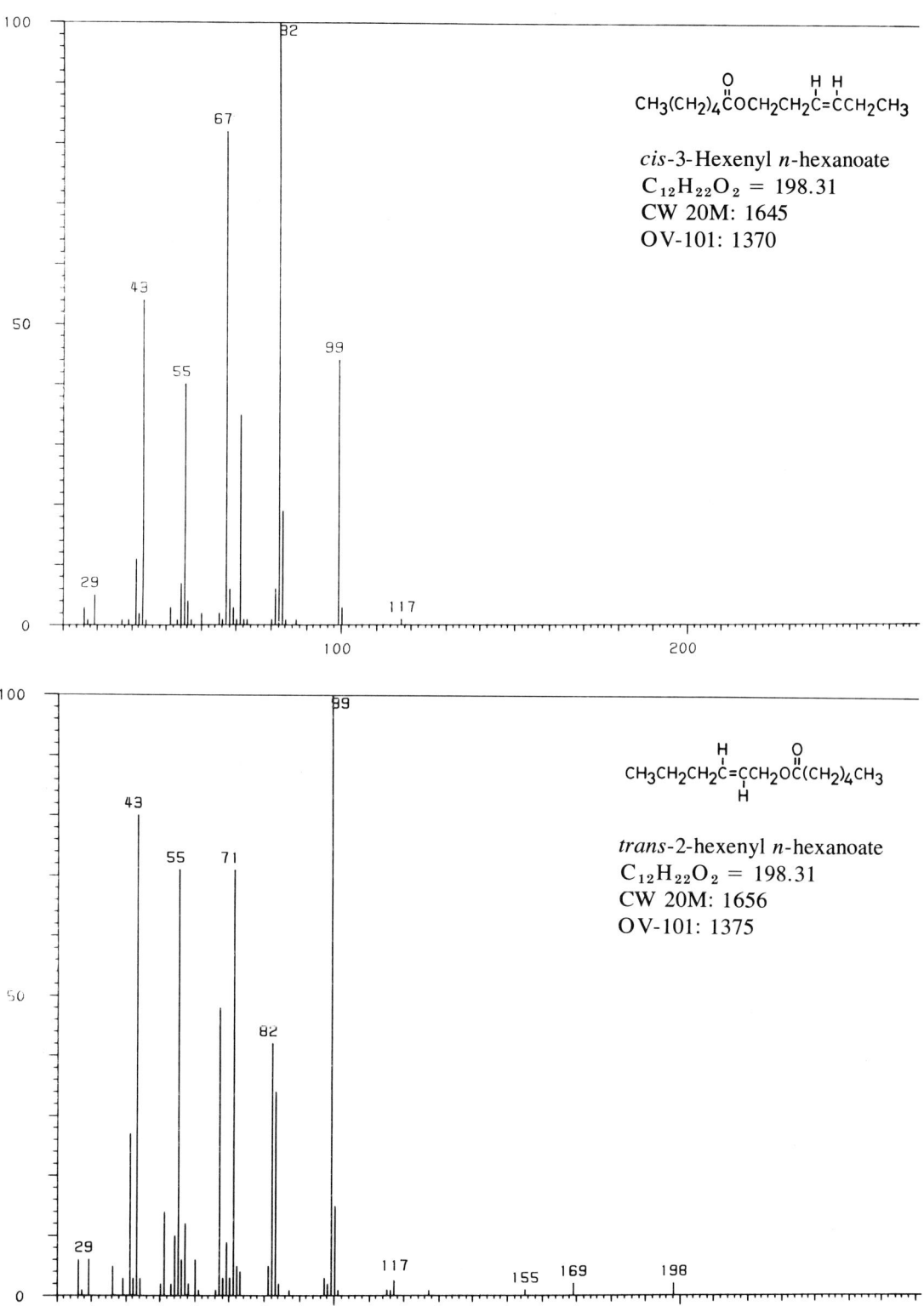

cis-3-Hexenyl *n*-hexanoate
$C_{12}H_{22}O_2 = 198.31$
CW 20M: 1645
OV-101: 1370

trans-2-hexenyl *n*-hexanoate
$C_{12}H_{22}O_2 = 198.31$
CW 20M: 1656
OV-101: 1375

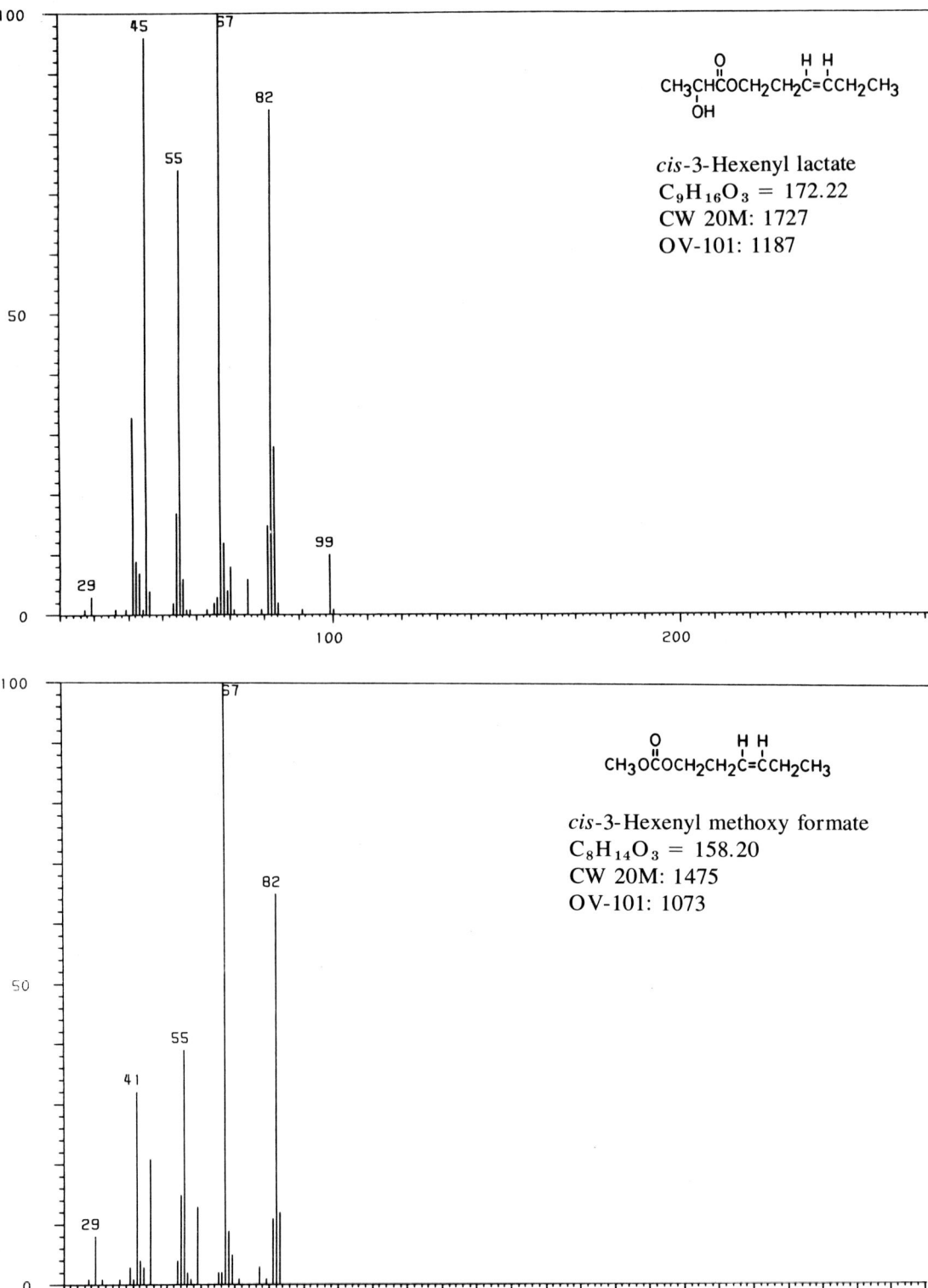

APPENDIX IV MASS SPECTRA OF INDIVIDUAL COMPOUNDS

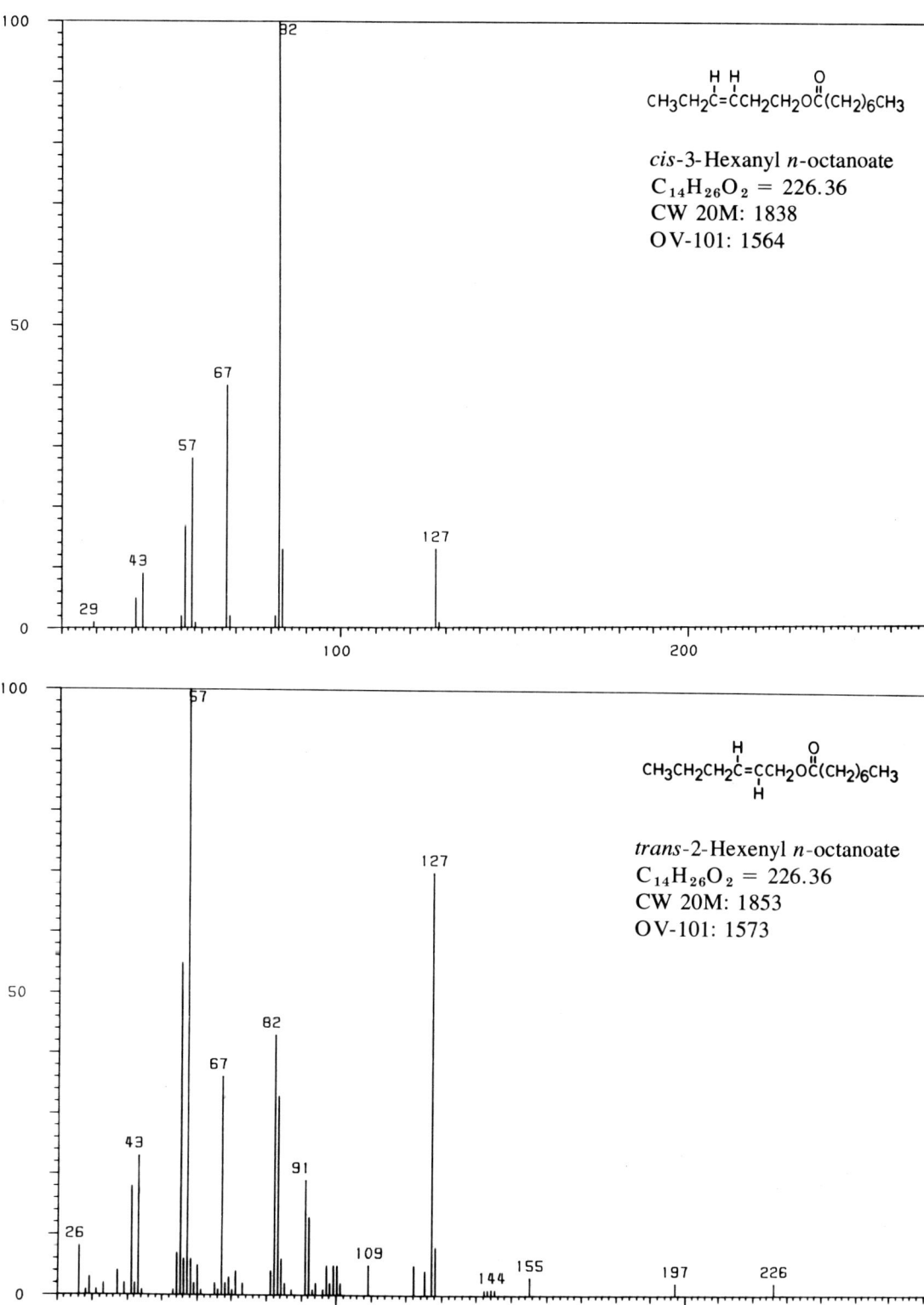

cis-3-Hexanyl *n*-octanoate
$C_{14}H_{26}O_2 = 226.36$
CW 20M: 1838
OV-101: 1564

trans-2-Hexenyl *n*-octanoate
$C_{14}H_{26}O_2 = 226.36$
CW 20M: 1853
OV-101: 1573

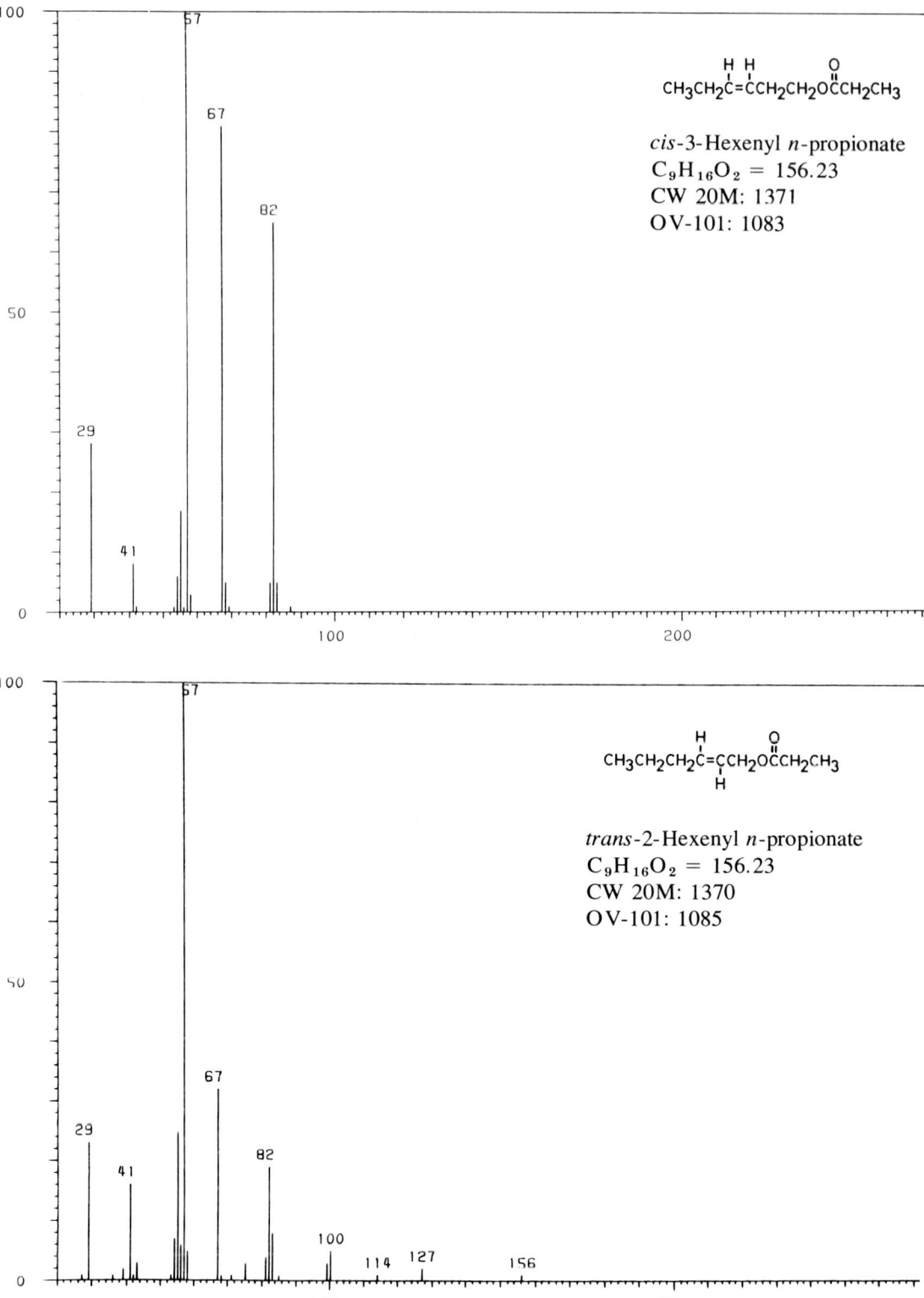

APPENDIX IV   MASS SPECTRA OF INDIVIDUAL COMPOUNDS   335

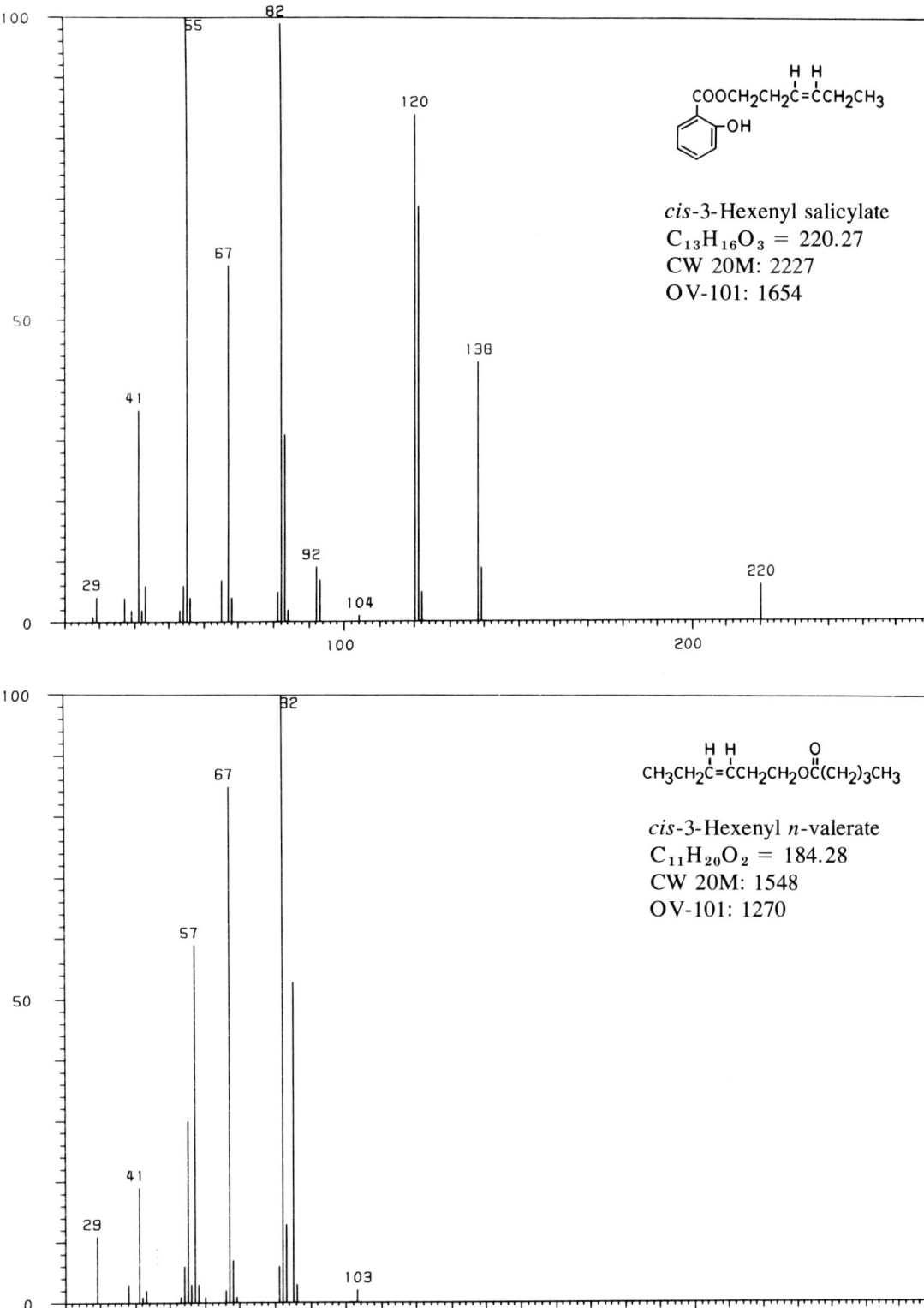

cis-3-Hexenyl salicylate
$C_{13}H_{16}O_3 = 220.27$
CW 20M: 2227
OV-101: 1654

cis-3-Hexenyl *n*-valerate
$C_{11}H_{20}O_2 = 184.28$
CW 20M: 1548
OV-101: 1270

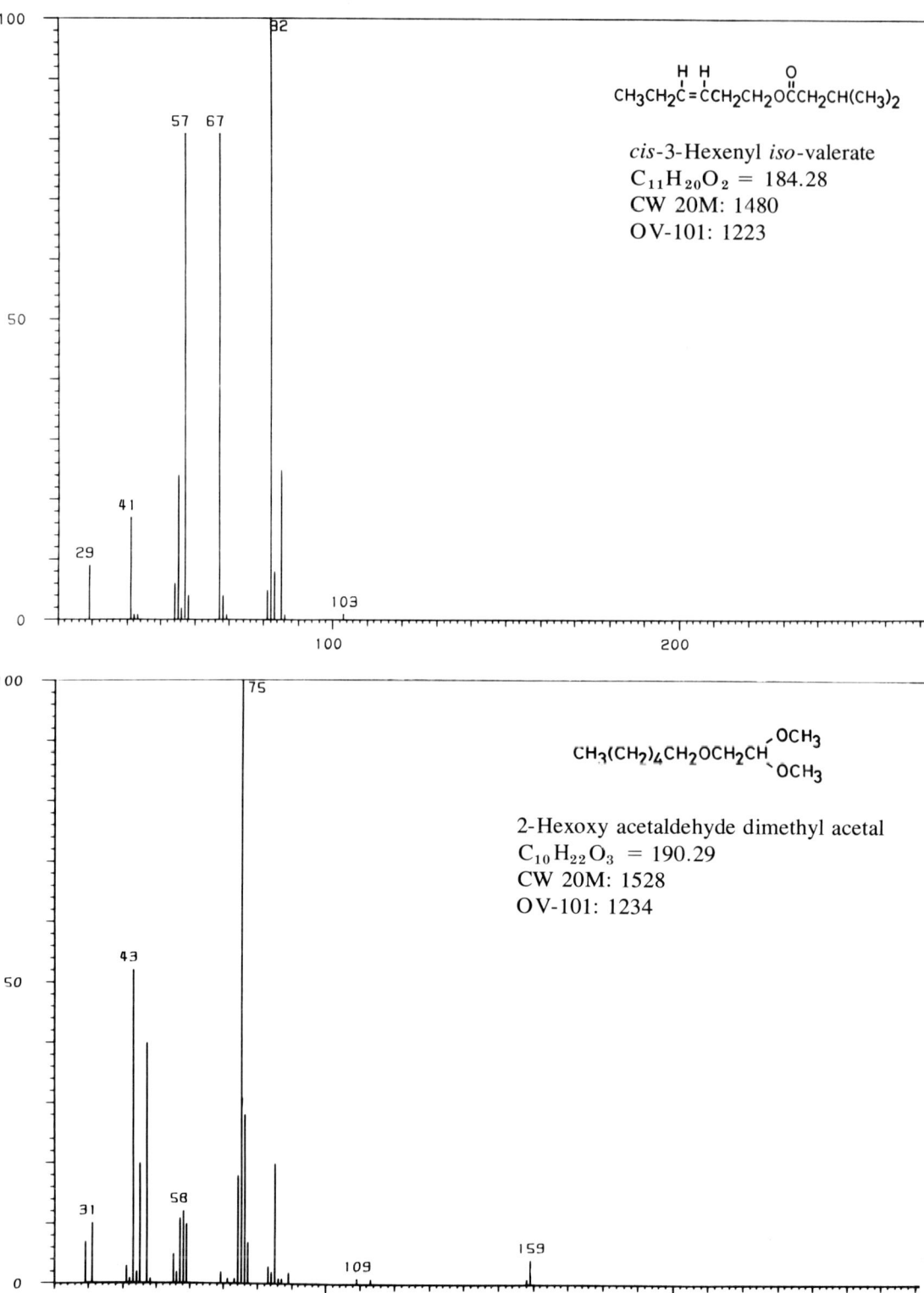

APPENDIX IV   MASS SPECTRA OF INDIVIDUAL COMPOUNDS   337

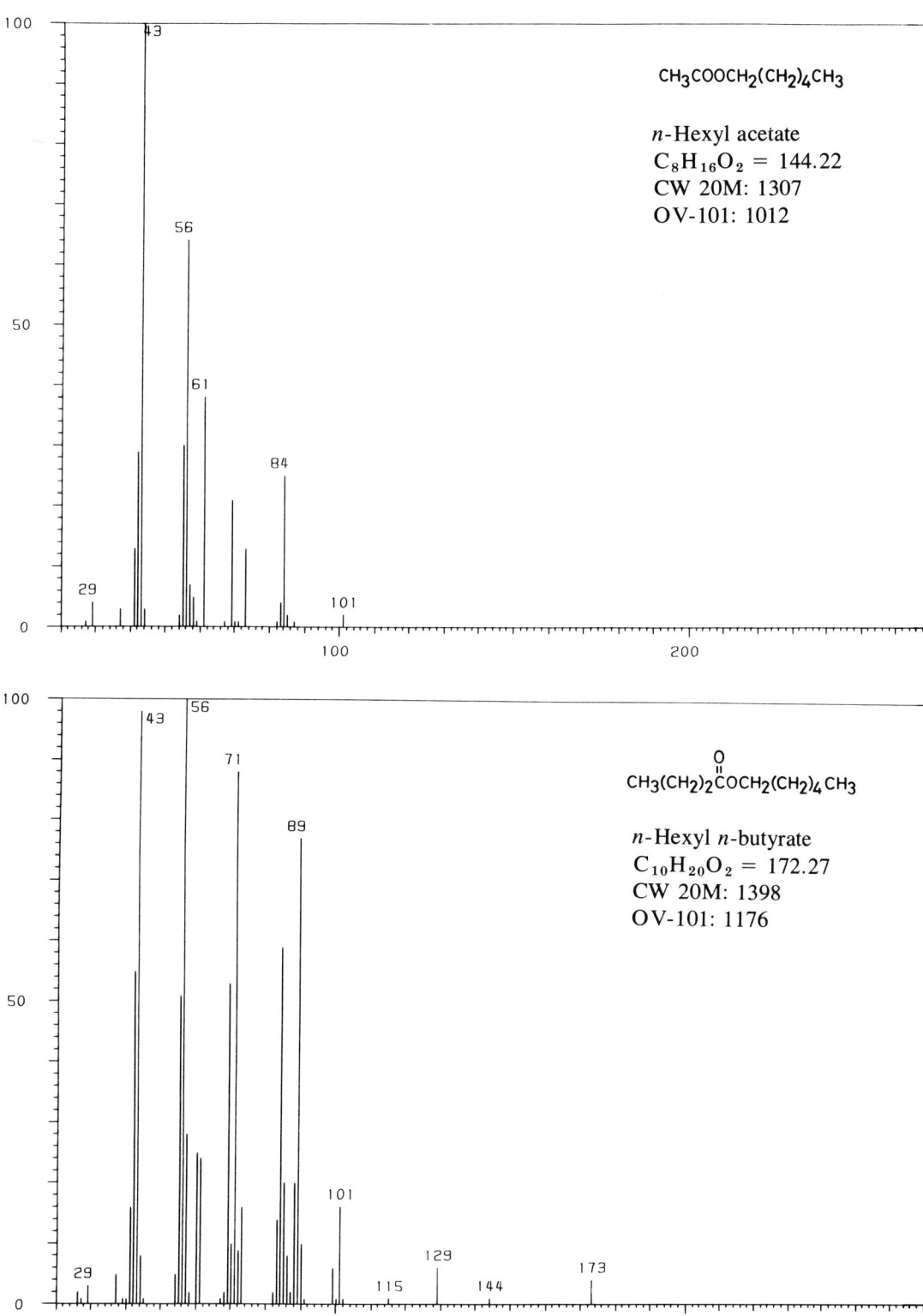

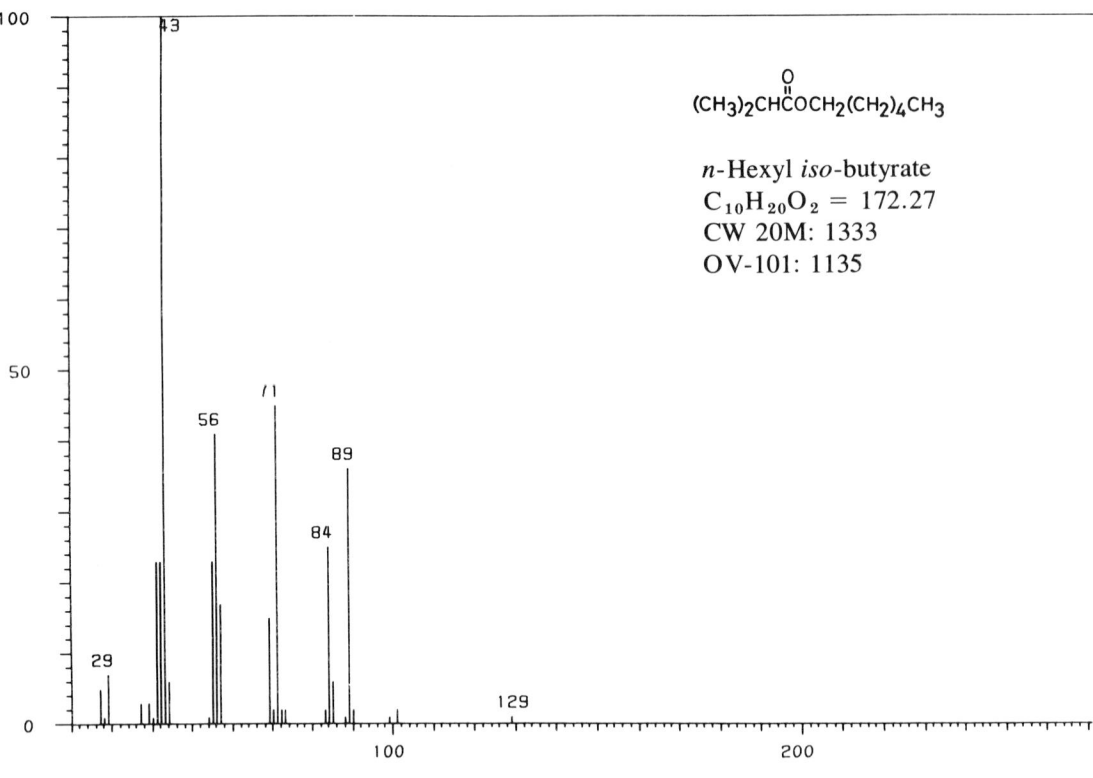

$(CH_3)_2CHCOCH_2(CH_2)_4CH_3$

n-Hexyl *iso*-butyrate
$C_{10}H_{20}O_2 = 172.27$
CW 20M: 1333
OV-101: 1135

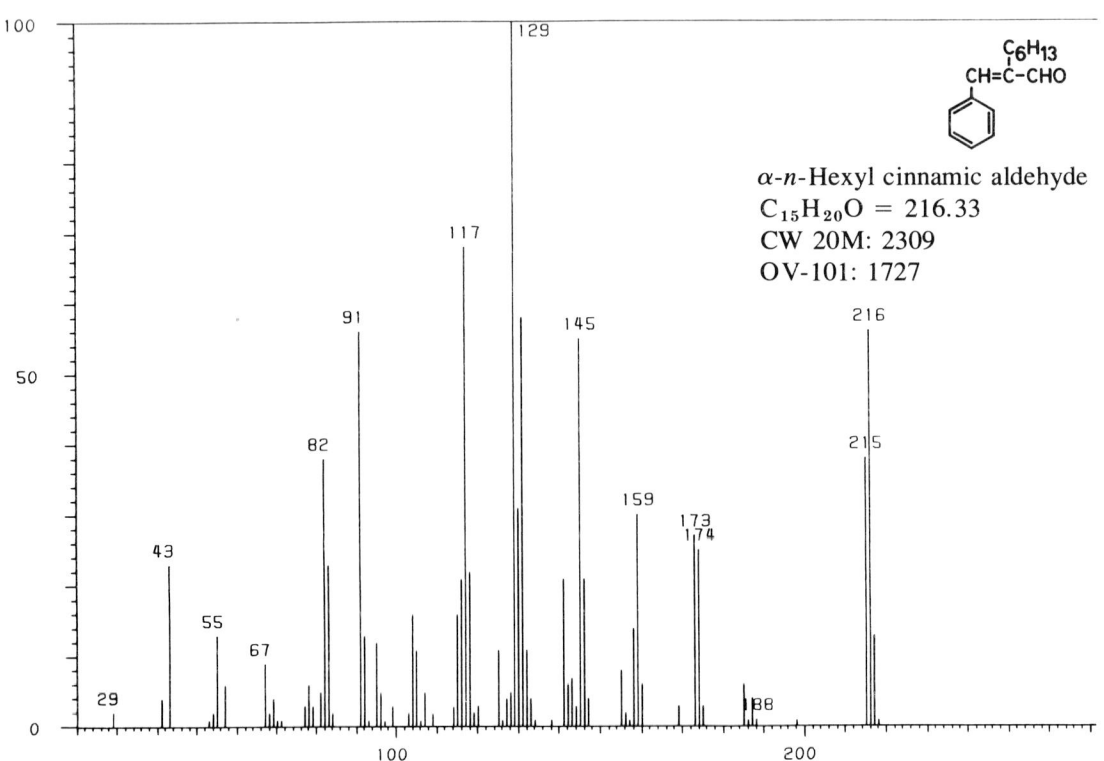

α-*n*-Hexyl cinnamic aldehyde
$C_{15}H_{20}O = 216.33$
CW 20M: 2309
OV-101: 1727

APPENDIX IV MASS SPECTRA OF INDIVIDUAL COMPOUNDS

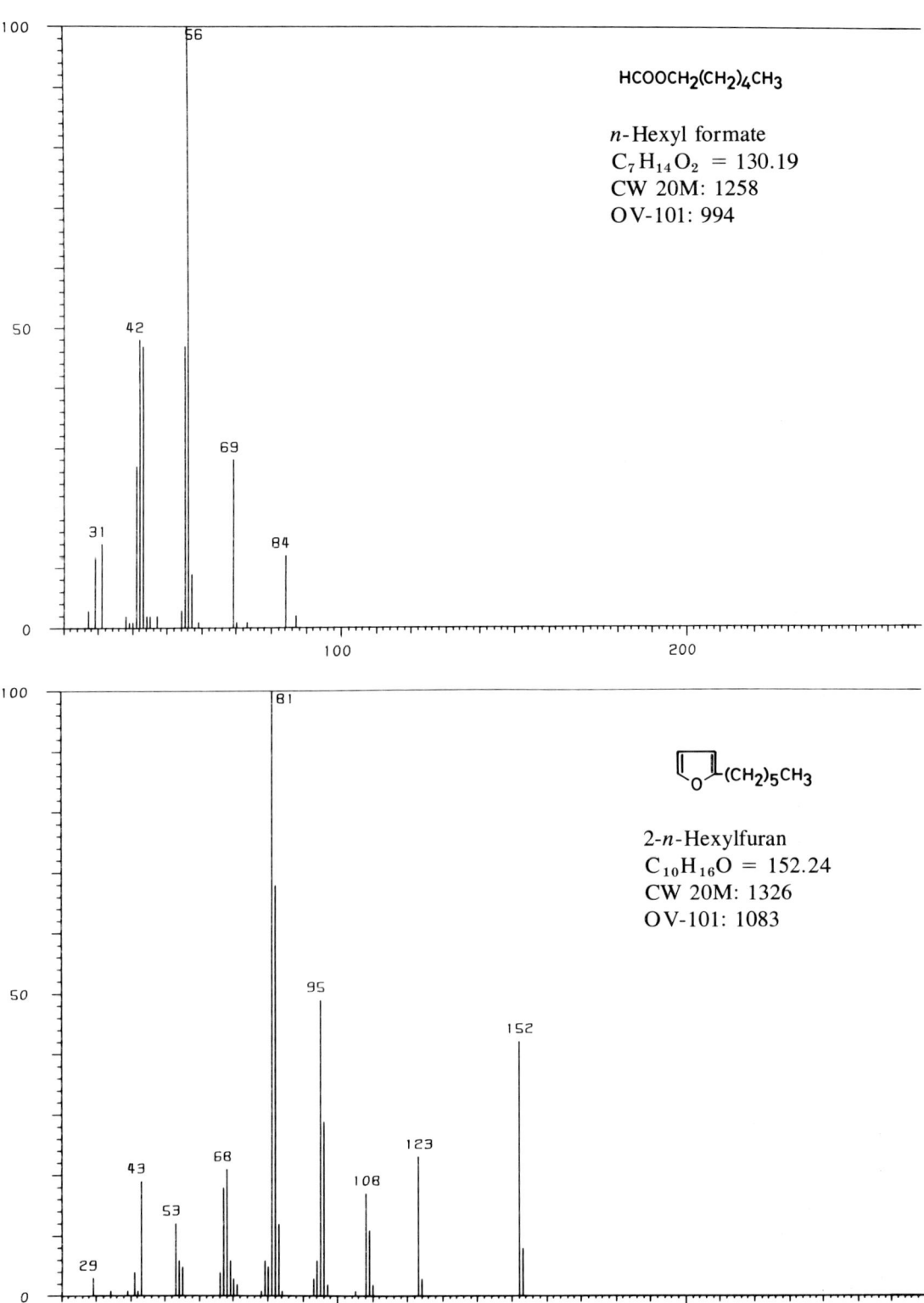

$HCOOCH_2(CH_2)_4CH_3$

n-Hexyl formate
$C_7H_{14}O_2 = 130.19$
CW 20M: 1258
OV-101: 994

2-*n*-Hexylfuran
$C_{10}H_{16}O = 152.24$
CW 20M: 1326
OV-101: 1083

n-Hexyl *n*-hexanoate
$C_{12}H_{24}O_2$ = 200.32
CW 20M: 1599
OV-101: 1371

n-Hexyl 2-methyl butyrate
$C_{11}H_{22}O_2$ = 186.30
CW 20M: 1418
OV-101: 1224

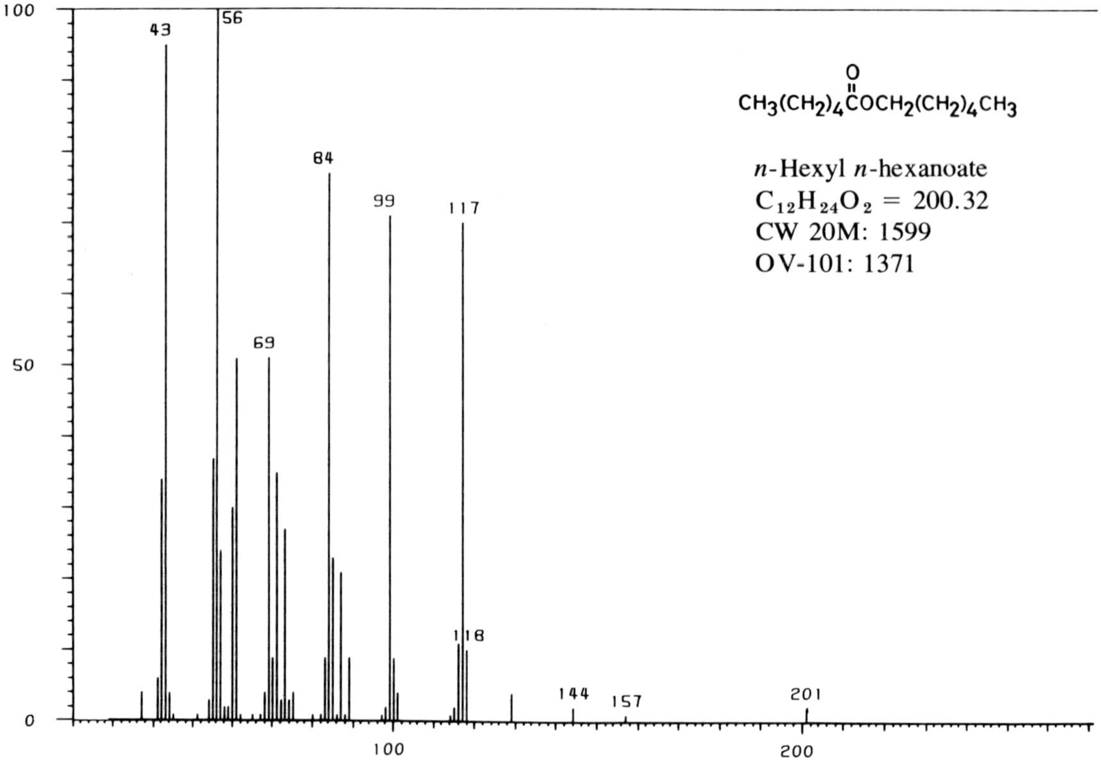

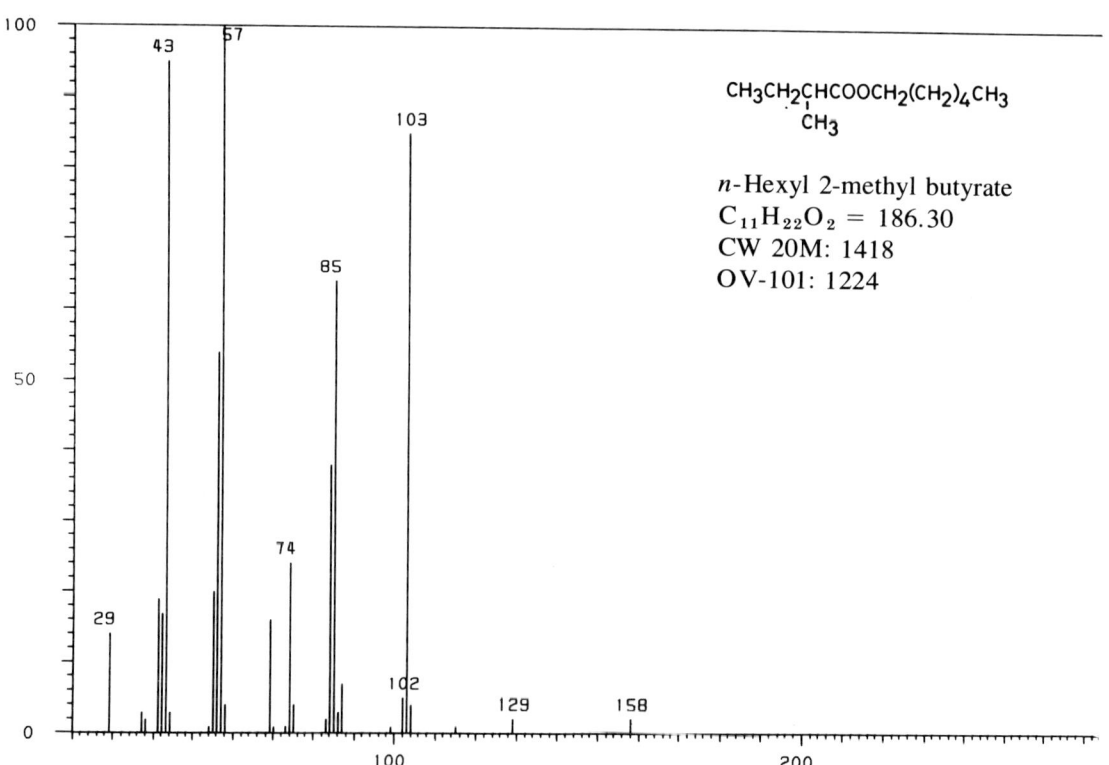

APPENDIX IV   MASS SPECTRA OF INDIVIDUAL COMPOUNDS

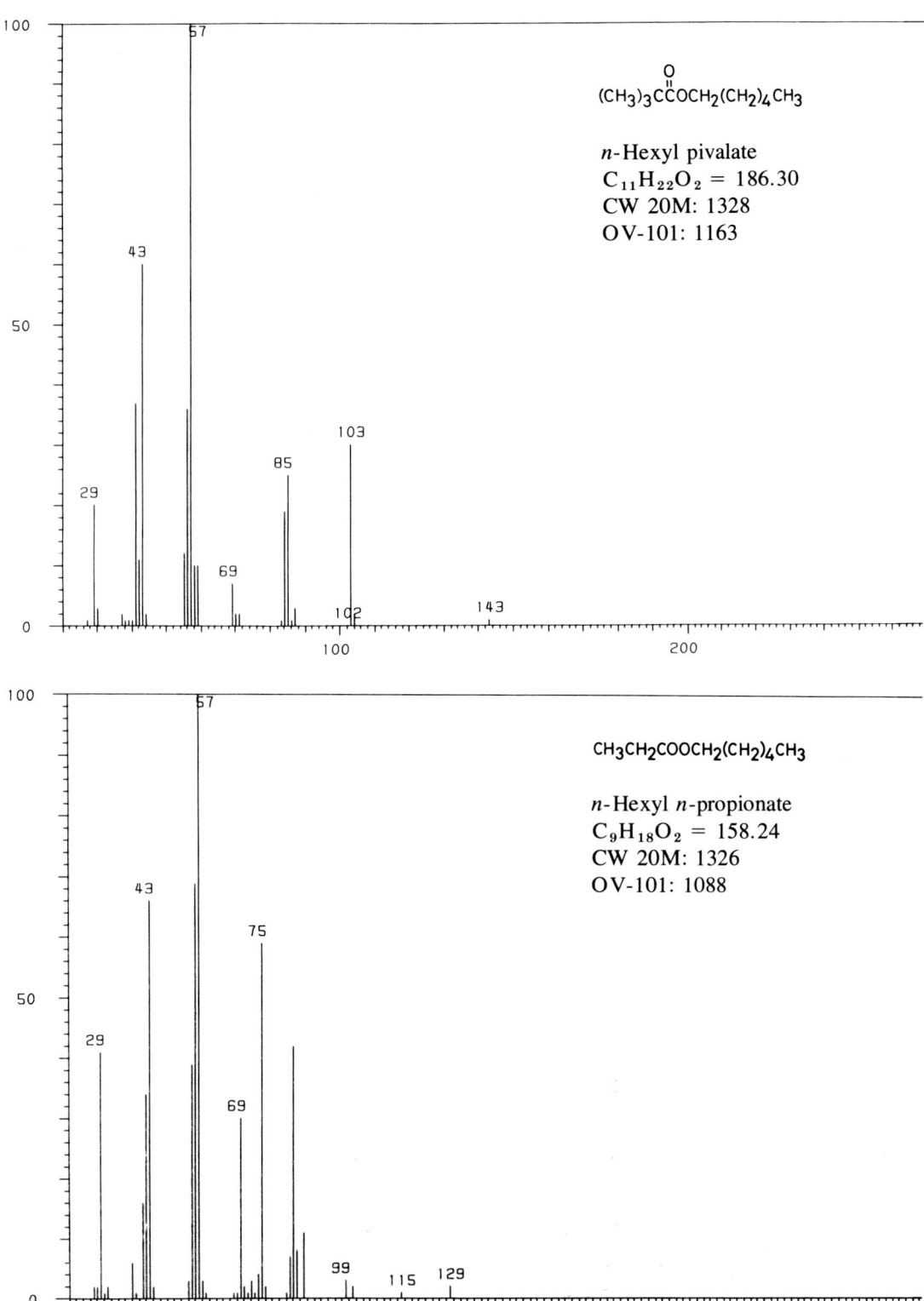

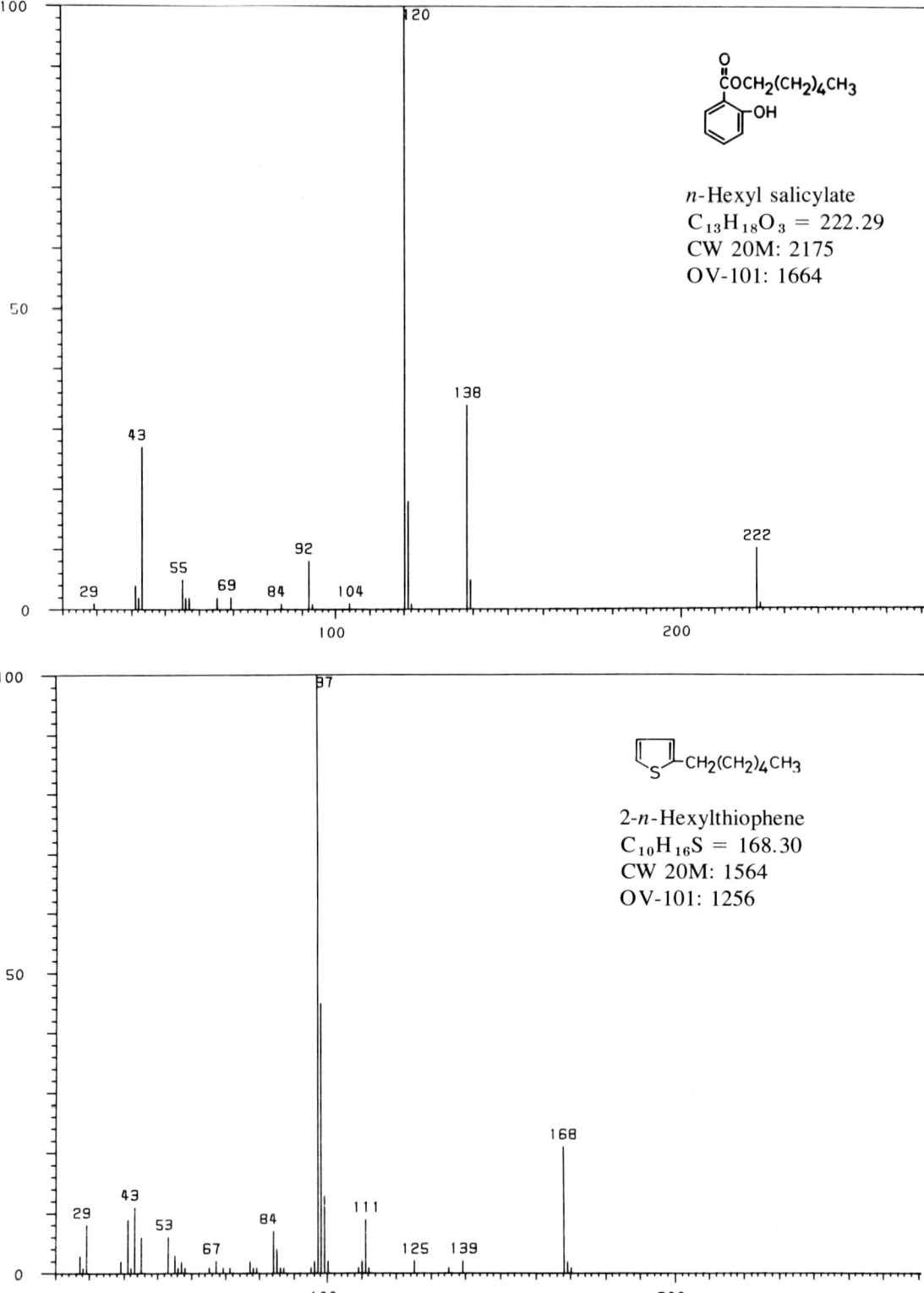

APPENDIX IV   MASS SPECTRA OF INDIVIDUAL COMPOUNDS

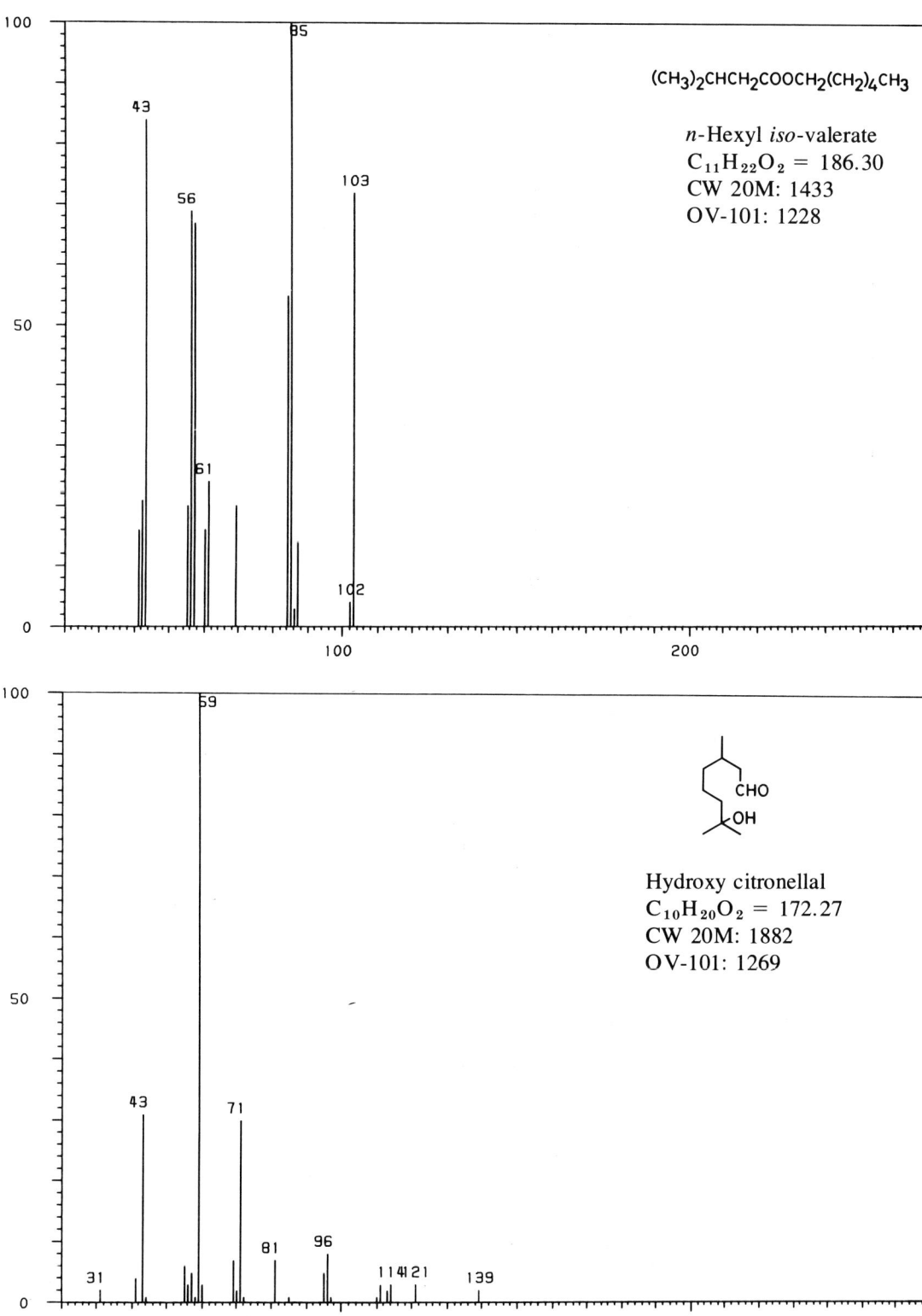

$(CH_3)_2CHCH_2COOCH_2(CH_2)_4CH_3$

n-Hexyl iso-valerate
$C_{11}H_{22}O_2 = 186.30$
CW 20M: 1433
OV-101: 1228

Hydroxy citronellal
$C_{10}H_{20}O_2 = 172.27$
CW 20M: 1882
OV-101: 1269

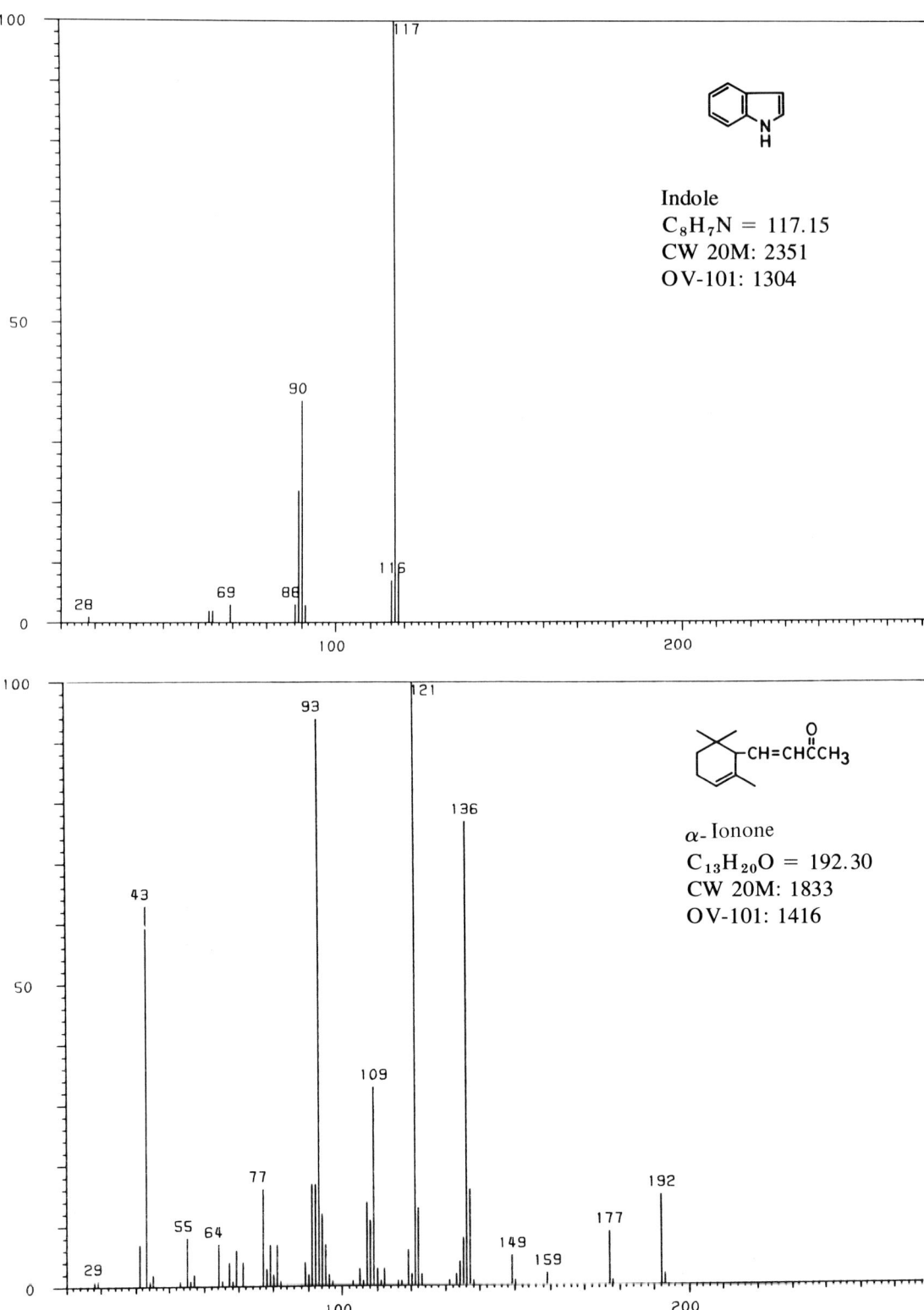

APPENDIX IV MASS SPECTRA OF INDIVIDUAL COMPOUNDS

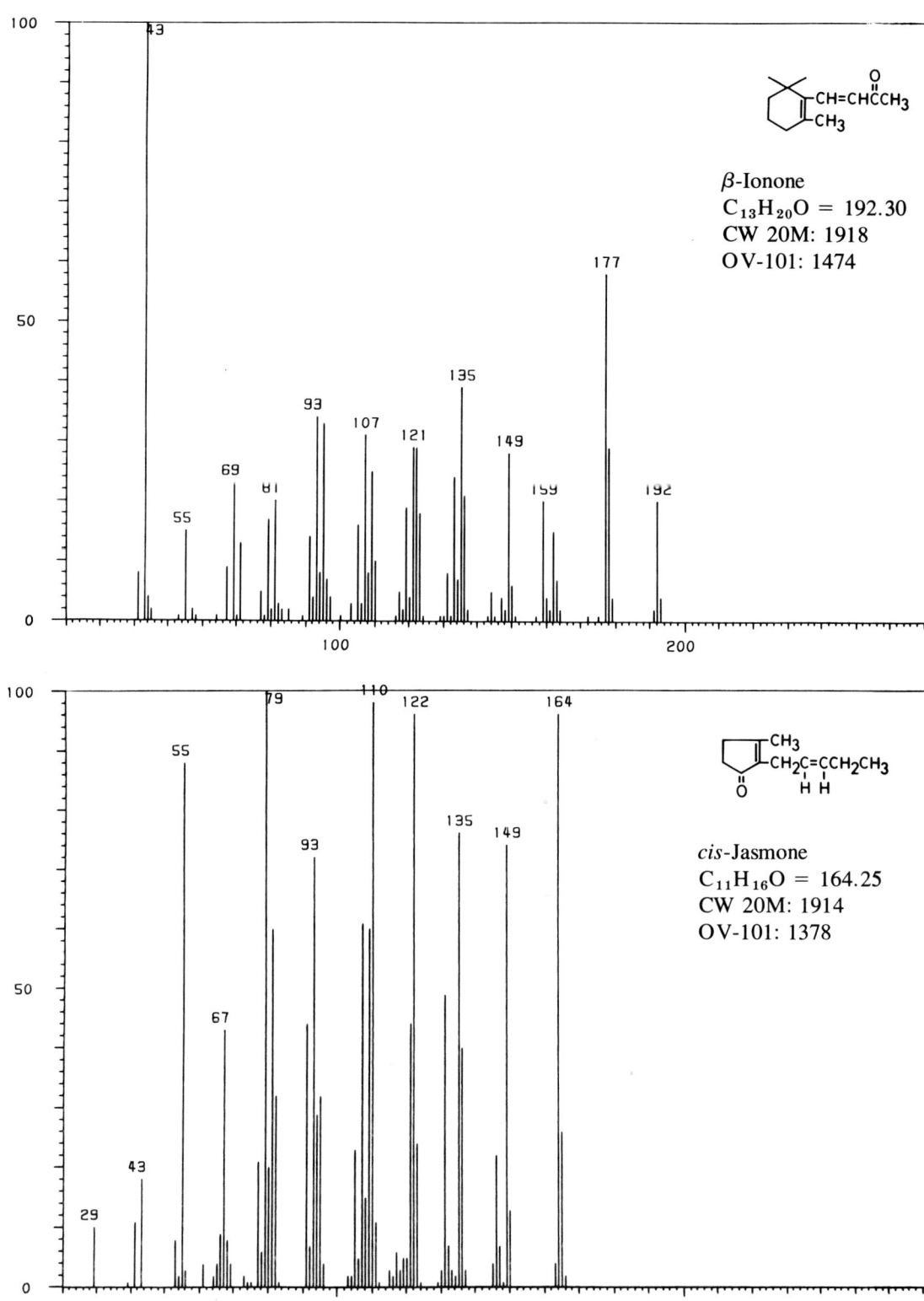

β-Ionone
$C_{13}H_{20}O = 192.30$
CW 20M: 1918
OV-101: 1474

cis-Jasmone
$C_{11}H_{16}O = 164.25$
CW 20M: 1914
OV-101: 1378

Lavandulyl acetate
$C_{12}H_{20}O_2$ = 196.29
CW 20M: 1597
OV-101: 1274

Limonene
$C_{10}H_{16}$ = 136.24
CW 20M: 1206
OV-101: 1030

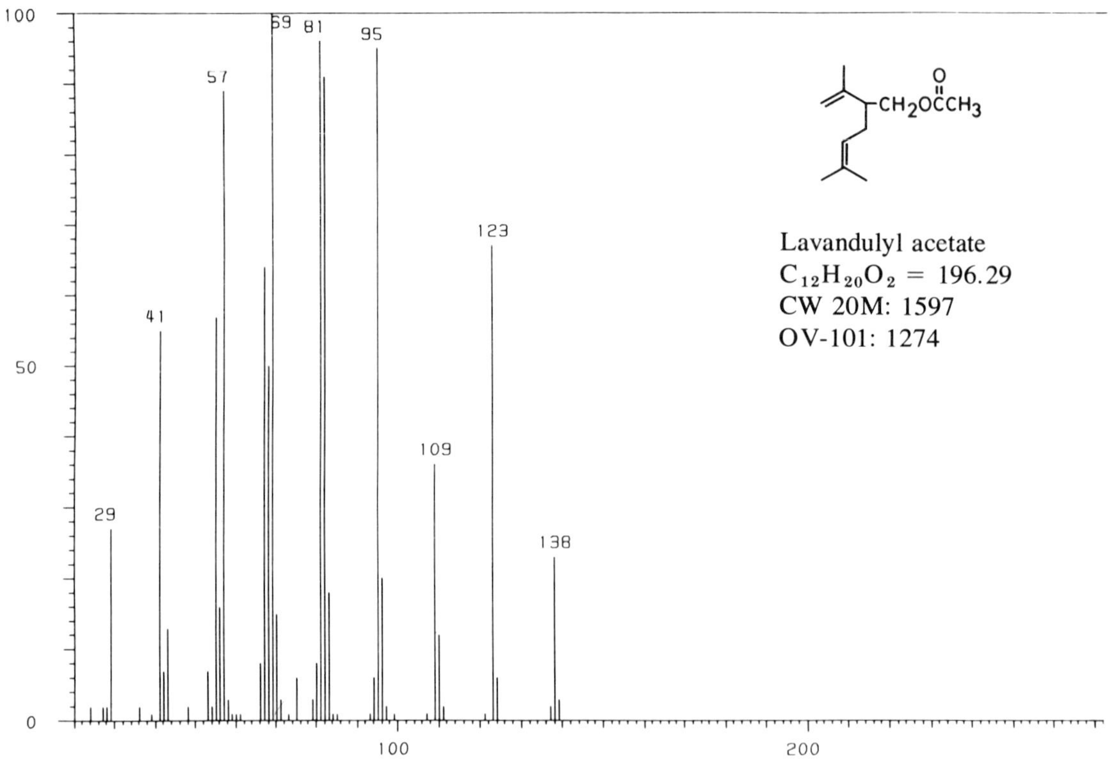

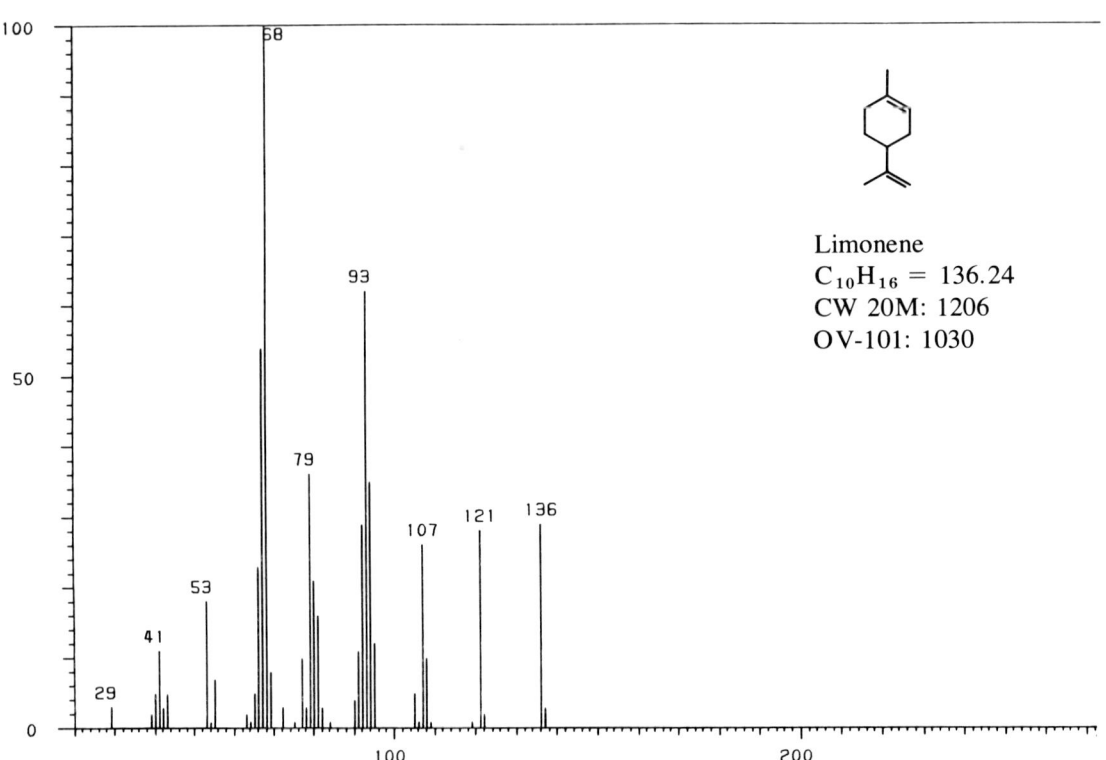

APPENDIX IV   MASS SPECTRA OF INDIVIDUAL COMPOUNDS

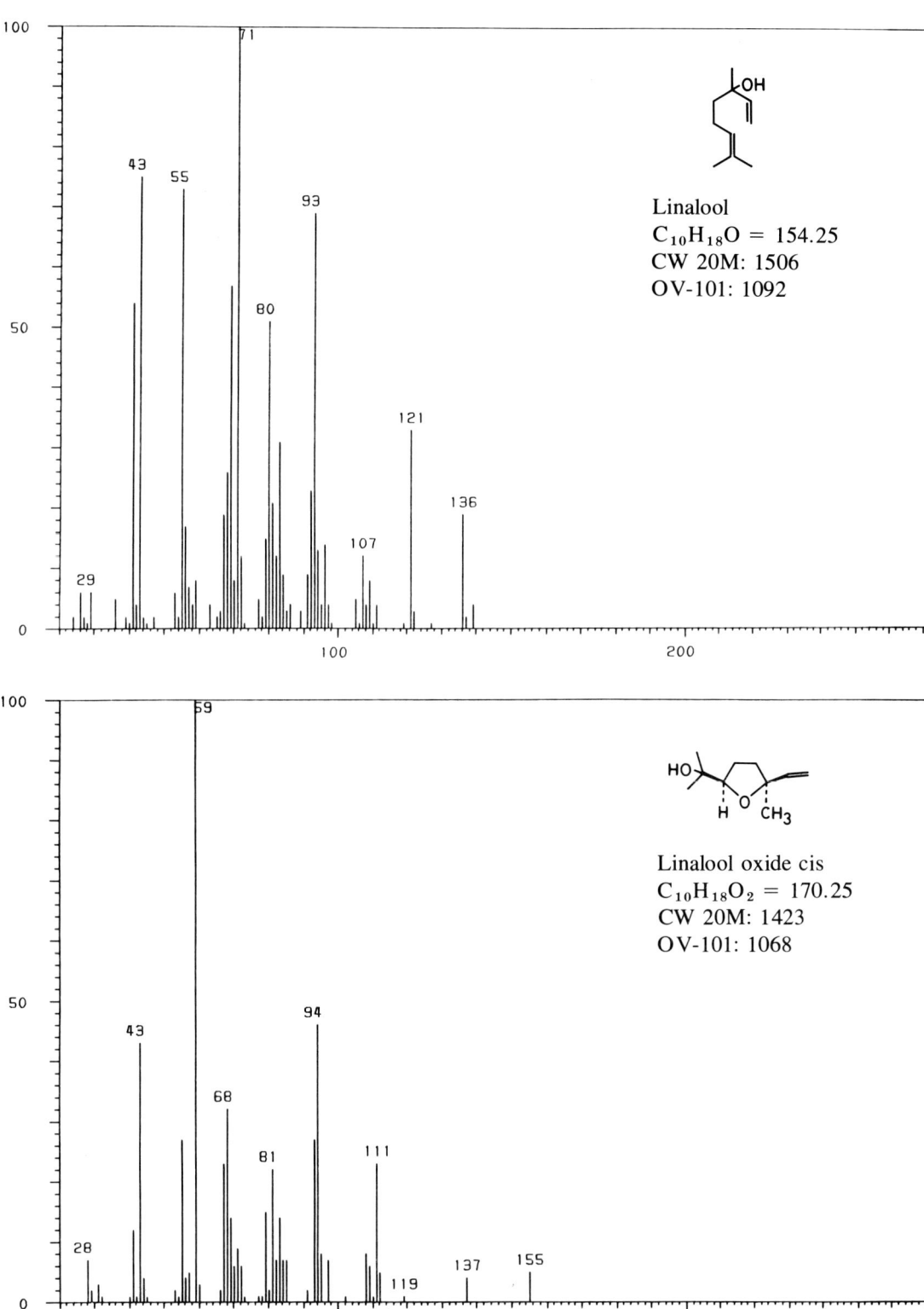

Linalool
$C_{10}H_{18}O = 154.25$
CW 20M: 1506
OV-101: 1092

Linalool oxide cis
$C_{10}H_{18}O_2 = 170.25$
CW 20M: 1423
OV-101: 1068

Linalool oxide trans
$C_{10}H_{18}O_2$ = 170.25
CW 20M: 1451
OV-101: 1082

Linalyl acetate
$C_{12}H_{20}O_2$ = 196.26
CW 20M: 1538
OV-101: 1246

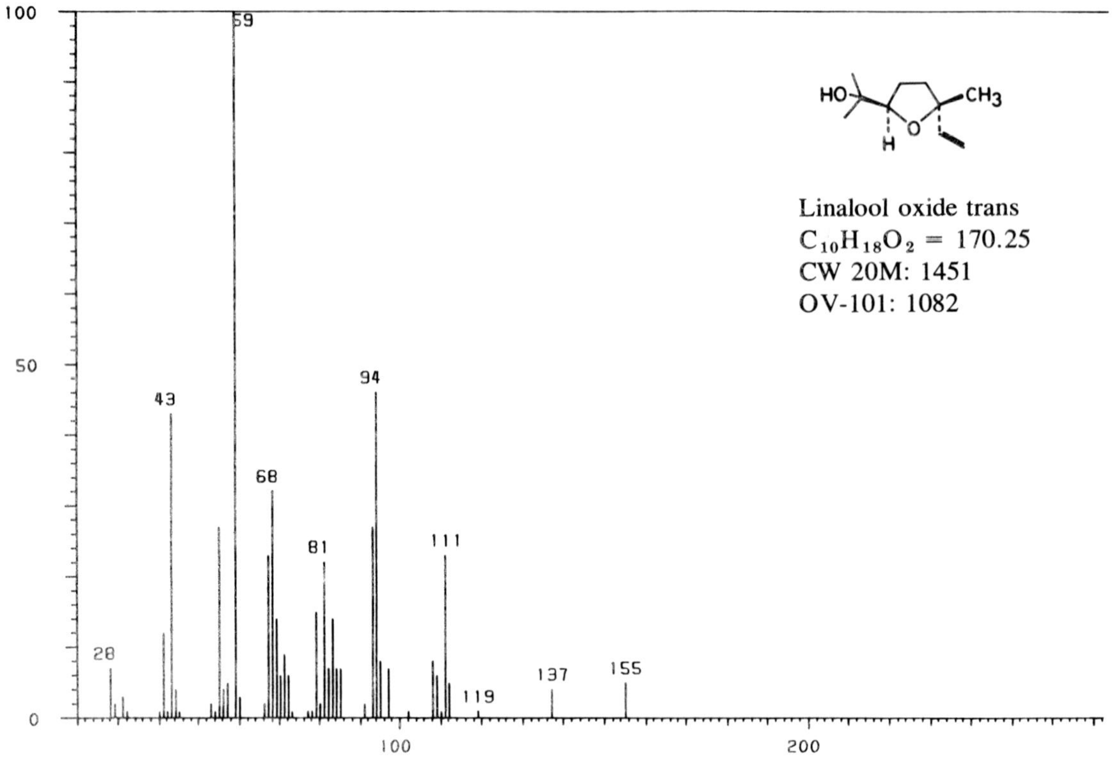

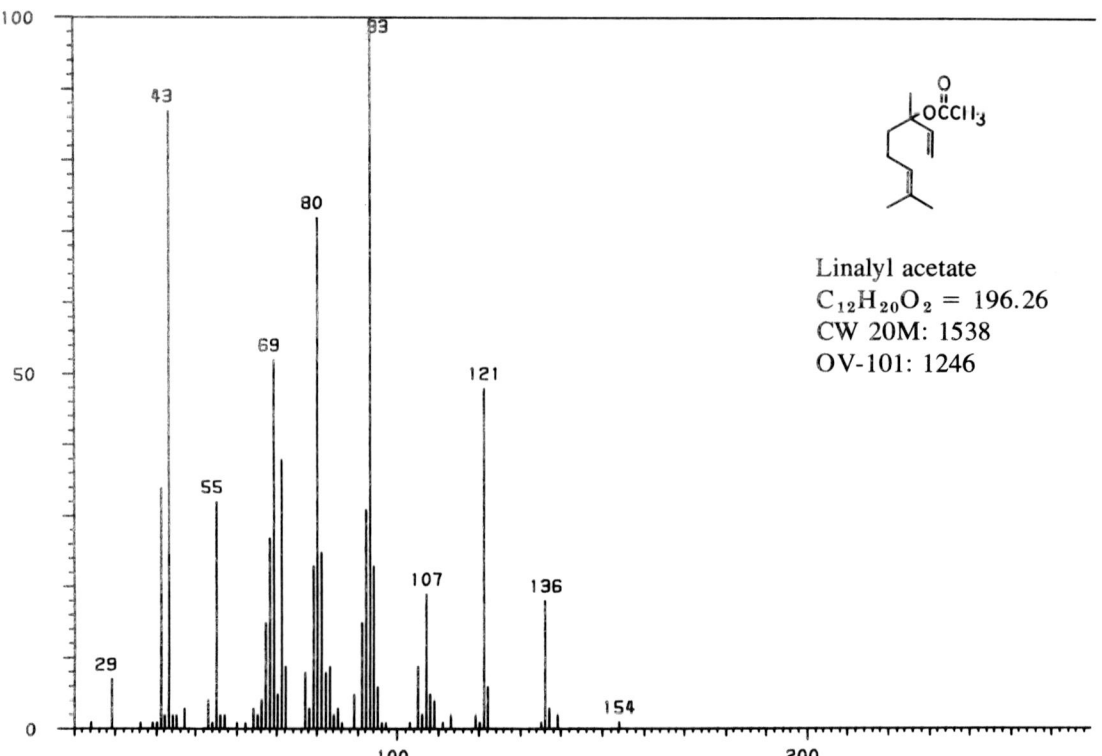

APPENDIX IV MASS SPECTRA OF INDIVIDUAL COMPOUNDS

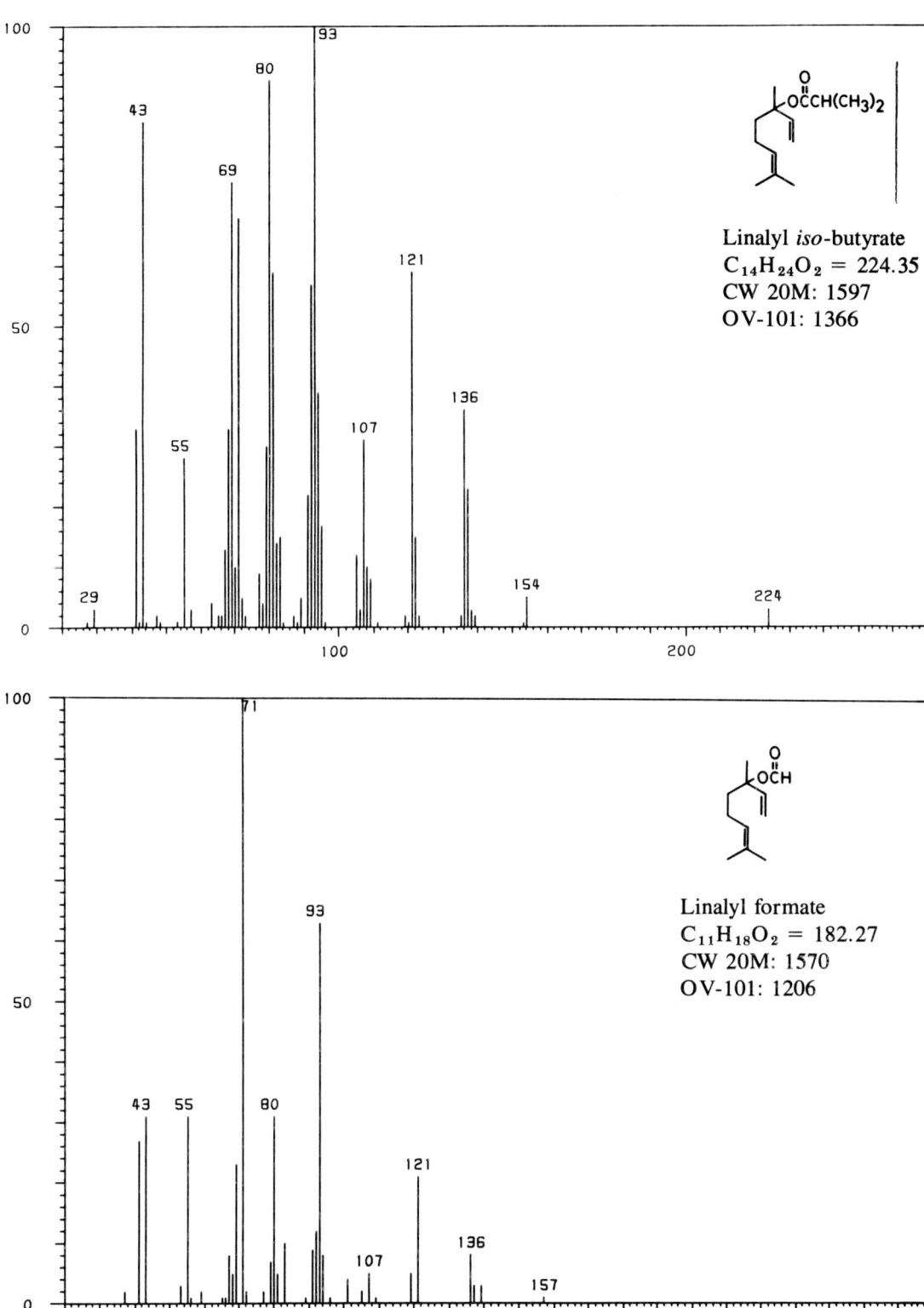

Linalyl *iso*-butyrate
$C_{14}H_{24}O_2 = 224.35$
CW 20M: 1597
OV-101: 1366

Linalyl formate
$C_{11}H_{18}O_2 = 182.27$
CW 20M: 1570
OV-101: 1206

Linalyl *n*-hexanoate
$C_{16}H_{28}O_2$ = 252.40
CW 20M: 1843
OV-101: 1582

Linalyl *n*-propionate
$C_{13}H_{22}O_2$ = 210.32
CW 20M: 1596
OV-101: 1324

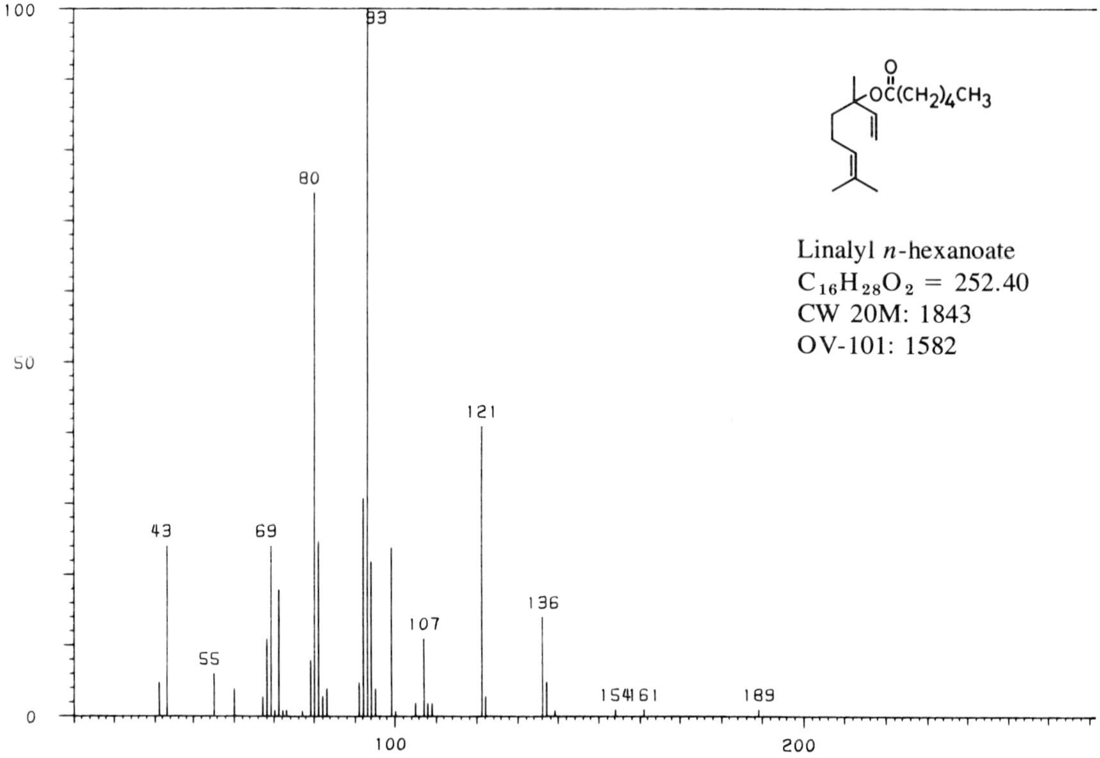

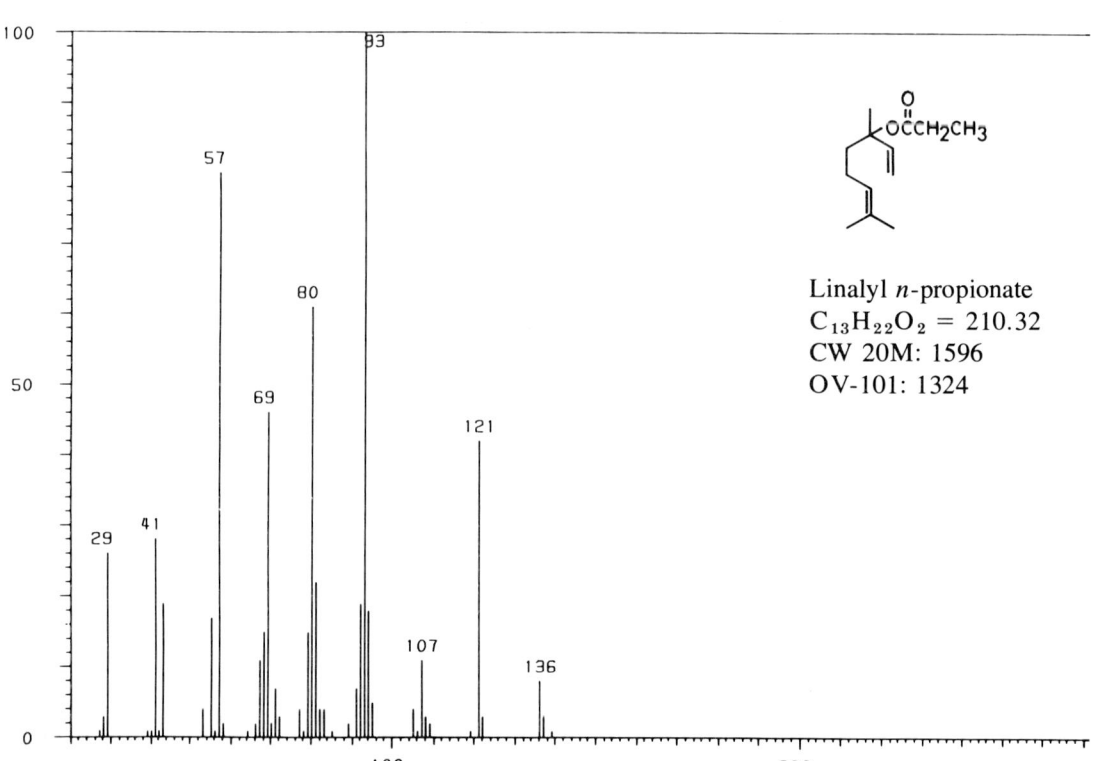

APPENDIX IV   MASS SPECTRA OF INDIVIDUAL COMPOUNDS

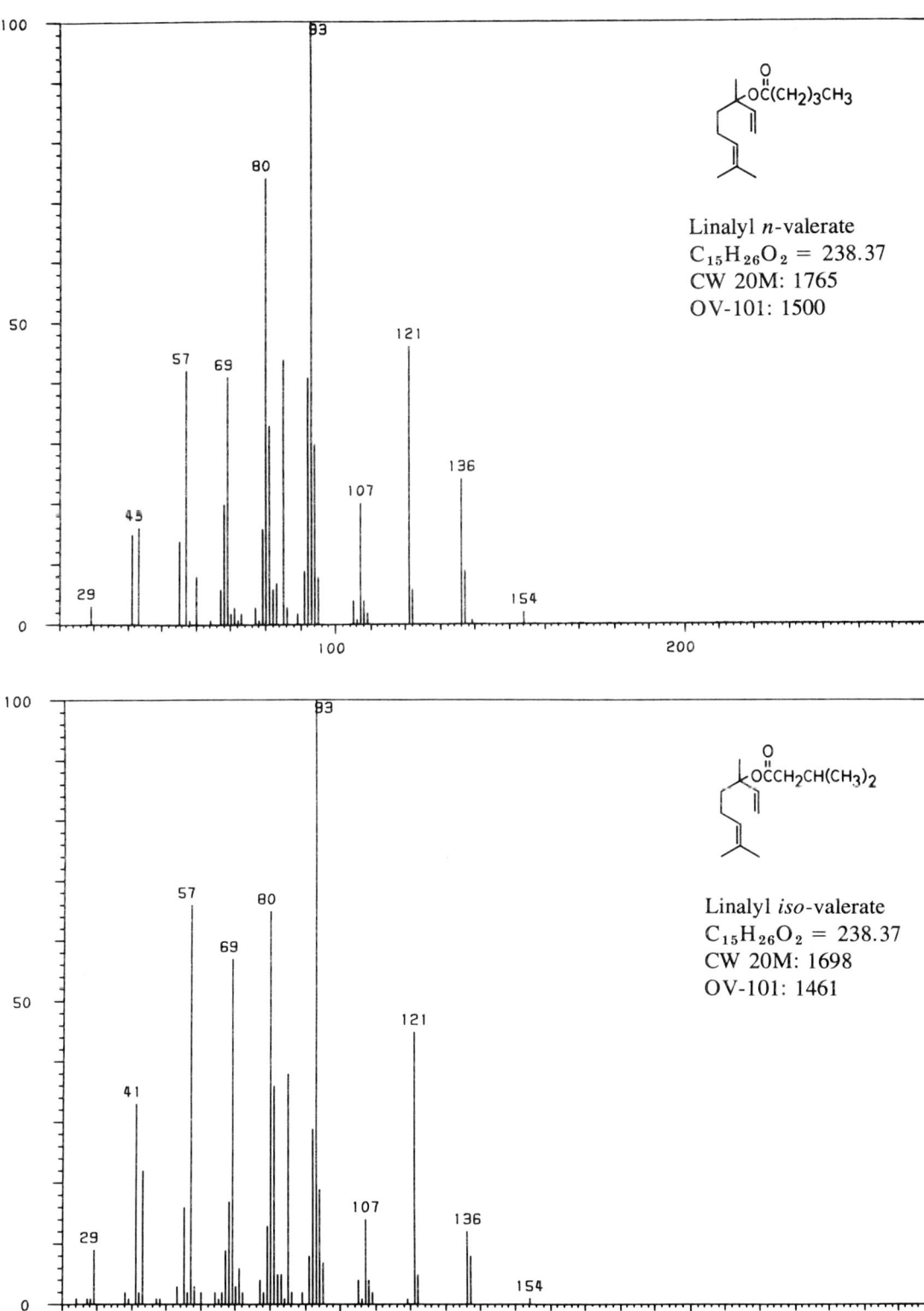

Linalyl *n*-valerate
$C_{15}H_{26}O_2 = 238.37$
CW 20M: 1765
OV-101: 1500

Linalyl *iso*-valerate
$C_{15}H_{26}O_2 = 238.37$
CW 20M: 1698
OV-101: 1461

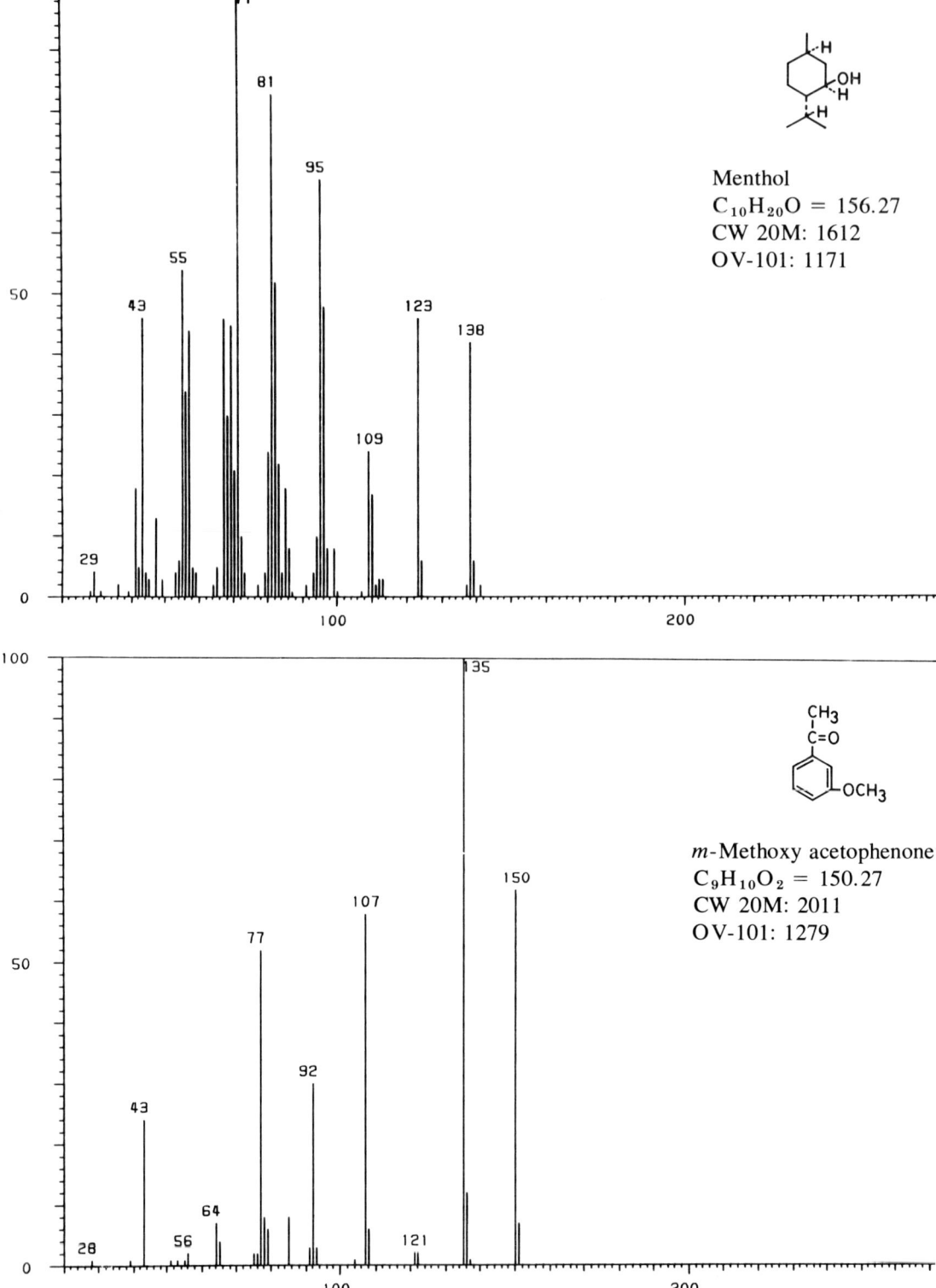

APPENDIX IV   MASS SPECTRA OF INDIVIDUAL COMPOUNDS

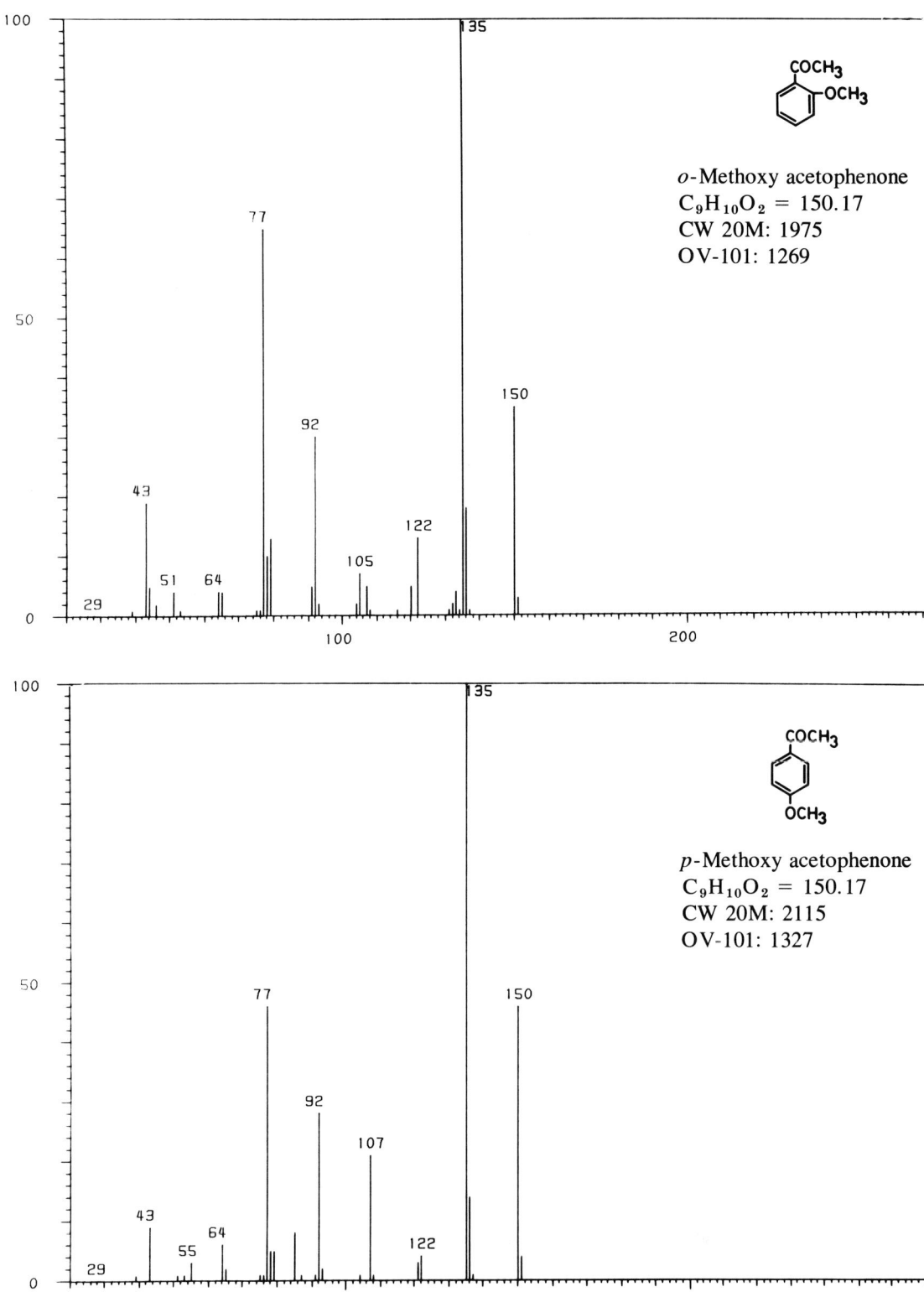

o-Methoxy acetophenone
$C_9H_{10}O_2 = 150.17$
CW 20M: 1975
OV-101: 1269

p-Methoxy acetophenone
$C_9H_{10}O_2 = 150.17$
CW 20M: 2115
OV-101: 1327

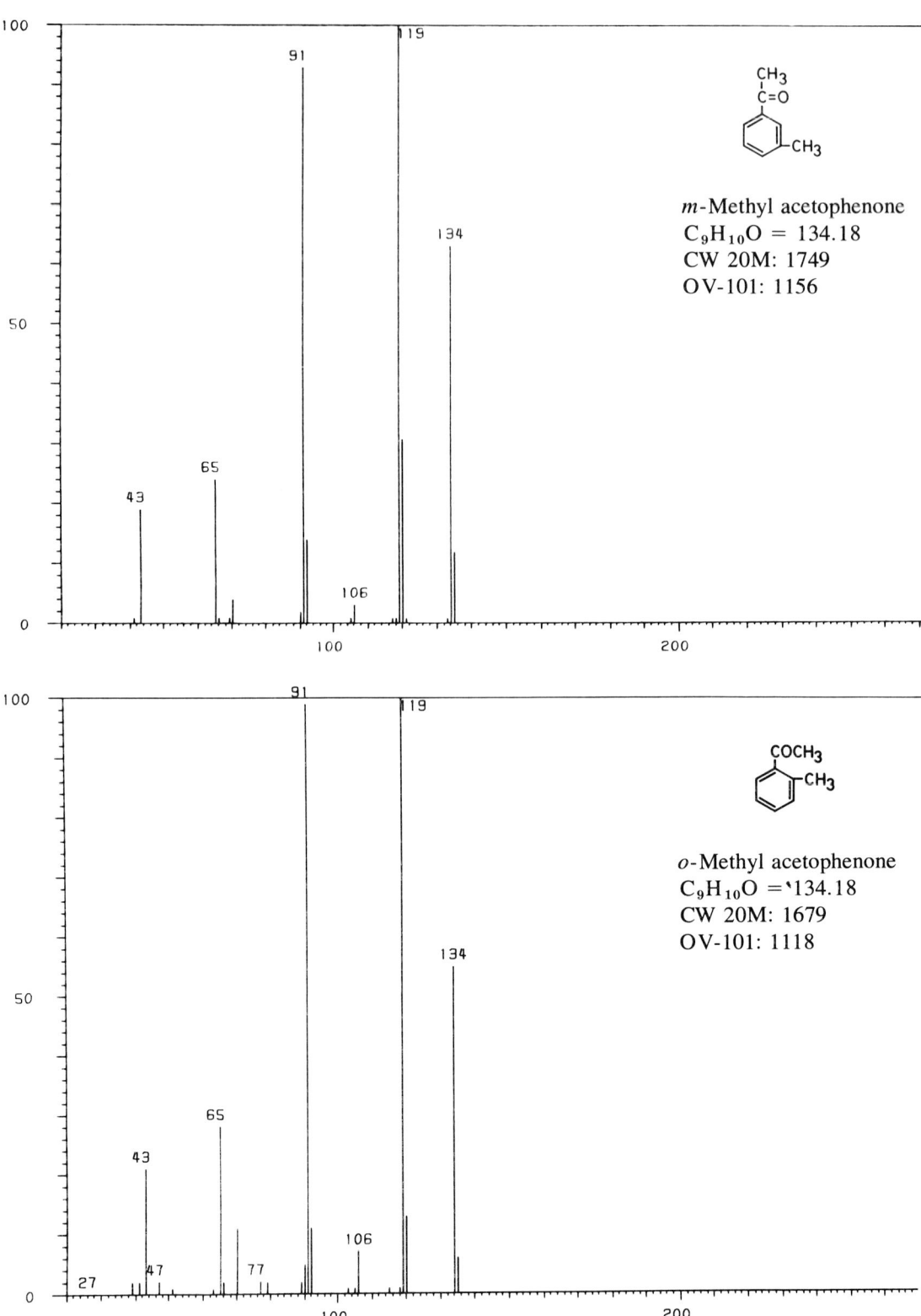

APPENDIX IV MASS SPECTRA OF INDIVIDUAL COMPOUNDS

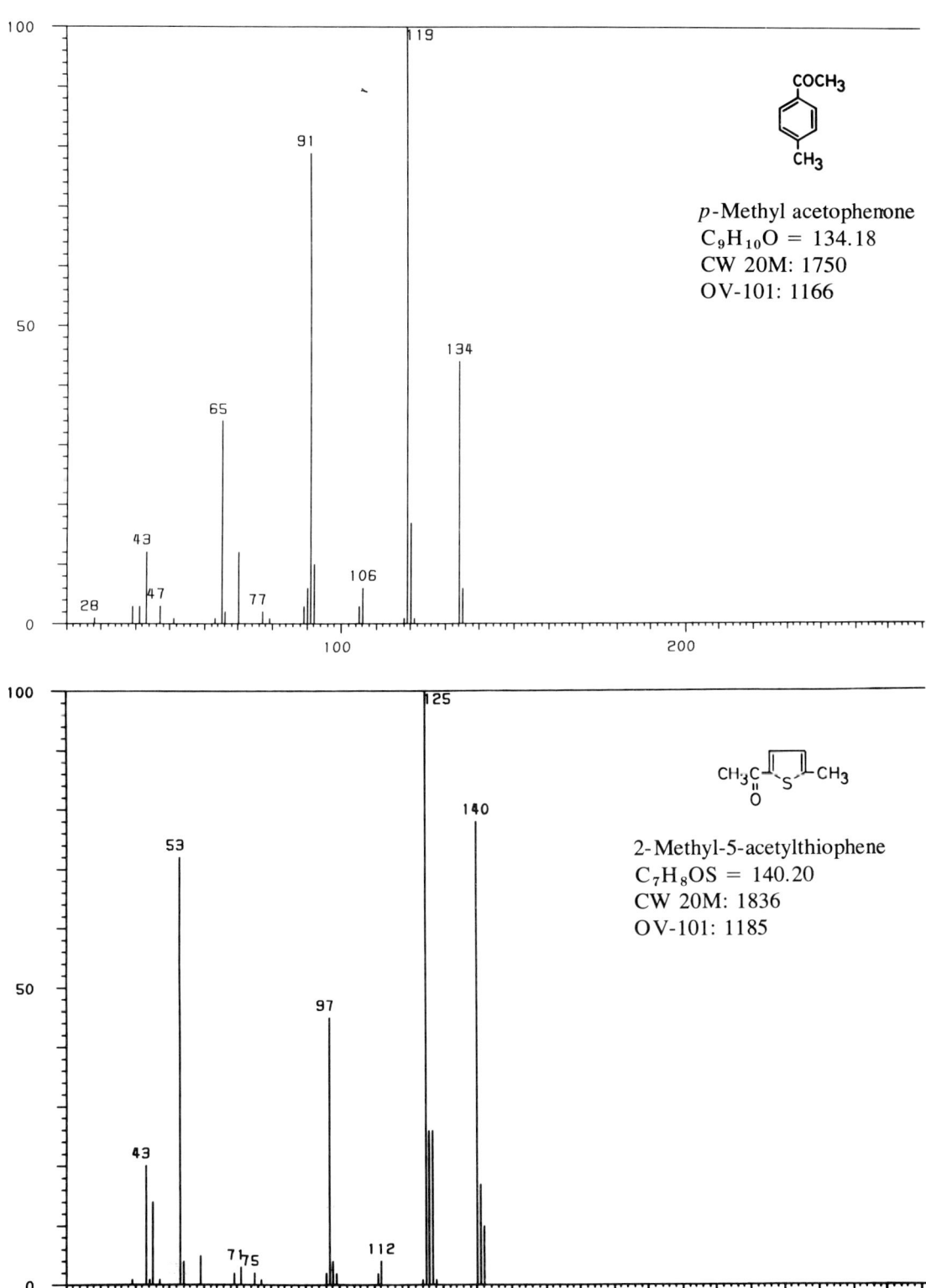

p-Methyl acetophenone
$C_9H_{10}O = 134.18$
CW 20M: 1750
OV-101: 1166

2-Methyl-5-acetylthiophene
$C_7H_8OS = 140.20$
CW 20M: 1836
OV-101: 1185

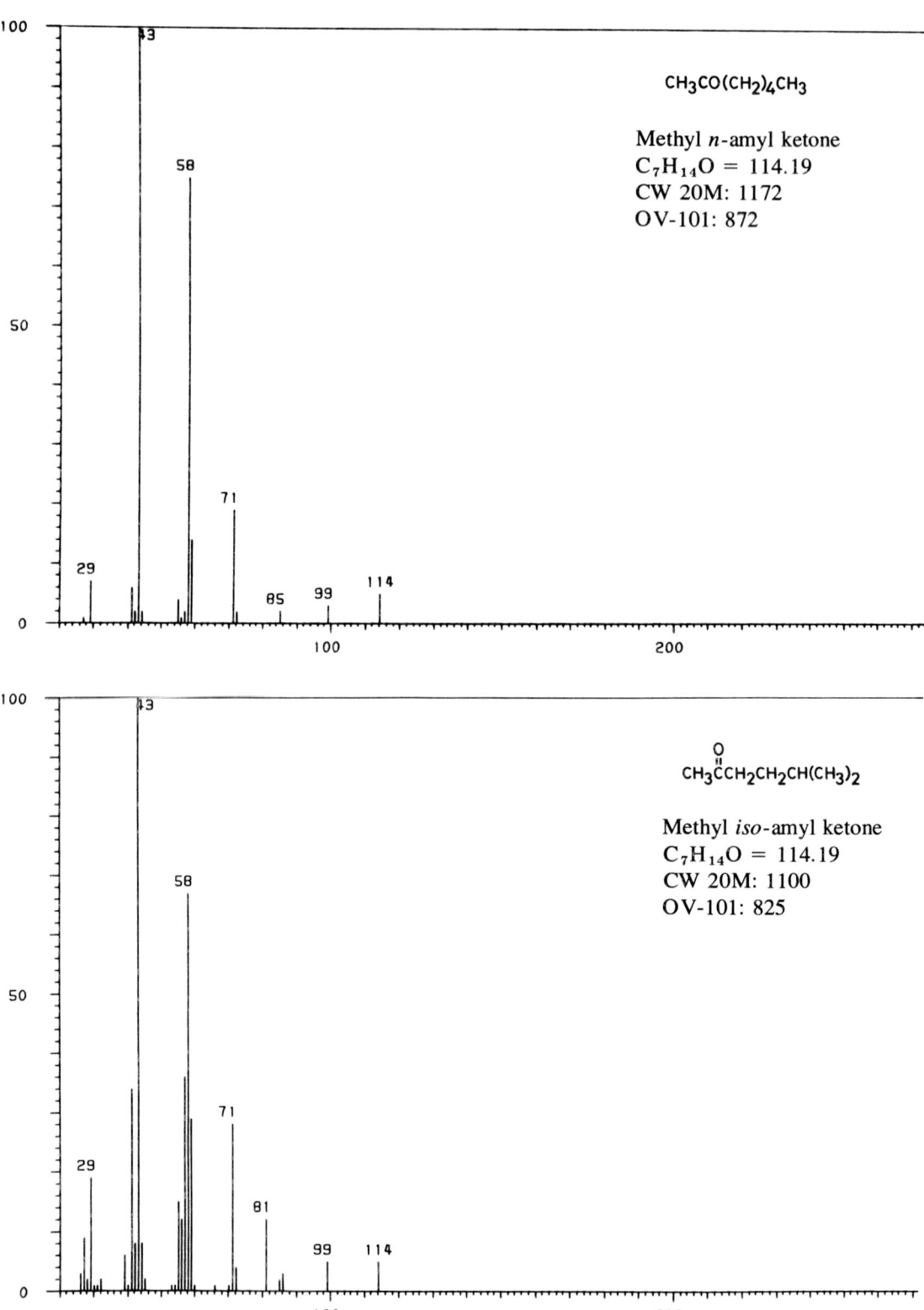

APPENDIX IV MASS SPECTRA OF INDIVIDUAL COMPOUNDS

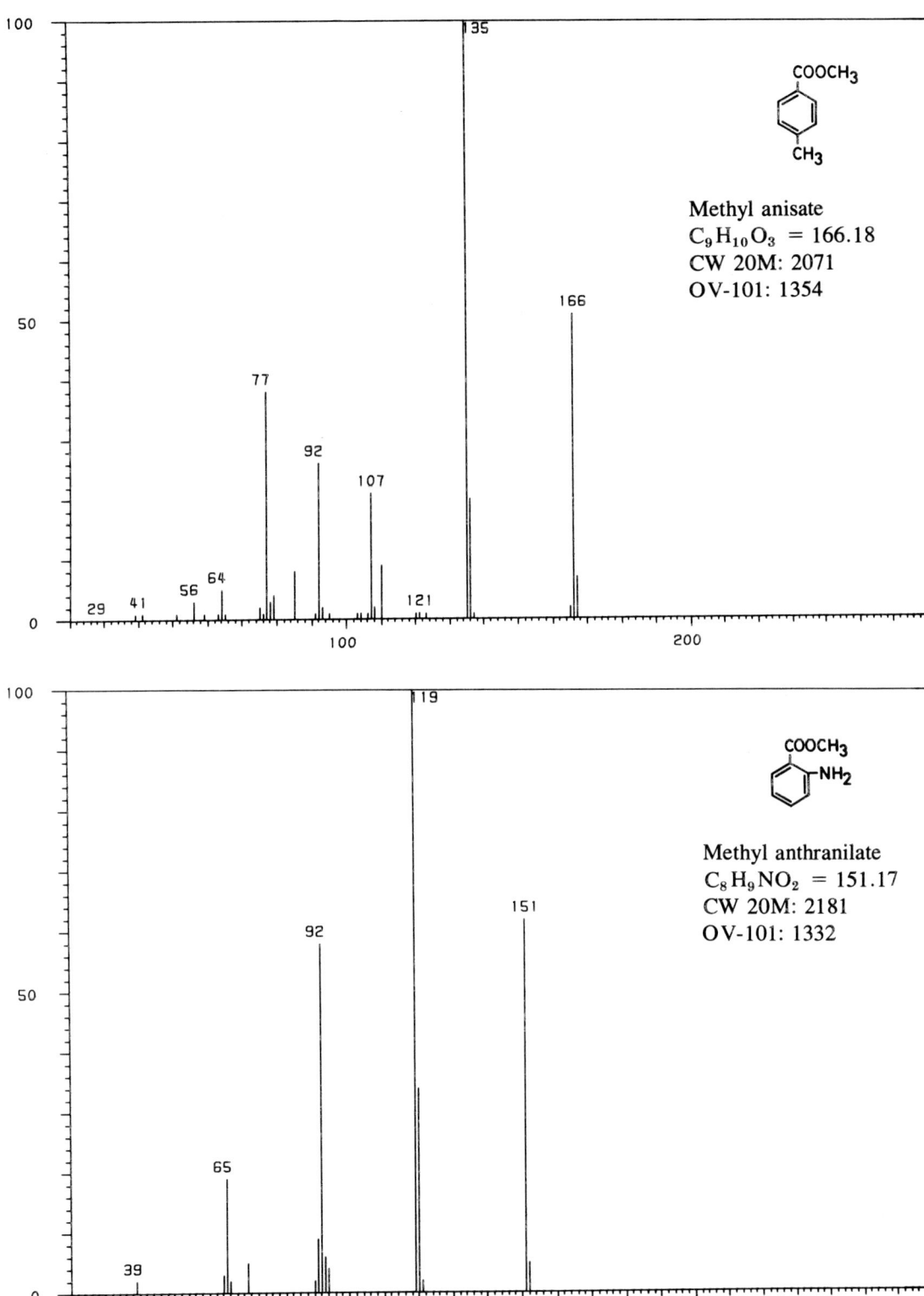

Methyl anisate
$C_9H_{10}O_3 = 166.18$
CW 20M: 2071
OV-101: 1354

Methyl anthranilate
$C_8H_9NO_2 = 151.17$
CW 20M: 2181
OV-101: 1332

Methyl benzoate
$C_8H_8O_2$ = 136.15
CW 20M: 1600
OV-101: 1078

3-Methyl butan-2-ol
$C_5H_{12}O$ = 88.15
CW 20M: 1052
OV-101: 671

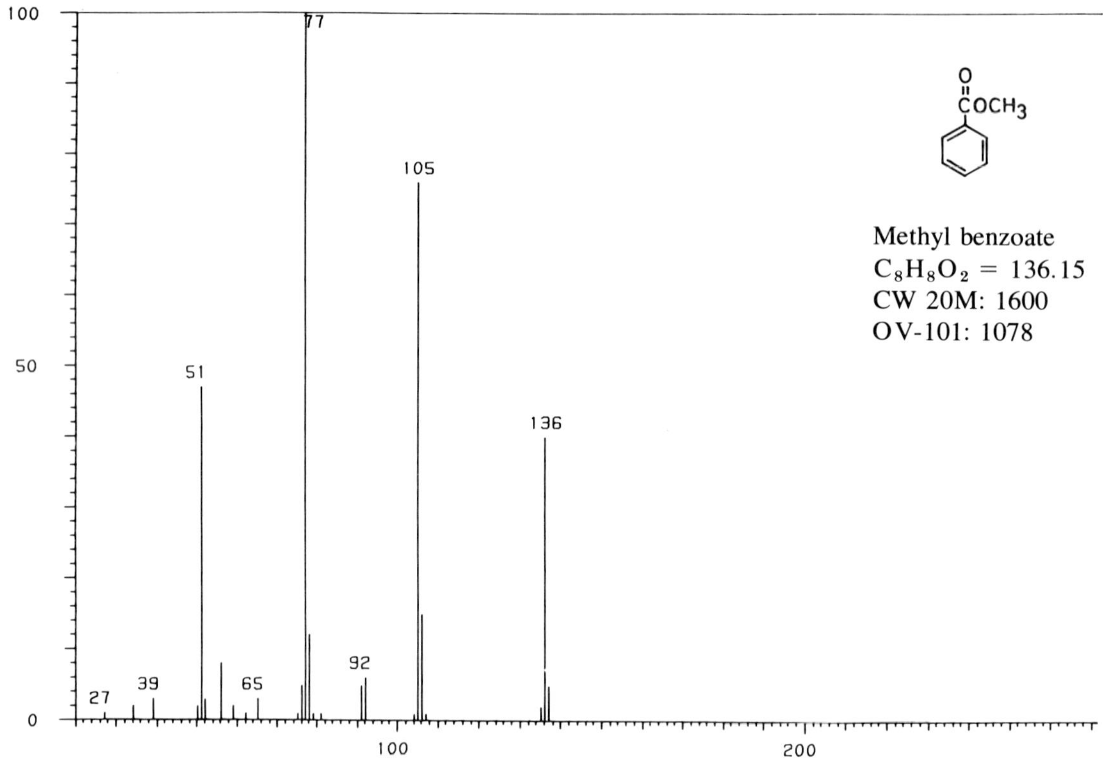

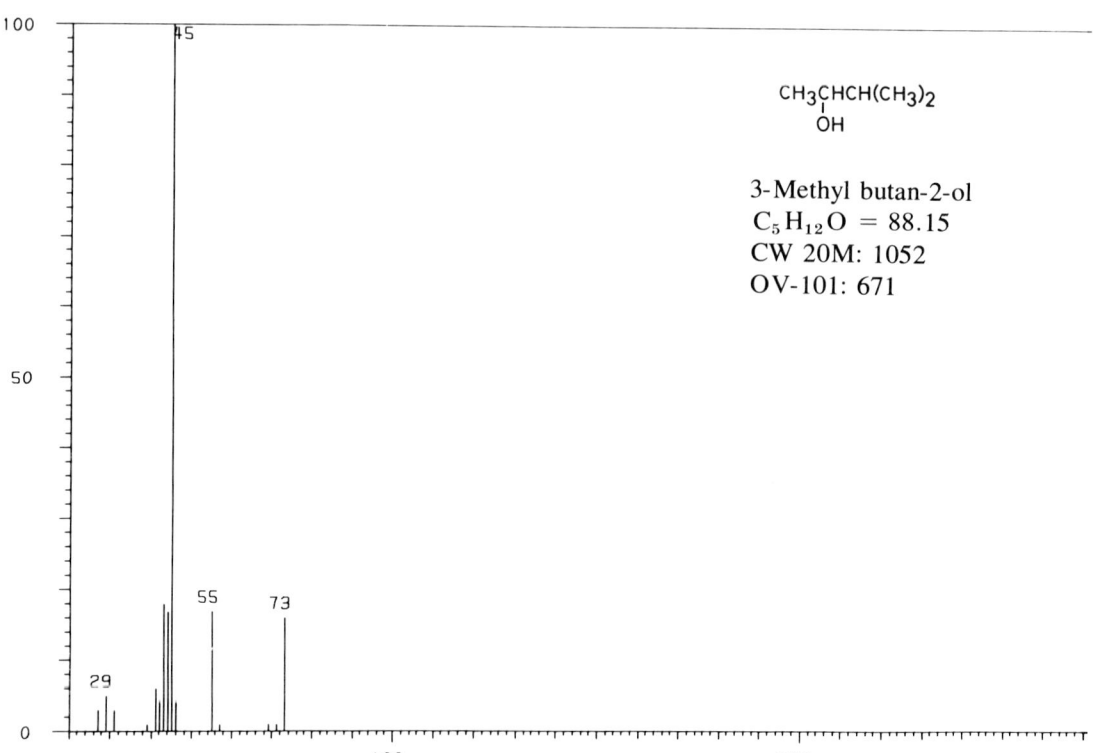

APPENDIX IV   MASS SPECTRA OF INDIVIDUAL COMPOUNDS

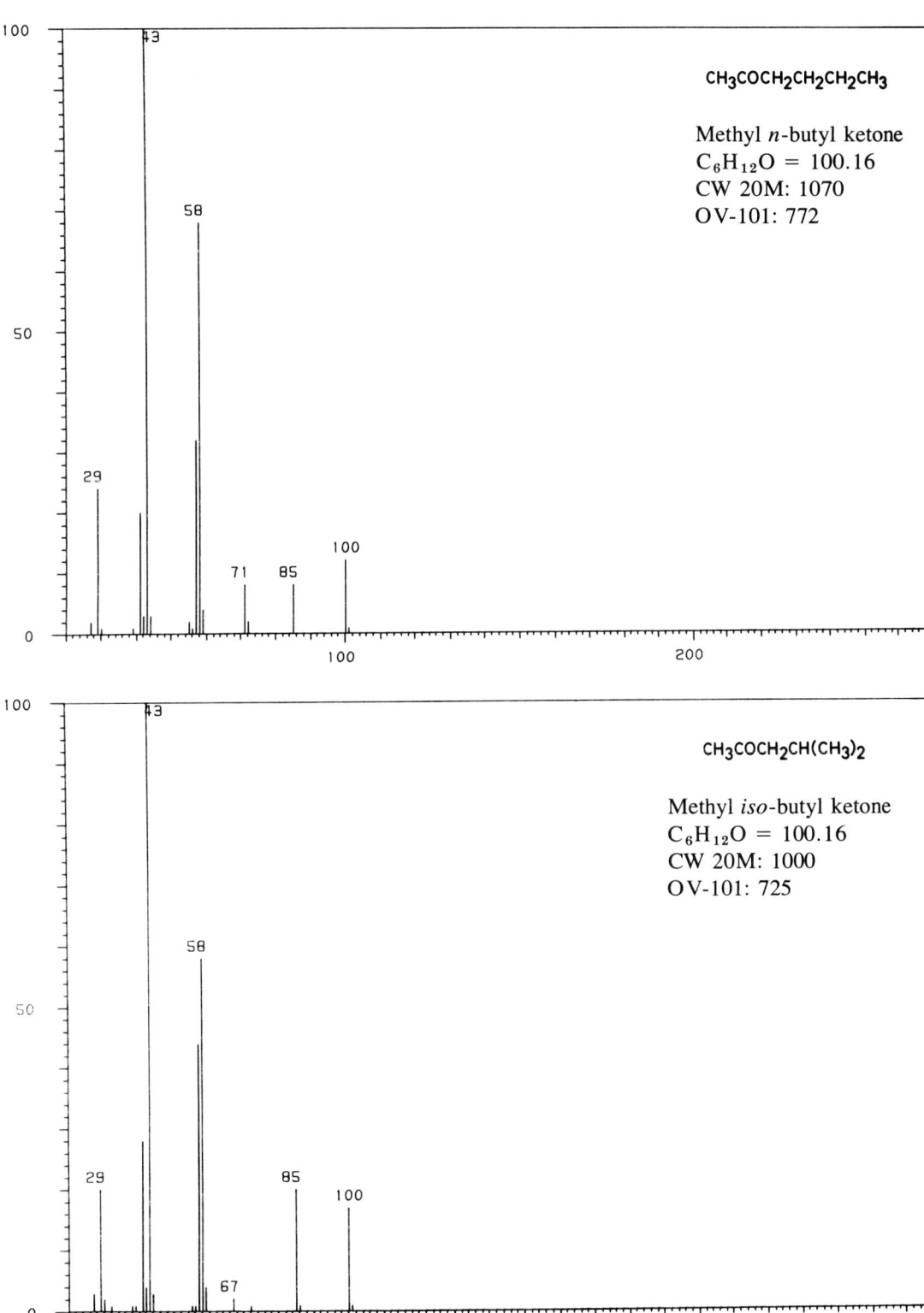

CH$_3$COCH$_2$CH$_2$CH$_2$CH$_3$

Methyl *n*-butyl ketone
C$_6$H$_{12}$O = 100.16
CW 20M: 1070
OV-101: 772

CH$_3$COCH$_2$CH(CH$_3$)$_2$

Methyl *iso*-butyl ketone
C$_6$H$_{12}$O = 100.16
CW 20M: 1000
OV-101: 725

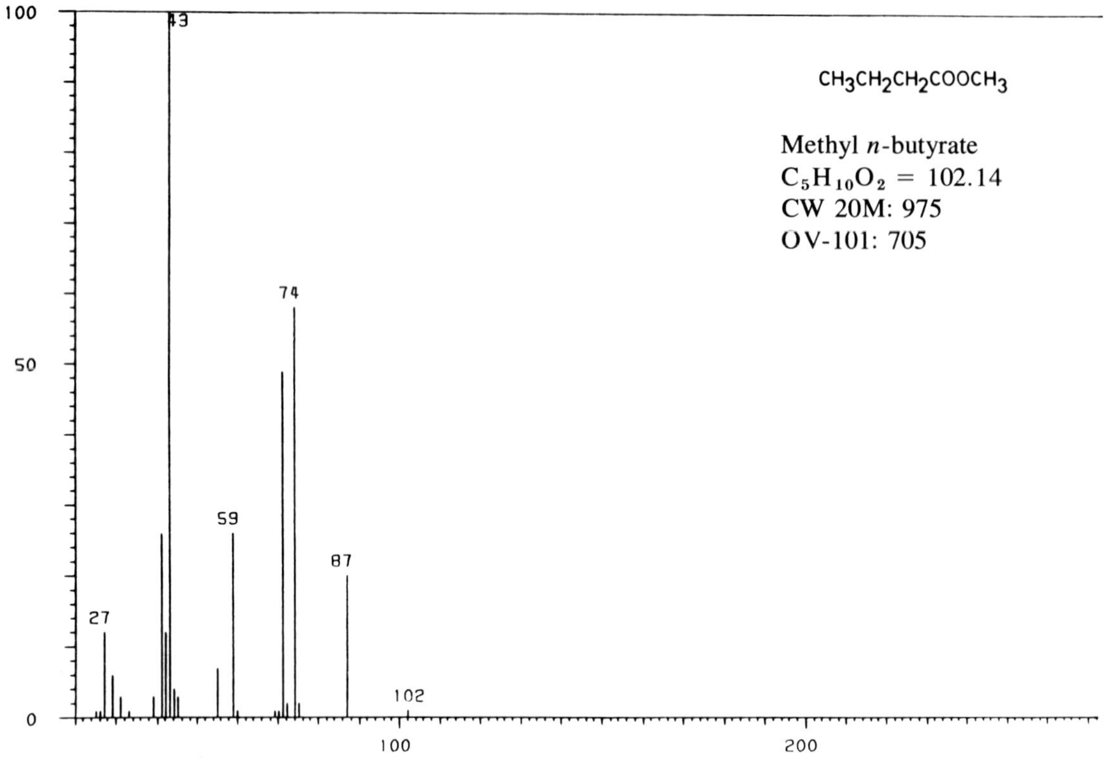

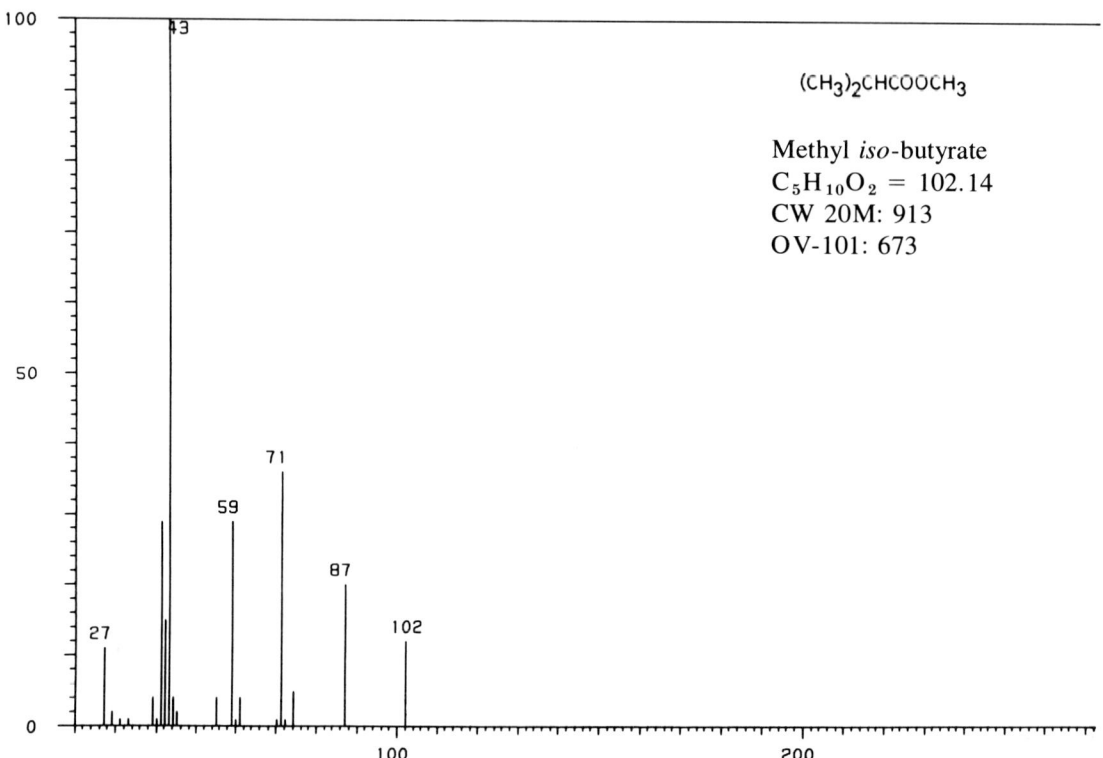

CH₃CH₂CH₂COOCH₃

Methyl *n*-butyrate
$C_5H_{10}O_2 = 102.14$
CW 20M: 975
OV-101: 705

(CH₃)₂CHCOOCH₃

Methyl *iso*-butyrate
$C_5H_{10}O_2 = 102.14$
CW 20M: 913
OV-101: 673

APPENDIX IV   MASS SPECTRA OF INDIVIDUAL COMPOUNDS

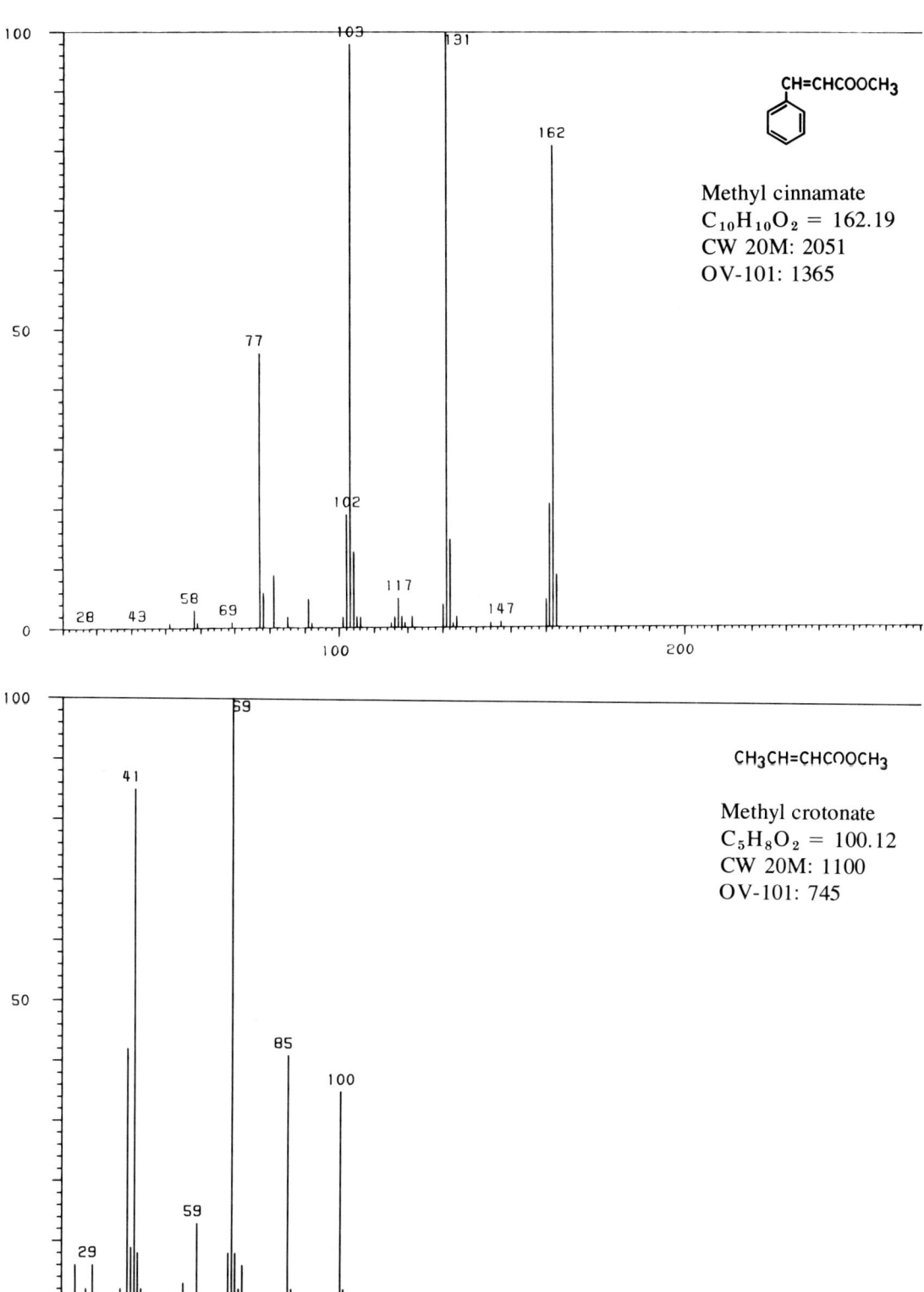

Methyl cinnamate
$C_{10}H_{10}O_2 = 162.19$
CW 20M: 2051
OV-101: 1365

Methyl crotonate
$C_5H_8O_2 = 100.12$
CW 20M: 1100
OV-101: 745

Methyl *n*-decanoate
$C_{11}H_{22}O_2$ = 186.30
CW 20M: 1581
OV-101: 1307

Methyl *n*-dodecanoate
$C_{13}H_{26}O_2$ = 214.35
CW 20M: 1785
OV-101: 1507

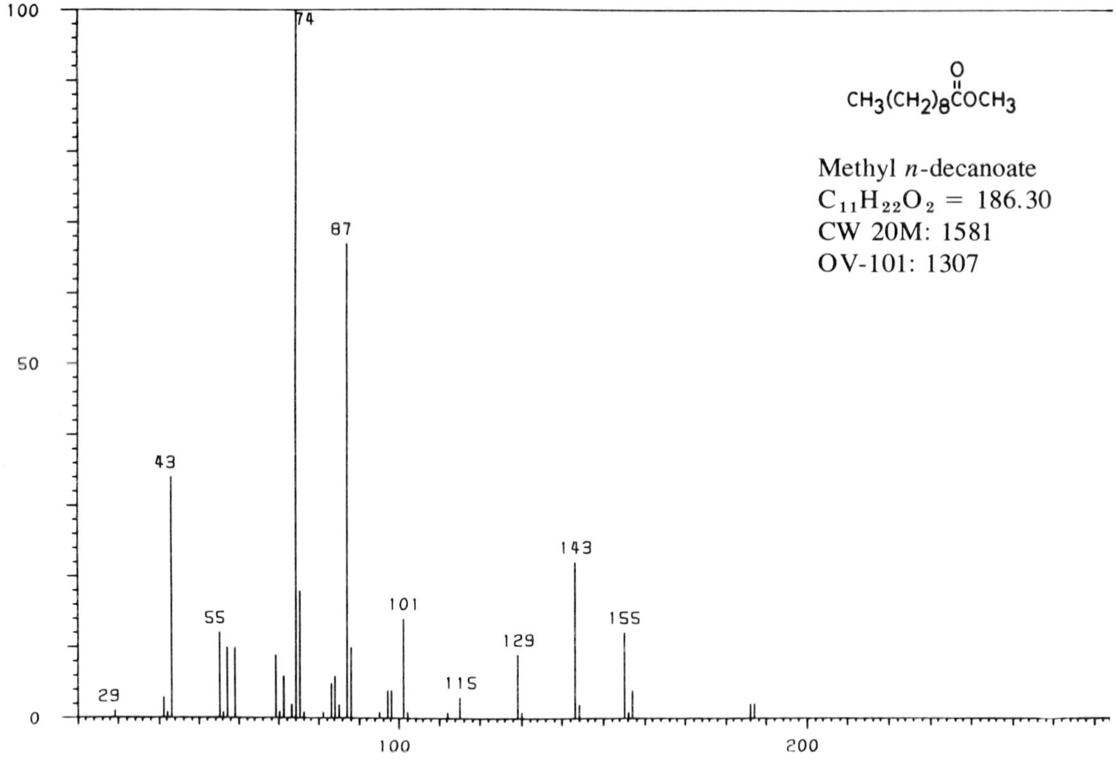

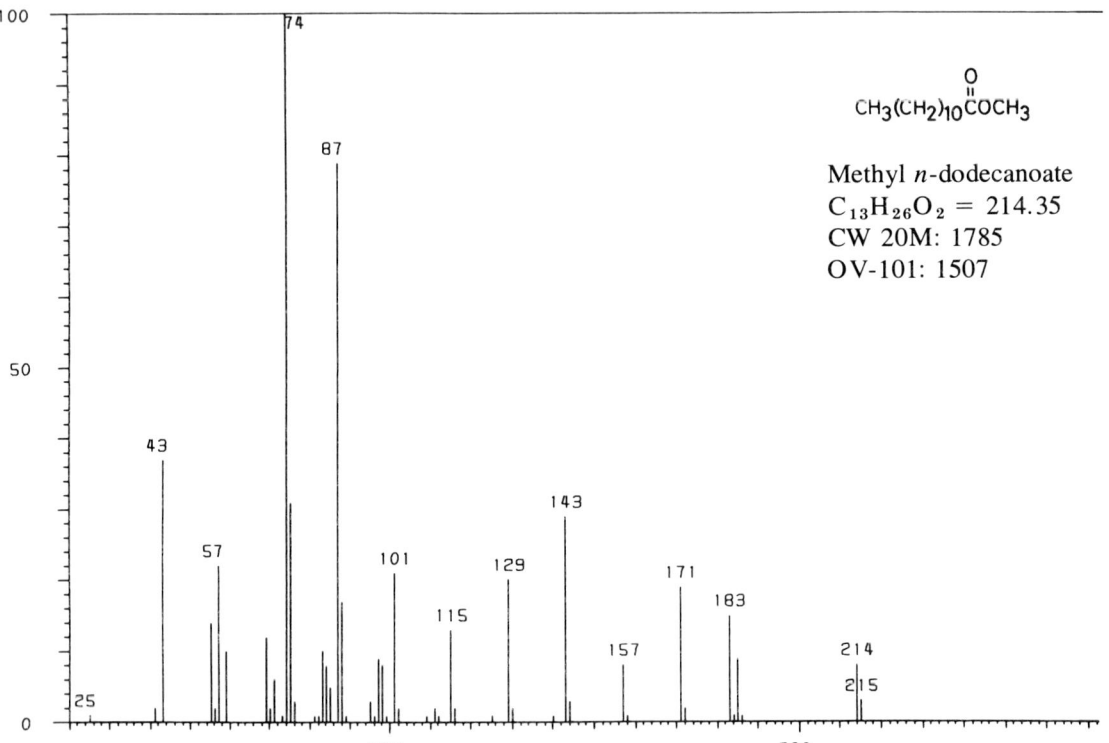

APPENDIX IV   MASS SPECTRA OF INDIVIDUAL COMPOUNDS   363

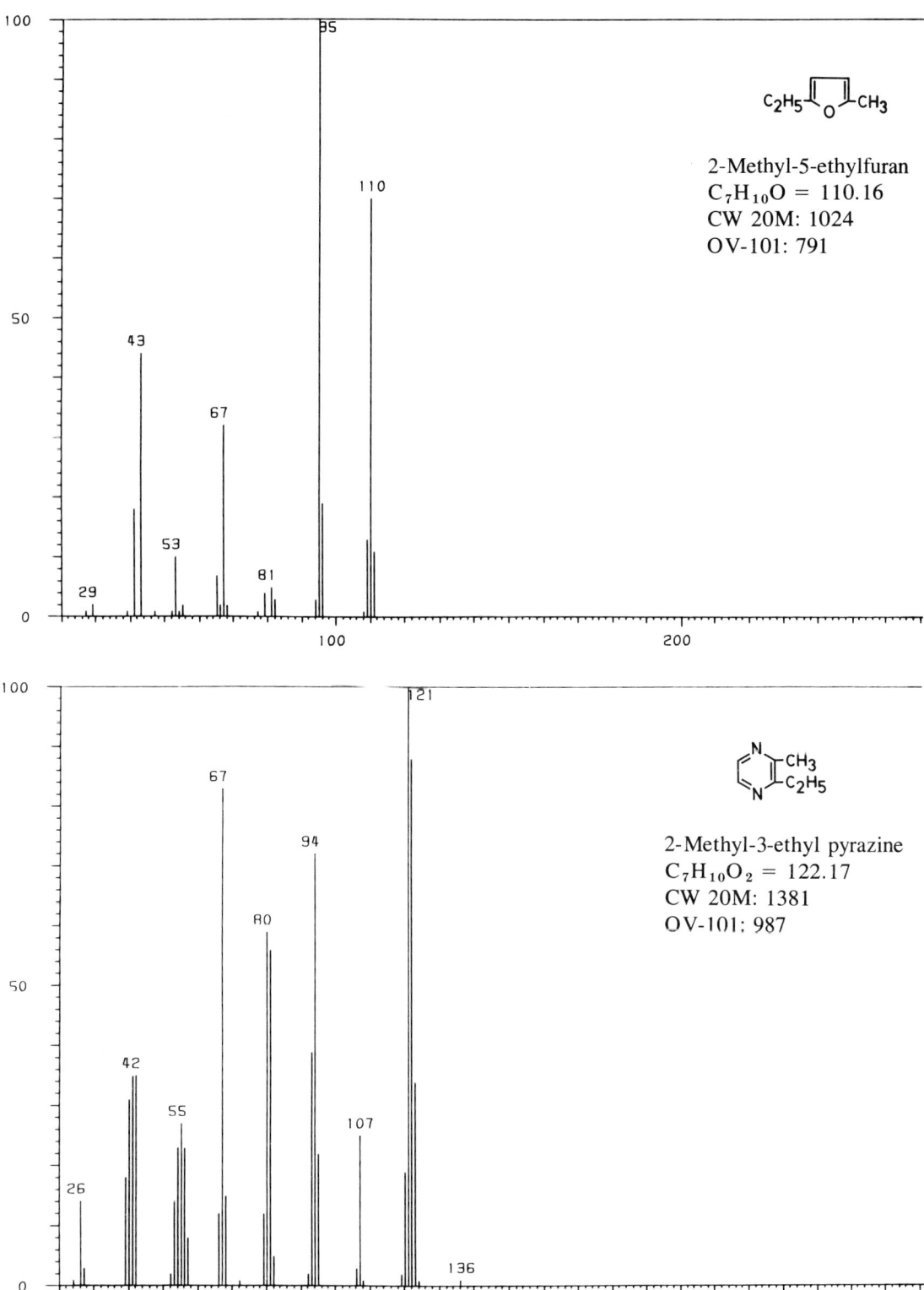

2-Methyl-5-ethylfuran
$C_7H_{10}O = 110.16$
CW 20M: 1024
OV-101: 791

2-Methyl-3-ethyl pyrazine
$C_7H_{10}O_2 = 122.17$
CW 20M: 1381
OV-101: 987

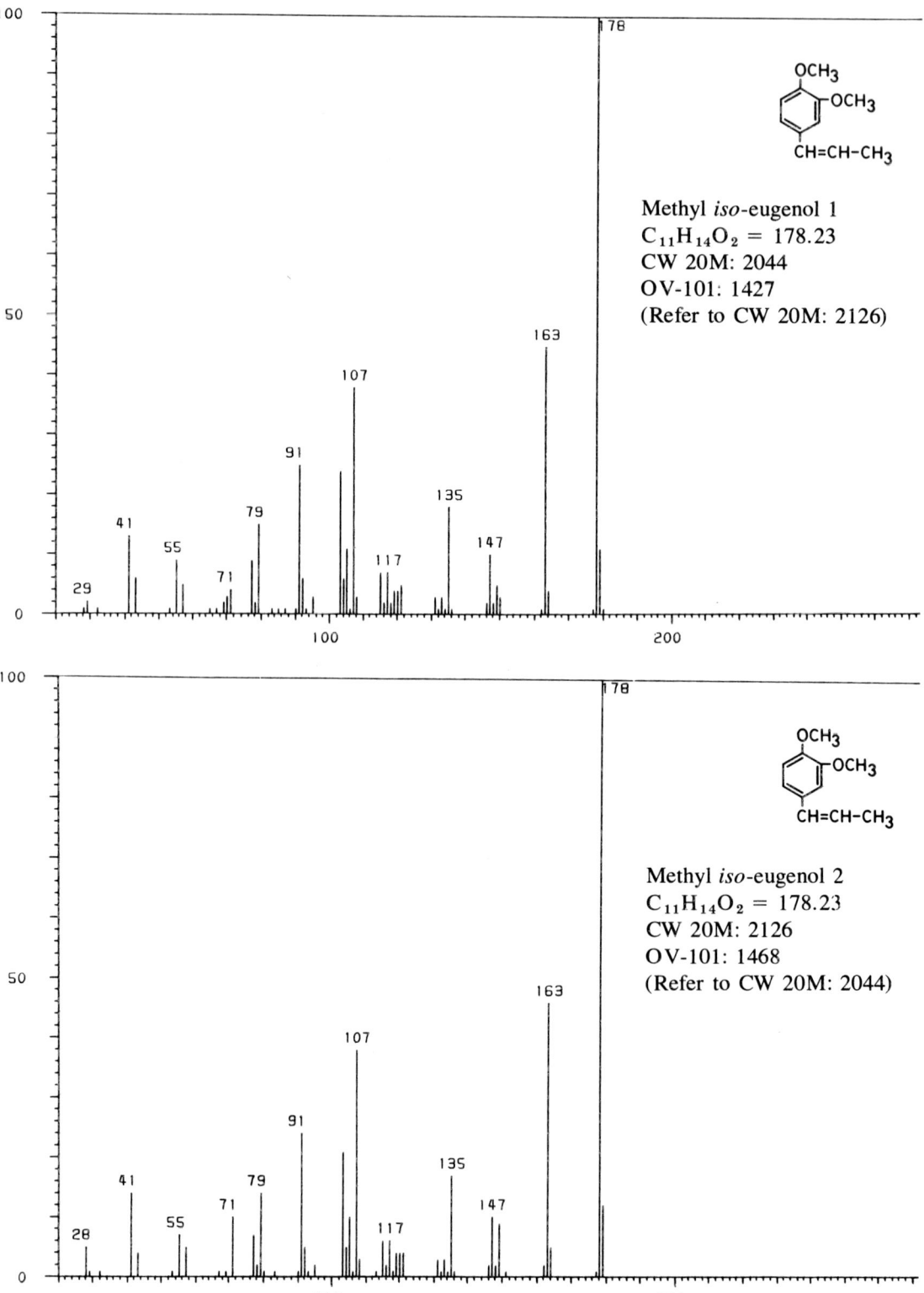

APPENDIX IV MASS SPECTRA OF INDIVIDUAL COMPOUNDS

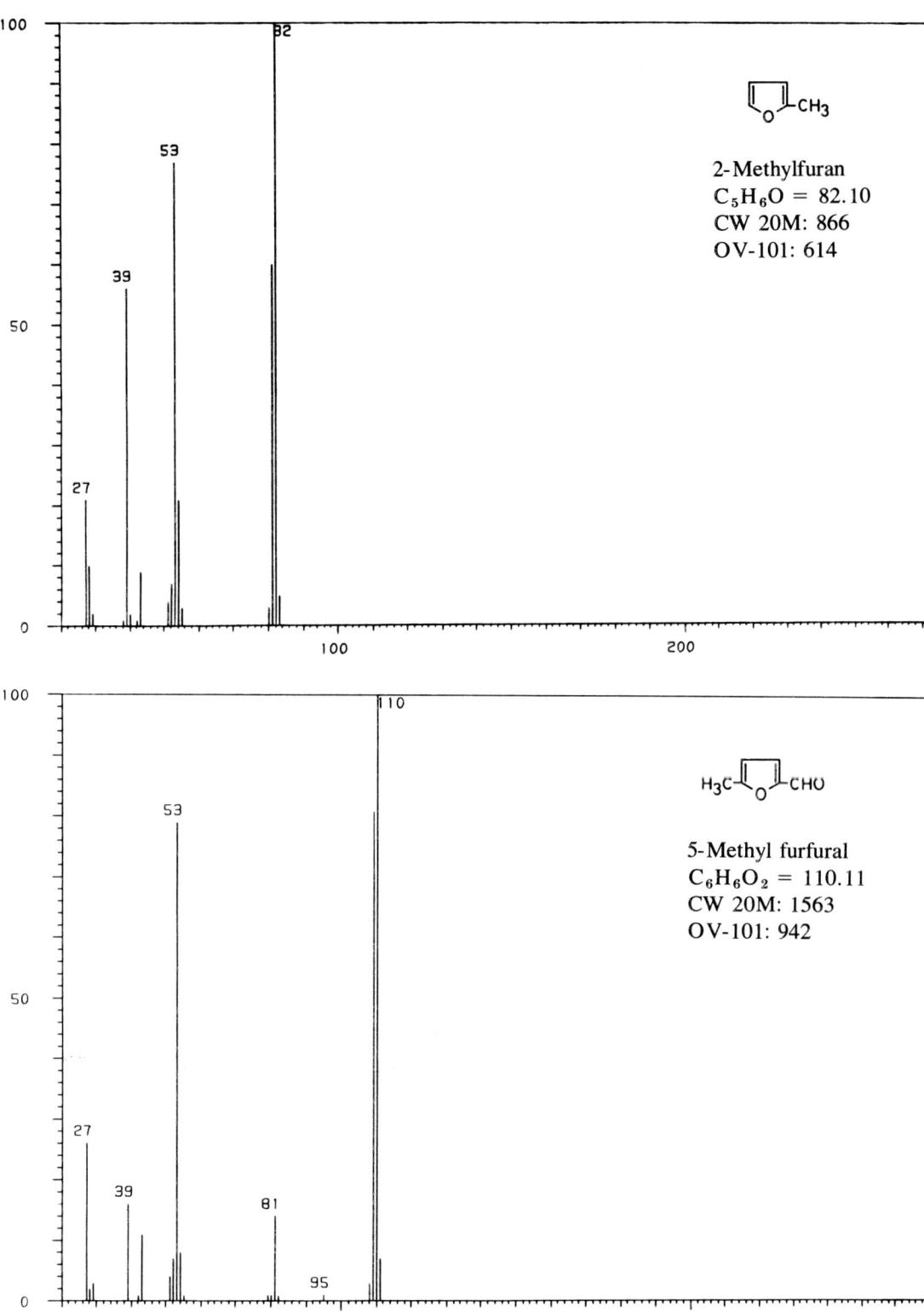

2-Methylfuran
$C_5H_6O = 82.10$
CW 20M: 866
OV-101: 614

5-Methyl furfural
$C_6H_6O_2 = 110.11$
CW 20M: 1563
OV-101: 942

Methyl furoate
$C_6H_6O_3$ = 126.11
CW 20M: 1561
OV-101: 956

Methyl *n*-heptanoate
$C_8H_{16}O_2$ = 144.22
CW 20M: 1276
OV-101: 1006

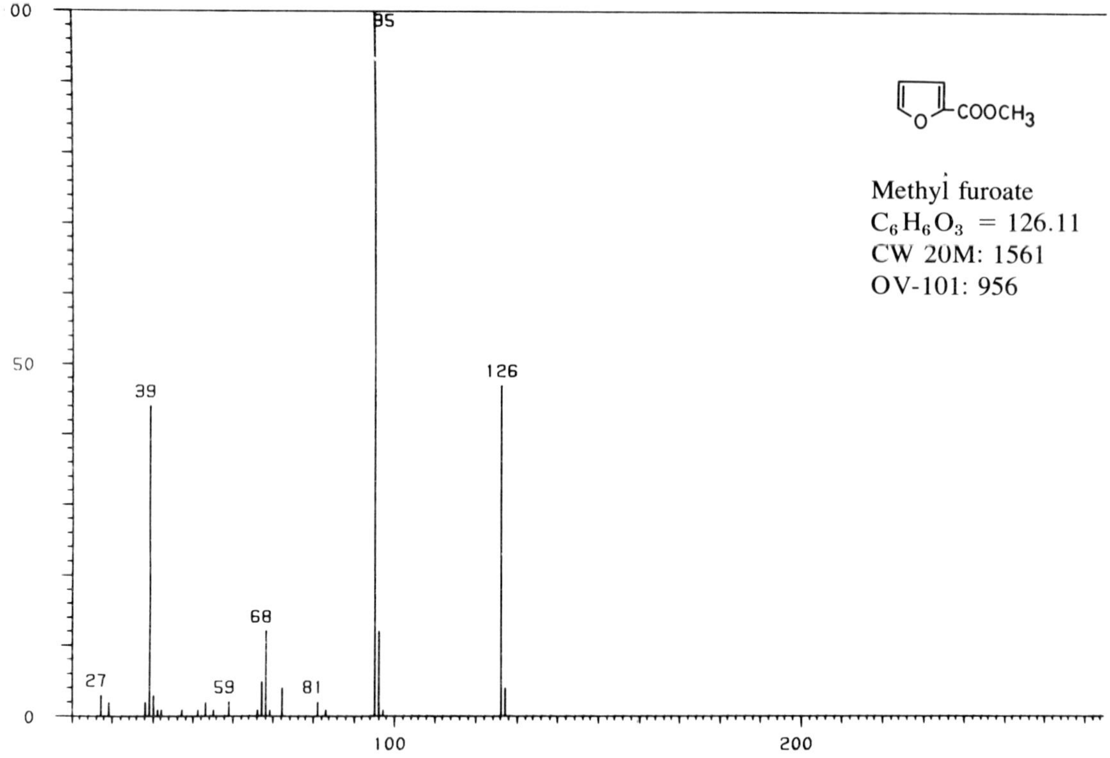

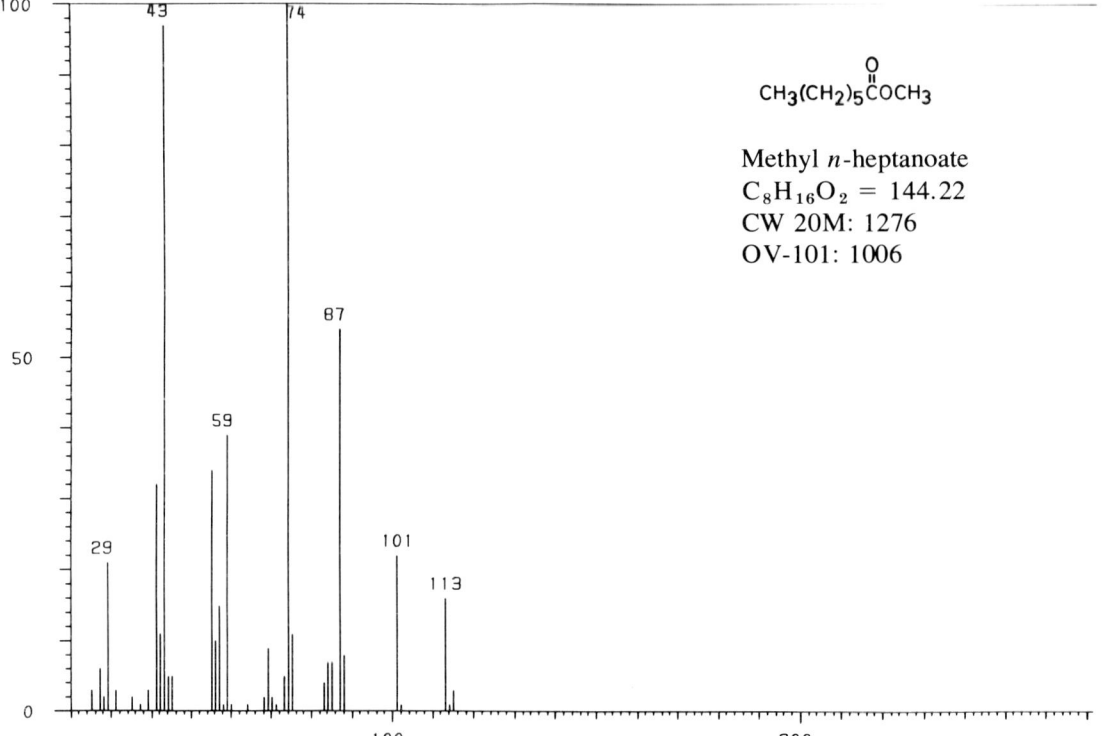

APPENDIX IV   MASS SPECTRA OF INDIVIDUAL COMPOUNDS

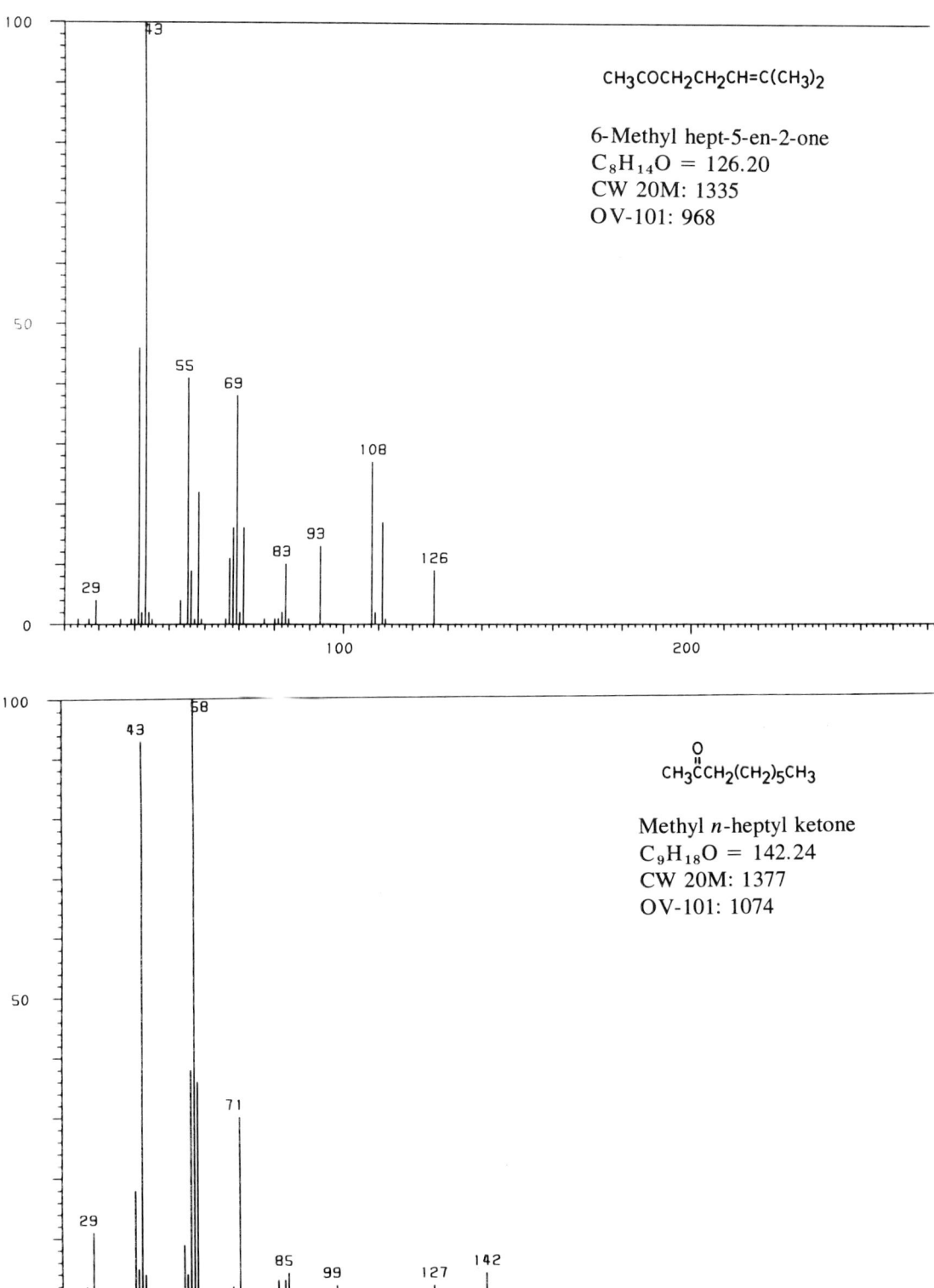

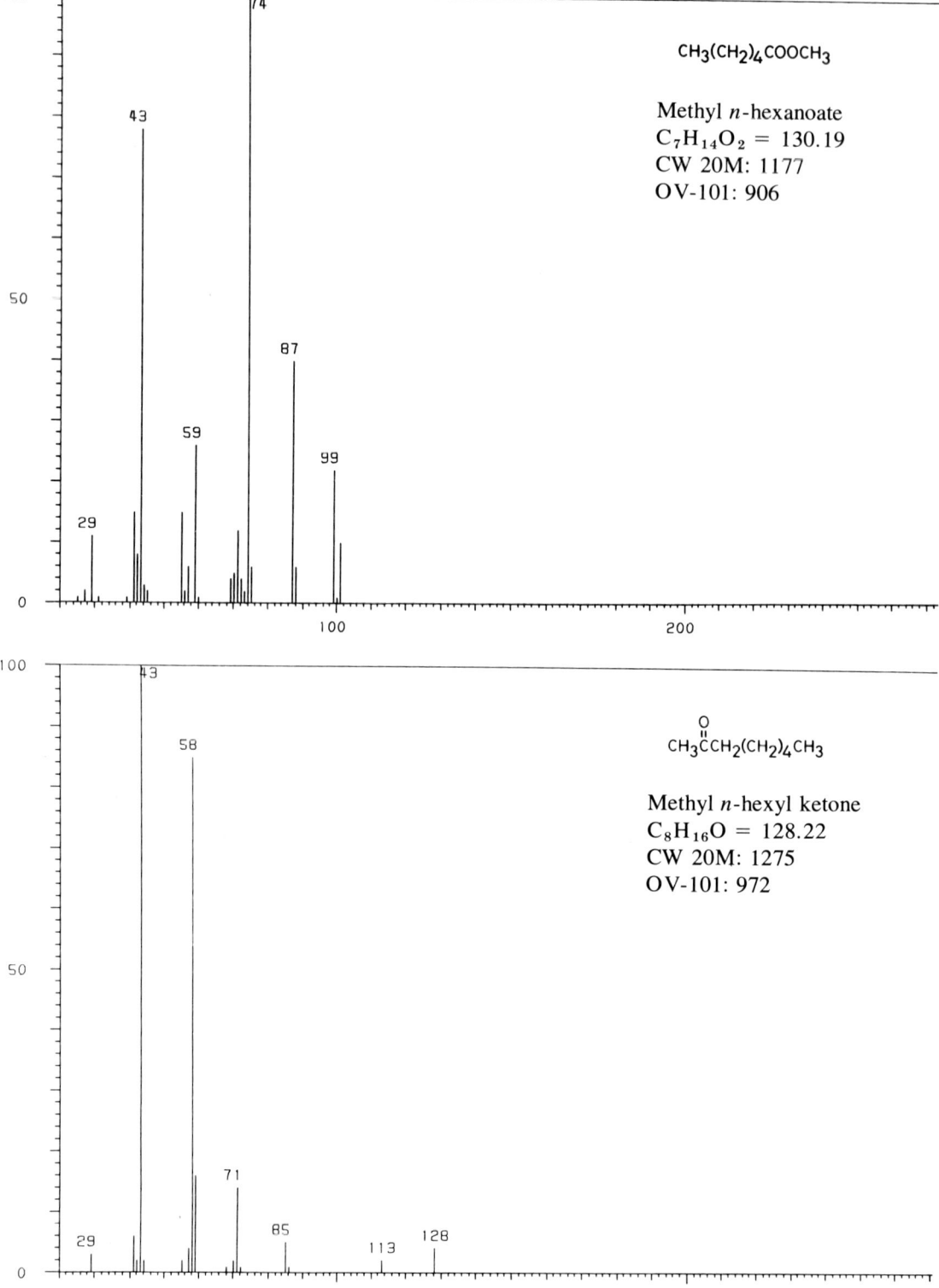

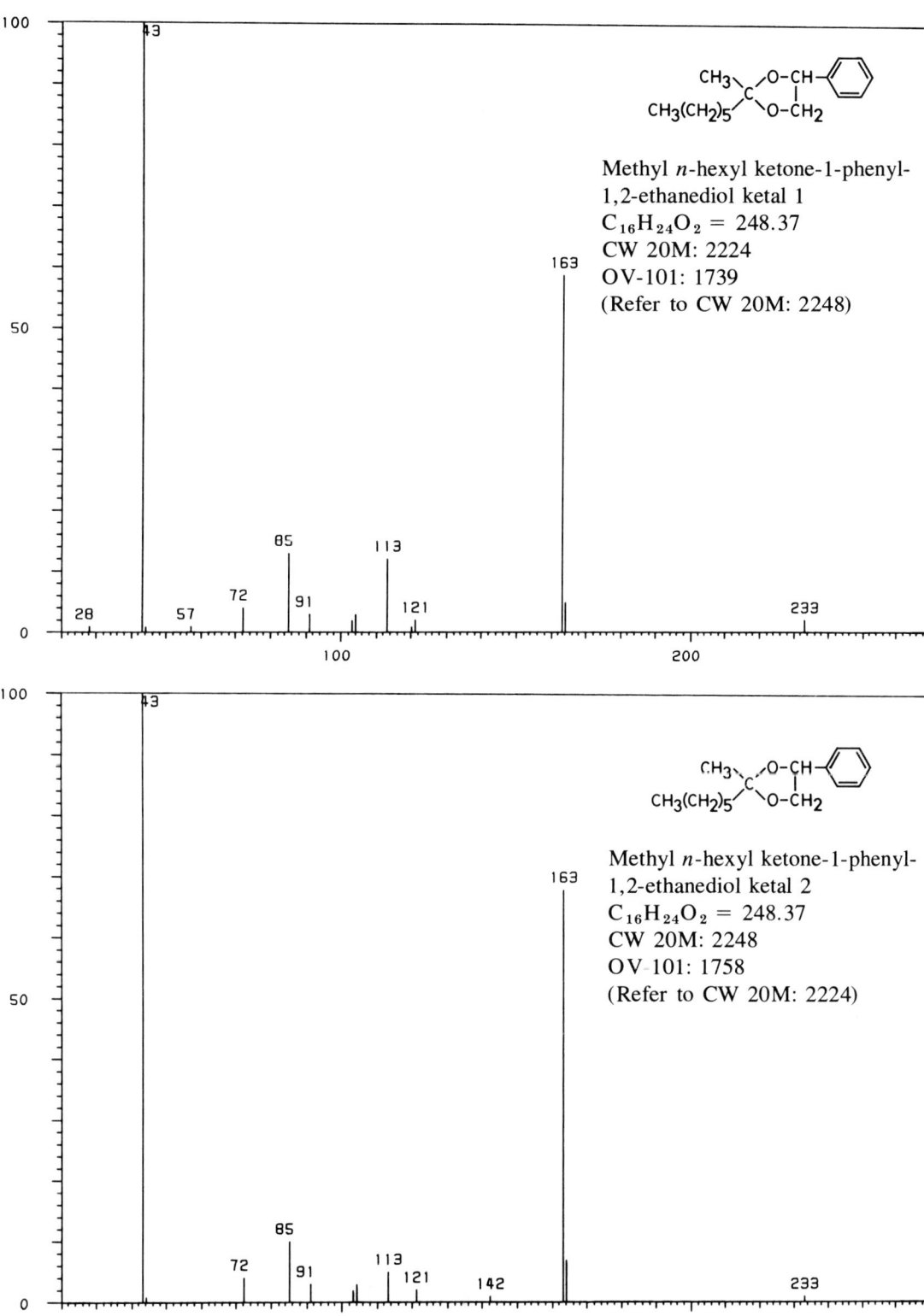

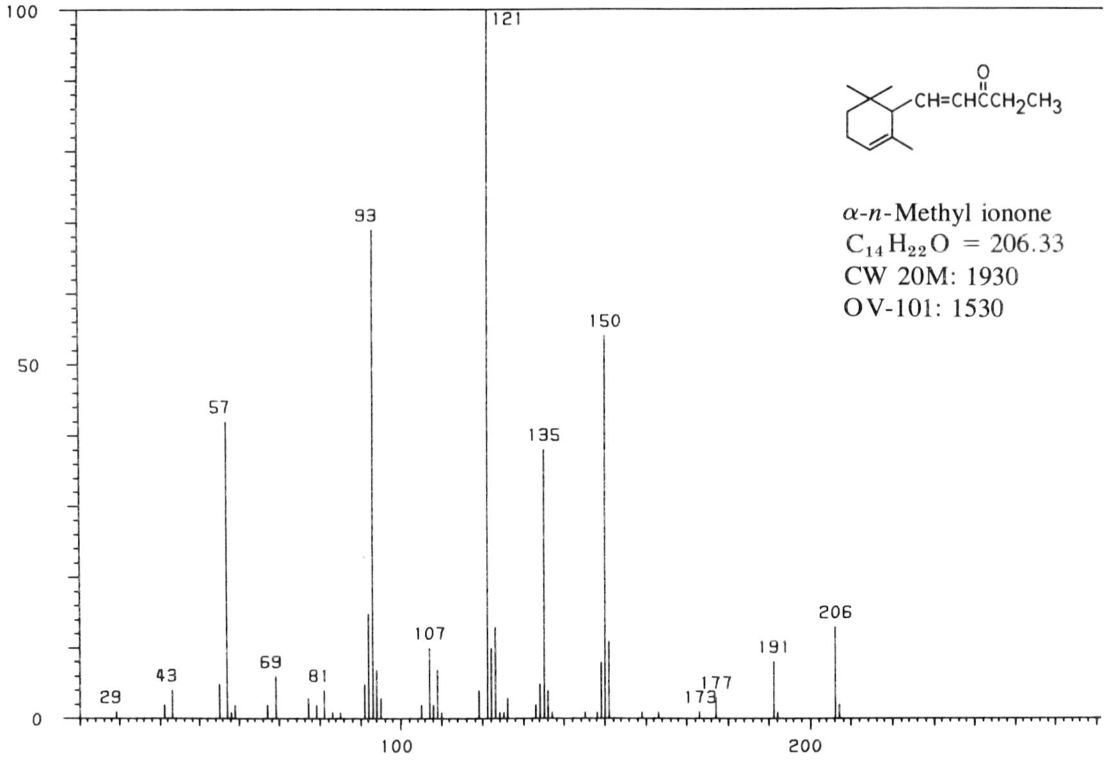

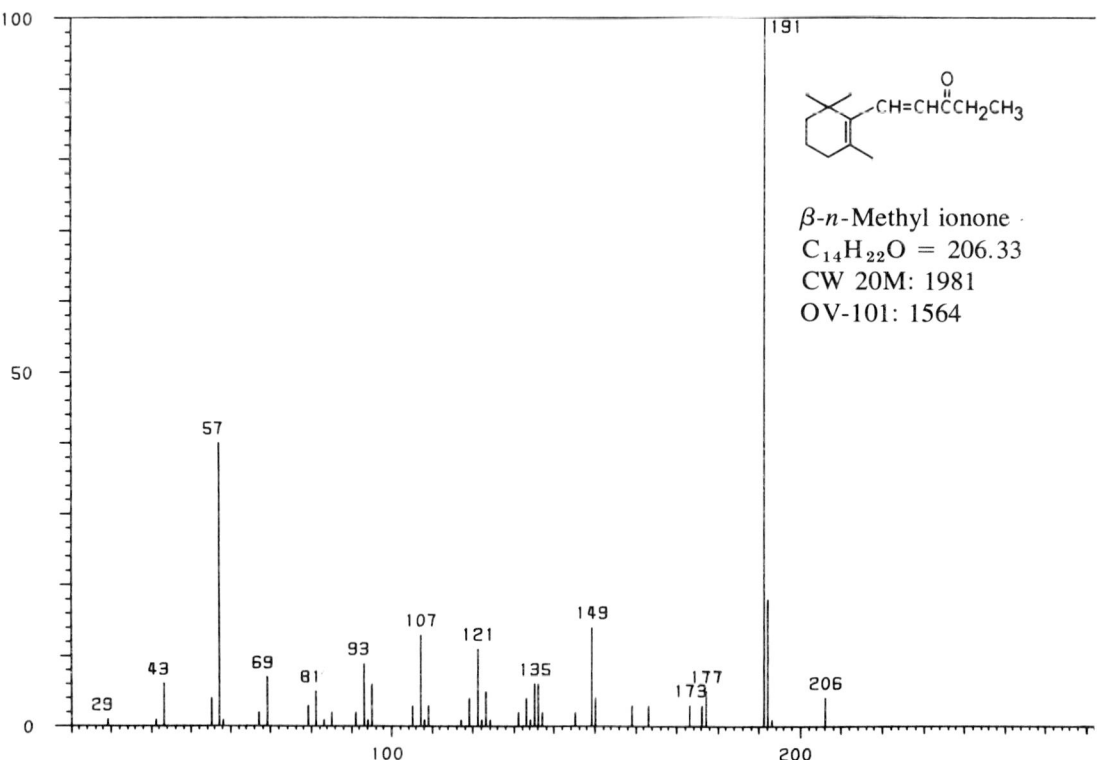

100
50
0
121
93
150
57
135
107
69
81
43
29
177
173
191
206
100
200
CH=CHCCH2CH3
O
α-n-Methyl ionone
C14H22O = 206.33
CW 20M: 1930
OV-101: 1530

100
50
0
191
57
149
107
121
93
69
81
43
29
135
177
173
206
100
200
CH=CHCCH2CH3
O
β-n-Methyl ionone
C14H22O = 206.33
CW 20M: 1981
OV-101: 1564

APPENDIX IV   MASS SPECTRA OF INDIVIDUAL COMPOUNDS

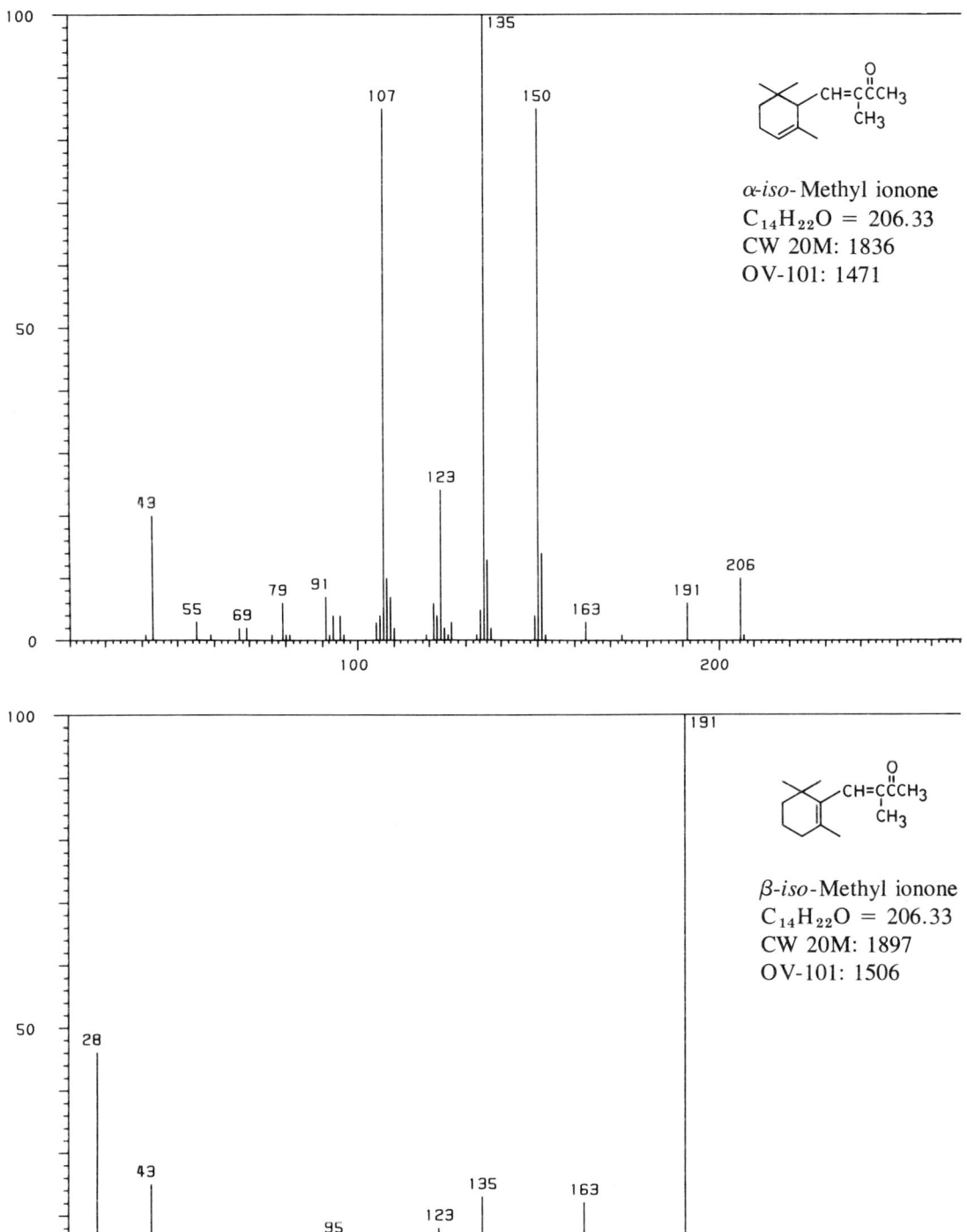

α-iso-Methyl ionone
$C_{14}H_{22}O = 206.33$
CW 20M: 1836
OV-101: 1471

β-iso-Methyl ionone
$C_{14}H_{22}O = 206.33$
CW 20M: 1897
OV-101: 1506

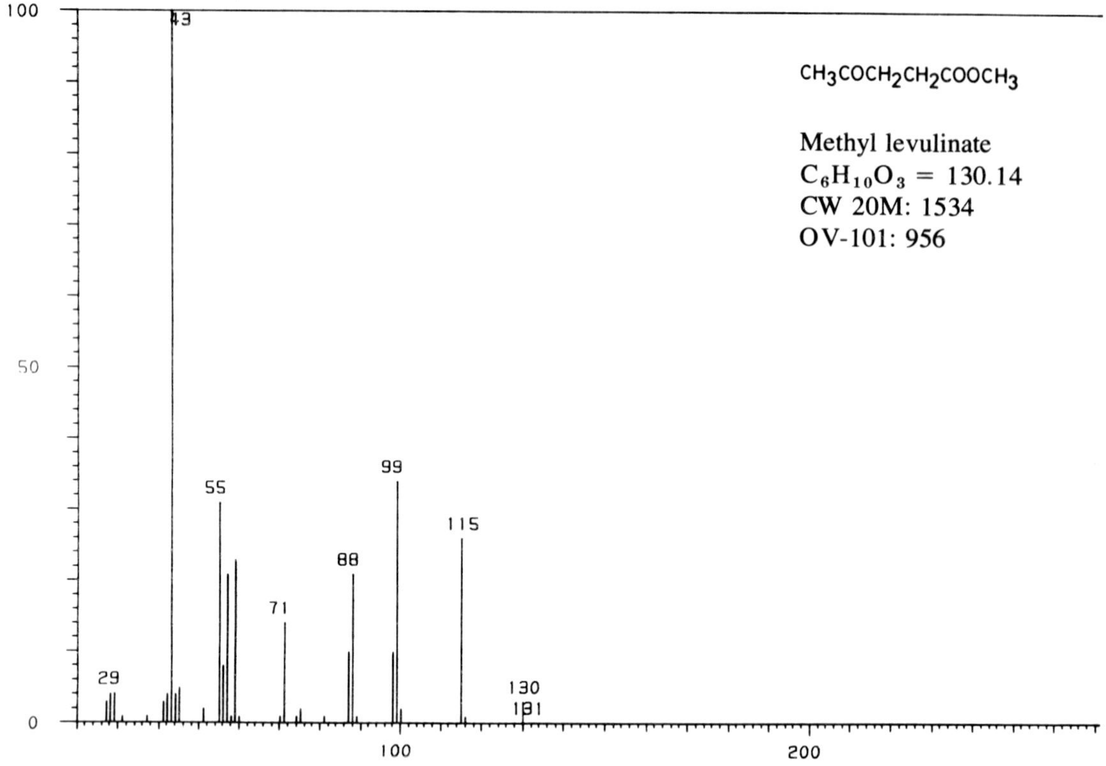

CH3COCH2CH2COOCH3
Methyl levulinate
$C_6H_{10}O_3$ = 130.14
CW 20M: 1534
OV-101: 956
100
50
0
43
29
55
71
88
99
115
130
131
100
200

COOCH3
NHCH3
Methyl *n*-methyl anthranilate
$C_9H_{11}NO_2$ = 165.20
CW 20M: 2042
OV-101: 1389
100
50
0
105
165
132
78
30
59
67
91
130
150
100
200

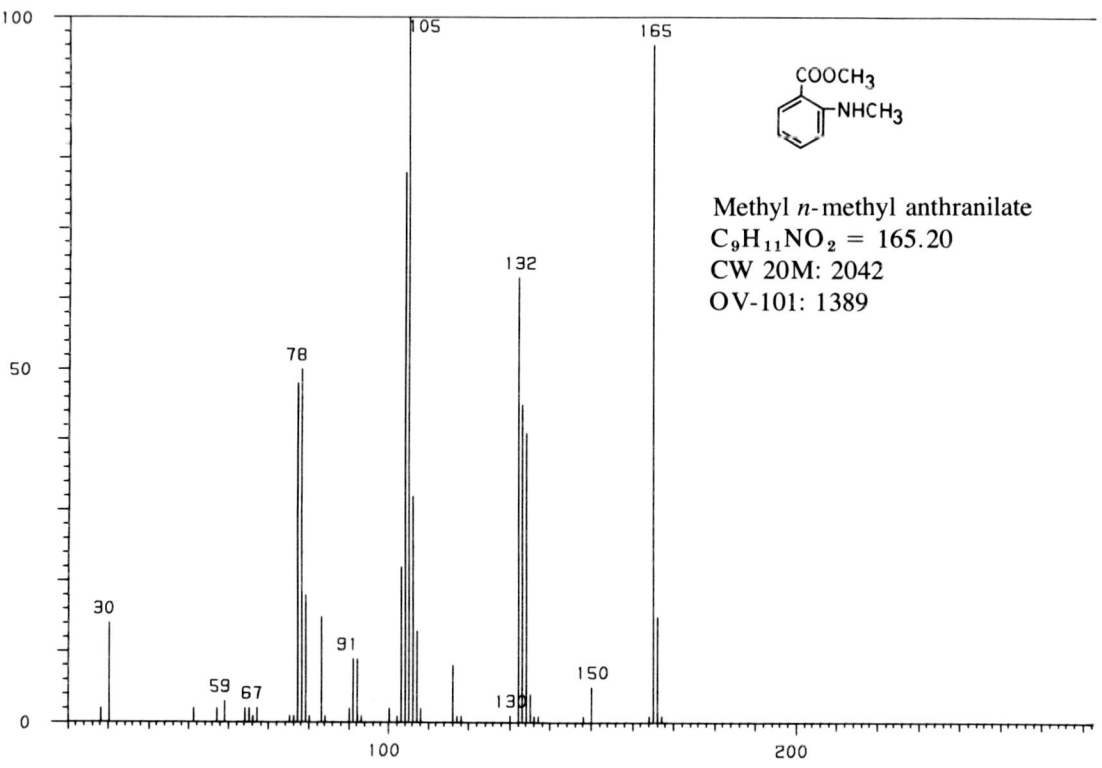

APPENDIX IV   MASS SPECTRA OF INDIVIDUAL COMPOUNDS

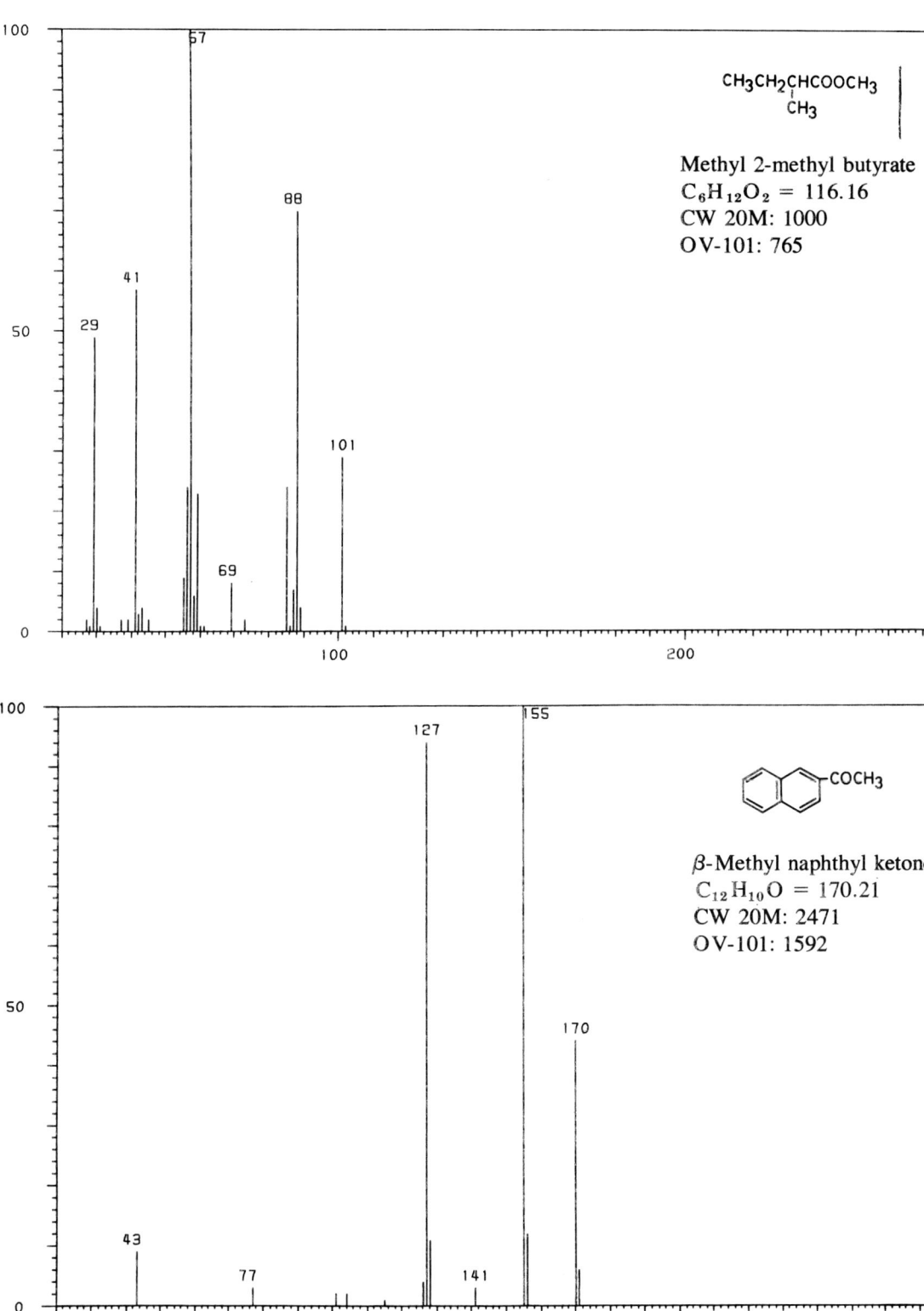

CH₃CH₂CHCOOCH₃
     |
     CH₃

Methyl 2-methyl butyrate
$C_6H_{12}O_2 = 116.16$
CW 20M: 1000
OV-101: 765

β-Methyl naphthyl ketone
$C_{12}H_{10}O = 170.21$
CW 20M: 2471
OV-101: 1592

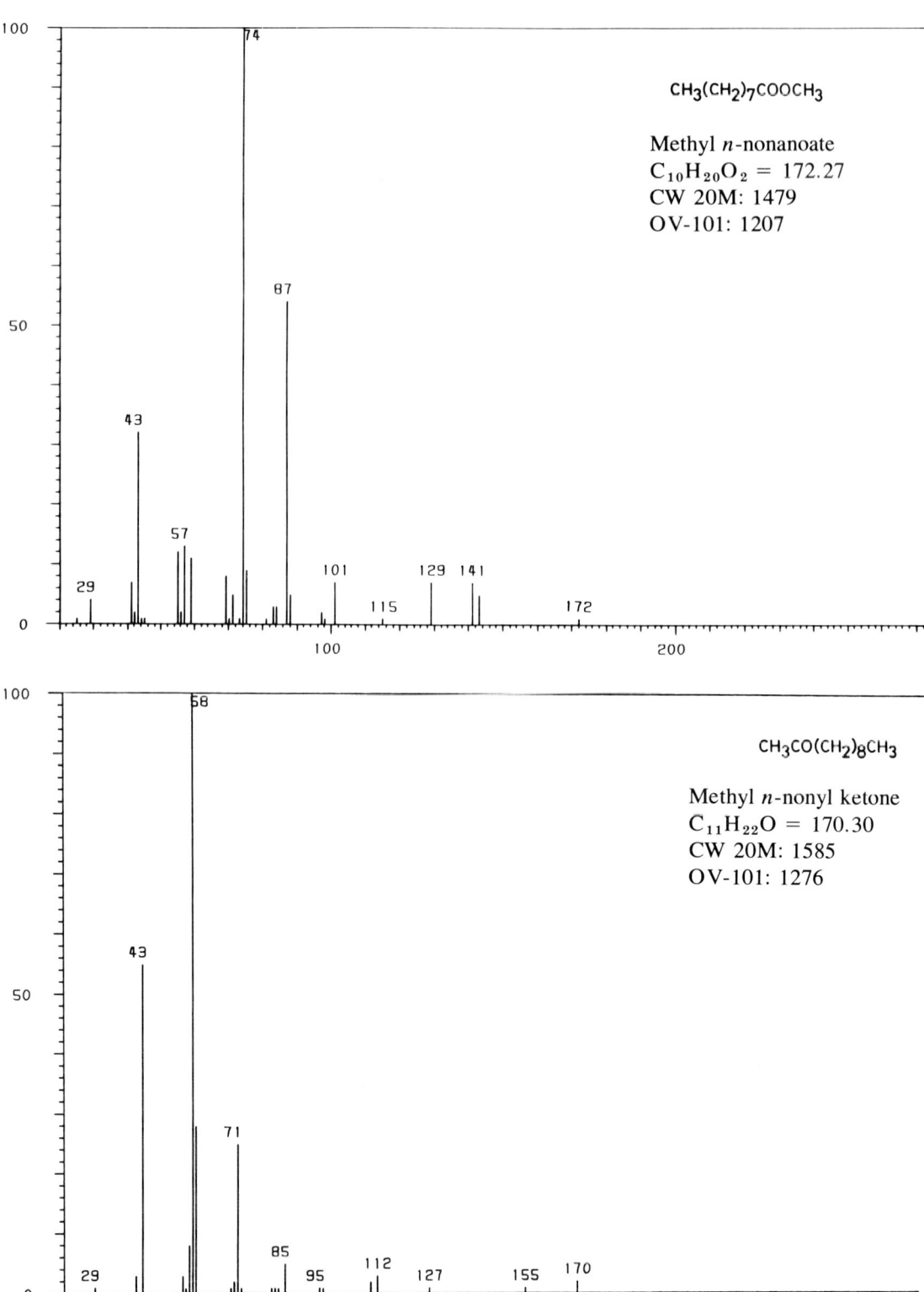

APPENDIX IV   MASS SPECTRA OF INDIVIDUAL COMPOUNDS

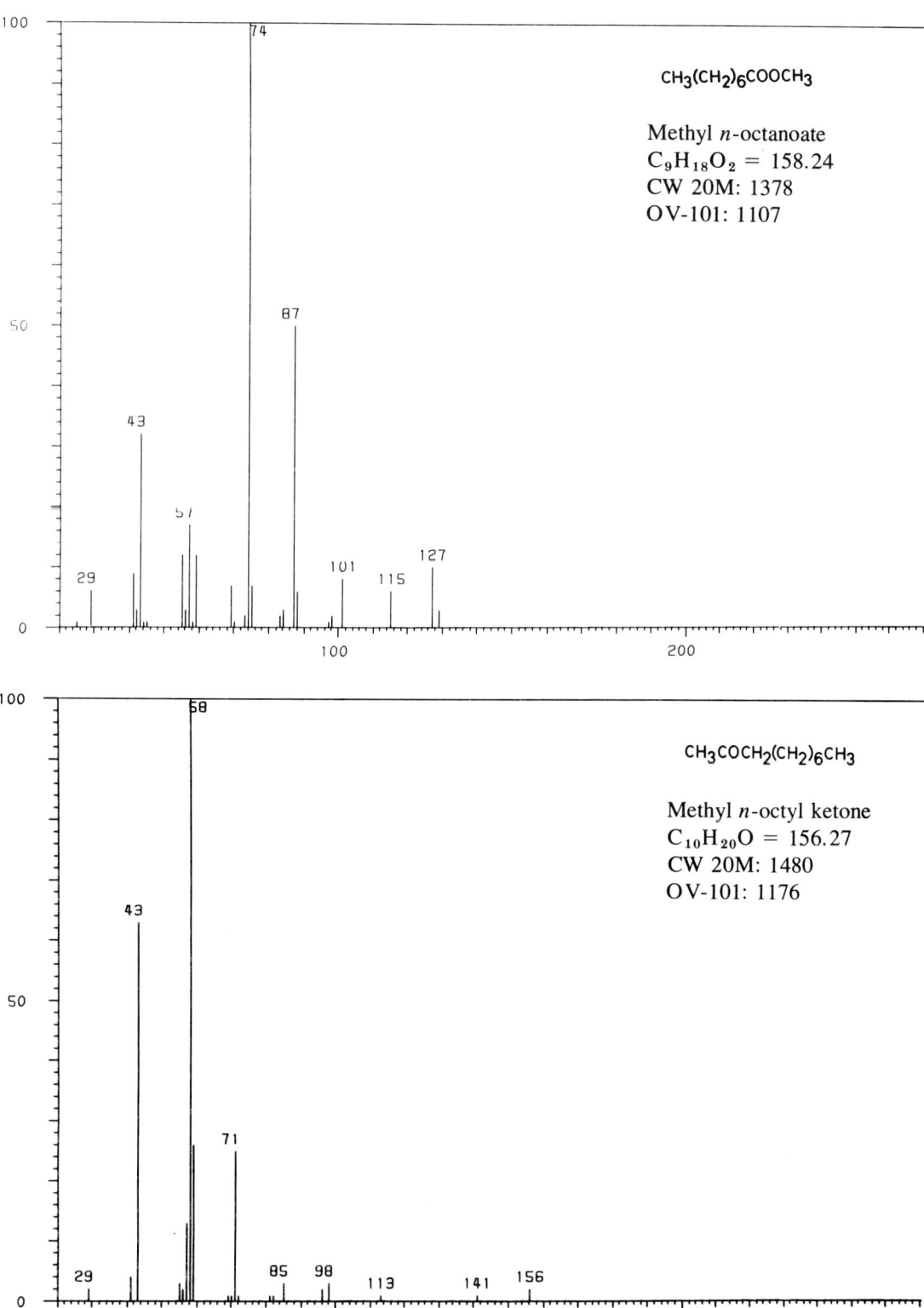

$CH_3(CH_2)_6COOCH_3$

Methyl n-octanoate
$C_9H_{18}O_2 = 158.24$
CW 20M: 1378
OV-101: 1107

$CH_3COCH_2(CH_2)_6CH_3$

Methyl n-octyl ketone
$C_{10}H_{20}O = 156.27$
CW 20M: 1480
OV-101: 1176

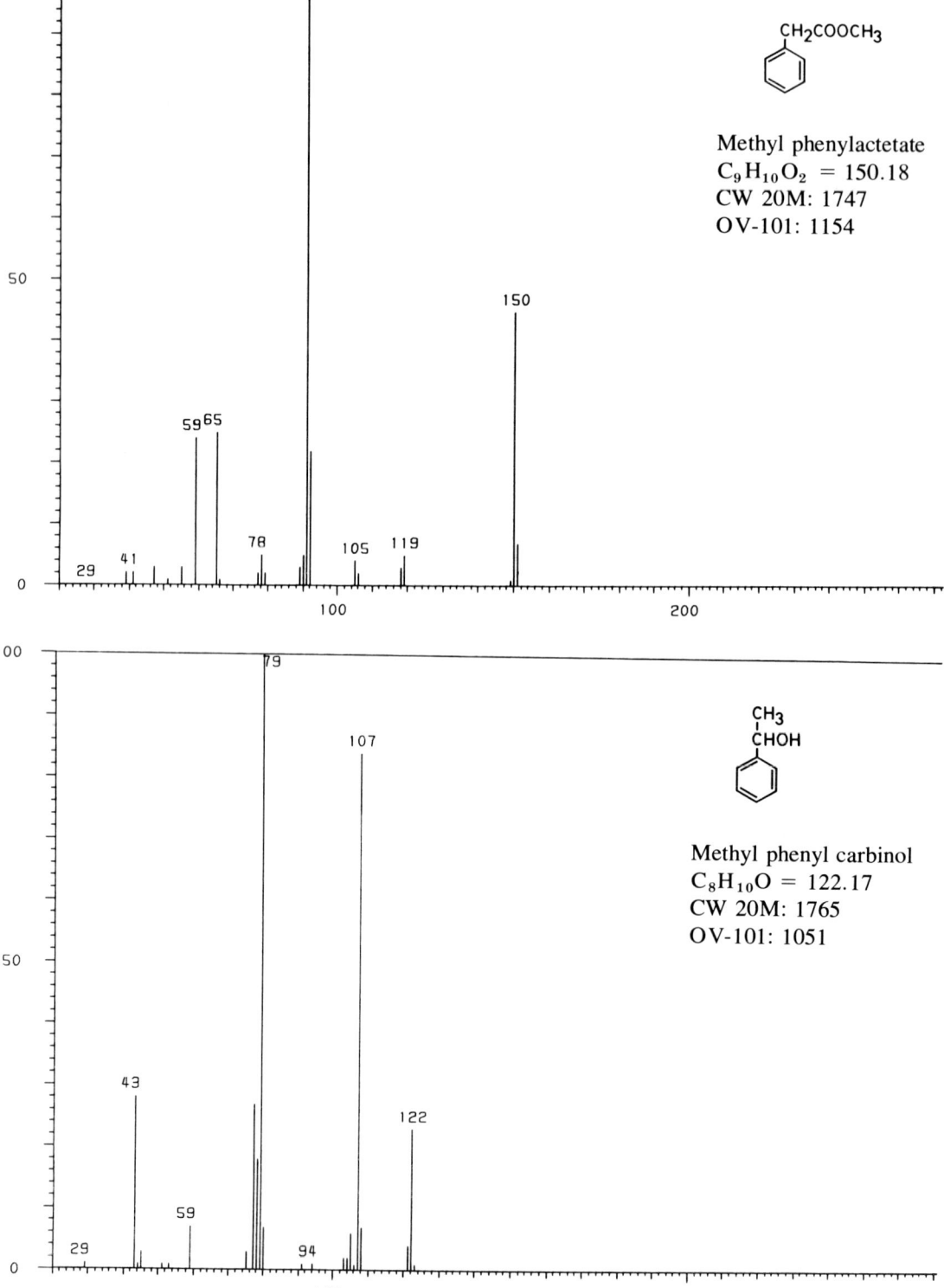

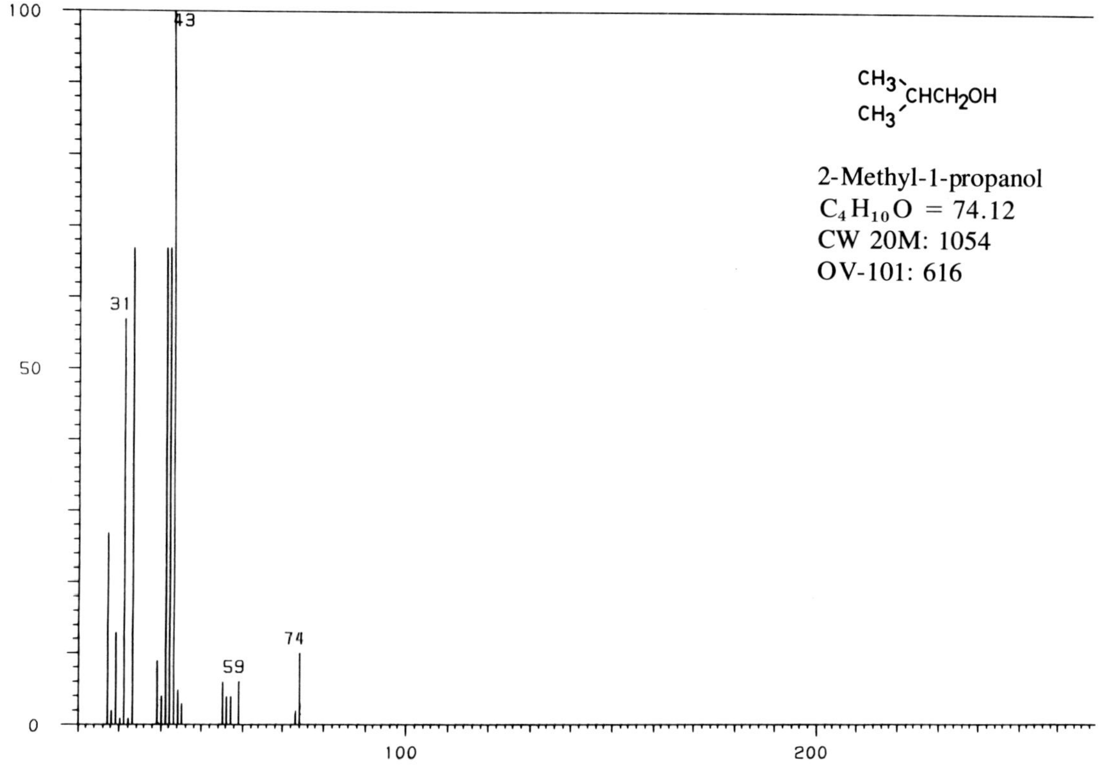

APPENDIX IV   MASS SPECTRA OF INDIVIDUAL COMPOUNDS

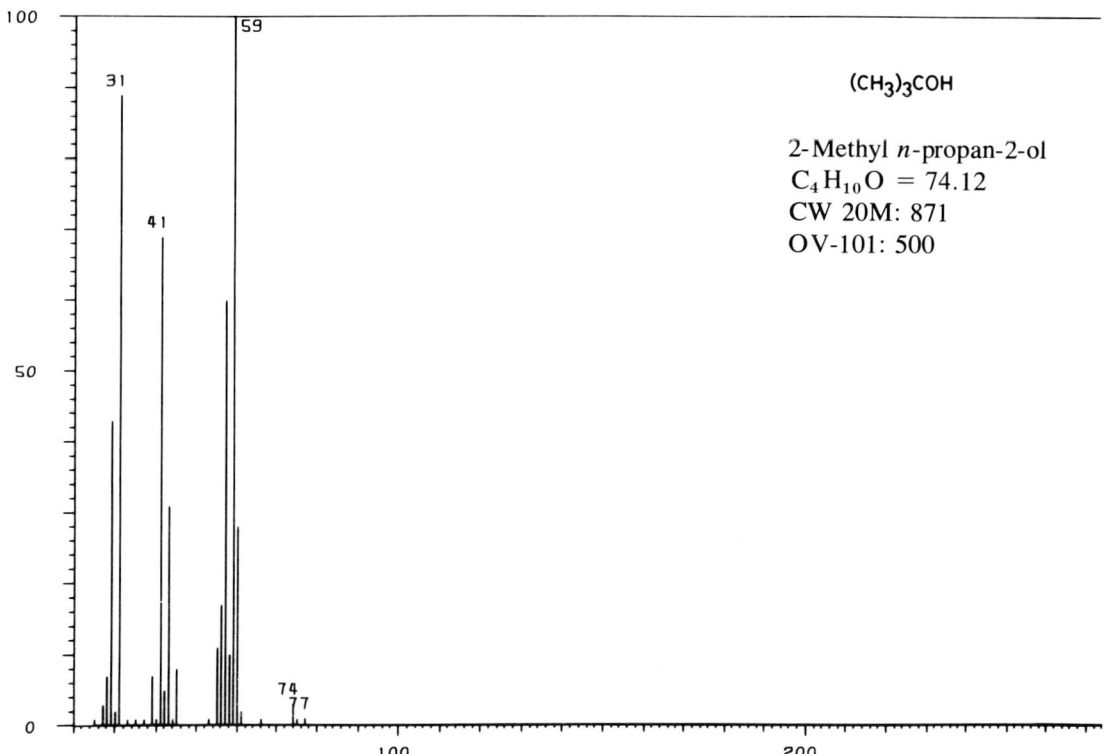

2-Methyl-1-propanol
$C_4H_{10}O = 74.12$
CW 20M: 1054
OV-101: 616

2-Methyl n-propan-2-ol
$C_4H_{10}O = 74.12$
CW 20M: 871
OV-101: 500

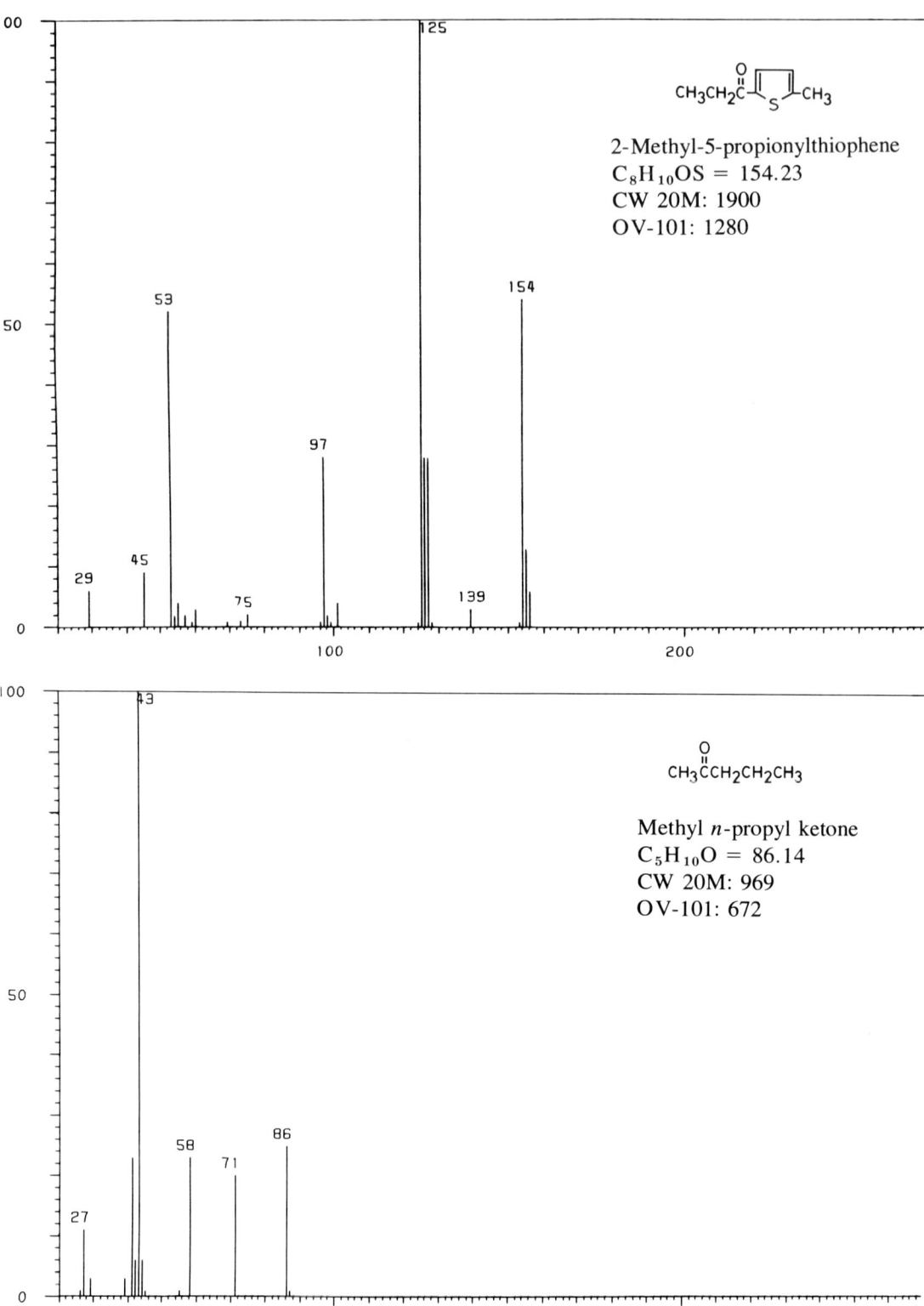

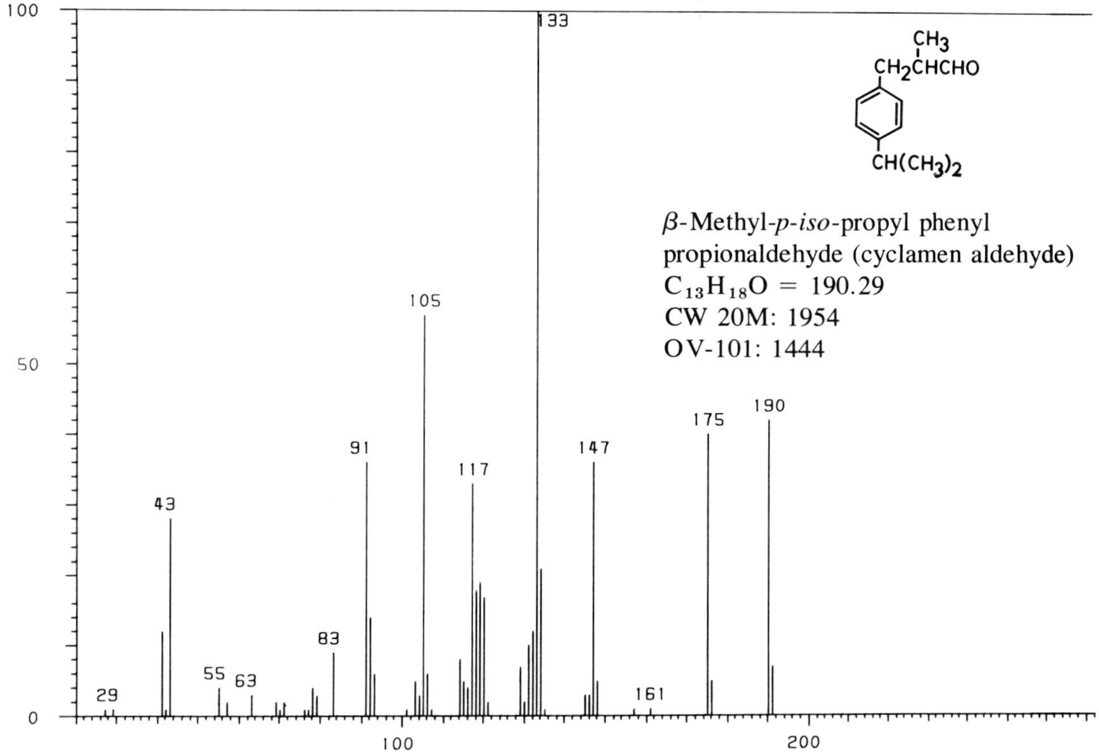

APPENDIX IV   MASS SPECTRA OF INDIVIDUAL COMPOUNDS

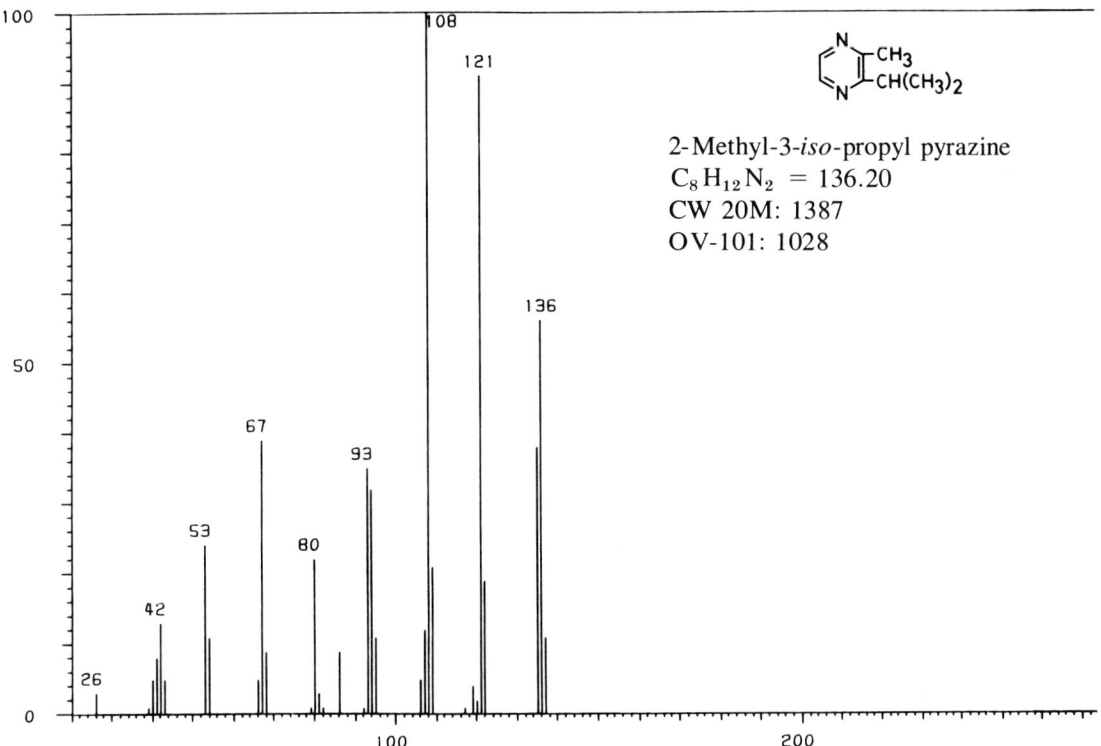

β-Methyl-*p*-*iso*-propyl phenyl propionaldehyde (cyclamen aldehyde)
$C_{13}H_{18}O = 190.29$
CW 20M: 1954
OV-101: 1444

2-Methyl-3-*iso*-propyl pyrazine
$C_8H_{12}N_2 = 136.20$
CW 20M: 1387
OV-101: 1028

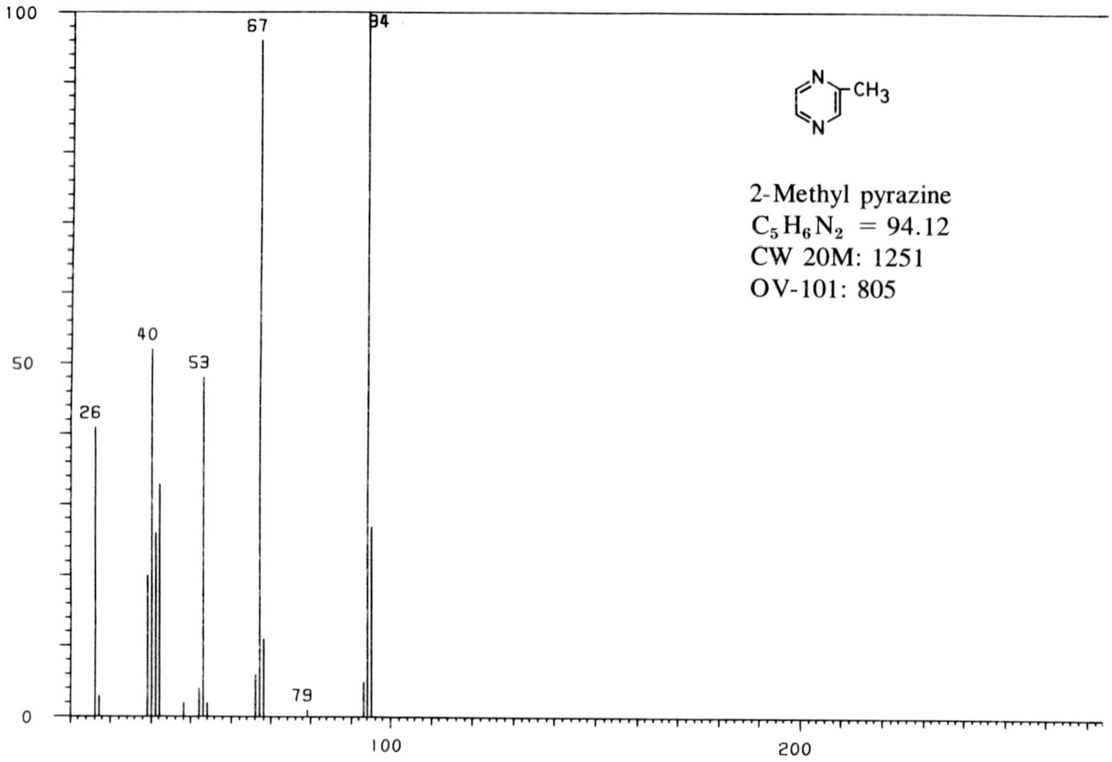

2-Methyl pyrazine
$C_5H_6N_2 = 94.12$
CW 20M: 1251
OV-101: 805

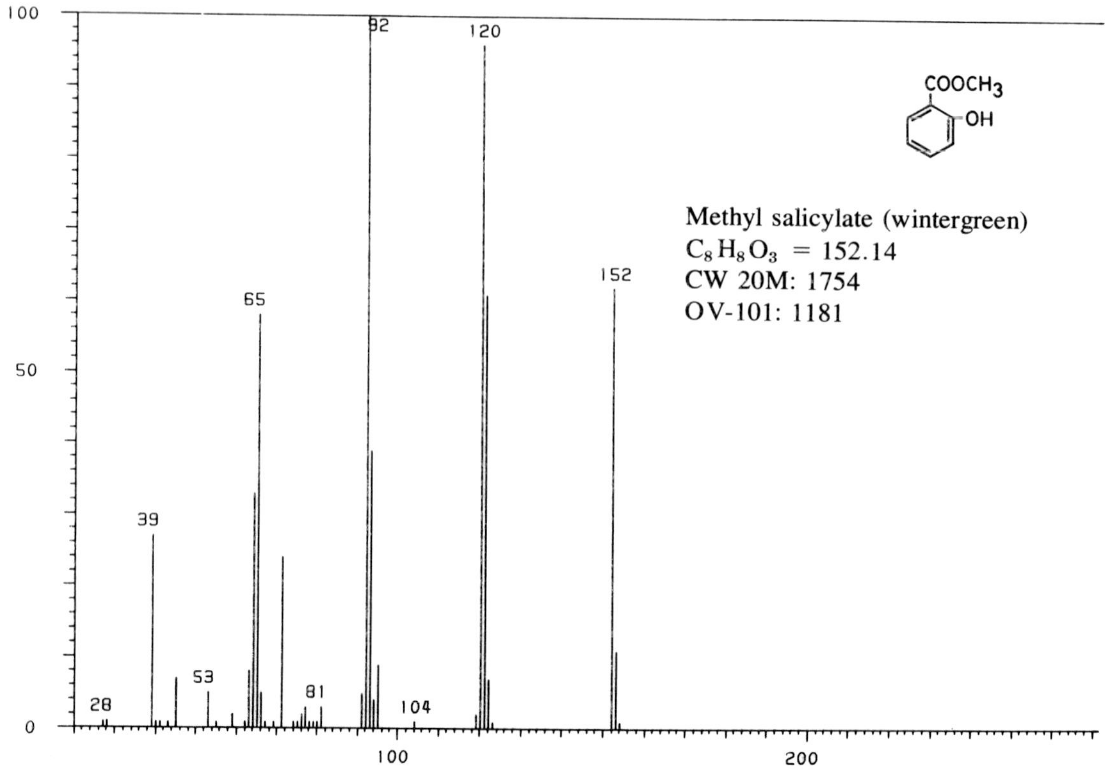

Methyl salicylate (wintergreen)
$C_8H_8O_3 = 152.14$
CW 20M: 1754
OV-101: 1181

APPENDIX IV MASS SPECTRA OF INDIVIDUAL COMPOUNDS

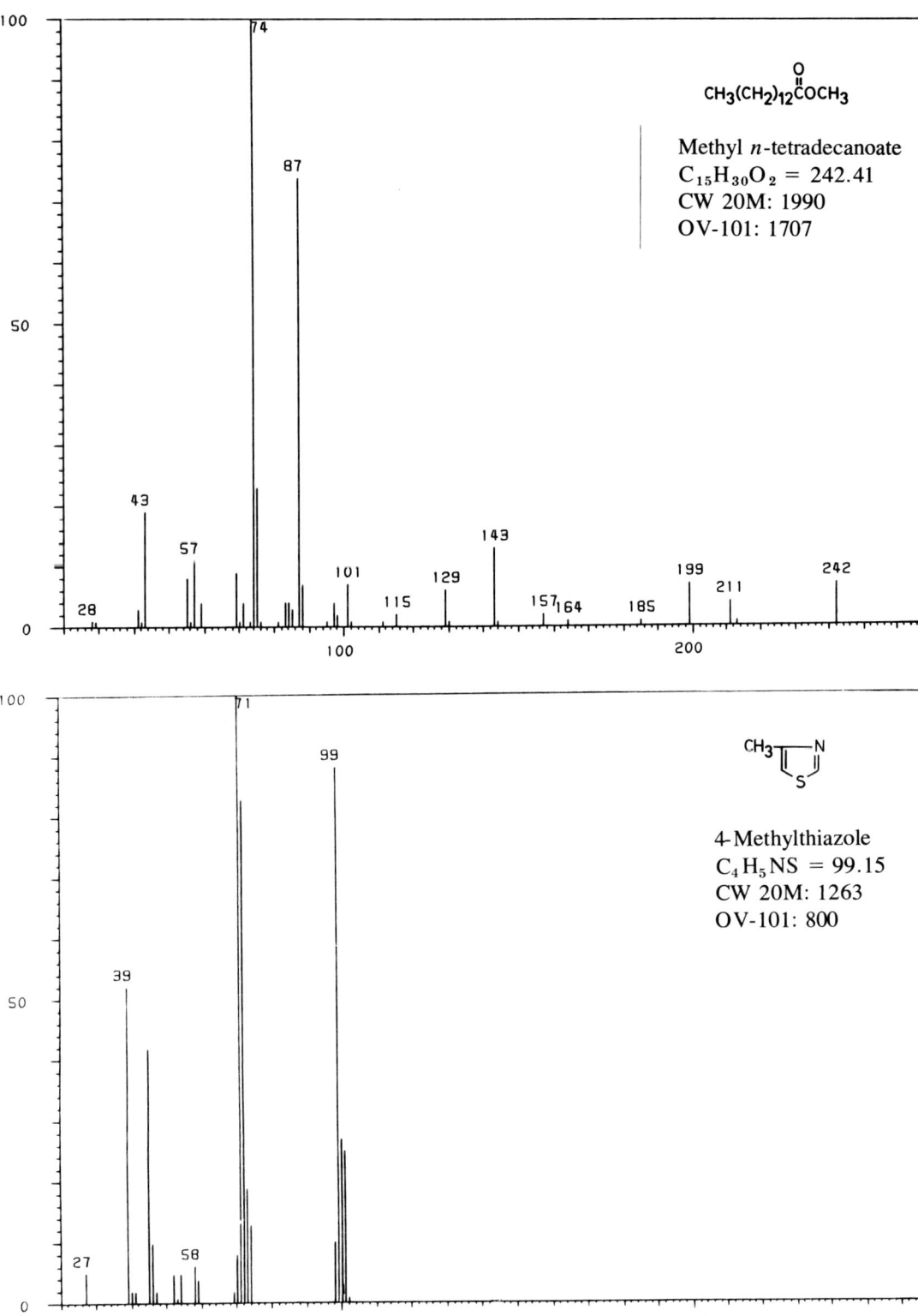

CH₃(CH₂)₁₂COCH₃

Methyl *n*-tetradecanoate
$C_{15}H_{30}O_2 = 242.41$
CW 20M: 1990
OV-101: 1707

4-Methylthiazole
$C_4H_5NS = 99.15$
CW 20M: 1263
OV-101: 800

4-methyl-5-thiazole ethanol
C_6H_9NOS = 143.20
CW 20M: 2216
OV-101: 1283

Methyl *p*-toluate
$C_9H_{10}O_2$ = 150.18
CW 20M: 1725
OV-101: 1194

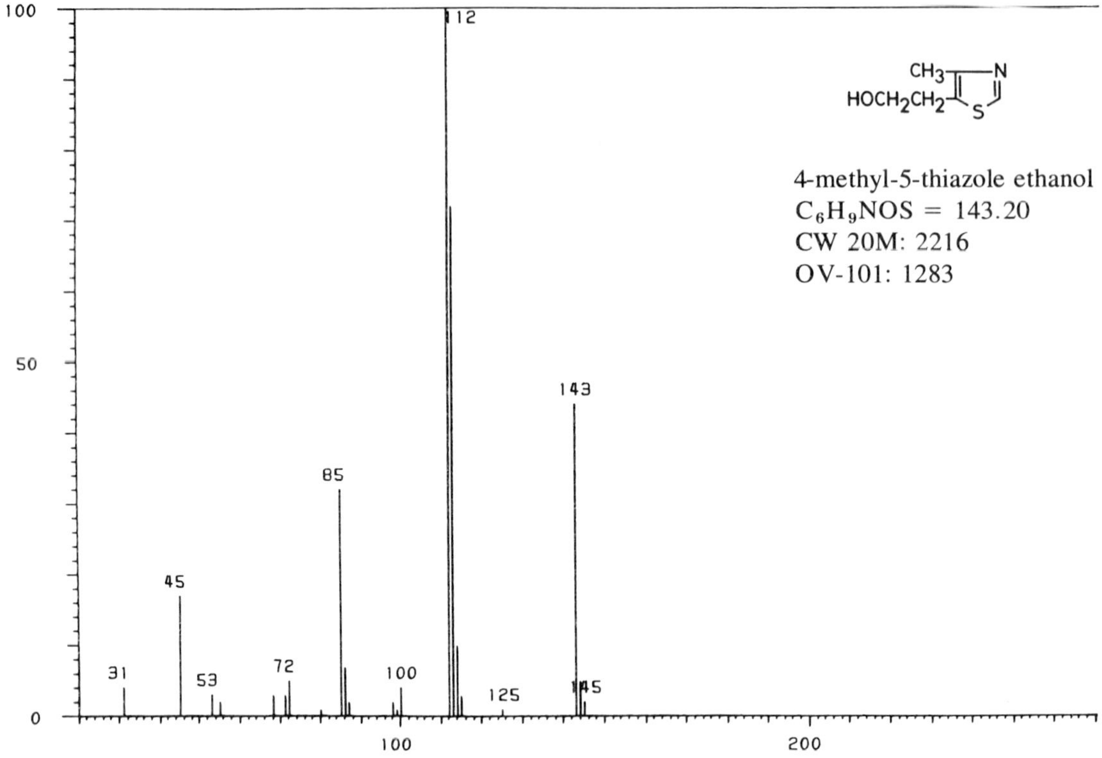

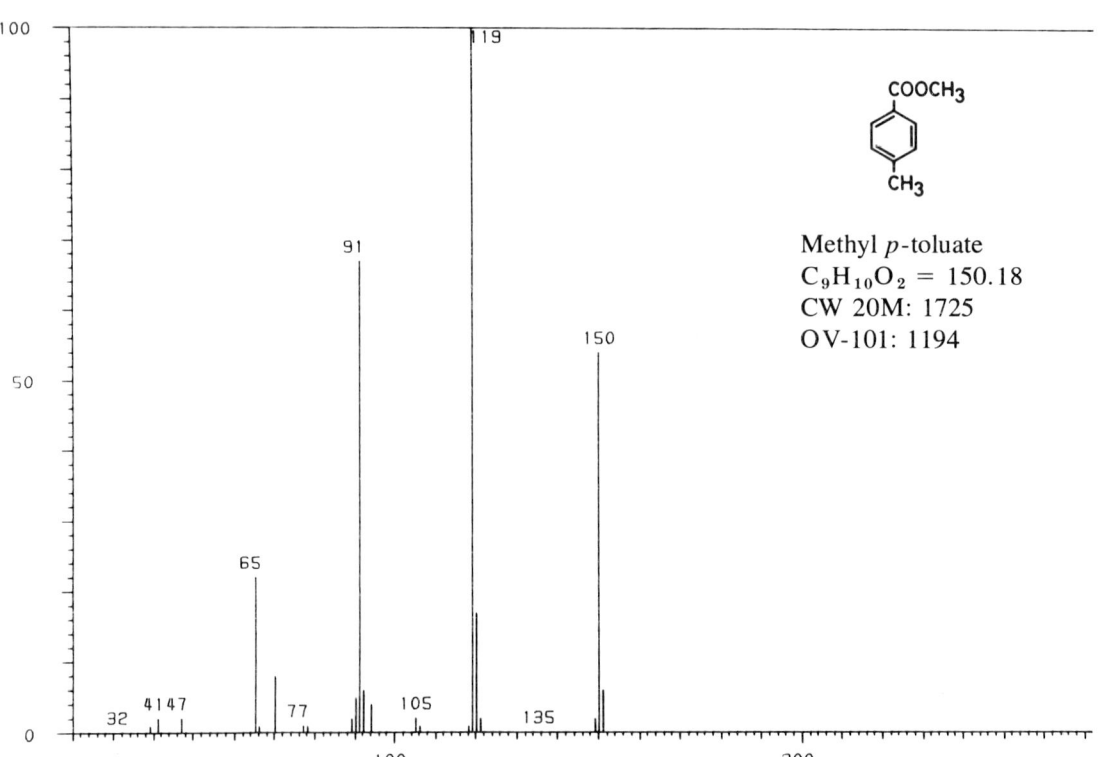

APPENDIX IV   MASS SPECTRA OF INDIVIDUAL COMPOUNDS

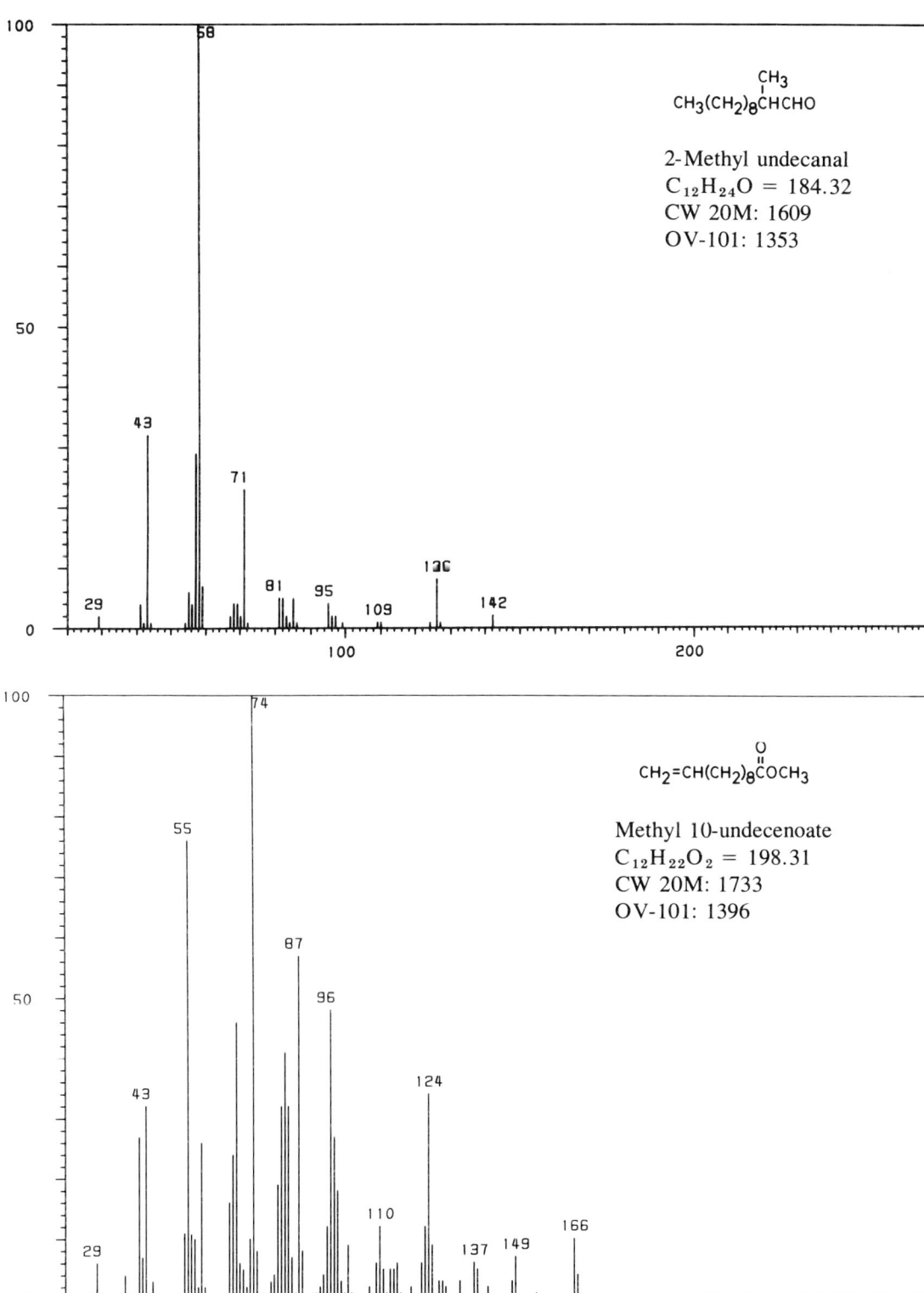

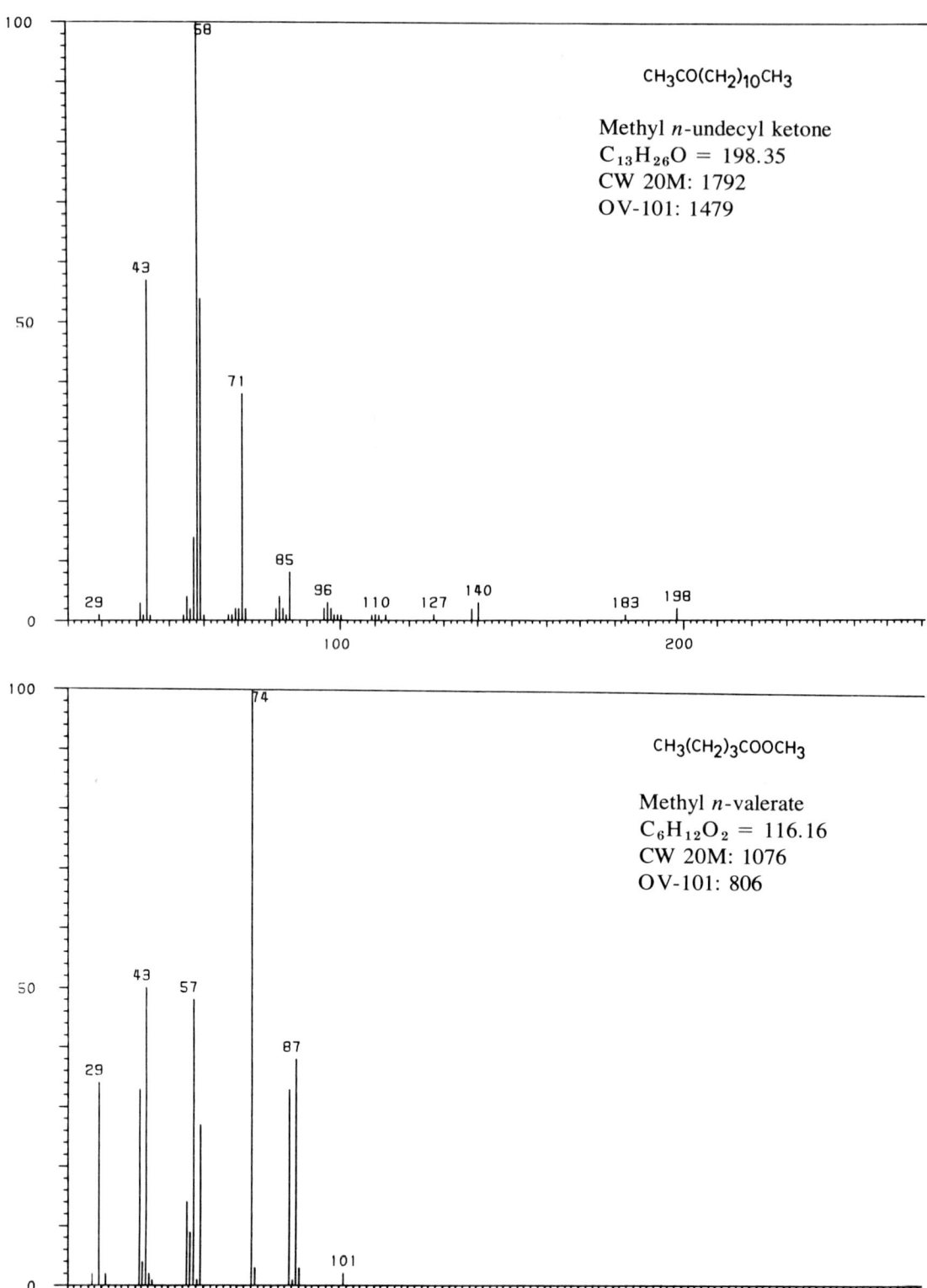

APPENDIX IV   MASS SPECTRA OF INDIVIDUAL COMPOUNDS   385

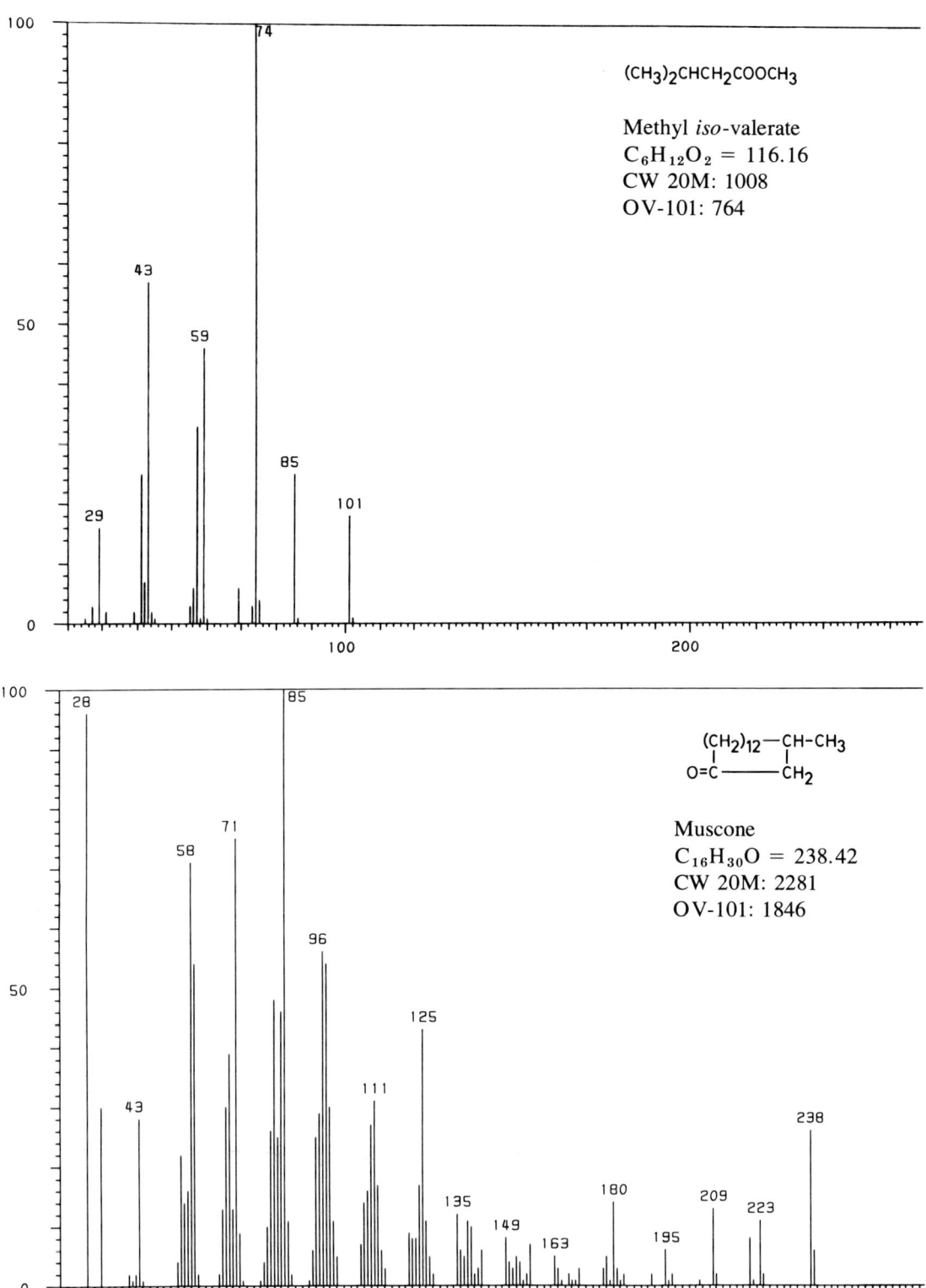

$(CH_3)_2CHCH_2COOCH_3$

Methyl *iso*-valerate
$C_6H_{12}O_2 = 116.16$
CW 20M: 1008
OV-101: 764

Muscone
$C_{16}H_{30}O = 238.42$
CW 20M: 2281
OV-101: 1846

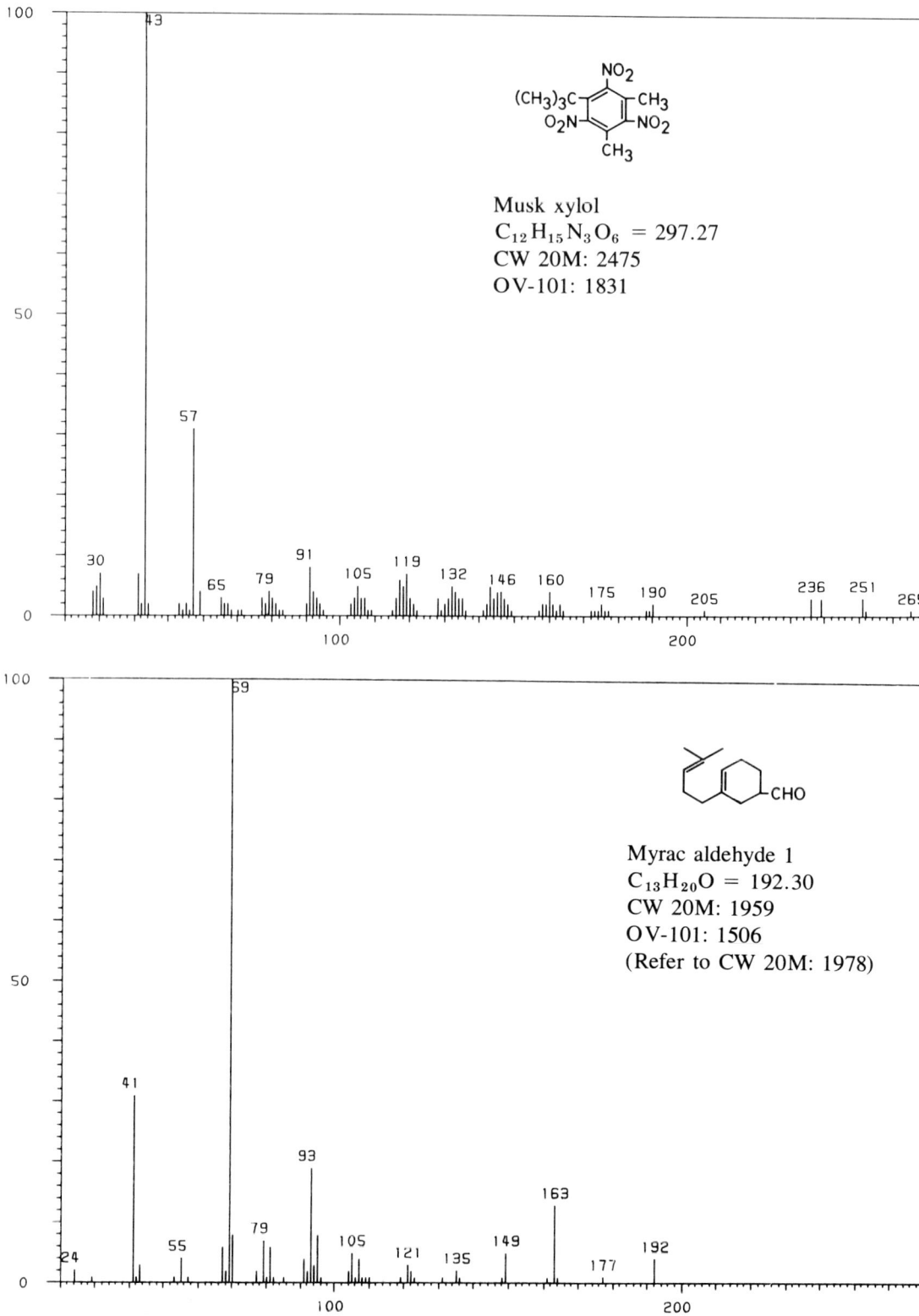

APPENDIX IV   MASS SPECTRA OF INDIVIDUAL COMPOUNDS

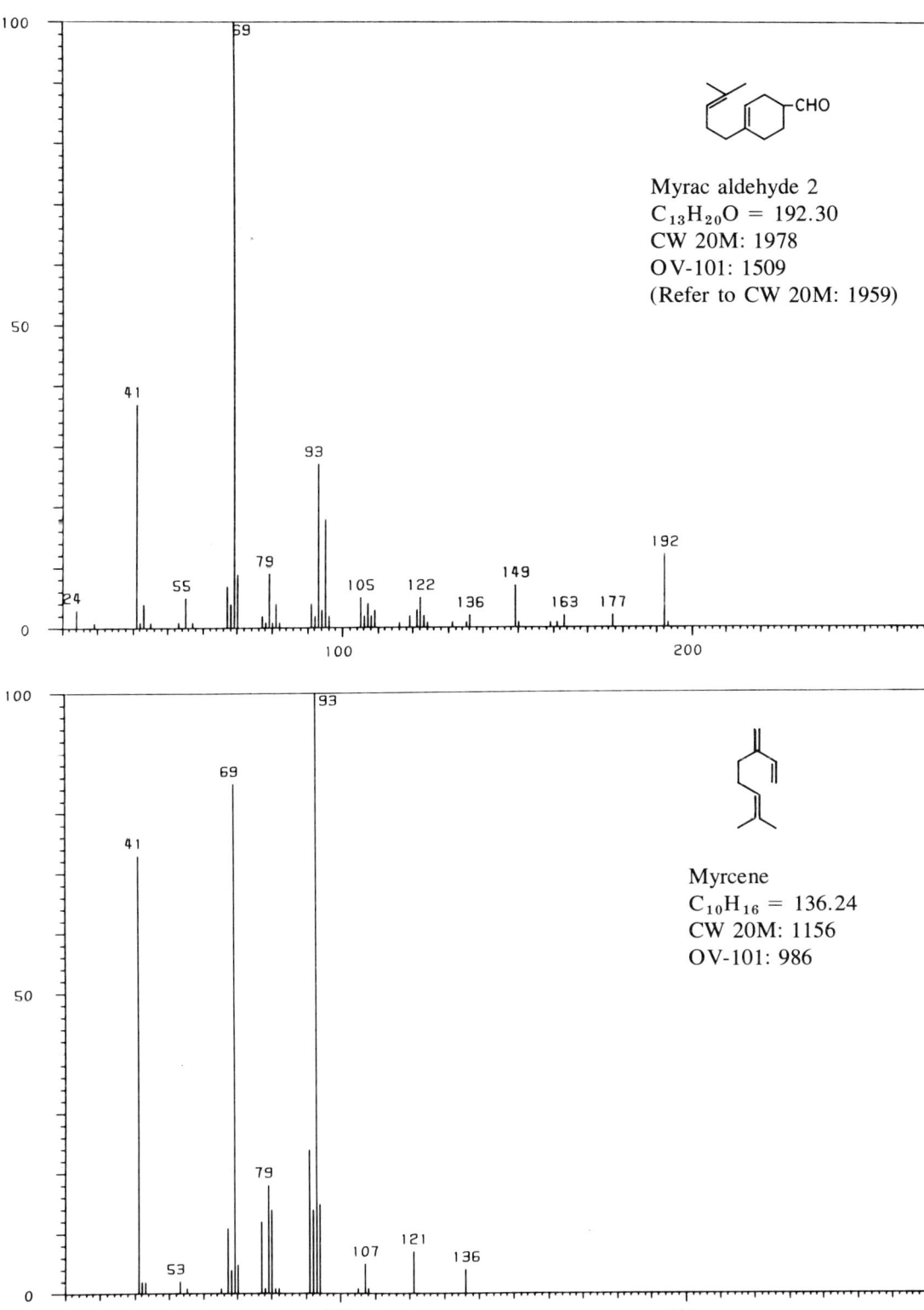

Myrac aldehyde 2
$C_{13}H_{20}O = 192.30$
CW 20M: 1978
OV-101: 1509
(Refer to CW 20M: 1959)

Myrcene
$C_{10}H_{16} = 136.24$
CW 20M: 1156
OV-101: 986

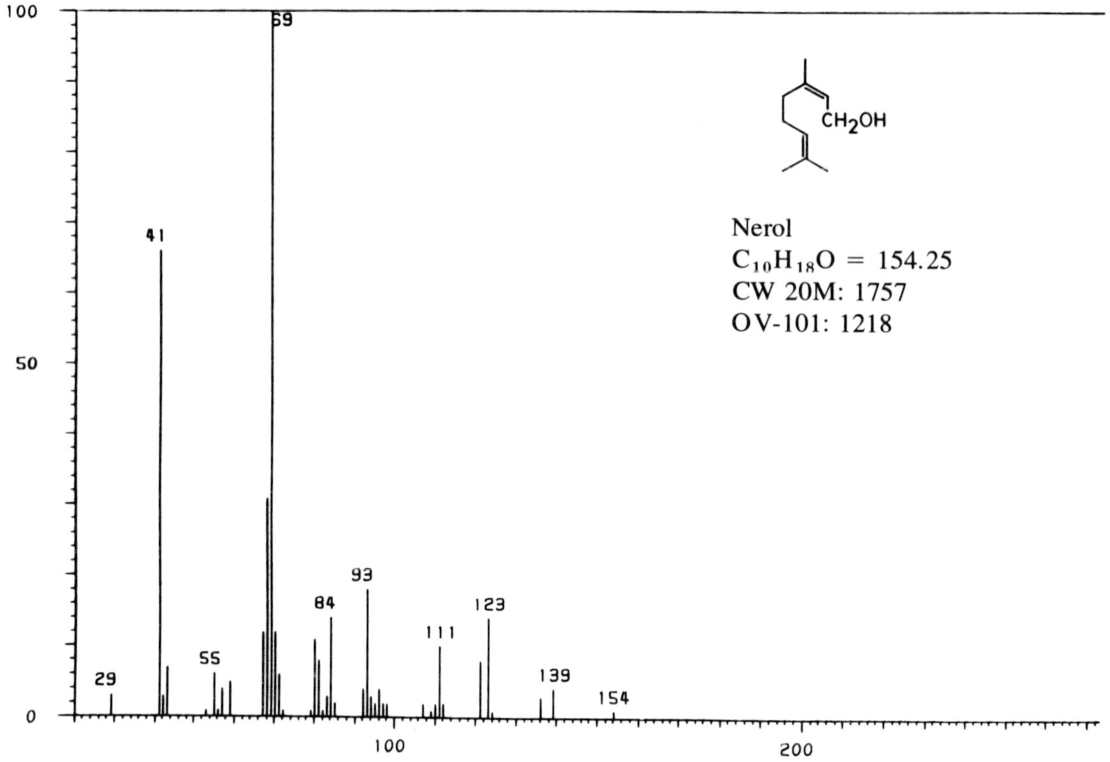

Nerol
$C_{10}H_{18}O = 154.25$
CW 20M: 1757
OV-101: 1218

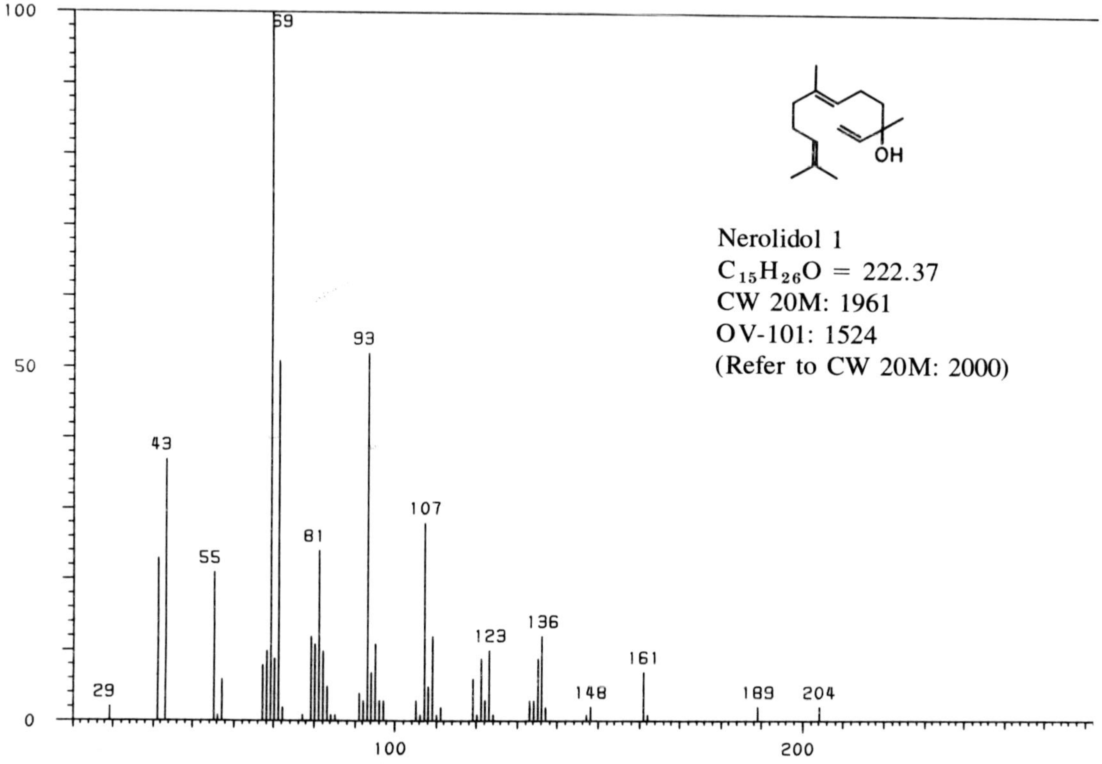

Nerolidol 1
$C_{15}H_{26}O = 222.37$
CW 20M: 1961
OV-101: 1524
(Refer to CW 20M: 2000)

APPENDIX IV   MASS SPECTRA OF INDIVIDUAL COMPOUNDS

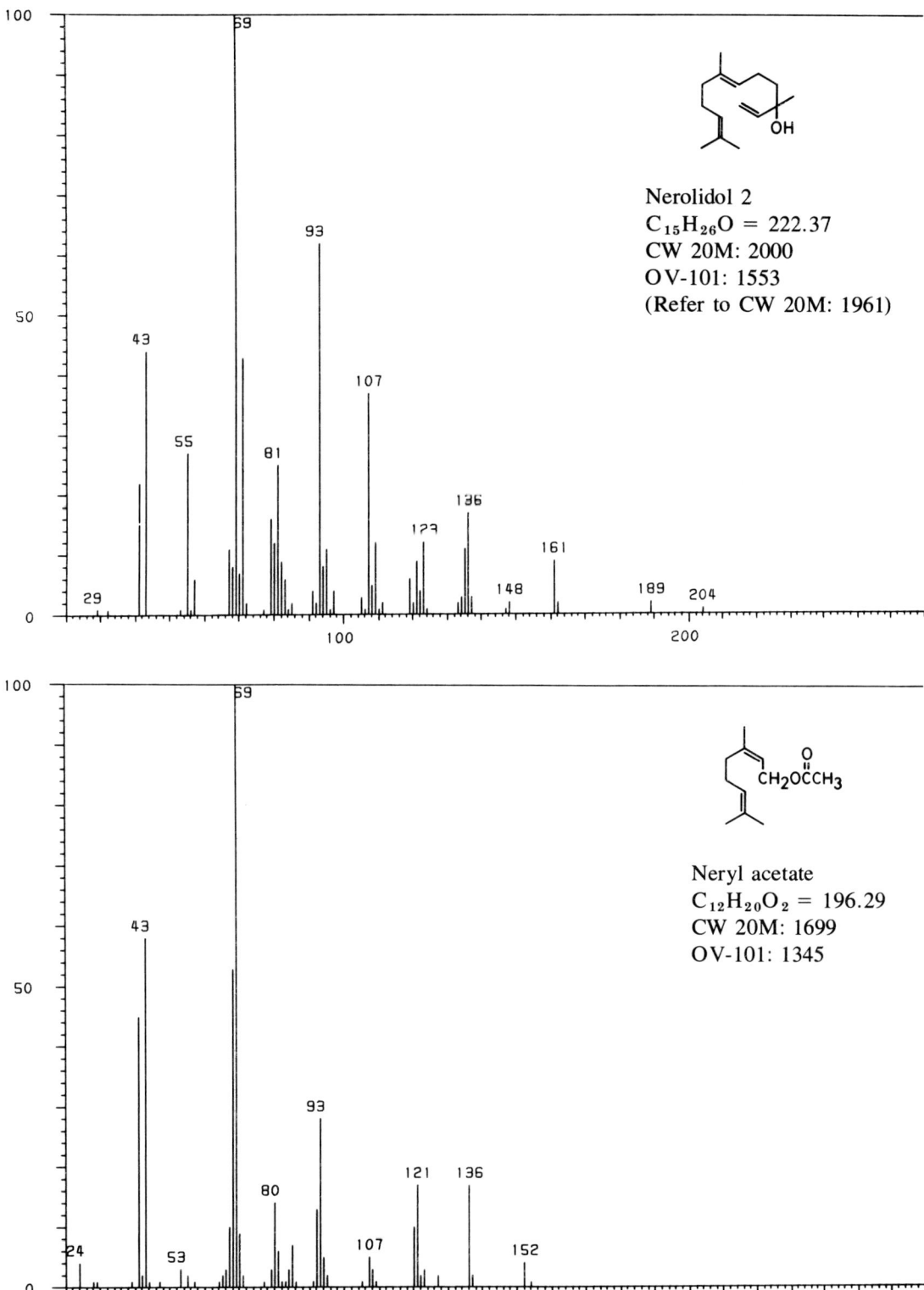

Nerolidol 2
$C_{15}H_{26}O = 222.37$
CW 20M: 2000
OV-101: 1553
(Refer to CW 20M: 1961)

Neryl acetate
$C_{12}H_{20}O_2 = 196.29$
CW 20M: 1699
OV-101: 1345

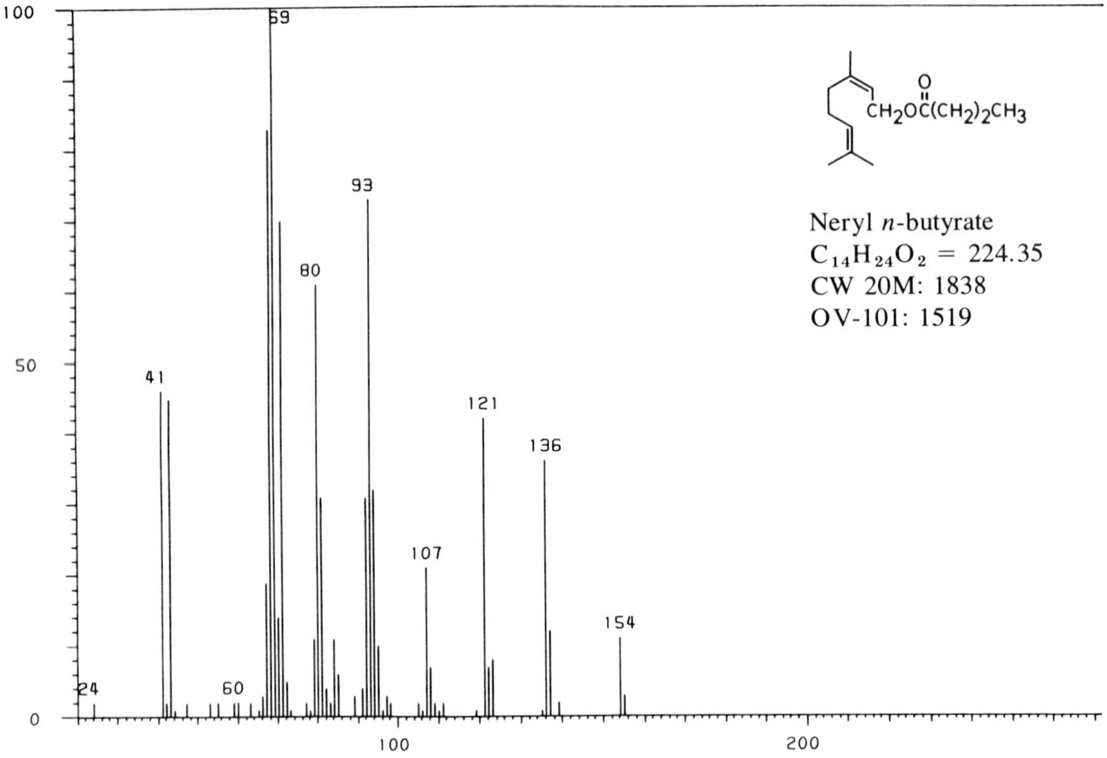

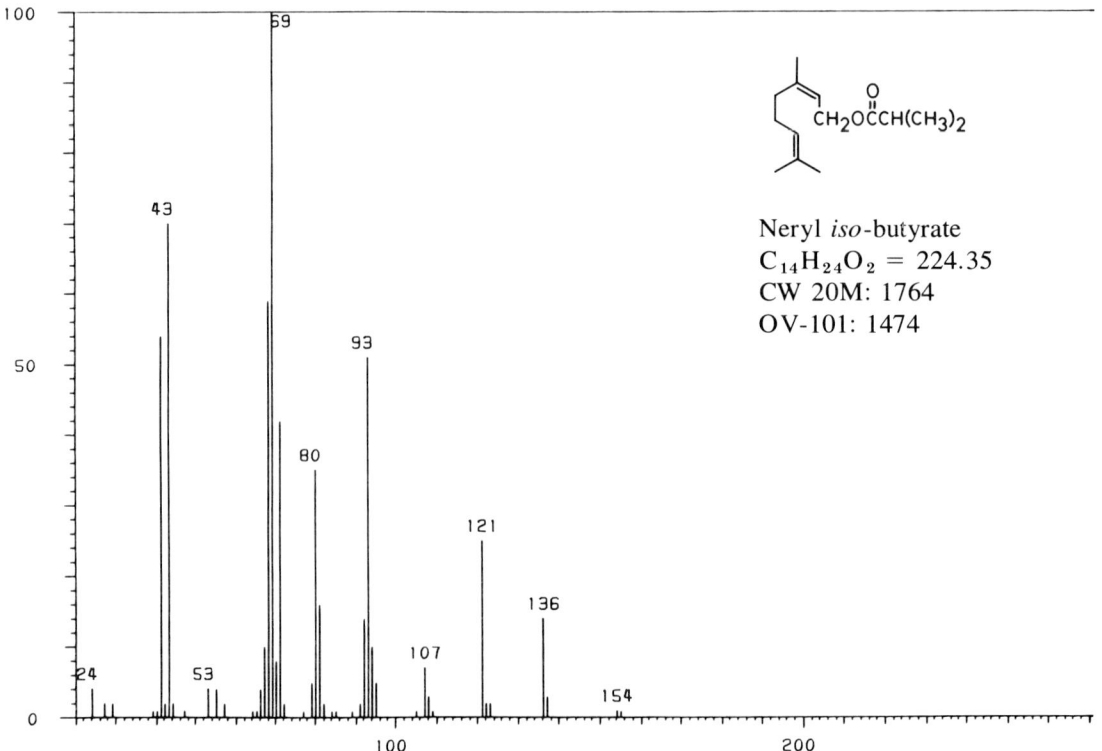

Neryl *n*-butyrate
$C_{14}H_{24}O_2 = 224.35$
CW 20M: 1838
OV-101: 1519

Neryl *iso*-butyrate
$C_{14}H_{24}O_2 = 224.35$
CW 20M: 1764
OV-101: 1474

APPENDIX IV   MASS SPECTRA OF INDIVIDUAL COMPOUNDS

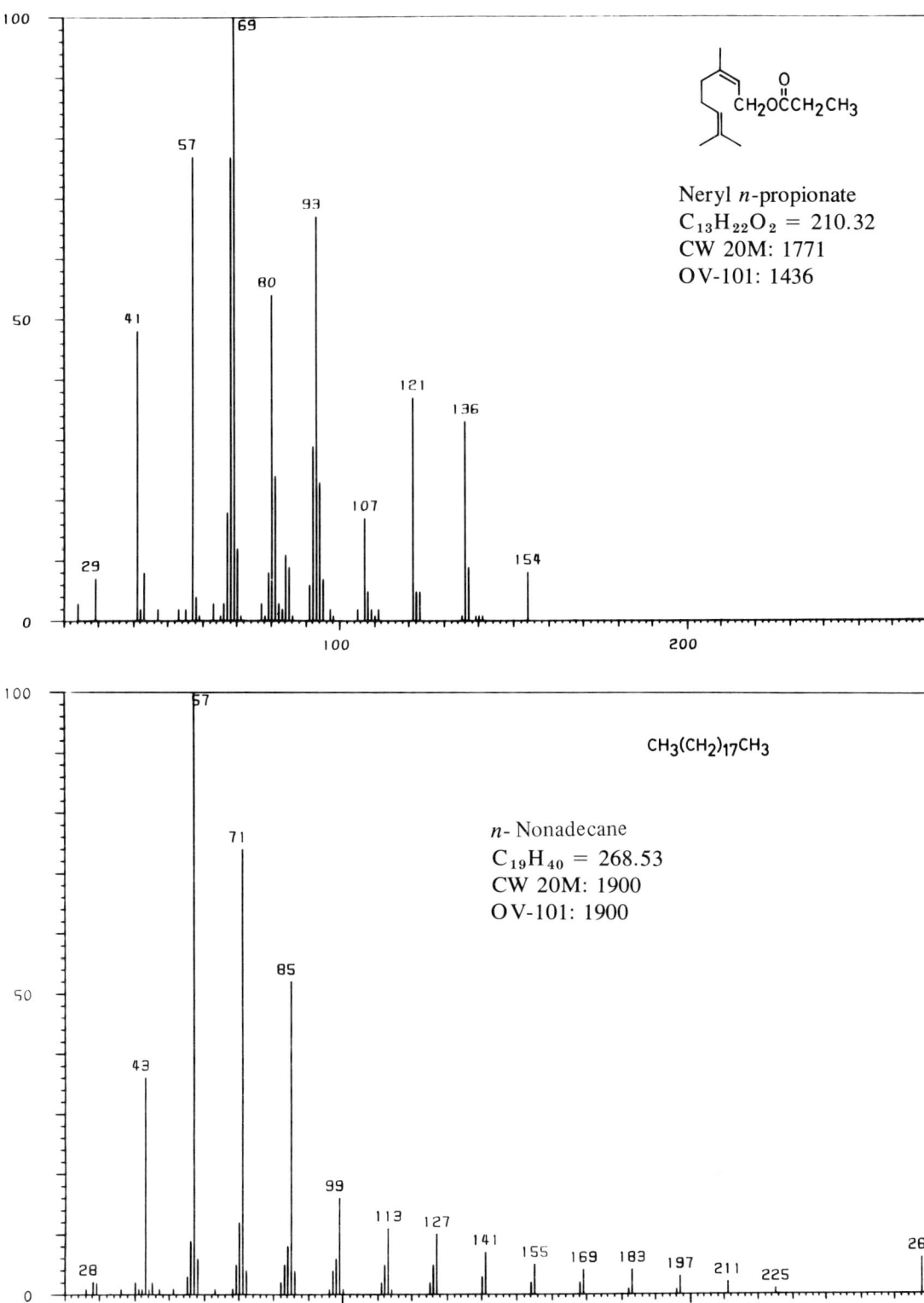

Neryl *n*-propionate
$C_{13}H_{22}O_2 = 210.32$
CW 20M: 1771
OV-101: 1436

n-Nonadecane
$C_{19}H_{40} = 268.53$
CW 20M: 1900
OV-101: 1900

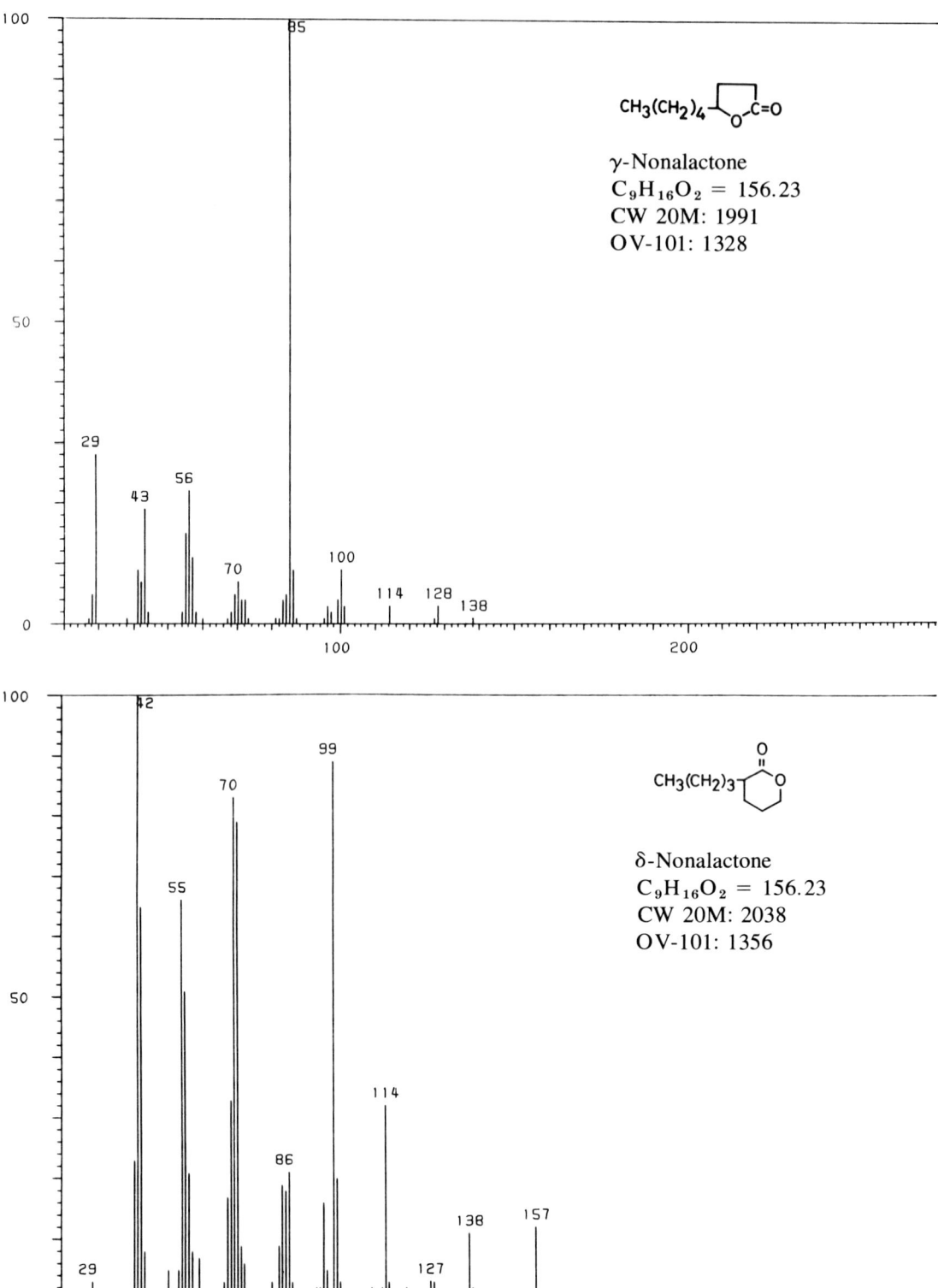

APPENDIX IV MASS SPECTRA OF INDIVIDUAL COMPOUNDS

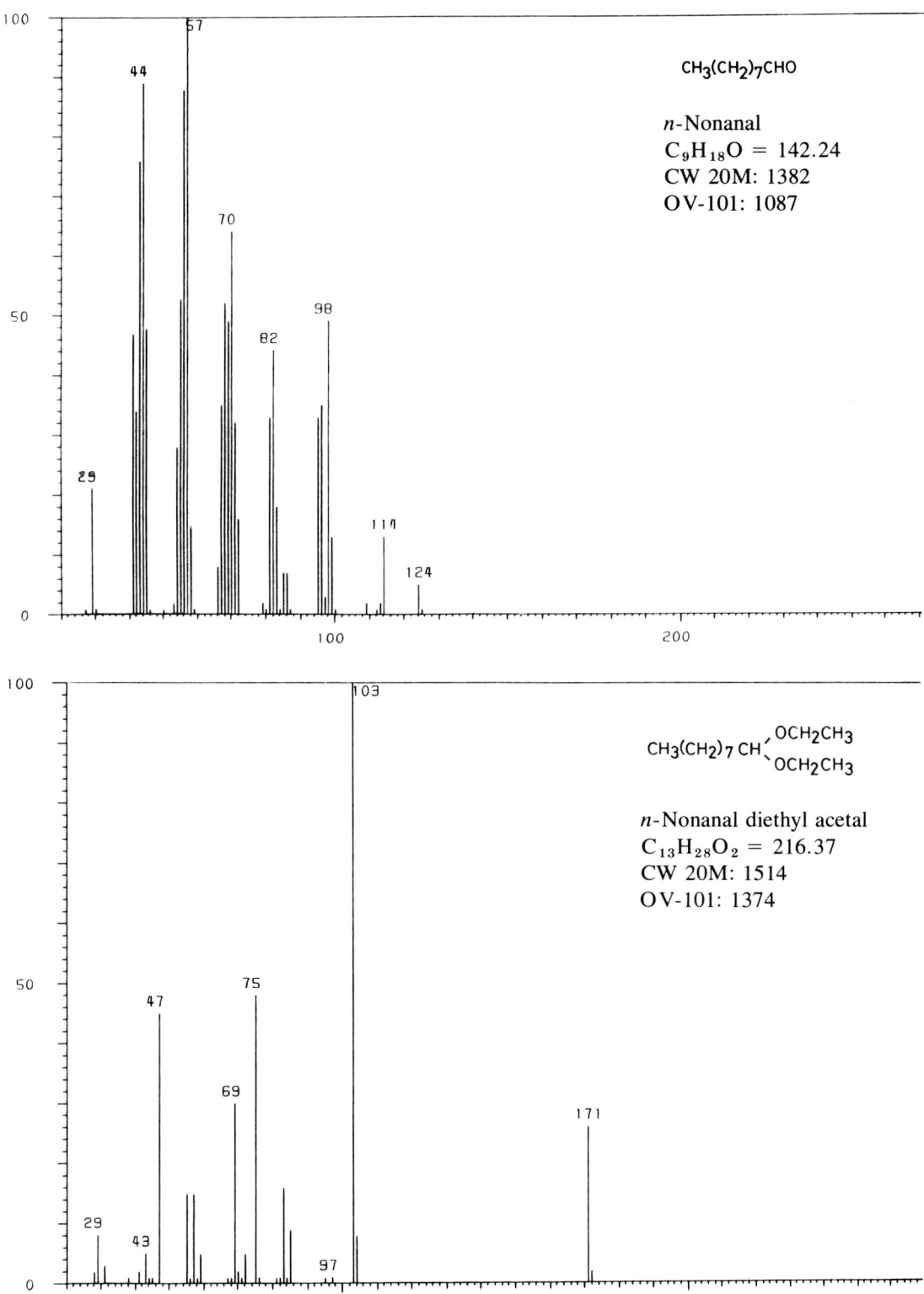

$CH_3(CH_2)_7CHO$

n-Nonanal
$C_9H_{18}O = 142.24$
CW 20M: 1382
OV-101: 1087

$CH_3(CH_2)_7CH\genfrac{}{}{0pt}{}{OCH_2CH_3}{OCH_2CH_3}$

n-Nonanal diethyl acetal
$C_{13}H_{28}O_2 = 216.37$
CW 20M: 1514
OV-101: 1374

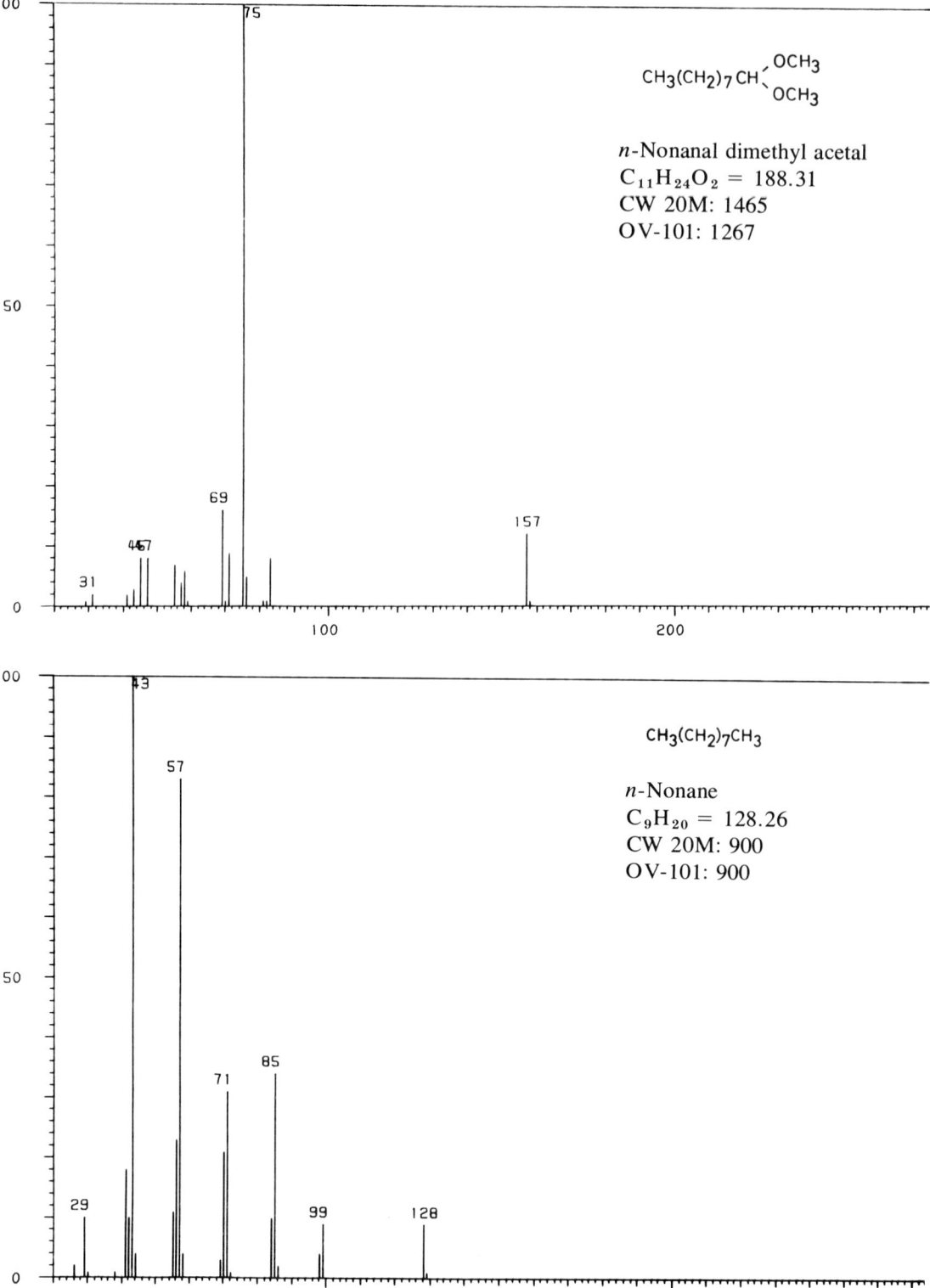

APPENDIX IV MASS SPECTRA OF INDIVIDUAL COMPOUNDS

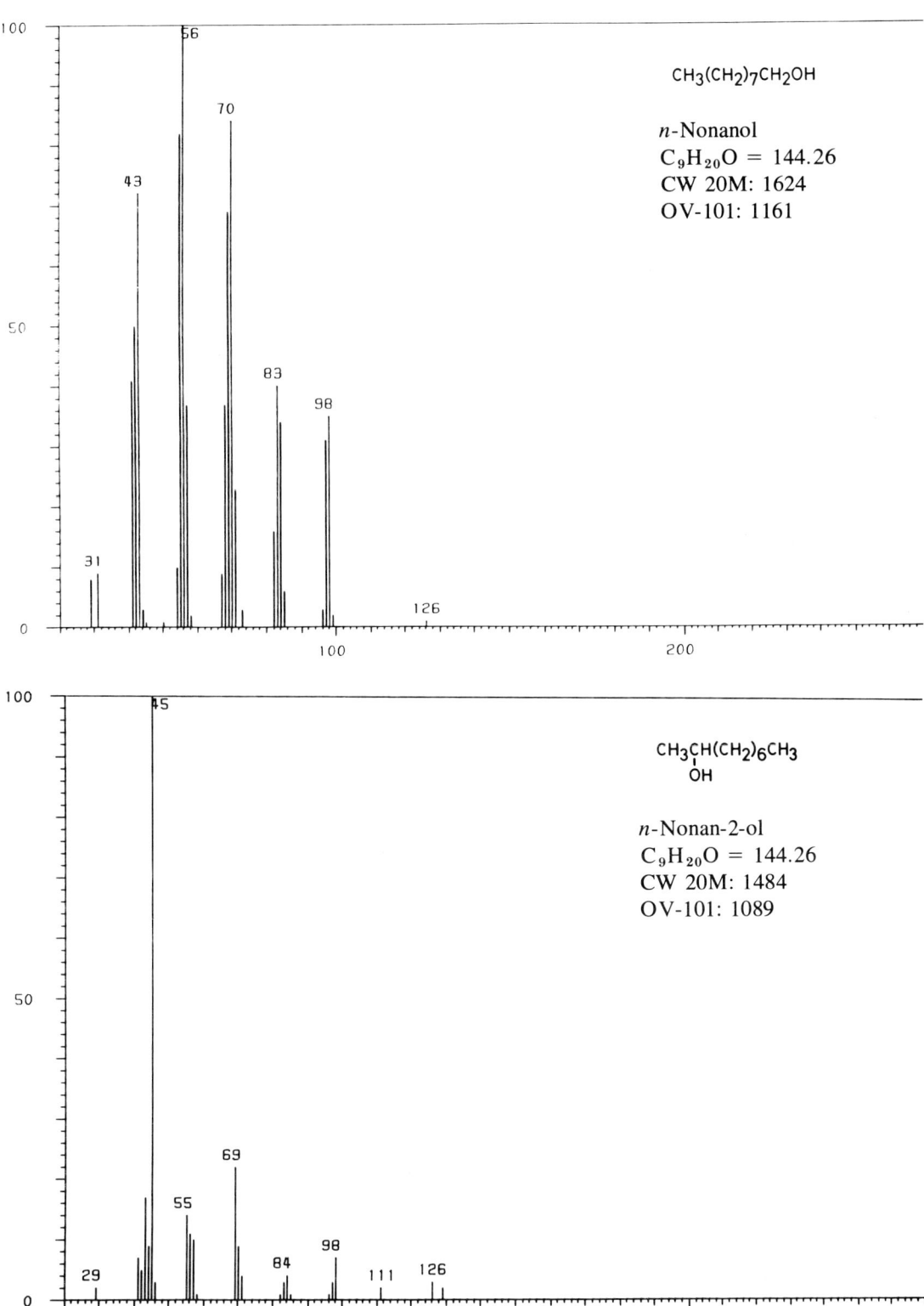

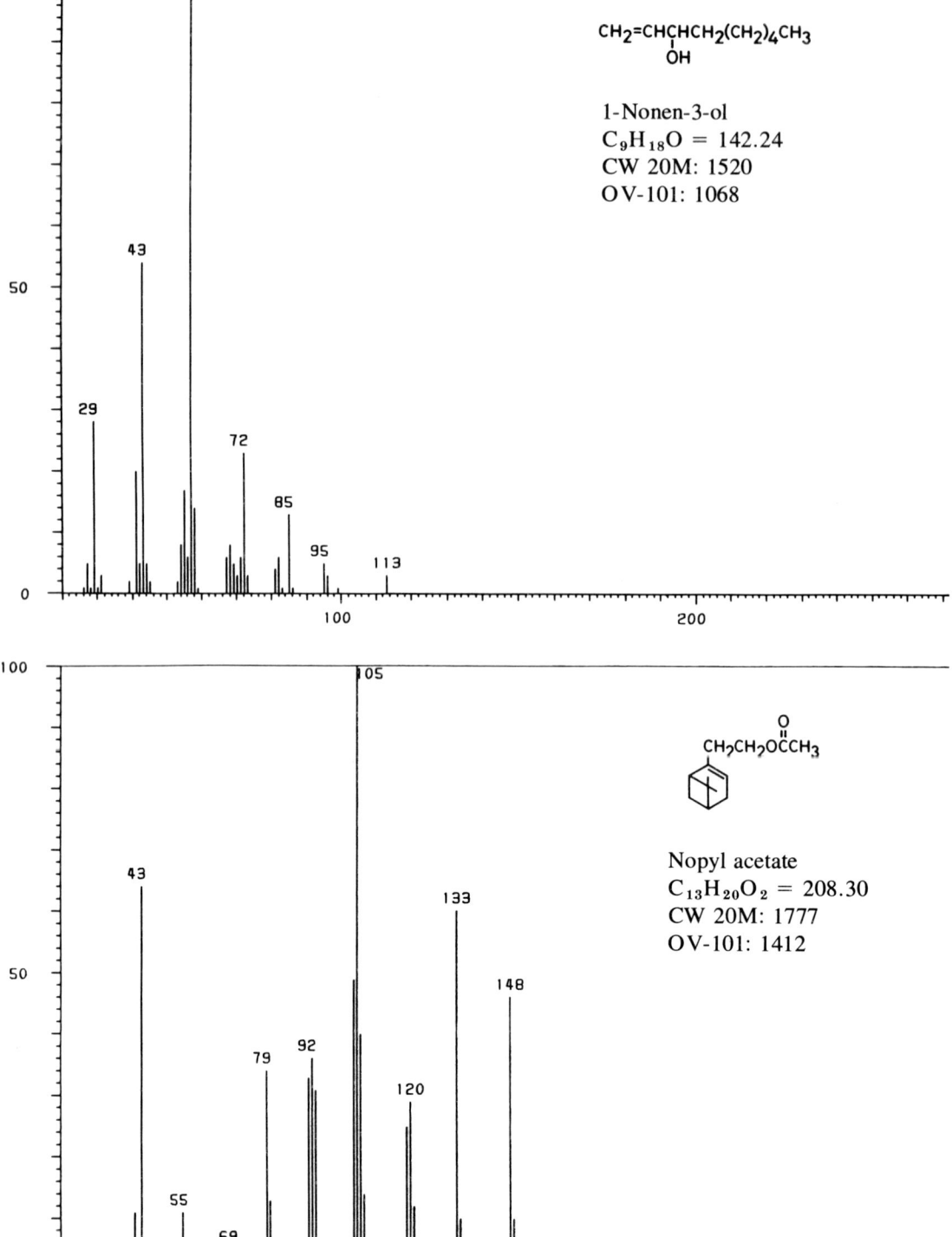

APPENDIX IV   MASS SPECTRA OF INDIVIDUAL COMPOUNDS

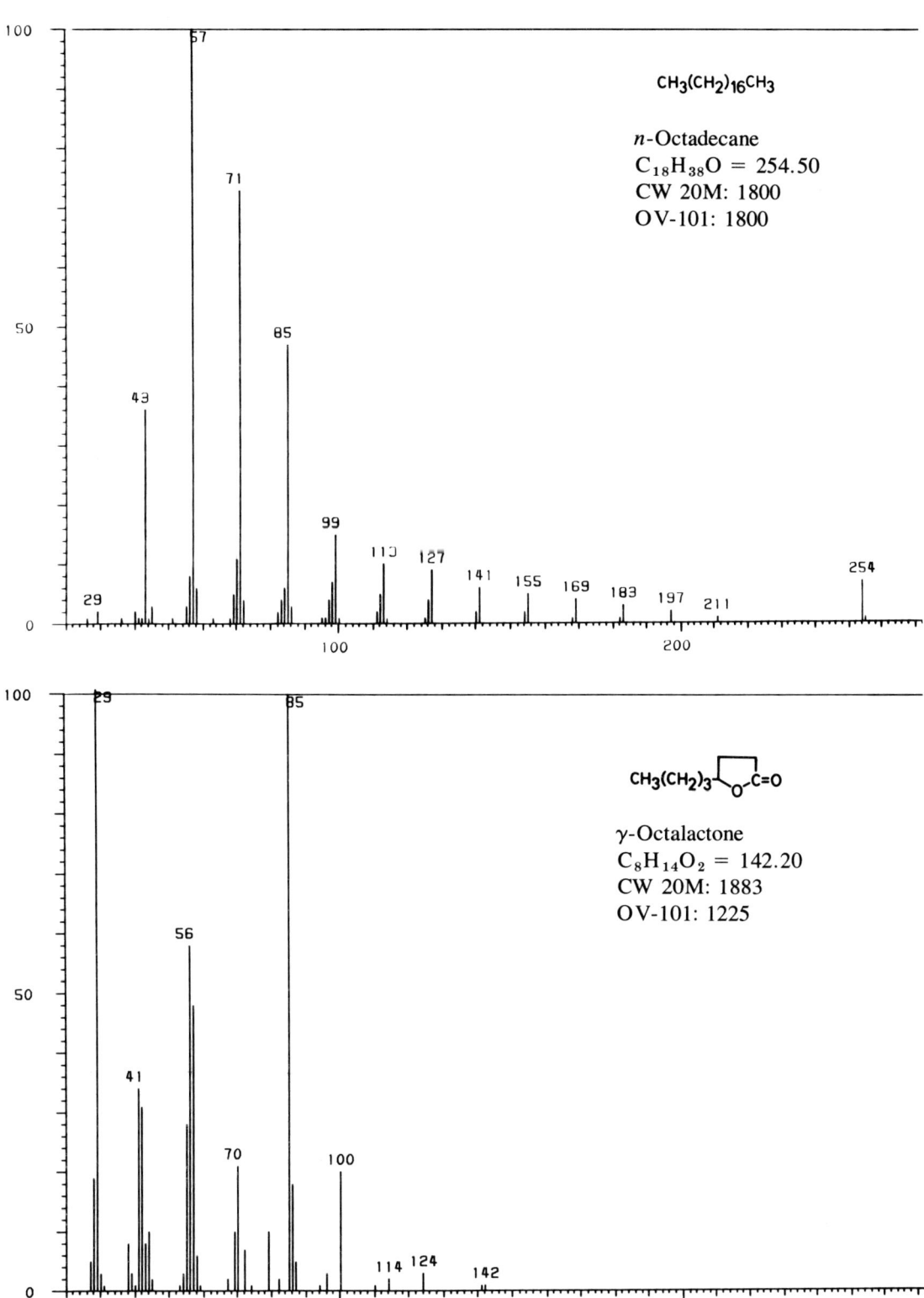

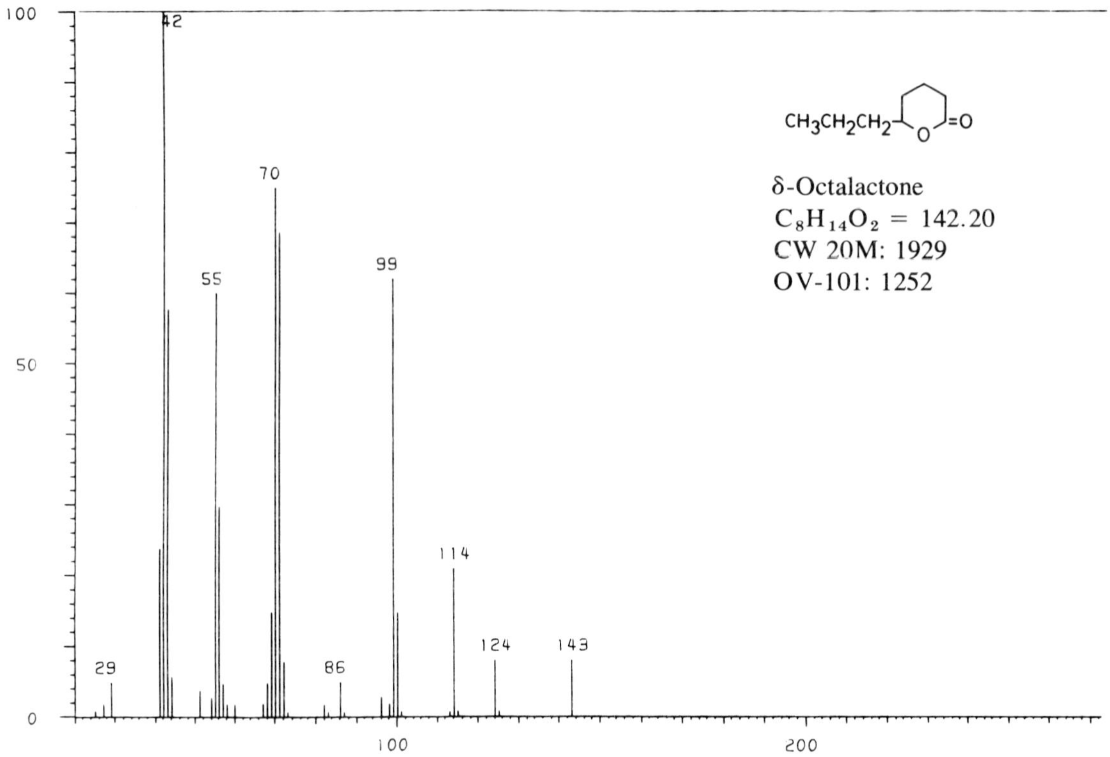

δ-Octalactone
$C_8H_{14}O_2$ = 142.20
CW 20M: 1929
OV-101: 1252

n-Octanal
$C_8H_{16}O$ = 128.22
CW 20M: 1278
OV-101: 985

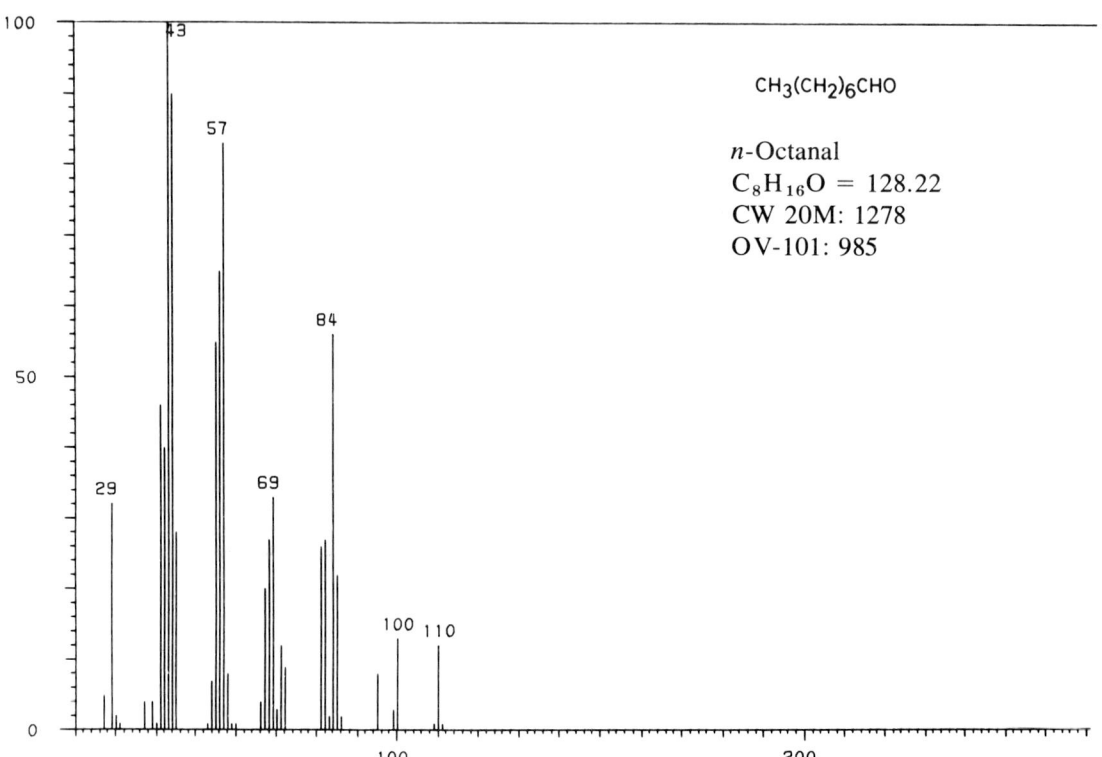

APPENDIX IV   MASS SPECTRA OF INDIVIDUAL COMPOUNDS

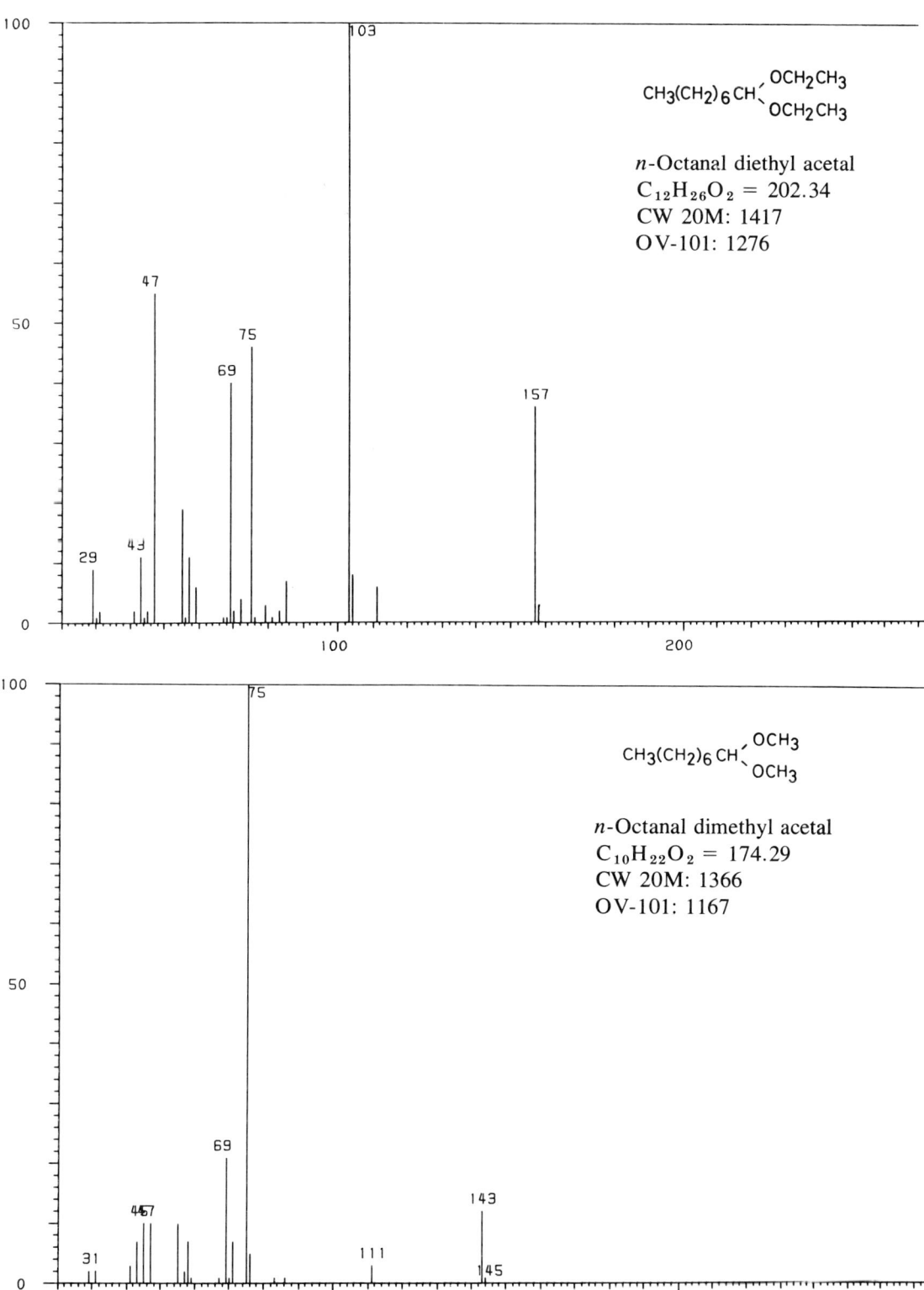

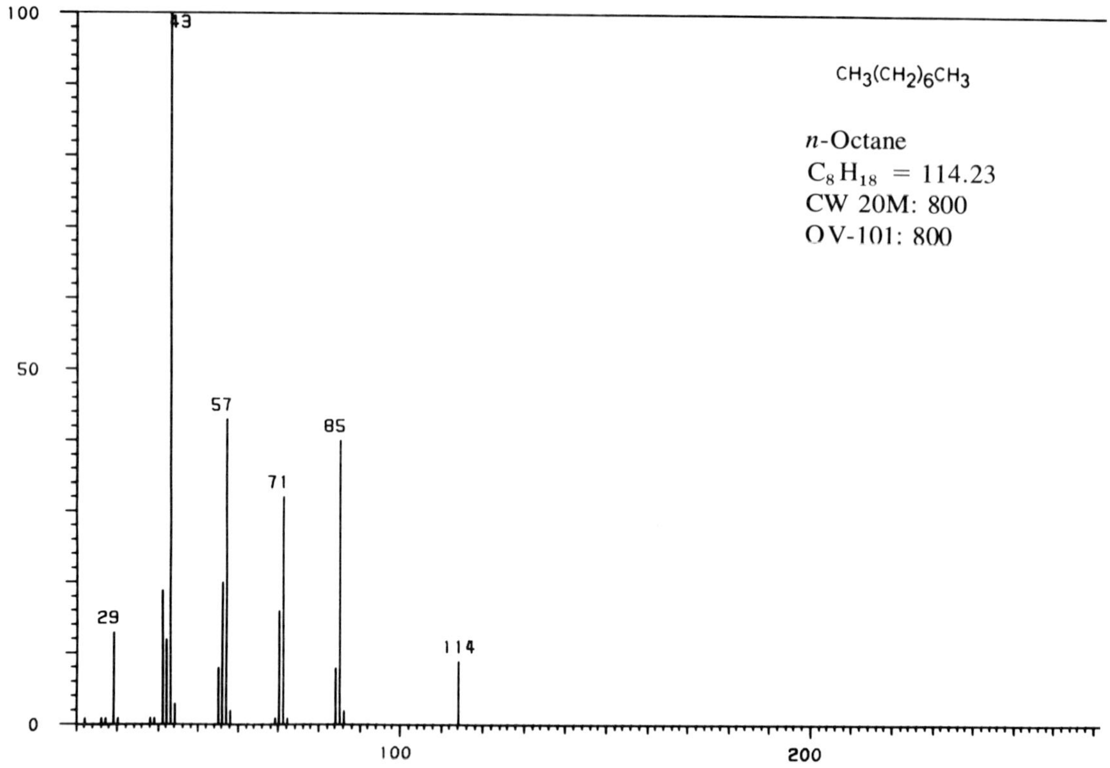

n-Octane
C_8H_{18} = 114.23
CW 20M: 800
OV-101: 800

n-Octanol
$C_8H_{18}O$ = 130.23
CW 20M: 1519
OV-101: 1061

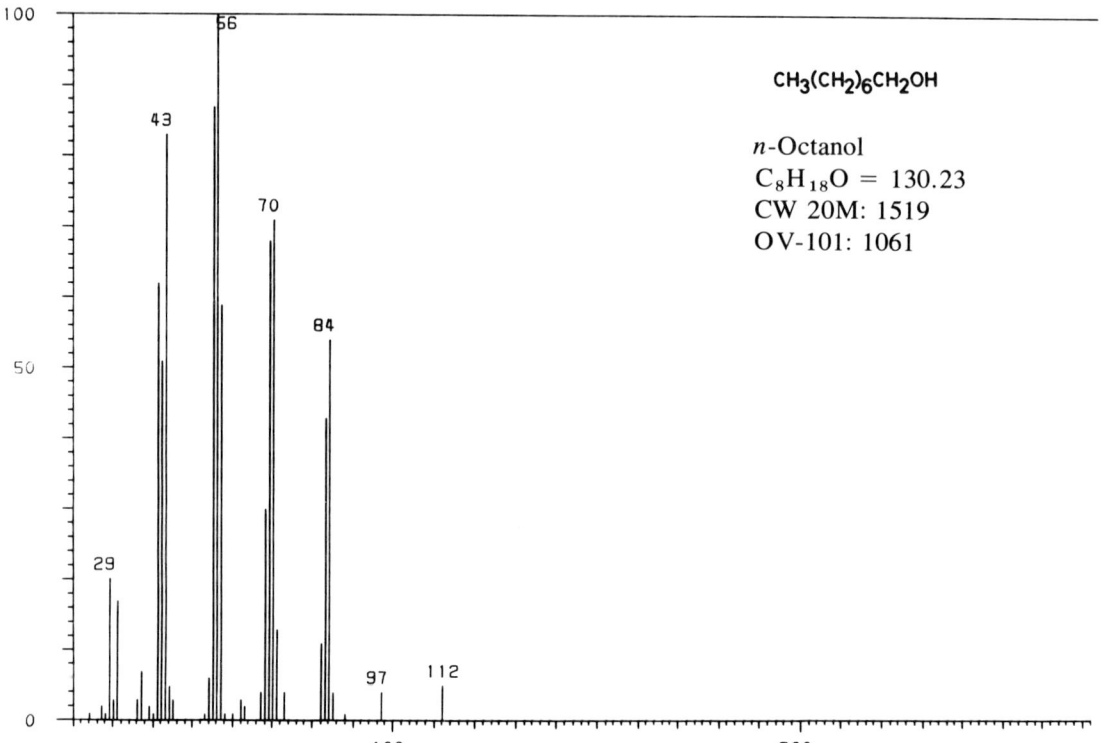

APPENDIX IV MASS SPECTRA OF INDIVIDUAL COMPOUNDS

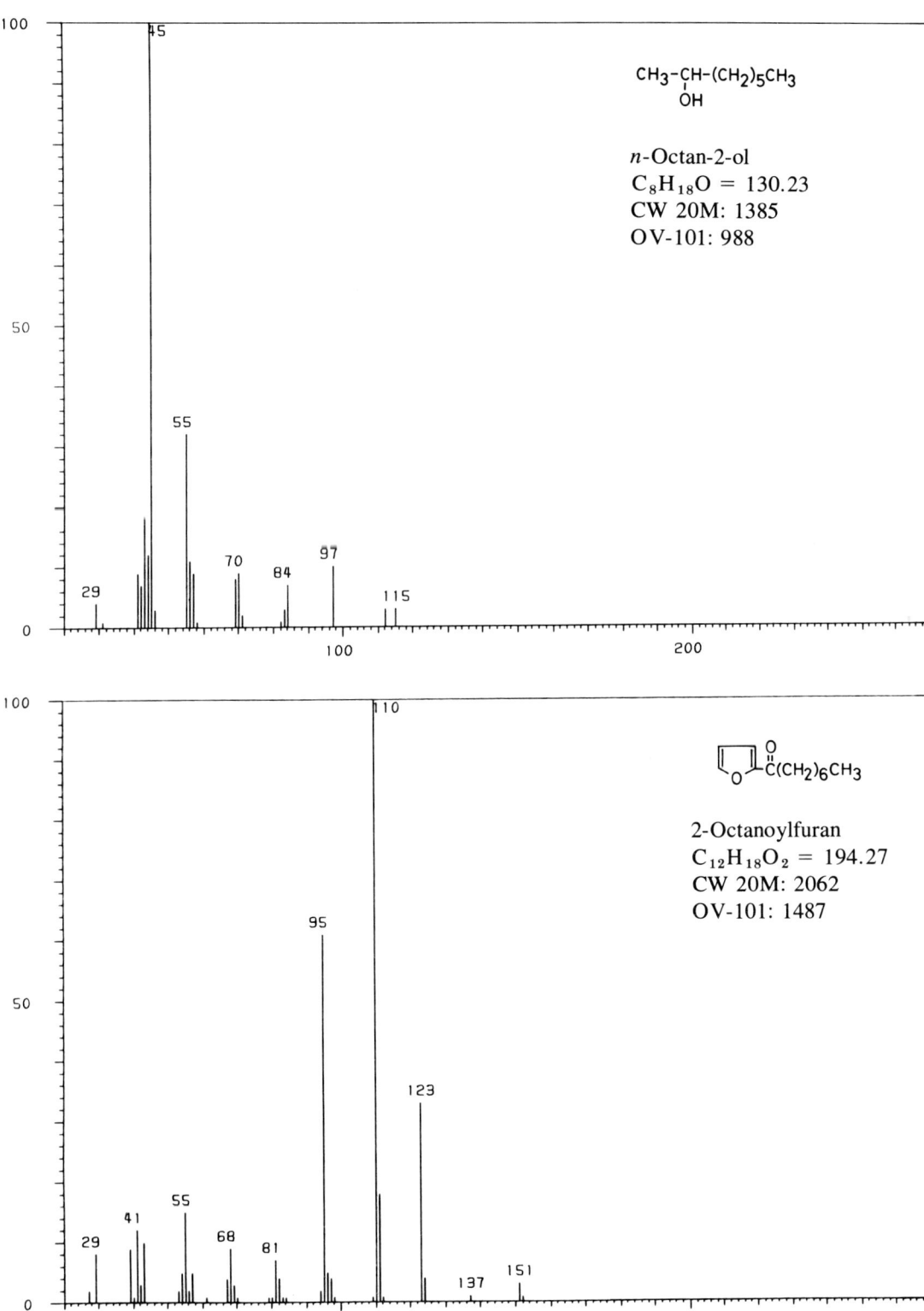

n-Octan-2-ol
$C_8H_{18}O = 130.23$
CW 20M: 1385
OV-101: 988

2-Octanoylfuran
$C_{12}H_{18}O_2 = 194.27$
CW 20M: 2062
OV-101: 1487

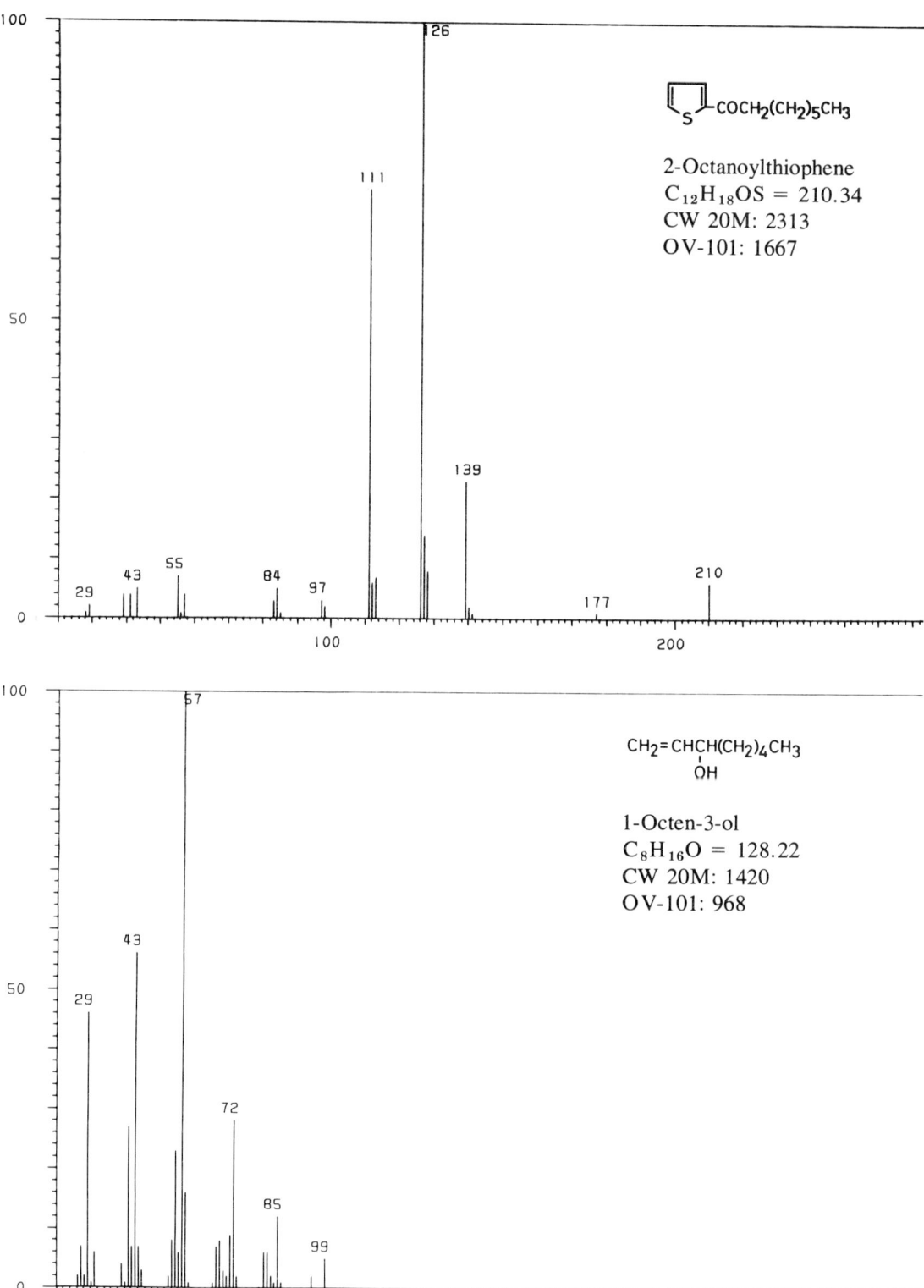

APPENDIX IV   MASS SPECTRA OF INDIVIDUAL COMPOUNDS

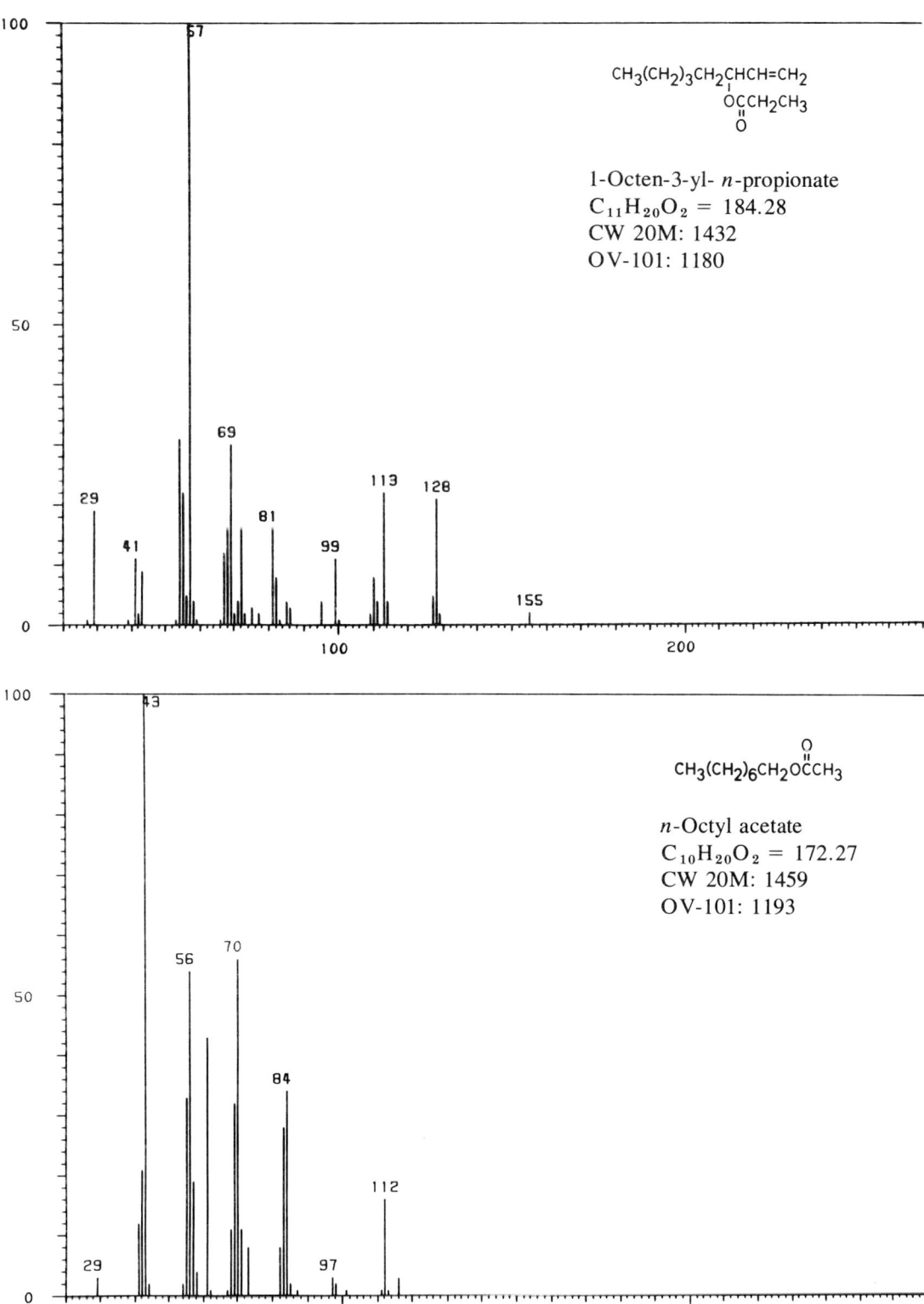

1-Octen-3-yl- n-propionate
$C_{11}H_{20}O_2 = 184.28$
CW 20M: 1432
OV-101: 1180

n-Octyl acetate
$C_{10}H_{20}O_2 = 172.27$
CW 20M: 1459
OV-101: 1193

$CH_3(CH_2)_2\overset{O}{\overset{\|}{C}}OCH_2(CH_2)_6CH_3$

n-Octyl *n*-butyrate
$C_{12}H_{24}O_2$ = 200.32
CW 20M: 1597
OV-101: 1373

$(CH_3)_2CHCOOCH_2(CH_2)_6CH_3$

n-Octyl *iso*-butyrate
$C_{12}H_{24}O_2$ = 200.32
CW 20M: 1529
OV-101: 1332

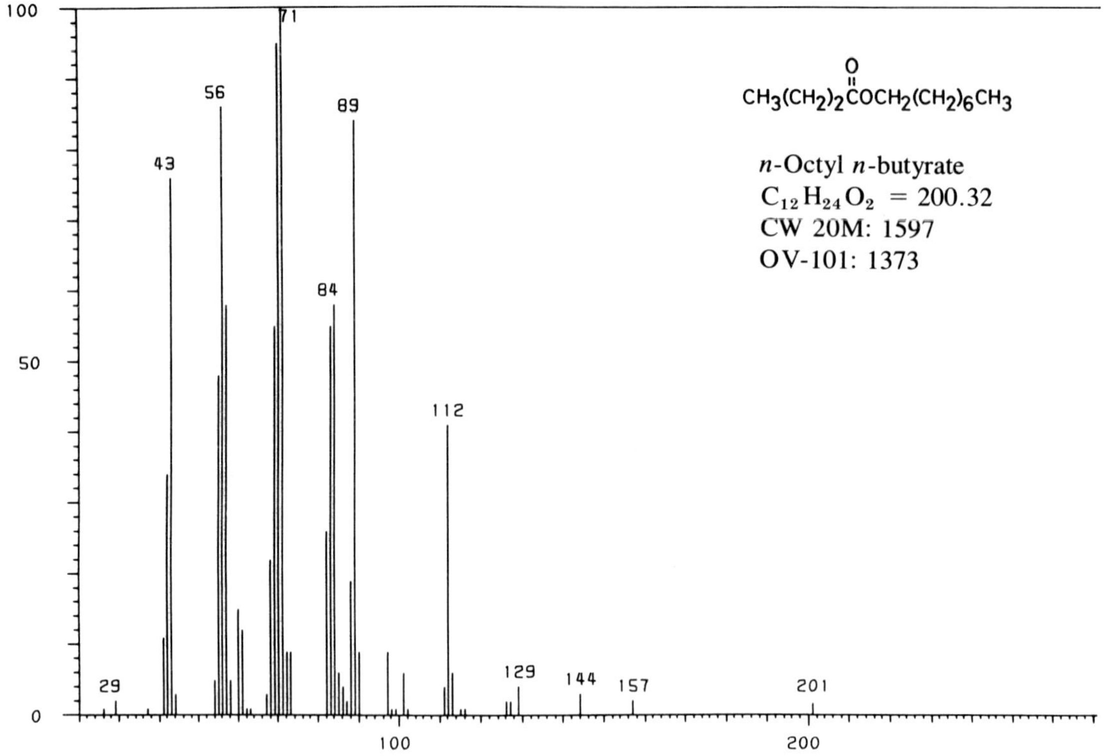

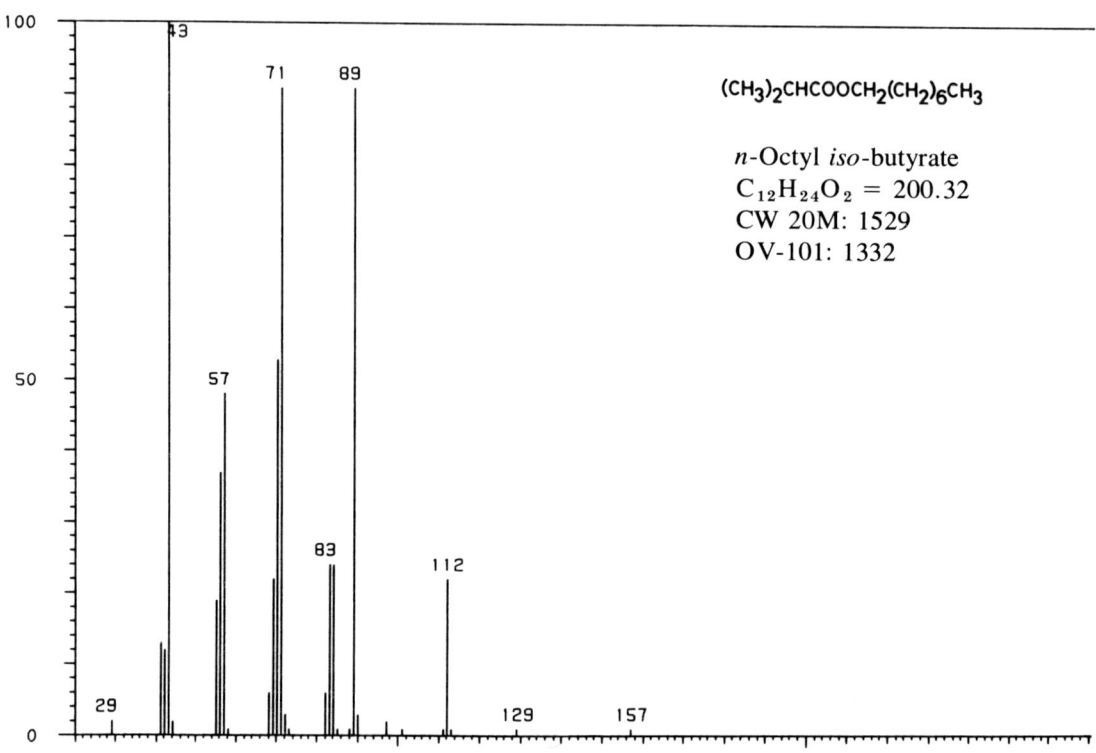

APPENDIX IV   MASS SPECTRA OF INDIVIDUAL COMPOUNDS

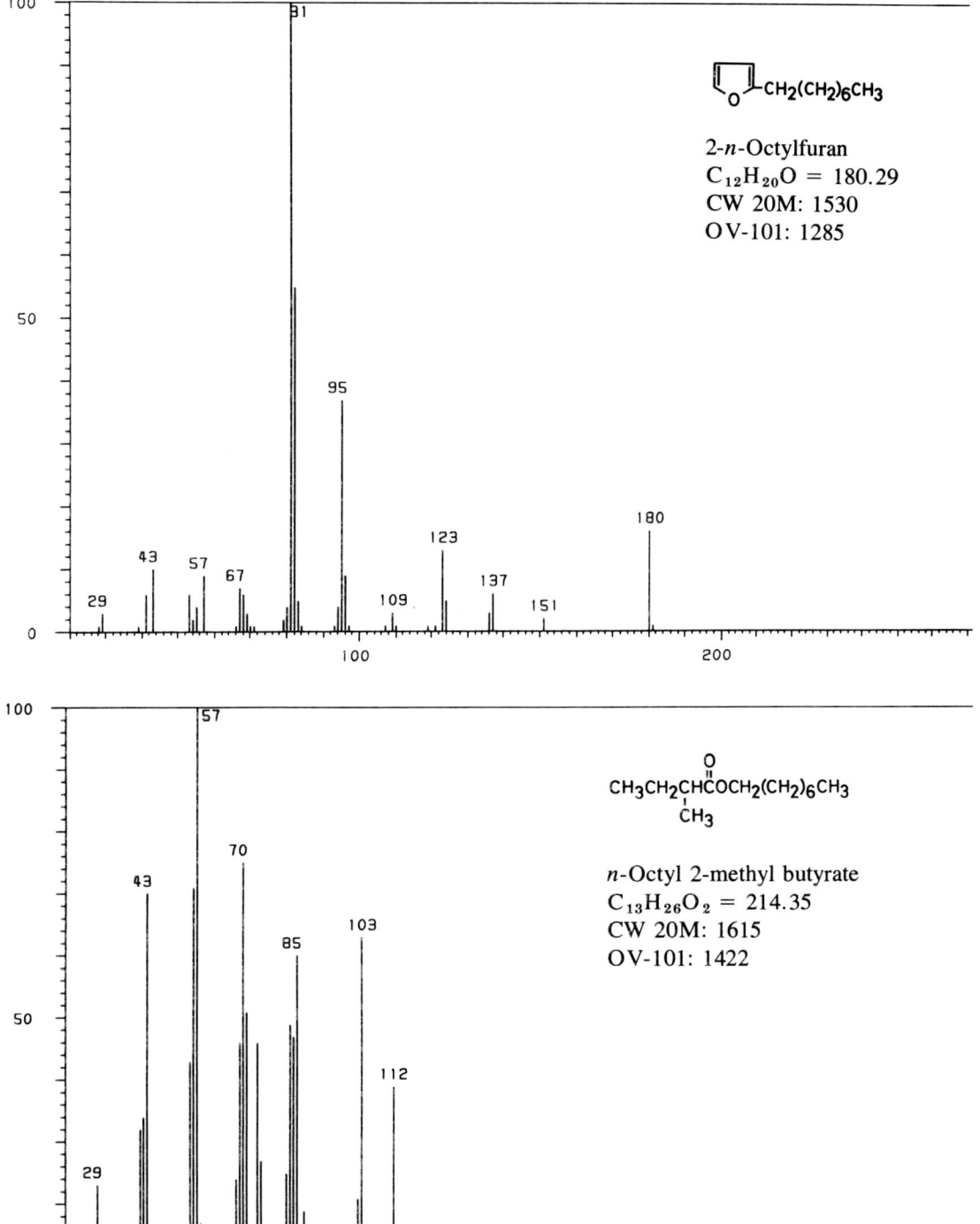

2-n-Octylfuran
$C_{12}H_{20}O = 180.29$
CW 20M: 1530
OV-101: 1285

n-Octyl 2-methyl butyrate
$C_{13}H_{26}O_2 = 214.35$
CW 20M: 1615
OV-101: 1422

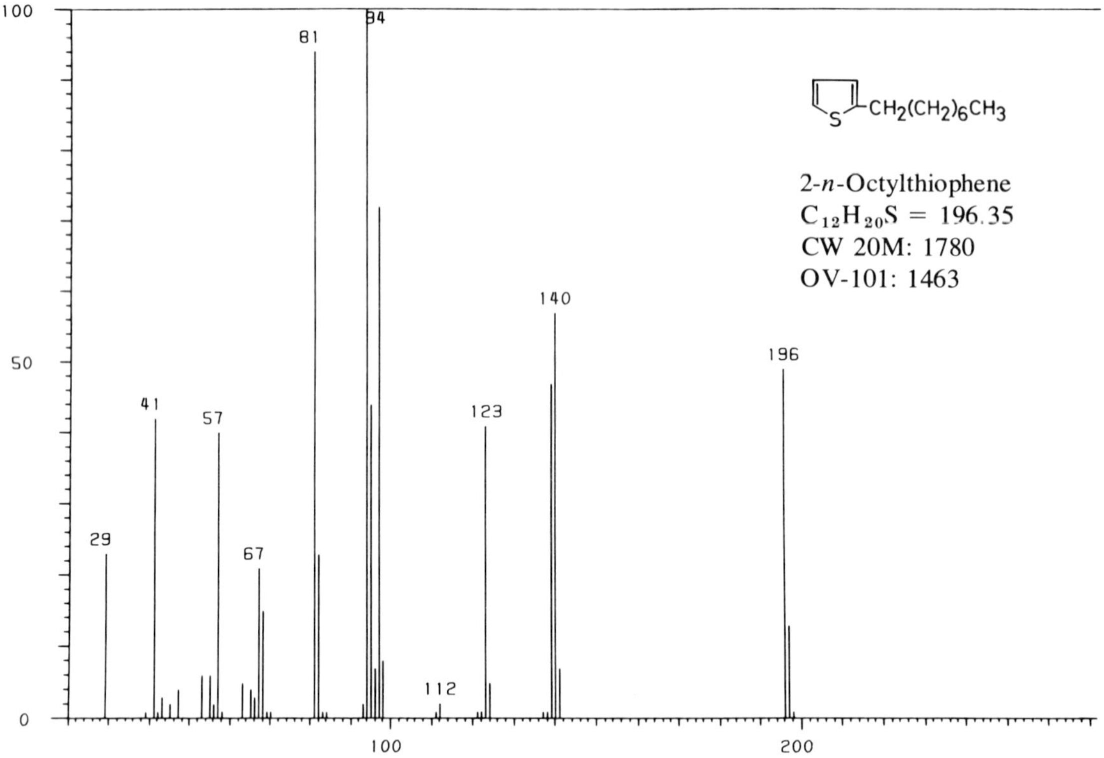

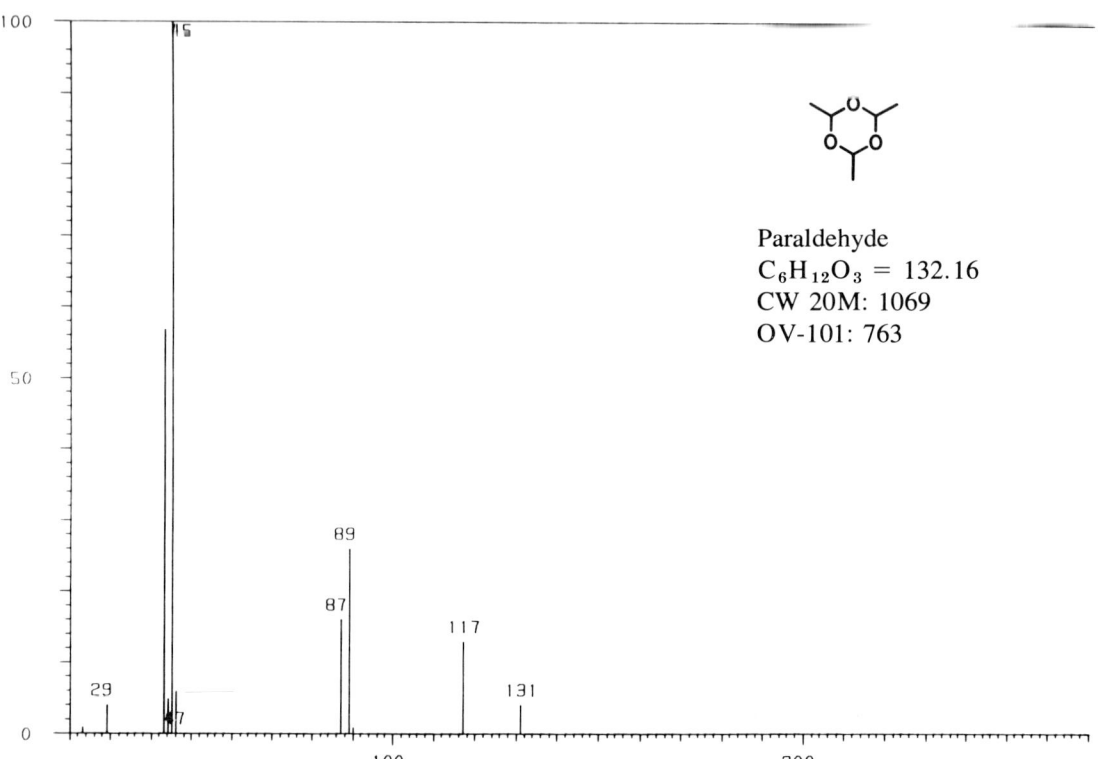

2-*n*-Octylthiophene
$C_{12}H_{20}S = 196.35$
CW 20M: 1780
OV-101: 1463

Paraldehyde
$C_6H_{12}O_3 = 132.16$
CW 20M: 1069
OV-101: 763

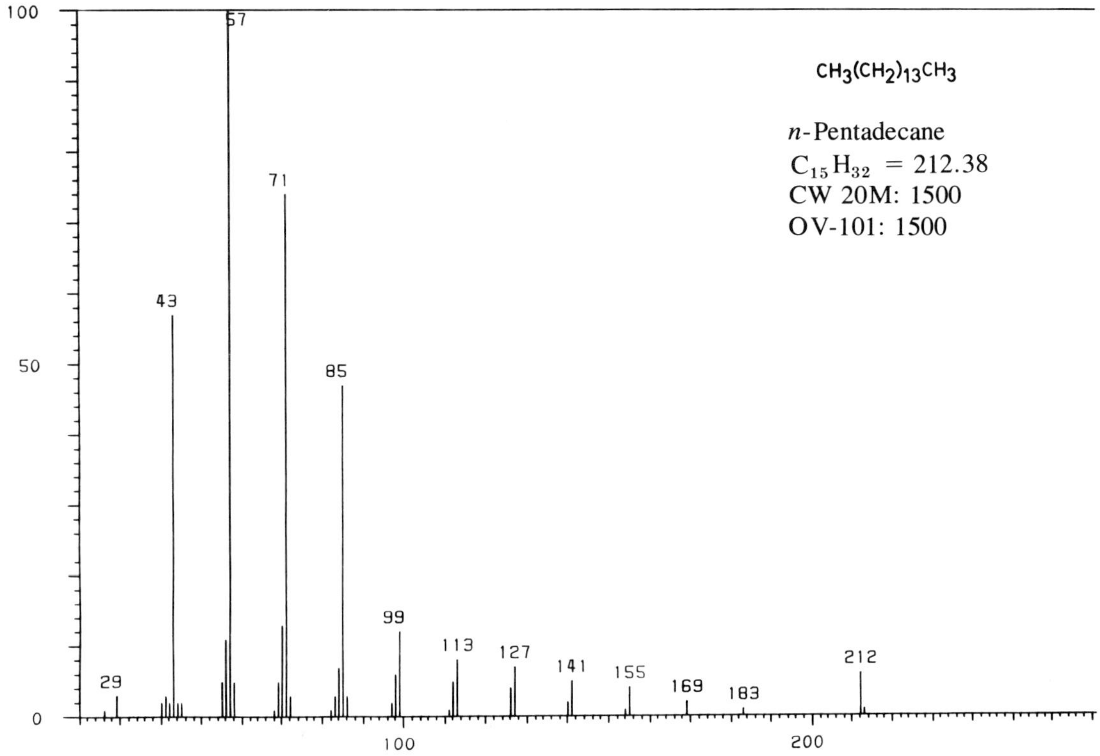

APPENDIX IV MASS SPECTRA OF INDIVIDUAL COMPOUNDS

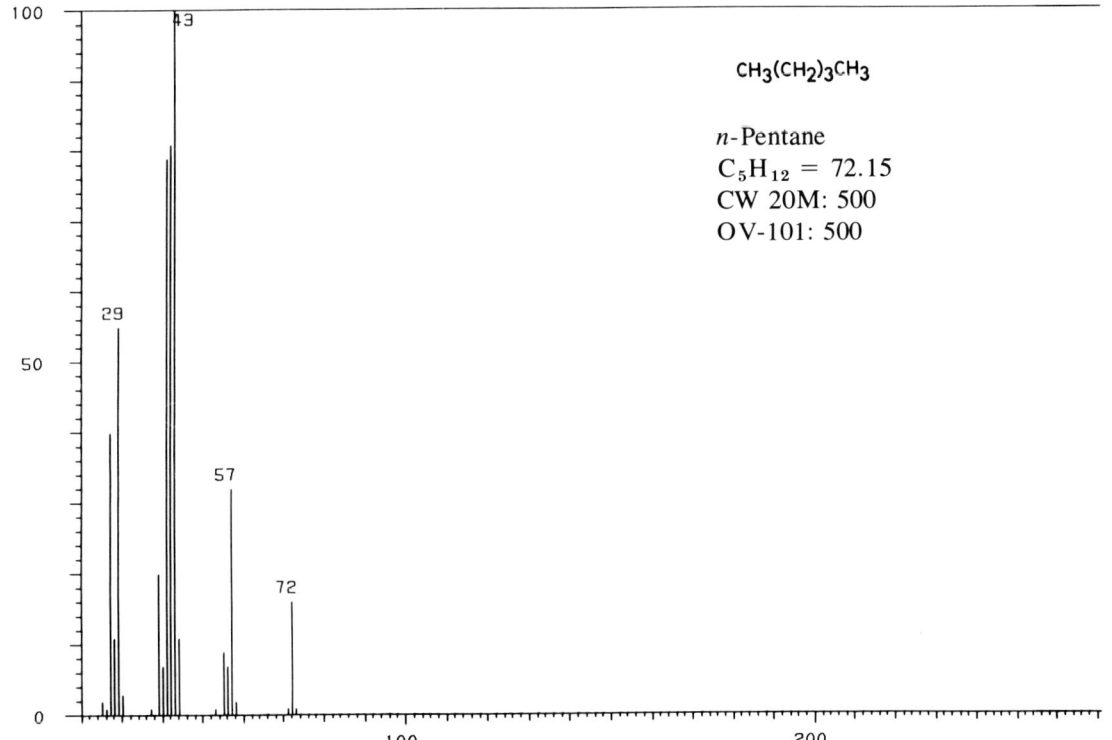

$CH_3(CH_2)_{13}CH_3$

n-Pentadecane
$C_{15}H_{32} = 212.38$
CW 20M: 1500
OV-101: 1500

$CH_3(CH_2)_3CH_3$

n-Pentane
$C_5H_{12} = 72.15$
CW 20M: 500
OV-101: 500

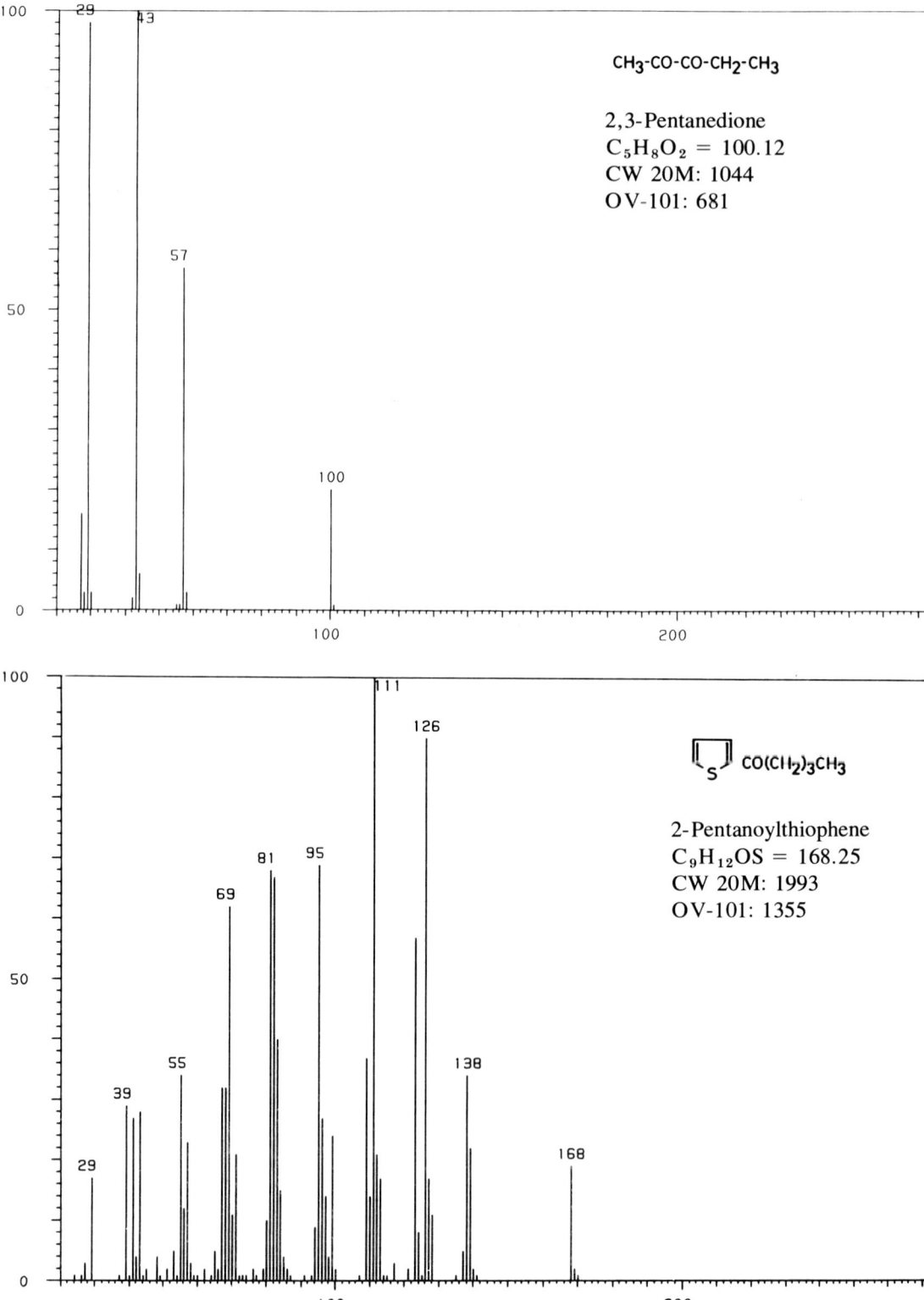

APPENDIX IV   MASS SPECTRA OF INDIVIDUAL COMPOUNDS

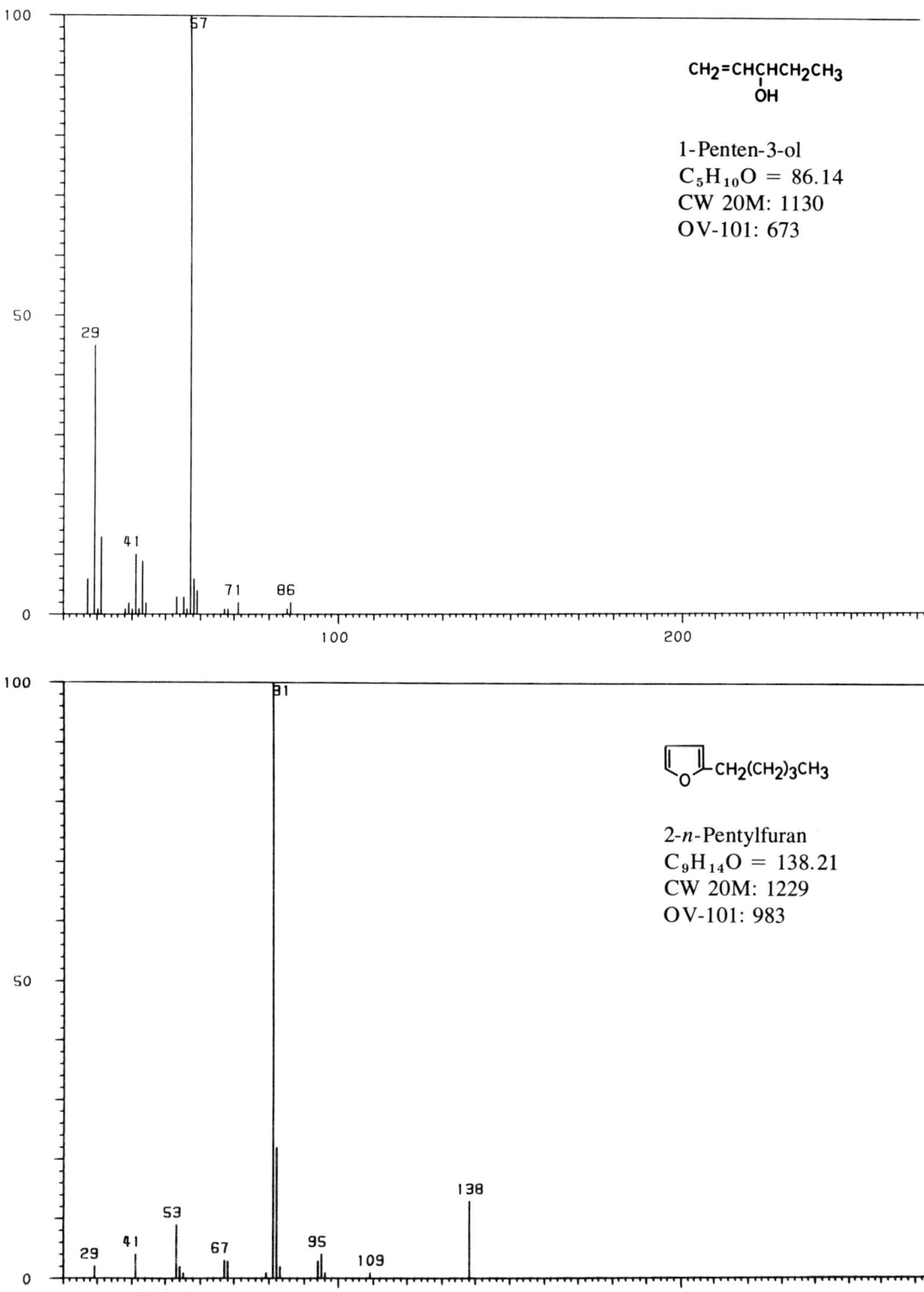

$CH_2=CHCHCH_2CH_3$
       $|$
       OH

1-Penten-3-ol
$C_5H_{10}O = 86.14$
CW 20M: 1130
OV-101: 673

2-n-Pentylfuran
$C_9H_{14}O = 138.21$
CW 20M: 1229
OV-101: 983

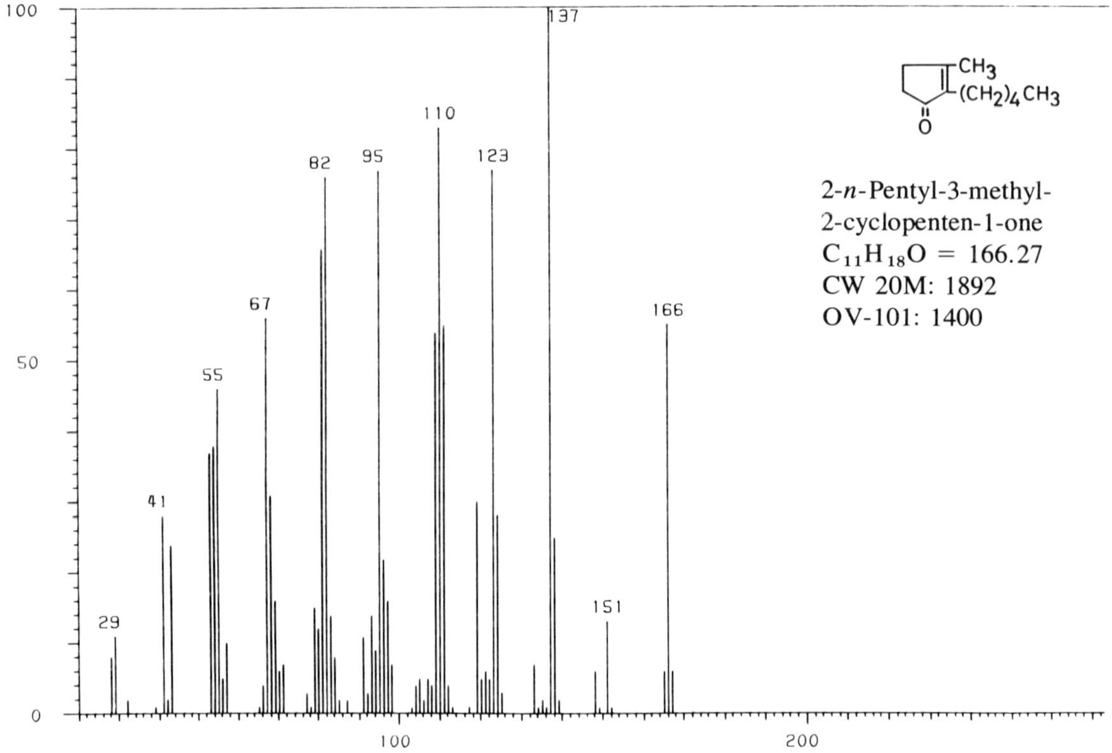

2-n-Pentyl-3-methyl-2-cyclopenten-1-one
$C_{11}H_{18}O = 166.27$
CW 20M: 1892
OV-101: 1400

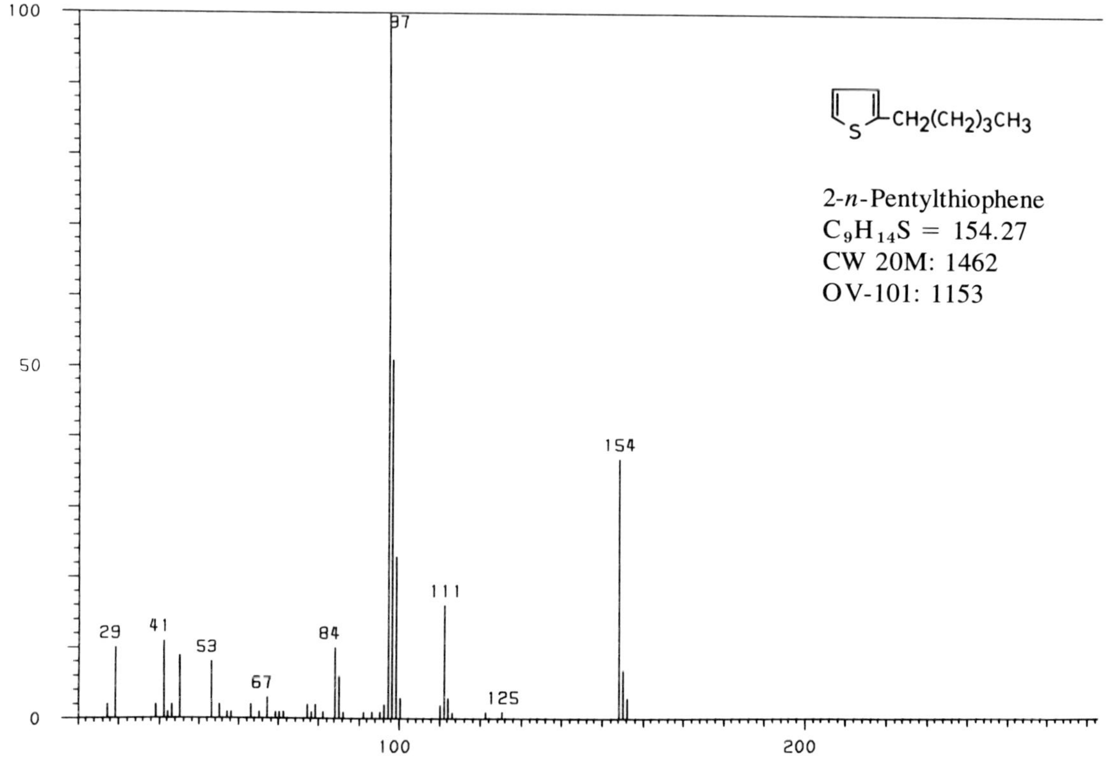

2-n-Pentylthiophene
$C_9H_{14}S = 154.27$
CW 20M: 1462
OV-101: 1153

APPENDIX IV   MASS SPECTRA OF INDIVIDUAL COMPOUNDS

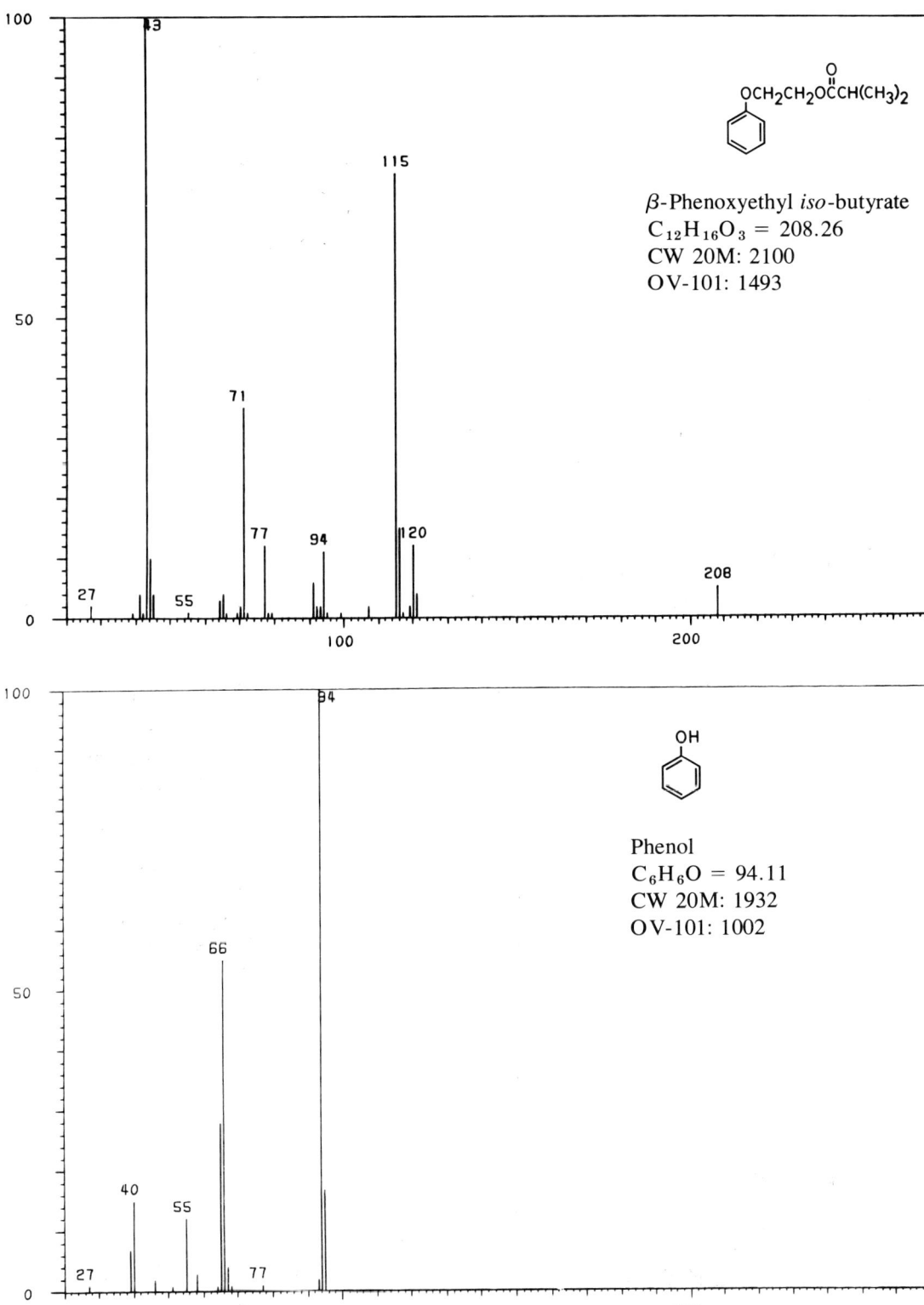

β-Phenoxyethyl *iso*-butyrate
$C_{12}H_{16}O_3 = 208.26$
CW 20M: 2100
OV-101: 1493

Phenol
$C_6H_6O = 94.11$
CW 20M: 1932
OV-101: 1002

Phenylacetaldehyde dimethyl acetal
$C_{10}H_{14}O_2$ = 166.22
CW 20M: 1665
OV-101: 1200

Phenyl *iso*-butyl methyl-carbinyl acetate
$C_{14}H_{20}O_2$ = 220.31
CW 20M: 1943
OV-101: 1459

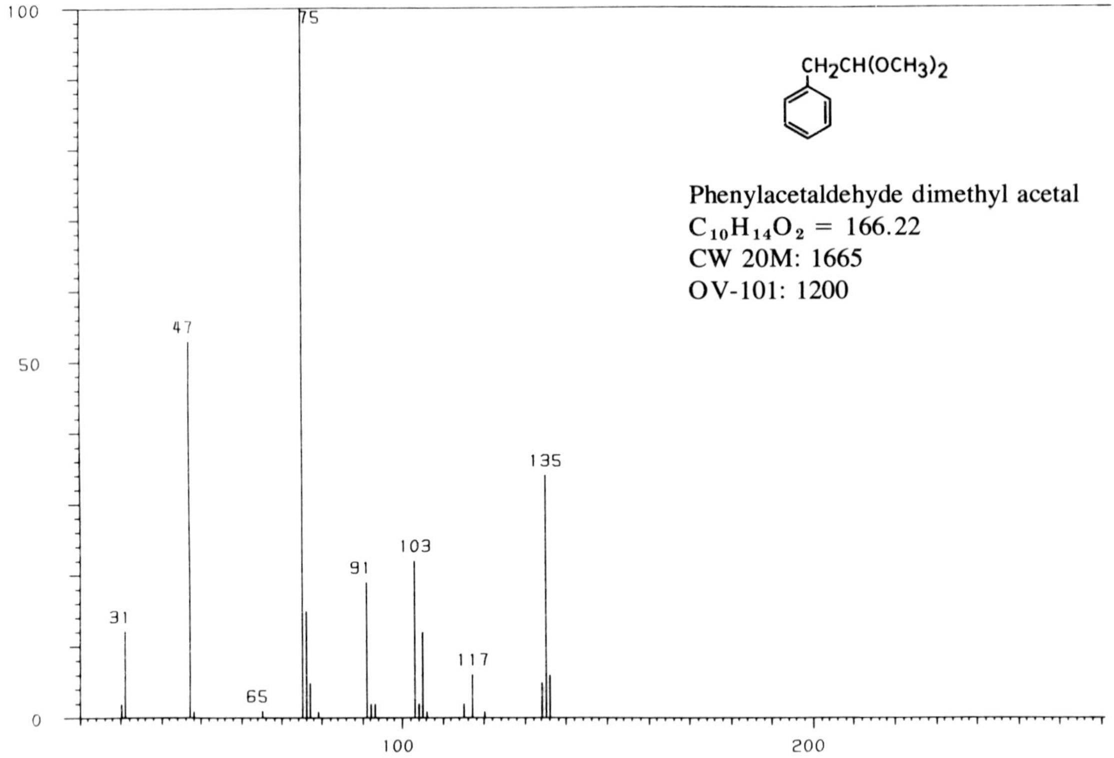

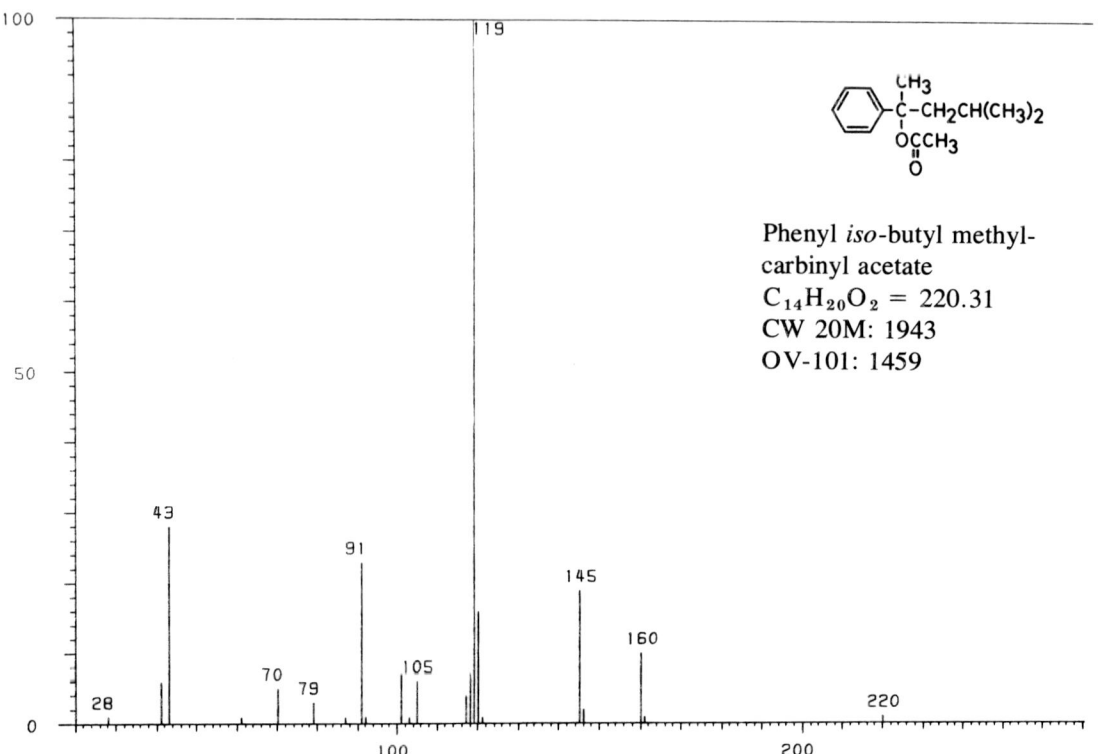

APPENDIX IV   MASS SPECTRA OF INDIVIDUAL COMPOUNDS   413

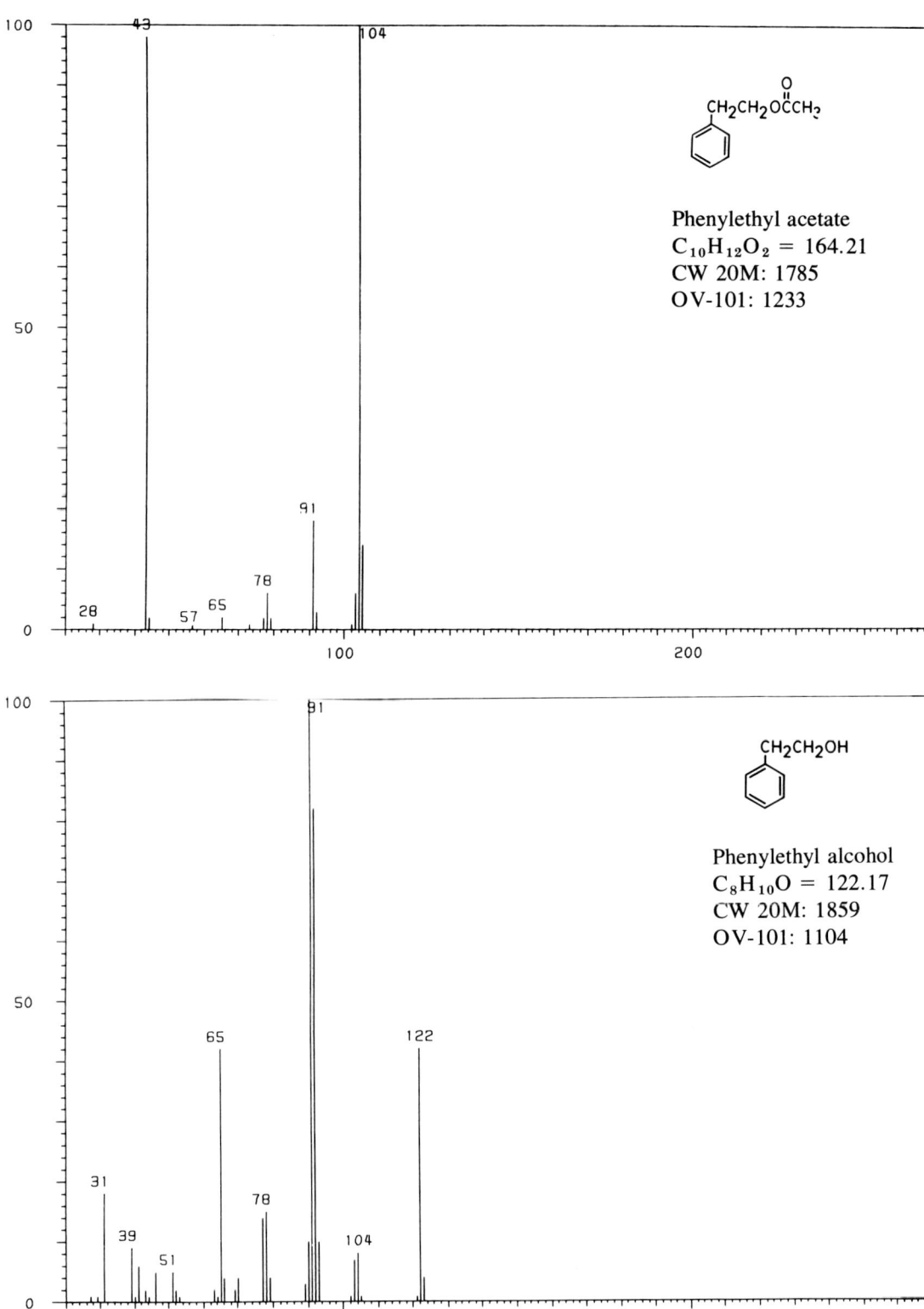

Phenylethyl acetate
$C_{10}H_{12}O_2 = 164.21$
CW 20M: 1785
OV-101: 1233

Phenylethyl alcohol
$C_8H_{10}O = 122.17$
CW 20M: 1859
OV-101: 1104

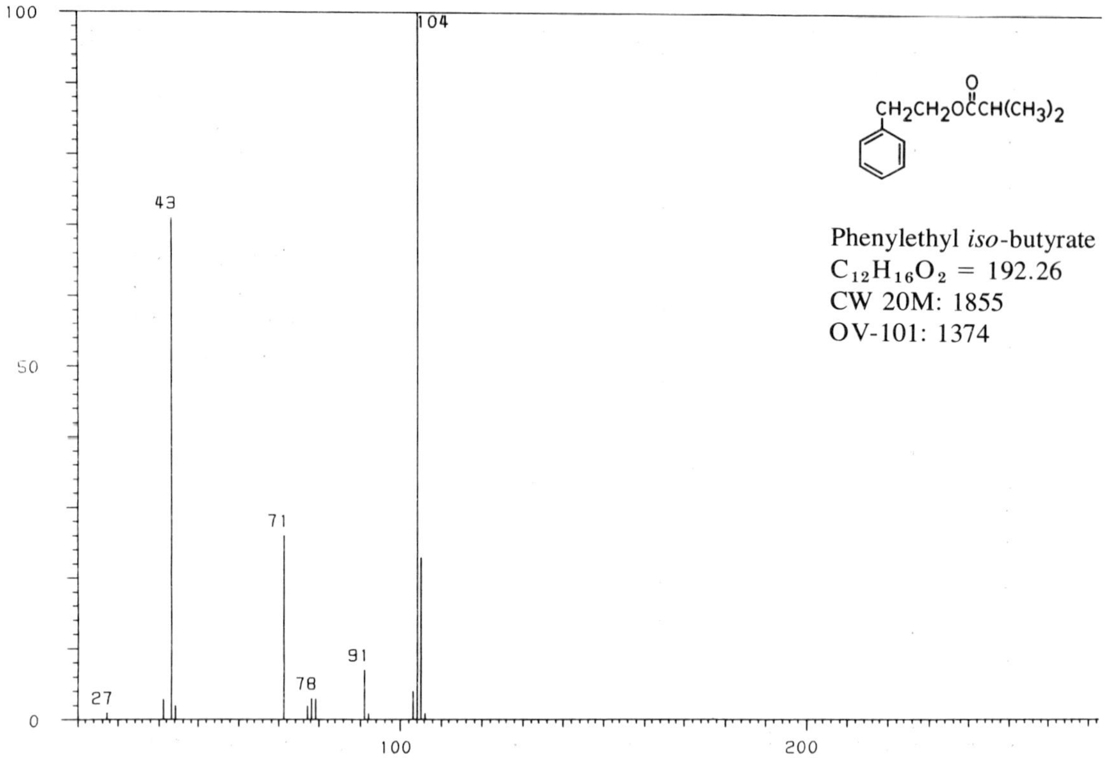

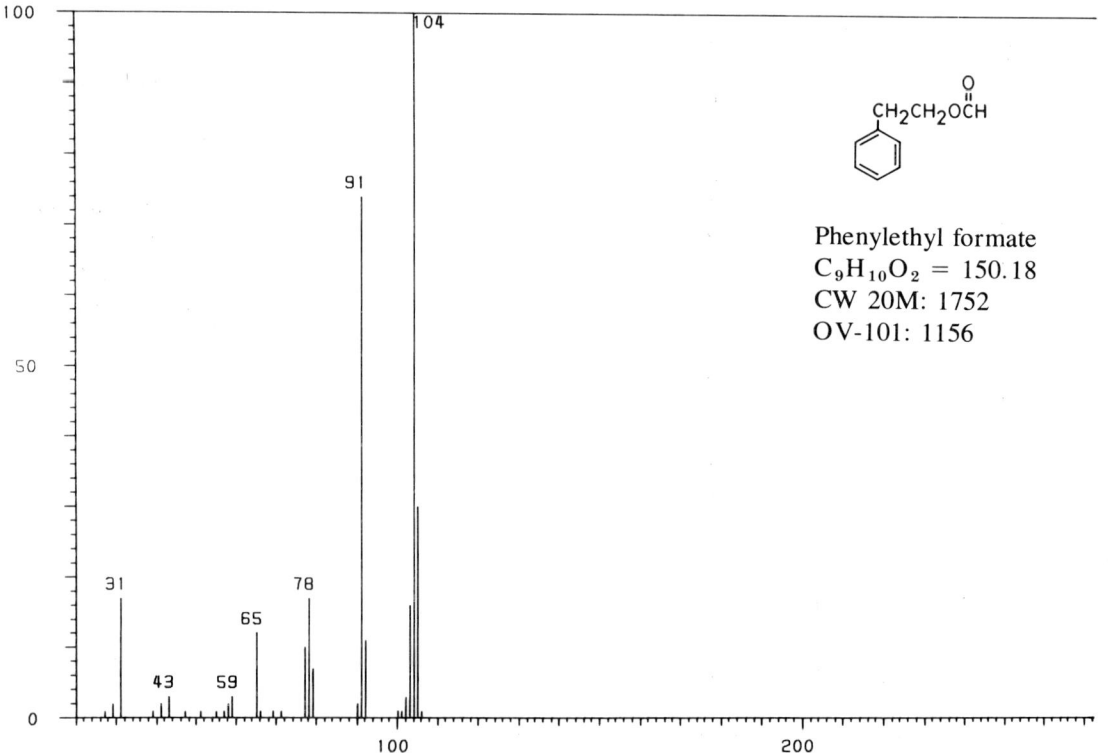

Phenylethyl *iso*-butyrate
$C_{12}H_{16}O_2 = 192.26$
CW 20M: 1855
OV-101: 1374

Phenylethyl formate
$C_9H_{10}O_2 = 150.18$
CW 20M: 1752
OV-101: 1156

APPENDIX IV   MASS SPECTRA OF INDIVIDUAL COMPOUNDS

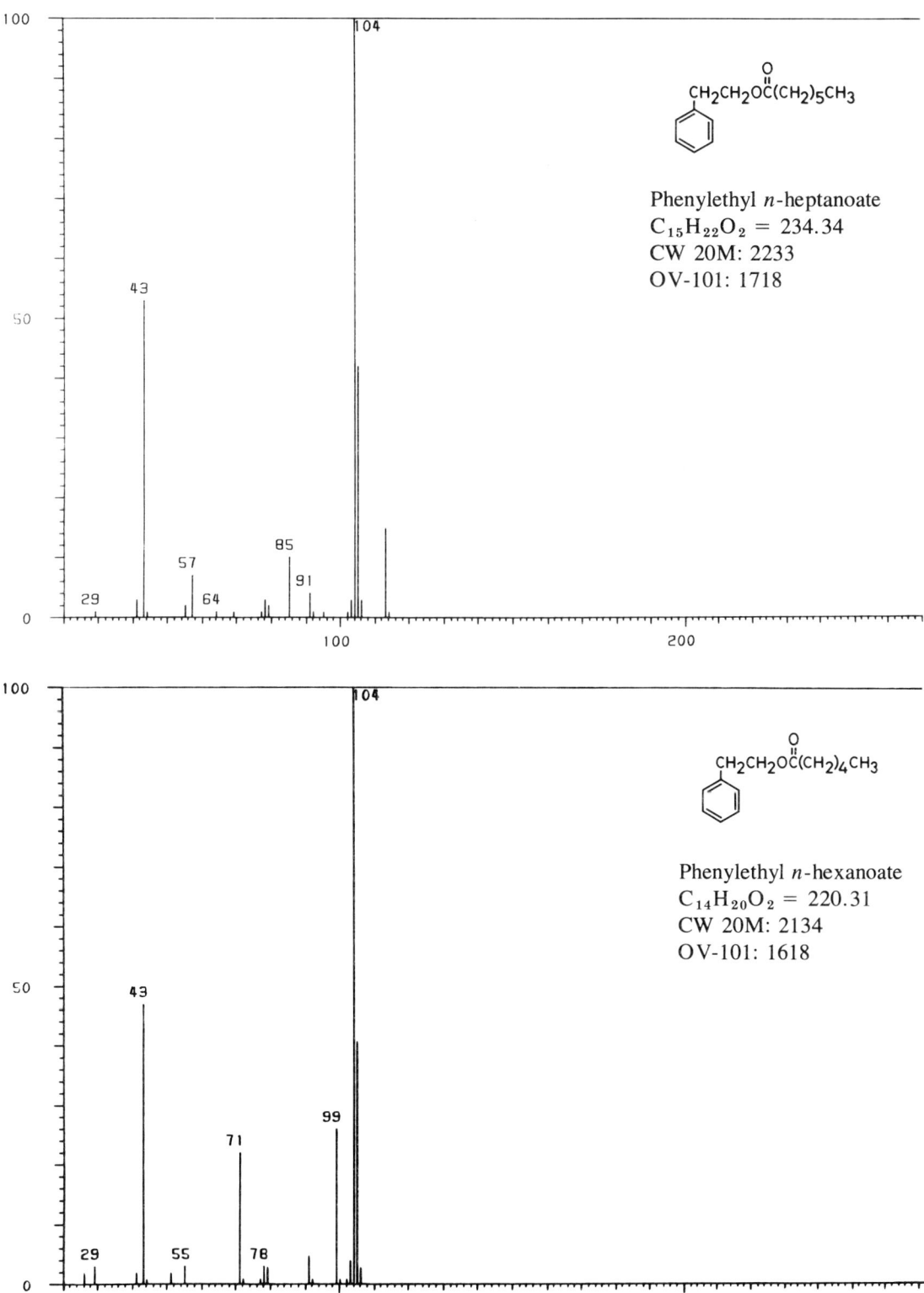

Phenylethyl n-heptanoate
$C_{15}H_{22}O_2 = 234.34$
CW 20M: 2233
OV-101: 1718

Phenylethyl n-hexanoate
$C_{14}H_{20}O_2 = 220.31$
CW 20M: 2134
OV-101: 1618

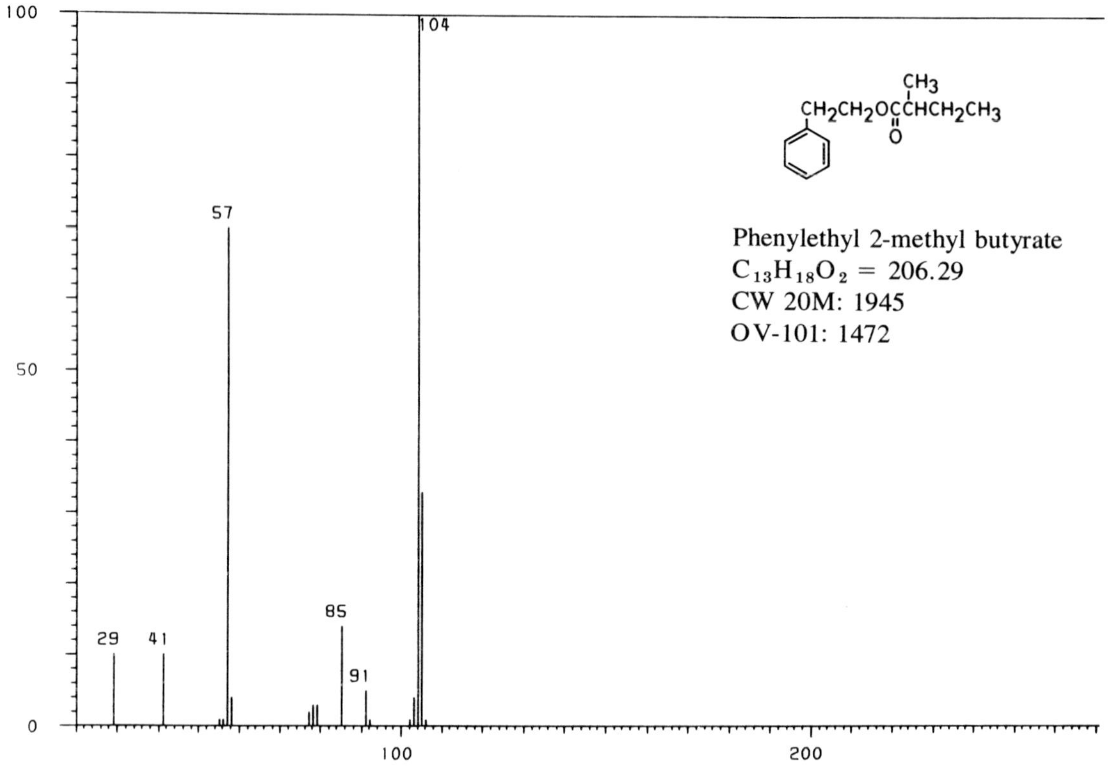

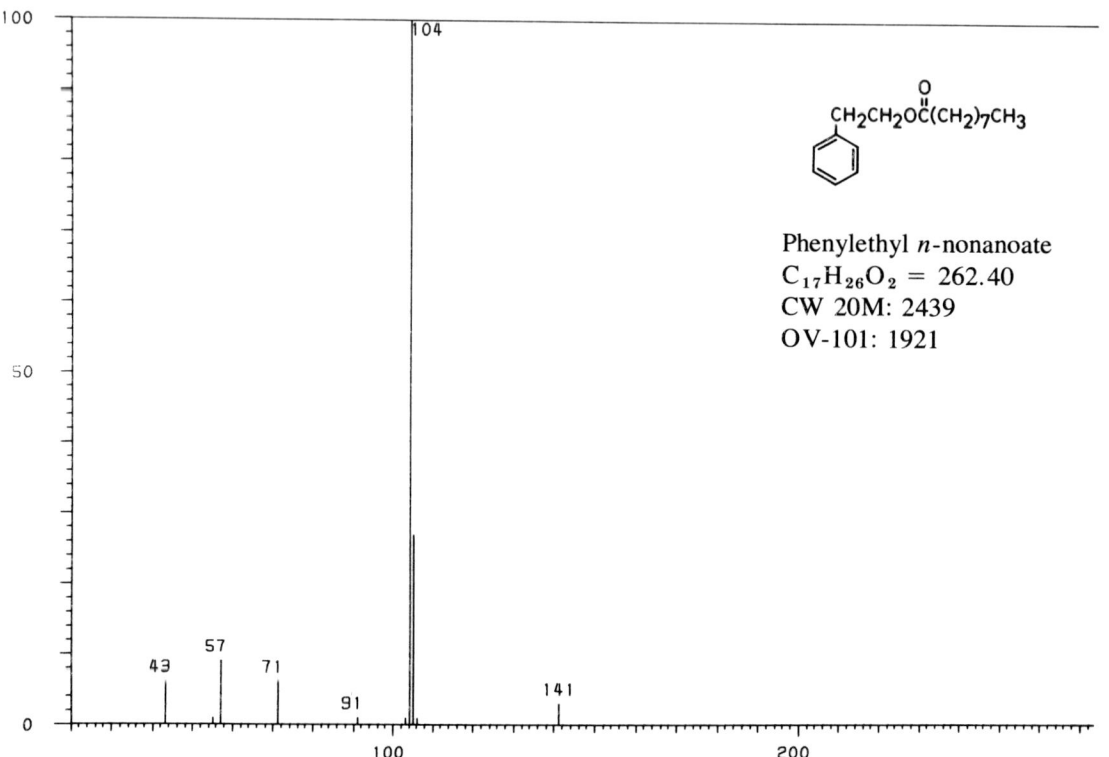

Phenylethyl 2-methyl butyrate
$C_{13}H_{18}O_2 = 206.29$
CW 20M: 1945
OV-101: 1472

Phenylethyl *n*-nonanoate
$C_{17}H_{26}O_2 = 262.40$
CW 20M: 2439
OV-101: 1921

APPENDIX IV MASS SPECTRA OF INDIVIDUAL COMPOUNDS

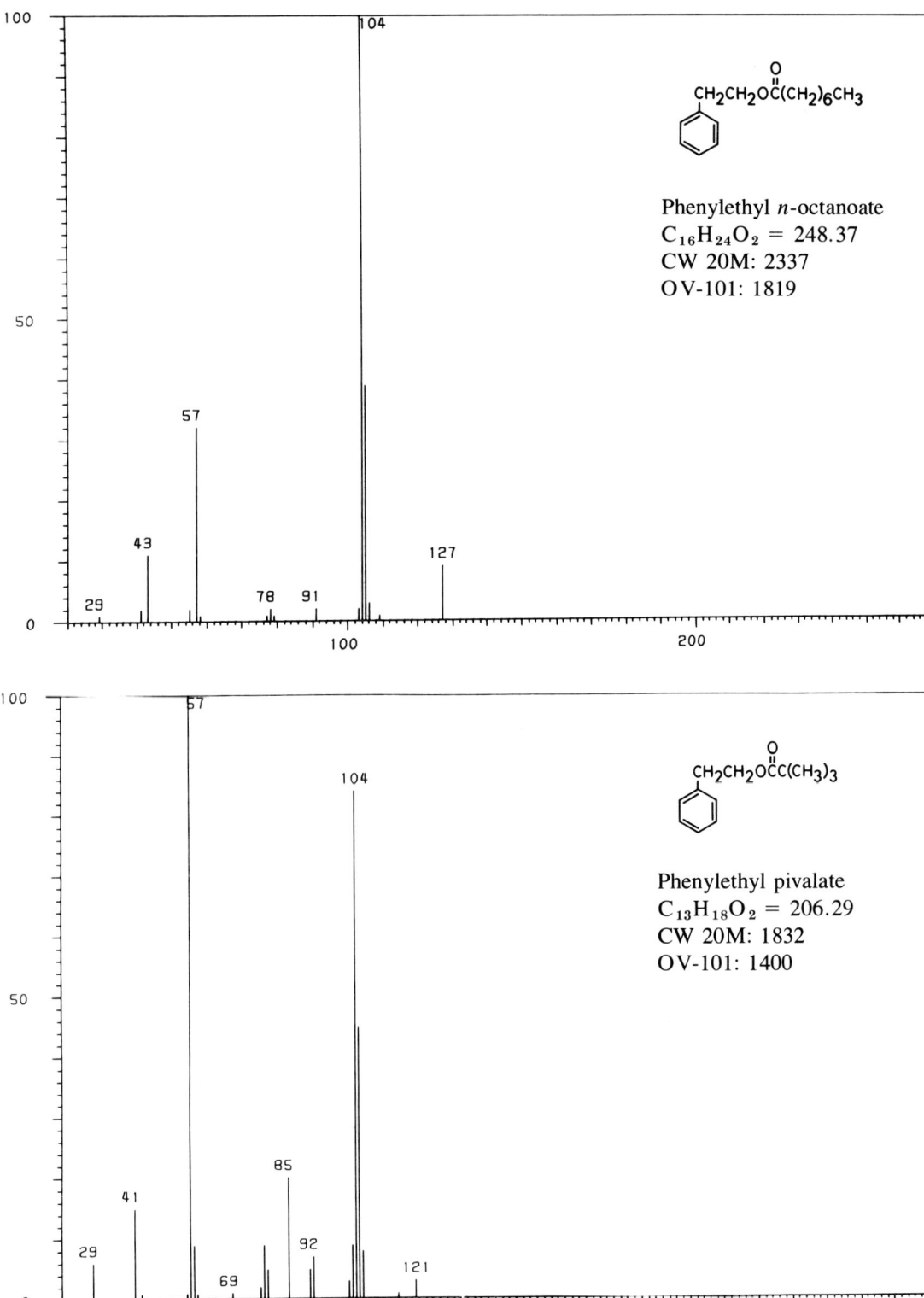

Phenylethyl *n*-octanoate
$C_{16}H_{24}O_2 = 248.37$
CW 20M: 2337
OV-101: 1819

Phenylethyl pivalate
$C_{13}H_{18}O_2 = 206.29$
CW 20M: 1832
OV-101: 1400

Phenylethyl *n*-propionate
$C_{11}H_{14}O_2$ = 178.23
CW 20M: 1855
OV-101: 1328

Phenylethyl tiglate
$C_{13}H_{16}O_2$ = 204.27
CW 20M: 2154
OV-101: 1562

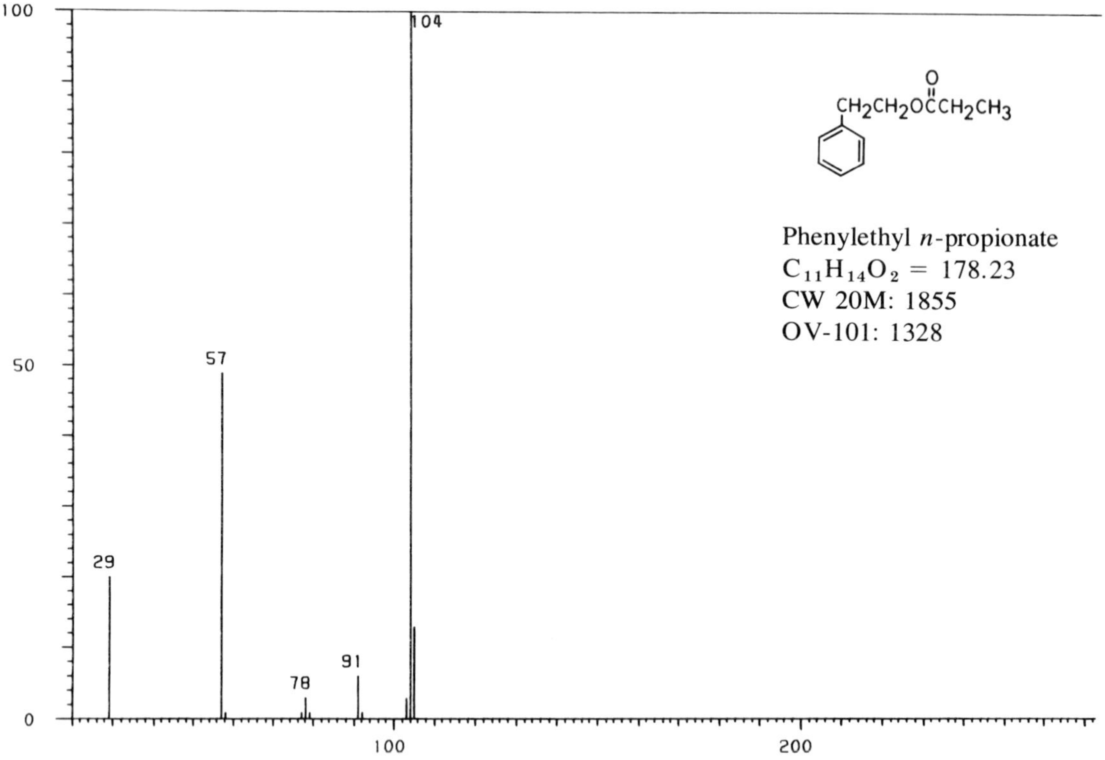

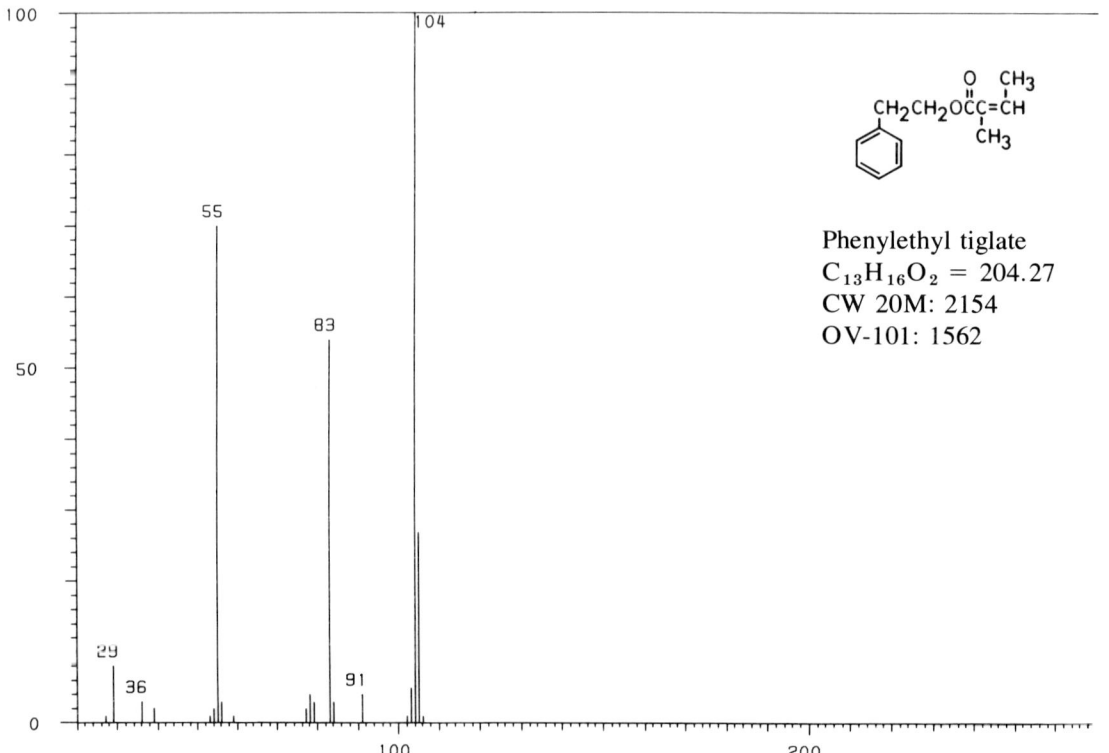

APPENDIX IV   MASS SPECTRA OF INDIVIDUAL COMPOUNDS   **419**

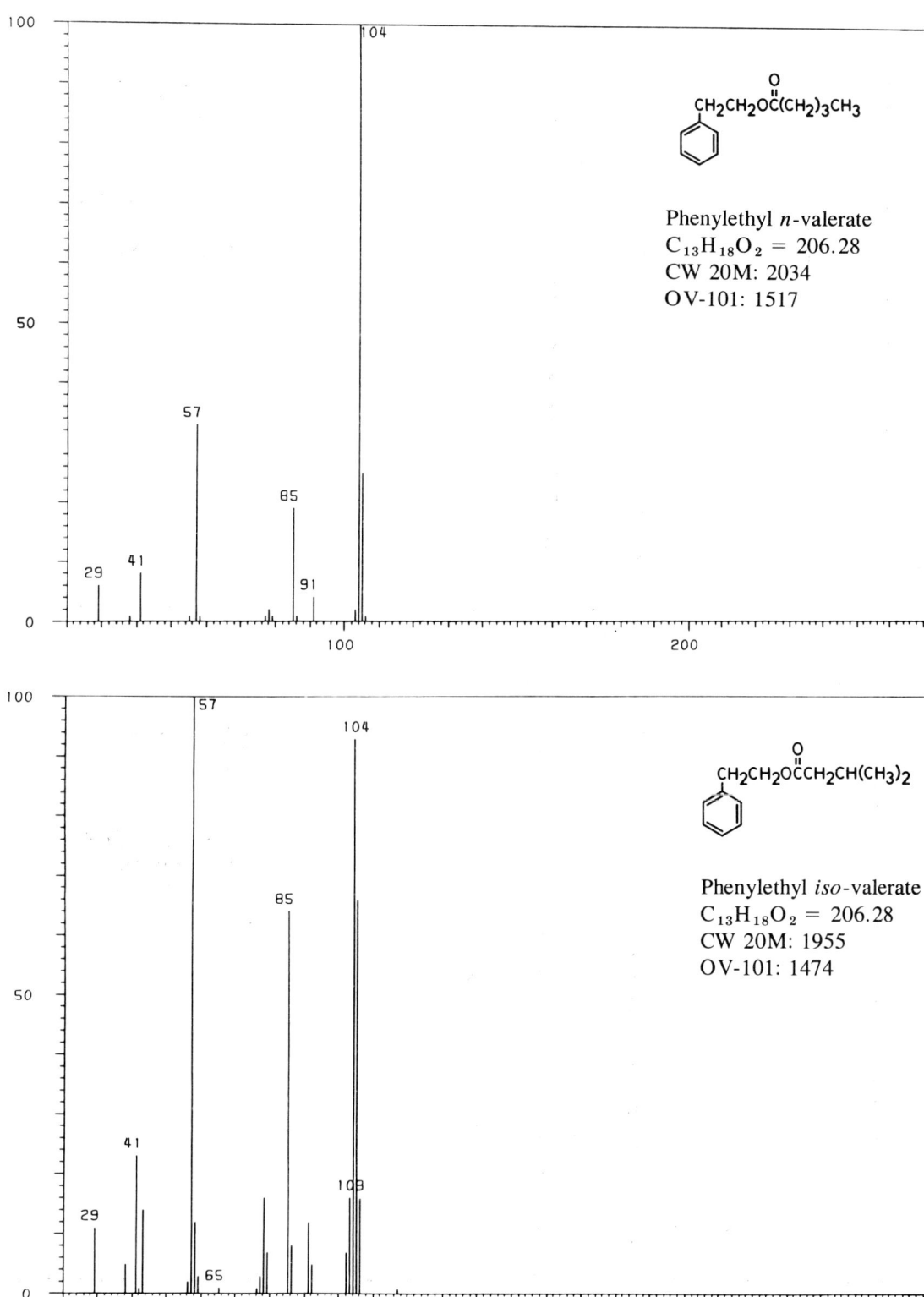

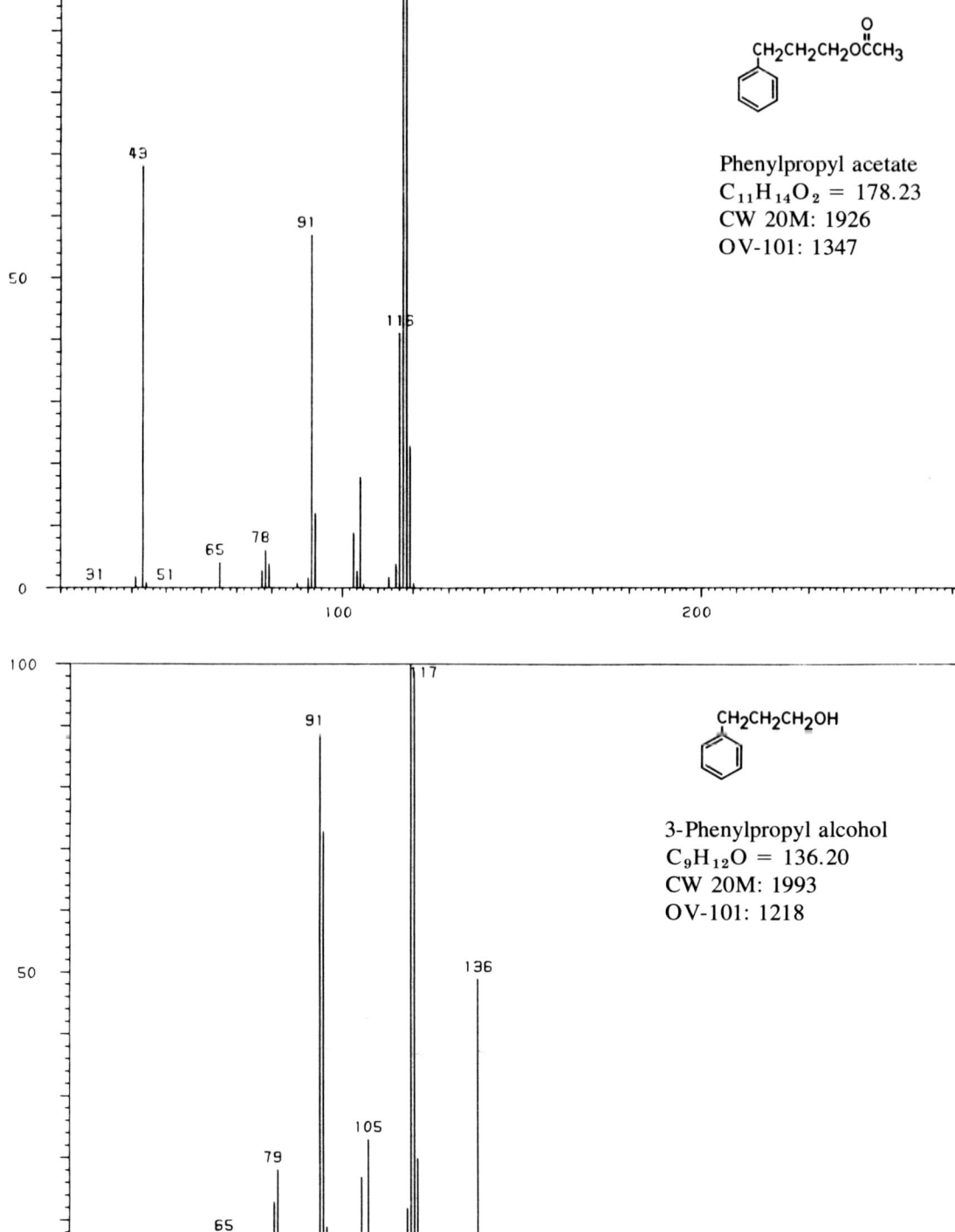

APPENDIX IV   MASS SPECTRA OF INDIVIDUAL COMPOUNDS

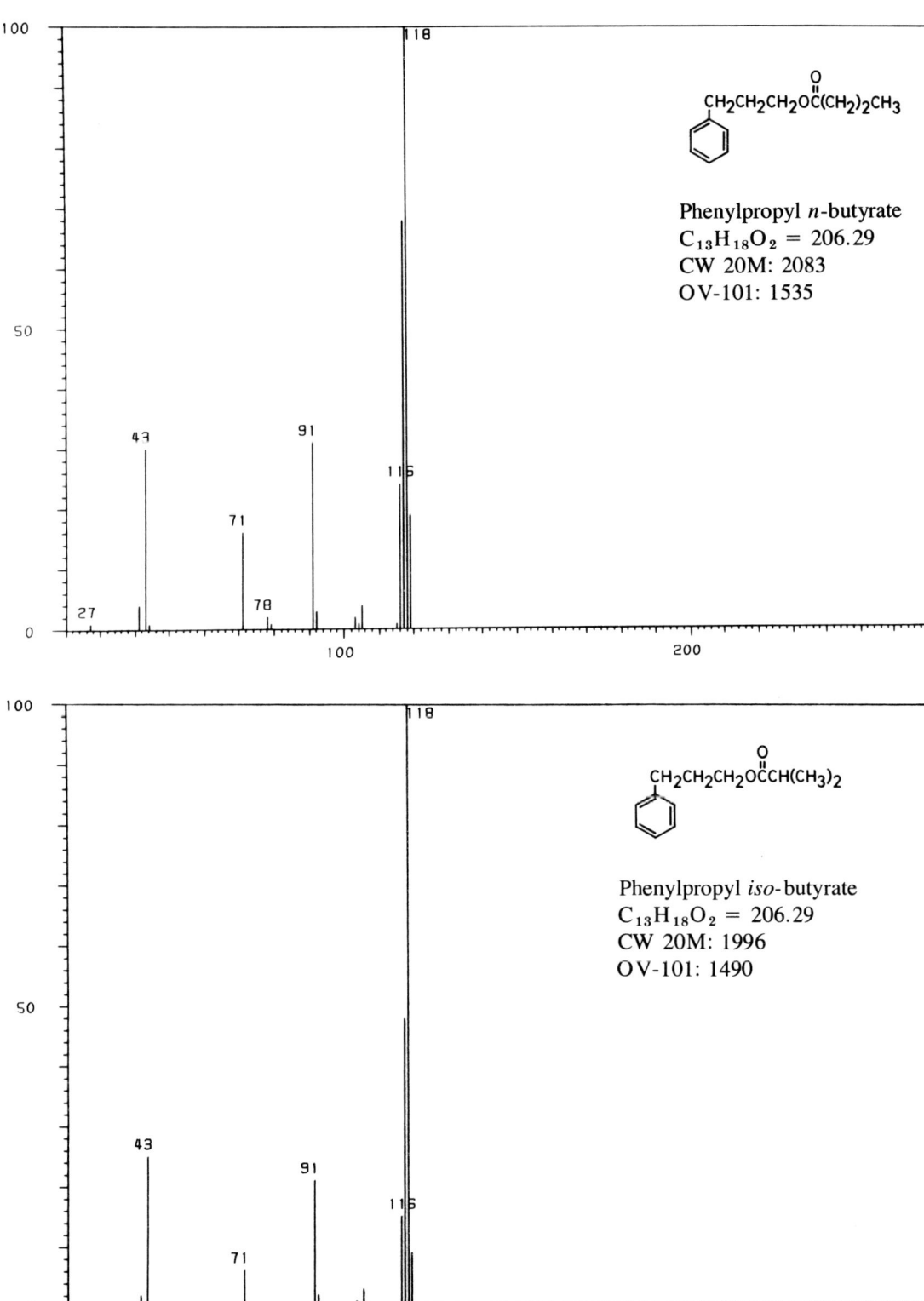

Phenylpropyl *n*-butyrate
$C_{13}H_{18}O_2 = 206.29$
CW 20M: 2083
OV-101: 1535

Phenylpropyl *iso*-butyrate
$C_{13}H_{18}O_2 = 206.29$
CW 20M: 1996
OV-101: 1490

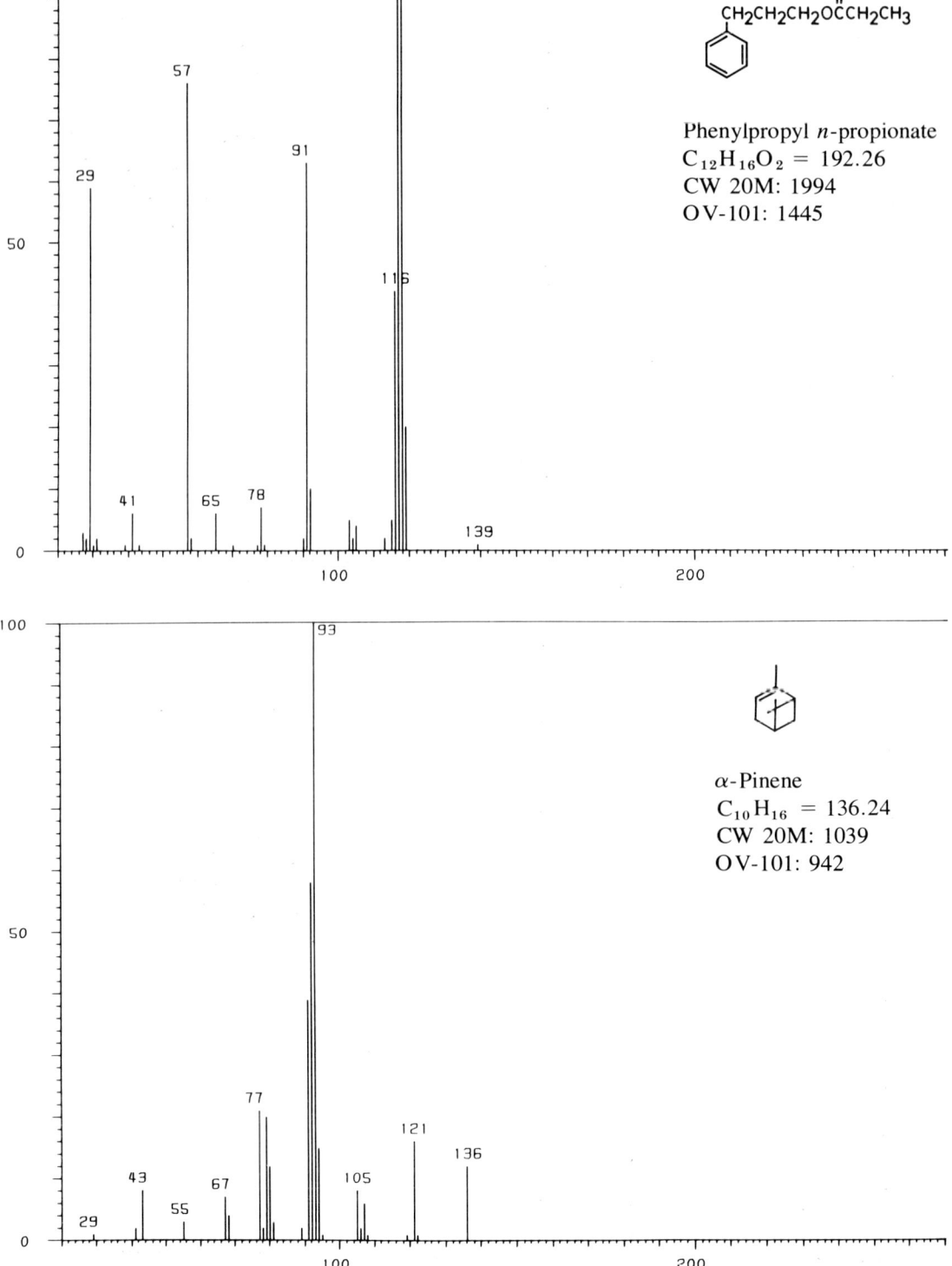

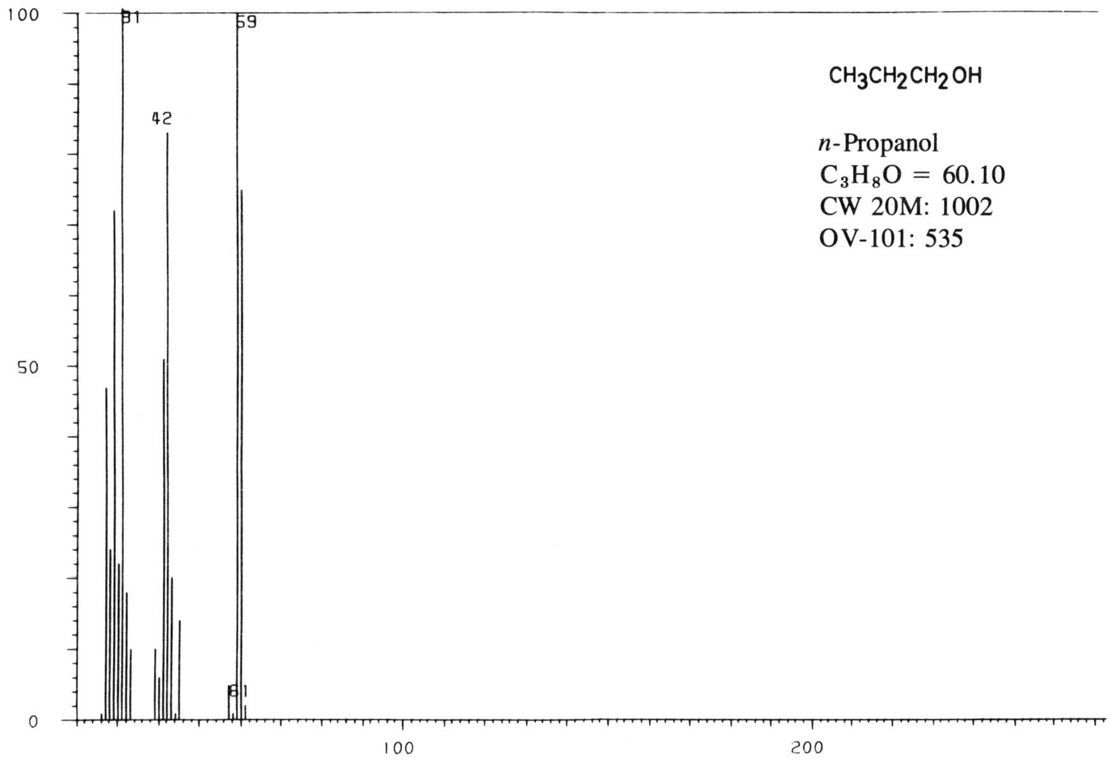

APPENDIX IV   MASS SPECTRA OF INDIVIDUAL COMPOUNDS

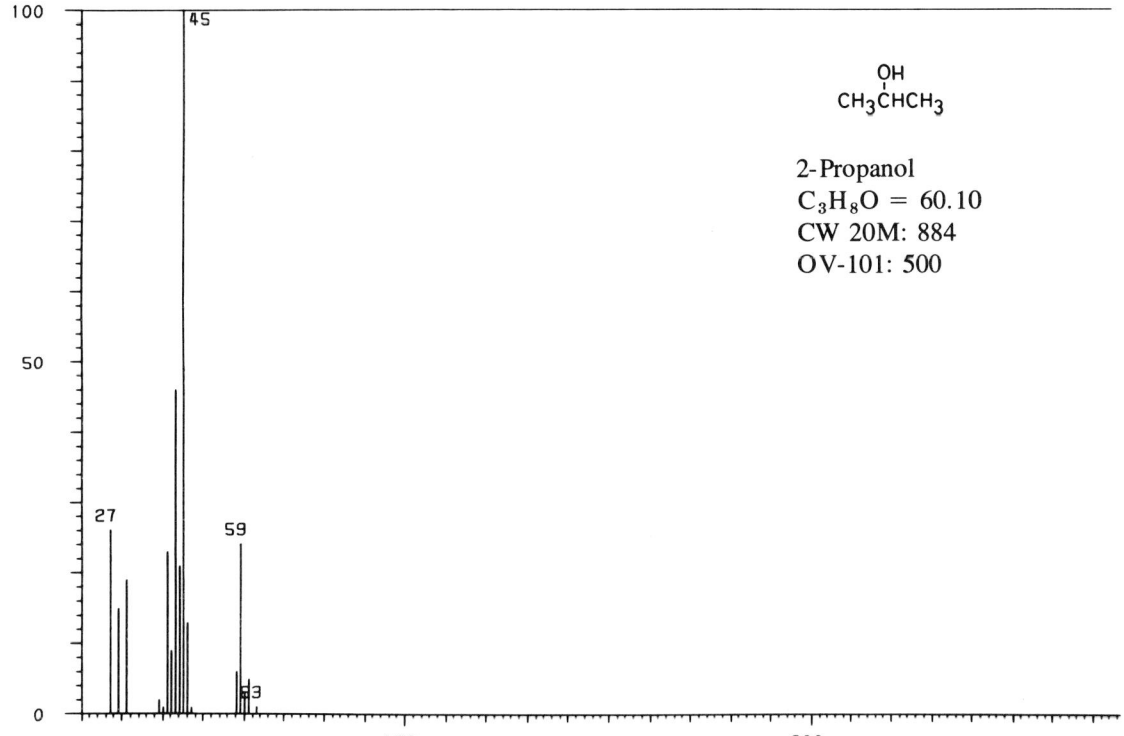

CH₃CH₂CH₂OH

n-Propanol
$C_3H_8O = 60.10$
CW 20M: 1002
OV-101: 535

OH
|
CH₃CHCH₃

2-Propanol
$C_3H_8O = 60.10$
CW 20M: 884
OV-101: 500

Propionaldehyde-1-phenyl-1,2-ethanediol acetal 1
$C_{11}H_{14}O_2$ = 178.23
CW 20M: 1871
OV-101: 1327
(Refer to CW 20M: 1880)

Propionaldehyde-1-phenyl-1,2-ethanediol acetal 2
$C_{11}H_{14}O_2$ = 178.23
CW 20M: 1880
OV-101: 1333
(Refer to CW 20M: 1871)

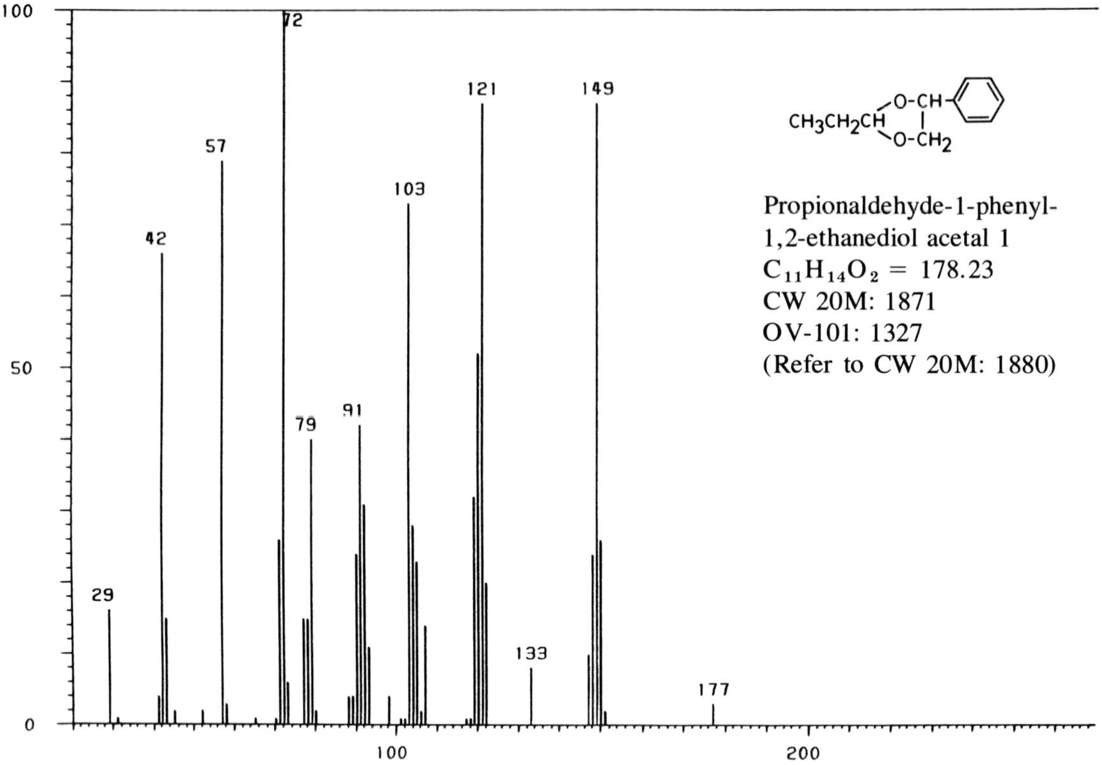

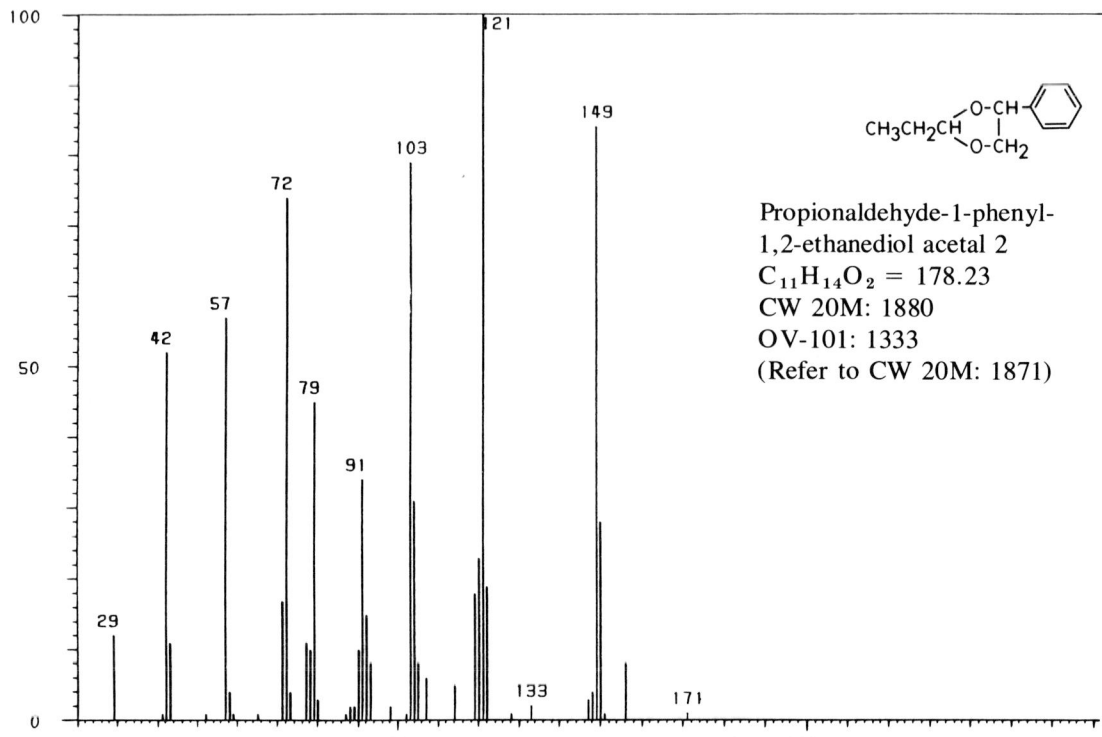

APPENDIX IV MASS SPECTRA OF INDIVIDUAL COMPOUNDS 425

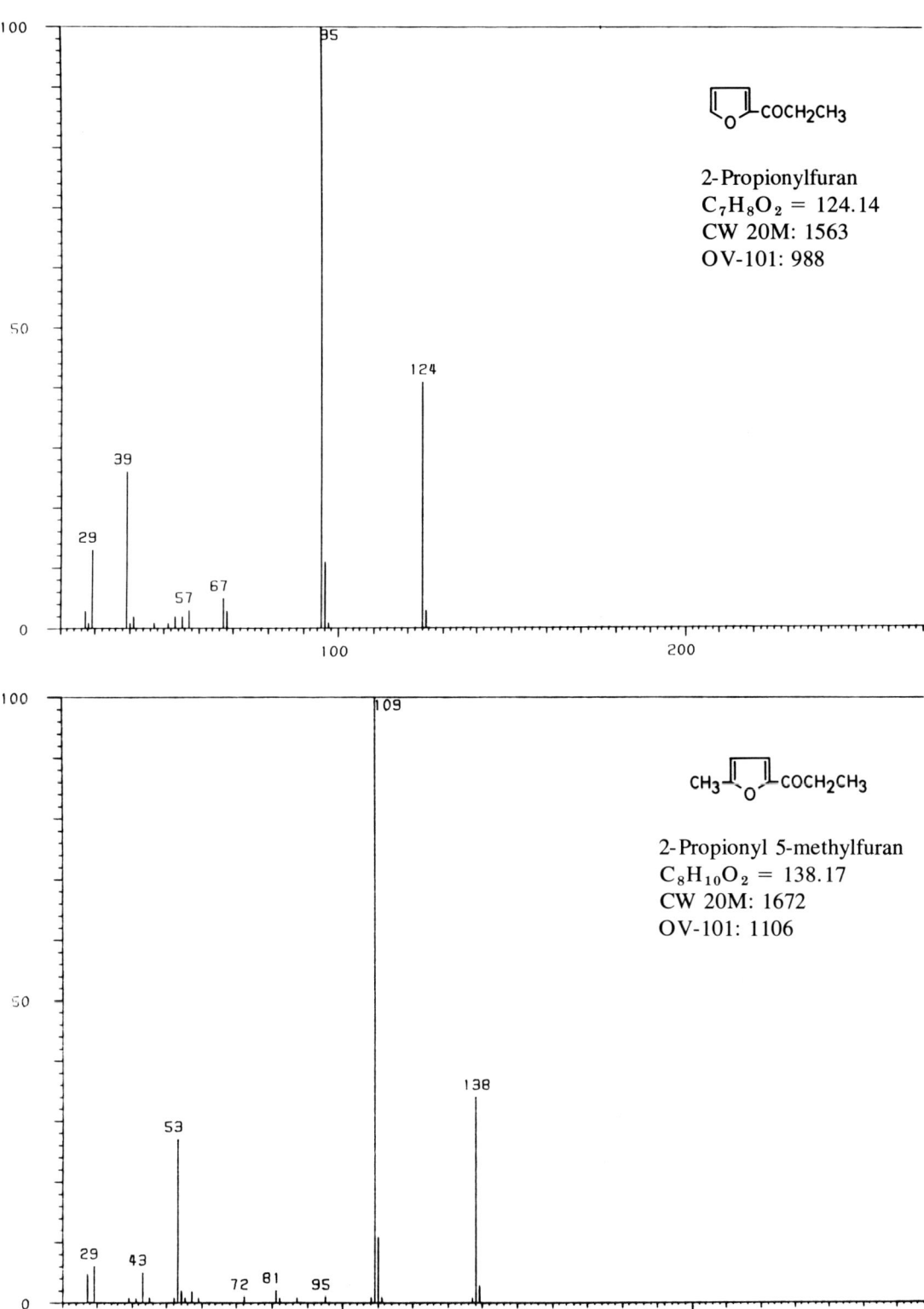

2-Propionylfuran
$C_7H_8O_2 = 124.14$
CW 20M: 1563
OV-101: 988

2-Propionyl 5-methylfuran
$C_8H_{10}O_2 = 138.17$
CW 20M: 1672
OV-101: 1106

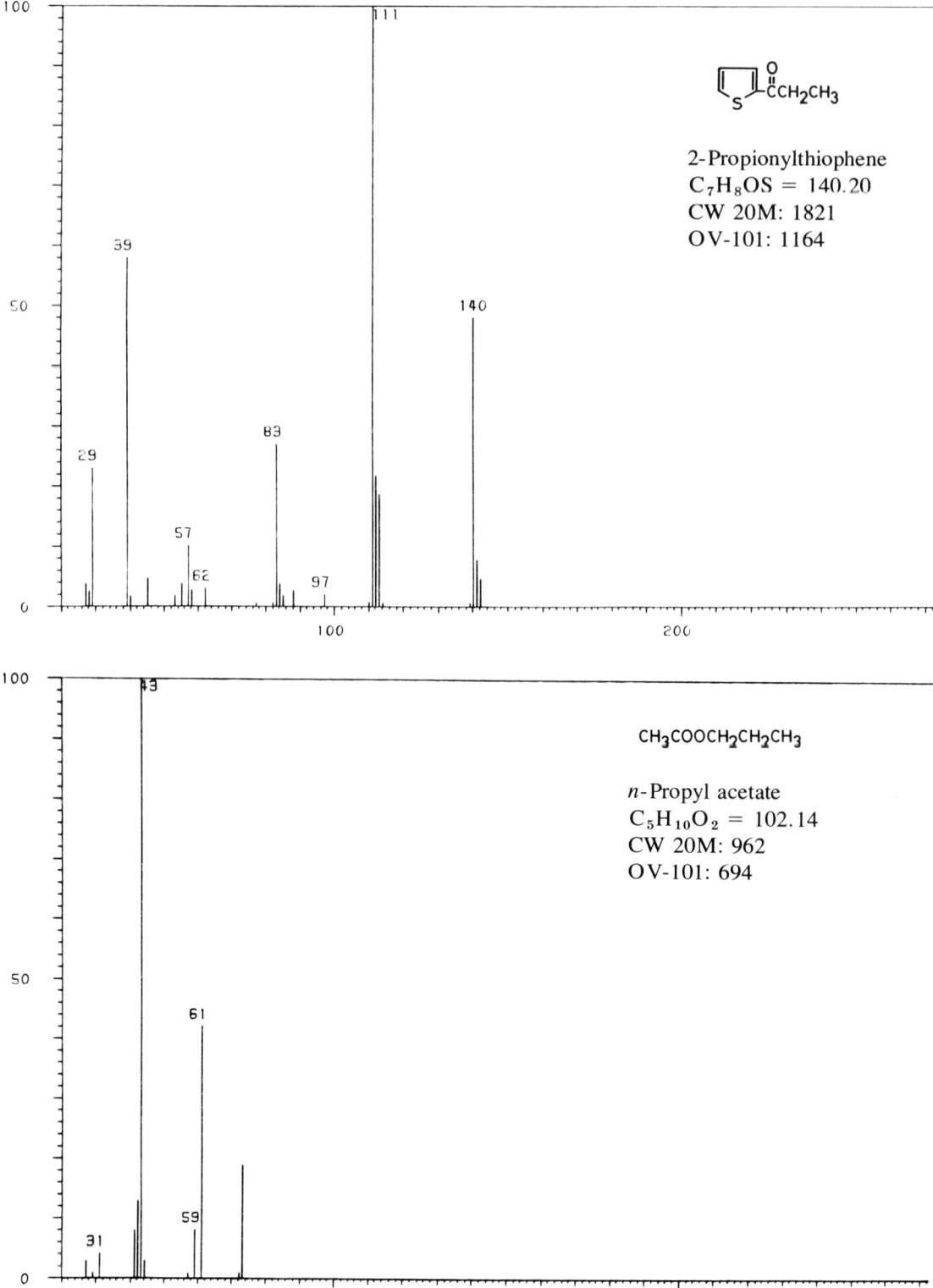

APPENDIX IV   MASS SPECTRA OF INDIVIDUAL COMPOUNDS

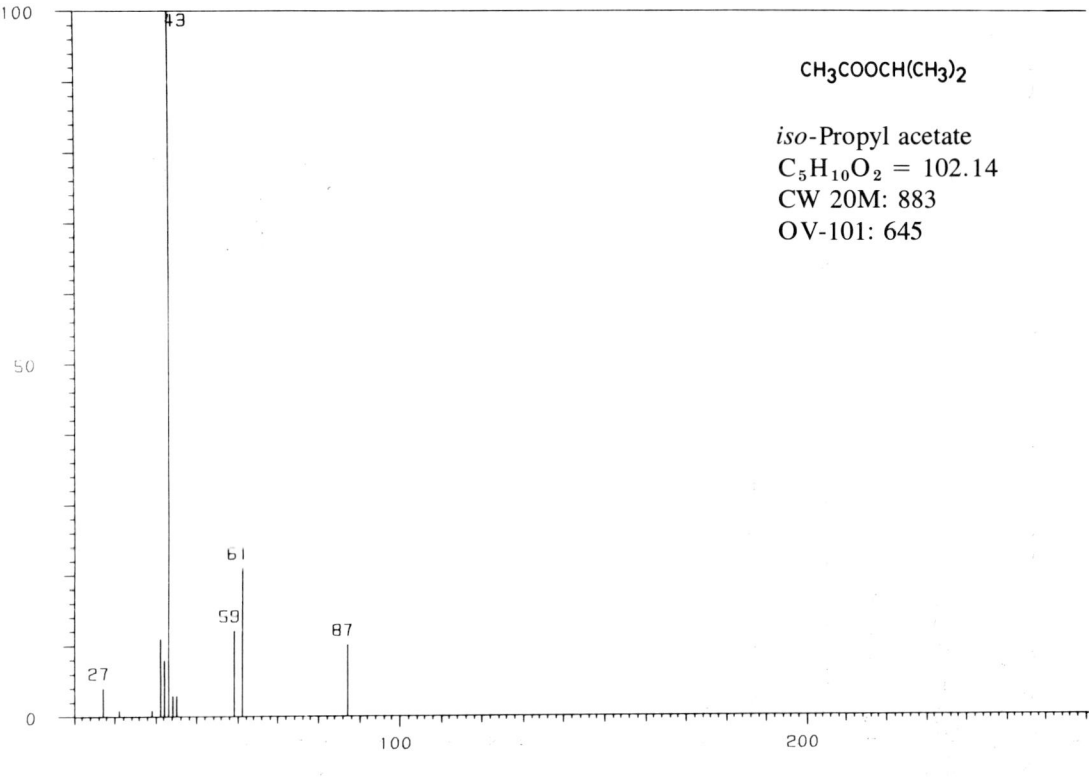

CH₃COOCH(CH₃)₂

iso-Propyl acetate
$C_5H_{10}O_2 = 102.14$
CW 20M: 883
OV-101: 645

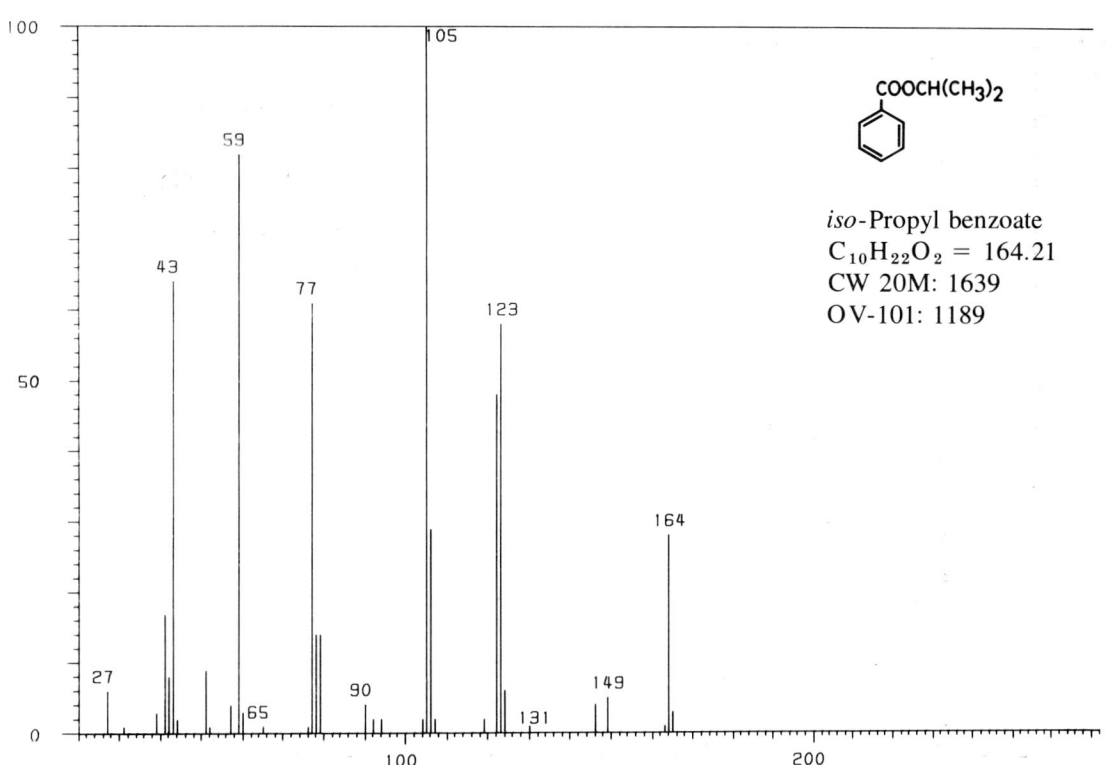

iso-Propyl benzoate
$C_{10}H_{22}O_2 = 164.21$
CW 20M: 1639
OV-101: 1189

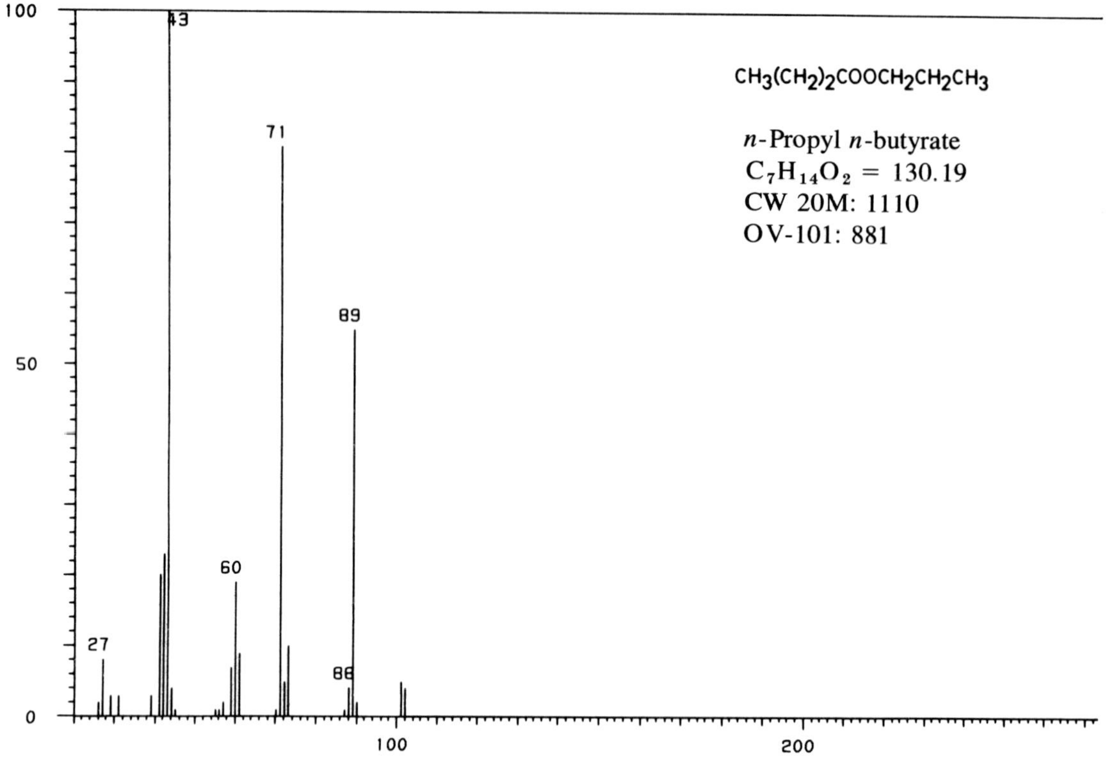
100
50
0
43
71
89
60
27
86
100
200
CH3(CH2)2COOCH2CH2CH3

n-Propyl *n*-butyrate
$C_7H_{14}O_2$ = 130.19
CW 20M: 1110
OV-101: 881

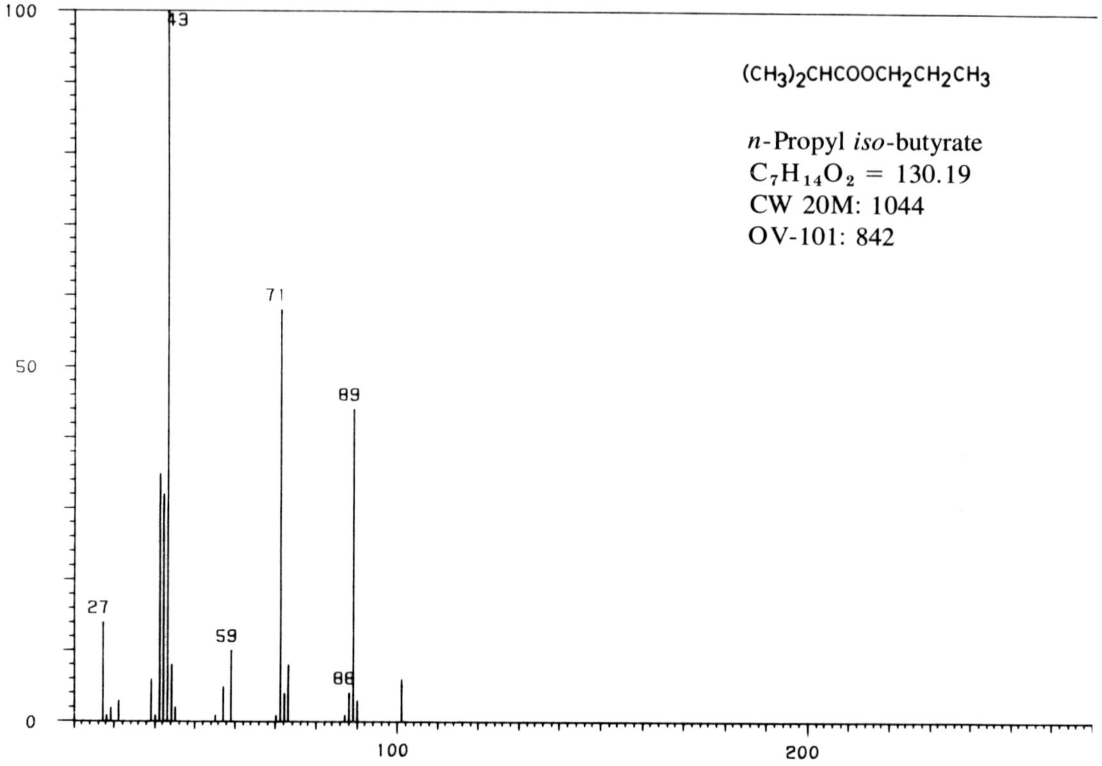
100
50
0
43
71
89
27
59
86
100
200
(CH3)2CHCOOCH2CH2CH3

n-Propyl *iso*-butyrate
$C_7H_{14}O_2$ = 130.19
CW 20M: 1044
OV-101: 842

APPENDIX IV MASS SPECTRA OF INDIVIDUAL COMPOUNDS

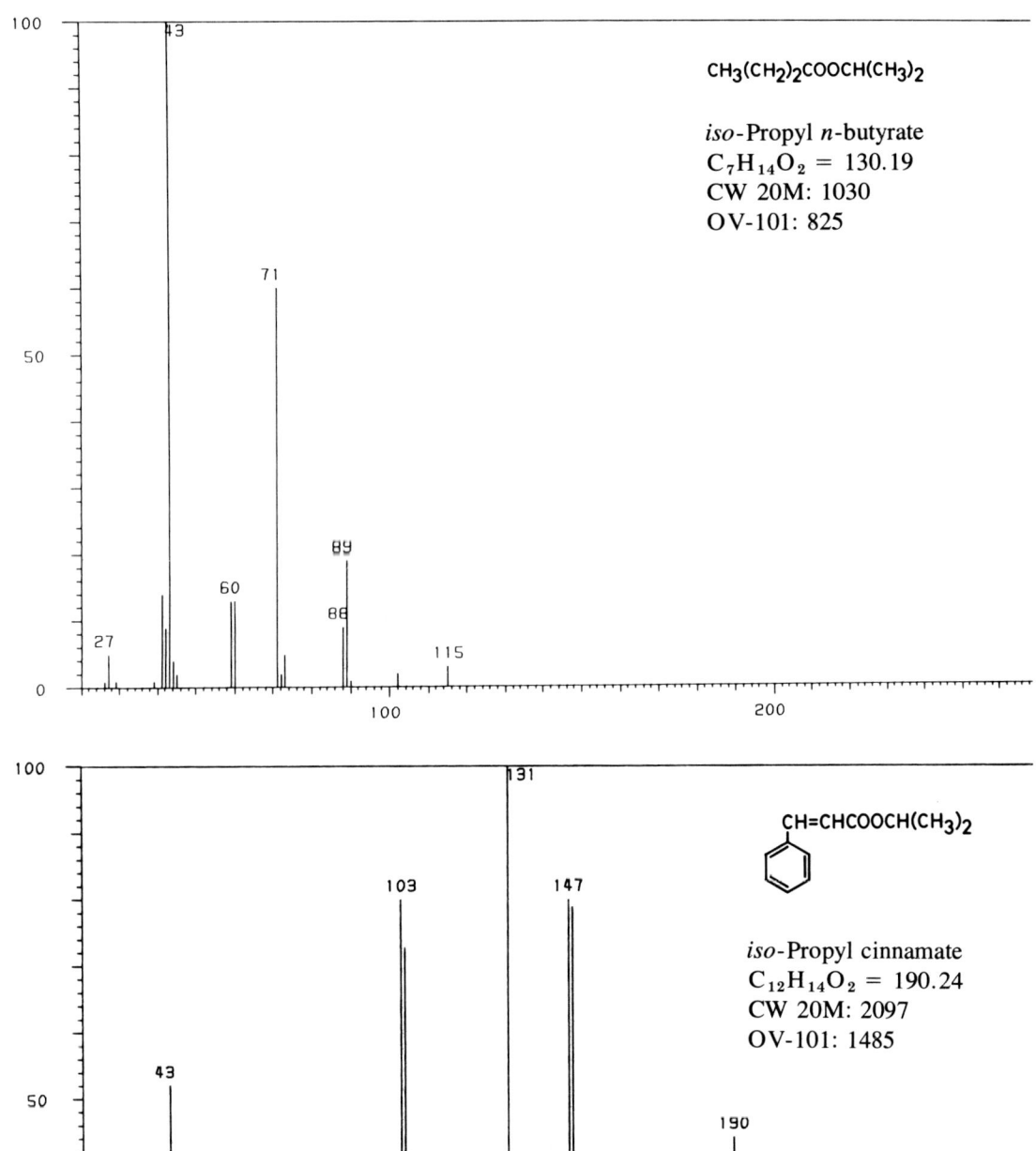

$CH_3(CH_2)_2COOCH(CH_3)_2$

iso-Propyl *n*-butyrate
$C_7H_{14}O_2 = 130.19$
CW 20M: 1030
OV-101: 825

iso-Propyl cinnamate
$C_{12}H_{14}O_2 = 190.24$
CW 20M: 2097
OV-101: 1485

2-*n*-Propyl-4,5-dimethyloxazole
$C_8H_{13}NO$ = 139.20
CW 20M: 1310
OV-101: 996

2-*iso*-Propyl-4,5-dimethyloxazole
$C_8H_{13}NO$ = 139.20
CW 20M: 1249
OV-101: 960

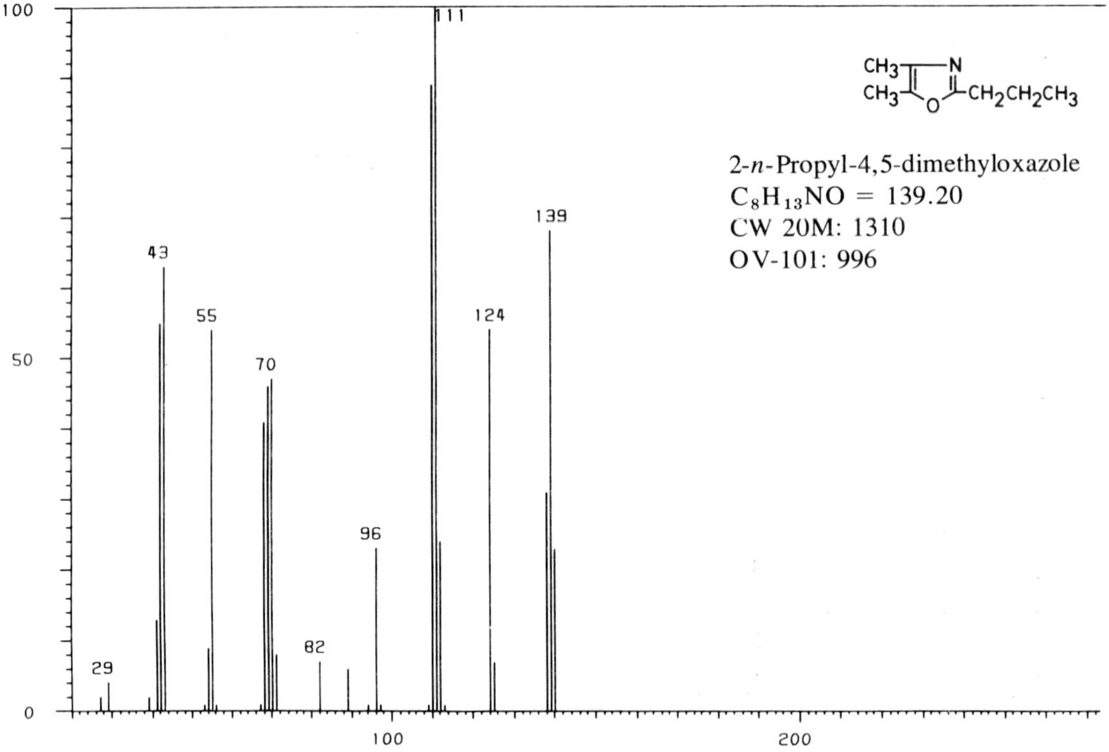

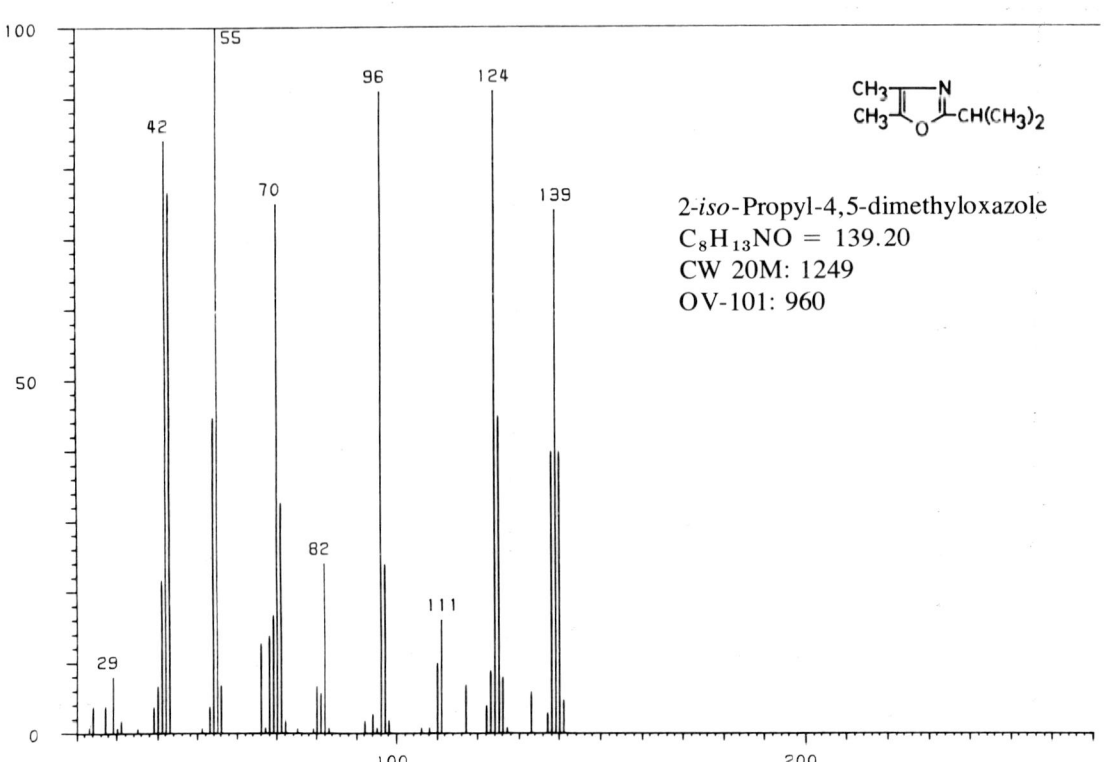

APPENDIX IV   MASS SPECTRA OF INDIVIDUAL COMPOUNDS   431

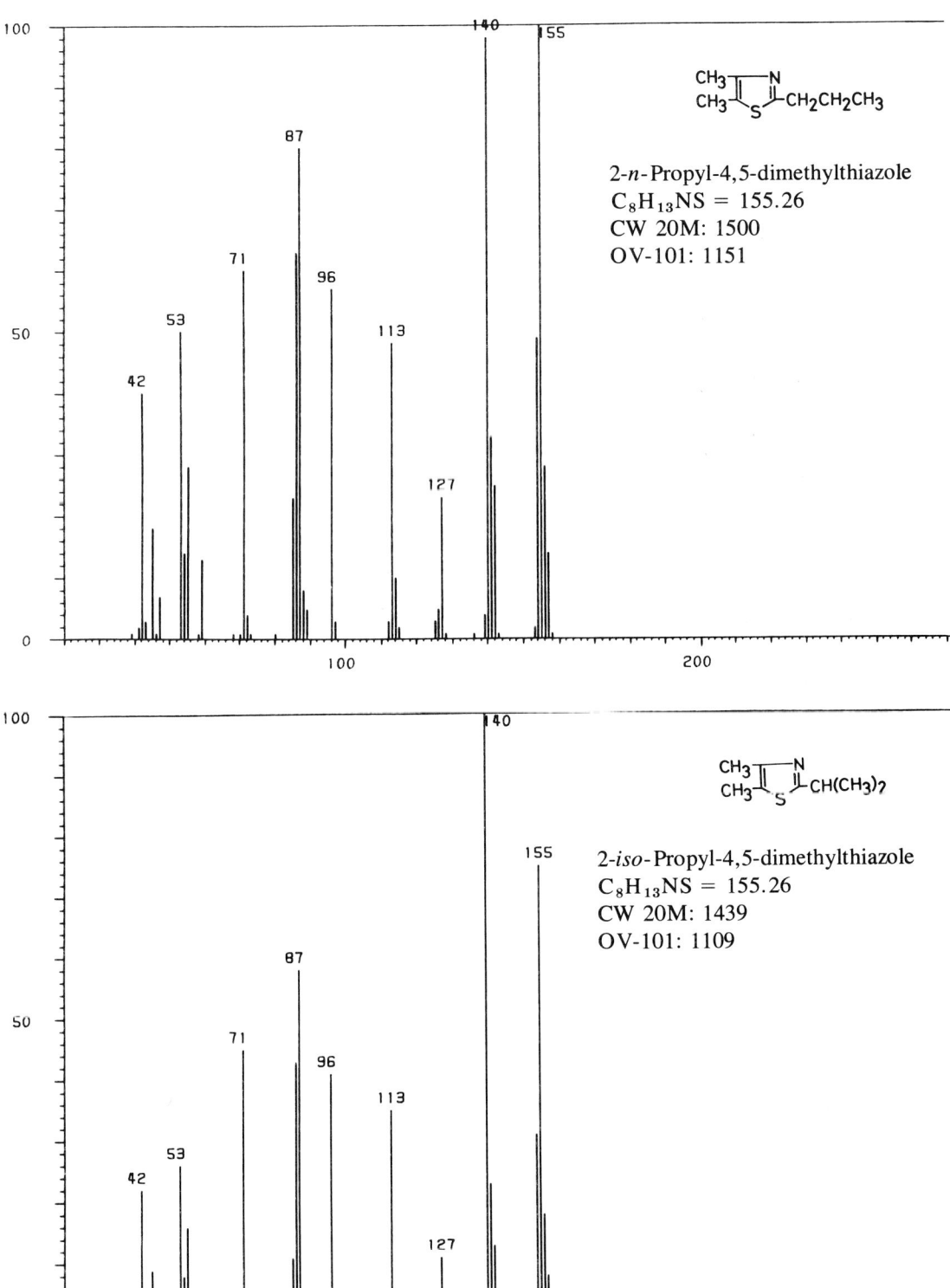

2-n-Propyl-4,5-dimethylthiazole
$C_8H_{13}NS = 155.26$
CW 20M: 1500
OV-101: 1151

2-iso-Propyl-4,5-dimethylthiazole
$C_8H_{13}NS = 155.26$
CW 20M: 1439
OV-101: 1109

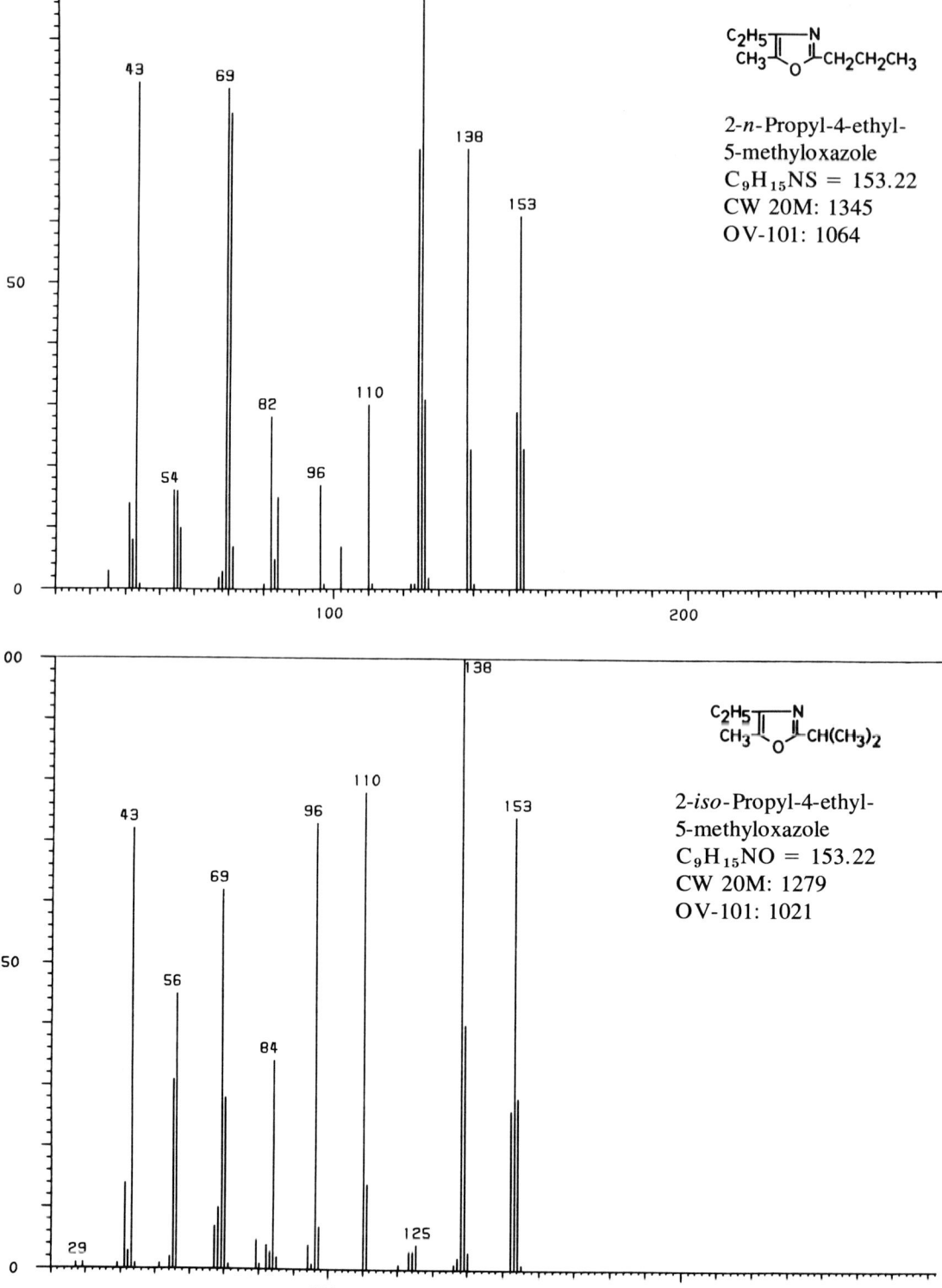

APPENDIX IV   MASS SPECTRA OF INDIVIDUAL COMPOUNDS   433

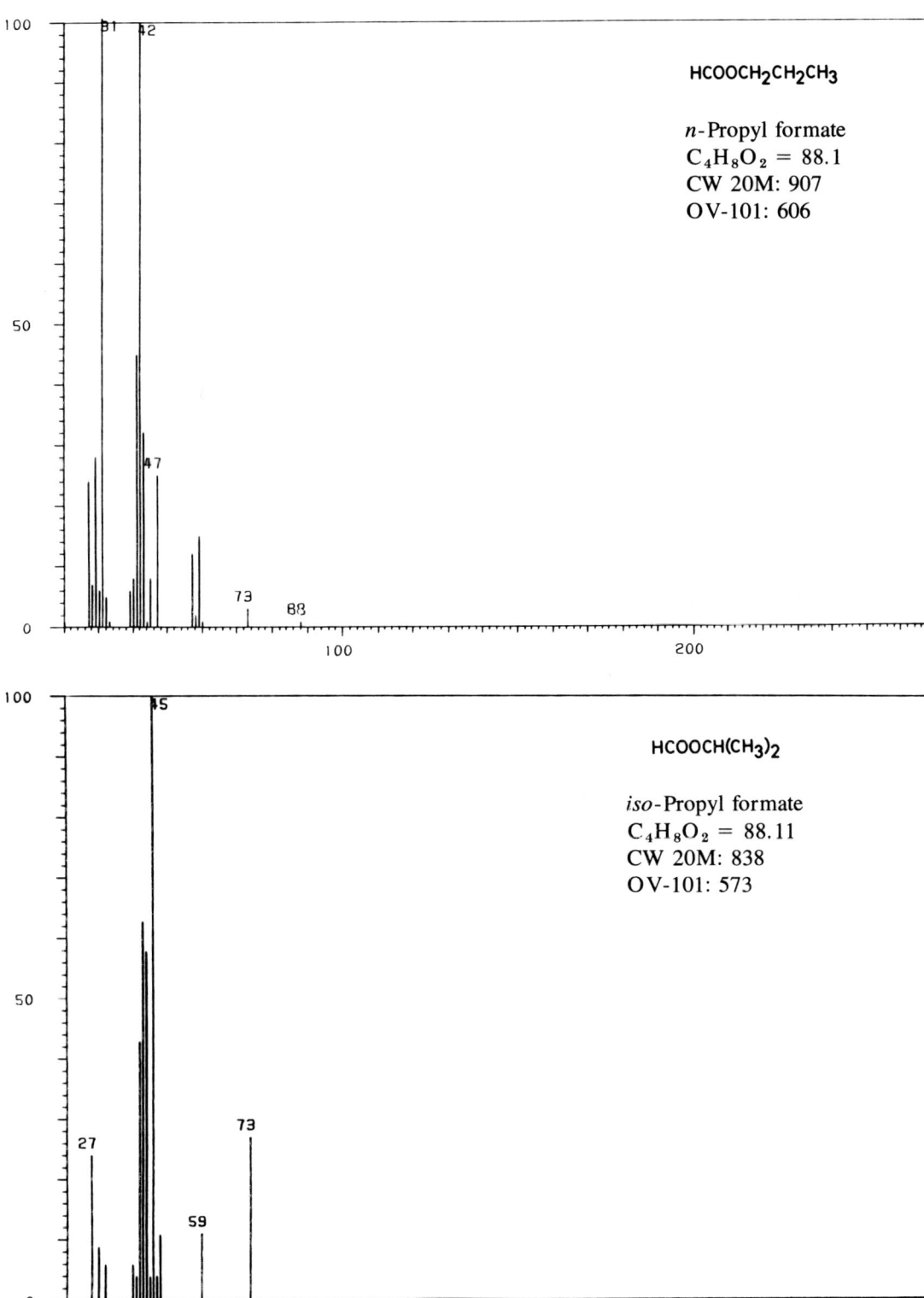

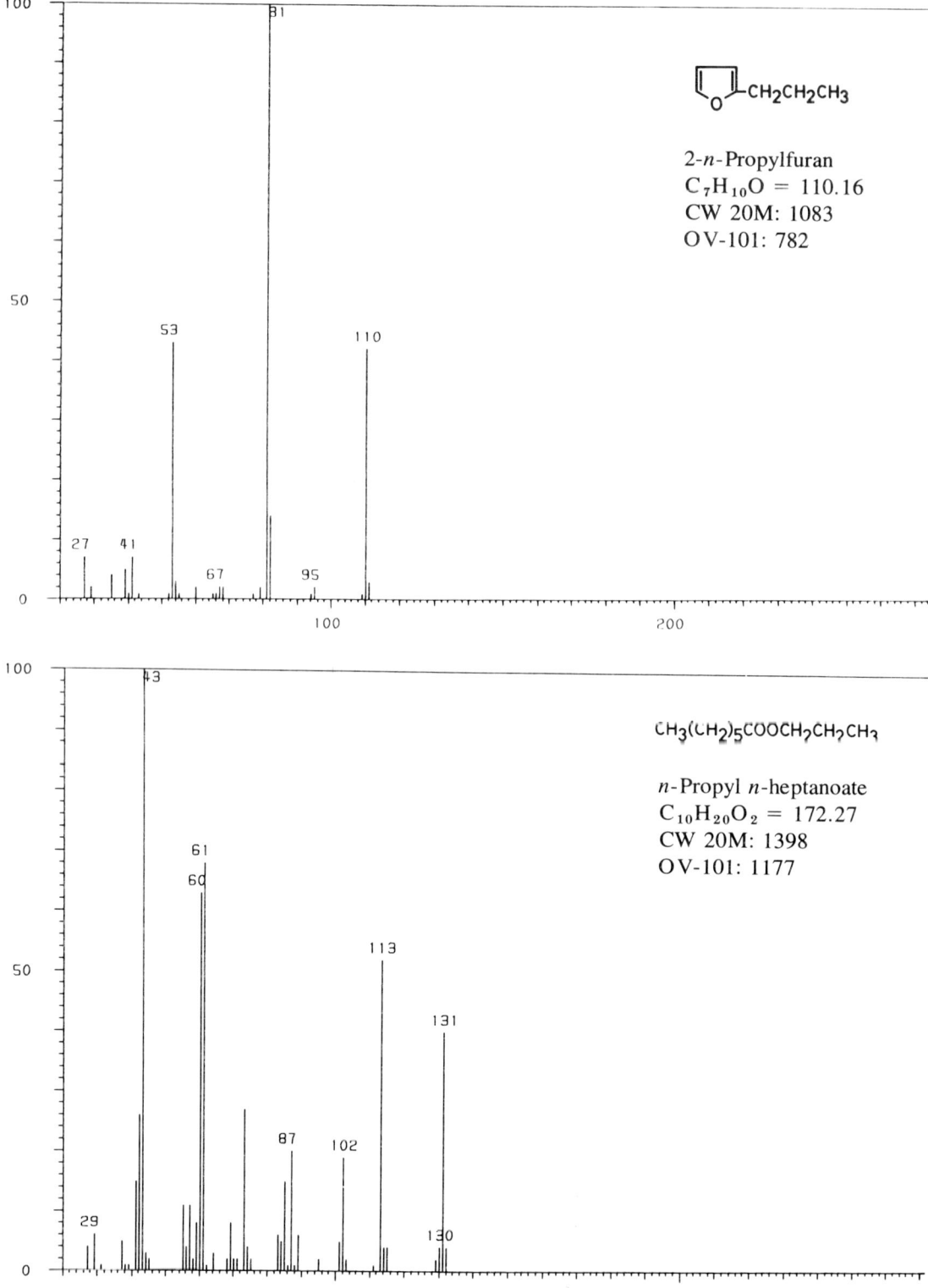

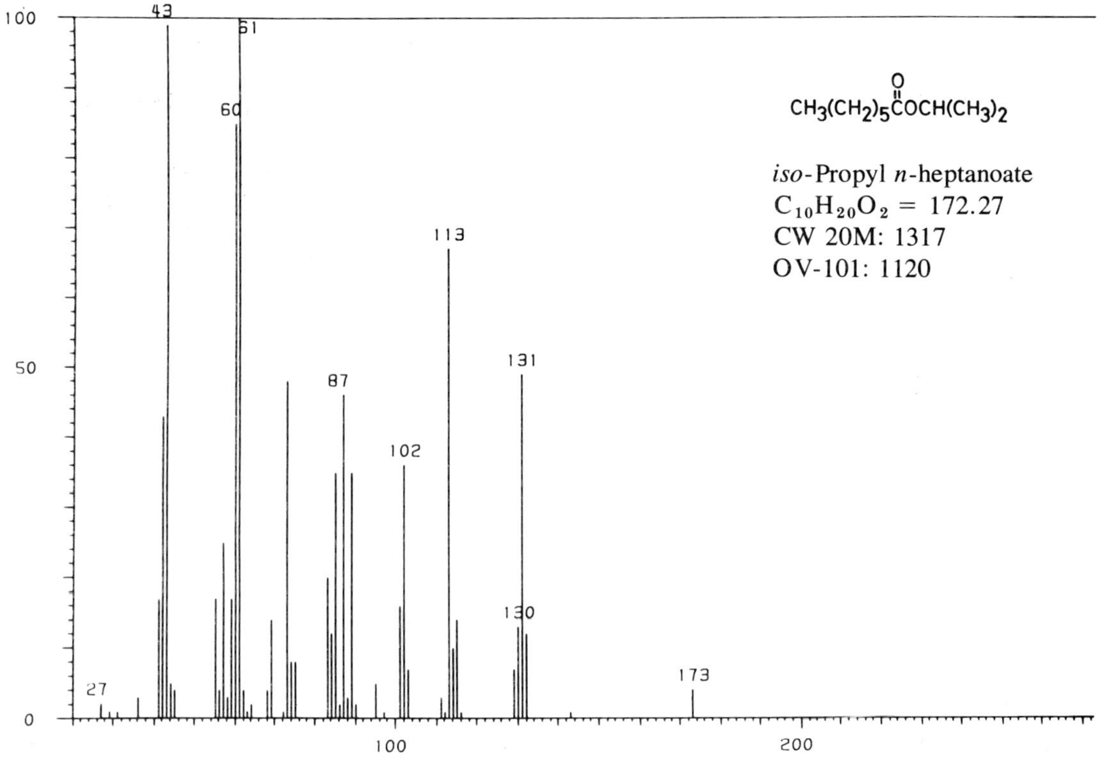

APPENDIX IV   MASS SPECTRA OF INDIVIDUAL COMPOUNDS   **435**

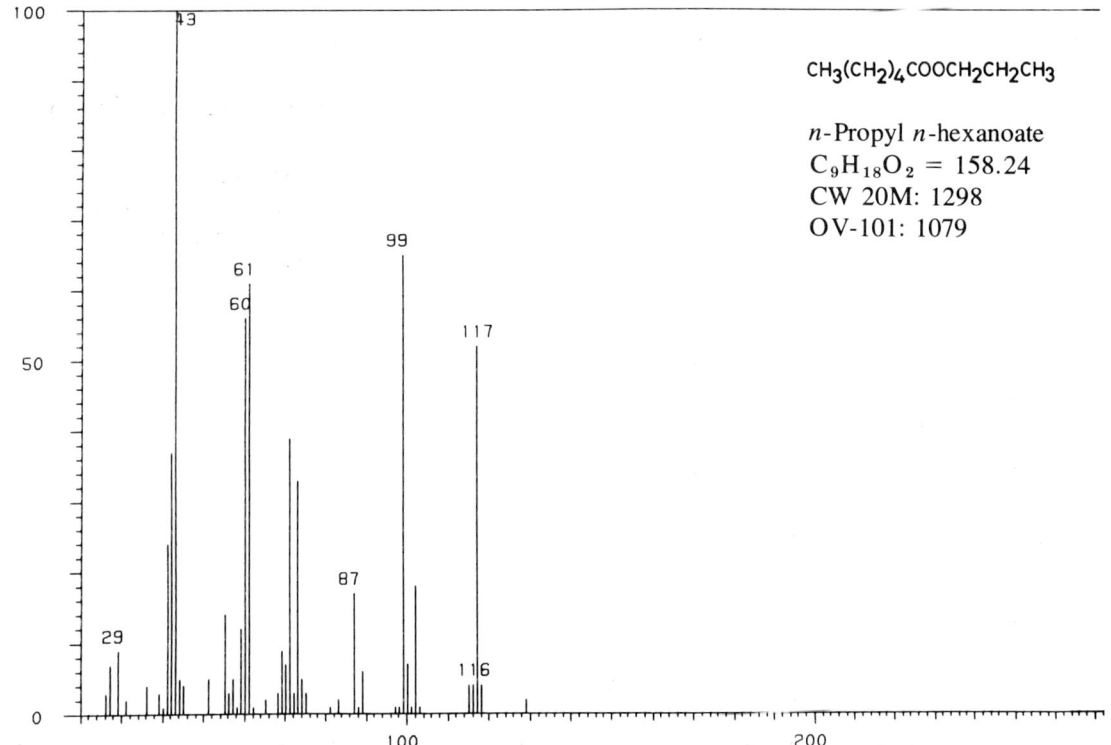

$CH_3(CH_2)_5\overset{O}{\overset{\|}{C}}OCH(CH_3)_2$

iso-Propyl *n*-heptanoate
$C_{10}H_{20}O_2 = 172.27$
CW 20M: 1317
OV-101: 1120

$CH_3(CH_2)_4COOCH_2CH_2CH_3$

n-Propyl *n*-hexanoate
$C_9H_{18}O_2 = 158.24$
CW 20M: 1298
OV-101: 1079

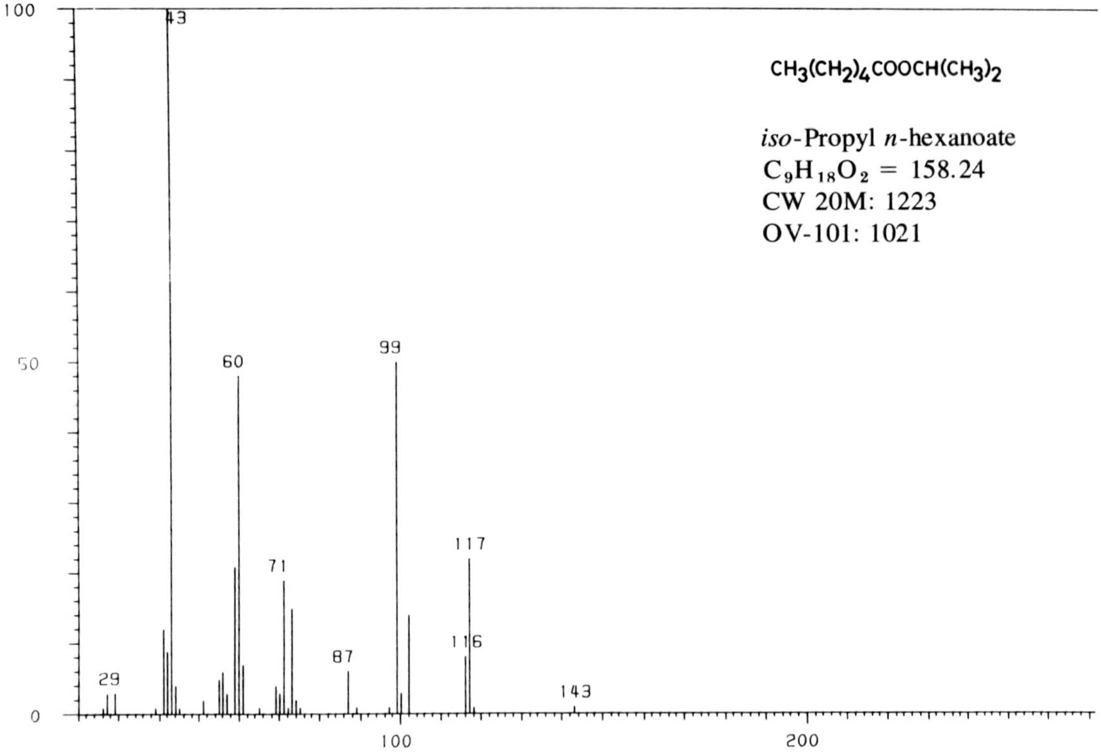

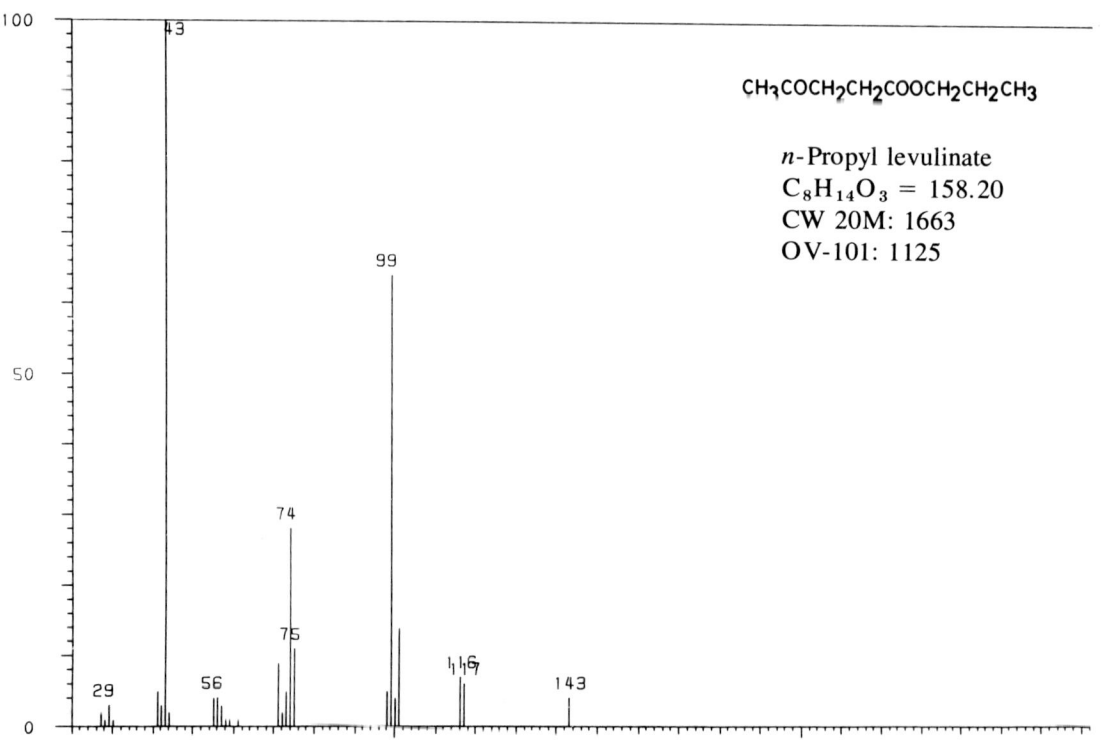

CH₃(CH₂)₄COOCH(CH₃)₂

iso-Propyl n-hexanoate
$C_9H_{18}O_2 = 158.24$
CW 20M: 1223
OV-101: 1021

CH₃COCH₂CH₂COOCH₂CH₂CH₃

n-Propyl levulinate
$C_8H_{14}O_3 = 158.20$
CW 20M: 1663
OV-101: 1125

APPENDIX IV   MASS SPECTRA OF INDIVIDUAL COMPOUNDS

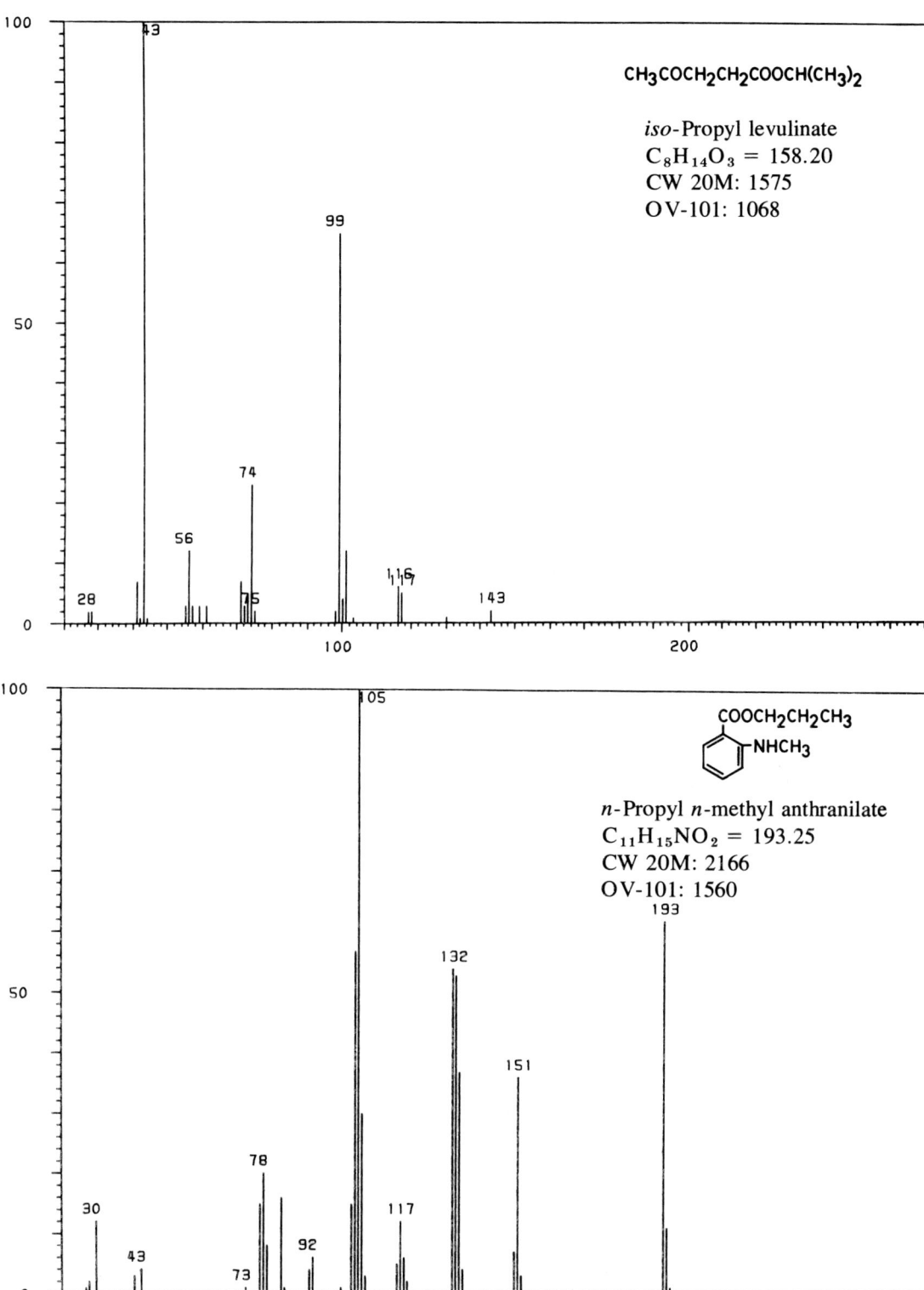

$CH_3COCH_2CH_2COOCH(CH_3)_2$

iso-Propyl levulinate
$C_8H_{14}O_3 = 158.20$
CW 20M: 1575
OV-101: 1068

n-Propyl *n*-methyl anthranilate
$C_{11}H_{15}NO_2 = 193.25$
CW 20M: 2166
OV-101: 1560

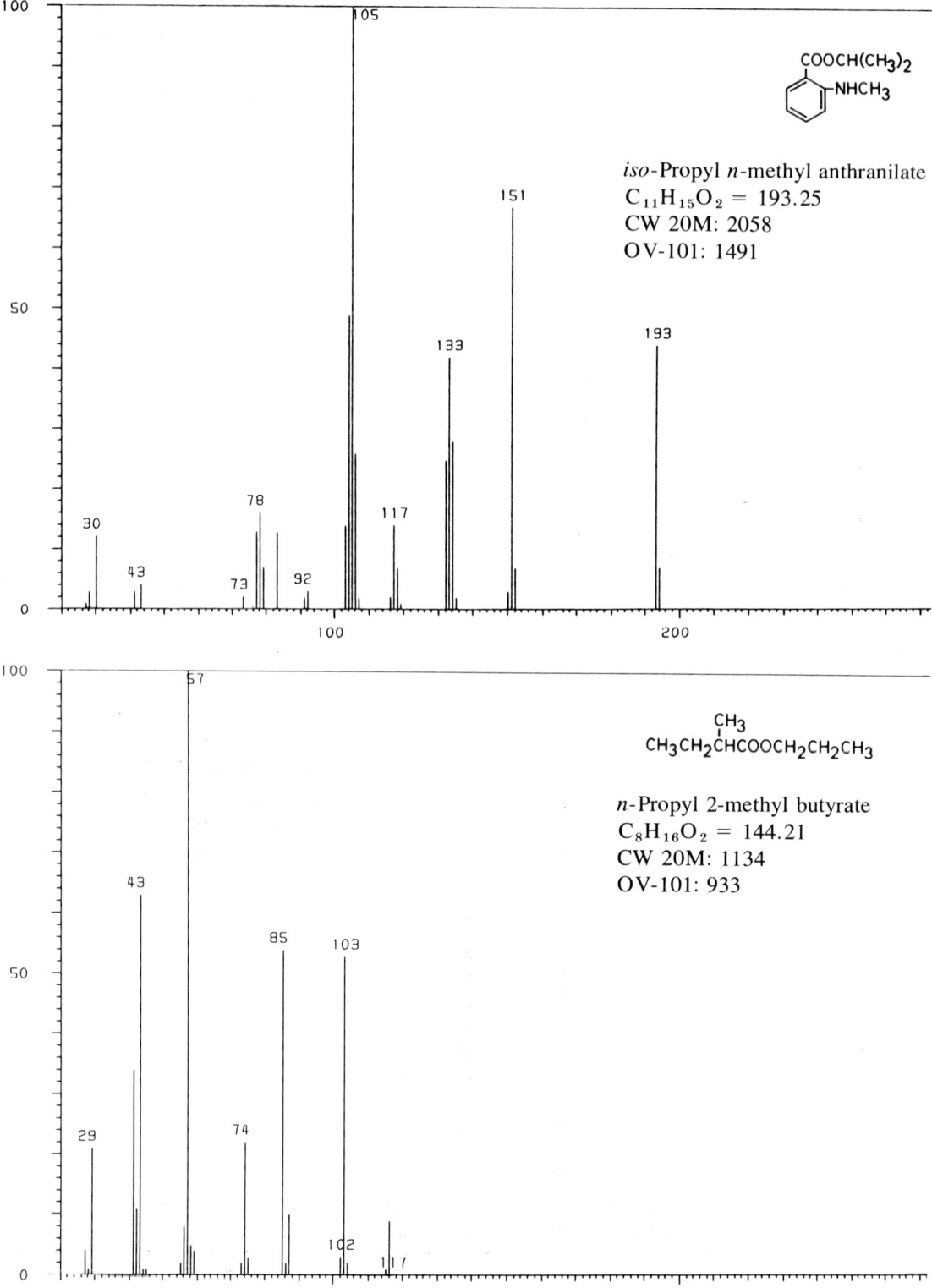

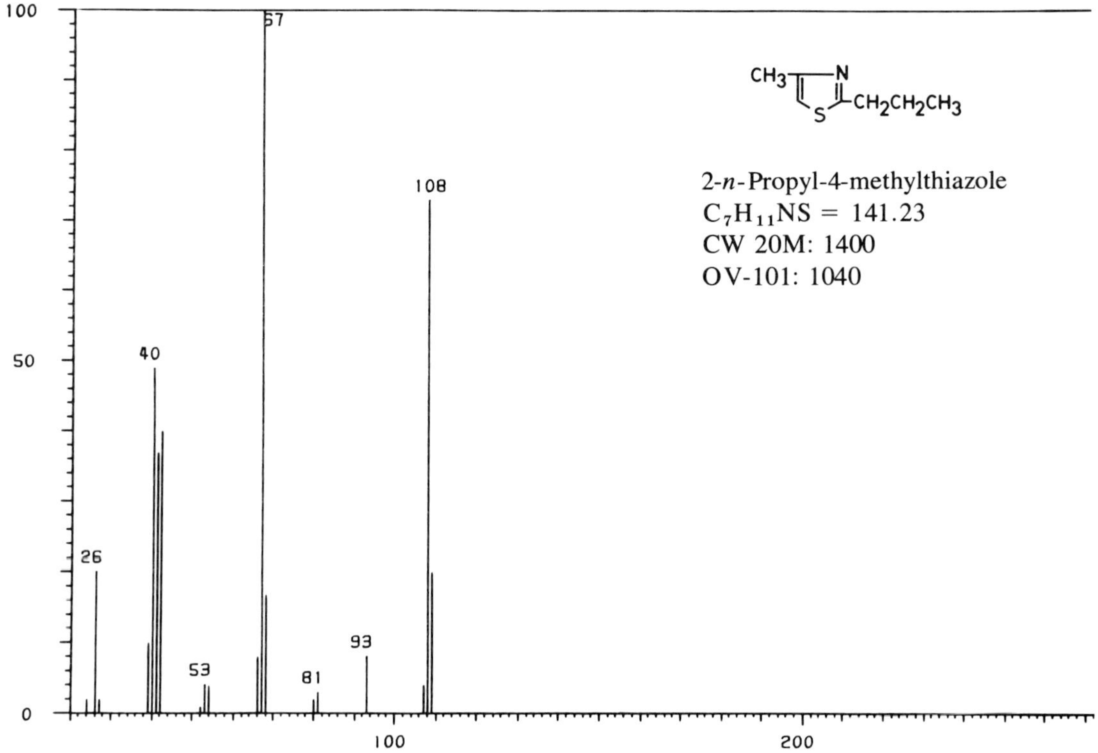

APPENDIX IV MASS SPECTRA OF INDIVIDUAL COMPOUNDS

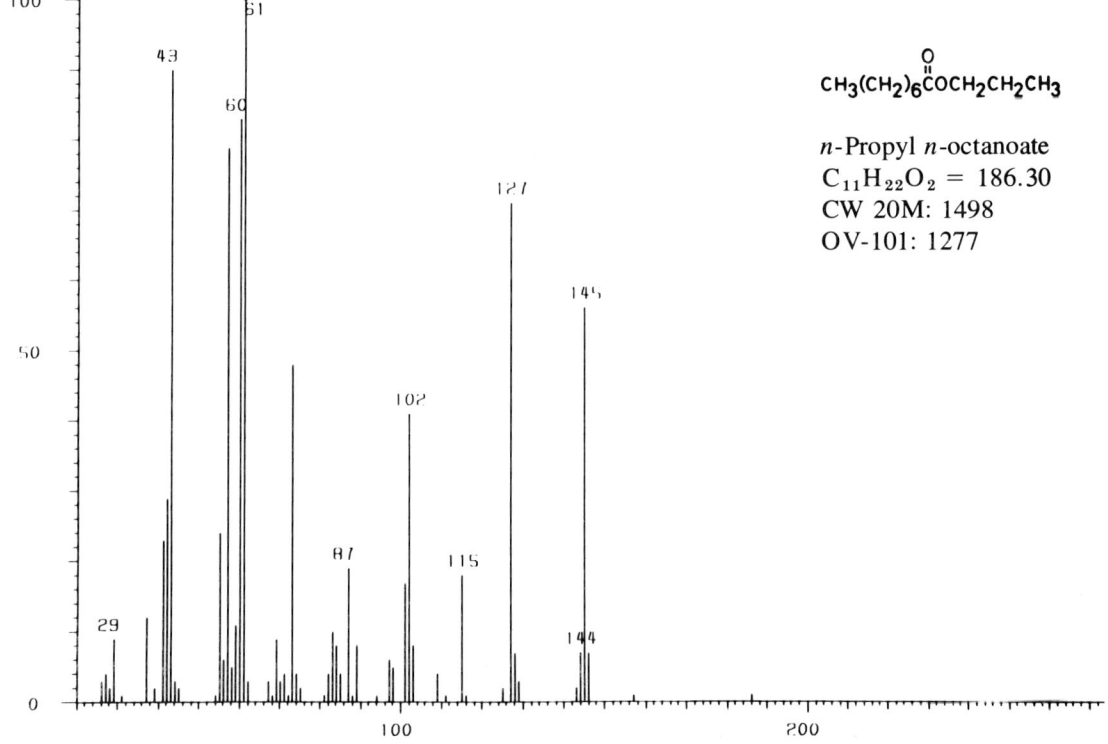

2-n-Propyl-4-methylthiazole
$C_7H_{11}NS = 141.23$
CW 20M: 1400
OV-101: 1040

n-Propyl n-octanoate
$C_{11}H_{22}O_2 = 186.30$
CW 20M: 1498
OV-101: 1277

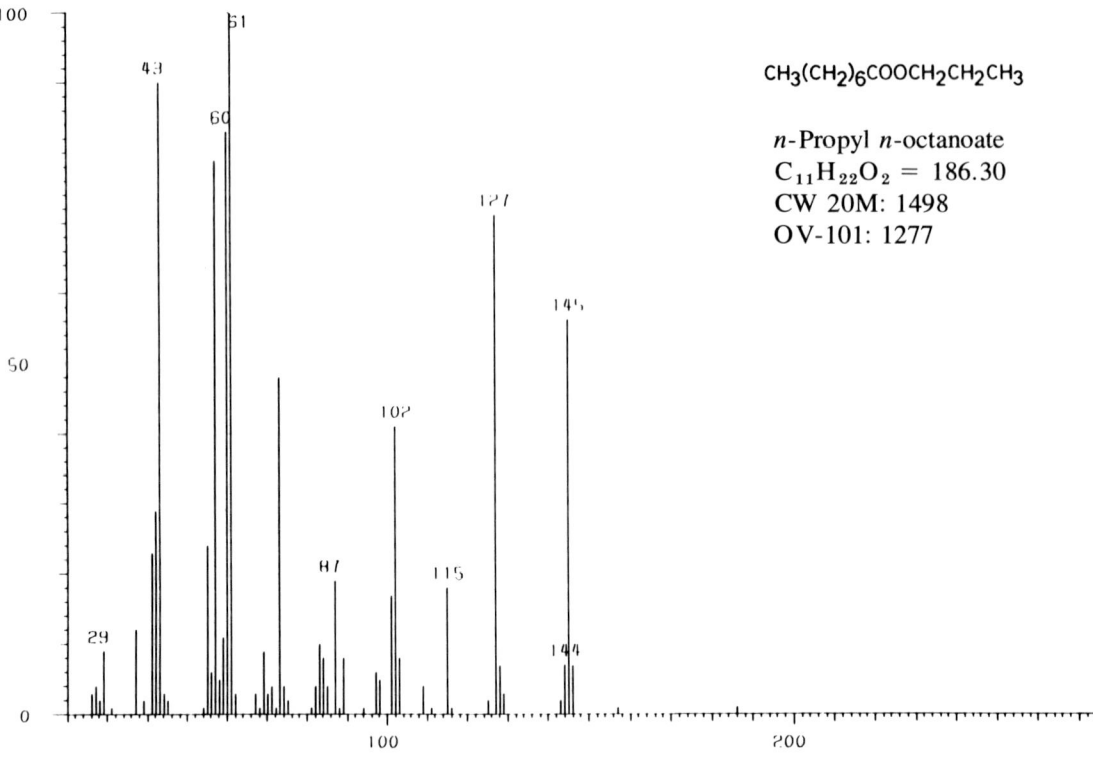

$CH_3(CH_2)_6COOCH_2CH_2CH_3$

n-Propyl *n*-octanoate
$C_{11}H_{22}O_2 = 186.30$
CW 20M: 1498
OV-101: 1277

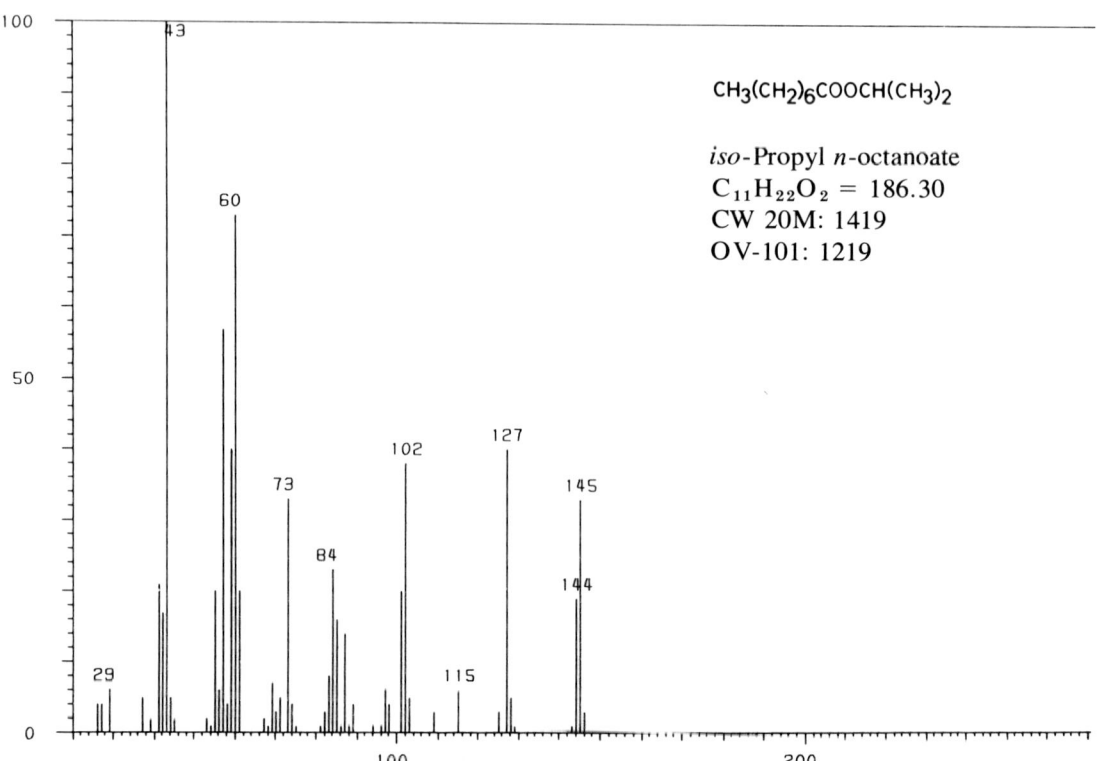

$CH_3(CH_2)_6COOCH(CH_3)_2$

iso-Propyl *n*-octanoate
$C_{11}H_{22}O_2 = 186.30$
CW 20M: 1419
OV-101: 1219

APPENDIX IV   MASS SPECTRA OF INDIVIDUAL COMPOUNDS

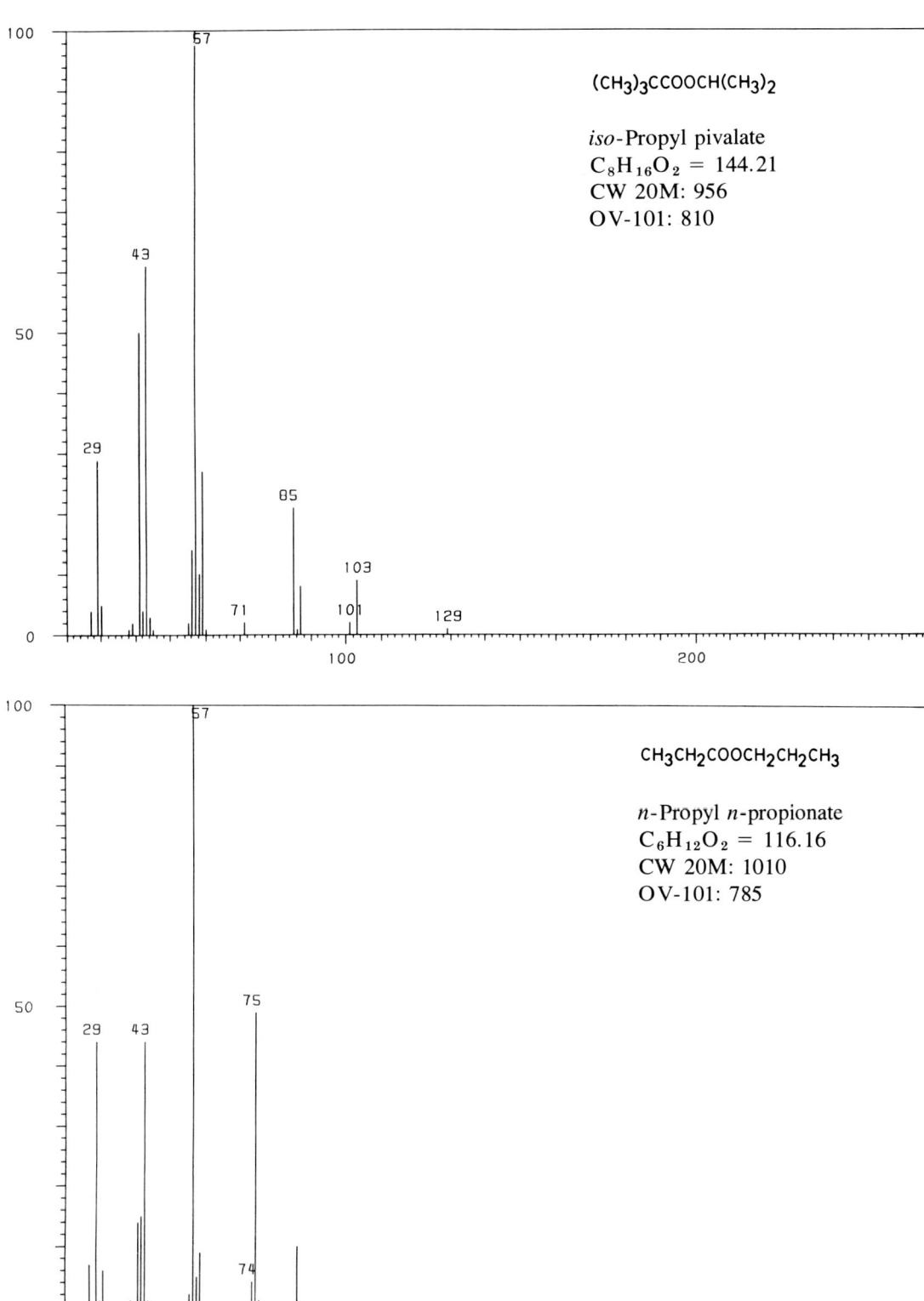

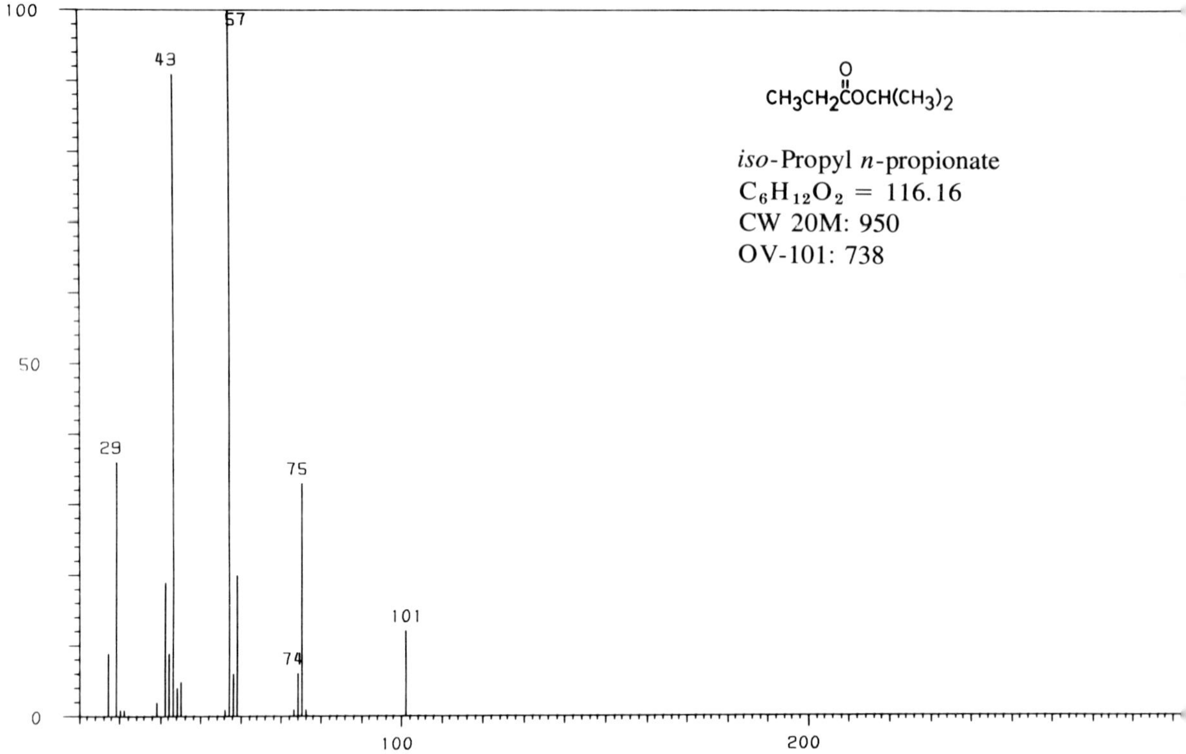

$$\underset{\text{iso-Propyl }n\text{-propionate}}{CH_3CH_2\overset{\overset{O}{\|}}{C}OCH(CH_3)_2}$$

iso-Propyl n-propionate
$C_6H_{12}O_2 = 116.16$
CW 20M: 950
OV-101: 738

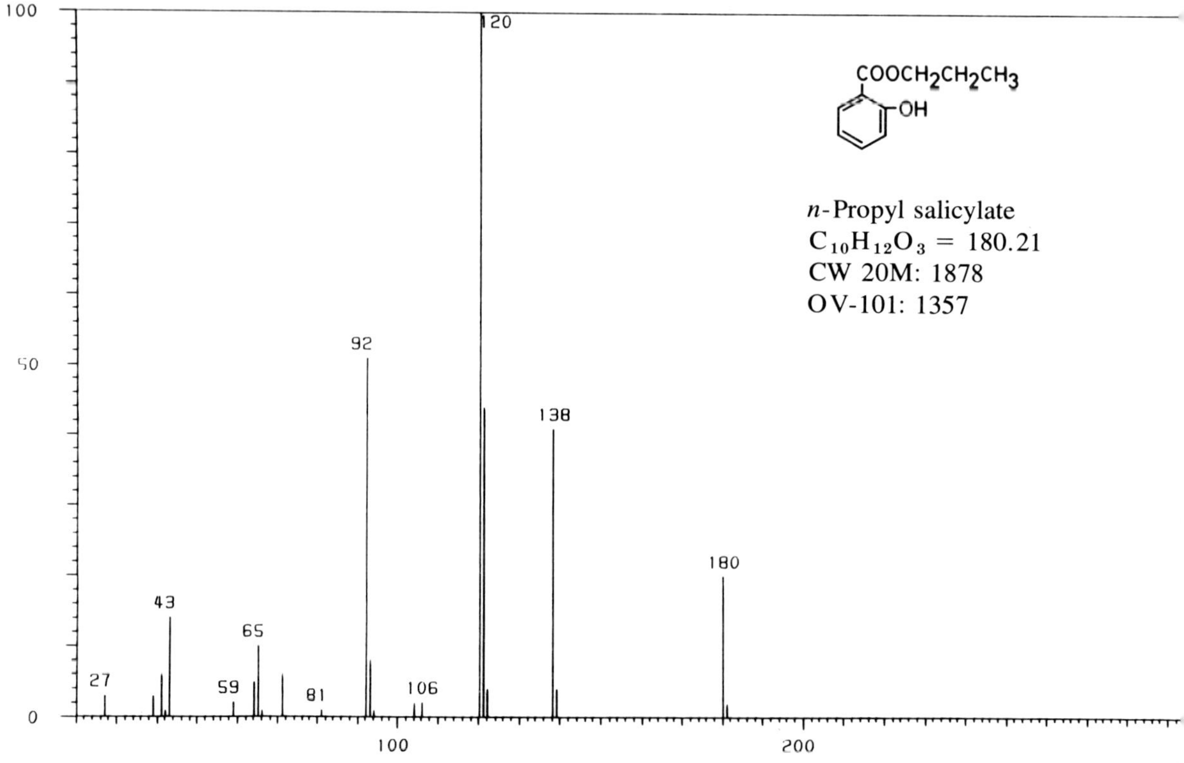

n-Propyl salicylate
$C_{10}H_{12}O_3 = 180.21$
CW 20M: 1878
OV-101: 1357

APPENDIX IV   MASS SPECTRA OF INDIVIDUAL COMPOUNDS   **443**

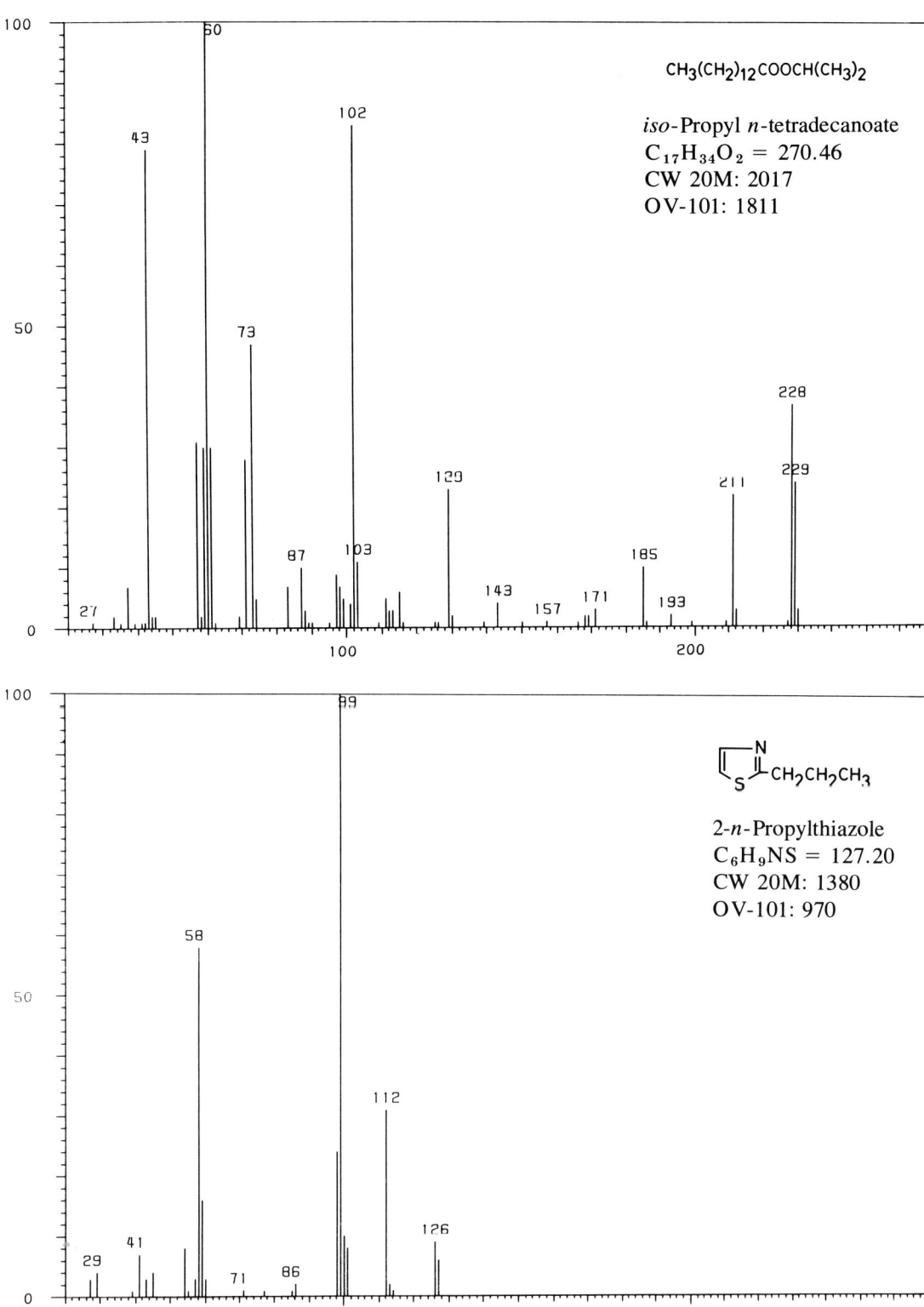

$CH_3(CH_2)_{12}COOCH(CH_3)_2$

iso-Propyl *n*-tetradecanoate
$C_{17}H_{34}O_2 = 270.46$
CW 20M: 2017
OV-101: 1811

2-*n*-Propylthiazole
$C_6H_9NS = 127.20$
CW 20M: 1380
OV-101: 970

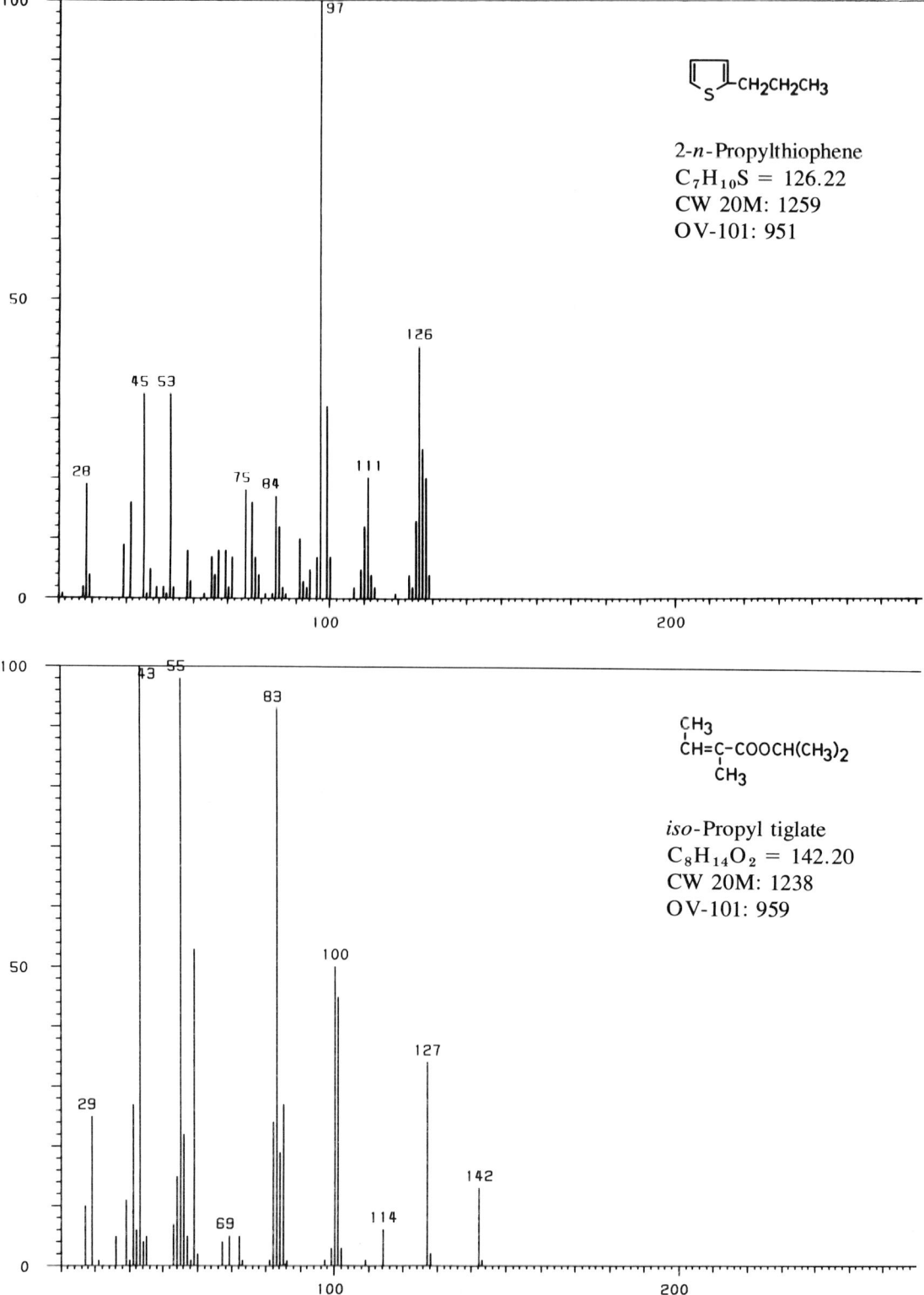

APPENDIX IV   MASS SPECTRA OF INDIVIDUAL COMPOUNDS   445

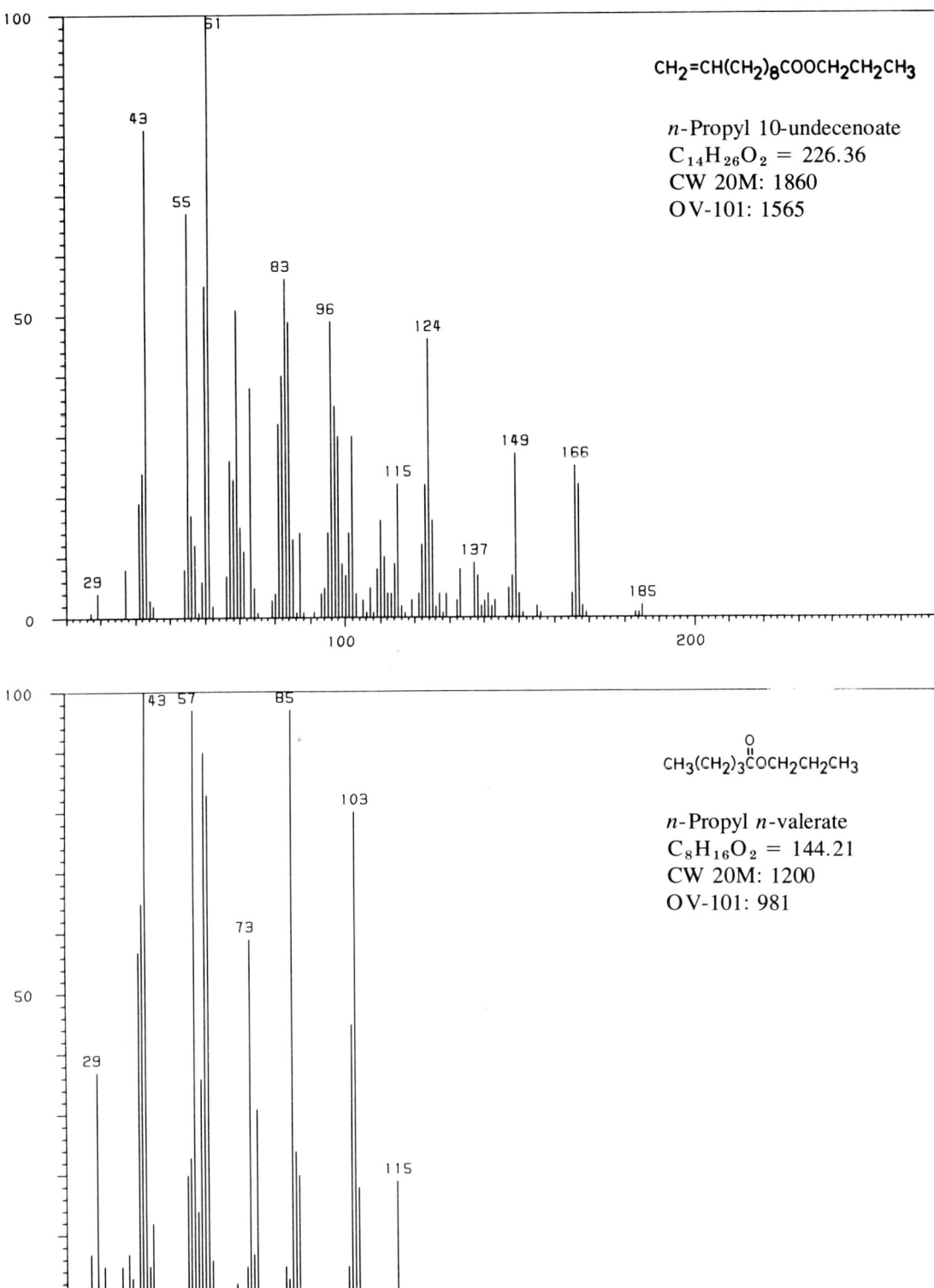

$CH_2=CH(CH_2)_8COOCH_2CH_2CH_3$

n-Propyl 10-undecenoate
$C_{14}H_{26}O_2 = 226.36$
CW 20M: 1860
OV-101: 1565

$CH_3(CH_2)_3\overset{\overset{O}{\|}}{C}OCH_2CH_2CH_3$

n-Propyl *n*-valerate
$C_8H_{16}O_2 = 144.21$
CW 20M: 1200
OV-101: 981

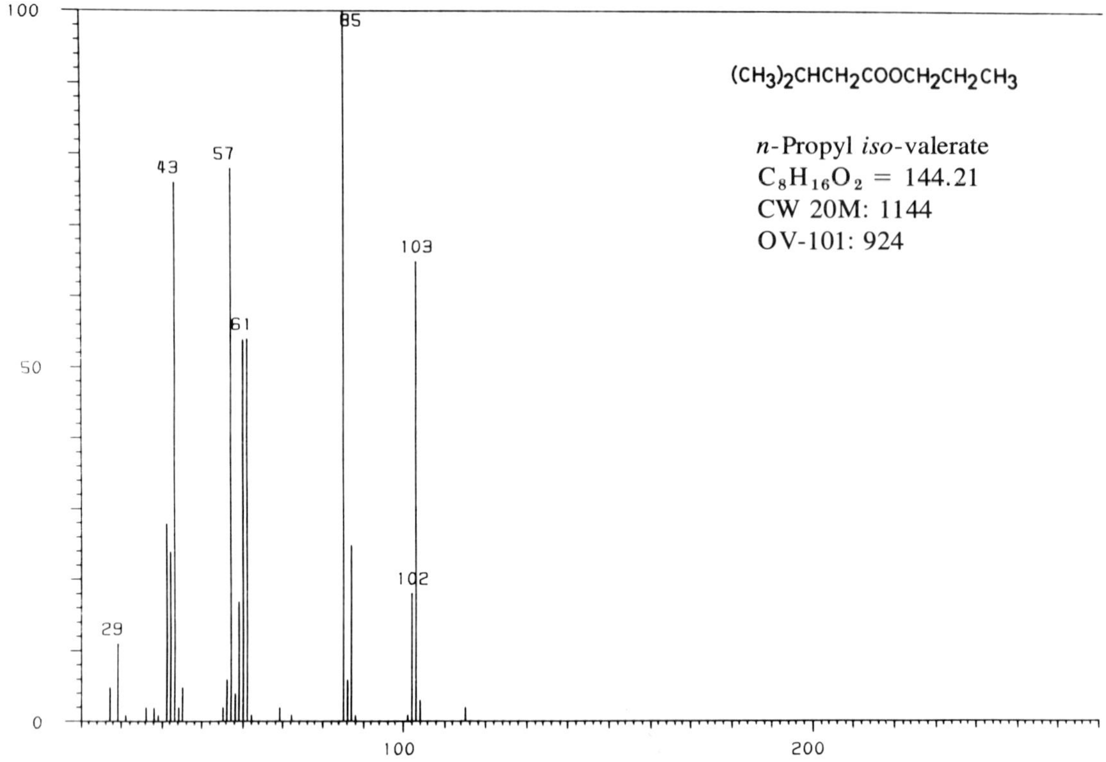

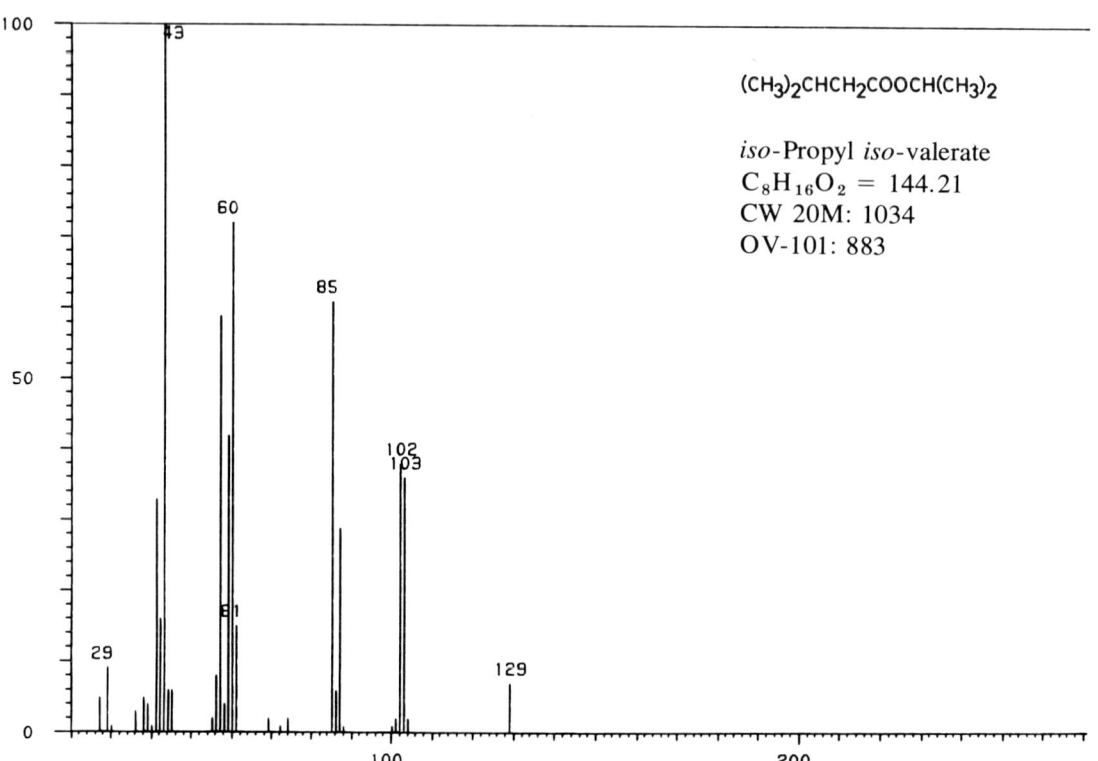

(CH₃)₂CHCH₂COOCH₂CH₂CH₃

n-Propyl *iso*-valerate
$C_8H_{16}O_2 = 144.21$
CW 20M: 1144
OV-101: 924

(CH₃)₂CHCH₂COOCH(CH₃)₂

iso-Propyl *iso*-valerate
$C_8H_{16}O_2 = 144.21$
CW 20M: 1034
OV-101: 883

APPENDIX IV MASS SPECTRA OF INDIVIDUAL COMPOUNDS 447

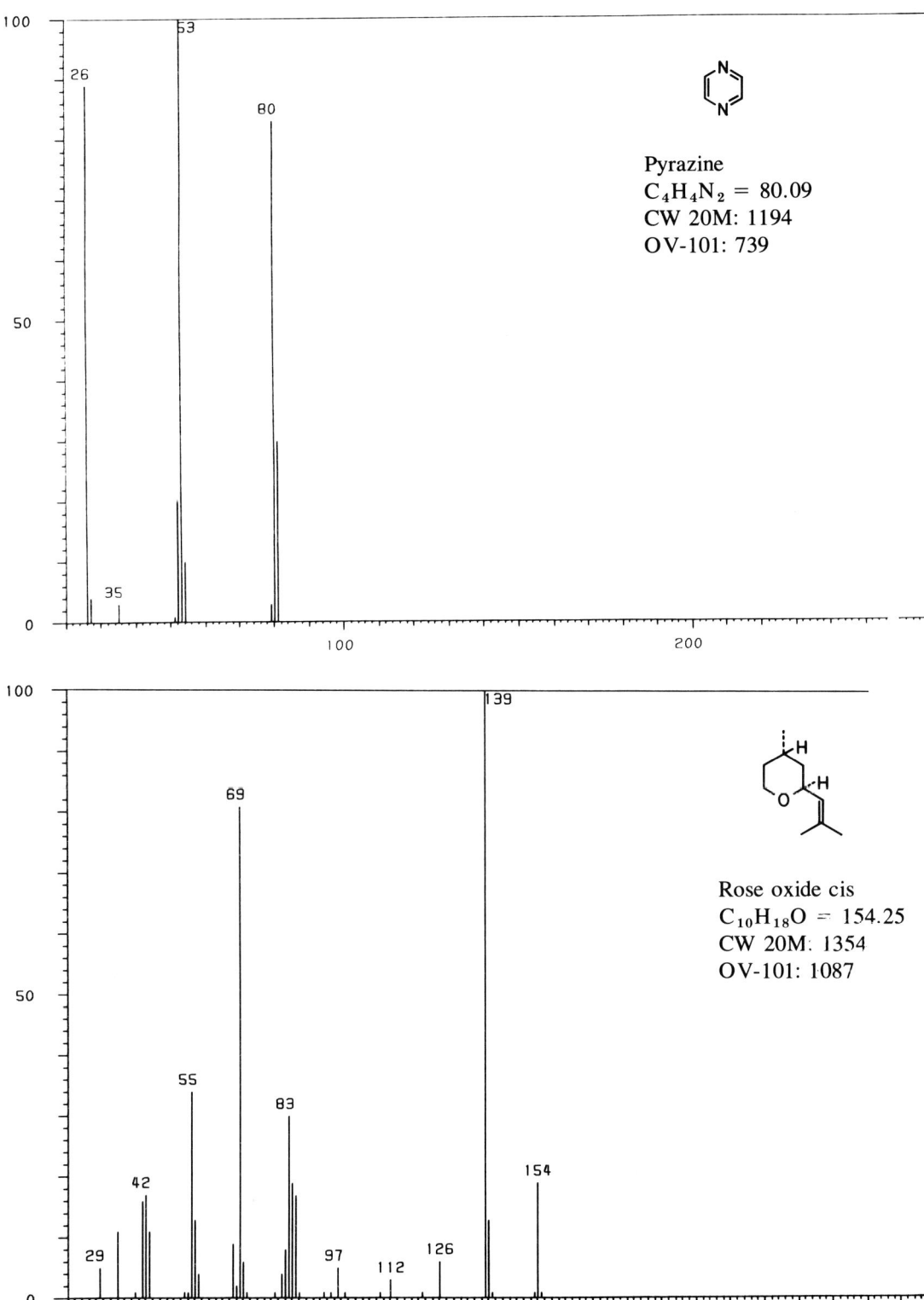

Pyrazine
$C_4H_4N_2 = 80.09$
CW 20M: 1194
OV-101: 739

Rose oxide cis
$C_{10}H_{18}O = 154.25$
CW 20M: 1354
OV-101: 1087

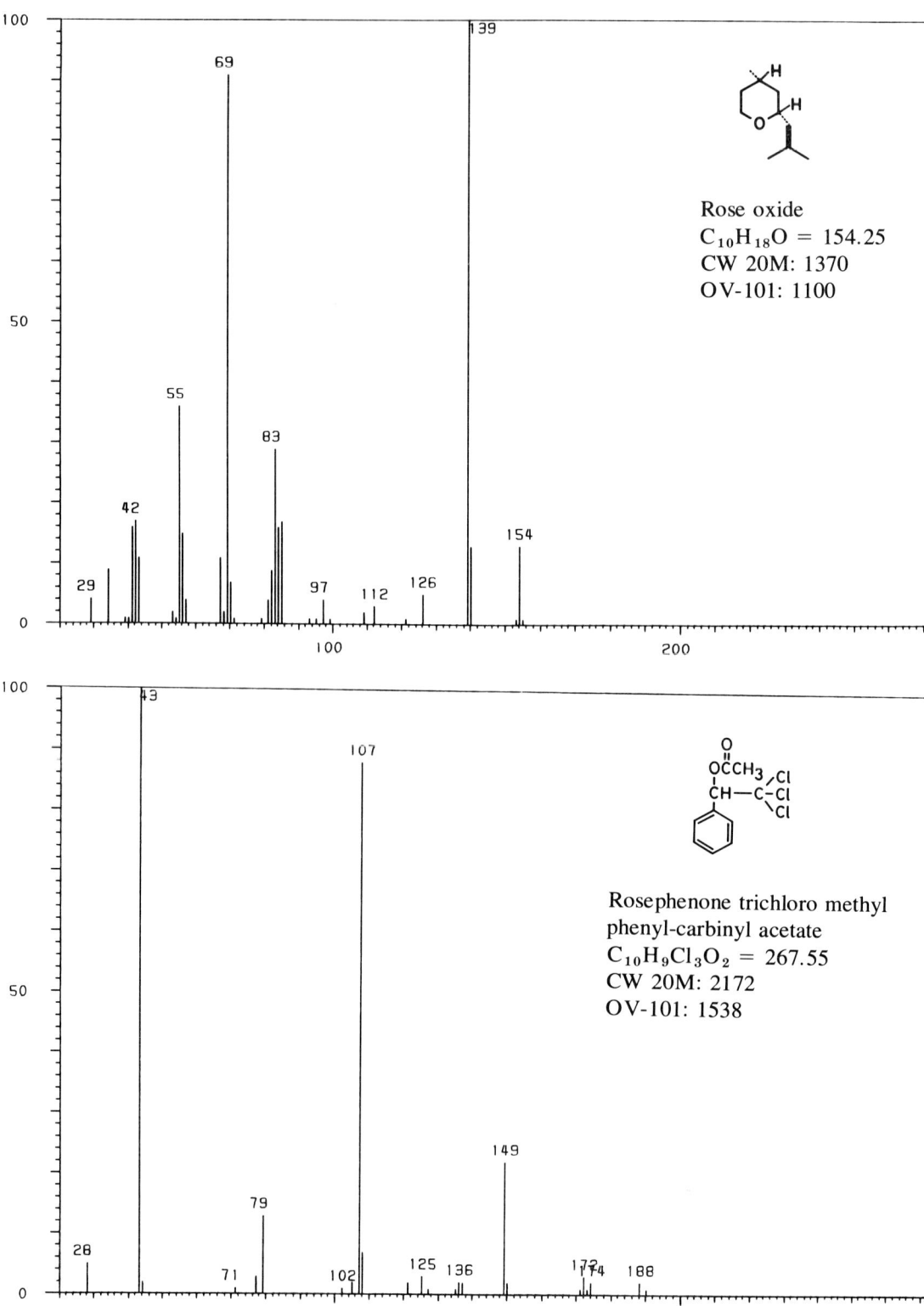

APPENDIX IV   MASS SPECTRA OF INDIVIDUAL COMPOUNDS

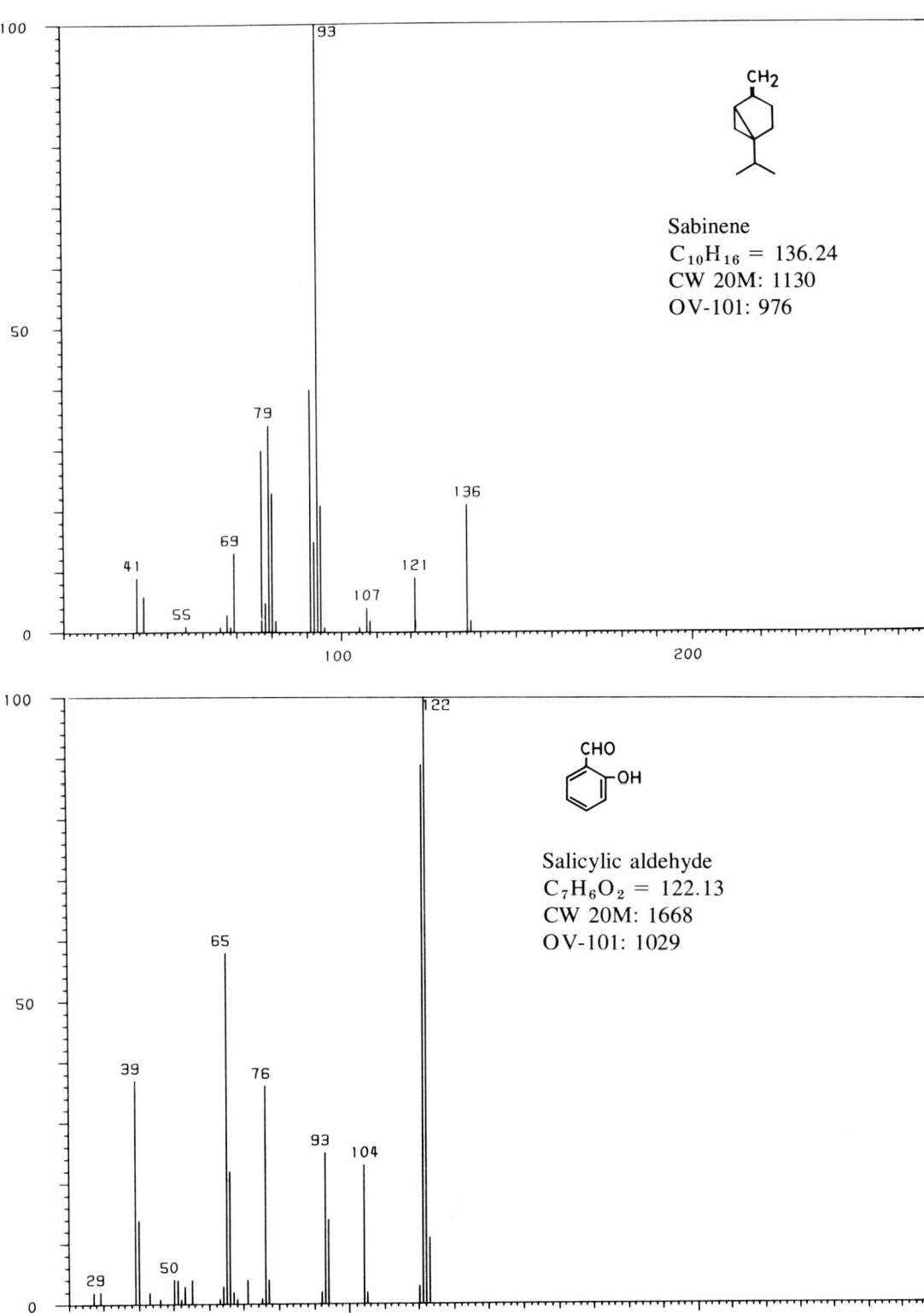

Sabinene
$C_{10}H_{16} = 136.24$
CW 20M: 1130
OV-101: 976

Salicylic aldehyde
$C_7H_6O_2 = 122.13$
CW 20M: 1668
OV-101: 1029

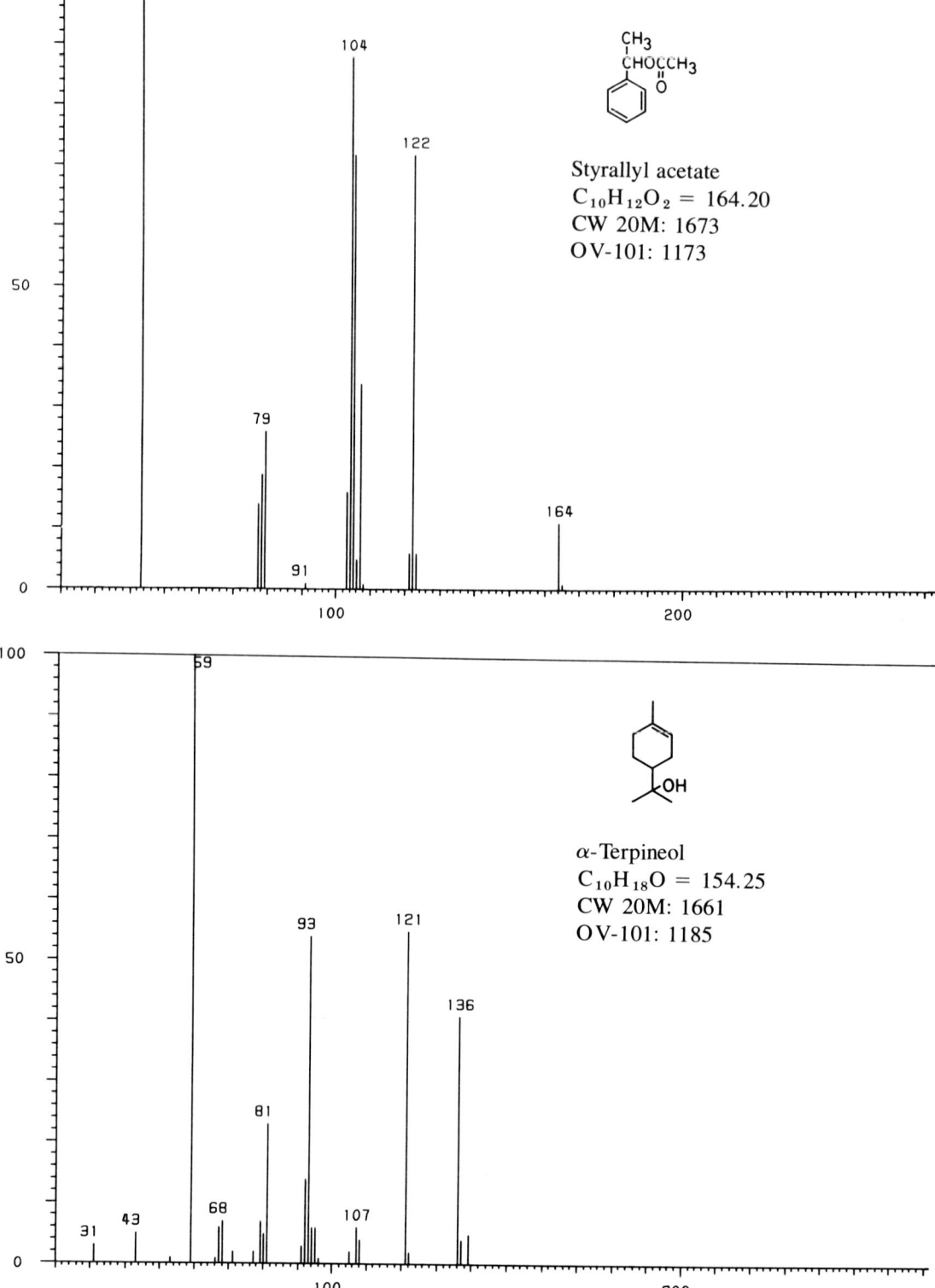

APPENDIX IV MASS SPECTRA OF INDIVIDUAL COMPOUNDS

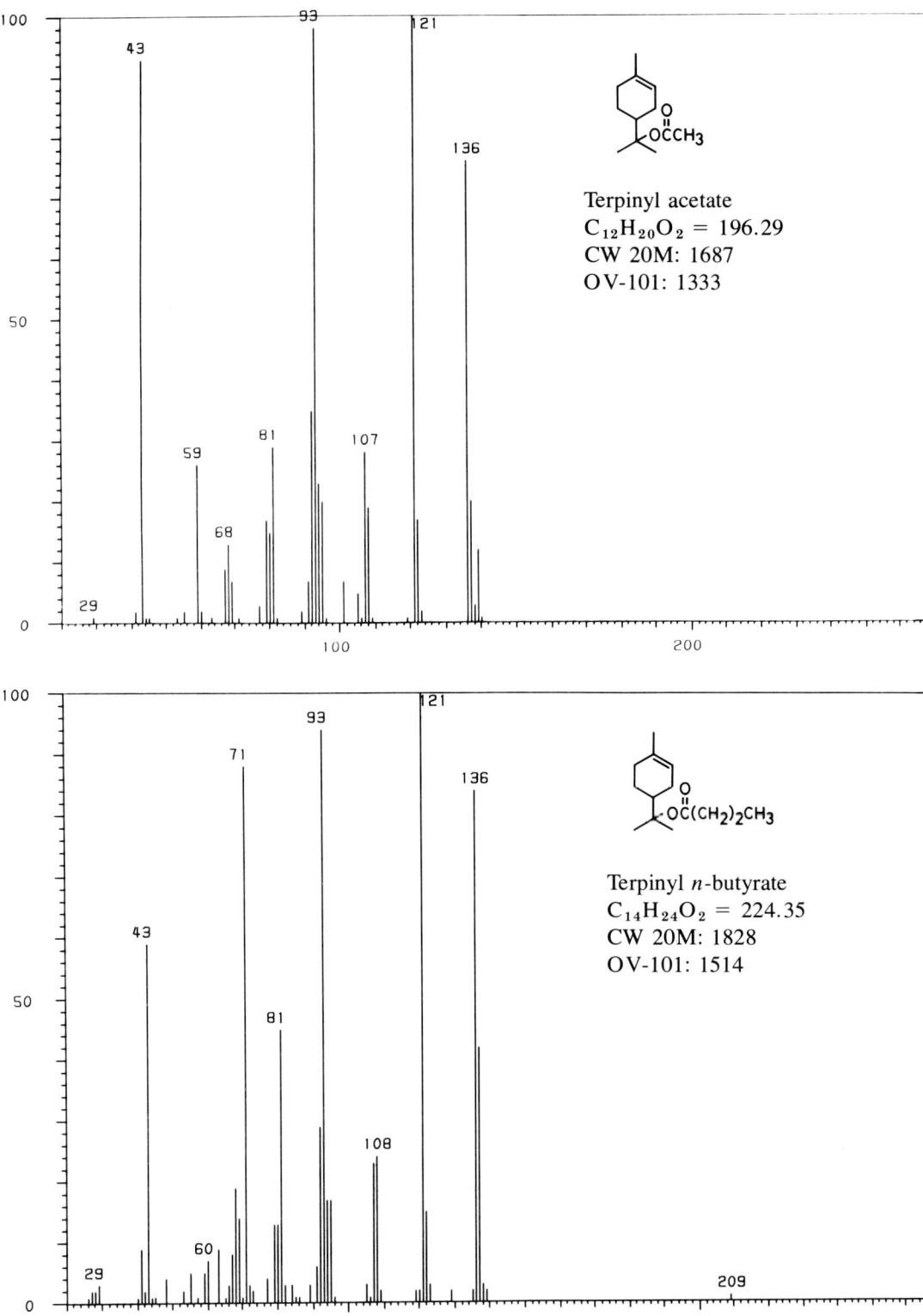

Terpinyl acetate
$C_{12}H_{20}O_2 = 196.29$
CW 20M: 1687
OV-101: 1333

Terpinyl n-butyrate
$C_{14}H_{24}O_2 = 224.35$
CW 20M: 1828
OV-101: 1514

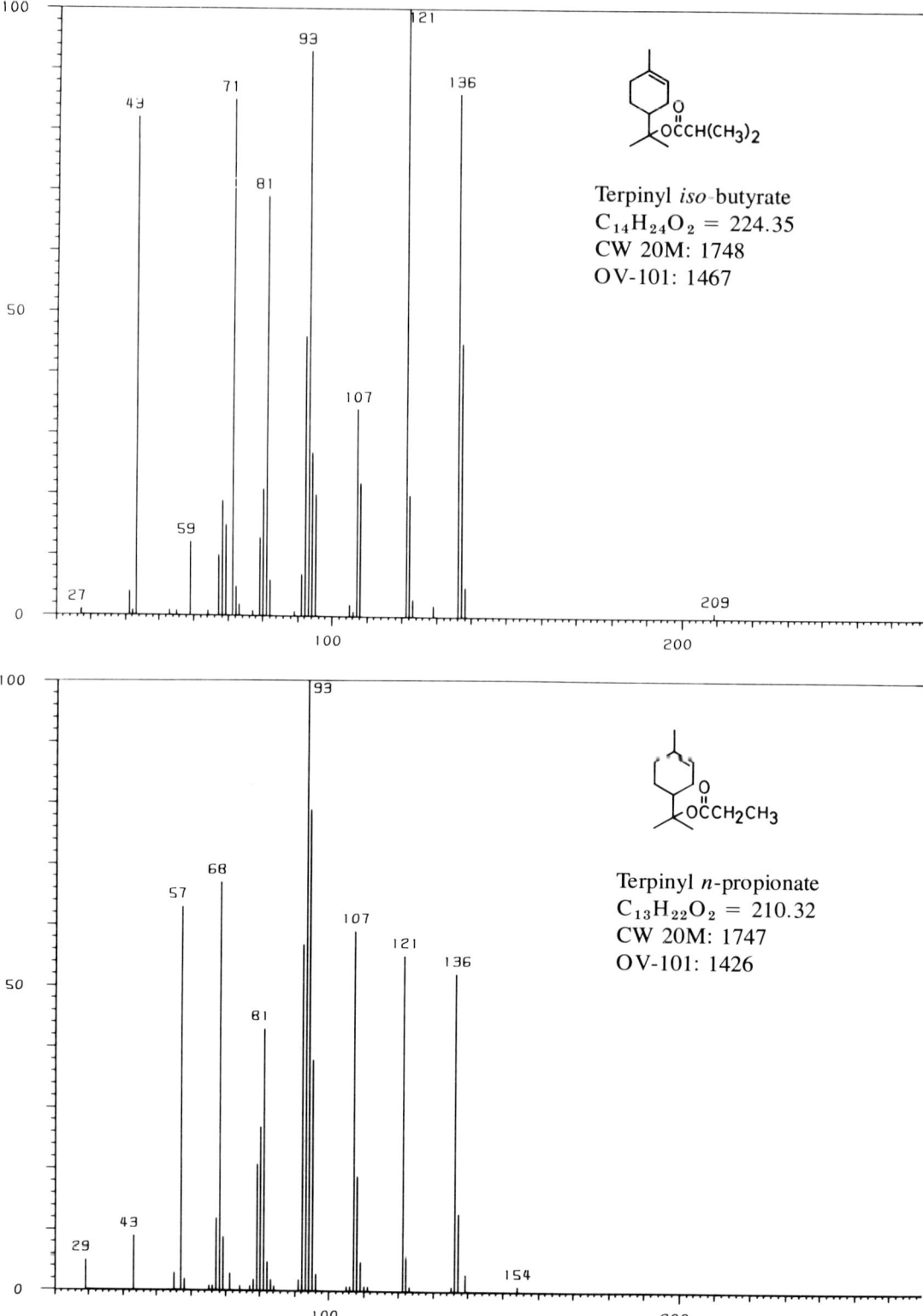

APPENDIX IV   MASS SPECTRA OF INDIVIDUAL COMPOUNDS

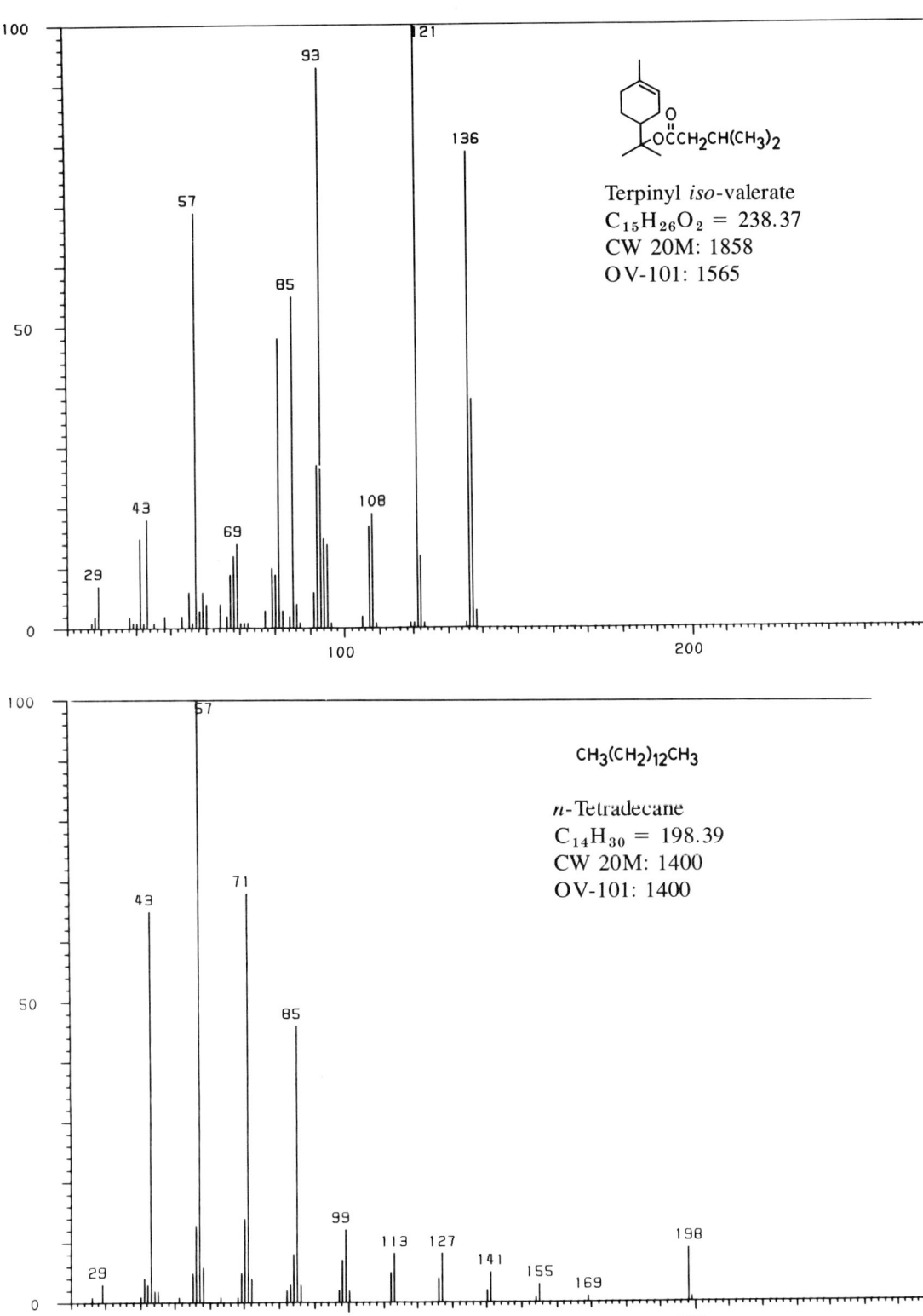

Terpinyl *iso*-valerate
$C_{15}H_{26}O_2 = 238.37$
CW 20M: 1858
OV-101: 1565

$CH_3(CH_2)_{12}CH_3$

n-Tetradecane
$C_{14}H_{30} = 198.39$
CW 20M: 1400
OV-101: 1400

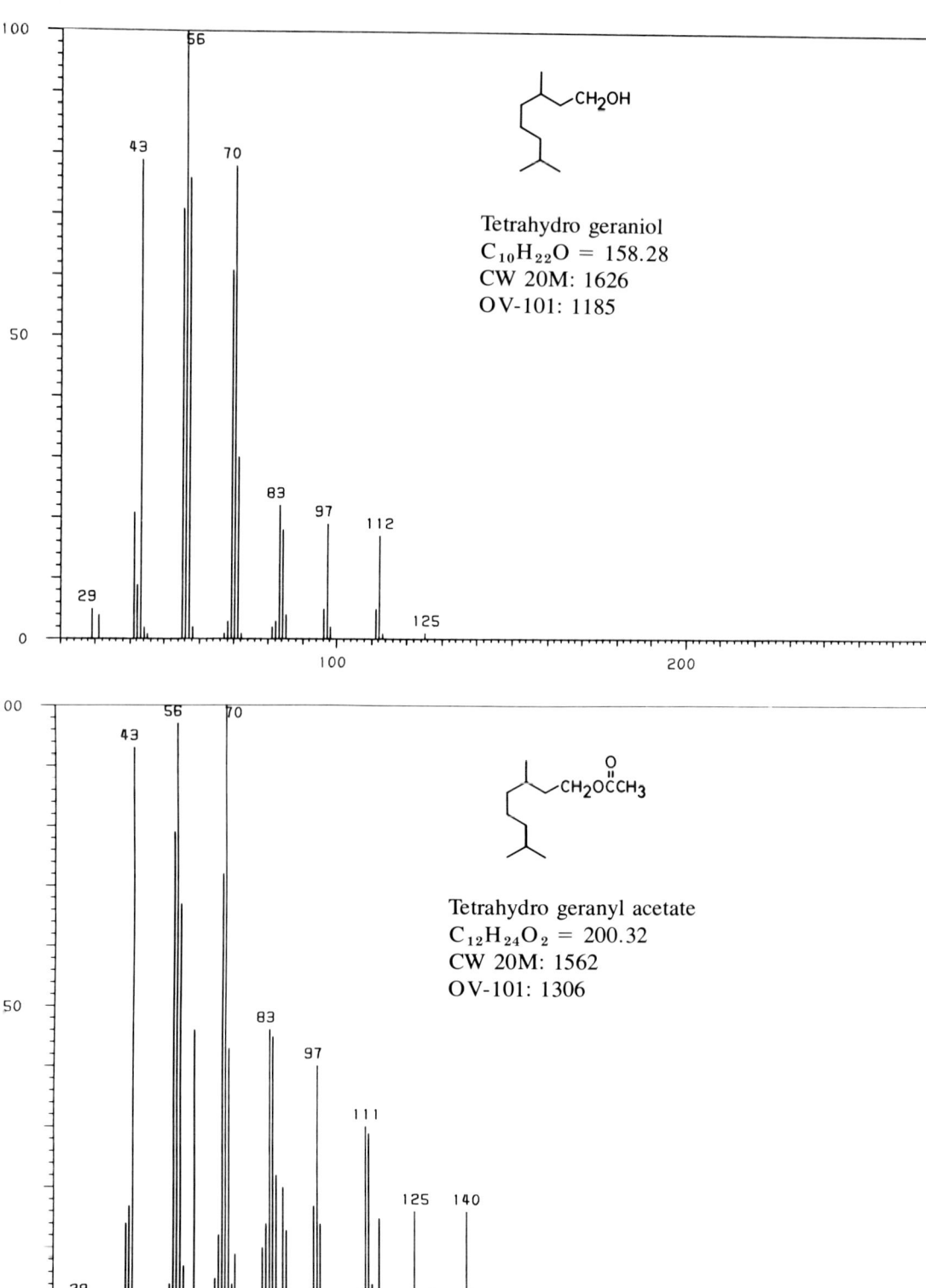

APPENDIX IV MASS SPECTRA OF INDIVIDUAL COMPOUNDS

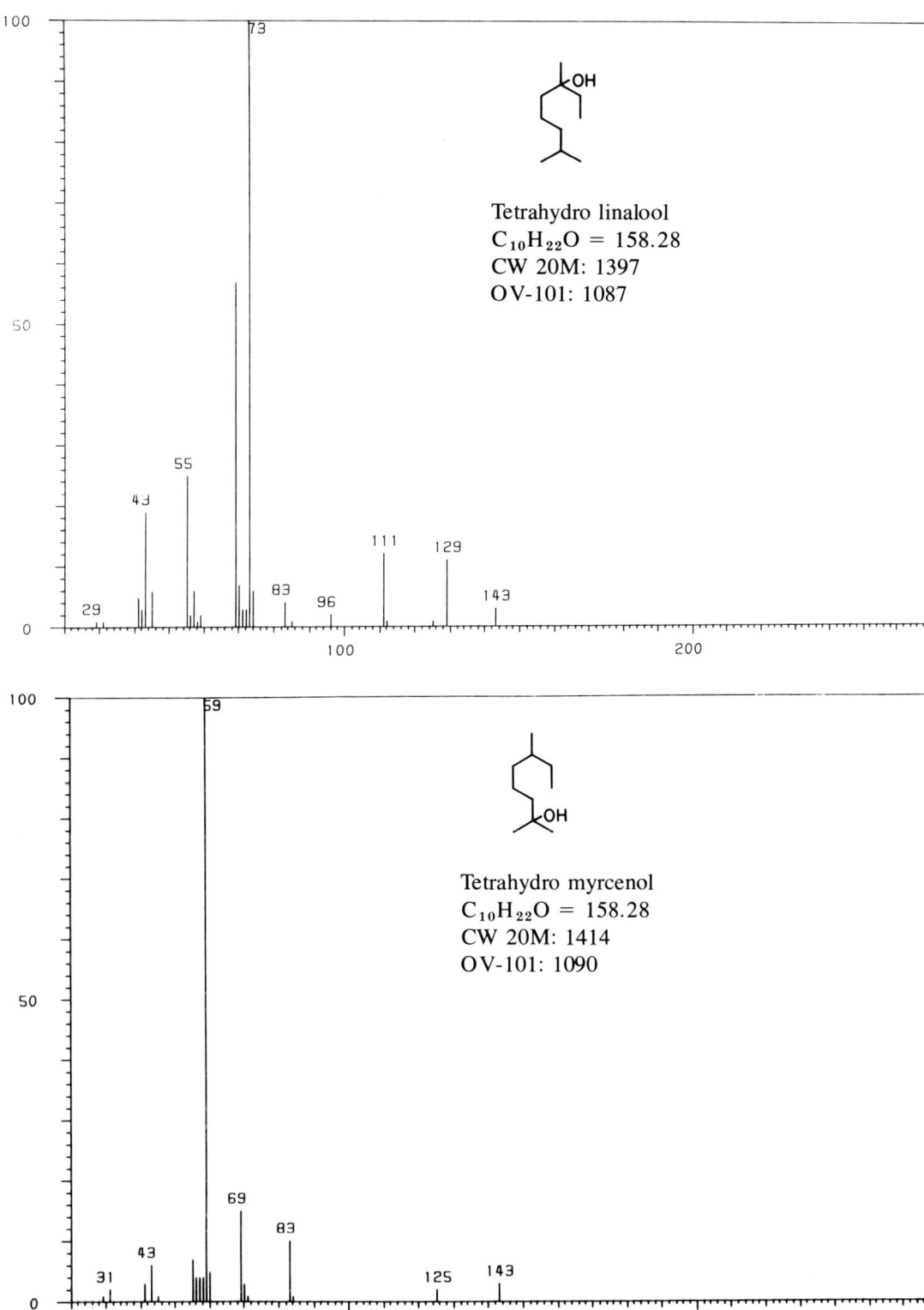

Tetrahydro linalool
$C_{10}H_{22}O = 158.28$
CW 20M: 1397
OV-101: 1087

Tetrahydro myrcenol
$C_{10}H_{22}O = 158.28$
CW 20M: 1414
OV-101: 1090

2,3,5,6-Tetramethyl pyrazine
$C_8H_{12}O_2$ = 136.20
CW 20M: 1458
OV-101: 1069

Thiophene
C_4H_4S = 84.14
CW 20M: 1035
OV-101: 650

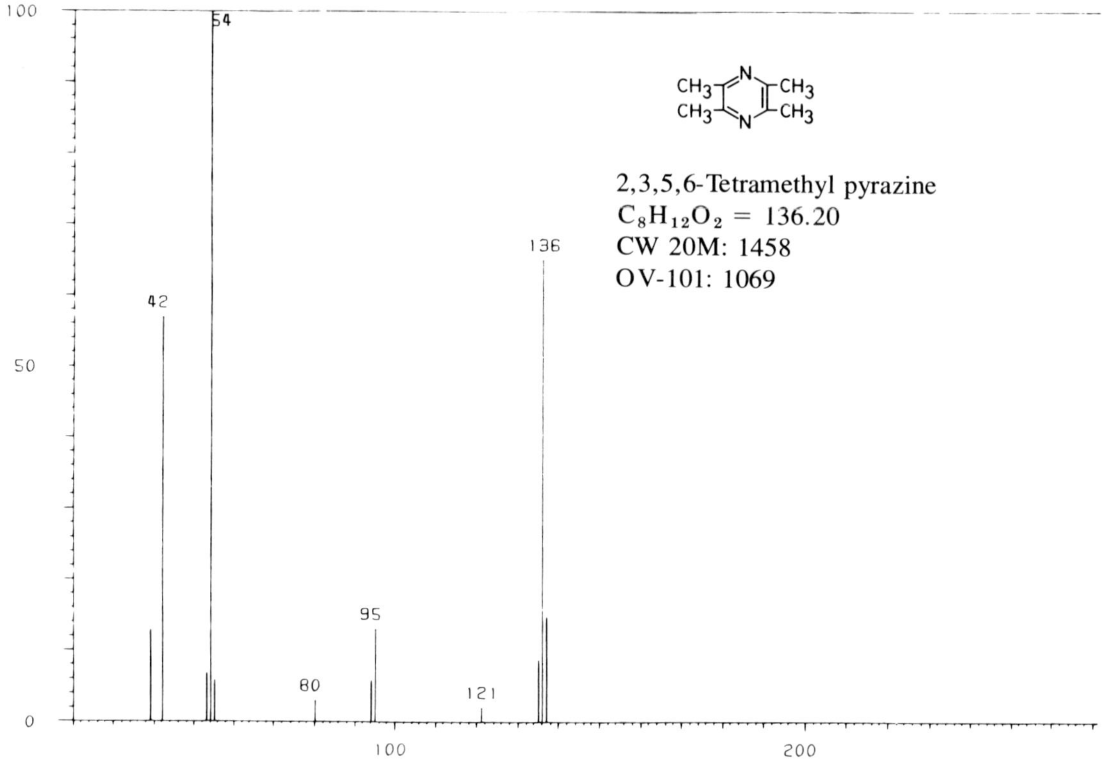

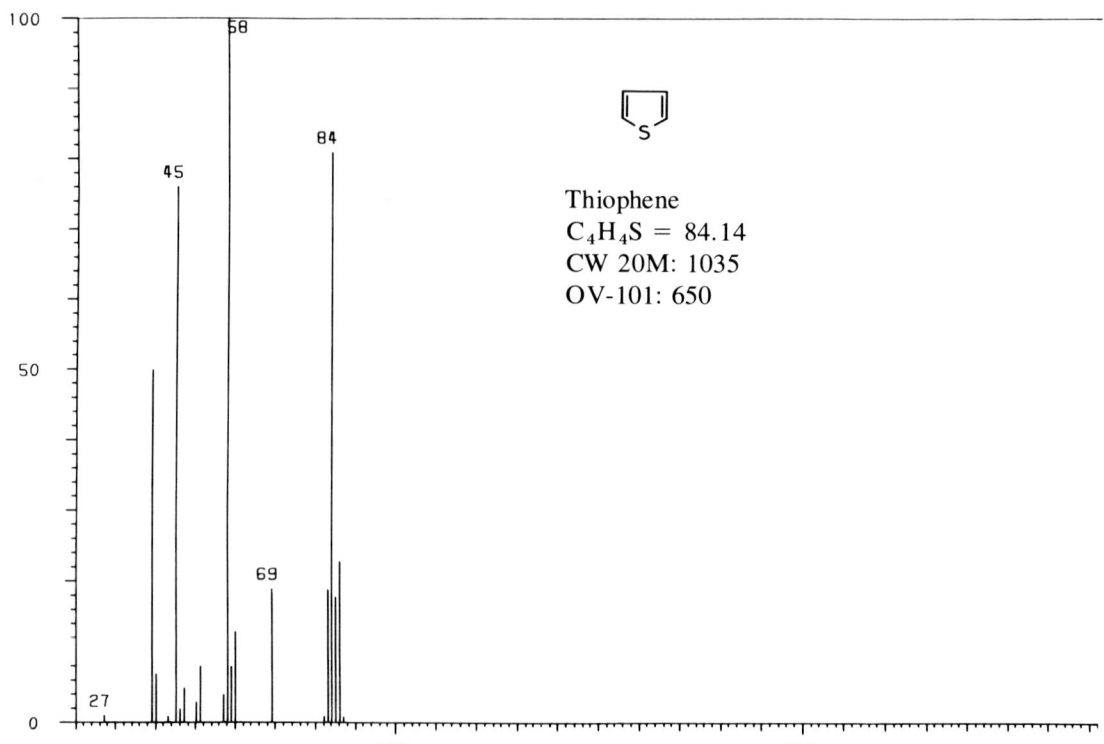

APPENDIX IV   MASS SPECTRA OF INDIVIDUAL COMPOUNDS   457

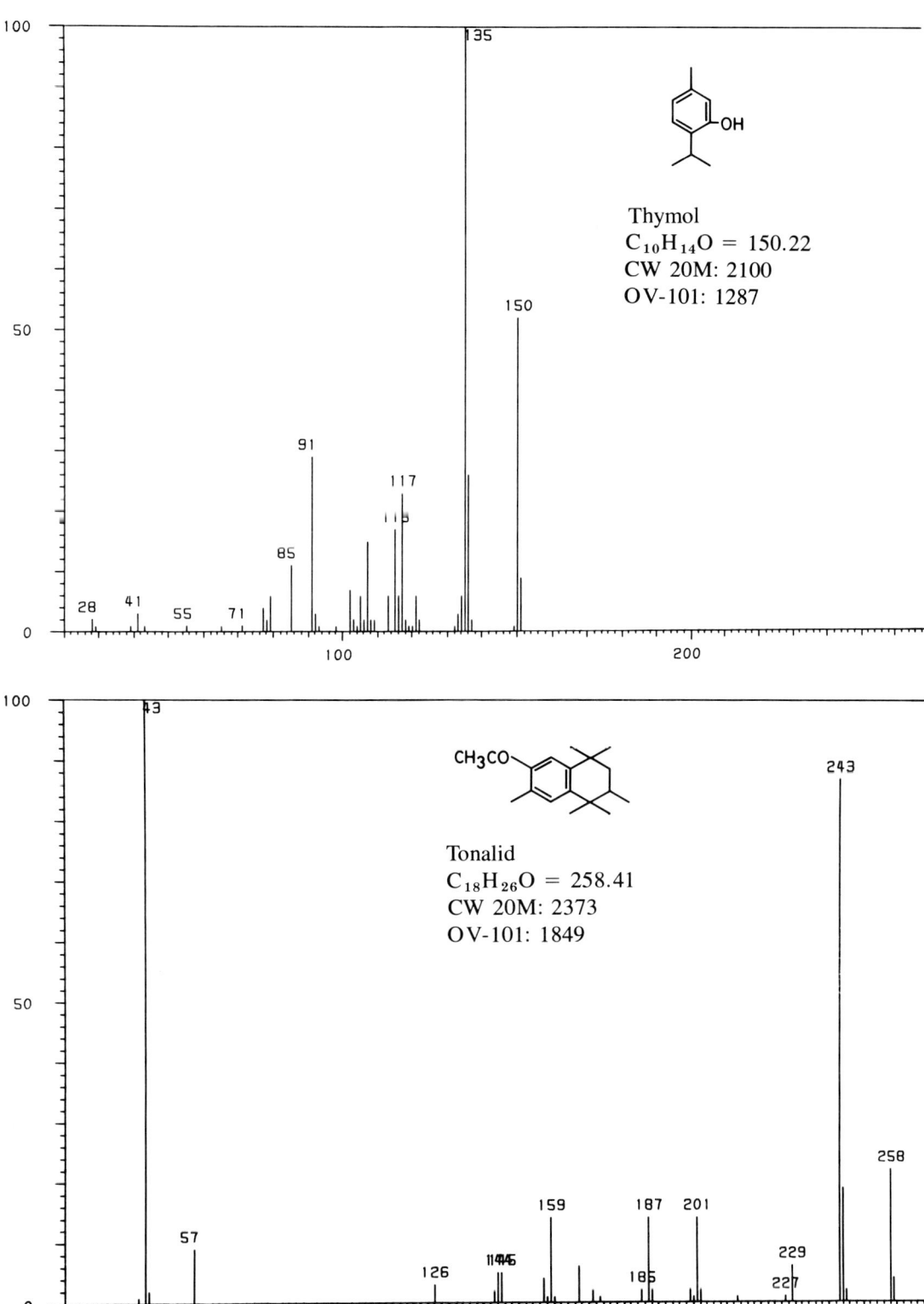

n-Tridecane
$C_{13}H_{28}$ = 184.37
CW 20M: 1300
OV-101: 1300

Triethyl citrate
$C_{12}H_{20}O_7$ = 276.29
CW 20M: 2386
OV-101: 1627

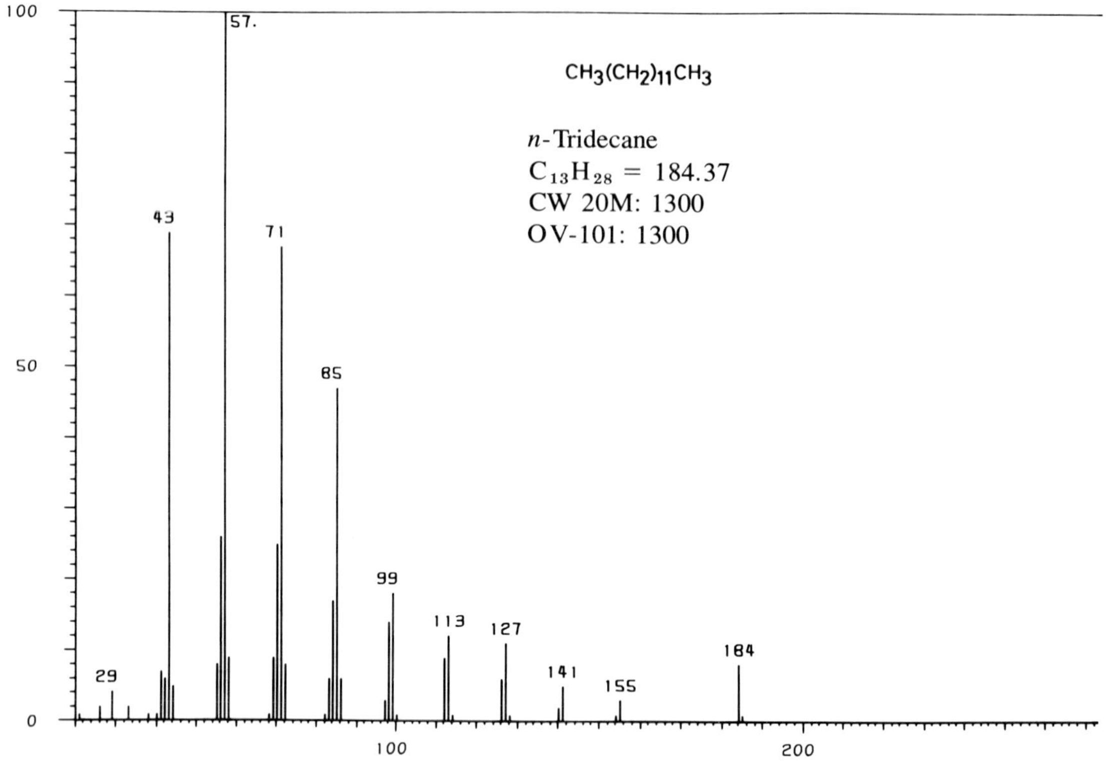

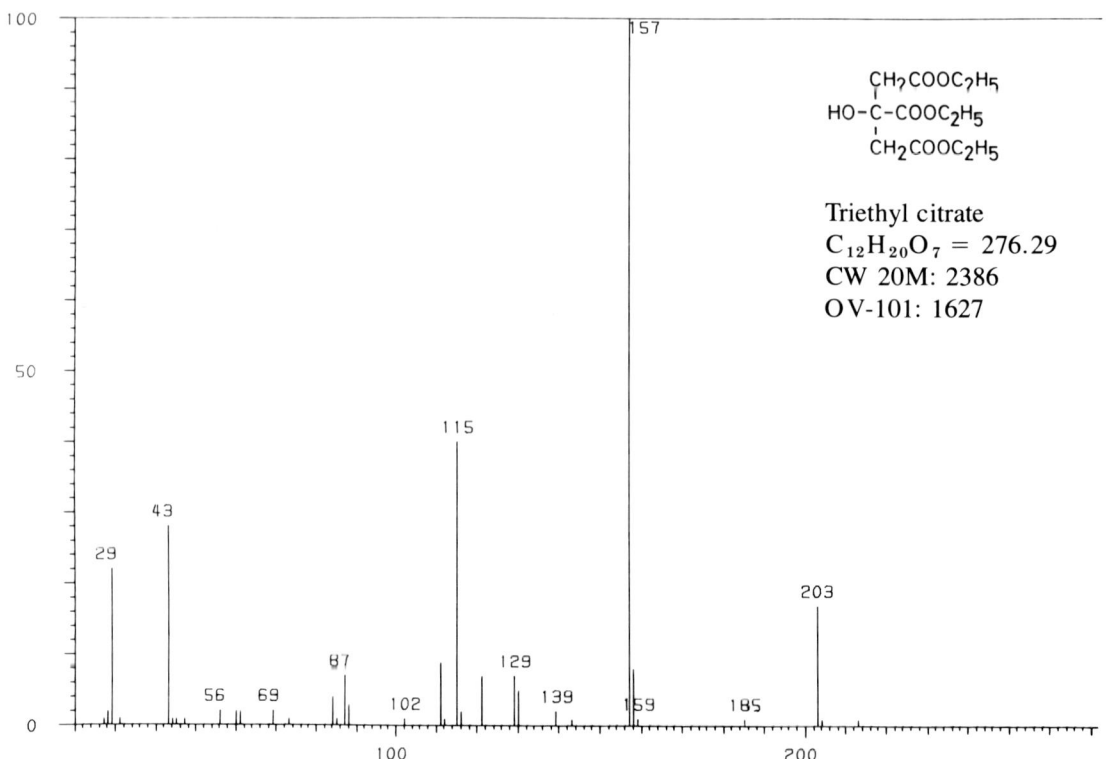

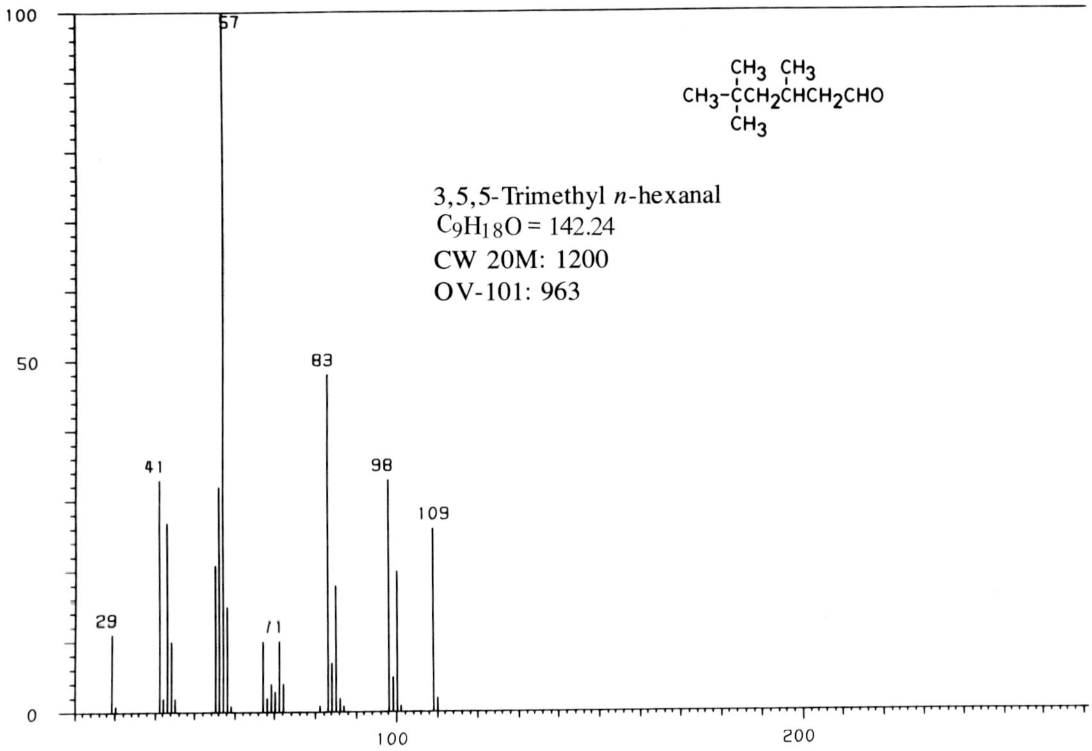

APPENDIX IV MASS SPECTRA OF INDIVIDUAL COMPOUNDS

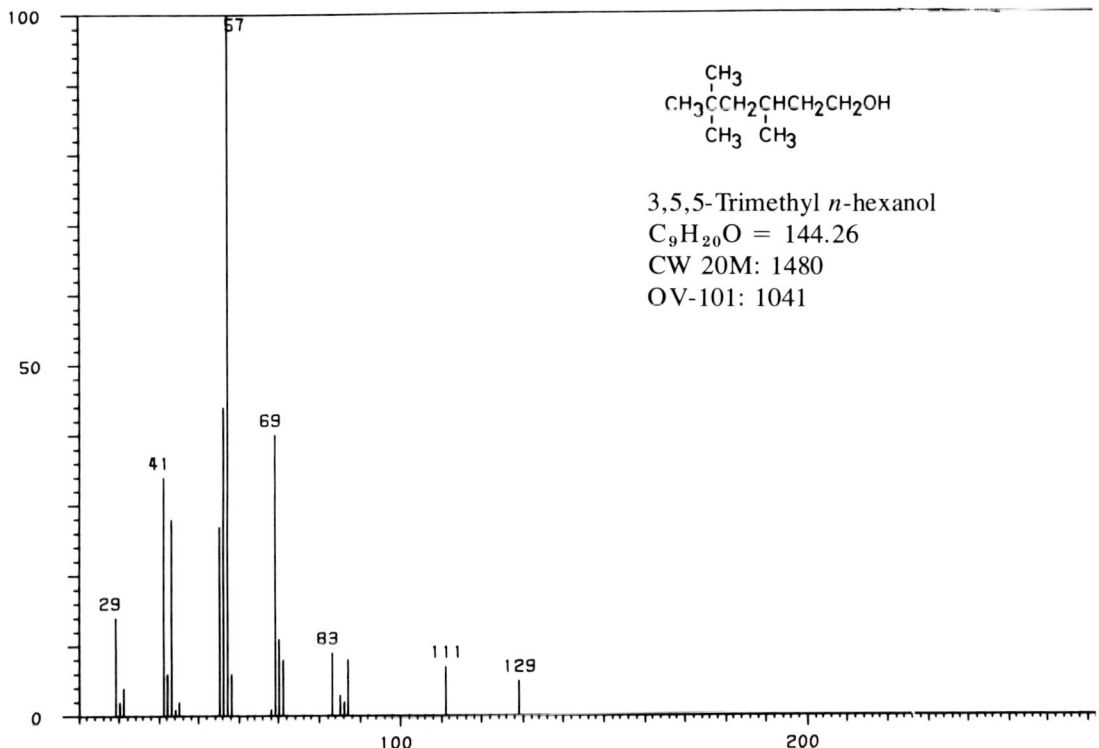

3,5,5-Trimethyl *n*-hexanal
$C_9H_{18}O = 142.24$
CW 20M: 1200
OV-101: 963

3,5,5-Trimethyl *n*-hexanol
$C_9H_{20}O = 144.26$
CW 20M: 1480
OV-101: 1041

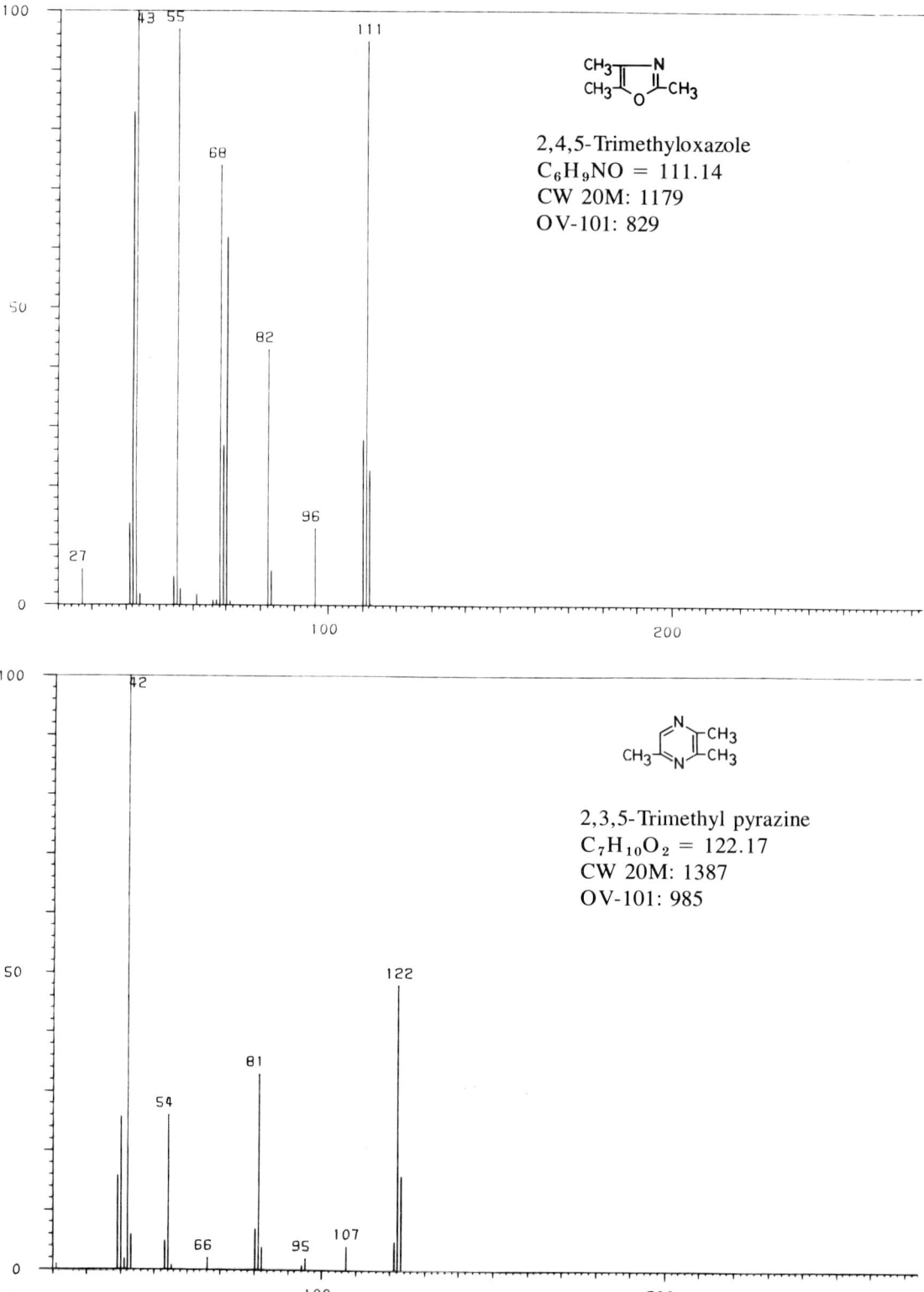

APPENDIX IV   MASS SPECTRA OF INDIVIDUAL COMPOUNDS

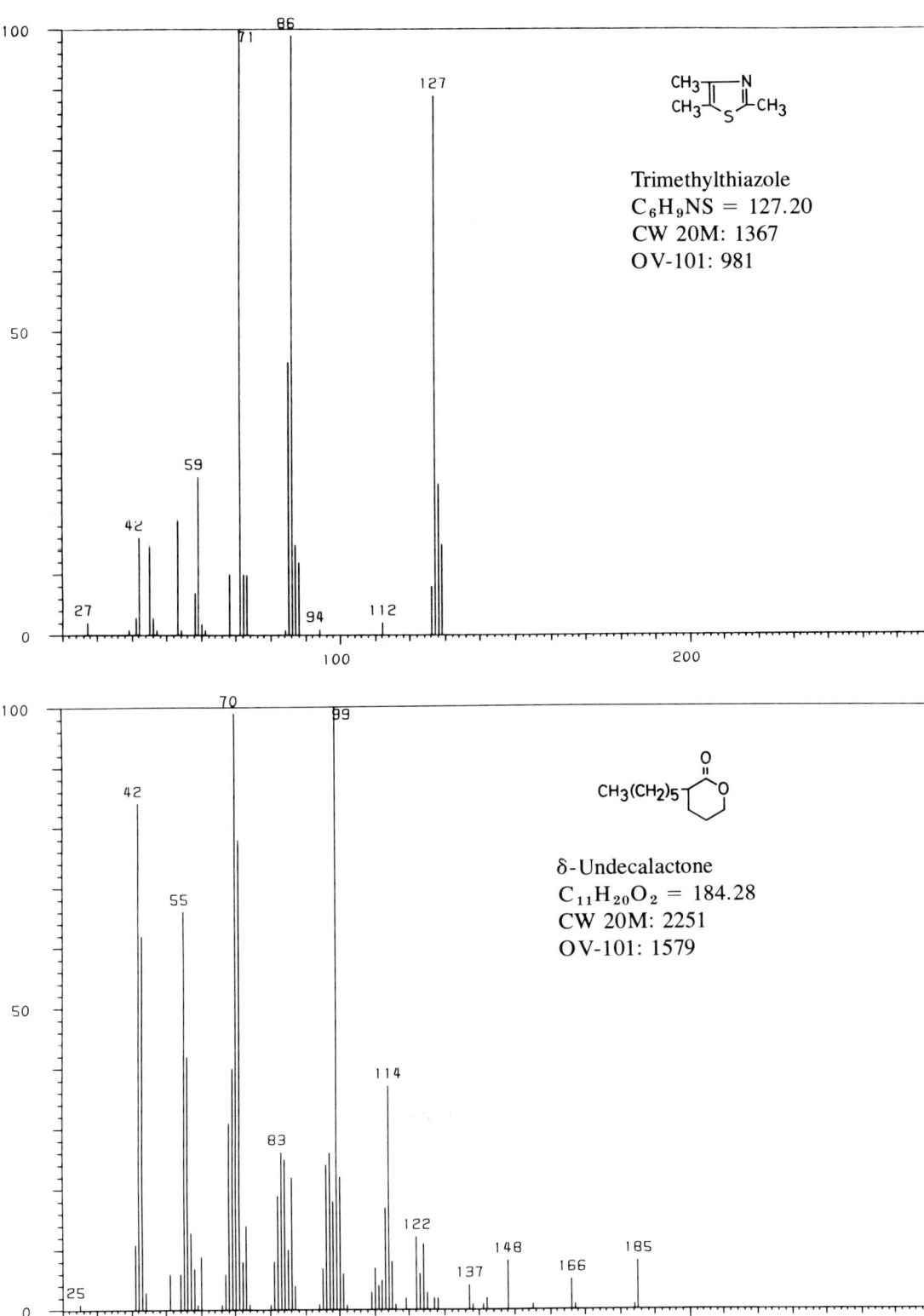

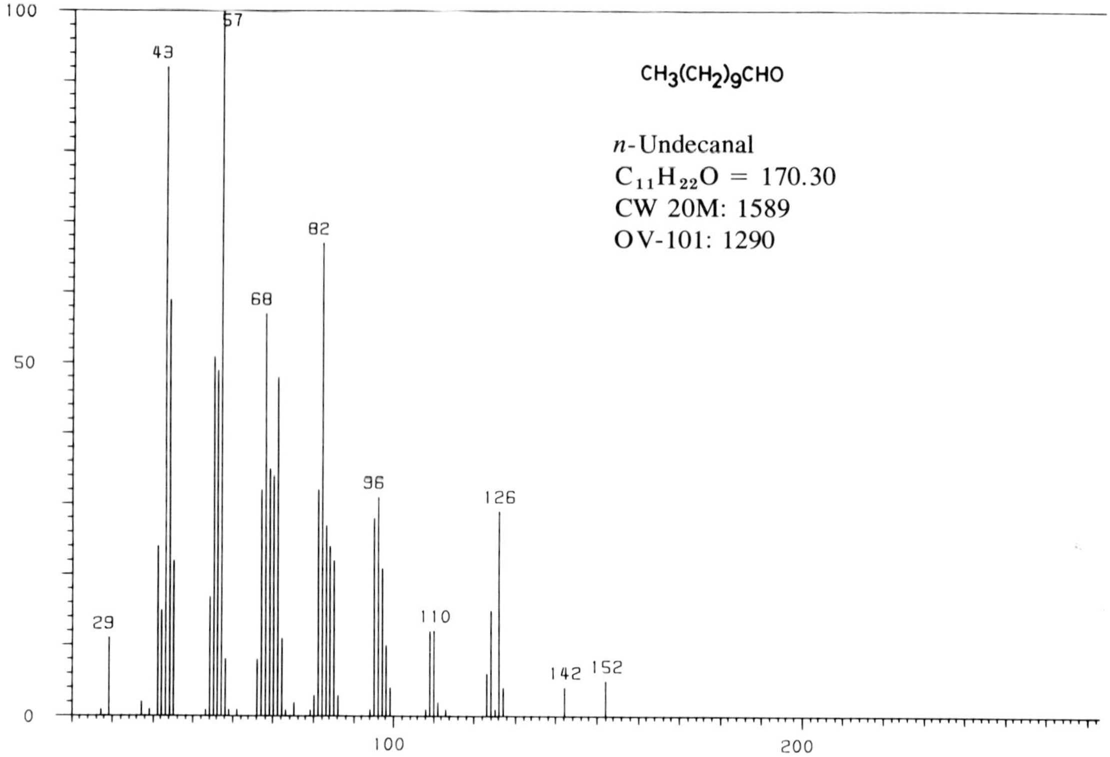

$CH_3(CH_2)_9CHO$

n-Undecanal
$C_{11}H_{22}O$ = 170.30
CW 20M: 1589
OV-101: 1290

$CH_3(CH_2)_9CH_3$

n-Undecane
$C_{11}H_{24}$ = 156.31
CW 20M: 1100
OV-101: 1100

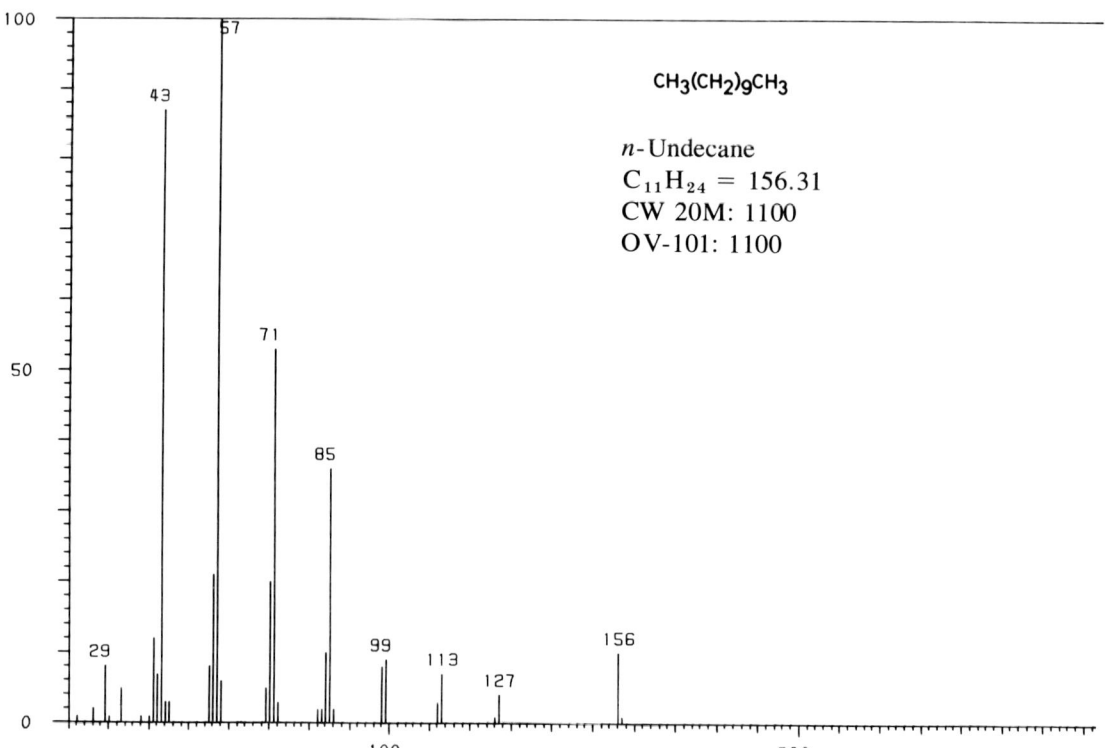

APPENDIX IV   MASS SPECTRA OF INDIVIDUAL COMPOUNDS   463

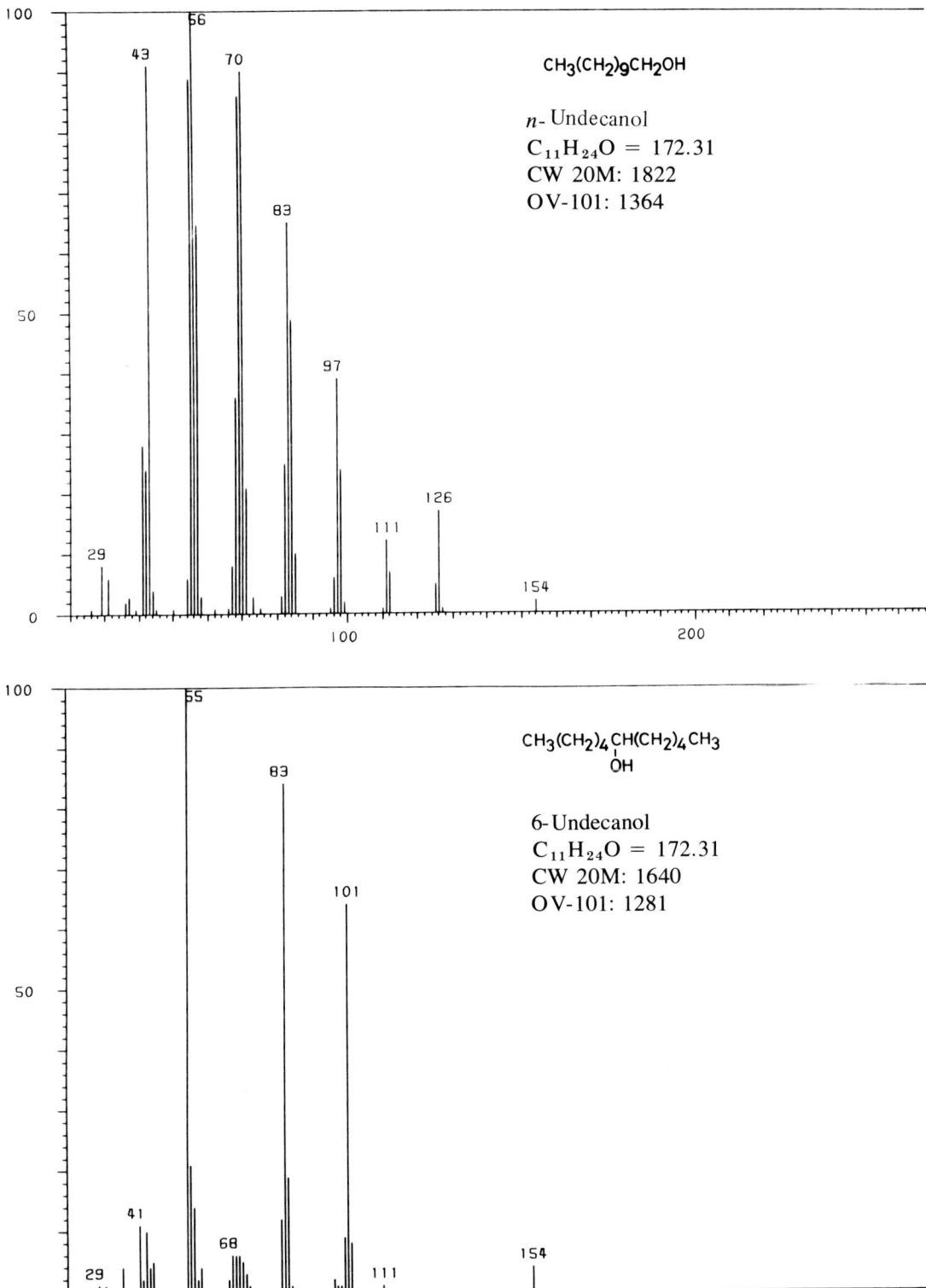

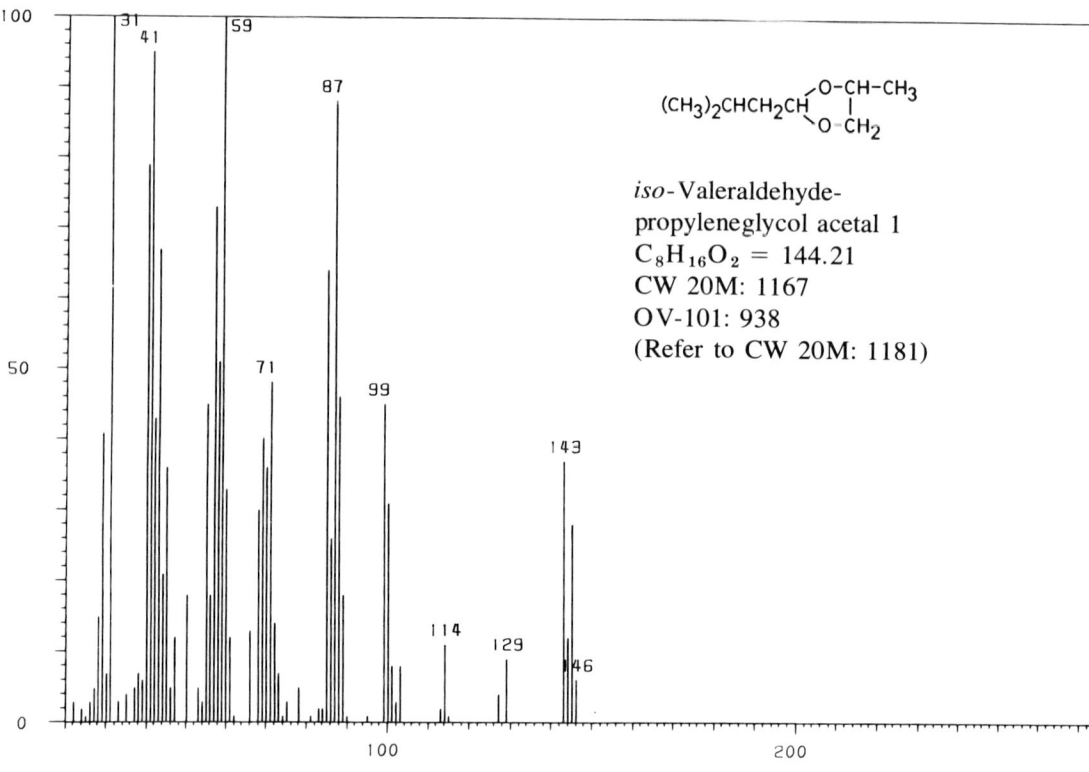

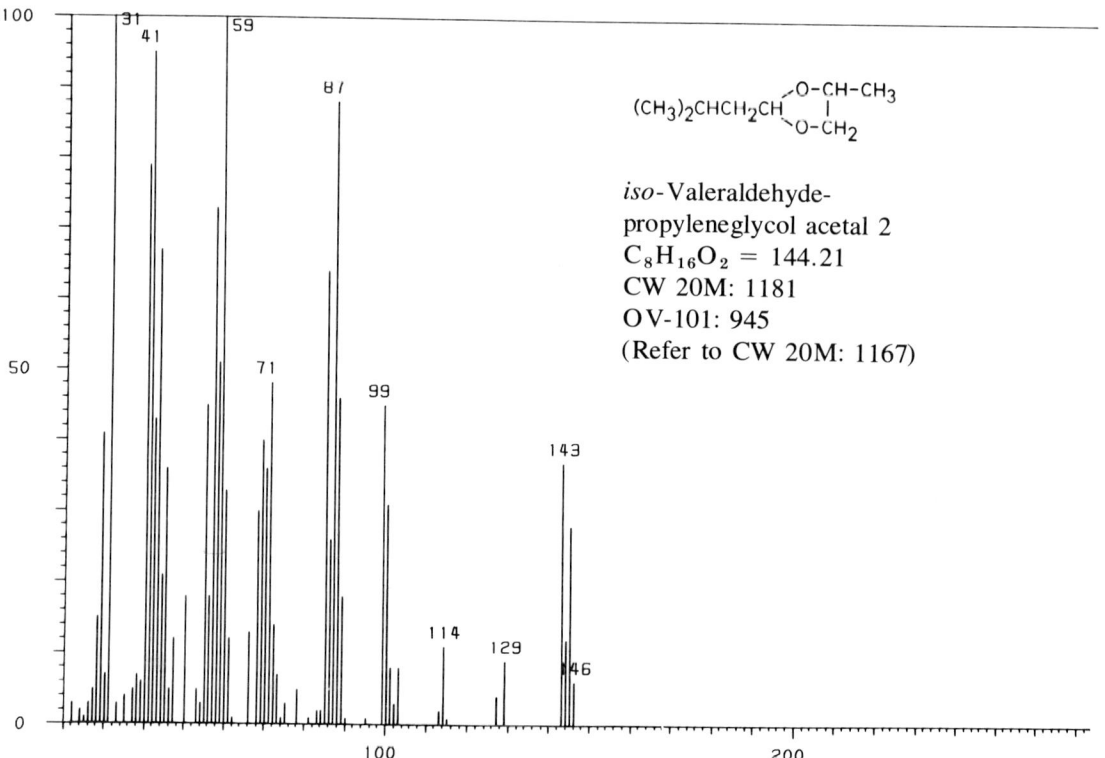

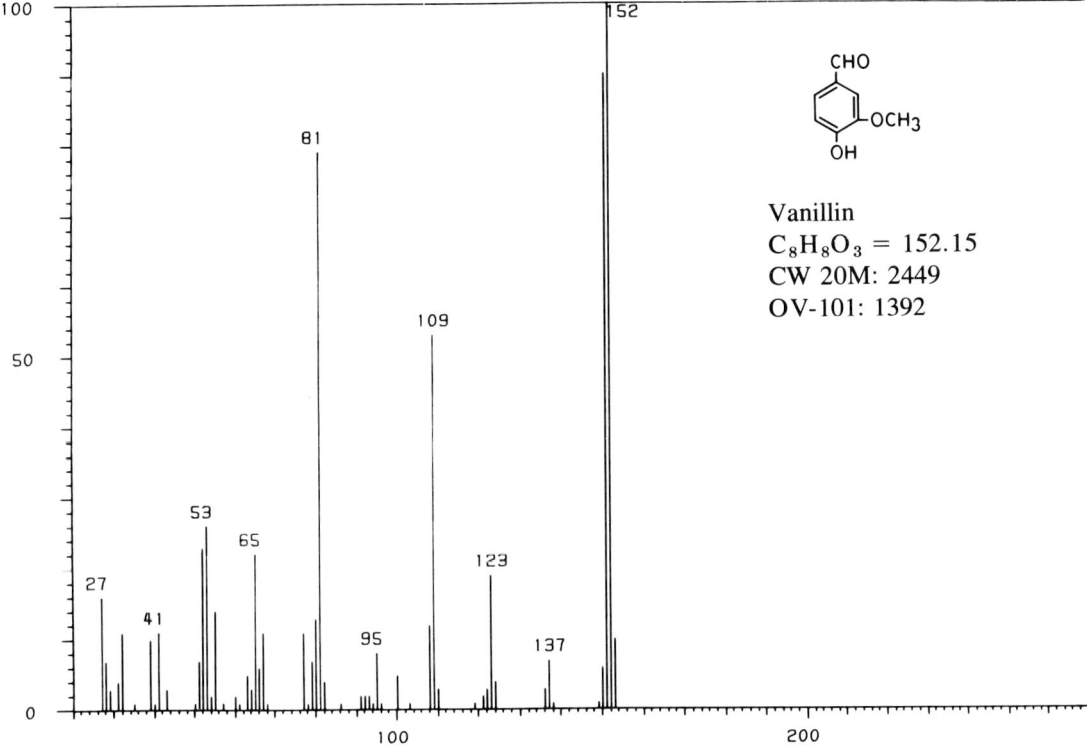

Vanillin
$C_8H_8O_3 = 152.15$
CW 20M: 2449
OV-101: 1392

COMMON SYNONYMS OF COMPOUNDS IN THE APPENDICES

(This listing is confined to commonly used synonyms of compounds for which another name has been used.)

A

Acetotoluene, *see* methylacetophenone
Acetyl benzene, *see* acetophenone
2-Acetylnaphthalene, *see* methyl napthyl ketone
Allyl α toluate, *see* allyl phenylacetate
    2-aminobenzoate, *see* allyl anthranilate
    caproate *see* allyl hexanoate
    caprylate, *see* allyl octanoate
    disulfide, *see* diallyl sulfide
    guaiacol, *see* eugenol
    trans-2-methyl-2-butenoate, *see* allyl tiglate
    pelargonate, *see* allyl nonanoate
    oenanthate, *see* allyl heptanoate
Amyl caproate, *see* amyl hexanoate
    caprylate, *see* amyl octanoate
    pentanoate, *see* amyl valerate

B

Benzazole, *see* indole
Benzodihydropyrone, *see* dihydro coumarin
2,3-Benzopyrrole, *see* indole
Bergamol, *see* linalyl acetate
Butyl 2-aminobenzoate, *see* butyl anthranilate
    caproate, *see* butyl hexanoate
    caprylate, *see* butyl octenoate
    γ-butyrolactone, *see* butyl levulinate
    2-hydroxypropanoate, *see* butyl lactate
    3-oxobutanoate, *see* butyl acetoacetate

C

Capric alcohol, *see* decanol
    aldehyde, *see* decanal
Capryl compounds, *see* corresponding octyl compound
Carbolic acid, *see* phenol
Cetone, *see* methyl ionone
Cocodescol, *see* 6-methyl coumarin
Coriandrol, *see* linalool
p-Cresyl methylether, *see* methyl anisole

D

Decanol, *see* amyl butyl carbinol
Diethyl butanedioate, *see* diethyl succinate
    decanedioate, *see* diethyl sebacate
Dimethyl butanedioate, *see* dimethyl succinate
    hydroquinone, *see* 1-4-dimethoxy benzene
3,7-Dimethyl-7-hydroxy octanal, *see* hydroxycitronellal
    ketone, *see* acetone
2-*cis*-3,7-Dimethyl-2,6-octadien-1-ol, *see* nerol
2-*trans*-3,7-Dimethyl-2,6-octadien-1-ol, *see* geraniol
3,7-Dimethyl-1,6-octadien-3-ol, *see* linalool
3,7-Dimethyl-1,6-octadien-3-yl esters, *see* corresponding linalyl ester
    phenethyl esters, *see* corresponding dimethylbenzylcarbinol ester
1,1-Dimethyl-2-phenylethanol, *see* dimethyl benzyl carbinol
    phenethyl alcohol, *see* dimethyl benzyl carbinol

1,1-Dimethyl-2-phenylethanol:
pyrazines, *see* corresponding dimethyl diazine
resorcinol, *see* 1,3-dimethoxybenzene
salicylate, *see* methyl anisate
Diphenyl ketone, *see* benzophenone

E

Ethanal, *see* acetaldehyde
Enanthaldehyde, *see* heptanal
3-Ethoxy-4-hydroxybenzalhyde, *see* ethyl vanillin
Ethyl-2-aminobenzoate, *see* ethyl anthranilate
   trans-2-butenoate, *see* ethyl crotonate
   caprate, *see* ethyl decanoate
   caproate, *see* ethyl hexanoate
   caprylate, *see* ethyl octanoate
   trans-2,3-dimethyl acrylate, *see* ethyl tiglate
   hendecanoate, *see* ethyl undecanoate
   2,4-hexadienoate, *see* ethyl sorbate
   o-hydroxybenzoate, *see* ethyl salicylate
   α-ketopropionate, *see* ethyl pyruvate
   ketovalerate, *see* ethyl levulinate
   p-methoxybenzoate, *see* ethyl anisate
   4-2-methoxyphenol, *see* ethylguaiacol
   β-methylacrylate, *see* ethyl crotonate
   trans-2-methyl-2-butenoate, *see* ethyl tiglate
   methyl ketone, *see* 2-butanone
   9-octadecenoate, *see* ethyl oleate
   pelargonate, *see* ethyl nonanoate
   propenoate, *see* ethyl acrylate
   toluate, *see* ethyl phenylacetate
Eugenyl acetate, *see* acetyl eugenol

G

Geranial, *see* also citral

H

Hendecanal, *see* undecanal
Heptyl caproate, *see* heptyl octanoate
Hexahydropyridine, *see* piperidine
Hexazane, *see* piperidine
Hexyl caproate, *see* hexyl hexanoate
   caprylate, *see* hexyl octanoate
   pentanoate, *see* hexyl valerate
Hyacinthin, *see* phenylacetaldehyde
Hydroquinone dimethyl ether, *see* 1-4-dimethoxy benzene
o-Hydroxyanisole, *see* guaiacol
2-Hydroxybenzaldehyde, *see* salicylaldehyde
Hydroxybenzene, *see* phenol
4-Hydroxy-3-methoxybenzaldehyde, *see* vanillin
3-Hydroxy-2-methyl-(1,4-pyran), *see* maltol
4-Hydroxy-3-pentenoic acid lactone, *see* angelica lactone
p-Hydroxytoluene, *see* *p*-cresol

I

2-Isopropyl-5-methyl-cyclohexanone, *see* menthone

M

Melonal, *see* 2,6-dimethyl-5-heptenal
p-Mentha-6,8-dien-2-ol, *see* carveol
8-*p*-Menthen-2-ol, *see* dihydrocarveol
cis-p-8(9)-one(2), *see* dihydrocarvone
p-Menth-8-(9)-en-2-yl-acetate, *see* dihydrocarvyl acetate
o-Methoxphenol, *see* guaiacol
Methoxy benzaldehyde, *see* anis aldehyde
   benzene, *see* anisole
   p-benzyl compounds, *see* anisyl compounds
Methyl 2-aminobenzoate, *see* methyl anthranilate
   caproate, *see* methyl hexanoate
   caprylate, *see* methyl octanoate
   catechol, *see* guaiacol
   p-cresol, *see* methyl anisole
   3,4-enedioxy-propylbenzene, *see* dihydrosafrole
   2-1,4-diazine, *see* 2-methyl pyrazine
   2,4-dimethylphenyl ketone, *see* di-methylacetophenone
   ethyl ketone, *see* 2-butanone
   hexyl ketone, *see* 2-octanone
   o-hydroxybenzoate, *see* methyl salicylate
   6-3-isopropenylcyclohexanol, *see* dihydrocarveol
   1-4-isopropenyl cyclohexan-2-one, *see* dihydrocarvone
   6-3-isopropenyl cyclohexyl acetate, *see* dihydrocarvyl acetate
   laurate, *see* methyl dodecanoate
   p-methoxybenzoate, *see* methyl anisate
   7-3-methylene-1,6-octadiene, *see* myrcene
   myristate, *see* methyl tetradecanoate
   pelargonate, *see* methyl nonanoate
   pentanoate, *see* methyl valerate
   3-2-(2-pentenyl)-2-cyclopenten-1-one, *see* jasmone
   phenyl ether, *see* anisole
   phenyl ketone, *see* acetophenone
   -3-phenylpropenoate, *see* methyl dihydrocinnamate
Myristaldehyde, *see* tetradecanal

N

Neral, *see* also citral
Niobe oil, *see* methyl benzoate
5-Nonanone, *see* dibutyl ketone

P

Peach aldehyde, *see* gamma-undecalactone
Pelargonicaldehyde, *see* nonanal
Pelargonyl acetate, *see* nonyl acetate
Pentamethylenimine, *see* piperidine
Pentanol, *see* amyl alcohol
4-Pentenoic acid, *see* allyl acetate
Pentyl compounds, *see* corresponding amyl compound
Phenethoxypropoxy ethane, *see* acetaldehyde, phenylethyl n-propyl acetal
Piperonal, *see* heliotropine
Piperonyl acetate, *see* heliotropyl acetate
2-propanone, *see* acetone
Propyl phenethyl acetal, *see* acetaldehyde, phenylethyl n-propyl acetal

R

Resorcinol dimethyl ether, *see* 1,3-dimethoxybenzene

T

Tetrahydro-4-methyl-2-(2-methylpropen-1-yl)-pyran, *see* rose oxide
Tolualdehyde, *see* phenylacetaldehyde
Toncair, *see* 6-methyl coumarin
Tuberyl alcohol, *see* dihydrocarveol

V

Vanillyl acetone, *see* zingerone
Verdural, *see* cis-3-hexen-1-yl acetate

INDEX

A

Alcohol subtraction, 21
Aldehyde subtraction, 21

C

Carbon skeleton analysis, 19–20
Cold trapping, 7
Columns, choice of type, 2 3
    connections, 3
    fused silica, 1, 3, 2–4
    glass capillaries, advantages of, 2, 3
    heat straightening of, 3
    inertness of, 1
    metal, 2
    PLOT, 3
    SCOT, 3
Column chromatography, 15–16

D

Detector, dual, 11–13
    electron capture, 14
    thermal conductivity, 13
    flame ionization, 4, 11–14
    flame photometric, 14
    nitrogen-phosphorus, 13–14
    photoionization, 8, 14
    selective, 11

E

Effluent splitters, 12

F

Fractionation, preliminary, 16–17
Functional group determination, 2, 15

G

GC/MS, 22
GC/MS interfaces, 23

H

Hydrogenation, 17, 20
HPLC, 16
Hold-up time, calculation of, 9

I

Inlet, 5
    all-glass, 4
    connection, 2, 3,
Inlet splitters, 5–7
    back pressure regulation of, 6–7
Inlet splitters, linearity checks of, 5–7

L

Lauric aldehyde, *see* dodecanal
Laurine, *see* hydroxycitronellal
Lauryl compounds, *see* corresponding dodecyl compound
Liquid phase film thickness, effect on partition rations, 5

M

Make-up gas, 3, 4, 13
   insulation of lines, 13

O

On-column injection, 8
Oven, 8, 9
Olefin subtraction, 21
Ozonolysis, 17

P

Platinium iridium connections, 23–24
Pre-column reactions, 19

R

Reaction GC, 18–19
Retention indices, 9–11
   calculation of, 9
   identification, criterion of, 1–2
   isothermal determination, 10–11
   liquid phase, effect of, 9–10
   program temperature determination, 11
   reliability as influenced by column efficiency, 1
   surface pretreatment, effect of, 9–10
   temperature, effect of, 10

S

Splitless injection, 7–8
Styralyl compounds, *see* corresponding methyl benzyl compound
Subtractive reactions, 20–22
Sulfur compounds, 3

T

Tailing, 3
Two-dimensional GC, 2
t_M, *see* hold-up time

TARRYTOWN TECHNICAL
Information Center
GENERAL FOODS CORP.